AF766872

The *Fusarium* Laboratory Manual

The *Fusarium* Laboratory Manual

John F. Leslie
and
Brett A. Summerell

Photographs by
Suzanne Bullock

Blackwell
Publishing

John F. Leslie is a professor in the Department of Plant Pathology at Kansas State University, Manhattan.

Brett A. Summerell is Director of Science and Public Programs at the Royal Botanic Gardens – Sydney (Australia). They have been organizing and teaching *Fusarium* Laboratory Workshops annually since 2000 and began the work on this manual in 2002 in Sydney under the sponsorship of the Australian-American Fulbright Commission.

Blackwell Publishing Professional
2121 State Avenue, Ames, Iowa 50014, USA

Orders: 1-800-862-6657
Office: 1-515-292-0140
Fax: 1-515-292-3348
Web site: www.blackwellprofessional.com

Blackwell Publishing Ltd
9600 Garsington Road, Oxford OX4 2DQ, UK
Tel.: +44 (0)1865 776868

Blackwell Publishing Asia
550 Swanston Street, Carlton, Victoria 3053, Australia
Tel.: +61 (0)3 8359 1011

Produced directly from camera-ready copy provided by the authors. Contribution no. 06-163-B from the Kansas Agricultural Experiment Station, Manhattan.

First edition, 2006

Library of Congress Cataloging-in-Publication Data

Leslie, John F., 1953-
Fusarium laboratory manual / John F. Leslie and Brett A. Summerell.
p. cm.
ISBN-13: 978-0-8138-1919-8 (alk. paper)
ISBN-10: 0-8138-1919-9 (alk. paper)
1. Fusarium-Laboratory manuals. I. Summerell, Brett A. II. Title.

SB741.F9L47 2006
632′.4677-dc22

2006004759

FSC www.fsc.org MIX Paper from responsible sources FSC® C013604

The last digit is the print number: 9 8 7 6 5

Table of Contents

Foreword

In 1753, Linnaeus placed all the fungi in the genus *Chaos*. In the case of the genus *Fusarium* things remained more or less chaotic until 1935 when Wollenweber & Reinking created some order in their monumental monograph, *Die Fusarien* (2359). The morphological differences that they used to differentiate between 65 species, 55 varieties and 22 forms were sometimes very small and variable. Subsequently two schools of *Fusarium* taxonomists developed – the "splitters" mainly in Europe and the "lumpers" in the USA where Snyder and Hansen reduced the genus to nine species in 1945. During the 1980s, collaboration between *Fusarium* taxonomists from Europe, the USA, Australia and South Africa resulted in a remarkable level of agreement between the taxonomic treatments of Gerlach & Nirenberg in Germany, Nelson, Toussoun & Marasas in the USA, and Burgess & Summerell in Australia. This relative uniformity shifted drastically again during the 1990s with the application of the phylogenetic species concept to DNA sequences of diagnostic genes. The resulting plethora of new species of *Fusarium* often cannot be distinguished morphologically and the genus *Fusarium* might appear to be some to be heading for chaos again.

Thus, the publication of this *Fusarium* Laboratory Manual by John Leslie and Brett Summerell is most timely because it integrates the morphological, biological and phylogenetic species concepts. The *Fusarium* Laboratory Manual includes detailed chapters on Techniques and Methods as well as on approaches to Taxonomy and Identification of *Fusarium* followed by Species Descriptions of 70 *Fusarium* species, arranged alphabetically from *F. acuminatum* to *F. verticillioides*. Each species description contains photographs and descriptions of key morphological characters, together with information on sexual stage, taxonomy, pathology, ecology, genetics and molecular biology. A comprehensive literature review is given for each species, including references to the most recent publications. This is a formidable undertaking for 70 species, considering that over 19,000 abstracts are available in Biological Abstracts for *Fusarium* with between 5000 and 6000 for *F. oxysporum* alone. This kind of information is provided for the first time for many of the 70 *Fusarium* species, particularly those described after 1983.

Many mycologists and plant pathologists who must identify isolates of *Fusarium* are extremely concerned about the proliferation of *Fusarium* species that are difficult, if not impossible, to identify morphologically. On the one hand they realize that many *Fusarium* taxa are heterogeneous and that in many cases the apparent limits of morphology have been reached for separating species. In short we have run out of morphological characters before we have run out of species that need to be separated. On the other hand they are alarmed that new species are being described based on molecular data from one or no more than a few isolates without due regard for the biology, pathology, toxicology and ecology of the taxa. The success or failure of implementing this "new" *Fusarium* taxonomy will depend on the successful integration of morphological characters, measures of cross fertility and DNA sequence data into a species concept based on the whole fungus. The methods and descriptions in this book provide a firm foundation from which such discussions and descriptions can begin.

The *Fusarium* Laboratory Manual is a milestone in the study of the genus *Fusarium* and will help bridge the gap between morphological and phylogenetic taxonomy. It will be used by everybody dealing with *Fusarium* in the Third Millennium.

W.F.O. Marasas
PROMEC Unit
Medical Research Council
Tygerberg
South Africa

Preface

Laboratory workshops, in which researchers spend a week with experts in the field to learn to identify various *Fusarium* species, are an established part of the *Fusarium* community. The best known of these were run by the late Dr. Paul Nelson at Pennsylvania State University in the 1980s and 1990s. It was at one of these workshops where one of us (jfl) was first introduced not only to *Fusarium* as a taxonomic object, but also to a cohort of other beginning *Fusarium* researchers, *e.g.*, Tom Gordon, Anne Desjardins, Jim Correll, and Marian Beremand, who along with the instructors of the class, T. A. Toussoun, W. F. O. Marasas, Carol Windels, Lester Burgess and Paul Nelson, have influenced much of the research in *Fusarium* over the last 20 years.

This manual springs from the tradition of *Fusarium* Laboratory Workshops and is in some ways a successor to both the manual of Nelson *et al.* (1551) and that of Burgess *et al.* (279), which had their roots in these teaching and outreach efforts. However, our goal was to take this tradition beyond morphological species descriptions to include additional techniques and identification processes that are widely used in the *Fusarium* research community. We also wanted to include some of the evolutionary biology and population genetic thinking that has begun to inform the understanding of agriculturally important fungal pathogens. The need for as many different techniques to be used when examining these fungi has never been clearer. In combination with the two recent volumes from APS Press (468, 2083), this volume presents a relatively current introduction to the genus *Fusarium*, the toxins these fungi produce and the diseases they can cause.

Much of the material in this manual has been used by participants who attended workshops held at Kansas State University (2000, 2001, 2003 and 2005), the University of Sydney (2002), and the University of Pretoria (2004). The material included in this manual is meant to serve as the basis for future workshops, as well as a guide to the field for those who are looking for a reference, a description of a common species, or a technique. The content also has been influenced by the presentations at these workshops of the other instructors – Lester Burgess, David Geiser, Antonio Logrieco, W. F. O. Marasas, John Rheeder, Baharuddin Salleh, Keith Seifert, Carol Windels, Brenda Wingfield, and Kurt A. Zeller. Their thoughts on what is important, and what is not, were passed through our filters as we developed this manual.

Thanks are due to many for their efforts both with the running of the workshops and the materials that have gone into this volume. First and foremost of these is Suzanne Bullock, who is responsible for the numerous photomicrographs in this manual and whose efforts should make the morphological species descriptions accessible even to those who have never before worked with these fungi. Second to Kurt Zeller, Amgad Saleh, Jim Jurgenson, and Brook van Scoyoc, for their help in developing many of the protocols included in this volume and distilling them to a form that is (hopefully) easily understood by those who have not used them before. Finally to all of those who have helped run the various workshops, but especially Amy Beyers, Anita Kesler, Ingelin Leslie, Brook van Scoyoc, and Celest McGowan, who have developed the protocols and the timing for both the scientific, fiscal, and social aspects of the workshops that the rest of us have followed with such great success.

Portions of this book were written while one of us (jfl) was on sabbatical leave at the University of Sydney and the Royal Botanic Gardens and Domain Trust (RBGDT) and sponsored by the Australian-American Fulbright Foundation and the RBGDT. We thank the American Phytopathological Society for permission to reprint Table 11-1 and Figures 4-2 and 11-1, the American Society for Microbiology for permission to reprint Figure 7-2, and the Genetics Society of America for permission to reprint Figure 7-3. Portions of chapter 9 are based on Leslie *et al.* (1235), and portions of chapter 11 are based on Summerell *et al.* (2087). Both the workshops and the manual have benefited from financial support from both our home institutions, and from INTSORMIL (the International Sorghum and Millet Collaborative Research Support Program, based at the University of Nebraska, Lincoln, Nebraska), IITA (the International Institute of Tropical Agriculture, Ibadan, Nigeria), and ABRS (the Australian Biological Resource Study).

Much of the book has been evaluated by others including Walter A. J. de Milliano, Anne Desjardins, Stanley Freeman, David Geiser, Cristiano Lima, Antonio Logrieco, W. F. O. Marasas, Randy Ploetz, Amgad Saleh, Baharuddin Salleh, Brenda Wingfield, and Kurt Zeller. Their eyes have found errors and omissions that we would have missed and their suggestions have only improved the book's accuracy, utility and readability. Errors that remain, however, are solely our own, and corrections and suggestions for topics to be considered for inclusion in future versions will be gratefully received.

A final, and perhaps the greatest note of gratitude is to our wives, Ingelin and Liz, and families for their encouragement, for sharing their homes and hospitality as early and final versions of the manuscript were first put to paper and then finally put to rest, and for tolerating our interest in this group of fungi, when they might have had other places they would like for us to have focused instead. They know better than we do that the idea of a book is grand and the finished product is even better, but that the time in between is not always one of great joy.

July 2005

John F. Leslie
Department of Plant Pathology
Kansas State University
Manhattan, Kansas, USA

Brett A. Summerell
Royal Botanic Gardens and Domain Trust
Mrs. Macquaries Road
Sydney, New South Wales, Australia

The *Fusarium* Laboratory Manual

1

Introduction

The genus *Fusarium* was introduced by Link in 1809 (1263), and is now approaching its third century as a genus that contains many plant-pathogenic fungi. The members of this genus can incite directly diseases in plants, humans, and domesticated animals, *e.g.*, Boonpasart *et al.* (221), Goldschmied *et al.* (718), Krcmery *et al.* (1136), Martino *et al.* (1373), Rabodonirina *et al.* (1760), Rebell (1783) and Vismer *et al.* (2263). The mortality rate for human patients with systemic *Fusarium* infections is > 70% (1136), and HIV-infected patients are susceptible to such *Fusarium* infections as well (562, 778, 1486). In addition, *Fusarium* spp. produce an intriguing array of secondary metabolites that are associated with plant disease, as well as with cancer and other growth defects in humans and domesticated animals. Some of these secondary metabolites are used commercially either directly or as the starting material for chemical syntheses of plant and animal growth promoters in both first world and third world settings (858, 1976, 2157). Allegations of the use of mycotoxins produced by some of these fungi as biological weapons also have been made (855, 1454, 1833). Two of these compounds, T-2 and diacetoxyscirpenol (DAS) are on the USDA/CDC select agent list and require special permits for study in the United States (www.cdc.gov/od/sap/docs/saist.pdf). Naturally occurring outbreaks of *Fusarium* toxicoses directly affecting humans have occurred historically, *e.g.*, Athens in the 5th century B.C. (1922), and in the Soviet Union during World War II (666, 982). Thus, *Fusarium* has always been a visible genus with many strains, species and metabolites of an importance that transcends just science or agriculture.

As a social phenomenon, *Fusarium* plant diseases have had several major impacts. One was the near devastation of the commercial banana industry in the 1960s by panama wilt caused by *Fusarium oxysporum* f. sp. *cubense* (1706). The recent losses of several billion dollars (2342) by many wheat and barley farmers to Fusarium head scab in the upper Midwest of the United States has shifted cropping strategies and bankrupted farmers in the region. At the same time, the causal agent of Fusarium head scab can be used in commercial fermentations to produce a precursor for one of the most widely used commercial cattle growth promotants (858). Recent problems caused by strains of *Fusarium* which may have originated from endophytes or pathogens of native *Gossypium* species (2286), are threatening the future of the cotton industry in Australia while simultaneously demonstrating the relatedness of native and agricultural populations and suggesting new avenues for understanding how these fungi evolve.

Many plants have at least one *Fusarium*-associated disease. A recent perusal of the plant disease list maintained by the American Phytopathological Society (www.apsnet.org/online/common/search.asp) revealed that over 81 of the 101 economically important plants on the list had at least one associated *Fusarium* disease. As these fungi also may grow as apparently symptomless endophytes under many conditions, the claim that, "If it is green, there is some *Fusarium* that can grow on it, in it, or with it" probably is not too far removed from the truth. The types of diseases induced are quite varied as is their severity, and may include root or stem rots, cankers, wilts, fruit or seed rots, and leaf diseases. Thus identifying the *Fusarium* strain present in a diseased plant sample, usually to species and sometimes further, has been and remains an important task in many plant diagnostic laboratories.

The need to identify strains and to attach names to them is as strong, or stronger, in *Fusarium* than it is in any fungal genus. That these names have a value that transcends the science from which they spring means that these names need to be assigned and changed with care and caution. Depending on the era and the identification scheme being followed (see Chapter 8), the number of *Fusarium* species could range from as few as nine to well over a thousand. Since the 1980s the number of recognized species has increased gradually, with the number of recognized species now > 80, of which 70 are described and illustrated in this text. If preliminary work from current molecular studies stands, then this number should increase, perhaps dramatically, during the coming years. As a genus, *Fusarium* lacks a large number of morphological characters that can be used to easily differentiate species, and it is not uncommon to "run out of characters before you run out of species." This problem has left numerous species definitions less-than-well defined, has complicated problems of identifying the "real" *Fusarium xxx*, and has resulted in spirited debates between individuals with quite different views of what a species is or should be. Although many morphological species concepts have changed, many have remained stable, even when challenged with genetic and molecular criteria.

The relatively large amount of work done on the morphological taxonomy of these fungi means that as a genus, *Fusarium* often has served as testing ground for new speciation concepts in fungi. The use of molecular approaches to differentiate species has been tried with a number of strains usually considered problematic, *i.e.* not clearly fitting within a given species, but not clearly distinguishable from it either. Studies of phylogenetic lineages based on multiple-gene genealogies, and the grouping patterns resulting from studies with amplified fragment length polymorphisms (AFLPs) both provide new means of evaluating relatedness. As these techniques are independent of one another, each can be used to test the hypothesis of distinctness generated by the other, to look for continuous gradations between representative types, and to identify putative inter-specific hybrids. Studies based on biological species concepts also have been used to delimit some species, most commonly within the *Gibberella fujikuroi* species complex, while also testing hypotheses of sameness/distinctness based on both molecular and morphological characters. Sub-specific differences could be indicative of "evolution in action", suggesting that these fungi are not evolutionary fossils, but entities that continue to change in response to both manmade and naturally occurring selective pressures and for the elucidation of evolutionary mechanisms. Thus *Fusarium* has been and remains an important genus for the testing and development of species concepts in fungi.

For a researcher or diagnostician not interested in state-of-the-art taxonomy and wanting primarily to attach a name to a culture, the apparent nomenclatural turmoil can be confusing and perhaps even threatening. Fortunately, many of the most important pathogens can be diagnosed, with some experience, on the basis of morphological characters. Even isolates not identified to a single species can be assigned to a group of usually closely related species. Accurate morphological characterizations require that the fungus be subcultured and handled with at least some care. Growing the fungus on media appropriate for the morphological features to develop takes time, sometimes several weeks, and means that a careful diagnosis often is a slow process. As molecular studies progress and the definition of common and important species solidify, the development of molecular diagnostics for many species also should be possible. These diagnostics should be much faster than the present morphological diagnoses, as at most limited culturing of the fungus is required. The molecular diagnostics will need to be carefully evaluated on a broad range of species as well as strains within the species to accurately define their diagnostic abilities and limitations. For laboratories that currently lack and are unable to acquire molecular expertise, morphological species definitions will remain the rule. Such scientists should seek partnerships with researchers with access to molecular technologies to confirm critical findings and conclusions, and to ensure that morphological similarities are not disguising significant genetic differences.

This book is designed as a laboratory manual and guide to techniques and species likely to be encountered. It is neither a mycological monograph (a number of species are missing) nor a comprehensive summary of work that has been done with various *Fusarium* spp. (there are > 19,000 citations to *Fusarium* in *Biological Abstracts* alone). It does contain recipes for many commonly used media, techniques that will work for common field and molecular applications, species descriptions of the most common species, and our thoughts on what a *Fusarium* species is and how the field might be evolving. We hope that you will find this manual a useful place to begin a serious study of a particular species or a *Fusarium*-induced disease, or to find a quick answer regarding these most interesting and important fungi.

Techniques and Methods

2

Media – Recipes and Preparation

2.1 Media for Growing and Identifying *Fusarium*

Various agar media have been used as standards on which to grow cultures for the identification of *Fusarium* species. *Fusarium* species have notoriously variable phenotypes when cultured on different agar media, *e.g.*, Burgess *et al.* (280). Morphological features that are common in cultures grown on some media may be absent or altered when the same strain is cultured on another medium. The basic principle is to use the same medium as that used by the authors of the identification guide to which comparisons are being made. Different media may have the same common name, *e.g.*, V-8 juice agar or potato dextrose agar, but vary in their composition. Thus, a standard recipe needs to be followed in preparing media to be used for species characterization. Adding a detergent to the media, *e.g.*, Triton X-100 or tergitol, can restrict colony size and make processes such as colony enumeration or replica plating simpler (1746, 2230). Carnation Leaf-piece Agar (CLA), Spezieller Nährstoffarmer Agar (SNA), and Potato Dextrose Agar (PDA) are the standard media used in the identification of *Fusarium* species in this volume.

Media usually are dispensed into Petri dishes or slants, sometimes termed an agar slope, in a test tube for media with agar, or in flasks for media without agar. Media for use in Petri dishes are made, autoclaved, and then dispensed into sterile Petri dishes after autoclaving. Media containing agar should be made in a flask that is twice the volume of the medium being made, *e.g.*, one liter of medium in a 2-L flask, with a 2-L flask generally the largest convenient size. Cooling the agar to ~50°C before it is dispensed is essential if a heat-sensitive solution, *e.g.*, an antibiotic, is to be added, and reduces the amount of bubbles on the medium surface and the condensate on the dish lid in all cases. The most commonly used Petri dishes are those with a 100 mm or 60 mm nominal diameter. One liter of medium usually suffices to fill 40-50 of the larger dishes (20-25 ml medium/dish), or 80-100 of the smaller dishes (10-12 ml medium/dish). If the medium is for sexual crosses, then more medium is dispensed per dish than if the medium is for a vegetative analysis (see section 2.6).

The same media used in Petri dishes also can be used in agar slants, but media that are viscous, *e.g.*, carrot agar (see section 2.6) or oatmeal agar, generally cannot be used in slants since even the slowest decompression cycle of most autoclaves results in the plugs capping the tubes being blown off or in the media boiling over the top tops of the tubes. A medium to be dispensed into slants is usually made up, melted, and dispensed into the slants with a repeating syringe before the medium is autoclaved. It is very important to ensure good mixing and complete melting of the agar before dispensing the medium; otherwise, some slants may receive too much agar and others too little, which can make the slants difficult to work with. The most commonly used test tube sizes for slants are 10×75, 13×100, and 16×150 mm nominal sizes, which receive 1.25, 2.5 and 6.0 ml of medium/slant respectively. After the medium has been added, the tubes are stoppered with cotton or foam plugs, or covered with commercially available plastic or metal caps. When autoclaving slants it is very important to use a slow decompression cycle to prevent the medium from boiling, as plugs wetted by boiling media render those cultures more susceptible to contamination. In practice, plugs wetted by media should be discarded and replaced with dry sterile plugs (made by autoclaving some tubes that contain no medium). After autoclaving, the tubes must be placed at a slant before they harden to provide additional surface area for culture growth. To get the media at a slant, the 10×75 and 13×100 mm tubes usually are left in the racks in which they were autoclaved, and the entire rack of tubes is slanted at one time. The larger slants, 16×150 mm size, may be slanted in the rack, but this often results in variation in the amount of slanting that actually occurs. A more uniform slant for these tubes is obtained by placing them individually on a slanting board to harden.

Carnation Leaf-Piece Agar (CLA). CLA is a natural substrate medium (625, 2017) prepared by aseptically placing sterile carnation leaf pieces, 3-5 mm², into a Petri dish and adding sterile 2% water agar (20 g agar in 1 L of H_2O). Usually one carnation leaf piece is added per 2 ml of medium. In 60 mm diameter Petri dishes, 5-6 pieces per dish and in 100 mm diameter dishes, 10-12.

The carnation leaf pieces are prepared from fresh carnation leaves free from fungicide or insecticide residue. Immediately after collection, the leaves are cut into 5-8 mm² pieces (they shrink when they are dried) and dried in a forced-air oven (~70°C) for 3-4 hours until brittle. Leaf pieces also can be dried in a microwave oven; the time required varies by oven. The dried leaf pieces are packaged in aluminum or polycarbonate containers and sterilized by gamma irradiation (2.5 megarads). Sterilized leaf pieces can be stored dry at room temperature for up to 12 months before use.

Most species of *Fusarium* sporulate on CLA in 6-10 days. Cultures grown on CLA produce macroconidia that are more uniform in size and form, than do cultures grown on carbohydrate-rich media such as PDA or Czapek's. Macroconidia form primarily in sporodochia, which usually develop on the leaf pieces. Macroconidia formed in sporodochia are preferred for identification as they are more consistent in shape and length than are macroconidia formed from more isolated phialides. Microconidia are more common on hyphae growing on the agar, often away from the leaf pieces. The mode of formation of microconidia, *i.e.* monophialides or polyphialides, the presence of false heads or chains of microconidia, and the presence of chlamydospores can be determined by direct examination with a compound microscope (100×) when strains are grown on small (60 mm diameter) plates of CLA. CLA also is suitable for the production of large numbers of conidia for experimental work, although many researchers also use the modified Czapek's medium (see section 2.5), and can be prepared for this purpose in large flat bottles or larger Petri plates, usually 150 mm nominal diameter.

Gibberella zeae (anamorph – *Fusarium graminearum*), homothallic strains of *Haemanectria haematococca* (anamorph – *Fusarium solani*), and appropriately mixed cultures of heterothallic *Fusarium* species form perithecia on CLA if incubated under light (see section 6.5). Older carnation leaves may give better results than younger carnation leaves for perithecia formation (2154). For genetic investigations, most researchers presently use carrot agar or V-8 juice agar (see section 2.6) to produce perithecia from which ascospores are collected.

Spezieller Nährstoffarmer Agar (SNA). SNA is a weak nutrient agar used for the identification and maintenance of strains of *Fusarium* and *Cylindrocarpon* (1566). SNA is prepared by autoclaving, in 1 L of distilled H_2O:

KH_2PO_4	1 g
KNO_3	1 g
$MgSO_4 \cdot 7H_2O$	0.5 g
KCl	0.5 g
Glucose	0.2 g
Sucrose	0.2 g
Agar	20 g

Placing 1-2 pieces of sterile filter paper (Whatman № 1), approximately 1 cm², on the agar surface after the medium has gelled can increase sporulation.

Culture degeneration, which is common on many synthetic media, usually does not occur on SNA. The medium promotes sporulation and good conidiogeneous cell development; however, because sporodochia formation is limited, macroconidial morphology is not as uniform or as reliable as seen with CLA. Cultures grown on SNA often are of value for examining microconidia, as this medium supports the formation of a range of different microconidia. In some cases microconidia formed on SNA will differ morphologically from those seen on CLA. Many strains also form chlamydospores more readily on SNA than they do on CLA, so for species in which microconidial morphology or the presence of chlamydospores are important identification criteria, cultures from both CLA and SNA should be examined. SNA is transparent, so cultures can be viewed directly with a compound microscope (up to 100×) or small agar blocks can be mounted on a slide with a drop of water and covered with a cover slip for observation at higher magnifications.

Potato Dextrose Agar (PDA). PDA is a carbohydrate-rich medium that contains 20 g dextrose, 20 g agar, and the broth from 250 g white potatoes made up to 1 L with tap water. The potatoes are unpeeled, washed, diced and boiled until soft (actual time varies with the size of the potato). Filter the boiled potatoes through a single layer of cheesecloth, which leaves some sediment in the broth. Commercially available preparations of PDA, *e.g.*, Difco, BBL, *etc.*, are often more convenient and usually are as suitable for identification purposes as home-made preparations.

Conidia formed on PDA are not as consistent in either size or shape as those formed on CLA or SNA, and thus are much less reliable for use for identification purposes. However, colony morphology, pigmentation and growth rates of cultures of most *Fusarium* species on PDA are reasonably consistent if the medium is prepared in a consistent manner, and if the cultures are initiated from standard inocula and incubated under standard conditions. These colony characteristics often are useful secondary criteria for identification.

PDA is used by some researchers for the isolation of *Fusarium* species. We do not recommend this medium for this purpose, as many saprophytic fungi and bacteria also can grow on the medium and interfere with the recovery of the *Fusarium* present. If PDA is used for the recovery of fungi from plant material, then the concentration of potato and dextrose should be reduced by 50-75%, and broad-spectrum antibiotics (see PPA Medium in section 2.3) included to inhibit bacterial growth.

2.2 Supplementary Identification Media

Water Agar (WA). WA (2%) consists of 20 g agar in 1 L of distilled H_2O. This medium is recommended for germinating conidia used to initiate *Fusarium* cultures (see sections 4.1-4.3). As hyphal growth is sparse on this medium, it is suitable for growing cultures from which individual hyphal tips are taken to initiate new colonies (see section 4.3). In some instances, the sparse growth on WA facilitates the isolation of *Fusarium* species from plant material, particularly roots.

WA (0.05%) consists of 0.5 g agar in 1 L of distilled H_2O. WA (0.5%) is used in the preparation of soil dilution series. The small amount of agar retards the sedimentation rates of fungal propagules. The water is heated until the agar dissolves, and the heated medium distributed into flasks, or, more commonly, McCartney bottles. Bottles are capped loosely during sterilization and then tightened after sterilization and cooling to room temperature.

Soil Agar (SA). Chlamydospore formation is enhanced on SA (1094), making this medium useful for the identification of some species of *Fusarium*. SA is prepared by placing 250-500 g of sieved dry soil into a flask and bringing the total volume to 1 L with tap water and autoclaving for 15 min. After the first autoclaving, add 15 g of agar and autoclave for a second 15 min. The amount of soil used varies with soil type. Abundant chlamydospore formation by various species has been observed on SA prepared with 250 g black clay soil, but other soil types also have been used successfully. While pouring plates, the autoclaved medium should be regularly mixed or swirled to ensure even distribution of the solids to all plates.

KCl Agar. When 4-8 g/L KCl is added to WA or to CLA the fungal cultures produce more and longer microconidial chains (626). The chains also are easier to see since there is less moisture on the agar surface and fewer moisture droplets in the aerial mycelium.

2.3 Media for Isolating *Fusarium*

A number of media have been developed for the specific isolation of *Fusarium* spp. Those with the longest history of use are PPA and Komada's, with Komada's usually preferred when the target is *F. oxysporum*, and PPA preferred for most other uses. MGA and RbGu media are variations on PPA that have been widely used by some researchers. MGA may become more prominent as the availability of PCNB decreases. New media are being continuously developed to meet particular needs, *e.g.*, SFM and SFA, or with formulations that are purportedly more selective for *Fusarium* species than were earlier medium, *e.g.*, the medium of Vujanovic *et al.* (2271) which contains myclobutanil as the selective agent.

Peptone PCNB Agar (PPA or Nash-Snyder Medium). The original medium was described by Nash & Snyder (1534), with modifications by Nelson *et al.* (1551). PPA is comprised of a sugar-free basal medium supplemented with antibiotics and fungicides. The base medium contains:

Peptone	15 g
KH_2PO_4	1 g
$MgSO_4 \cdot 7H_2O$	0.5 g

PCNB (Pentachloronitrobenzene)	750	mg
Agar	20	g
H_2O	to 1	L

PCNB is usually added as 1 g of Terrachlor, which contains 75% PCNB (w/w). The pH should be adjusted to 5.5-6.5, if necessary. The streptomycin stock solution is 5 g of streptomycin in 100 ml distilled H_2O, and is used at the rate of 20 ml/L of medium. The neomycin stock solution is 1 g of neomycin sulfate in 100 ml distilled H_2O, and is used at the rate of 12 ml/L of PPA. Streptomycin is effective against Gram-negative bacteria, and neomycin against Gram-positive bacteria.

The medium may be poured into plates onto which soil or diseased plant material is placed, or dispensed into vials into which the material is placed directly. Vials can be particularly useful if samples are taken immediately from field material and then returned to the lab for processing, and as a means to avoid problems associated with the importation of diseased plant material that are usually not encountered with cultures. If plates are being used in soil dilutions (see section 3.3), then it is best if they are allowed to sit for several days to dry before use. This drying time allows the water in the soil suspensions to be absorbed more quickly into the agar gel.

PPA is highly inhibitory to most other fungi and bacteria but allows slow growth of *Fusarium*. It enables the selective isolation of *Fusarium* species from soil dilutions and from rotting plant material that often is infested with other fast-growing microbes. Most species of *Fusarium* do not form distinctive colonies on PPA. Often, sporulation is poor and conidial morphology abnormal. This means that colonies usually must be subcultured before a definitive identification can be made. Although the sporulation is poor, there may be enough spores so that a subculture originating from a single spore can be generated from colonies growing on PPA. Cultures should not be left on PPA for more than 2-4 weeks. The sole nutrient source in the medium is peptone, and its breakdown generates relatively high levels of ammonia that eventually kill the colonies.

Komada's Medium. Komada's medium was developed for the selective isolation of *F. oxysporum* from soil (1115). The basal medium contains:

D-Galactose	20	g
L-Asparagine	2	g
K_2HPO_4	1	g
KCl	0.5	g
$MgSO_4{\cdot}7H_2O$	0.5	g
PCNB (Pentachloronitrobenzene)	750	mg
Fe_3Na EDTA	10	mg
Distilled H_2O	to 1	L

PCNB is usually added as 1 g of Terrachlor, which contains 75% PCNB (w/w). The pH is adjusted to 3.8 ± 0.2 with 10% phosphoric acid, if necessary, prior to autoclaving. The basal medium is autoclaved and cooled to ~50°C before adding filter-sterilized supplemental stock solutions. The streptomycin stock solution is 5 g of streptomycin in 100 ml distilled H_2O, and is used at the rate of 6 ml/L of medium. The Oxgall stock solution contains 5 g Oxgall and 10 g $Na_2B_4O_7{\cdot}10H_2O$ in distilled H_2O, and is used at the rate of 10 ml/L of medium.

Colonies of *F. oxysporum* are distinctly pigmented on this medium, and usually separable from other *Fusarium* species on this basis (272). However, there is enough overlap with some related species, *e.g.*, those in the *G. fujikuroi* species complex, that simple counts of colonies with different morphologies need not be an accurate measure of the *F. oxysporum* levels in the sample. Growth of other *Fusarium* species may be suppressed by the medium, and this medium often is not a good choice for the recovery of *Fusarium* communities that contain species other than *F. oxysporum*.

Malachite Green Agar (MGA). MGA was developed in 1997 by Castellá *et al.* (326) as an alternative to PPA or Komada's medium, and has performed well in some comparison studies with other media (232). This medium contains:

Peptone	15	g
KH_2PO_4	1	g
$MgSO_4{\cdot}7H_2O$	0.5	g
Malachite green oxalate	2.5	mg
Agar	20	g
H_2O	to 1	L

After autoclaving, streptomycin and chloramphenicol are added as for PPA. This medium has not been used as widely as has PPA or Komada's. The only difference between this medium and PPA is that the PCNB is replaced by 2.5 ppm malachite green. This medium probably will become more important as the availability of PCNB (due to its carcinogenicity) declines. MGA also is reported to be more inhibitory of common contaminants, *e.g.*, *Aspergillus* spp. and *Penicillium* spp., without reducing the number of colonies of *Fusarium* spp. recovered than is PPA. The malachite green level in the medium is important, as the level of malachite green (15-50 ppm) used in the Singh-Nene medium (1991), from which this formulation was derived, prevents germination of many *Fusarium* spores.

Another variant on PPA (712) is to replace the PCNB with 1 g/L 2-benzoxazolinone, an antimicrobial compound produced by many maize lines. This medium is particularly effective for selecting for maize pathogens such as *F. graminearum*, *F. subglutinans*, and *F. verticillioides*.

Selective Fusarium Agar (SFA). SFA was developed for the selective isolation of *Fusarium* species from soil debris. SFA is a modified Czapek-Dox medium (see section 2.5) that contains antimicrobial agents (2151).

Glucose (Dextrose)	20	g
KH_2PO_4	0.5	g
$NaNO_3$	2	g
$MgSO_4 \cdot 7H_2O$	0.5	g
Yeast extract	1	g
1% $FeSO_4 \cdot 7H_2O$ (aqueous)	1	ml
Agar	20	g
H_2O	to 1	L

The basal medium is autoclaved and allowed to cool to ~50°C before adding filter-sterilized antibiotic stock solutions. The streptomycin and neomycin stocks are the same as those for PPA, and are used at the same rate – streptomycin stock is 5 g of streptomycin in 100 ml distilled H_2O, and used at the rate of 20 ml/L of SFA, and the neomycin stock is 1 g of neomycin sulfate in 100 ml distilled H_2O, and used at the rate of 12 ml/L of SFA. Dichloran (2,6-dichloro-4-nitroanaline) is prepared as 50 mg of dichloran (or 100 mg Allisan or Botran) in 100 ml ethanol, and used at the rate of 13 ml/L SFA. If dichloran is not available, then PCNB, as used in PPA – 750 mg or 1 g Terrachlor – can be incorporated into the medium prior to autoclaving.

SFA permits the slow growth of *Fusarium* species from plant roots and soil debris, and is less inhibitory than PPA to most fungi. Colonies of different species developing from a single root fragment or piece of debris are easier to differentiate on SFA than on PPA. Although SFA was developed for isolation of *Fusarium* from soil debris, it is not suitable for the isolation of *Fusarium* species from soil dilutions.

Rose Bengal-Glycerine-Urea Medium (RbGU). This medium was developed by van Wyk *et al.* (2224) for isolations from both soil and plant debris.

Glycerol	10	g
Urea	1	g
L-Alanine	0.5	g
PCNB (Pentachloronitrobenzene)	1	g
Rose Bengal	0.5	g
Agar	15	g
H_2O	to 1	L

The basal medium is autoclaved and allowed to cool to ~50°C before adding a filter-sterilized streptomycin solution that is the same as that used for PPA – 5 g of streptomycin in 100 ml distilled H_2O. The medium is similar in selectivity to both PPA and Komada's medium but with the advantage that colony morphology is somewhat easier to distinguish on this medium.

Specific Screening Media (SSM). SSM was developed for the selective isolation of *Fusarium pseudograminearum* and other fungal pathogens from the crown region of wheat plants (2077). SSM contains 10 g dextrose, 20 g agar, and the broth from 125 g white potatoes made up to 1 L with tap water. The potatoes are unpeeled, washed, diced, and boiled until soft. Filter the boiled potatoes through a single layer of cheesecloth, which leaves some sediment in the broth. Commercially available preparations of PDA, *e.g.*, Difco, BBL, *etc.*, may be used at half of the recommended strength, but must be supplemented with an additional 10 g/L agar. After the basal medium has been autoclaved and cooled to ~50°C supplemental stock solutions are added. The streptomycin stock solution is 5 g of streptomycin in 100 ml distilled H_2O, and is used at the rate of 3 ml/L of medium. The neomycin stock solution is 1 g of neomycin sulfate in 100 ml distilled H_2O, and is used at the rate of 6 ml/L of SSM. Dichloran is prepared as 50 mg of dichloran (or 100 mg Allisan or Botran) in 100 ml ethanol, and used at the rate of 13 ml/L of SSM.

SSM permits the formation of distinctive colonies of *F. pseudograminearum*, and suppresses the growth of mucoraceous fungi and *Trichoderma* species. This medium is particularly well-suited for studies of crown rot of wheat in which *F. pseudograminearum* is known to form a morphologically distinguishable colony on the medium. In such studies, *e.g.*, Burgess *et al.* (268), the medium is prepared in large trays to enable a very large number of samples to be processed rapidly.

2.4 Media for the Preparation of Natural Inocula

Chaff-Grain Medium. Inoculum suitable for addition to soil in pathogenicity tests can be prepared by using colonized grain-chaff as a substrate (we have successfully used wheat, barley and oats, but expect that maize, sorghum and rice also would work). Cereal chaff and grain are mixed together in an approximately 5:1 ratio (often a commercial horse feed will suffice).

To a 2-L beaker, add approximately 500 ml of the chaff-grain mixture, and then add tap water to approximately the 1-L level. Mix thoroughly to release any air bubbles and to wet the entire mixture; add additional water to bring the total volume back to 1 L, if necessary. Place the beaker at 5°C for at least overnight, but not more than 24 hours, to leach phenolic compounds. After leaching is complete, cover the mouth of the beaker with a layer of cheesecloth or Miracloth, and invert the beaker on a drainboard to drain the water. After 5-10 min, wrap the cereal-chaff mixture in the cheesecloth and squeeze until no more water can be released.

The drained mixture is distributed into glass jars or Erlenmeyer flasks, which are filled to a depth of approximately 5 cm. The containers are sealed with a large cotton wool plug and autoclaved for 15 minutes on each of two successive days. Containers are inoculated with a conidial or mycelial suspension ($\geq 10^5$ cfu/ml) at the rate of 2 ml of the fungal suspension per 250 ml chaff-grain mixture.

The inoculated material is shaken vigorously to evenly distribute the inoculum throughout the medium. Inoculated material is incubated at 25°C until the material is completely colonized. The length of the incubation depends on the rate of growth for the fungus. Usually 7-14 days of incubation is sufficient. These cultures should not be maintained for more than 21 days, as viability begins to decrease. Cultures should be shaken daily for the first 3-4 days to encourage more rapid and uniform colonization of the substrate. Once the substrate is completely colonized, it is removed from the container and air dried at room temperature (20-25°C) overnight (perhaps somewhat longer if the relative humidity is high). When dry, then material is crushed to the required size (usually particles pass through a 2 mm aperture sieve) for addition to soil. The dried crushed substrate may be stored for up to 12 months at 2-5°C.

In pot cultures or disease nurseries, the inoculum often is mixed with soil at a rate of 1-2% of the final total volume before planting. In other cases, *e.g.*, Liddell *et al.* (1255), the inoculum is placed as a layer above the seed and then covered with soil, and the seedlings grow through the inoculum layer. Infection may occur at this time, or it may be delayed by withholding moisture from the layer that contains the inoculum.

This inoculum is particularly appropriate for species that do not form chlamydospores, but that persist in soil as hyphae in plant residues. This procedure has been used with numerous *Fusarium* species, *e.g.*, Liddell (1253) and Liddell *et al.* (1255), including *F. crookwellense*, *F. culmorum*, *F. oxysporum*, *F. pseudograminearum*, *F. solani*, and *F. verticillioides*. It also is suitable for studies with a number of fungal genera, *e.g.*, Summerell *et al.* (2081), including *Cylindrocarpon*, *Pythium*, *Rhizoctonia*, and *Sclerotium*.

2.5 Synthetic and Semi-synthetic Media

Synthetic and semi-synthetic media often are used for studying physiological characters, for isolating DNA, and for genetic studies. The media in this section are all based on the Czapek's Dox medium recipe, as modified by Correll *et al.* (402). The basal medium contains:

KH_2PO_4	1 g
$MgSO_4 \cdot 7\,H_2O$	0.5 g
KCl	0.5 g
Agar (Bacto)	20 g
Trace element solution	0.2 ml
Distilled H_2O	to 1 L

It is critical to use distilled water in this medium to prevent unintentional contamination of the medium with salts that may adversely affect growth characteristics or trace element levels. Agar should be at least of bacteriological grade. Higher levels of agar purity occasionally are warranted, *e.g.*, see Nitrite medium below, but are not necessary for routine work. Lower grades of agar may have trace amounts of vitamins, nucleic acids, and/or amino acids that may make it difficult to score auxotrophic mutations clearly. For work with liquid cultures, the agar is omitted completely.

The trace element solution supplies critical nutrients required for some enzyme activities, *e.g.*, Mo is essential for nitrate reductase activity, and should not be omitted. Note that all *Fusarium* species require iron, zinc, manganese, copper, molybdenum and boron for proper vegetative growth and sporulation, but that excessive amounts of these trace elements may inhibitory (1310). The trace element solution contains:

Citric acid	5 g
$ZnSO_4 \cdot 6H_2O$	5 g
$Fe(NH_4)_2(SO_4)_2 \cdot 6H_2O$	1 g
$CuSO_4 \cdot 5H_2O$	250 mg
$MnSO_4$	50 mg
H_3BO_3 (Boric Acid)	50 mg
$Na_2MoO_4 \cdot 2H_2O$	50 mg
Distilled water	95 ml

This recipe makes a nominal 100 ml. The citric acid should be added to the water first and dissolved completely to prevent other salts from precipitating. The solution may be filter-sterilized, although this is not necessary. More commonly this solution is stored in a capped bottle unsterilized at 4°C indefinitely following the addition of approximately 1 ml of $CHCl_3$ to the stock solution. The $CHCl_3$ forms a small ball at the bottom of the container and it should be possible to detect its presence by smelling the solution in the bottle. If during storage the smell disappears or the ball of $CHCl_3$ shrinks significantly or is no longer visible, then the stock solution is probably no longer saturated with $CHCl_3$, and additional $CHCl_3$ should be added. A monthly check usually suffices.

Additional ingredients are added to the basal medium (see below), depending upon the experimental protocol, prior to autoclaving and use. Adding a drop or two of vegetable-dye based food coloring, available at the local grocery store, to media before autoclaving allows media types to be distinguished based on color rather than by relying on stripes on the side of a plate or differences in the plugs for slants to make these distinctions. The food coloring does not affect the morphology or growth rates of *Fusarium* spp.

Minimal Medium (MM). This medium is the standard Czapek's-Dox formulation as modified by Correll *et al.* (402). This medium is made by adding 2 g/L $NaNO_3$ and 30 g/L sucrose to the basal medium prior to autoclaving. Most *Fusarium* species grow well vegetatively on this medium, but sporulation characters vary, which makes it unsuitable for routine taxonomic studies. As with PDA, the high level of sugar can cause problems with character degeneration in some strains. In studies of vegetative compatibility, this medium functions as the NO_3 test medium. This medium works well to identify strains with any nutritional auxotrophies (they will grow poorly if at all), and is suitable as a liquid medium for studies of enzyme activity and for the isolation of nucleic acids.

Complete medium (CM). This medium is a semi-synthetic version of the Czapek's-Dox medium that we routinely use in our laboratories. To the basal medium the following additions are made prior to autoclaving:

Sucrose	30	g
$NaNO_3$	2	g
N-Z Amine	2.5	g
Yeast extract	1	g
Vitamin stock solution	10	ml

N-Z Amine is an amino acid digest that is used as a source of amino acids and fixed nitrogen. Other hydrolytic protein digests can be substituted for N-Z Amine at a similar level of supplementation. We routinely use Bacto (Difco, Detroit, Michigan) Yeast Extract, but other yeast extracts should perform similarly. Yeast extracts vary somewhat from lot to lot. This variation can be particularly important for methionine and adenine auxotrophs, and samples from several lots should be tested to make sure that an adequate level of the supplement is present before embarking on extensive studies of such mutants with this medium. The vitamin stock solution is composed of:

Inositol	4	g
Ca pantothenate	200	mg
Choline•Cl	200	mg
Thiamine	100	mg
Pyridoxine	75	mg
Nicotinamide	75	mg
Ascorbic acid	50	mg
Riboflavin	30	mg
p-aminobenzoic acid	5	mg
Folic acid	5	mg
Biotin	5	mg
50:50 ethanol:H_2O	to 1	L

The vitamin solution may be filter-sterilized, although this is not necessary. More commonly though, this solution is stored indefinitely in the dark (dark bottle or bottle wrapped in aluminum foil) in a tightly capped bottle unsterilized at 4°C following the addition of 1-2 ml of $CHCl_3$ to the stock solution. The $CHCl_3$ forms a small ball at the bottom of the container and it should be possible to detect its presence by smelling the solution in the bottle. If during storage the smell disappears or the ball of $CHCl_3$ shrinks significantly or is no longer visible, then additional $CHCl_3$ should be added to maintain a saturated solution. A monthly check usually suffices. Storage in the dark is important as some of the vitamins decompose when exposed to light. Choline and pantothentic acid must be added as salts to maintain pH and to increase solubility. Nicotinic acid is added as nicotinamide for the same reason. Some researchers add additional vitamins to this mixture, but we have not observed any cases in which such additions have significantly affected experimental results.

Virtually all *Fusarium* species grow well vegetatively on CM, but sporulation characters vary, which makes it unsuitable for growing cultures for routine taxonomic studies. As with PDA, the high level of sugar can cause problems with character degeneration in some strains. The availability of fixed nitrogen means that *nit* mutants (see section 5.3) can revert without a noticeable change in phenotype. *nit* mutants isolated from sectors on chlorate-containing medium usually must be grown on CM for preservation or to obtain large numbers of spores, but in general their culture on this medium should be as limited as possible. Auxotrophic mutants that require amino acids not represented in casein usually grow poorly on CM, as they rely solely on the amino acids in the yeast extract for growth. Arginine is almost always limiting, and, if required, must be added as a separate supplement. Virtually all other *Fusarium* strains, both prototrophs and auxotrophs, grow well on this medium.

Chlorate medium. This medium was designed for the recovery of nitrate non-utilizing (*nit*) mutants for use in vegetative compatibility tests (402, 1745). To the basal medium the following additions are made prior to autoclaving:

Sucrose	30 g
$NaNO_3$	2 g
L-asparagine	1.6 g
$KClO_3$	15 g

A common level of chlorate in the medium is 1.5% (the level reflected in the recipe). This level is usually the minimum chlorate level that is effective for generating *nit* mutant sectors. This recipe has numerous variants. If too few NitM sectors are recovered, then changing the fixed nitrogen source from asparagine to proline may help (1090). If colonies seem to grow slowly across the plate without generating sectors, then omitting the asparagine often helps to restrict colony growth sufficiently to permit the identification of *nit* sectors. The chlorate level also can be increased, with 2-3% chlorate commonly used in many labs. The maximum chlorate level is 6% (60 g $KClO_3$/L). Plates with high levels of chlorate dry easily and have a feathery pattern of crystal formation in the medium. Such plates should be discarded. A final alternative is to drop the pH of the medium, but not so far as to prevent the agar from gelling. Conditions developed for generating *nit* mutants in *Fusarium* often, but not always, work with other ascomycete fungi *e.g.*, *Colletotrichum* (250), *Aspergillus* (156, 557), and *Verticillium* (1126). The media developed for studies in *Aspergillus* can be particularly useful for strains that sector relatively poorly on the chlorate media described here.

Phenotyping Medium. Phenotyping medium was developed by Correll *et al.* (402) to distinguish the mutations that affect different portions of the nitrate assimilation pathway (see section 5.3). There are three forms of phenotyping medium that are named according to the nitrogen source that is added to basal medium before autoclaving: ammonium (1.6 g/L ammonium tartrate), hypoxanthine (0.2 g/L hypoxanthine), and nitrite (0.5 g/L $NaNO_2$). All contain 30 g/L sucrose as the carbon source. The ammonium medium is the control. Strains that will not grow on this medium have another nutritional requirement that needs to be diagnosed. Hypoxanthine medium is used to differentiate NitM mutants, as these mutants are unable to utilize hypoxanthine as a sole nitrogen source due to a defective molybdenum co-factor that is shared by nitrate reductase and purine dehydrogenase (1376). Nitrite medium can be problematic. Nitrite is potentially toxic to the cells. If they are supplied with too much NO_2 they will not grow due to NO_2 toxicity, but if there is not enough NO_2 then they will not grow due to N insufficiency. In some cases, replacing the bacteriological agar in nitrite medium with Noble agar will make it easier to score the growth on this medium.

Media to Restrict Radial Growth of Colonies. Minimal medium + tergitol and sorbose (MMTS) and Complete medium + tergitol and sorbose (CMTS) were modified by Bowden & Leslie (230) from the original recipe of Puhalla & Spieth (1746) to restrict radial growth of colonies on agar plates, and to enable the isolation of numerous colonies from a single plate following a selective treatment, *e.g.*, a mutagenesis experiment. These media are made by substituting 30 g/L sorbose for the 30 g/L sucrose in the minimal or complete recipes given above, and adding 0.5 ml/L tergitol (type NP-10). If colonies in a more "pelleted" form are desired from a liquid culture, then similar changes can be made to the liquid media (1216). The sorbose level used can range from 2-6% (20-60 g/L), and tergitol from 0.05-0.2% (0.5-2 ml/L). Either compound can be used alone, but a higher concentration of the single compound is required to obtain the same net effect on colony diameter. In some cases a small amount of other sugars, *e.g.*, 0.2% dextrose and 0.2% fructose, also may be added to the medium to support initial growth and spore germination. Both interspecific and intra-specific variation for sensitivity to tergitol and sorbose are known, and the level of sorbose and/or tergitol used in the medium may need to be adjusted to reflect this variation.

2.6 Media for Sexual Crosses

Obtaining a sexual stage is essential for analyses of biological species in some heterothallic species. Many species will form perithecia and mature ascospores on CLA. Some investigators also have used cultures grown on various natural substrates, *e.g.*, rice straw (905) or mulberry twigs (469) on water agar or PDA, to obtain the sexual stage. For studies of development or for genetic studies, however, more perithecia with a larger number of ascospores usually are required. Two media commonly are used for this purpose, Carrot agar and V-8 Juice agar. Fertility on either medium occasionally can be increased by placing a piece of sterile filter paper (Whatman №. 1) over a portion of the plate prior to inoculation, and perhaps impregenating it with a drop or two of linoleic acid (535).

Carrot Agar. Carrot agar was developed for fertility studies of *Fusarium* species with a *Gibberella* sexual stage (1090). The medium is prepared by washing, peeling and dicing 400 g of fresh carrots. Place the carrots in a flask in 400 ml of water and autoclave for 20 minutes.

Blend the carrots in a blender until the mixture appears smooth and with no large lumps. Add an additional 500 ml of H_2O, using some of the water to rinse the blender, and 20 g of agar. Autoclave for 30 minutes. Both autoclaving steps and the length of the second one are important to avoid contamination of plates by soilborne bacteria with heat-resistant spores. Plates should be poured thick (15-17 ml per 60 mm diameter dish), as they may be incubated for up to 6 weeks. Plates that are a normal thickness usually dry, the agar curls, and meaningless, if any, results are obtained. Some species, *e.g.*, *G. nygamai* (1085), are reported to be more fertile on a half strength medium (200 g carrots) than on the full strength medium, but there has been no systematic testing of this effect for most of the species with known sexual stages.

V-8 Juice Agar. This medium commonly is used to elicit the formation of the sexual stage of many *Fusarium* species. In some cases this medium also is used for routine maintenance of vegetative cultures. The medium is commonly made by mixing 300 ml of V-8 Juice (Cambell Soup Co., Camden, NJ) with 700 ml water, and 20 g of agar. Before addition of the agar, the pH of the juice should be adjusted to 5.5-6.5 using 1 M NaOH. There are many variants of this basic formula, see Tuite (2189) for numerous examples. These variations usually have to do with whether (and how much) $CaCO_3$ is added to the medium, and whether and how the pH is adjusted. We have found that for fertile crosses, V-8 juice agar usually gives good results, but for crosses that are less-fertile, carrot agar usually is more consistent and more productive.

2.7 Sterilization of Media and Materials

Sterilization of media and utensils is essential to destroy or kill all the living organisms. The method of sterilization depends primarily mainly on the physical nature of the material to be sterilized.

Heat Sterilization. The temperature and time required for killing are inversely related, *i.e.* the higher the temperature the shorter the period of time for which the material must be exposed to it. Both the temperature and the length of the exposure (Table 2-1) also depend on the nature of the heat applied – moist, *e.g.*, with an autoclave, or dry, *e.g.*, in an oven. Exposing materials to these conditions is not a guarantee of sterility, but as a general guideline moist heat, applied in an autoclave at 121°C for 15 minutes, and dry heat, applied in an oven at 160°C for one hour, usually suffices. More heat-resistant microbes, however, may require longer exposures before they are killed. Such microbes usually are rich in organic substances, with fat composition especially important in resisting killing due to moist heat.

Table 2-1. Times of exposure to moist and dry heat that result in sterilization of media and equipment.

Temperature °C	Moist Heat	Dry Heat
100	20 hr	
110	150 min	
115	50 min	
121	15 min	8 hr
125	6.5 min	
130	2.5 min	
140		150 min
160		1 hr
170		40 min
180		20 min

Moist Heat. Moist heat generally is applied along with an increase in pressure to keep liquid materials from boiling and is usually administered in a pressure cooker, autoclave, or other containment vessel that can withstand the additional pressure generated. Moist heat kills microorganisms by coagulating and denaturing enzymes and structural proteins, and requires that the steam contact the microorganism. If the microbes are protected from the steam by a layer of grease or oil, then a longer exposure is required for effective sterilization. All culture media are sterilized with moist heat, with those containing large amounts of lipids usually being autoclaved for a more extended period of time. Items exposed to saturated steam usually reach the equilibrium temperature much more quickly than do those exposed to dry heat. Although 121°C for fifteen minutes in the autoclave is the most common treatment, some materials are not amenable to the temperature, the pressure, or the setting, *e.g.*, large lots of soil or plant material, and often can be effectively sterilized following exposure to a normal steam for 30-45 min on each of several (usually at least three) successive days.

Autoclaves must be adjusted to prevent the chamber pressure from falling too rapidly as this will result in media boiling and wetting the plugs (and to contamination of the media before or during its use). Leaving media in the autoclave for 5-10 min after it has returned to atmospheric pressure is usually a good safety precaution, as superheated liquids can boil over causing both burns and large messes when disturbed. Materials should not be left in an autoclave for an extended period following the completion of the cycle, however, because a vacuum usually builds up that both makes the autoclave difficult to open and leads to loss of the liquid volume. Sterilizing large and small volumes of media in the same run also should be avoided wherever possible, as the extra time required for a large volume of liquid to reach the sterilization temperature may result in too long of an exposure for some of the media present only in a small volume.

pH also is an important variable when moist heat is applied since the resistance of bacterial spores is highest at pH 7.0 and decreases as either the acidity or alkalinity increases. On a practical level, this means that metal instruments boiled in water containing 2% Na_2CO_3 for 10 min. have been as effectively treated as those boiled in plain water for several hours.

Dry Heat. Dry heat kills microbes by oxidation and is preferred for dry glassware such as test tubes, Petri dishes, flasks, pipettes, syringes and instruments such as forceps, scalpels, and scissors. Dry heat also is used to sterilize dry materials in sealed containers and powders, fats, oils and greases that are impermeable to moisture. These materials may exhibit significant mass transfer effects with respect to heat transfer, *i.e.*, it may take a relatively long time for all of the material to reach the ambient temperature, and they are, therefore, best sterilized as several smaller amounts, rather than as one large one. For example, an oven packed with glassware may require 2-3 hours to reach 160°C, and this length of time must be added to the time required for sterilization.

Sterilization by Filtration. Microbes also can be removed from liquids by means of filters with very small pores that trap contaminating yeasts, bacteria and spores. This method is used for sterilizing items such antibiotic and vitamin solutions that are heat labile. Most filter sterilization is done with membrane filters. Membrane filters are thin, porous sheets composed of cellulose esters or similar polymeric materials. All particles exceeding the pore size, which range from 0.22-14 μm depending on the type of filter purchased, are retained by the filter. The fibers defining the pores generally are resistant to water, dilute acids and alkalis, aliphatic or aromatic hydrocarbons, and non-polar liquids of interest to most microbiologists.

Normally a 0.22 μm membrane filter should be used for filter sterilization, as these filters will retain all known bacteria. Reusable membranes that are sterilized, usually by autoclaving after mounting in an appropriate holder, may be purchased; however, disposable units that come sterile and ready to use are common and convenient. The liquid to be sterilized usually is allowed to flow through the filter by gravity, or by positive pressure, *e.g.*, when a solution is forced through a filter that is mounted on a syringe. If the suspension being sterilized contains a great deal of suspended particulate matter, then filtration through glass wool or some other prefilter may be necessary.

Sterilization of Instruments During Use. Forceps, inoculating needles and other instruments must be sterilized before and after contact with a culture to avoid cross-contamination. Inoculating needles are readily sterilized by heating to red hot in a flame. The needle must be cooled to room temperature before being used to make a transfer. A needle that is too hot is one of the most common reasons for unsuccessful subculturing. Forceps and scalpels are sterilized by standing them in a beaker or similar container of alcohol. Before use, the alcohol is burnt off by passing the instrument through a flame and igniting the alcohol. The instrument should not be held in the flame immediately before use as this will heat it up too much. Do not place a hot or flaming instrument in or near the alcohol, since it is possible to ignite all of the alcohol in the container. As a safety measure, have a watch glass or other item near at hand that can be used to cover the mouth of the container containing the alcohol in case the alcohol catches fire.

Sterilizing Work Surfaces. Trays, benches and other surfaces may be sterilized with any commercially available liquid disinfectant. Alcohol also is commonly used and works best as a sterilant when mixed with water; a solution of 70% ethanol is suitable. Hypochlorite (bleach) solutions, *e.g.* 1% sodium hypochlorite, also are very effective, but may have an offensive odor or leave a solid residue and are harmful to clothing and many other common laboratory materials.

3

Techniques for Recovering *Fusarium*

Fusarium is common in both agricultural and native environments, but before strains can be identified, characterized and studied, they must be recovered from the environment and subcultured in such a manner that other microbes that might accompany them have been eliminated. Following that, a single-spore subculture (Section 4.3) is needed to ensure that only a single strain is being examined, as many natural sources of *Fusarium* contain genetically distinct strains of the same and different species. The collection strategy and recovery techniques used depend on the question(s) being asked. A strategy suitable for identifying the causal agent of a plant disease could be quite different from one in which the goal was an inventory of the *Fusarium* species present on a native grassland or determining if there were *Fusarium* strains associated with a "sick building." This chapter provides a few general guidelines towards the design of collecting strategies, but is neither a complete nor all-inclusive guide to these strategies, which are heavily dependent upon the question being addressed.

3.1 Collecting Strategy

This section presumes a sampling strategy in which a number of samples are needed from the same location for an evaluation of some sort. The most important properties of such a sample usually are its size and its representtativeness of the *Fusarium* species/strains present in the original source.

Sample size is an important character. Normally the sample should be large enough to ensure that all of the most common species/strains were recovered with some statistically significant degree of confidence. Assuming a random distribution of the species/strains within the sample, *i.e.*, that the sample is a representative one, these values are calculated from a binomial/multinomial equation.

If a particular strain is present at frequency a in the sample, then the probability that this strain would be selected on any individual draw from the sample is a, and the probability that it would not be selected is $1 - a$. If x draws are made from the sample, *i.e.*, the sample size, then $(1 - a)^x$ is the probability that none of the draws will be of the strain in question. Thus, the lower the value for a the larger the sample will need to be to be certain that a strain representing the group has been recovered. Statistically significant frequencies for items in a sample often are considered to be either 5% or 1%. The certainty of recovery, $1 - (1 - a)^x$, usually is set at 95% or 99%. Given a value for a or x, the equation can be easily solved for the other variable. To be 95% certain of seeing a strain present at a frequency of 5%, the sample size should be 59, and to be 99% certain the sample size should be 90. To be 95% certain of seeing a strain present at a frequency of 1%, the sample size should be 299, and to be 99% certain the sample size should be 459. The application of these numbers requires some discretion. For example in a sample of 59, which has 95% certainty of identifying all species/strains present at the 5% level, the probability of missing at least one of the species present at the 1% level is 50%.

The target of the sampling also helps determine the sample size. For example, suppose the goal of the sampling is to obtain 59 cultures of a particular species. If that species is the only one present, then a sample size of 66 should suffice for analysis at the 5% level with 95% confidence, assuming that 90% of the items sampled will yield a fungal culture. If cultures of the target species can be recovered from only 50% of the items sampled, however, then the sample size will need to be proportionately increased to ensure that enough cultures of the target fungus have been obtained for the results to be statistically meaningful.

Comparing samples from different locations often is a goal of a project that requires sampling. In such cases the samples are most useful if they are taken in a hierarchical manner. The smallest unit usually is a single plant or a single lesion on a plant. From there the progression usually is to a "field", to a town, village or county, to a state or other larger district, etc. The number of layers in the hierarchy usually is determined by the known or suspected distribution of the organism of interest and the practical limitations on the total number of samples that can be analyzed. As a general rule, more samples are generally preferred to fewer, but once a sample is "large enough", increasing it further may add little additional statistical power to the argument. If only a relatively few samples can be analyzed, then the locations to be sampled need to be chosen with special care to ensure that the results obtained can support the inevitably broad inferences that will be made from them. Unless the questions asked include, "Is there more than one species/strain per plant?", we recommend that only a single sample be taken per plant. Many plants are infected with multiple strains of *Fusarium* from the same or different species, *e.g.*, Kedera *et al.* (1029), and Leslie *et al.* (1226). In such cases if a second recovery is made of the same strain/species its independence from the first recovery cannot be critically assessed. Thus, multiple isolates from the same plant should not be assumed to be independent events and probably are better handled in a qualitative, rather than a quantitative manner.

When collecting in a more-or-less uniform setting, *e.g.*, a monoculture agricultural field, the spatial pattern associated with the sample may be of interest as well. A rectangular grid often is used in settings where there are a series of parallel rows. The sample often is taken in a diagonal manner beginning at an arbitrary point in the field, and then moving over one row, and down the row the width of the row. The process can be continued for the width of the field, the length of the row, or until the sample size is large enough. If the samples at the sampling site are not arranged in rows, then setting up a grid based on rectangular co-ordinates can be time-consuming and at least somewhat frustrating. Under these conditions, using a polar grid for sample collection can be much more effective and time efficient.

To set up a polar grid, a central stake is needed, as is a protractor (or some other means of measuring an angle), a compass, and a string or rope in which knots have been tied at the distances from the centerpoint that are to be sampled. By sampling from the centerpoint, and from 12 points on each of five concentric circles around that centerpoint, a set of 61 isolates can be obtained. The central stake is driven into the ground, and a sample is taken immediately adjacent to it. Using the compass, the string is extended out sufficiently far to the north and samples are taken where the knots fall on the ground. The string is moved clockwise an angle of 30 degrees and another series of five points sampled. The process continues until the entire circle (with 12 samples per circle) is complete. We have used circles with a radius of 1, 4, 8, 10 and 12 m in our studies, but any five distances could be used.

3.2 Isolation Techniques – Plants

Isolation protocols have a significant impact on the recovery of the *Fusarium* species from diseased plants (1411). It is common to recover more than one species of *Fusarium* from individual pieces of diseased plant material, but some species may be pathogens, while others are saprophytes or endophytes that may have a role as secondary pathogens or which may have no role in the disease process whatsoever. In plant pathology settings, the isolation technique selected should maximize the recovery of the "true" pathogen while restricting the recovery of "weed" fungi. This problem can be particularly severe in intensively cultivated soil (1258). A number of *Fusarium*

Table 3-1. Species of *Fusarium* regularly recovered from various parts of diseased plants as saprophytes.

Plant Part	*Fusarium* Species
Roots & stem bases	*F. acuminatum, F. avenaceum, F. compactum, F. equiseti, F. proliferatum, F. oxysporum, F. solani*
Leaves & aerial parts	*F. proliferatum, F. semitectum*
Flowers	*F. semitectum*
Seed & grain	*F. chlamydosporum, F. equiseti, F. poae, F. semitectum*

species commonly are recovered as "weeds" (Table 3-1), both from plant parts and soil, so the identification of the causal agent always requires some care. If in doubt, and where possible, Koch's postulates should be repeated with questionable strains, but such analyses usually are far beyond the scope of routine, utilitarian laboratory diagnoses. Note also that strains that can synthesize mycotoxins or other noxious metabolites may be important economically or from a health standpoint even if they cause no plant disease. If mycotoxin contamination risk assessment is part of the objective of the study, then surface sterilizing all of the plant material from which isolates are recovered may lead to erroneous conclusions.

Working from fresh and freshly diseased plant material usually reduces the number of secondary pathogens and saprophytes that are recovered from a sample. The longer the plant has been diseased, and the longer that the sample has been awaiting analysis the more likely it is that fungal species other than the pathogen, including other *Fusarium* species, will have invaded the diseased area. This problem is particularly acute for soilborne diseases, such as wilt diseases and root rots. Samples from plants with these diseases should be expected to yield a range of fungi other than the original causal organism Amongst *Fusarium* species, *F. equiseti* and *F. semitectum* are saprophytic colonizers regularly recovered as secondary invaders of diseased tissue. *F. oxysporum* and *F. solani* are common soilborne fungi and frequently present in samples from diseased roots even if they do not cause the reported disease. It is not unusual for these fungi to be reported as pathogens or causal agents of diseases based simply on their recovery from diseased material. Repeating Koch's postulates with such strains is essential to prevent the confusion that can result when a disease is attributed to the wrong causal agent. The frequent reporting of species such as *F. equiseti* and *F. semitectum* as pathogens of a broad range of plant species is consistent with the hypothesis that strains of these species have limited pathogenic capabilities but are widely distributed saprophytes. Thus, reports of the pathogenicity of these species should be viewed with caution unless supported with data that go beyond their recovery from diseased plant material.

The type of disease often dictates the isolation procedure used. The most obvious part of the plant from which to recover the fungus is the part of the plant affected by the disease. This task, however, must be approached with an open mind, as some symptoms can be a response of the plant to infection at another point. For example, the bakanae disease of rice (2298) is caused by excessive gibberellic acid production by *F. fujikuroi*, and the pathogen may be found anywhere in the plant (not necessarily at the crown or the growing tip). As a general rule, however, if there is a lesion or disease focus it is best to attempt to isolate the fungus from the leading edge of this region. Often this point is the one at which the fungus is most likely to be actively growing and easily recovered. Careful examination of the diseased plant also may indicate the best isolation procedure. Examine plant material both before and after washing it, as a thorough wash can dislodge diagnostic characters, *e.g.*, sporodochia or perithecia, on the outside of the stem. Cut stems/leaves/lesions and examine cross sections or cuts through the lumen of the stem to determine if vascular browning has occurred or if hyphae are present in the lumen. Such symptoms can be diagnostic for vascular wilts caused by *F. oxysporum*, or for crown and stem rots of wheat and barley caused by *F. pseudograminearum* or *F. culmorum*.

In some situations the pathogen causing the disease is well-defined, and the question being asked is one of quantity, *i.e.*, how frequently is the pathogen recovered, rather than quality, *i.e.*, to which species do the pathogenic isolates belong. In such cases, specialized isolation techniques often are developed that permit recovery of a pathogen from numerous samples, while still providing enough information to correctly identify the pathogen. These techniques usually are developed in response to a need to be able to quickly identify and quantify the amount of a particular species or set of species. For example, with the wheat crown rot pathogen, *F. pseudograminearum*, there is a selective medium (SSM, Section 2.3) on which many fungi and bacteria cannot grow, but the fungal pathogen can be identified tentatively by colony morphology. These identification/recovery schemes always must be validated through a more detailed examination of a randomly selected subset of isolates recovered on a medium such as PPA that does not bias recovery of one or more *Fusarium* species. These identification processes are essential for the relatively large scale experiments needed for some bioclimatic modeling studies, *e.g.*, Backhouse & Burgess (113).

Once cultured, the value of the culture must be determined, *i.e.*, is the isolate cultured of sufficient import that its analysis is worthy of additional time and material resources? The quality and quantity of the plant samples available usually guide this decision. A culture from a fresh, representative diseased plant usually is much more valuable than a single isolate from degraded tissue or other suspect source. It is not uncommon, unfortunately, for a client to expect a plant diagnostician to perform a miracle by recovering only the causal agent and producing an unquestionable diagnosis on the basis of a limited sample of very suspect material. In such cases, experience plays an important role in the diagnostic decision, and the formal tasks of exhaustive isolation, strain recovery and subculturing often are replaced by a more pragmatic decision-making process grounded in the familiarity of the diagnostician with hosts, diseases, and the frequency with which proven pathogenic organisms have been recovered from similar samples. The combination of host plant, the part of the plant from which the strain was recovered, climatic region, local weather data, and isolation protocol often greatly reduce the number of *Fusarium* species expected to be recovered. In at least some cases a tentative diagnosis can be made on this information alone, with examination of cultural morphology serving to confirm the inferred result, rather than as the primary means of identification.

3.3 Isolation Techniques – Soil

Fusarium species can be directly isolated from soil with a dilution plate technique, by isolation from organic debris extracted from soil, or by baiting from soil using a living root system bait or some form of sterile organic bait, *e.g.* sterile cereal straw. None of these techniques are definitive in recovering all species of *Fusarium* from a soil and it is important to select a technique that is likely to provide answers to the question(s) being asked.

Sampling and collecting soil. There are numerous variables to consider when sampling soil. The distribution of *Fusarium* hyphae and spores in soil usually is irregular both spatially and temporally. The spatial distribution usually varies across a particular site and down the soil profile. In general, propagule abundance decreases with soil depth, with samples from depths > 30 cm from the surface yielding relatively little fungus or information. For many species propagule distribution reflects the distribution of vegetation, especially for those species that have a close association with a particular plant species. Seasonal variations in abundance also may occur, with temperature and moisture (usually estimated by the amount of recent rainfall) two commonly important environmental factors. We usually collect soil from plant rhizospheres because these areas often contain a wider range of species. In arid and semi-arid regions, the difference between the rhizospheres and barren soil may be even more prominent.

It is not possible to control the amount of moisture in field soil when collecting, so this parameter must be standardized when samples are returned to the laboratory. In most cases we store soil samples in paper bags. Such storage allows the soil samples to dry prior to processing, however if the soil sample is too wet then a plastic bag may have to be used. We do not like to use plastic bags for soil samples because these bags do not allow the soil sample to dry and may encourage bacterial growth that can complicate the recovery of *Fusarium* isolates. Our standard technique is to air-dry the soil sample in a laminar flow hood for 24-48 hrs, and then to grind the sample in a mortar with a pestle. The objective of the grinding is to generate soil particles of a consistent size for use in soil dilutions, but not to alter recovery *per se*.

Soil dilution plate technique. This technique involves the uniform distribution of a soil suspension across a medium selective for the isolation of *Fusarium* species (section 2.3). The concentration of the soil suspension depends on the relative number of *Fusarium* isolates in the soil, but usually is between 1:50 and 1:2000. The dilution medium is 0.05% WA (section 2.2), which while still liquid is viscous enough to be evenly distributed across a plate of selective medium. A common protocol is to use 1 ml of the soil suspension for a 100-mm diameter Petri dish. The suspension is transferred with a pipette and spread evenly across the agar surface. Even distribution of the soil suspension may be accomplished by carefully manipulating the plate with gentle agitation to move a wetting front across the entire agar surface or by using a glass "hockey stick" applicator to spread the suspension. Petri dishes containing agar to be used with soil dilutions should be allowed to dry for 3-5 days before use. The drier media more quickly soaks up excess water in the suspension, which helps minimize bacterial contamination. The number of colonies per plate depends on the number of propagules in the soil and the dilution factor. An ideal number is 20-30 colonies per plate, which allows for relatively easy differentiation and subculture of the isolates. Plates should be incubated under lights to increase color formation and the development of other diagnostic characters. Colonies usually must be subcultured from the isolation medium before they can be identified as most isolation media, but especially PPA, are somewhat toxic to the fungus and the resulting colonies have atypical and/or inconsistent spore morphology and other diagnostic features. For example, most strains of *Fusarium* will die on PPA before they have a chance to form chlamydospores.

Soil dilution techniques will only recover a subset of any fungal group to which it is applied. Species that do not form long-term survival structures, *e.g.*, chlamydospores (section 12.3), are much less likely to be isolated from soil with this technique unless they are present in small pieces of suspended debris. Thus, species that form chlamydospores are more likely to be recovered than those that do not, and species that produce numerous chlamydospores are more likely to be recovered than those that produce relatively few.

Debris isolation technique. This technique uses the same media as used for soil dilutions (Section 2.3), but pieces of plant debris are removed from the soil prior to grinding in the mortar and pestle. The soil sample is washed through a sieve, or nest of sieves (4 mm, 2 mm, and 0.5 mm aperture). The length of time for washing varies and depends on the soil type. The debris remaining on the sieves is dried over silica gel or in a laminar flow hood. Once dry, the debris is transferred to a selective medium, usually 10-12 pieces per plate. The debris, especially the smaller pieces, should not be surface sterilized with either ethanol or a dilute beach solution because the debris pieces usually are so small and porous that the surface-sterilization technique kills most or all of the fungi associated with the debris – both on the surface and in the interior of the debris pieces. Recovering isolates from soil debris usually expands the spectrum of species recovered by including species that are endophytes or live within the plant tissue but that do not form chlamydospores. If the objective of an isolation is to determine the range of species found in a soil sample, then both soil dilution and debris plating techniques should be used together to ensure that the entire spectrum of recoverable fungi has been sampled.

3.4 Isolation Techniques – Spore Trapping and Air Sampling

Air sampling techniques are not commonly used for the isolation of most *Fusarium* species, as the unequivocal identification of a *Fusarium* spore to a species based on the microscopic observation of just a single spore is all but impossible. The techniques that have been used can be broadly divided into passive spore trapping, *e.g.*, selective media in an exposed Petri dish, or active spore trapping, *e.g.*, an Anderson spore sampler (65); a Burkard (Hirst) spore sampler (www.burkard.co.uk/ instmts.htm, 868), or a Rotorod (www.multidata. com). Spore samplers require experience to ensure that the spores recovered are reflective of those actually present (263, 1192, 1603). Identifications that go beyond the genus level usually require that representative spores be cultured for identification or that there be independent knowledge that a particular species is expected at a high level, *e.g.*, during a disease epidemic.

Studies of the mycology of airborne samples from diverse locations such as hospitals (25), libraries (264), and subway stations (100). *Fusarium* rarely is the dominant spore type found in these settings, however. Spore frequency in indoor air often is reflective of that found in the outdoor air (816, 1955), although air management systems, *e.g.*, air conditioning can make significant differences (25). There are reports of allergenic response to environments containing *Fusarium* spores, *e.g.*, Dixit *et al.* (509) and Fadda *et al.* (594), but the presence of other fungi in the samples confound the results and make the data difficult to interpret. There also are reports of recovery of airborne *Fusarium* toxins, *e.g.*, Lappalainen *et al.* (1186), but these reports probably are due to toxins associated with dust particles rather than with the spores.

In a plant pathological context, air sampling techniques have usually been used more to explore epidemiological and/or ecological questions than they have been for surveying the diversity of *Fusarium* in the atmosphere, *e.g.*, Fernando *et al.* (620), Lukezic & Kaiser (1301), and Schmale & Bergstrom (1912). Rainfall and season both affect the number of spores recovered, *e.g.*, Cosentino *et al.* (406), Ooka & Kommedahl (1629), and Rossi *et al.* (1835). *Fusarium* spores often are recovered at higher numbers at night than during the day, *e.g.*, Bai *et al.* (123), Bandyopadhyay *et al.* (138), Paulitz (1662) and Schmale *et al.* (1914). This result is not unexpected since the higher evening humidity and the lack of ultraviolet light would both help prolong the life of most *Fusarium* spores, which often have relatively thin walls and lack protective melanins and other pigmentation. The study of airborne spores of *Fusarium* is an area that probably has been understudied and that would benefit from more carefully designed investigations.

3.5 Seed Disinfestation

Seeds of many cereals, *e.g.*, maize and sorghum, often carry *Fusarium* isolates as endophytes even in apparently healthy kernels. Determining the relative effects of various fungal strains in seedling and other whole plant assays is not possible if the strain applied is competing with an endogenous strain(s) that need not be constant for all of the kernels in a seed lot. In such scenarios it also is difficult to determine whether the effects observed are due to the endogenous strain or to its exogenously applied counterpart. Similar problems also occur with some cowpeas and some flower bulbs, *e.g.*, *Gladiolus*.

For maize, treatments with NaOCl (63, 439) have been reported to reduce the infestation incidence, but not to eliminate it completely. A benomyl in acetone treatment (438) was reported to be effective for asparagus, but was not effective for maize seed (439). Soaking seeds in hot water, up to 60°C, sometimes accompanied by presoaking in KSO_3 (1471) or a post-soaking treatment with $HgCl_2$ (1867) also have been reported to be effective.

The protocol of Daniels (439) in which maize seeds are placed in water at 60°C for 5 min. has been generally successfully applied to both maize and sorghum (585,

958, 959, 1233), although in some cases the length of time in the hot water is extended beyond five minutes. Treatment in hot water is a trade off between killing the endogenous *Fusarium* and killing the seed and preventing germination. Above 60°C, treatments of even five minutes generally prevent germination, and treatments at < 50°C usually do not kill enough of the endogenous fungi to be worth employing. Seeds treated at 60°C suffer a reduction in germination of about 10% relative to an untreated control, which is comparable to the declines observed following treatment with NaOCl. The sensitivity to temperature means that the temperature of the water in the water bath in which the seeds are treated must be very carefully controlled. Even the addition of too many seeds at one time may reduce the effectiveness of the disinfestation process, as the additional seeds may decrease the temperature in the water bath. A similar technique has been tested for *Fusarium* spp. from cowpeas (1993).

The protocol recommended by Daniels (439) is the one usually employed for cereals. Seeds are put into lots of ~15 g (50 seeds) and each lot is wrapped in 1-2 layers of cheesecloth. Do not use lots of seed larger than this size, since the seeds in the center of a larger package will not warm as quickly as those near the edges and might not get hot enough to kill all of the endogenous *Fusarium* mycelium and spores. The seeds are soaked in distilled water at room temperature for four hours to allow them to hydrate. Following soaking, the seed packages are transferred to a water bath at 60°C for five minutes. The temperature of the water bath should be monitored closely and additional warm water should be added to bring the temperature back to 60°C if necessary. After removal from the hot water bath, the seeds should be rinsed in sterile distilled water and then removed from the cheesecloth packets and allowed to dry for 24 hours on sterile paper toweling in a laminar flow hood. Treated seeds should be used within a few days of the treatment and should not be stored for any significant period of time due to their high moisture content.

For *Gladiolus* the combined efficacy of hot water temperature and exposure time have been combined in a single mathematical model (1820). Note that the size of the item being disinfested with hot water is important, probably due to heat mass transfer effects), and that heat treatments of items such as cantaloupe melons (322) have not been successful.

4

Techniques for Growing and Maintaining *Fusarium*

Fusarium strains will grow under remarkably broad conditions, but the utility of cultures for various purposes often depends on the conditions under which they were grown. In this chapter we provide general information on conditions for vegetative propagation of *Fusarium*, and methods to deal with commonly encountered cultural problems. We describe a recommended set of culture conditions that should lead to the expression of the commonly observed morphological characters, protocols that can be used to recover cultures that originate from single conidia, and some of the equipment that we have used to implement these protocols. We also include a protocol for a commonly used technique for recovering auxotrophic mutants induced in the laboratory that has been adapted to *Fusarium*. Finally there is a section on the preservation of *Fusarium* cultures for long-term storage – via lyophilization, in silica gel, and as frozen suspensions at -70°C – and for the preparation of cultures that can be dried to meet the standards of the International Code for Botanical Nomenclature (770, Article 8) for the description of new species.

4.1 Vegetative Propagation

Incubation conditions. Sporulation and pigmentation are favored by light, including ultra-violet wavelengths, and fluctuating temperatures (1198, 2014, 2017). Commonly used conditions that accommodate the vegetative growth of many *Fusarium* species are alternating 12-hour periods of light (25°C) and darkness (20°C). We commonly incubate cultures approximately 40 cm below a light bank, 75 cm wide, that contains four 40W cool white fluorescent tubes and one 36W black light tube (Philips TL 36W/80 RS F40 BLB) on custom-made shelving, and in lighted growth chambers, *e.g.*, a Percival CU-22L or CU-36L (www. percivalscientific.com). Culture dishes for vegetative propagation usually are labeled on the back side of the plate for identification and incubated upside down, *i.e.*, with the lid side down. Dishes may stacked two to three high without any noticeable effects on morphology. If a series of lighted shelves are placed on top of one another, then there should be some distance from the light box of the lower shelf and the incubation surface of the upper shelf, so that cultures on the shelf above the lights are not warmer than those that are further away. The formation of condensate on the lid of the plate often is indicative of this problem. If condensate forms on the lid of the plate and seals it, the morphological characters may differ from those presented here. Plates that are placed too close to the lights also may dry prematurely.

Measurements of vegetative growth. The rate of vegetative growth is sometimes used as a secondary character in species identifications and descriptions. Cultures that have begun to visibly degenerate morphologically should not be used for these measurements as their growth rates are not necessarily the same as those of strains with normal morphology. Vegetative growth rate measurements usually are made in constant (25 or 30°C), rather than fluctuating, temperatures, and in complete darkness. These data are not available for all *Fusarium* species (Table 4-1), but when available usually are based on radial expansion of a colony initiated from a single conidium on a water agar inoculum block placed at the center of the plate. Measurements usually are taken after 72 hours of growth. Two measurements are taken at 90° angles (perpendicular) per plate, with 3-5 plates measured per strain.

An alternative to measuring radial diameters on a plate is to use race tubes. This technique was first developed by Ryan *et al.* (1852) for use with *Neurospora*, and modified for use with plant pathogens by Pearson *et al.* (1665). Race tubes are constructed by bending 50 cm sections of 16 mm O.D. Pyrex (or equivalent) glass tubing to a 45° angle 4 cm from each end. Tubes are filled with 20 ml of molten non-sterile agar medium, which covers the approximately 40 cm bottom of the tubes between the bends. We usually plug the ends with cotton, although other closures can be used. Note that the plastic and stainless steel caps that are commonly used to cover the outside of slant cultures are not satisfactory for this purpose, as they tend to fall off. The cotton plugs are relatively easily wetted during autoclaving, and having extra dry plugs on hand for replacement of wet plugs after the medium in the autoclaved race tubes has hardened is a good precaution. Tubes are autoclaved for 20 min at 121°C with the bent ends up. Taping several tubes together prior to autoclaving often makes this task easier. Tubes should be cooled on a level surface so that the medium is evenly distributed along the bottom of the tube. It is important to let race tubes dry for several days before inoculating them to allow the medium to dry and to allow any condensate present to evaporate. Hyphae will grow more rapidly through the water on the surface, typically where the agar meets the outer glass, than they will across the solid surface, which leads to irregular growth patterns and meaningless growth measurements. For this reason race tubes should not be incubated too close to the lights. Race tubes also are suitable for studies of circadian rhythms (527, 1272, 1773), and incubation characters such as the type and amount of light and temperature that affect the circadian patterns (733, 1242, 1796). Although these rhythms have not been studied as such in *Fusarium*, they are known to affect conidiation, mitochondrial activity, and mycelial growth in a number of other fungi including *Neurospora*, *Aspergillus*, and *Paecilomyces* (249, 765, 768, 1885).

Race tubes are inoculated 0.5-1.0 cm from one end with a 1-2 mm^3 agar plug taken from the edge of a growing colony. Three to five replicate race tubes usually are

Table 4-1. Growth rate (in mm) of representative cultures of *Fusarium* spp. on PDA at 25 and 30°C after three days in complete darkness. Data are from Burgess *et al.* (279), unless otherwise noted.

Species	25°C	30°C
F. acuminatum	25-35	5-28
F. andiyazi [a]	27-33	27-40
F. anthophilum	25-40	20-45
F. armeniacum	44-58	37-54
F. avenaceum	28-40	5-25
F. aywerte	40-45	32-42
F. babinda	25-37	15-22
F. beomiforme	30-39	36-46
F. camptoceras [b]	23-29	20-29
F. chlamydosporum	34-46	37-55
F. compactum	41-54	42-58
F. crookwellense	54-66	15-25
F. culmorum	55-68	15-25
F. decemcellulare	15-25	11-22
F. dimerum	4-10	5-12
F. equiseti	34-46	28-44
F. graminearum	47-61	5-20
F. heterosporum	28-41	8-30
F. hostae [c]	24-29	20-24
F. konzum [d]	21-34	21-38
F. lateritium	8-20	5-15
F. longipes	37-54	44-61
F. merismoides	4-10	5-12
F. musarum [b]	50-59	44-49
F. napiforme [e]	20-35	20-32
F. nelsonii [b]	24-39	26-41
F. nurragi	32-40	6-22
F. nygamai	25-35	32-42
F. oxysporum	25-40	25-40
F. poae	42-54	24-39
F. polyphialidicum	23-40	20-38
F. proliferatum	25-35	25-32
F. pseudograminearum	39-51	10-25
F. pseudonygamai [d]	24-30	24-29
F. redolens [c]	32-37	33-39
F. sambucinum	24-35	11-21
F. semitectum	35-45	16-33
F. scirpi	36-48	36-49
F. sporotrichioides	51-61	32-42
F. solani	21-29	26-36
F. subglutinans	23-37	11-34
F. thapsinum [d]	19-28	14-27
F. tricinctum	29-39	2-15
F. verticillioides [d]	21-30	22-35

[a]Marasas *et al.* (1346); [b]Marasas *et al.* (1347); [c]Geiser *et al.* (689); [d]Zeller *et al.* (2412); [e]Marasas *et al.* (1344).

inoculated for each strain. Incubation conditions, constant 25 or 30°C and continuous darkness, as for the radial growth measurements, commonly are used. Growth measurements for race-tube cultures begin when the colony is large enough to fill the width of the tube (1-3 days after inoculation depending on rate of growth). For some of the very slow growing *Fusarium* sp., *e.g.*, *F. dimerum* and *F. merismoides*, standard race tubes are not suitable for making growth measurements, as the colonies take too long to become large enough to cover the entire front. The growing front is marked as a line on the back of the tube with a laboratory marker at this time, and marked again daily for the time period desired (usually no more than 1-2 weeks). Use the thinnest tip marker available to mark the growing front as a thin line improves the accuracy of the measurements. The distance between the marks is used to calculate the average growth rate and its day-to-day variation.

Data from race tubes and radial growth measurements are not directly comparable, even if the same medium and incubation conditions were used. Both the time of measurement initiation (when a conidium germinates or when a colony reaches a threshold size) and the starting material (a single conidium or a vigorously growing colony) are quite different. With radial growth measurements, 72 hours is the normal termination, while for race tubes growth measurements may continue almost indefinitely. The surface area covered by a colony with a particular linear measure of radial growth and the same linear measure of growth in a race tube radial are not the same. A colony in a race tube that grows 10 mm along the tube, will cover ~160 mm^2 (10 mm linear growth × 16 mm race tube width), and one that has grown 20 mm approximately 320 mm^2. The area covered daily by a radialy growing colony depends on the diameter of the colony at the start of the day period, and is significantly larger for older colonies (with larger diameters) than it is for young ones (with smaller colony diameters). For example, a colony with a diameter of 10 mm would have a surface area of ~79 mm^2 and one with a diameter of 20 mm of ~314 mm^2.

Degenerate Cultural Variants. The use of standard culturing procedures enables degenerate cultural variants to be recognized quickly and discarded (2168). Strains of some *Fusarium* species, *e.g.*, *F. armeniacum*, *F. compactum*, and *F. longipes*, are much more susceptible to cultural degeneration than are strains from some other species, *e.g.*, most species in section *Liseola*. There are two types of degenerate cultural variants. The pionnotal type results in a flat slimy colony that lacks aerial mycelium, but produces abundant macroconidia, which need not be typically shaped for the species The mycelial type consists of sterile, usually white, mycelium. Only rarely can wild-type cultures be recovered from either type of degenerate variant. Degenerate strains usually lose the capacity to serve as the female parent in a sexual cross. Fertility as a male usually is not altered in the pionnotal mutants, as the conidia can still serve to spermitize the female, but degenerate variants that no longer sporulate usually function poorly, if at all, as a male parent.

Although cultural degeneration is a major problem in studies on *Fusarium*, little is known of the mechanism(s) that underlie the process. For strains prone to cultural degeneration, its occurrence is promoted by frequent subculturing on carbohydrate-rich media, *e.g.*, PDA, MM and CM, via mass transfers of mycelium. Similar problems have been seen in other fungi, *e.g.*, *Podospora anserina* and *N. crassa* (31, 773), but *Fusarium* species are particularly well known for this problem. The frequency of degenerate cultures can be minimized by using natural substrate or low nutrient media, *e.g.*, CLA or SNA, and by using a single germinated conidium or a hyphal tip to initiate subcultures.

Degenerate cultures differ from wild-type cultures both morphologically and physiologically. Degenerate cultural variants of pathogenic species often are avirulent and should be avoided in studies of disease. In contrast, mycotoxin production often is unaffected in degenerate cultures (2349), and in some cases may even be increased, *e.g.*, Hidy *et al.* (858) and Keith (1031). Identification of degenerate cultures, if it can be done at all, usually is based on DNA sequence data (Chapter 7).

Culture Mites. If plant material is brought into the laboratory for the recovery of *Fusarium* isolates, then at some time culture mites will be a problem. As they move from plate to plate in an incubator or a laboratory the mites can contaminate the cultures they traverse with bacteria, yeasts, or other filamentous fungi. They also can cross-contaminate cultures by carrying spores from one plate to the next leading to mixed fungal cultures that may be morphologically uniform but genetically mixed. These pests also eat the spores and aerial hyphae of the growing cultures, and it is possible to watch, over a period of several days, a colony with many spores go from a fluffy appearance to one that is flat, slimy and has relatively few spores. These types of changes should not be confused with the cultural degeneration described elsewhere. As soon as a mite infestation is discovered it is important to take steps to end it. If infestations are infrequent, then quarantining infected material (until experiments are complete) followed by autoclaving it, and disinfesting the incubator surfaces usually suffices. If infestations are frequent then more proactive steps should be taken.

As soon as a mite infestation is discovered, everything in the incubator should be considered contaminated. Older cultures should be discarded (preferably autoclaved immediately), and younger essential cultures should be "quarantined", analyzed, and then autoclaved. Experiments based on a (potentially) mite-infested culture should be repeated and any data based on the mite-infested culture(s) discarded. Subcultures from cultures in a mite-infested incubator should generally not be made, because of the risk that the subculture may propagate the mites, or be mixed genetically or otherwise contaminated. Those isolates that must be subcultured should be repurified from a single conidium and checked closely to be certain that the mites have not been subcultured along with them.

Disinfestation in our laboratories has two components. As soon as an infestation is discovered, a piece of a chlorovos-impregnated vinyl strip is placed in the infested incubator. (These strips are sold under a variety of trade names for use in buildings, *e.g.*, barns, to reduce insect problems.) A small portion of a standard strip (10-20% depending on incubator size) is sufficient, and too large a piece is to be avoided due to the toxicity of the chlorovos. Wear gloves when handling the strip and cut it with a sharp pair of scissors that is washed immediately after use. Leave the strip in the incubator until the last of the contaminated cultures is discarded. Once the cultures are discarded, then wipe all of the surfaces down with a dicofol (Kelthane), or other commercially available miticide, solution made up according to the manufacturer's instructions. Alternatively, moving cultures to an atmosphere with a high level of CO_2 at room temperature is effective in stopping mite infestations of critical infected cultures of *Neurospora* (1433), but to our knowledge has never been tried with *Fusarium* cultures.

If mite infestations are relatively frequent and more proactive measures to reduce problems are required, then mites can be excluded from slants by gluing rice paper over the mouth of the test tube (2016), after the medium in the slant has been sterilized. The glue contains 20 g gelatin, 5 g Cu_2SO_4 (to inhibit microbial growth in the glue), and water to 100 ml. The mixture is heated to dissolve the gelatin then poured into Petri dishes to gel and stored at 5°C. A tube is capped by warming the mouth in the flame of an alcohol or Bunsen burner. The warmed test tube mouth is inserted first into the glue and then placed onto a small piece of rice paper that covers the mouth of the tube. The glue is allowed to set and the excess paper around the outside of the mouth is touched to a flame to burn away, leaving a cap covering solely the mouth of the tube. Tubes are capped with a plastic or metal cap to maintain sterility; do not stopper these tubes with cotton or foam plugs.

4.2 Preparing Cultures for Identification

Morphological characteristics are basic for the identification and taxonomy of *Fusarium*; however, the species in the genus vary significantly both morphologically and physiologically. At least some of this variation is in response to environmental differences, including differences in media, temperature, and light. Accurate use of the descriptions in this book as a guide to species identification requires that cultures be grown on specific media under particular conditions. Deviations from these cultural conditions can result in significant errors in identifications based on morphological characters. Cultures from which DNA is to be isolated are much less sensitive to differences in cultural conditions, with DNA yield and the lack of carryover of inhibitors of DNA manipulation enzymes usually the most important parameters.

In this manual, morphological identifications generally are based on characters observed on one of three media – CLA, PDA, or SNA. The standard for inoculation is a single germinated conidium (Fig. 4-1, see also section 4.3). After germination the agar block carrying the germinated conidium is transferred to the test medium. With SNA and PDA, the agar block usually is placed in the center of the plate, or the center of the surface in an agar slant in a test tube. With CLA, the agar block should be placed near a carnation leaf piece so that both leaf pieces and water agar are relatively quickly colonized. Sporodochia containing macroconidia are formed more readily on the carnation leaf pieces and microconidia are usually more abundant in the portions of the colony growing on the water agar.

Chlamydospore production occurs on CLA, SA and SNA. Chlamydospores form on CLA, usually within 2-4 weeks of inoculation, but are more readily formed on SA (in 10-14 days when incubated in the dark at 25°C) and SNA (in 2-3 weeks when incubated in the light or the dark at 25°C). Cultures for chlamydospore production growing on CLA and SNA usually are initiated from single germinating conidia, but cultures on SA may be started with a mass transfer from a colony growing on CLA (water agar portion) or SNA.

Cultures for DNA extraction usually are started with a mass transfer of spores and hyphae from a vigorously growing culture. PDA, MM and CM all have been used successfully. Most researchers use cultures growing in Erlenmeyer flasks that contain liquid medium as 25-40% of the total volume of the flask (Section 7.1), but culture

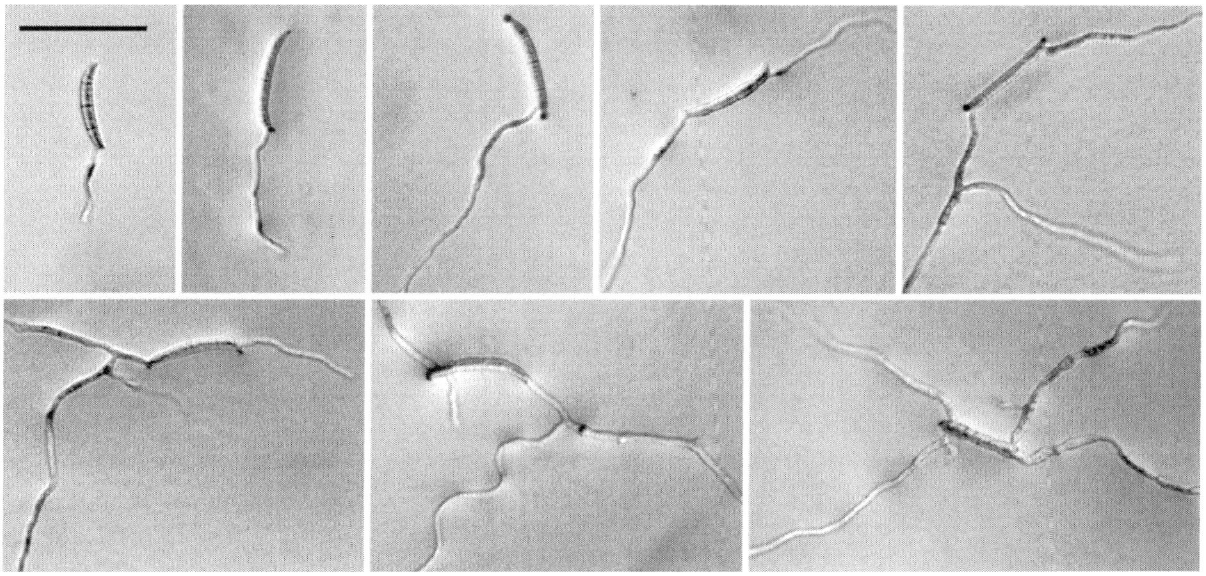

Figure 4-1. Germinating macroconidia of *Fusarium oxysporum*

material that has been scraped from the surface of plates or agar test tube slants also have been used successfully. Erlenmeyer flask cultures may be placed on the bench top and incubated at room temperature without agitation for 2-5 days days. More commonly these cultures are incubated on an orbital shaker at 150-200 rpm for 2-3 days. Cultures that have had the opportunity to grow for an extended period of time, sometimes as little as 4-5 days, may accumulate complex carbohydrates. These compounds often co-precipitate with the DNA and make DNA extractions more difficult and less efficient. To avoid this problem, heavier inocula, *e.g.*, 10^5 spores/ml in the medium after inoculation, often are used to shorten the incubation period and the time in which these carbohydrates can accumulate.

4.3 Single Spore Subcultures

Soil and many plants often harbor numerous *Fusarium* spp. In some instances these species can be distinguished on PPA, but in general colonies must be subcultured to other media before identification. If the subcultures from PPA are made with single conidia (preferably of uninucleate origin) or from hyphal tips, then the variation in morphological characters of the resulting characters is may be severely reduced. For genetic studies of both field populations and laboratory crosses, an ultimately uninucleate origin of the culture is critical, since significant genetic differences are not necessarily morphologically obvious. The time invested to obtain clean cultures that originate from a single conidium usually is more than repaid in greater certainty and clarity of identifications, and the reduced need to re-evaluate results or rerun experiments when the results obtained could be explained by mixed cultures. For example, until recently *Fusarium xylarioides* was commonly maintained as a mixed culture of two quite different species (690). Conidia are separated either by dilution plating or by micromanipulation and the spores allowed to germinate overnight.

Dilution plating. Dilution plating can be done with soil (Section 3.3) or with spores from cultures that are already growing. Cultures from plant material usually must be subcultured on some medium before single spores can be isolated. For *Fusarium*, most workers follow the general protocol of Hansen & Smith (811). To 10 ml of sterile water in a test tube is added a small scrape of culture material (either macroconidia, microconidia or both depending on the species). For spore suspensions, spore concentrations can be reduced with additional water, until 1-10 conidia are seen in a drop viewed at low power, usually 100×. Alternatively, a series of plates from increasing dilutions of the spore suspension can be made If the dilutions are plated to media in a series of 100-mm diameter Petri dishes, the "ideal" plates usually have 50-200 germinating spores per plate. Soil dilutions (Section 3.3) usually are made with sterile 0.05% WA (section 2.2). The small amount of agar in the solution makes it easier to disperse the soil particles. For both soil dilutions and spore suspensions, the best results are obtained if the plates have had a chance to "dry" for a few days to a week before use. The liquid applied to the plates is adsorbed more readily, and the colonies appear somewhat more rapidly and are more easily distinguished. Plates usually are incubated right-side up, *i.e.*, with the lid up, overnight at 25°C. Cultures with bacterial contaminants may be rid of these contaminants at this stage by including antibiotics in the medium onto which the spores/soil particles are plated, or by adding a drop of 25% lactic acid to the conidial suspension. Either of these techniques may slow germination, and up to 48 hr may be required for high levels of germination and growth to be easily detected.

Micromanipulation. Micromanipulation of spores is more technically demanding, but has the advantage that the subcultured colonies almost certainly result from a single spore. Micromanipulation is usually done only with cultures that have been freed of non-*Fusarium* contaminants. It also is a common way to separate ascospores resulting from sexual crosses. For micromanipulation, a small amount of culture material is picked up with an inoculating loop or needle and spread across one edge of an agar slab, usually 3% water agar, formed on a cover slip in an agar-layer casting block (802). If a loop is used, then a "loopful" of liquid, usually 0.2% Tween 80, is used when the spores are taken from the colony to help reduce clumping on the agar micromanipulating slab. Individual spores are pulled perpendicular to the streaked edge across the agar surface with the glass micromanipulating needle. Usually this process scores the agar surface resulting in a visible line that can be used to locate the germinated spore the following day. Spores usually are pulled half to two-thirds of the distance to the opposite side of the agar slab. This distance clearly separates the germlings of the manipulated spores from the colonies that result from the germination of the spores that remain in the initial unresolved spore streak. Twenty to 30 spores can be separated on a single agar slab. Although this many ascospores may be recovered from a single perithecium for genetic analyses, for purification of vegetative cultures usually only a few conidia are separated unless problems with spore germination are anticipated. Slabs on which spores have been separated are placed in sterile, usually glass, Petri dishes that contain no media, and are incubated right side up at 25°C over night.

Isolation of germlings. Germinated spores, also termed germlings, are identified by examination in a dissecting microscope, usually 30-70×. The hyphae of the germinated spores make them appear as small threads on the agar surface. Some spores may have only one germ hyphae per spore, but others may have several (Fig. 4-1). Individual germlings are excised with a small sterile "spatula" made from nichrome or platinum-iridium wire mounted in a standard needle holder. The tip of the wire is flattened by hammering it with a conventional construction hammer on a smooth hard surface; a small anvil works best. Once flattened, the wire can be trimmed with a sharp pair of scissors to a spade-like shape. The edges of the spade can be used to cut the agar surface around the germling, and the flat surface to cut the bottom of the piece away and to lift the agar block out and transfer it to the target medium. The spade tip of the wire is sterilized by flaming and then usually cooled in the agar before beginning the cutting process. Nichrome wire, which is much cheaper than platinum-iridium wire, succumbs to this heating and cooling process relatively quickly, and the tip must be regularly flattened and trimmed to retain its utility. The platinum-iridium wire is little affected by the repeated heating and cooling and the tip usually retains its shape and its rigidity unless it is dropped point first onto the lab bench or the floor. The platinum-iridium wire also is easier to flatten and to trim than is the nichrome wire.

Hyphal tip transfer technique. This technique usually is used only when spores are unavailable, or when colonies that develop from germinated conidia frequently and quickly degenerate, *e.g.*, *F. longipes*. Cultures from which hyphal tips are to be taken should be growing on a weak medium, *e.g.*, CLA, WA or SNA, and should not have completely covered the agar surface; pouring relatively thin plates also helps reduce the amount of growth. The objective is to recover a hyphal tip from the margin of a sparsely growing colony. Individual hyphal tips are identified under the dissecting microscope and are excised from the medium with a platinum-iridium or nichrome wire spatula in a manner similar to that used for excising germlings originating from spores.

4.4 Mutagenesis

Mutants may be induced in *Fusarium* spp. with a number of different mutagens, both chemical and physical. Commonly used mutagens are ultraviolet light (1216) and nitrosoguanidine (96, 97), but the list of possibilities is much more extensive, *cf.*, Burnett (286), Esser & Kuenen (589), and Moore & Frazer (1469) for more comprehensive lists. Uninucleate spores are preferred as the targets for mutagenesis; for *Fusarium* spp. microconidia should be used whenever they are available. Physical mutagens, *e.g.*, ultraviolet light or X-rays, are preferable for many experiments as they usually leave no toxic residue to dispose of and are relatively easy to handle.

Chemical mutagens usually must be washed from the spores after the mutagenesis period is complete. The dose of chemical mutagen applied often is more difficult to assess due to differences in uptake and solubility. Chemical mutagens also may need to be treated as hazardous waste, and special treatments may be required for their detoxification. In general, do not begin a mutagenesis experiment with a chemical mutagen unless a detoxification and/or disposal protocol is already in place.

In many experiments, the goal is to kill 90-95% of the cells and then screen the survivors for a desirable deviation from wild type. Although 90-95% killing is the target, there are instances when the target killing rate may be as low as 50% or as high as 99%. In general, the higher the rate of killing the larger the number of mutations induced, and the greater the chance for recovery of a mutant with the desired phenotype. With a higher number of mutations, however, the possibility that the number of mutations carried by a survivor will be > 1 increases. Strains with multiple mutations often are physiologically inferior and there is a risk that a strain with a desired mutation may be unusable for further experimentation due to poor growth under any conditions. If more than one mutation is present in the strain, then conclusions based solely on the phenotype of the mutagenized strain can be questioned. The possibility of cryptic, undetected mutations altering results also can limit the utility of strains that may carry multiple mutations. Increasing the proportion of strains that survive the mutagenesis treatment reduces the frequency of strains with multiple mutations, but increases the number of survivors that have no mutations at all. These unaltered strains must then be removed in the screening process that follows the mutagenesis. The presence of unmutated strains results in a lower frequency of mutants with the desired phenotype, and usually lengthens the screening process.

For species in which it is possible to do classical genetics, *e.g.*, *G. zeae* and many members of the *G. fujikuroi* species complex, and show that phenotypes co-segregate, higher killing rates (90+%) usually are desirable, as genetic crosses can be made to demonstrate that a single gene (or at least a small chromosomal region) is responsible for the observed phenotype. In species where such crosses are not possible, which includes most other *Fusarium* spp., a lower killing rate, *ca.* 50%, is recommended to reduce the possibility of multiple mutations.

Mutagenesis with ultraviolet light. Ultraviolet light is a clean, relatively safe, readily available, and easy to use mutagen that can induce a wide variety of mutations due to both its direct action and to the manner in which the lesions induced in the DNA background are repaired (463, 826, 834, 1158, 2376). The protocol described below was developed for *Gibberella zeae* (1216) and has been used with other *Fusarium* spp.

Make a relatively dense spore suspension, a concentration of 10^7-10^8 spores/ml is commonly used. Addition of a small amount of detergent, *e.g.*, 0.2% Tween 60 or Tween 80, increases spore dispersal, but is not essential. Microconidia are preferable if they are produced, but macroconidia also can be used successfully. Approximately 10 ml of the spore suspension is placed in a sterile glass Petri dish (100 mm diameter) with a very thin stir bar, and placed on a stir plate covered in black cloth (so that the ultraviolet light is absorbed and not reflected). The suspension should be stirred gently with the stir bar.

An ultraviolet light source with an emission wavelength of 254 nm should be mounted in a stand and placed over the Petri dish. The ultraviolet light source should be allowed to warm up before use (the warm up period may take 10-30 minutes depending on the source) so that the UV emission is constant and stable before it is used to irradiate the culture. Older bulbs typically emit less UV light, and may require a longer exposure or that the culture be placed closer to the light source for a sufficient dose to be received by the culture. An ultraviolet light meter may be required to precisely calibrate the dose of mutagen applied.

For *G. zeae*, 80-90% killing results from a 60-90 second exposure to 44 ergs/sec/mm^2. Following the mutagenesis, the culture must be kept in the dark to prevent photoreactivation and repair of the lesions induced by the ultraviolet light. Screening for desired mutant phenotypes should not begin for at least 24 hours following mutagenesis to allow time for mitotic DNA replication, which "fixes" the induced mutations, and allows time for segregation of the mutants into separate cells. Selection pressure applied before this time can kill all of the survivors, not just those that lack the desired phenotype.

Filtration enrichment. Filtration enrichment is a common means of selecting for auxotrophic mutations that was developed for use with *Neurospora crassa* (77, 2363) and *Cochliobolus heterostrophus* (1199, 2391), and adapted for *Fusarium* (1216). The strategy is that cells that have mutations in genes that do not affect nutritional requirements will be able to grow normally on a minimal medium, while those that require an additional nutritional supplement(s) will not. Cells with no nutritional requirements form pellets that can be caught in a filter, while those that have such requirements grow slowly, if at all, and therefore can pass through the filter. The flask contents are filtered repeatedly for several days, and care must be exercised not to contaminate the contents during the filtering process. At the end of the filtration, most wild-type material should have been removed, thus upping the percentage of desirable mutants amongst the survivors (Fig. 4-2). The last filtration usually retains relatively little material on the filter.

This protocol was optimized for *G. zeae* strain ATCC 20273 (1216), but should be adaptable to any *Fusarium* strain by changing the timing and perhaps the mesh of the filters used in the filtration steps. Ten milliliters of a heavy, irradiated spore suspension (~10^7 spores/ml) is added to a 500 ml Erlenmeyer flask containing 250 ml of medium with 0.2% glucose, 0.2% fructose and 6% sorbose in place of the normal sugar. This set of carbon sources results in more restricted colonies, a result that also can be achieved for some *Fusarium* spp. by supplementing the medium 0.01-0.1% tergitol (section 2.5). The flask is wrapped in aluminum foil to prevent the entry of light for the first 24 hours of incubation and stoppered with a cotton plug. The flask is incubated on a rotary shaker (100-150 rpm). Incubations are made at the "non-permissive" temperature, usually 28-32°C for most *Fusarium* spp., if temperature sensitive mutants are sought or at a temperature that yields good growth, usually 20-25°C, if temperature-sensitive mutants are not desired.

After 4-6 hours, the contents of the flask are filtered through a sterile ~105-μM-mesh polypropylene filter. The filtrate is returned to the flask and the incubation continued. This filtration and any others up to 24 hours post mutagenesis (usually no more than one) should be done in a dark room with a red light to avoid photoreactivation repair of lesions induced by the ultraviolet light. For the next 72-120 hours, the contents of the flask are filtered through a 30-μm mesh nylon screen approximately once every 12 hours. Strains that grow particularly rapidly may need to be filtered more often and may need to be filtered through the 105 μm mesh a second, or perhaps even a third, time before changing to the finer mesh. Strains that grow more slowly may need to have longer times for growth between the filtration steps, and to be filtered for a longer period of time. The amount of material retained on the filters should be examined. If the filters retain too much material, then the non-growing/poorly growing propagules may be trapped in the mat formed by the larger colonies when they are retained on the filter.

Following the final filtration, the contents of the flask are mixed with double strength agar on which the desired mutant class(es) will grow, commonly CM or supplemented MM that contains 0.1% tergitol to restrict colony growth. The fresh agar should be barely molten (48-52°C). As soon as the flask contents have been mixed with the fresh agar it should be poured into sterile Petri dishes, the agar allowed to gel, and the plates incubated at a "permissive" temperature at which all of the remaining survivors are capable of growing. Colonies usually appear 2-6 days after inoculation at which time they should be transferred to fresh media (usually CM or supplemented MM) and purified by single-spore subculturing (section 4.3).

Figure 4-2. Filtration enrichment efficacy in *Gibberella zeae* [from Leslie (1216); reproduced with the permission of the American Phytopathological Society]. A wild-type strain (○ or □, ATCC 20273) and an adenine auxotroph (● or ■, *ade-1*; ATCC 48063), were inoculated in 10:1 (○ or ●) or 100:1 (■ or □) ratios and subjected to the filtration enrichment protocol described in section 4.4. Note that wild-type has been effectively removed from the suspension within 72 hours of the commencement of filtration for either the 10:1 or the 100:1 mixture.

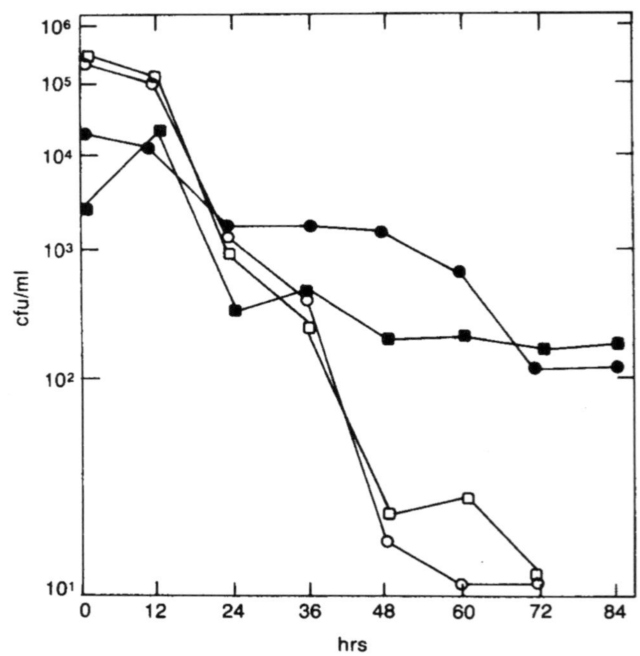

4.5 Preservation of Cultures

A number of preservation techniques have been used with *Fusarium* cultures including lyophilization, freezing of spore suspensions at -70°C or in liquid nitrogen, and storage on filter paper, silica gel or soil at 4 or -20°C. None of these techniques suffices for preparing cultures as herbarium specimens required for the description of a new species under the International Code of Botanical

Nomenclature (770). We describe techniques with which we have experience, but the filter paper preservation technique, which we have not worked with extensively, also appears to work well and might be a suitable alternative to preservation on silica gel (635).

Freezing at -70°C. Preservation by freezing at -70°C is the easiest way to preserve cultures of *Fusarium* for future study. Cultures are grown on an appropriate medium in slants made in 10×75 mm test tubes (1-1.25 ml medium per tube) until they have sporulated extensively. A spore suspension is made with sterile 15% glycerol (15:85 glycerol:distilled H_2O). We normally use a Pasteur pipette, with a full pipette barrel providing approximately 2 ml of the 15% glycerol solution. The Pasteur pipette tip is used to work the surface of the culture to free spores (and some hyphal fragments) that might be attached. The spore suspension is removed from the tube and transferred immediately to a 2-ml cryovial. The cryovial may left on the bench for 30 min to an hour, if necessary, before being placed into a box in the ultra-low temperature freezer. If the vial has been sitting for some time, then it should be inverted several times or otherwise mixed, *e.g.*, with a vortexer, before freezing to insure that the spores and hyphal fragments have not settled to the bottom of the vial.

Cultures can be recovered by scraping some of the ice from the top of the vial. Cultures that have thawed completely should be remade, but if no significant thawing has occurred, then the contents of the vial may be reused for years as long as the contents remain sterile.

Cultures stored in this manner remain viable for at least two years for all of the species that we have tested. Strains from the *G. fujikuroi* species complex and of *F. oxysporum* have been kept in this manner for at least ten years without significant loss in viability. Strains of *G. zeae* begin to lose viability after 4-5 years of storage in this manner. Whenever viability seems to be dropping, we empty the entire contents of a vial onto an appropriate medium, subculture it, and then freeze it again. Reisolation from a single spore may be done at this time, but is not necessary if the original frozen culture was derived from a single-spore purified isolate and if the culture morphology does not appear to be degenerating.

Lyophilization. Freeze drying of cultures is a traditional method for preservation that works well with most strains of *Fusarium* for an extended period of time (20+ years). This method is relatively time consuming, and once opened lyophilized cultures must be remade and cannot simply be restored. Cultures preserved in this manner usually are stored at 4°C, and several copies of the same culture usually are preserved at the same time. There are several variations on this method. The preservation process for the different technique variations all begin with colonized carnation leaf pieces from CLA. Strain revival requires plating the lyophilized contents onto an appropriate medium. As a control, one vial or ampoule of each strain lyophilized often is checked for viability and purity immediately after lyophilization.

Lyophilization technique number 1 [after Tio et al. *(2151)].* Small glass ampoules (60 mm length × 5 mm inside diameter) are plugged with cotton and autoclaved. Three to five well-colonized carnation leaf pieces are put into the ampoule, a cotton plug inserted, and the ampoule pulled into an hour-glass shape, with the cotton plug and the carnation leaves on the side of the constriction with the closed end of the ampoule. The open end of the ampoule is attached to a nipple on the lyophilizer, and the contents of the ampoule dried under vacuum (10^{-1} – 10^{-2} Torr) for 18 hours. After drying, but before removing from the lyophilizer, the ampoule is cut into two pieces with the flame from a torch that simultaneously seals the closed end of the ampoule completely while retaining the vacuum within it. Ampoules are stored at 4°C. To revive the cultures, the ampoule is surface sterilized, wrapped in a layer of facial tissue or paper toweling, placed on a flat surface and hit gently with a hammer or other blunt object to crack the ampoule. The dried leaf pieces are removed from the remains of the ampoule with a pair of sterile forceps, transferred to an appropriate medium, and incubated for growth.

Lyophilization technique number 2. This technique is that of Fisher *et al.* (625), and is the one most commonly cited for preservation of *Fusarium* spp. Screw-cap vials (5 ml volume) are plugged with split rubber stoppers and autoclaved. After cooling, several well-colonized carnation leaf pieces are transferred to each vial followed by 1 ml of sterile skim milk [autoclave skim milk for 10 min at 10 p.s.i. (= 6.9 kPa) instead of the usual 15 p.s.i. (= 10.4 kPa) to sterilize it], and the plug loosely reinserted (do not cap the vials at this time) such that air can be evacuated from the vial. Vials are flash frozen in a tray containing liquid N_2. A Lucite plate that is slightly larger than the tray is placed on top of the vials, and the tray, vials and cover moved to the precooled (-35°C) drying chamber of a VirTis (Gardiner, NY) model no. 10-MR-SA on a refrigerated lyophilizer. After 10 min, a vacuum (10 µm Hg) is pulled, and the internal shelf is heated to and then maintained at 15-20°C for 16-20 hours. Vials are sealed under vacuum by inflating a rubber diaphragm in the chamber above the cover plate. Inflation of the diaphragm puts downward pressure on the cover plate and presses all of the split rubber stoppers firmly into the mouth of the vial. After being sealed, vials are capped and stored at -20 to -30°C, although higher temperatures, *e.g.*, 4°C also have been used by some investigators. Cultures are recovered by placing the dried skim milk/carnation leaf piece pellet onto a suitable medium after removing it aseptically from the vial. In many cases, it is necessary to break the pellet into pieces with a sterile scalpel, spatula, or other sterile instrument, before the pellet can be removed from the vial, as the diameter of the vial mouth usually is somewhat smaller than that of the rest of the vial.

Silica gel. Several researchers and the Fungal Genetics Stock Center (University of Missouri at Kansas City, Kansas City, Missouri) preserve *Fusarium* cultures on silica gel by using a variation of a procedure that has long been used for *Neurospora crassa* (330, 1682). Silica gel cultures are easier to prepare than lyophilized cultures, but not as simple as freezing at -70°C in 15% glycerol. Like the glycerol cultures, however, they can be used a little bit at a time by shaking out a few silica gel crystals onto an appropriate medium, and then replugging the tube. To be successful, cultures should conidiate profusely. Cultures also need to be checked for viability with some regularity as cultures preserved on silica gel tend to have a shorter storage life span than do cultures preserved either by freezing at -70°C or by lyophilization (2343, 2344).

Silica gel (non-indicating) is dispensed into culture tubes (13 × 100 mm) and sterilized for at least 90 min at 180°C. This process both dries and sterilizes the silica gel. Do not autoclave the silica gel to sterilize it! The silica gel should fill 60-75% of the tube. Both screw cap and regular tubes plugged with cotton can be used for this process. Tubes containing sterile silica gel are cooled in an ice bath immediately prior to use. A heavy conidial suspension is made by adding a "Pasteur-pipette full", ~2 ml, of sterile skim milk to a 10 × 75 mm slant culture. The pipette tip is used to work the surface and free as many spores as possible. Some mycelial fragments and small agar chunks also will be freed from the surface. The agar chunks and mycelial fragments do not usually affect the preservation process unless they are so large that they block the tip of the Pasteur pipette. The spore suspension is sucked back up into the pipette, and then spread up and down across the silica gel in a single tube. The tube should be vigorously vortexed and then placed immediately back into the ice bath. The heat generated by the addition of a liquid to the dried silica gel is sufficient to kill the *Fusarium* spores if they are subjected to anything other than a brief exposure. Do one tube/culture at a time. After sitting on ice for a few moments the tube can be vortexed again. The skim milk acts as a protectant, but not as a nutrient source for the spores. Dried cultures are stored at 4°C either loosely capped, if a screw cap tube is used, or plugged with cotton and covered lightly with a small piece of ParaFilm M (American National Can, Neenah, WI), if a regular tube is used. Cultures are recovered by aseptically shaking a few crystals onto an appropriate medium, and then replugging or recapping the tube used and returning it to the 4°C storage. When viability begins to decrease, the remaining contents of the tube are dumped onto an agar plate, and a fresh culture made and preserved.

Soil preservation. Soil commonly is used to preserve *Fusarium* cultures, but the practice is not recommended. Both Booth (222) and Windels *et al.* (2343, 2344) found that the fungus can colonize the soil and grow during storage, which increases the possibility of contamination and of recovering strains in which the morphology has been altered. Gaylarde & Kelley (687) found that *F. merismoides* stored in soil appeared to senesce as evidenced by the appearance of additional DNA restriction fragments that they associated with a plasmid.

Preservation for herbarium records. Descriptions of new species of *Fusarium* and their teleomorphs requires that dried specimens be dried and deposited in a recognized herbarium. For most *Fusarium* species these specimens will be dried laboratory cultures, with CLA or SNA (depending upon features desired) generally used as the substrate for vegetative cultures and V-8 juice or carrot agar used for the sexual structures.

An appropriate culture, usually 2-3 weeks old for vegetative cultures or just as a cirrus of ascospores begins to appear for teleomorphs, is killed by exposing the plate to formalin in a closed container for three days. Water agar (2%) is dispensed in 6 ml aliquots into test tubes and sterilized. The lid of the Petri dish with the killed culture is removed, inverted and covered with 1.5-1.75 ml of sterile glycerol. The glycerol is overlaid with 6 ml of molten water agar, and the culture to be preserved is removed aseptically from the Petri dish with a large sterile spatula and floated on the surface of the water agar. Cultures are covered with a piece of paper (to keep out dust, but not to prevent air circulation), and stored in a cabinet or drawer for several days to a week to allow them to dry. Once dry, the culture will have a rubber-like texture and can be removed from the Petri dish lid for more permanent herbarium storage.

5

Vegetative Compatibility Groups

Vegetative compatibility, also known as heterokaryon compatibility, has a long history of both mycological and genetic studies (1220), and has been documented in numerous ascomycetes other than *Fusarium* (Table 5-1). In studies with *Fusarium*, vegetative compatibility (or incompatibility) is the term virtually always used, with heterokaryon compatibility or heterokaryon incompatibility more commonly used in some other systems, *e.g.*, *Aspergillus nidulans*, *Neurospora crassa*, and *Podospora anserina*. At its simplest, vegetative compatibility means that two hyphae can anastomose and fuse to form a stable heterokaryon. Those strains that can form a stable heterokaryon (Fig. 5-1A) are said to be vegetatively compatible and to be in the same vegetative compatibility group (VCG), while those that cannot form such heterokaryons (Fig. 5-1B) are vegetatively incompatible and are in different VCGs. The vegetative compatibility phenotype has a multigenic basis, and can be used as a means of identifying a set of isolates that all share common alleles at those loci. As the number of loci governing the trait is large (10-15 in most ascomycete fungi), the number of possible VCGs also is large (thousands), and chance identities, although possible, should be proportionately few in a randomly mating population. The hypothesis that all strains in the same VCG are clonal, however, is not always true and should be made with caution unless the hypothesis has been tested and verified with an independent set of markers, *e.g.*, Chulze *et al.* (372) and Elmer (565). The *vic* genes also play no clear role in interspecific heterokaryon formation (1314) and the action of the *vic* genes apparently can be overridden whenever auxotrophic mutants are used to force heterokaryons (20, 1462).

In an asexual population, such as those thought to exist for *F. oxysporum*, there should be no sexual recombination, and isolates with similar pathogenicity that arose as part of a clonal lineage should all be in the same VCG. Thus, if there was a quick way to identify VCG, these identifications might serve as a substitute for time-consuming pathogenicity tests, and provide a sufficiently rapid analysis to enable the evaluation of populations of strains. These ideas were first formulated in a testable manner by John Puhalla and his colleagues (404, 405, 1745). The major impediment was identifying mutants that could be used to force heterokaryons and show that heterokaryosis had occurred. This problem was solved by pairing strains that formed different pigments and looking for complementation with respect to pigment production (1744). The current strategy for identifying VCGs is to generate *nit* (NO_3 non-utilizing) mutants as spontaneous ClO_3-resistant sectors (402, 1090, 1745), and then differentiating them by their growth on media containing

Table 5-1. Fungal genera in which there is published evidence for a vegetative compatibility system.

Acremonium	*Amylostereum*
Ascochyta	*Aspergillus*
Beauveria	*Botryosphaeria*
Botrytis	*Ceratocystis*
Cochliobolus	*Colletotrichum*
Cryphonectria	*Cryptosporiopsis*
Cryptostroma	*Diaporthe*
Diplodia	*Discula*
Eutypa	*Fusarium*
Gaeumannomyces	*Glomus*
Graphis	*Hypoxylon*
Leptographium	*Leucostoma*
Leucocytospora	*Magnaporthe*
Metarhizium	*Monilinia*
Morchella	*Neurospora*
Ophiostoma	*Paecilomyces*
Penicillium	*Pestalotiopsis*
Phaeographis	*Phaeoramularia*
Phomopsis	*Plectosphaerella*
Podospora	*Pseudocercosporella*
Rhizoctonia	*Rosellinia*
Sclerotinia	*Septoria*
Sphaeropsis	*Stagonospora*
Trichoderma	*Typhula*
Verticillium	

diverse nitrogen sources (see section 5.2). Mutants could be paired on an appropriate medium, usually MM with NO_3 as the sole nitrogen source, and vegetatively compatible strains detected relatively easily. From this premise have come a large number of studies of vegetative compatibility in numerous ascomycete species. By far and away, however, most such studies have been conducted with *F. oxysporum* (see section 5.1).

In this chapter we describe some of the history of and genetic basis for the vegetative compatibility trait both in *Fusarium* and in other fungal species (section 5.1), describe the general strategy for determining if *Fusarium* strains are vegetatively compatible (section 5.2), provide protocols for recovering and identifying *nit* mutants (sections 5.3 and 5.4), identify some potential trouble spots in the VCG identification process (section 5.5), and provide strategies for use when screening a population for the presence of previously identified VCGs (section 5.6), or when evaluating a population for overall VCG diversity (section 5.6).

5.1 History of and Genetic Basis Underlying Vegetative Compatibility

The first studies of vegetative compatibility were conducted by Garnjobst (684) with *Neurospora crassa*. The *Neurospora* experiments included classical genetic studies, which identified five loci that were involved in the process including mating type (*A*/*a*). The involvement of mating type in the vegetative compatibility process is now known to be limited to *N. crassa* (946, 1683), and is mediated through the *tol* gene (1973), which has no impact on any of the other genes involved in vegetative compatibility in *N. crassa* (1230). By using chromosome rearrangements that result in partial genomic duplications, Mylyk (1509) increased the number of loci known to be involved in vegetative compatibility in *Neurospora* to 10.

Initial molecular studies indicated that the molecules mediating the vegetative incompatibility process were proteins (2336). More recently some of *het* genes have been cloned and molecular studies of their mode of action begun (709, 1902). The proteins encoded by the *het* genes of *Neurospora* cloned thus far have no clear similarity to one another, and one locus (*het-6*) apparently encodes two very different proteins (2006). These proteins may be involved in signaling processes, and it now appears unlikely that the proteins encoded by the *het* loci are not directly responsible for the killing that occurs when strains in different VCGs attempt to form a heterokaryon. Multiple alleles are confirmed at at least one *het* locus in *N. crassa*, *het-c* (1903), are likely at a second, *het-8* (903), and numerous, poorly characterized, loci other than the *het* loci also may play a role in determining the reactions attributed to the *het* loci (78).

Although some basic genetic studies of *vic* loci have been made in *Fusarium*, most research has focused on the utility of the phenotype in assessing genetic variation. The genetic and molecular basis for the function of the *vic* loci in *Fusarium* usually is inferred from what is known about these loci in other fungi. All known *vic* genes in *Fusarium* are of the "allelic" form. In these interactions, the alleles at *vic1* interact with other *vic1* alleles, the alleles at *vic2* interact with other *vic2* alleles, *etc.*, but the *vic1* and *vic2* alleles do not interact with each other. The number of *vic* loci in different *Fusarium* species usually is estimated at 10-15, approximately the known number of chromosomes for many of these species. When two hyphae fuse to form an initial heterokaryotic cell, a mutual assessment is made by each nucleus of the *vic* alleles in the other nucleus. If both nuclei have identical sets of *vic* alleles, then the heterokaryon is stable. If the alleles at one (or more) of the *vic* loci are

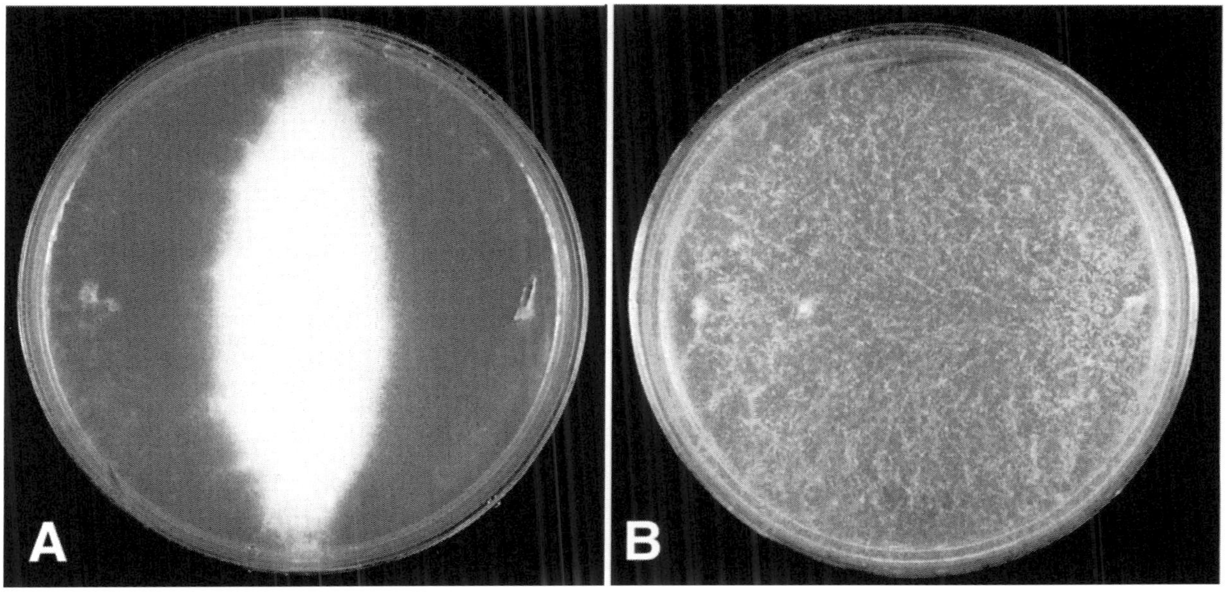

Figure 5-1. A. Prototrophic heterokaryon formed on MM + $NaNO_3$ by *nit* mutants derived from vegetatively compatible strains of *F. verticillioides*. **B.** Lack of heterokaryon formation on MM + $NaNO_3$ by *nit* mutants derived from vegetatively incompatible strains of *F. verticilliodes*.

different, then the heterokaryon is transitory and unstable, and the initially formed heterokaryotic cell is walled off on both sides of the heterokaryotic cell and dies. The killing reaction may be apoptotic in nature. The speed with which killing occurs depends on the number of loci with allelic differences, and occurs more rapidly if there are numerous differences. Strains that differ at a single locus may have heterokaryons that persist long enough for weak nutritional complementation to occur. These weak complementation events often last long enough for organelles, *e.g.*, mitochondria, and other cytoplasmic constituents to be exchanged by the two participating strains (740). The killing process as well as the stability of the heterokaryon may depend on numerous genes other than the *vic* loci, that either are not well studied or that have yet to be conclusively linked to the heterokaryon formation process.

Vegetative compatibility can be detected in several different manners. In *Fusarium* and *Neurospora*, vegetative compatibility is detected as the ability of auxotrophic strains to form a prototrophic heterokaryon. In this type of interaction, the positive event is the formation of the heterokaryon. Strains for which no prototrophic heterokaryon is formed may be due to vegetative incompatibility or to the physical inability of one (or both) of the strains to form a heterokaryon with any other strain. Strains that lack the ability to form heterokaryons, including even between mutants derived from the same strain have been termed heterokaryon self-incompatible (403). In this type of system, the scoring error that is usually made is in scoring a compatible reaction as incompatible due to a lack of physiological complementation.

Another type of vegetative compatibility interaction is observed in *Cryphonectria parasitica*, as incompatible wild-type strains form a barrage when paired (58). In such cases, the incompatibility of two strains is indicated by the positive reaction, *i.e.*, the formation of the barrage. Strains that do not form a barrage are termed vegetatively compatible, and the scoring error that is made is to score strains as compatible, because no barrage was formed, when they are not. Barrages have been reported in *Fusarium* (1397, 2052), but this phenotype has not been widely used to score VCGs in population studies.

VCGs provide a crude marker for population genetic studies in that the numerous underlying genes together produce a single result when two strains are compared – the same or different. Isolates in different VCGs are not clones, but isolates in the same VCG could be. If there are 10 segregating *vic* genes in a population, each with two distinguishable alleles, ~1000 (2^{10}) different VCGs, can result. With 15 segregating loci, the number of potential VCGs increases to ~32,000 (2^{15}). If there are more than two alleles per locus, then the number of possible VCGs increases even more rapidly. For example, with three alleles per locus there are ~59,000 (3^{10}) possible VCGs with 10 segregating loci, and > 14,000,000 (3^{15}) possible VCGs with 15 segregating loci.

These VCGs are not expected to be present with equal frequency within even a randomly mating population. The expected frequency for any given VCG is the product of the frequencies of the alleles that constitute it. Only if all of the alleles are present at the same frequency, *e.g.*, 50% for each allele at loci with two alleles, will all of the VCGs in the population occur with equal frequency.

The number of VCGs resulting from a cross depends on the number of heterozygous *vic* loci, and is simply two raised to that power. For example with two segregating *vic* loci each with two alleles there are four possible VCGs, while with six heterozygous *vic* loci, the expected number of VCGs amongst the progeny is 64. If the segregating *vic* loci are unlinked to one another, then the different VCGs in the progeny should all be equally frequent. If there is linkage of two (or more) *vic* loci, or if there is selection either for or against a marker linked to a *vic* locus, then the relative number of progeny in each VCG will not be a constant. Note that strains in the same VCG will be relatively more common if there are relatively few *vic* loci segregating in the cross, and that these progeny, although in the same VCG, will not be clonally related to one another.

Thus if two isolates are clones of a common strain they will always be in the same VCG, but isolates in the same VCG need not be clones. In this sense, the number of VCGs identified in a sample is the minimum number of genetic individuals, *i.e.*, haplotypes, in the sample.

The inability to discern different degrees of relatedness by using VCGs – only whether the strains are identical or not – limits the types of analyses that can be performed. VCGs are most effective when they are used to determine if a particular strain is present in a population. For the most effective analyses, the question usually is of the form, "Are strains belonging to VCG '__' present in a sample?" The question has greatest meaning in asexual populations, but if a particular strain type has a heavy selective advantage or disproportionate presence, then this question may have some meaning in the analysis of a sexual population as well. Comparing VCGs of strains from different populations probably makes sense only if the VCG of interest is reasonably frequent, *e.g.*, > 5% of all strains, in at least one of the populations being analyzed. Similarly, requests for "all VCGs" of some type are meaningful only if the number of VCGs of that type are relatively few, or of great economic or scientific interest.

In *Fusarium*, VCG studies usually have described one of two patterns. In one pattern there are many strains, and many VCGs, with one or perhaps a few strains per VCG. Conclusions from such studies usually are that most strains are different, that there is a great deal of genotypic diversity within the population, and that a mechanism for generating the genotypic variation (commonly sexual recombination) needs to be identified. Further population-based studies of such strains require markers other than VCGs, because the questions usually are: "How related are various strains to one another?" or, "How related are two populations to one another?" rather than, "Are these strains identical?" Sexually reproducing *Fusarium* spp., *e.g.*, *G. moniliformis* (372), *G. fujikuroi* (2088) and *G. zeae* (229), usually have few strains per VCG. Soil populations of *F. oxysporum* are similarly diverse (75, 404, 560), raising the possibility of cryptic sexual reproduction in these populations.

The alternative pattern is for there to be one, or a few, VCGs, encompassing the range of pathogenic strains. It is not uncommon for the isolates of an *F. oxysporum* form species to belong to a relatively few VCGs. Some of the best examples of a relatively limited number of VCGs within a form species include *F. oxysporum* f. sp. *cubense* (167, 1104), *F. oxysporum* f. sp. *cyclamenis* (2365), *F. oxysporum* f. sp. *melonis* (950), *F. radicis-lycopersci* (502, 1014, 1015, 1834), and *F. oxysporum* f. sp. *canariensis* (787, 1713). In general, the number of VCGs increases as the number of isolates and the diversity of the geographic origin of the isolates increases. Isolates within a form species are not always monophyletic in origin (105) or genetically uniform (211).

A numbering system was first proposed for various VCGs by Puhalla (1745). This system has been formalized (1008, 1076) and the assignment of new VCG numbers is being coordinated by Dr. Talma Katan (Volcani Institute, Bet Dagan, Israel).

5.2 Overall Strategy for Determining if Strains are Vegetatively Compatible

A standard strategy for identifying and working with *nit* mutants to determine vegetative compatibility has been developed (Fig. 5-2). This overall strategy is discussed briefly here, and in more detail in the succeeding sections of this chapter.

For making heterokaryons in *Fusarium*, complementary auxotrophic mutants are needed. In principal, any complementary auxotrophic mutations can be used to force the heterokaryons, see the Perkins *et al.* (1686) listing of mutant loci in *Neurospora* for examples of the diverse mutations that can be induced in filamentous fungi. In practice, NO_3 non-utilizing (*nit*) mutants usually are used to force heterokaryons for VCG analyses in *Fusarium* and numerous other species. *nit* mutants arise spontaneously in most strains of *Fusarium* when isolates are cultured on a medium containing $KClO_3$, a toxic analog of $NaNO_3$.

Several classes of *nit* mutants can be distinguished based on their ability to grow on media containing different nitrogenous compounds as the sole source of nitrogen. Mutants in the *nit1*, *nit3* and NitM mutant classes all can be used to force heterokaryons. Mutants in the *crn* class, however, must be identified and discarded.

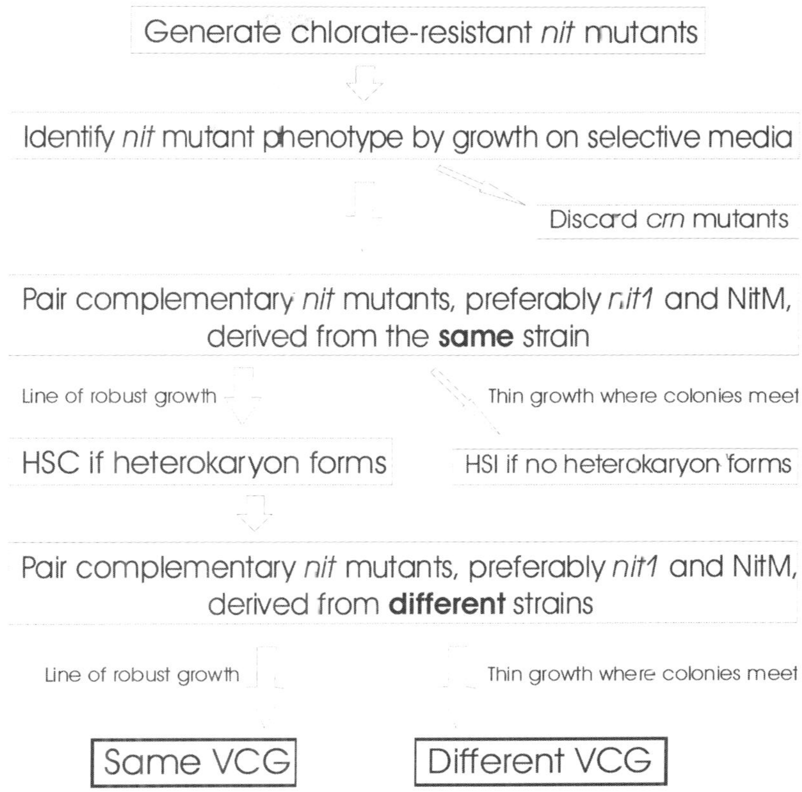

Figure 5-2. Flow chart with general strategy for generating *nit* mutants and determining whether strains of *Fusarium* are vegetatively compatible. HSI – Heterokaryon self-incompatible; HSC – heterokaryon self compatible.

Heterokaryon formation is a complex process that depends on more than just the ability of strains to complement physiologically. Not all field strains are capable of making such heterokaryons. These strains, which are termed HSI – heterokaryon self-incompatible (403) – would be scored, incorrectly, as incompatible with all others if not eliminated from the process before making determinations of vegetative compatibility and assigning isolates to VCGs. These strains are identified by pairing complementary *nit* mutants generated from the same source isolate with one another. Failure of these mutants, which must be in the same VCG if the culture is pure, to produce a prototrophic heterokaryon indicates that the wild-type strain from which they were derived was heterokaryon self-incompatible, and results in the elimination of that wild-type strain from further testing. Strains in which complementary mutants can form heterokaryons are termed "heterokaryon self-compatible" and are included in further testing.

The final step in the overall vegetative compatibility testing process is to make pairings between *nit* mutants derived from different strains. Not all mutants are equally useful in pairing. We usually try to use two (occasionally more) *nit* mutants derived from each strain of interest to definitively determine compatibility. Pairings are made between mutants in different *nit* classes, with *nit1* by NitM pairings generally preferred. A line of heavy growth where the colonies meet on the agar surface indicates complementation. ***N.B.*** The pairings used to determine vegetative compatibility are vegetative and asexual and should not be referred to as "crosses" since that term is reserved for descriptions of sexual interactions.

5.3 Recovering and Identifying *nit* Mutants

The protocol in this section presumes that cultures have been purified through the subculturing of a single conidium (section 4.3). If such subculturing is not done, then mixed cultures can result in extremely confusing results that cannot be meaningfully analyzed.

Inoculating plates for recovery of ClO_3-resistant sectors. Place the isolate from which *nit* mutants are to be generated on MM, PDA, or other medium that results in dense fungal growth. After the colony begins to grow vigorously, usually 2-4 days, take small plugs (1-2 mm^3) from the edge of the colony. Transfer these inoculum blocks to Petri dishes (100-mm diameter) containing MM + ClO_3. Three equally-spaced inoculum blocks per plate (in the form of an equilateral triangle) work well for most isolates. If an isolate grows slowly, then as many as six inoculum blocks may be placed on a single plate. The risk is that a colony that produces ClO_3-resistant sectors relatively quickly will overgrow one that produces these sectors relatively later, and that the sectors produced later

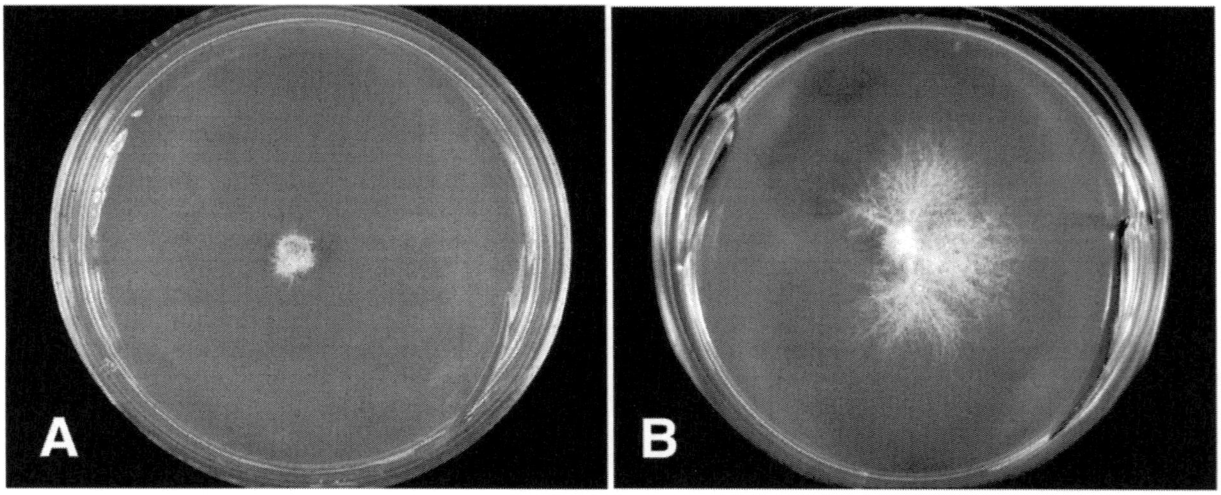

Figure 5-3. Colony of *F. verticillioides* on a MM + $KClO_3$ plate prior to (**A**) and following (**B**) sectoring to produce a ClO_3-resistant, NO_3-non-utilizing (*nit*) mutant.

may not be recoverable. The number of plates per isolate varies with the objective of the experiment, but 10-15 plates per isolate is not uncommon, although smaller numbers may suffice for some studies.

The initial level of ClO_3^- in the medium (section 2.5) usually is 1.5%, but this level may be raised to as much as 6%. The 2-3% level is relatively common. Isolates that are sensitive to $KClO_3$ may neither sector nor grow if the $KClO_3$ level is too high. Thus, it is important to begin screens for *nit* mutants at lower levels of $KClO_3$ and work up to higher levels as necessary, rather than routinely using a higher level for all work. Above the 4% $KClO_3$ level, the ClO_3^- may precipitate as the medium dries, and care must be taken to ensure that the plates remain moist enough to avoid this complication.

Colonies usually grow slowly initially and often are limited to the inoculum block and the area very near to it (Fig. 5-3A). Colonies begin to sector within 3-5 days after inoculation (Fig. 5-3B). These sectors usually are relatively thin and grow much more rapidly than the rest of the colony. If only dense, slow-growing colonies that seem to "creep" across the plate are seen, then the desired *nit* mutants are not being generated, and a change in media usually is needed. Sectoring frequency depends on the isolate, with some sectoring relatively quickly and others relatively slowly (Fig. 5-4). By the end of two weeks, most ($\geq$ 95%) isolates of either *Fusarium oxysporum* or members of the *G. fujikuroi* species complex usually have produced enough *nit* mutants for most studies. As a practical matter, we usually subculture sectors 7, 10 and 14 days after inoculation, and discard the plates after the 14-day subcultures have been taken.

If the number of ClO_3-resistant sectors is low. If ClO_3-resistant sectors occur only rarely, then it may be necessary to alter the medium being used. The first strategy usually is to increase the amount of ClO_3 in the medium. Standard increments are by 0.5%, until the 4% $KClO_3$ level is reached. Beyond this level additional $KClO_3$ usually is of marginal value. Changing the nitrogen source occasionally increases the number of *nit* mutant sectors, but complete removal of the $NaNO_3$ or the asparagine from the medium usually is the most efficient, as also has been documented for *Aspergillus flavus* (156). Researchers who work with other fungi have developed techniques, usually specialized media, for these fungi that often can be used for *Fusarium*, but we have not had extensive experience with any of these media. These media

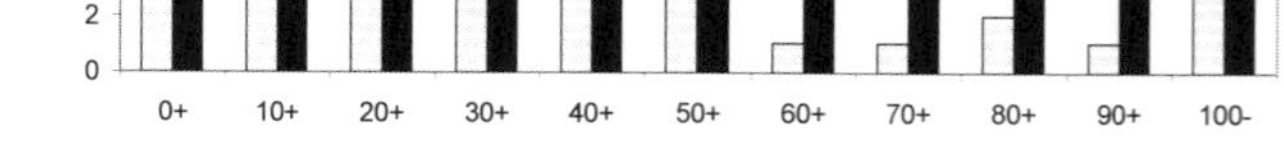

Figure 5-4. Proportion of colonies of *F. verticillioides* producing *nit* mutants on minimal medium + 1.5% $KClO_3$ after one week (stippled) or two weeks (solid black) of incubation at 25°C; after Klittich *et al.* (1089).

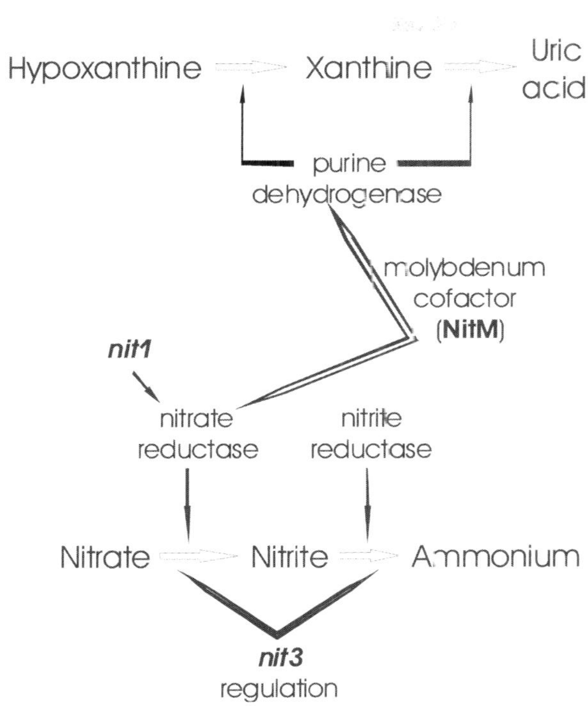

Figure 5-5. NO_3 metabolism pathway as it relates to the generation and classification of *nit* mutants for VCG testing.

include: (i) a water-agar-ClO_3 medium for *Verticillium dahliae* (1126), and (ii) a rose-bengal supplemented medium for several fungal species (557).

Initial characterization of ClO_3-resistant sectors. Sectors usually are subcultured to MM with $NaNO_3$ as the sole nitrogen source as soon as the sector is detected. Strains that grow thinly on the minimal medium are *nit* mutants and are saved for later analysis. Strains with heavy growth that resembles wild type are either the wild-type parent that has escaped the screen, strains that are heterokaryons for more than one *nit* mutation, or *crn* mutants that can both utilize $NaNO_3$ and are resistant to ClO_3. Regardless of their genetic basis (and all three problems are known to occur), cultures with heavy growth on minimal medium with $NaNO_3$ as the only nitrogen source should be discarded. *nit* mutants usually are not purified through a single conidium subculture (section 4.3) unless the *nit* mutant will be used for more than a single test, or as a standard reference. Unusual or questionable results, however, usually necessitate the single spore subculture of *nit* mutants to rule out problems.

Nitrate assimilation. NO_3 utilization in *Fusarium* follows a pattern that is similar to that seen in other ascomycete fungi (Fig. 5-5). NO_3 is reduced to NO_2 by nitrate reductase. This enzyme also can reduce ClO_3 to ClO_2, a toxic process. The NO_2 is then reduced to NH_4 by nitrite reductase. The NH_4 can be used by the cells in many different ways. The loci in *Fusarium* have functional homologues in other filamentous fungi, *e.g.*, *Aspergillus nidulans* and *Neurospora crassa*, but the name assigned to the same locus in the different organisms is not always the same (Table 5-2).

The nitrate reductase enzyme is an octamer of a single polypeptide holoenzyme that is encoded by *nit1*. Strains carrying mutations at *nit1* cannot use NO_3 as a nitrogen source, but are unaffected in their ability to use other nitrogenous compounds as nitrogen sources. In addition to the NIT1 protein, a molybdenum-containing cofactor is part of the active nitrate reductase enzyme. There are five loci that encode for a piece of this cofactor in *Fusarium* (*nit2*, *nit4*, *nit5*, *nit6* and *nit7*), but identification of mutants to the individual locus usually is unimportant in the context of VCG tests. Collectively these five loci are termed NitM mutants as they all confer the same phenotype. Mutants at different NitM loci can complement one another, and the use of strains with mutations at different NitM loci is an important part of the testing protocol described in section 5.5. This co-factor also functions as a part of purine dehydrogenase. Neither nitrate reductase nor purine dehydrogenase can function if one of the pieces of the co-factor has been altered. Thus, NitM mutants can use neither NO_3 nor hypoxanthine as nitrogen sources, but can use other nitrogenous compounds, as nitrogen sources for growth.

There are two regulatory loci that impact NO_3 utili-

Table 5-2. Names used in other filamentous fungi for genes of importance in NO_3 assimilation.

Function	*Fusarium* spp.	*Aspergillus* spp.	*Neurospora crassa*
Nitrate reductase holoenzyme	*nit1*	*niaD*	*nit-3*
Molybdenum co-factor	NitM (*nit2*, *nit4*, *nit5*, *nit6*, & *nit7*)	*cnx* (*cnxABC*, *cnxE*, *cnxF*, *cnxG*, & *cnxH*)	*nit-1*, *nit-7*, *nit-8*, & *nit-9*
Nitrate pathway specific regulator	*nit3*	*nirA*	*nit-4*
Global nitrogen regulator	*nnu*	*areA*	*nit-2*

Table 5-3. Utilization of nitrogen sources in a standard phenotyping screen by *nit* mutants that commonly result after growth on MM + $KClO_3$.

Strain Type	Medium Supplement				
	NH_4	NO_3	NO_2	Hypoxanthine	ClO_3
Wild type	+	+	+	+	-
nit1	+	-	+	+	+
nit3	+	-	-	+	+
NitM	+	-	+	-	+
crn	+	+	+	+	+

zation in *Fusarium*. The first is a global nitrogen regulator, *nnu* (1070, 1218), which is known to be similar to the global nitrogen regulators in *Neurospora* and *Aspergillus* (505). This locus also has been called *AreA* in *G. fujikuroi*, probably *F. fujikuroi* (1445, 2185), but the *nnu* name has priority for *Fusarium*. Mutants in the gene encoding the global nitrogen regulator usually are not recovered as ClO_3-resistant sectors following growth on minimal medium containing $KClO_3$. This protein is a positive regulatory protein that turns on the genes that encode the proteins used to catabolize and assimilate many secondary nitrogen sources (649, 2340). A second regulatory protein, encoded by the *nit3* locus, is a pathway-specific regulatory protein that affects only the NO_3 utilization pathway. Mutants in *nit3* can be recovered following growth on MM + $KClO_3$. These mutants synthesize neither nitrate nor nitrite reductase, and can utilize neither NO_3 nor NO_2 as nitrogen sources.

Phenotyping ClO_3-resistant mutants. The combined abilities of the different classes of *nit* mutants to use various nitrogen sources can be used as a means to distinguish them as well (Table 5-3). Wild type strains can utilize all of the nitrogen sources, but are sensitive to ClO_3. *nit1* strains can grow on all of the media except for those with NO_3 as the sole nitrogen source. *nit3* mutants can grow on all media except for those with NO_3 or NO_2 as the nitrogen source. NitM mutants can grow only on media with NO_2 or NH_4 as the nitrogen source. *crn* mutants (see section 5-4) can grow on all of the media and are both ClO_3-resistant and NO_3-utilizing. In essence the medium containing NH_4 serves as a positive control, and the medium with NO_3 as a negative control.

To identify the phenotype of *nit* mutants begin with cultures that have been growing on minimal medium slants for 3-4 days at 25°C. Although freshly growing cultures are best, slant cultures that have been kept at 5°C or spores from frozen stocks (-80°C) also can be used. Four phenotyping media that differ in their nitrogen source are used for the test – MM + $NaNO_3$, MM + $NaNO_2$, MM + hypoxanthine, and MM + ammonium tartrate. Media should be dispensed into standard Petri dishes (100 × 15 mm). Label the plates on the reverse side, and mark four rows of evenly spaced dots. The top and bottom row should each be three dots, and the middle two rows four dots each. The three dots on the bottom row are used for a *nit1*, a *nit3*, and a NitM control (do not omit the controls!). The remaining 11 dots are inoculated with small (≤ 1 mm^3) inoculum blocks of the *nit* mutants whose phenotype is to be identified. Strains are cultured in the light or the dark at 25°C for 3-4 days.

Interpreting growth on phenotyping media. Results from phenotyping media should result in typing of the strains to the categories in Table 5-3. The control strains should fit into their proper categories. If the controls cannot be scored correctly, then the scoring of the unknown strains should be questioned and the experiment may need to be repeated. Such problems usually are associated with the MM + $NaNO_2$ medium or with the size of the inoculum blocks.

MM + ammonium tartrate is the positive control. All of the strains should grow well. Those strains that do not grow well should be discarded, and should not be scored further.

MM + $NaNO_3$ is the negative control. All strains should grow very thinly, with little aerial mycelium. The colonies may appear to be almost transparent. If vigorous growth occurs, then the culture should be discarded. Vigorous growth may indicate that the mutant has reverted, that a bit of the wild-type colony was subcultured instead of the mutant, that the culture is a mixture of two (or more) *nit* mutants, or that a *crn* (ClO_3-resistant NO_3-utilizing) mutant has been generated. Colonies growing vigorously should be discarded, and should not be scored further.

MM + $NaNO_2$ is diagnostic for *nit3* mutants. *nit3* mutants should grow poorly on this medium (similar to growth on MM + $NaNO_3$). All other strains should grow well on this medium.

MM + hypoxanthine is diagnostic for NitM mutants. NitM mutants should grow poorly on this medium (similar to growth on the $NaNO_3$ medium). All other strains should grow well on this medium.

5.4 Typical pairing protocols

Pairings are done on MM + $NaNO_3$ with NO_3 as the sole nitrogen source. A *nit* mutant will grow thinly on this medium unless it forms a heterokaryon with another mutant. Tests can be made on small (60 × 15 mm) Petri dishes, in 24-well plates, or in 96-well plates.

Pairings in 60×15 mm Petri dishes. Tests conducted in this manner usually are the easiest to read and to score, and usually are relatively easy to document photographically. However, these tests also require more media, incubation space, and time. If a large number of isolates are being examined, then either the 24-well or 96-well microtiter plates are preferable as they require less time and materials than does this method.

Plates are inoculated with small blocks of mycelial (≤ 1 mm^3) cultures that have been growing on minimal medium. Inoculated plates are incubated at 25°C; light is not required. The colonies usually meet after 3-4 days and heterokaryons can be observed beginning 1-2 days later. Heterokaryon formation usually is complete by 14 days.

Usually two strains, A and B, are tested for compatibility in this configuration. Four inoculum blocks are used per plate, with a *nit1* A mutant in the upper left, a NitM B in the upper right, a NitM A in the lower left, and a *nit1* B in the lower right. This combination results in a positive control pairing for each strain, and two tests of the pairing combination. There are three possible outcomes. If a heterokaryon forms between each of the four colony pairs, then the A and B strains are in the same VCG. If a heterokaryon forms only when the A *nit1* is paired with the A NitM, and when the B *nit1* is paired with the B NitM, then the positive controls have both worked, but the A and B strains are in different VCGs. Other results can occur, *e.g.*, only one heterokaryon formed, or 3/4 possible pairs forming heterokaryons. These patterns are relatively rare (≤ 5%), and usually result from the presence of heterokaryon self-incompatible strains (see section 5.5).

Pairings in 24-well plates. This format is the easiest to use if a moderate number of cultures need to be tested. These tests are more rapid and require less medium and incubator space than do the pairings on 60×15 Petri dishes. Persons with relatively little experience with VCG pairings can usually obtain good results from this protocol.

Grow *nit* mutants to be used on small (10×75 mm) slants of complete medium for 2-3 days. Prepare 24-well tissue culture plates by adding 1.7 ml minimal medium per well, usually with a repeating syringe. Plates can be used immediately, but results often are easier to score if plates are allowed to dry for a day or two before use so that the liquid from the spore suspension is absorbed more quickly into the gel.

Make a spore suspension of a *nit* mutant by adding ~1 ml of 2% Tween 60 to the slant culture with a Pasteur pipette. Use the tip of the Pasteur pipette to work the surface of the colony and release the conidia. Use the Pasteur pipettes to dispense small drops of each of the strains involved in the pairing on opposite sides of the well. Try to avoid having the liquid from the two drops come into contact with one another. After inoculation, allow the plates to sit until the liquid is absorbed before moving them to the incubator.

Incubate plates at 25°C; light is not required. Positive reactions appear as lines across the middle of the well if small inoculum drops are used, and as larger fuzzy colonies if the colonies came in contact prematurely (usually because the drops used to inoculate the wells were too large). Scoring for vegetative compatibility (positive reactions) can begin as early as four days after inoculation, with all results usually being clear within seven days following inoculation. The best strategy for the arrangement of strains in these plates depends on the objective of the experiment (see section 5.6).

Pairings in 96-well plates. This format is the most desirable if large numbers of pairings need to be made. These tests are the least demanding in terms of time and materials, but are the most demanding in terms of the precise technical expertise of the person performing the experiment. The general approach and the design of the pairing grid follows that used for the 24 well plates.

Grow *nit* mutants to be used on small (10×75 mm) slants of complete medium for 2-3 days. Prepare 96-well tissue culture plates by adding 200-250 µl MM + $NaNO_3$ per well. The amount of agar used per well is important, since wells with too little agar may dry up before the end of the test and be unscorable, while wells with too much agar may overflow when the spore suspensions are added and cross-contaminate adjacent wells. Plates can be used immediately, but results often are easier to score if plates are allowed to dry for a day or two before use so that the liquid from the spore suspension is absorbed more quickly into the gel.

Make a spore suspension of a *nit* mutant by adding approximately 1 ml of 0.2% Tween 60 to the slant culture with a Pasteur pipette. Use the tip of the Pasteur pipette to work the surface of the colony and release the conidia there. Use the Pasteur pipettes to dispense small drops of each of the strains involved in the pairing in the well. Placing drops on the opposite sides of the wells, as is done with the 24-well plates, usually is the goal, but the relatively small surface area in the well makes this goal difficult to achieve. After inoculation, allow the plates to sit until the liquid is absorbed before moving them to the incubator.

Incubate plates at 25°C; light is not required. Positive reactions occasionally appear as lines, but more com-

Table 5-4. Distribution of ClO_3-resistant sectors into *nit* mutant classes from *F. verticillioides* following growth on minimal medium with the indicated nitrogen source and 1.5% $KClO_3$ [after Klittich & Leslie (1090)].

Nitrogen source	Relative (%) frequency of *nit* mutants	Distribution of *nit* mutants		
		nit1	*nit3*	NitM
L-Asparagine	100	69	19	13
L-Aspartic acid	100	50	31	19
Ammonium tartrate	69	45	36	18
L-Glutamine	106	71	24	6
L-Glutamic acid	75	75	8	17
Hypoxanthine	113	67	17	17
$NaNO_3$	69	64	18	18
L-Ornithine	75	83	17	0
L-Proline	106	65	24	12
L-Threonine	94	40	13	47
L-Tryptophan	88	36	43	21
Urea	6	0	0	100
Uric acid	15	33	47	20

monly appear as dense colonies covering the entire surface of the colony. Results usually are scored after 3-4 days of incubation. Allowing plates to sit for too long before they are scored may result in erratic results since some prototrophic heterokaryons may grow well enough to cross-contaminate adjacent wells. The best strategy for the arrangement of strains in these plates depends on the objective of the experiment (see section 5.6).

5.5 Common trouble spots – HSI, *crn*, and NitMs

No technique is completely trouble-free, and VCG pairings are no exception. Below we identify some of the areas that can commonly cause problems and for which an explanation is available. Studies of the molecular basis for the formation of a stable heterokaryon are currently in progress and should provide a better understanding of the mechanism(s) responsible for the vegetative compatibility phenomenon and its application to the analysis of VCGs in populations.

No or few NitM mutants. The medium and the nitrogen source used affect the proportion of the different *nit* mutant classes that are recovered. The standard is to use minimal medium with asparagine as the nitrogen source, but other nitrogen sources can be substituted (Table 5-4). Other media also can be used, with $KClO_3$-supplemented PDA or Corn meal agar the alternatives most commonly found in the literature. Culturing isolates on PDA + $KClO_3$ usually results in a preponderance of *nit1* mutants. When screening for known VCGs (section 5.6), this property can be very useful. When characterizing a novel population with VCGs (section 5.6), however, this preponderance of *nit1* mutants can complicate the analysis (as mutants in different classes are needed for these tests) and this medium should be avoided. Changing the nitrogen source in the medium, usually by replacing the asparagine with threonine (Table 5-4), often suffices to solve this problem.

Nothing grows on MM + $NaNO_2$. NO_2 can be used as a nitrogen source at low levels, but is toxic at higher levels. The recommended levels (section 2.5) are a balance between having enough NO_2 to be able to grow, and not having so much as to be toxic. Thus, great care must be exercised when making this medium. Most of our problems in this area have been associated with improperly made media, usually too much or too little $NaNO_2$. If nothing grows on MM + $NaNO_2$ when the *nit* mutants are being phenotyped, then *nit1* mutants will be incorrectly scored as *nit3* mutants, and NitM mutants will not have any of the common phenotypes. The controls on each plate are particularly important for detecting this problem whenever it occurs.

Chlorate-resistant, nitrate-utilizing (*crn*) mutants. This class of mutants is a common problem in VCG analyses (1091). In general these mutants are discarded as they cannot be used to force heterokaryons growing on a minimal medium.

There are several mechanisms that could explain the observed phenotype. The simplest is that the activity of nitrate reductase is reduced such that there no longer is enough ClO_3 being converted to ClO_2 to be toxic, but that there is still enough NO_3 being converted to NO_2 for the cell to be viable (Fig. 5-6). Of the five *crn* loci identified, three (*crn1*, *crn3* and *crn5*) have phenotypes that are consistent with this hypothesis. *crn2* and *crn4* both have normal or higher than normal levels of nitrate reductase activity (Fig. 5-6). These loci are not allelic with *nit1* or any of the NitM loci, although *crn2* is closely linked to *nit5*. Thus, it is unlikely that these phenotypes are due to a change in the nitrate reductase enzyme that results in decreased sensitivity to ClO_3. Instead, these mutants probably are similar to the *crnA* mutants of *Aspergillus nidulans* (413) in which the NO_3-uptake mechanism is changed and ClO_3 uptake is reduced or blocked, while enough NO_3 uptake occurs to permit visibly normal growth.

Revertants to wild type. Most investigators do not purify *nit* mutants by single-spore isolations, and there is no guarantee that sectors are genetically pure. The most common problem seen is when a *nit* mutant reverts to the wild-type phenotype. In such cases the mutant should be discarded. The genetic basis for such reversions has not been studied in any detail, but is commonly due to the presence of both wild type and mutant nuclei in a heterokaryon. Such heterokaryons are unbalanced, with the nucleus carrying the *nit* mutant selected for on medium containing $KClO_3$, and the wild-type nucleus selected for on MM + $NaNO_3$. Revertants are readily detected on MM + $NaNO_3$, as they grow as wild type strains on this medium, whereas the *nit* mutants have relatively thin and wispy growth. Thus we recommend that *nit* mutants be routinely maintained on minimal medium, and that they are cultured on media on which wild-type growth can occur only when necessary for the preparation of inocula for VCG tests, or when a large number of spores are needed for further studies. Repeated subculture on complete medium can result in the loss of the *nit* mutant. *nit* mutants usually can be recovered from colonies with both wild-type and *nit* sectors on minimal medium by subculturing from the *nit* sector, but colonies resulting from such subcultures should be purified by a single spore isolation before being used in any further VCG tests.

Irregular pairing patterns. In an ideal setting, VCG pairings are consistently either clearly positive or clearly negative, and all strains from the same VCG behave in the same manner, *i.e.*, pair with all other strains in the VCG and with none of the strains in any other VCG. In most cases these general expectations are met, however exceptional isolates and pairings can be found that do not meet these general expectations. There are two general classes into which irregular pairings fall. In the first case a strain, or set of strains, appears to form heterokaryons with more strains than expected.

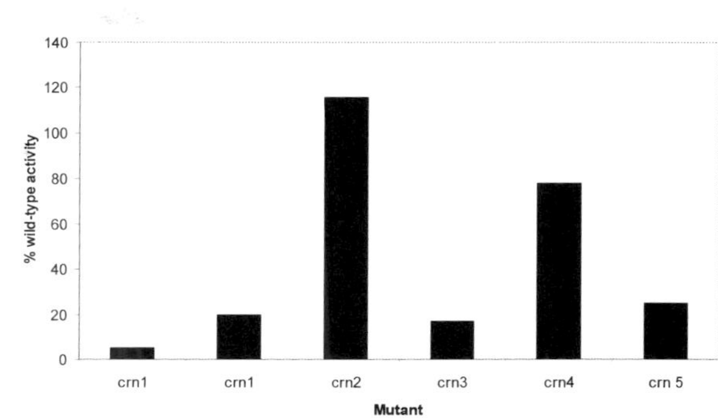

Figure 5-6. Nitrate reductase activity of *crn* mutants relative to their wild-type (wt) progenitors [after Klittich & Leslie (1091)]. Each bar represents the mean of three independent assays of a single mutant.

Heterokaryon formation with "too many" other strains. Vegetative compatibility usually is attributed solely to interactions of alleles at the *vic* loci. As Leslie & Zeller (1231, 1232) have proposed, however, the heterokaryon formation and maintenance pathway (Fig. 5-7) is likely to be much more complex, and recognition factors used by these fungi to distinguish themselves from members of other VCGs also may be used to distinguish themselves from other species or pathogenicity determinants (1143). Natural variants at these loci can result in strains that have unusual pairing interactions, and usually can pair with strains that would otherwise be placed into different VCGs. Our experience is that these types of strains are not common in field populations, and are generally best handled by not being "forced" into any single VCG, but instead should be discussed separately in the text.

In irregular pairings in which strains pair with members of more than one VCG, the resulting heterokaryon usually is not as robust as that formed between members of the same VCG. If these weak pairings occur between

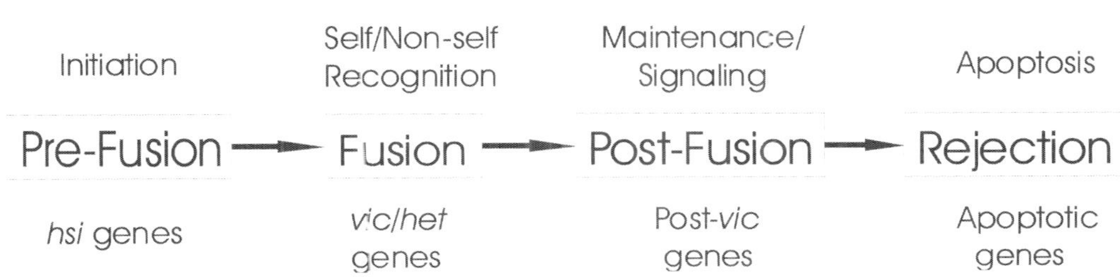

Figure 5-7. Four step model for VCG activity [after Leslie & Zeller (1231). Step one is regulated by loci that result in heterokaryon self-incompatible mutations. Step two is regulated by *vic* loci. Mutants that affect steps three and four are known (Leslie & Zeller, unpublished), but not well characterized.

all of the members of two otherwise distinct VCGs, then the two VCGs likely differ at only one *vic* locus. Such partial heterokaryon compatibility loci have been described in *Aspergillus* (387), and the killing reactions associated with various *het* loci in *Neurospora* also are known to vary (1509). Preliminary genetic evidence (2411) suggests that similar loci exist in *F. verticillioides*. This phenotype probably occurs across the genus and could be responsible for some of the "bridging" reactions that have been reported that can result in some members of apparently distinct VCGs being capable of pairing with one another, *e.g.*, Gunn & Summerell (787), Kerényi *et al.* (1038), and Vakalounakis & Fragkiadakis (2208). If the exceptional pairings are limited to a single strain, then it is likely that the vegetative compatibility results from a mutation in that strain that has disabled part or all of the heterokaryon formation and maintenance pathway.

Heterokaryon formation with "too few" other strains. The most common reason for this observation is that a strain carries a mutation from the *hsi* (heterokaryon self-incompatible or vegetatively self-incompatible) mutant class. These mutants occur in strains from species from the *Liseola* section and from *F. oxysporum* (403, 947). These strains usually are identified when complementary *nit* mutants derived from the same wild-type strain are unable to form a heterokaryon. These strains commonly are segregated from the rest of the strains in a study as soon as the *hsi* diagnosis is made, as each *hsi* strain usually appears to constitute a unique VCG of which the *hsi* strain is the only member. In general, *hsi* strains are reported as a single value, usually a percentage of the total number of strains examined, for the entire study. Occasionally, a *hsi* strain will form a heterokaryon with a *nit* mutant derived from other strains, and tests for heterokaryon formation between *hsi* strains and representative *nit* mutants of the more common VCGs usually should be attempted.

A final type of pairing irregularity occurs when *nit* mutants derived from the same wild-type strain behave differently in pairing reactions with *nit* mutants derived from other wild-type strains. Usually the "abnormal" mutant pairs with only some of the *nit* mutants derived from other strains in the same VCG. When doing large surveys in which strains are assigned to one of a relatively few VCGs, *e.g.*, many studies of individual form species of *F. oxysporum*, it is important to check at least two *nit* mutants of each strain that do not appear to belong to a known VCG before finalizing an assignment. When we identify such problematic *nit* mutants, we usually discard them as soon as a replacement can be generated.

5.6 Characterizing a Population with VCGs

VCGs are commonly used in two general manners to characterize a set of strains. In one case, VCGs are used to characterize strains from a novel population for which no information is available. Such populations may be subsets of larger populations that are uncharacterized but thought to be composed of only one or a few VCGs, *e.g.*, Harveson & Rush (823). In the other case, there is reason to think that strains will belong to one of a relatively limited number of VCGs, *e.g.*, when screening strains from a previously characterized form species of *F. oxysporum*.

Characterizing a novel population with VCGs. In these characterizations, *nit* mutants should be generated on MM + $KClO_3$, as both *nit1* and NitM mutants will be needed for every strain in the population. Mutants need not be purified by subculturing single spores, unless they are selected for use as representatives for their VCG. At least one, and preferably 2-3, *nit1* and NitM mutants should be selected for each strain (these numbers provide back-up mutants should a strain revert during the analysis). We commonly use 10 regular-size (100×15 mm) $KClO_3$ plates, each with 3-4 colonies, per isolate to obtain a series of *nit* mutants over a two-week period. If a NitM strain cannot be obtained for all of the wild-type strains, then a *nit3* mutant may be substituted, but *nit1* + *nit3* pairings usually do not give results that are as clear as those obtained from *nit1* + NitM pairings. These pairings are made most conveniently in 24-well microtiter dishes.

Mutants are arranged in a grid, with *nit1* mutants along one axis and NitM mutants on the other. The number of pairings to be made will be the square of the number of wild-type strains from which the *nit* mutants were derived, plus a control for each mutant in which a microtiter plate well contains only the single mutant (to identify revertants). Tests are inoculated and scored as described above (section 5.4). The grid arrangement results in a positive control for each wild-type strain, and for two tests, between different mutants, for each pair of wild-type strains. Repeating the pairing once is recommended, and questionable results should be evaluated more closely in pairings on MM + $NaNO_3$ in 60×15 mm Petri dishes.

Screening for known VCGs. This technique commonly is used when screening strains from a single form species of *F. oxysporum*, but also is applicable to any other setting in which a only a relatively limited number of VCGs is expected to occur. For each established VCG, two representative complementary NitM mutants, *i.e.*, NitM mutants that will form a heterokaryon when paired with each other, should be generated and purified by single-conidial subculture. For the field isolates, *nit* mutants should be generated on PDA + 1.5% $KClO_3$. These mutants need not have their phenotype determined, and need not be purified by single-conidial subculture. Two *nit* mutants should be generated for each field isolate.

Strains again are screened in a grid arrangement, with the strains representing the established VCGs along one axis, and the strains representing the uncharacterized field isolates along the other. Depending on the phenotype of the *nit* mutants from the field strains, 2-4 of the pairings with the representative of the established VCG to which the unknown strain belongs will be positive. Such strains have had their VCG definitively identified and need not be analyzed further. Strains that are positive when paired with none of the representatives of the established VCGs need to be treated as though they were part of a novel, uncharacterized population, and the studies described in the preceding section should be conducted with some or all of the strains that paired with the representatives of none of the established testers. Some of these strains may be HSI, while others probably belong to rarer VCGs, or to more common VCGs that had not previously been identified. In general, an established VCG will contain enough isolates to constitute at least 5% of the field population being examined.

Thus, the strategy for the analysis of a population with VCGs will depend on how the genetic variation for the *vic* loci is distributed in the population. If the population is primarily clonal, or if particular clonal haplotypes are important from a disease or other perspective, then VCG analyses can be quite useful once the correlations between VCG and the trait of interest have been validated. In populations in which genetic exchange and recombination occur, VCG analyses will probably provide relatively little information other than that every strain is different. In general, VCG studies of these genotypically diverse populations should be avoided.

6

Fertility Concepts

Sexual fertility is an important practical parameter when biological species concepts are being applied, as these species concepts usually require some evidence of sexual cross-fertility before two strains are assigned to a common species. In this chapter, we discuss the basics of sexual cross-fertility as it is understood in *Fusarium* spp., and provide some background based on work done in other ascomycete fungal species. Fungal species may be heterothallic, homothallic, or pseudohomothallic in terms of their sexual reproductive strategies, with only heterothallic and homothallic styles known so far in *Fusarium*. Mating type is a critical parameter for the heterothallic species, and molecular methods of determining mating type now make the identification of this character possible without requiring sexual cross-fertility. Beyond mating type there are a number of different traits that can affect the ability of a strain to complete the sexual portion of the life cycle. Finally we include protocols that can be used to make sexual crosses under laboratory conditions for some of the commonly studied *Fusarium* spp., and provide lists of available standard strains for making crosses for routine diagnostics. Protocols for the amplification of diagnostic DNA fragments from the mating type idiomorphs are given in the following chapter (section 7-2).

6.1 Heterothallic, Homothallic and Pseudohomothallic

Homothallic and pseudohomothallic. The most commonly encountered terms referencing fungal mating systems are heterothallic and homothallic. Homothallic cultures, as originally defined by Blakeslee in 1904 (196), are cultures that produce sexual spores from a culture arising from a single spore. Notice that although homothallic cultures may self, they also may outcross. Within the homothallic fungi there are two groups, homothallic and pseudohomothallic. The true or primary homothallics, *e.g.*, *Aspergillus nidulans*, *Gibberella zeae*, and numerous *Nectria* species that can complete the life cycle with cultures that are initiated from a spore or mycelial fragment that contains a single nucleus. For pseudohomothallic fungi, a culture that originates from a single sexual spore can complete the life cycle. Usually pseudohomothallic fungi grow vegetatively as a heterokaryon and maintain one or more copies of two genetically distinct nuclei in each cell. Pseudohomothallic fungi have all of the genetic mechanisms and controls to function as heterothallics, but have instead developed mechanisms, usually in the form of spatial controls of meiosis (1771) and blocks to recombination, *e.g.*, Gallegos *et al.* (670) that permit them to function as homothallics. There are no known pseudohomothallic species in *Fusarium*, but in many cases studies of life cycles have not been conducted in sufficient detail for the homothallic and pseudohomothallic life styles to necessarily be distinguished.

Homothallic and pseudohomothallic strains are by definition self-fertile. Even though they are potentially self-fertile, it does not mean that they always, or perhaps even commonly, are. Certainly diversity in field populations suggests that a significant amount of outcrossing occurs in *Gibberella zeae* (229, 230, 2409, 2410), and a classical genetic map has been established for this homothallic fungus as well (994). Heterozygosity in crosses with homothallic fungi can occur both accidentally and intentionally under laboratory conditions. The most common ways of forcing crosses in homothallic fungi include:

- Crossing self-sterile mutants that can make perithecia only when crossed with a complementary self-sterile mutant.
- Crossing strains with at least one autonomous spore-color mutant and identifying heterozygous asci which are then the only ones used in the genetic analysis.
- Making crosses between strains carrying complementary nutritional mutants, *e.g.*, *nit* mutants (section 5.3), which usually are unlinked, and analyzing only the prototrophic recombinants (or spores known to have originated from a heterozygous meiosis). This technique has the disadvantage that regions linked to the alleles being selected against will be under-represented in the progeny set as a whole due to their linkage. This technique is described in more detail below (section 6.5) for *G. zeae*, and is the only technique that has been used to force crosses of homothallic wild-type strains of *Fusarium*.
- Partially disable the mating-type locus so that one (or both) of the parents is no longer self-fertile, *e.g.*, Lee *et al.* (1206). If only one parent is disabled, then the cross must be made in such a manner that the self-sterile parent is the only one that has an opportunity to produce perithecia, *i.e.*, the mating-type-disabled parent should be the female parent for the cross. If the crosses are made by using the standard protocols defined below (section 6.5), then the mating-type disabled strain serves as the female parent and the strain with the wild-type mating type is the male parent. Strains with partially disrupted *MAT* loci are available for *G. zeae* (1206).

In heterokaryons of non-*Fusarium* homothallic fungi, whether naturally occurring or forced, karyogamy to form the pre-meiotic diploid nucleus does not occur at random between nuclear types. For example, in *Sordaria fimicola* (1618) and *Aspergillus nidulans* (875, 1719) in heterokaryons with 1:1 nuclear ratios, if nuclear fusions were random with respect to nuclear genotype, then one quarter of the diploids should be one kind of homozygote, one quarter the other kind of homozygote, and one half heterozygotes. If the heterokaryon is unbalanced, then the number of homozygous diploids of the dominant type should increase at the expense, primarily, of the heterozygous diploids. When heterokaryons were constructed and the asci observed however, there was an excess of heterozygous asci, suggesting that not only were nuclear fusions not random, but that they were enriched for heterozygous fusions. Comparable data are not presently available for any homothallic *Fusarium* species.

Heterothallic. Species whose life cycle requires a sexual interaction between two morphologically (usually) indistinguishable, but physiologically distinct, strains arising from separate spores are termed heterothallic. The physiological difference between heterothallic strains that is of importance usually is the difference in mating type. Mating type alleles can be functional or non-functional, with non-functional alleles often lacking some or all of the wild-type activity. Strains carrying non-functional alleles often are sterile and may be blocked in meiosis or in the development of perithecia. Strains that carry different functional mating type alleles and that can interact to potentially form the sexual stage are termed compatible. Strains that share all or part of a mating type allele and that cannot interact to form the sexual stage are termed incompatible. Most of the species in *Fusarium* are heterothallic.

Gibberella sacchari. Sexuality in *G. sacchari* is somewhat unusual and merits special mention. In general, this species is heterothallic in its mating interactions, but occasional strains may make fertile perithecia with all of the resulting ascospores identical to the female parental vegetative mycelium (243). The importance of these events in a field setting is not clear, and probably depends on the relative advantages of sexually and asexually produced spores, and the phenotypic costs associated with their production. In a laboratory setting, these occasional homothallic perithecia can lead to false positives when crosses are scored for fertility as the tester strains for these species, FGSC 7610 and FGSC 7611, can occasionally make these "homothallic" perithecia. When these homothallic perithecia are formed, there usually are relatively few of them in the culture, and even fewer visibly ooze ascospores. Homothallic ascospore formation under these conditions is not consistent, and fertile perithecia often are not seen in every replication. Homothallic perithecia usually are detected only when crosses are made between isolates that normally would be incompatible, *e.g.*, the strains are in the same species and are the same mating type, or the strains belong to different species. Although homothallic perithecia also may form on plates in which heterothallic perithecia occur, the homothallic perithecia presumably are greatly outnumbered by their heterothallic counterparts, would be difficult, if not impossible to identify, and would have little affect on the experimental results unless spores from a homothallic perithecium were analyzed in relatively large numbers.

The presence of homothallic perithecia can be confirmed by analyzing ascospores from putative homothallic perithecia. These ascospores should be genetically identical to the maternal parent and to one another, and should be different from the paternal parent. Any set of genetic markers can be used to make such determinations. A quick solution is to generate *nit* mutants from both parents and from 8-10 progeny, and to check all of them for vegetative compatibility (section 5.6). If the parents are in different VCGs and a sexual cross has occurred, then the ascospore progeny should generally be in VCGs that are different from those of either of the parents (and probably will be in different VCGs from each other as well). If the ascospores were formed in a homothallic manner, however, then the ascospore progeny should all belong to the same VCG. The VCG containing the ascospores should be the same as that of the female parent, and different from that of the male parent.

Thus, instances in which only a relatively few perithecia are observed, especially if a large number normally would be expected, should be evaluated very carefully. The viability and biparental origin of the progeny from such perithecia should be confirmed before concluding that a cross has occurred. The molecular basis(es) for the occasional production of homothallic perithecia by *G. sacchari* is unknown, as is the degree to which this character is found in other *Fusarium* spp. Given what is known of the molecular basis of mating type in *Fusarium*; however, the mating type switching mechanisms know in the yeasts *Saccharomyces cerevisiae* (853), *Schizosaccharomyces pombe* (437), and some other filamentous ascomycetes (1684) probably do not suffice to explain the production of homothallic perithecia by *G. sacchari*.

6.2 Mating type

Mating types commonly are viewed as the master regulatory genes for successful crosses between strains of filamentous fungi. Fungal mating systems range from the simple to the complex. Some are controlled by complex molecules, such as carotenes or steroids (317, 759, 1417), and some by molecular signaling pathways that are still being unraveled (324, 399).

***Mating Type* Nomenclature.** The terminology used to describe a fungal mating system is based on the number of genetic loci, and the number of alleles at those loci, involved in the process. These loci were initially defined on the basis of classical genetics, and most encode multiple transcripts and proteins. The genetically simplest fungal mating system is the dimictic system found in most heterothallic ascomycetes, including all heterothallic *Fusarium* species. In the dimictic mating system, there is a single mating type locus, termed *MAT* in *Fusarium*, that has two functional alleles, termed *MAT-1* and *MAT-2* in *Fusarium*, not *MAT1-1* and *MAT1-2* as in some other plant pathogenic ascomycetes (2192). These alleles are "idiomorphs", *i.e.*, the alleles occur at the same relative position on a chromosome but have no sequence similarity to one another (and may even be of different sizes). Although there are only two functional "alleles" at *MAT*, the number of proteins encoded by an allele is not limited to just one, and the genes necessary for successful mating and perithecial development are not limited to those at the mating type loci, *cf.*, Leslie & Raju (1228) and Raju & Leslie (1769).

In a dimictic mating system, compatible strains have different, often termed the opposite, mating type alleles. Until DNA sequences of the *MAT* alleles became available, the two *MAT* alleles were designated arbitrarily as either "+" or "-". With the sequencing of many of these genes, their molecular homologs have been identified, and the names converted from the old +/- naming system to the newer *MAT-1/2* terminology (Table 6-1). The *MAT-1/2* terminology is now the preferred form and should be used in place of the earlier +/- system.

Although a dimictic mating system is found in most ascomycetes, the genetic basis and organization of the *MAT* idiomorphs can vary considerably. The mating type alleles of *S. cerevisiae*, for example, are quite different from any of those known in *Fusarium*. In addition to controlling the mating process, in some fungal systems, the *MAT* alleles also are responsible for proteins by which nuclei of different mating type can distinguish themselves from one another in the sexual heterokaryon that occurs after fertilization but before karyogamy. There are some general areas of sequence conservation across many filamentous ascomycete genera, with *Fusarium* *MAT* alleles retaining more similarity to those from *Neurospora* and *Podospora*, than to those from *Saccharomyces*, *Schizosaccharomyces*, or *Cochliobolus* (Fig. 6-1).

Table 6-1. Mating type tester strains and published names for mating type idiomorphs in strains from the *Gibberella fujikuroi* species complex [after Kerényi *et al.* (1039), Steenkamp *et al.* (2041), and Zeller *et al.* (2412)].

Fusarium Species	*Gibberella* Species	*G. fujikuroi* Mating Population	+/- Mating Type Designation	*MAT-1/2*[1] Mating Type Designation	KSU[2] Strain Number	FGSC[3] Strain Number
verticillioides	*moniliformis*	A	-	*MATA-1*	A-00149	7600
			+	*MATA-2*	A-00999	7603
sacchari	*sacchari*	B	-	*MATB-1*	B-03853	7611
			+	*MATB-2*	B-03852	7610
fujikuroi	*fujikuroi*	C	+	*MATC-1*	C-01993	8931
			-	*MATC-2*	C-01995	8932
proliferatum	*intermedia*	D	-	*MATD-1*	D-04854	7615
			+	*MATD-2*	D-04853	7614
subglutinans	*subglutinans*	E	-	*MATE-1*	E-00990	7616
			+	*MATE-2*	E-02192	7617
thapsinum	*thapsina*	F	+	*MATF-1*	F-04094	7057
			-	*MATF-2*	F-04093	7056
nygamai	*nygamai*	G	+	*MATG-1*	G-05111	8934
			-	*MATG-2*	G-05112	8933
circinatum	*circinata*	H	-	*MATH-1*	H-10847	9022
			+	*MATH-2*	H-10850	9023
konzum	*konza*	I	n.a.	*MATI-1*	I-11616	8910
			n.a.	*MATI-2*	I-11615	8911

[1]The letter following the *MAT* designation should be used only if referring to the mating population of the *G. fujikuroi* species complex. For example, either *G. fujikuroi MATA-1* or *G. moniliformis MAT-1*.
[2]Strain number from the Department of Plant Pathology, Kansas State University, Manhattan, Kansas.
[3]Strain number from the Fungal Genetics Stock Center, Department of Biological Sciences, University of Missouri – Kansas City, Kansas City, Missouri. FGSC is the preferred source for these strains, and their strain numbers should be included in all publications that use these strains.

These conserved regions have been used to develop degenerate PCR primers that can be used to amplify diagnostic portions of both *MAT* idiomorphs from many *Fusarium* species (section 7.2). Inactivation or deletion of the coding sequences in the *MAT* alleles in most fungal systems usually results in sterility, even though most *MAT* mutants retain some of the components of the mating function(s). The differences in coding sequence prevent the conversion of one idiomorph to another via mutation. Unlike *Neurospora crassa*, the *MAT* alleles in *Fusarium* are not known to have any vegetative compatibility functions.

The mating type alleles in *N. crassa* were cloned in the late 1980s (708, 2032), and those from *Fusarium* some 10 years later (2402). Although sequenced from a few *Fusarium* species, much additional work remains to be done with the *MAT* alleles to determine the genes that they regulate, and how these genes facilitate the sexual reproduction process. The *MAT* region in *F. verticillioides* is flanked relatively closely by the RFLP marker 6E75 (1.7 map units) on one side and by the AFLP marker EAGMAT0087C (0.8 map units) on the other (995). These markers should enable the orientation of the genes in this region. Phylogenies based on the sequences from the *MAT* alleles, sometimes (1601), but not always (2041), are congruent with those based on other gene sequences.

Gibberella zeae. The organization of transcripts in the homothallic *G. zeae* is unusual, in that all of the transcripts encoded by both mating types of the heterothallic fungi are located in a single chromosomal gene cluster in the *G. zeae* genome (Fig. 6-1). The intervening sequences in the heterothallic and homothallic species also differ, although the significance of these differences is not known. This organization for the *MAT* region is quite different from that found in homothallic *Neurospora* species (157, 710), which may have as little as the *A-1* portion of the *A* (*MAT-1*) idiomorph and still be fertile. PCR primers used for diagnostic amplification of the *MAT* region from strains in the *G. fujikuroi* species complex cannot be used to amplify similar sequences from the *G. zeae* genome (1039, 2041), but the more degenerate primers of Kerényi *et al.* (1037) appear to work for most of the *Fusarium* species tested. The *G. zeae MAT* region resembles the organization found in *Sordaria macrospora* (1751) in that all of the *MAT* alleles are at a single locus, but resembles the heterothallic *Gibberella* species, in the relative orientation of the coding sequences for the *MAT-1* related transcripts. The *S. macrospora* mating type locus has been examined in more detail that has its counterpart in *G. zeae*, and has several interesting features including: (i) synthesis of the Smta-1 and SmtA-3 proteins from a dicistronic mRNA, (ii) unusual splicing of the *SmtA-3* coding region to generate several different proteins, and (iii) the chimeric nature of SmtA-3, which has regions of sequence similarity with proteins encoded by both the *A* and *a* idiomorphs of *N. crassa*. Whether these unusual features also carry over to *G. zeae* is not known.

MAT-2. The *MAT-2* idiomorph sequence has a single open reading frame termed *MAT-2-1*, which encodes a protein that has conserved regions that are similar to mating type encoded proteins in several other fungal species including a-1 from *N. crassa*, mat-Mc from *Schizosaccharomyces pombe*, FPR1 from *Podospora anserina*, and MAT-2 from *Cochliobolus heterostrophus*. In both *Cochliobolus* and *Neurospora*, there are two transcripts associated with this open reading frame that differ from one another in terms of how the mRNA is spliced. The number of transcripts associated with this open reading frame in *Fusarium* is not known. The a-1 protein from *N. crassa* is known to bind to DNA sequences that have an HMG-box binding sequence, and may play a role in ascosporogenesis in addition to its role in determining mating specificity. A similar role has been postulated for this protein in *Fusarium*.

MAT-1. The *MAT-1* idiomorph is significantly more complicated than is the *MAT-2* allele (Fig. 6-1). This idiomorph encodes at least three transcripts (*MAT-1-1*, *MAT-1-2* and *MAT-1-3*), with orientation similar to that seen in both *Podospora* and *Neurospora*. In *Neurospora*, but not *Podospora*, it appears that it is critical for these genes to be located in the mating type region if they are to be expressed correctly, with the situation in *Fusarium* currently unresolved. The *MAT-1-1* gene is the only one with homology to *C. heterostrophus*, and it also has detectable similarity with the α1 protein from *S. cerevisiae*. The *MAT-1-2* gene has some similarity to the SMR1 protein from *P. anserina* and the A-2 protein of *N. crassa*. The *MAT-1-3* gene has some similarity to the SMR2 protein of *P. anserina*, to the A-3 protein of *N. crassa*, and contains HMG-like sequences that suggest that it is a DNA-binding protein.

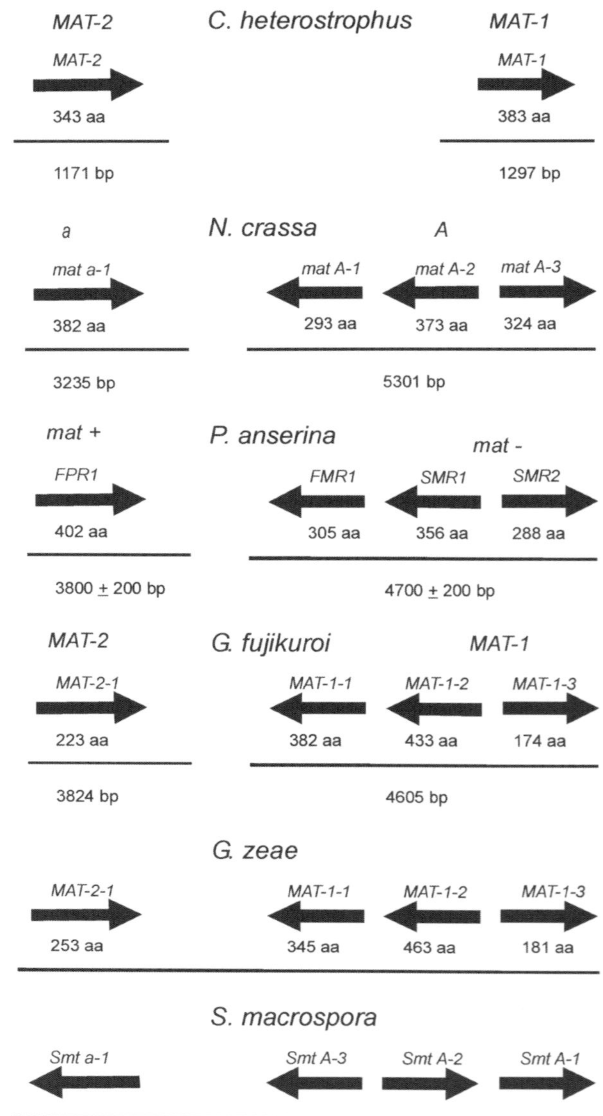

Figure 6-1. Comparative structural organization of the mating type idiomorphs of *C. heterostrophus, N. crassa, P. anserina, G. fujikuroi, G. zeae*, and *S. macrospora* [after Coppin *et al.* (399), Pöggeler & Kück (1751), and Yun *et al.* (2402)]; positions and sizes of introns are not indicated. Sizes of proteins are given in amino acids, and sizes of DNA in bp; lengths are not drawn to scale.

In *Neurospora*, both the A-2 and A-3 transcripts are constitutively produced during vegetative growth and sexual reproduction. Mutants in either *A-2* or *A-3* mate normally and the resulting perithecia have numerous asci and ascospores. A double mutant in both *A-2* and *A-3*, can mate, but the resulting perithecia contain only a few asci and ascospores, suggesting that these proteins have at least some redundant function(s) and are involved in the regulation of sexual development that occurs after fertilization.

Although not examined intensively in *Fusarium*, the mating type loci in other fungi often can be replaced or supplemented with those from another species and the resulting strains still are fertile. The degree of fertility remaining varies depending upon the substitution made, the source of the foreign mating type, whether the native mating type allele is deleted, disabled or functional, and the location and orientation of integration of the foreign mating type into the recipient's genome. In many cases, mating type alleles from species with no known sexual stage are known to be transcribed (80) and can function when introduced into a closely related species. Conversely, the introduction of mating type alleles from a sexually fertile species into strains of a species for which no sexual stage is known usually does not result in sexual fertility of the recipient strains. Thus, mating type sequences and function *per se* appear unlikely to be serving as barriers to inter-specific sexual reproduction, and there is more required to produce a sexually fertile cross than simply two strains with functional, complementary mating type alleles.

6.3 Mating types and their effects on population structure

The biggest effect that mating types have on populations is their influence on the amount of inbreeding that can occur within a population. In general, mating types impose a disassortative mating structure on the population, *i.e.*, an individual is more likely to mate with another individual to whom it is not closely related than it is to an individual to whom it is closely related. Contrast this situation with that found commonly in higher eukaryotes where like tends to mate with like, *i.e.*, assortative mating.

The dimictic system in *Fusarium* has one locus with two functional alleles. This system serves primarily to prevent self-fertilization. Each spore can mate with 2/3 of its sister spores in species with four spores per tetrad or with 4/7 of its sister spores in species with eight spores per tetrad. In addition, each spore can mate with 50% of the "world at large". The expectation for randomly mating populations is that both *MAT* alleles will be present at equal frequencies, but this expectation is not always found, suggesting that asexual reproduction is quite important in many populations of *Fusarium*, even if the members of that population are capable of sexual reproduction.

The *MAT* locus is a classic case of "balanced" selection, in which there is pressure to maintain both alleles at the locus in the population. Such pressure means that the genes linked to the *MAT* region are unlikely to be under heavy selection pressure, as selection in this region could result in an imbalance in the mating type ratio in the population, perhaps even the elimination of a mating type, and an accompanying overall reduction in fertility.

In most other organisms, "male" and "female" are sufficient to define the different types required for a successful cross. In fungi, the "male" and "female" differences (section 6.4) may be important, but the differences in mating type are the critical differences for determining cross-fertility. In a population context, this means that mating type often can be used as a substitute for "male" and "female" in equations derived for other organisms. For example in calculating N_e, the effective population number, for *Fusarium* populations substituting mating type frequencies for the frequencies of the two sexes results in informative equations (1224). N_e is largest when the two *MAT* alleles are present at the same frequency, *i.e.*, 50%, but even highly significant deviations from the expected 1:1 ratio often do not reduce N_e by more than 10% relative to the N_e expected if the *MAT* alleles are present at the equal frequencies.

6.4 Male, female, and hermaphrodite

In some fungi, *e.g.*, the yeasts *Saccharomyces cerevisiae* and *Schizosaccharomyces pombe* and most basidiomycetes, mating type is the only crossing criterion. In the yeasts two cells fuse to form a diploid, with potential fusion partners determined solely by mating type. In most basidiomycetes, haploid mycelia fuse to form a specialized heterokaryon, the dikaryon, that may persist indefinitely but that eventually differentiates the sexual structures. In most filamentous ascomycetes, including *Fusarium*, however, there are clearly defined male and female roles. The female structure, *i.e.*, a protoperithecium, usually is differentiated, but remains immature until it is fertilized by a nucleus carrying an allele of the opposite mating type. In some ascomycetes, a special structure, the trichogyne (194), is used to retrieve the male nucleus that is used for fertilization, but the existence of trichogynes has not been documented for any *Fusarium* species. The protoperithecium (and after fertilization the perithecium) is a specialized structure that is used solely for sexual reproduction. Male structures are not specialized, with ascospores, asexual macroconidia or microconidia, or hyphal fragments all capable of serving as the male nucleus. It is important not to confuse differences attributable to male/female differences with those due to mating type as these functions are distinct in these fungi even though they may be combined in other orgnisms.

Plants can be either monoecious or dioecious, depending on whether an individual produces both male and female sexual structures (monoecious) or only one of the other (dioecious). Attempts have been made to extend this terminology to fungi. In these extensions, a monoecious fungus is one in which an individual can serve as both a donor and a recipient of nuclei for karyogamy. The ability to donate the nucleus is taken as analogous to the male function, and the ability to receive nuclei as analogous to the female function.

All *Fusarium* species are monoecious based on this definition, with an idealized strain in a heterothallic species best described as a "self-sterile" hermaphrodite. This idealized strain can serve as a male through the production of conidia, and as a female through the production of perithecia, but cannot fertilize itself because different mating types are required for a cross to be successful. Perithecial differentiation and development is a complicated process (section 10.3) in many fungi, *e.g.*, *A. nidulans* (2092), in which many genes play a role. A mutation in any one of these numerous genes can result in a female-sterile strain, and female-sterile strains may carry more than one mutation that confers female sterility.

Field strains of *Fusarium* often lack the ability to serve as a female strain in crosses made under laboratory conditions (1224). Thus, strains that are highly fertile as female parents need to be identified or synthesized for use as tester strains if fertile sexual crosses are used as the basis for the assignment of previously unidentified strains to a biological species. The development of female-fertile tester strains may require one to several generations of crosses between strains of marginal fertility, usually as backcrosses or sib crosses, before suitable tester strains are identified (section 6.6).

Male functions are less well understood than are female functions. To date we have detected no strains of *Fusarium* that are female fertile that are not also male fertile. Presumably this "female only" class is missing because strains that are female sterile can still produce conidia, while those that cannot produce conidia do not survive the strain purification process in which strains are subcultured from single spores. Whether there are conidiating male-sterile strains, *i.e.*, strains that produce conidia that cannot satisfy the male function, remains to be determined. Such strains would be detected simply as sterile in a standard set of identification crosses. To be analyzed definitively, such strains need to be assigned to a species on the basis of some other character, *e.g.*, morphology, DNA sequence of a diagnostic gene(s) or AFLP fingerprint, and the inability to cross with the standard tester isolates for that species then rigorously examined. Some cases of perceived male sterility also could be due to homozygosity of genes that block meiosis or other developmental processes, such as the recessive meiotic mutants that are common in field populations of *Neurospora crassa* (1228, 1769).

The ability of a field isolate to serve as a male or a female parent in a sexual cross made under laboratory conditions should not be assumed. In general, male-fertile/ female-sterile strains are relatively common under field conditions. Strains should first be tested for male fertility by using them as the male parent in a cross with a female-fertile tester strain. Mating type may be determined on the basis of a diagnostic PCR reaction (section 7.2), but this *a priori* determination is not essential. If mating type has been identified, then crosses need be made only with the tester strain of the opposite mating type. If mating type is unknown, then crosses must be made with both testers, with only one of the two crosses expected to be fertile. Once male-fertility is established the male/female roles in the crosses are reversed, and the field isolate is used as the female strain to determine female fertility. Tests for female fertility are necessary only if this information is required, and are not necessary if the experimental objective is the identification of a strain to species or the determination of the strain's mating type.

Population significance of "male-only" strains. Male-fertile, female-sterile strains have an unusual role in populations of fungi such as *Fusarium*. Whenever sexual reproduction occurs, the hermaphrodites in the population will contribute something more than half of the gametes to the next generation. Specifically, these female-fertile strains will contribute all of the female gametes and some of the male gametes. The male-only strains will contribute the remainder of the male gametes, but none of the female gametes. Due to the inequality of the contributions to the next sexual generation, the male-only strains are selected against every time sexual recombination occurs. If sexual recombination occurs frequently, then the proportion of male-only strains in the population is small due to the regular selection that occurs every time the population reproduces sexually. If sexual reproduction is not frequent, however, then the proportion of male-only strains in the population will be higher. An important conclusion from this reasoning is that if a significant number of male-only strains is found in the population, then the population is reproducing both sexually and asexually at a significant rate, *i.e.*, there is a mixed mode of reproduction. The frequency of male-only strains can be used to estimate the relative frequency of sexual and asexual reproduction in field populations (1224).

The relative proportion of male-only and hermaphroditic strains in a population provides an alternative to *MAT* allele frequencies for the calculation of N_e (section 6.2). This process has no equivalent in classical population genetics of diploids (1224). N_e based on fertility is at its maximum, *i.e.*, 100% of the number of individuals counted, when all of the strains are hermaphrodites and at its minimum when all of the strains are male-only and no sexual recombination can occur. In the *Fusarium* populations that have been evaluated, the lack of female-fertile isolates usually reduces N_e much more than does any observed imbalance in mating type frequencies (246, 372, 1224). For example, *F. verticillioides*, which is viewed as relatively fertile, has a reduction in N_e attributable to mating type imbalance (237:446) of 9.4%, and a reduction of N_e attributable to female sterility of 11% (341/683 strains female fertile). *F. thapsinum*, which is viewed as relatively infertile, has a reduction in N_e attributable to mating type imbalance (53:40) of only 1.9%, but a reduction in N_e attributable to female sterility of 68% (9/93 strains female fertile).

6.5 Crossing protocols

Sexual stages are known for a number of *Fusarium* species (section 10.5). Some teleomorphs, are known only from natural substrates, while others are known only from laboratory cultures. Recently, sexual studies have been most important for differentiating species in the *Gibberella fujikuroi* species complex, *e.g.*, Leslie (1219, 1221, 1222) and for genetic analyses of *G. moniliformis* (995, 2373) and *G. zeae* (994). We present detailed crossing protocols for these groups. These techniques should be suitable for use with many strains from species known or expected to have a *Gibberella* idiomorph. There are other techniques that have been used successfully as well. These techniques include making crosses on carnation leaf agar – usually with spontaneous formation of perithecia by homothallic strains – and making crosses on natural substrates, *e.g.*, mulberry twigs (469) and rice straw (905), incubated on water agar, or other defined or semi-defined media (506).

Incubation conditions. A number of factors influence the fertility of a cross. The importance of media (section 2.6), mating type (section 6.2), and of male- and female-fertility (section 6.4) are discussed elsewhere in this volume. In addition, light, temperature, and gas exchange also are important factors that can reduce fertility or prevent sexual crosses from forming or developing properly. The addition of exogenous fatty acids, *e.g.*, linoleic acid, also can increase the fertility of some crosses (535).

Light is required for most crosses. The mixture of cool-white and near-UV light that suffices for adequate sporulation (section 4.1) also suffices for sexual development and differentiation. Strains to be used as the male parent have no light requirements other than those necessary for adequate sporulation. Strains to be used as the female parent can be incubated in complete darkness until fertilized. Once fertilized, light is required for perithecium formation.

Temperature plays a very important role in determining the fertility of crosses, and temperatures that are either too cold or too warm can prevent perithecium formation in what should otherwise be a fertile cross. A temperature that is off by just a few degrees can have a major impact on a fertility study, *e.g.*, Britz *et al.* (246) and Covert *et al.* (414). In our experience the maximum for perithecium formation for crosses in most species is 25-27°C. Some species have an optimum in this range, but others grow only asexually at these temperatures. The lowest temperature usually used for crossing is 15°C, but many species form perithecia poorly if the temperature is below 20°C. If the temperature optimum for a species is unknown, then 22-23°C usually is a reasonable range at which to begin.

The importance of gas exchange for perithecial development has not been studied as such in *Fusarium*, but can be important in other fungi where excessive CO_2 can reduce or prevent fruiting body formation (18, 1859, 2424). The importance of gas exchange for perithecium formation by *Fusarium* can be inferred anecdotally however. We have found that crosses growing in Petri dishes sealed with plastic wrap or Parafilm M usually form few, if any, perithecia. Similarly, Petri dishes that are incubated with the lid down can accumulate significant amounts of water as condensate. If this water "seals" the plate, then the fertility of the cross in that dish cannot be adequately assessed. Thus, Petri dishes containing crosses should be incubated with the lid up, and the amount of condensate on the lids monitored (or removed) as necessary.

***Gibberella fujikuroi* species complex.** Crosses have been used to define biological species within this species complex. Although commonly referred to by letter as mating populations or biological species in the form "*Gibberella fujikuroi* mating population __", all of these mating populations are now associated with unique anamorph and teleomorph names (Table 6-1) that in general are preferable and should be used instead. The crossing protocol described below is, in general, that of Klittich & Leslie (1090).

In the laboratory, crosses with these fungi are made by inoculating the strain serving as the female parent on a Petri dish containing carrot agar and allowing the inoculated strain to colonize the substrate. Once the female parent has colonized most or all of the agar surface, the male parent is applied as a spore suspension and the fertilized cross allowed to develop. The validity of this interpretation of the roles of the different strains was demonstrated in *G. moniliformis* by using *pale* (*pal*) mutants that produce perithecia with reduced pigmentation (333). *PAL* is a nuclear gene whose phenotypic expression is limited to strains serving as the female parent in a cross. If the *pal* mutant was the female parent and a wild-type strain the male, then the perithecia were all pale in phenotype, *i.e.*, the genotype of the female parent was expressed. If the roles were reversed and the wild type strain was the female and the *pal* strain the male, then the perithecia were the more common blue-black color.

For a sexual cross to occur both strains must be in the same biological species. One strain must carry the *MAT-1* allele and the other the *MAT-2* allele, although it does not matter which parent (male or female) carries which mating type allele. No other combinations are fertile. At least one of the strains must be female fertile (see section 6.4), and a female-fertile strain must be used as the female parent in the crossing protocol. For routine screening the strain used as the female parent commonly is one of the tester strains for the mating population.

Strains of the female parent are inoculated into Petri dishes, usually 60×15 mm, containing carrot agar, and the strain serving as the male is inoculated on a slant (6-8 ml of medium in a 16×150 mm test tube) of complete medium on the same day. It is very important that the carrot agar plates be poured relatively thick, as thin plates often dry out during the incubation and before the cultures have a chance to develop mature perithecia. The strains are allowed to grow for one week at an appropriate temperature. Light is not essential for either of these incubations. Crosses also may be made in 100×15 mm Petri dishes. Crosses made in the larger Petri dishes require a similar amount of time to mature as do those in the smaller dishes, but require more media and lighted incubation space.

After seven days, conidia from the male parent are suspended in 3-5 ml of a 0.2-0.25% Tween 60 solution. Approximately one ml of the spore suspension is spread on the surface of the female culture, and worked into the mycelia of the female colony with a bent glass rod. We use rods bent 90° in the approximate shape of a hockey stick that can be purchased as "Petri dish spreaders" from Bellco Glass (www.bellcoglass.com). At the end of this process the mycelia usually appear to be wet and appressed. Fertilized plates should be incubated at an appropriate temperature with the lid up and should be no more than two layers deep in an incubator with a mix of cool-white and near-UV fluorescent light bulbs. The length of time until perithecia can be seen depends on the species and the cultural conditions, but generally ranges from 7-35 days post-fertilization. Ascospores usually can be seen oozing from the perithecia of fertile crosses 1-2 weeks after perithecium formation is observed, but probably begin to eject ascospores well before the ascospore cirrhi are observed.

To score a cross as positive, we usually require that a cross be fertile in crosses made on at least two separate occasions. For a cross to be scored as negative, we require that sterile crosses be observed on at least three separate occasions. It is important to include both positive and negative controls (usually intercrosses of tester strains for the species in question) in these experiments. The positive control indicates that the media and cultural conditions were adequate for perithecium formation. The negative control ensures that the female-fertile tester strains were not cross-contaminated.

Gibberella zeae. This species is homothallic and individual strains can form perithecia without a partner strain on both CLA and carrot agar (230). The usual question with this species is how can heterozygous perithecia, or spores with a biparental origin be increased in number and/or distinguished from those formed homothallically. Recombinant progeny have been selected by crossing two mutant strains (often *nit* mutants), and then identifying prototrophic progeny. This process is technically simple, but results in a relative lack of progeny with genetic material in the regions of the parents that carried the marker alleles that were selected against. Alternatively, ascospores can be recovered only from perithecia known to be of biparental origin. In such cases all of the spores can be analyzed, and not just those that are recombinant for the markers being tracked.

Three crossing procedures have been tested for efficacy in *G. zeae*: (i) a spermitization technique, (ii) a mycelial plug crossing method, and (iii) a mixed inoculum crossing method. The spermitization technique is the same as that described above for the members of the *G. fujikuroi* species complex. This technique works well if it is important to be able to identify the male and female parents of a cross, otherwise one of the other techniques is preferred. The spermitization technique also works well if the strain serving as the female parent carries a partially inactivated *MAT* allele (1206). Both the mycelial plug and the mixed inoculum methods are effective means of generating heterozygous perithecia. Heterozygous perithecia in the mixed inoculum technique generally are scattered across the plate while heterozygous perithecia resulting from the interaction of the two colonies in the mycelial plug technique usually are concentrated at the colony interface.

In the mycelial plug method, two small (1 mm^3) mycelial blocks, one from each parent, are placed equidistant from each other and the edge of a Petri dish containing carrot agar. Plates may be incubated under lights or in the dark at 23-25°C. After one week of incubation, one ml of a sterile 0.2-0.25% Tween 60 suspension is added to the culture. The solution is worked into the mycelia with a bent glass rod while rotating the plate. The "fertilized" cultures are incubated in the light as described above for members of the *G. fujikuroi* species complex, with perithecia appearing 2-3 weeks after fertilization.

In the mixed inoculation technique, the initial inoculum is 0.1 ml of a spore suspension containing approximately 10^4 conidia/ml of each of the two parents. The spore suspension is spread evenly across the surface of the carrot agar. The plates are then incubated and treated with Tween® 60 as described for crosses being made via the mycelial plug method.

6.6 Developing Female-Fertile Tester Strains

A critical question has been whether genes other than mating type also play a major role in the crossing process. In some cases relatively equal numbers of mating type alleles are found in a population, but intercrosses of strains known to be of opposite mating type results in only a small percentage of the total possible crosses being fertile. In many cases strains can be backcrossed or intercrossed to identify fertile tester isolates for use in laboratory tests, or as testers for assigning strains to biological species. The ease of generating such testers varies by species, but usually no more than one or two generations of backcrossing or intercrossing is necessary to increase fertility, and is the basis for the standard testers used for *G. sacchari*, *G. intermedia*, *G. thapsina*, *G. nygamai*, and *G. konza* (Table 6-1; 1085, 1092, 1236, 1229, 2412).

The strategy for developing these testers depends on whether there is a trait of importance that must be maintained as a heterozygote or made homozygous in the tester strains, and whether there is a particular genetic background into which fertility needs to be inbred. If a highly fertile field strain has been identified, then the common technique is to make a cross between that strain and another with some degree of female fertility. Progeny are collected and tested for mating type and female fertility. If a sufficiently female fertile strain of the opposite mating type is identified amongst the progeny, then it is used as the tester along with the highly fertile field isolate. If no suitable strain is identified after the first cross, then the progeny strain with the highest level of female fertility is backcrossed to the highly fertile field isolate and the progeny analysis continued until a suitably fertile strain of opposite mating type is identified.

If no highly fertile field strains are available (the usual situation), then the strains to be intercrossed are the field isolates with the highest levels of female fertility detected. Again progeny are evaluated for mating type and female fertility. The most highly fertile progeny of each mating type are selected. If either of these strains is insufficiently fertile, then these progeny are crossed with each other, and the progeny again screened for mating type and female fertility. Successive rounds of progeny (sib) mating continue until sufficiently female-fertile strains are identified.

Usually one or two rounds of sib-mating or backcrossing are sufficient to generate female fertile tester strains. Although these strains will be genetically similar they are not the near-isogenic lines required for some genetic studies. Synthesizing near-isogenic tester strains is much more difficult, and requires a larger number of generations of backcrossing or intercrossing, than does mere identification of a strain pair of opposite mating type with relatively high fertility. See Leslie (1215) for the theory underlying such isogenization processes, which usually requires ten or more generations of backcrossing to a fixed parent, and Klittich & Bronson (1088) for an applied fungal example.

6.7 Species Identification Through Sexual Crosses

Biological species can be identified by making crosses with tester isolates and identifying crosses that produce fertile perithecia. Conceptually, the easiest method is to simply cross every field isolate with every tester isolate (now up to 18 for nine biological species) and then score the results. As noted above, conclusions that a cross is positive should be based on the formation of perithecia on at least two occasions, and the conclusions that a cross is negative should be based on the lack of formation of perithecia on at least three occasions. In practice, not all of the possible crosses with the tester isolates generally are made. Instead crosses are made first with testers for species for which the results are most likely to be positive. For example, isolates recovered from maize should be tested first with the testers of *F. verticillioides*, *F. subglutinans*, and *F. proliferatum*, as these species are those most commonly recovered from maize. If mating type is identified on the basis of PCR amplification, then the number of crosses performed can be reduced by half, as the unidentified strain need be crossed only with the tester for each species that is of the opposite mating type. If AFLP or other DNA sequencing information has been used to make a tentative identification, then the only crosses that need be made are with the tester with which the strain is expected to be fertile. If these crosses are positive, then there is no need to repeat them, as the DNA data serve as the duplicate result to indicate repeatability in terms of species identification.

7

Nucleic Acid Analyses

Analyses of DNA sequences, either directly or as fragments from a genomic DNA fractionation protocol, are becoming common. These analyses have different objectives, and the technique employed varies with the objective. For analyses of taxonomic relationship between species, genomic sequences of one or several genes, *e.g.*, β-tubulin (*tub-2*), translation elongation factor 1-α (*tef-1*), histone H3, or portions of the nuclear or mitochondrial ribosomal RNA coding region, commonly are used. For distinguishing species, particular DNA sequences may be critical, or a pattern of bands on a gel resulting from one of many processes may be used. Some PCR primers that can be used to amplify a specific fragment from a particular species have been developed, but they have yet to be widely tested, and their reliability for analyses of strains from various crops and/or geographic locations is unproven. PCR primers also have been developed for the mating type loci in most species, and can be used for identifying mating type in strains that are not known to be fertile, and to simplify crossing grids in population analyses (section 6.7). Amplified Fragment Length Polymorphisms (AFLPs, 2266) have been used to identify variability within populations and to distinguish species from one another (1346, 2412). The overall patterns resulting from the AFLP process are too variable for use as a phylogenetic tool, but individual bands within a pattern may be of use in such analyses.

Due to the numerous uses for nucleic acid analyses it is very difficult to give a complete listing of protocols that might be used with *Fusarium*. Many of these protocols vary relatively little once the DNA has been isolated from the organisms, whether the original source was bacterial, fungal or human. Instead of attempting to detail all of the possible protocols, we have opted to give a number of protocols that have some fungal and/or *Fusarium* specificity – DNA isolation and purification, AFLP amplification, and mating type allele identification. Other techniques that are either unique to a particular usage or that are more generally applied across molecular biology, *e.g.*, Southern, Northern and Western blots, DNA sequencing, construction of genomic cDNA and EST libraries, and plasmid design and construction are not included.

7.1 DNA extraction and purification

The first step in any analysis of nucleic acids is to extract the DNA from the fungus and purify it. The amount of DNA and the quality needed for the reactions usually determines the extraction protocol employed. In general studies that rely on PCR amplifications for analyses need relatively little DNA. We provide protocols for working with materials from pure cultures. Techniques for extracting DNA from other sources, *e.g.*, soil (1161) are dependent on the substrate being analyzed and the amount of DNA that needs to be recovered.

We include three laboratory protocols – one for relatively large scale preparations that is based on the cetyl trimethyl ammonium bromide (CTAB) protocol of Murray and Thompson (1505), another developed for "mini-preps", and a third for a very quick "micro-prep" that yields only enough DNA for one or two PCR reactions. The quality of the DNA from the micro-extraction is variable. There also are numerous kit and column purifications available. These kits usually work well (make sure to get one that is specific for fungi or for plants and not for bacteria or animal cells), but they are more expensive per isolation than are the preparations outlined below. If only a few isolations are needed or if experience with molecular biological techniques is generally lacking, then the kits usually are worth the extra expense. If a large number of strains need to be extracted at more-or-less the same time, however, then these protocols usually are more cost-effective.

For expression studies and the preparation of libraries, the "maxi" preparation usually is used. For AFLPs and most other PCR analyses, the "mini" preparation or any of the commercially available kits will yield enough DNA for tens to hundreds of reactions. If only a few (< 5) PCR reactions need be run, then the "micro" preparation protocol can be used.

Obtaining material from which to extract DNA. For both the "maxi-prep" and "mini-prep" DNA extraction protocols, liquid cultures must be grown from which DNA is then extracted. The size of the culture depends on the amount of DNA desired. In all cases the culture to be used should be clean (no mites or bacterial or fungal contaminants) and purified through a single-spore subculture (section 4.3).

For the "maxi-prep" protocol dispense 100 ml of minimal (or other appropriate) medium into a 500-ml Erlenmeyer flask, and inoculate with 5-10×10^6 spores. For the "mini-prep" protocol, dispense 25 ml of minimal (or other appropriate) medium in a 125-ml Erlenmeyer flask, and inoculate with 1-2×10^6 spores. Incubations usually last 2-7 days, but shorter is almost always better as less complex carbohydrate is formed by the fungus and accumulates in the medium. This complex carbohydrate can reduce yield and purity of the DNA recovered. If problems with carbohydrate accumulation are encountered, then the inoculum level should be increased and the length of incubation reduced.

Flasks can be incubated static on a bench top at room temperature, but growth generally is much better if incubated on a shaker at 150-200 rpm. Room temperature usually is adequate, but controlling the temperature at 25°C can give somewhat more reproducible results, especially if the room has a tendency towards higher (> 30°C) or lower (< 20°C) temperatures during part or all of the incubation period. Light conditions do not affect the quality or quantity of DNA extracted, and incubating cultures in total darkness is acceptable for these purposes.

Harvest mycelia from the culture(s) by pouring the liquid media containing fungal growth through a non-gauze milk-filter (KenAG, Ashland, OH) – alternatives include Miracloth (CalBiochem cat. no. 475855) or several layers of cheesecloth. Allow the culture filtrate to drain briefly for a minute, then wash the filtered mycelia while still in the filter/funnel twice with sterile distilled water. Remove the washed mycelia from the funnel with a clean spatula onto a pad of clean paper towels, and blot the mycelia dry. At this stage the mycelia can be used immediately, or frozen and stored unground, wrapped in aluminum foil at -20°C for at least several months.

"Maxi-prep" DNA extraction protocol [after Kerényi *et al.* (1039)].

1. Prepare buffers.

   CTAB buffer solution (2% CTAB, 100 mM Tris-HCl pH 8.0, 20 mM EDTA and 1.4 M NaCl). The following recipe is for 500 ml of buffer.

   - Place 10 g of dry CTAB [Sigma (St. Louis, Missouri), cat. no. H-6269] in a 1-L flask. Add 350 ml distilled H_2O.
   - Stir on a stir plate for a few minutes, then heat in a microwave oven on high for one minute.
   - Return to the stir plate and mix for an additional 1-2 minutes, and then heat in the microwave for an additional minute.
   - Return to the stir plate and add the Tris (50 ml of 1 M Tris-HCl pH 8.0 stock or 6.06 g solid). Allow the Tris to dissolve and mix before continuing.
   - Add the EDTA ($Na_4EDTA \cdot 2H_2O$; 20 ml of 0.5 M EDTA stock), and continue mixing. The pH must be ≥ 8 for the EDTA to stay in solution.
   - Add the NaCl (40.9 g). Continue mixing until everything is dissolved.
   - Transfer the solution to a graduated cylinder and adjust the final volume to 500 ml by adding distilled water.
   - Store the solution in a capped bottle.
   - Add 2-mercaptoethanol (β-mercaptoethanol) at 1% (v/v) immediately before use. Once the mercaptoethanol has been added, the solution should be kept tightly capped and opened only in a chemical hood or other very well ventilated area, preferably with an outdoor exhaust.

   1× TE Buffer (pH 8.0) (= 10 mM Tris Base, 1 mM Na_2 EDTA). Make as a 50× stock solution and dilute as needed for use. Recipe is for 1 L of the 50× stock.

   - To a 2-L beaker add 700 ml of double-distilled water.

- Add 60.6 g of Tris base (500 mM) and 18.6 g of Na_2 EDTA (50 mM) to the water.
- Stir vigorously on a stir plate.
- After the Tris and the EDTA are both in solution, adjust the pH to 8.0 with concentrated HCl.
- Adjust the final volume to 1 L with double-distilled water and autoclave.

2. Approximately 1-2 hr prior to beginning a set of extractions, place the extraction buffer into a 65°C incubator or water bath to warm it. Alternately, the buffer can be heated to approximately 65°C in a microwave oven immediately prior to use.
   - Turn on the centrifuge and adjust the temperature to 4°C. Place the correct rotor in the centrifuge and allow it to cool.
3. Remove frozen mycelial sample (1.5-2 g wet weight) from -20°C freezer. **N.B.** The total number of samples that can be processed at one time is limited by the number of slots in the centrifuge rotor.
   - Grind to a fine powder with liquid N_2 in a chilled mortar and pestle. Use a separate mortar and pestle for each extraction to avoid cross contamination.
   - Transfer ground mycelia to a sterile, labeled 25- or 40-ml screw cap centrifuge tube.
   - Store at -70°C (for up to several months if necessary) until ready for the next step.
4. Remove ground mycelial samples from -70°C freezer.
   - Add 8 ml of hot CTAB buffer and 80 µl of 2-mercaptoethanol (if not already in the buffer) to each tube. Cap tightly. Invert to mix. (Note: It is easiest to prepare enough buffer with 2-mercaptoethanol for all of the tubes in a flask or beaker and then dispense it in aliquots to the tubes.) Beginning at this step, bench work should be conducted in a chemical hood to minimize exposure to 2-mercaptoethanol.
   - Incubate at 65°C for 30 min. on a shaker with gentle agitation.
5. Add 8 ml of chloroform:isoamyl alcohol (24:1 v:v) to each tube.
   - Mix with gentle rocking for an additional 5-10 min.
   - Centrifuge at ~6500 × *g* for 10 minutes.
   - Transfer the top (aqueous) 6 ml to a fresh sterile screw cap 25-ml centrifuge tube.
   - Discard the remaining liquid in a chloroform:isoamyl alcohol waste disposal bottle. Do not pour down the sink!
6. Add 6 ml of isopropanol (2-propanol) to each tube.
   - Mix with gentle rocking for 5-10 min.
   - Centrifuge at ~9,200 × *g* for 10 min.
   - Drain the supernatant and the invert tubes on a paper towel to dry pellet for 5-10 min (or overnight).
7. Resuspend crude pellets in 1 ml of TE buffer (10 mM Tris-HCl, 1 mM EDTA, pH 7.4). Use a 1000-µl pipetter (or equivalent), and a fresh tip for each pellet. Be patient. A large pellet may take some time to dissolve. Working the pellet up and down in the pipette tip can help the pellet dissolve, but excessive pipetting can shear the DNA into relatively short fragments.
   - Allow the solution to stand until the pellet dissolves completely.
   - Transfer to a labeled 1.5-ml microcentrifuge tube.
8. Add phenol:chloroform:isoamyl alcohol (25:24:1 v:v:v) to fill the microcentrifuge tube (~500 µl).
   - Mix well and vortex briefly.
   - Centrifuge at ~9,200 × *g* for 5 min in a microcentrifuge.
   - Transfer the top (aqueous) approximately 800 µl to a fresh, labeled microcentrifuge tube. Use a 1000-µl pipetter and a fresh tip for each tube.
   - Discard remaining liquid in a phenol:chloroform waste disposal bottle. Do not pour down the sink!
9. Add 500 µl chloroform:isoamyl alcohol (24:1, v:v), but do not overfill the microcentrifuge tube.
   - Centrifuge at ~9,200 × *g* for 5 min in a microcentrifuge.
   - Transfer the top (aqueous) ~500 µl to a fresh, labeled microcentrifuge tube. Use a 1000-µl pipetter and a fresh tip for each tube.
   - Discard the remaining liquid in a chloroform:isoamyl alcohol waste disposal bottle. Do not pour down the sink!
10. Add 1.5 µl of RNase A (Sigma cat. no. R-4642, purchase as a solution in 50% glycerol and use without diluting, concentration 2 mg/ml) to each tube. Use a 10 or 20-µl pipetter and a fresh tip for each tube.
    - Incubate at 37°C for 30 min.
11. Add 1 ml isopropanol to each microcentrifuge tube.
    - Mix each tube by repeatedly inverting it until the DNA begins to precipitate.
    - Centrifuge at ~9,200 × *g* for 5 min. in a microcentrifuge.
    - Drain off the supernatant.

- Add 1 ml ice cold 70% ethanol to each tube.
- Invert each tube gently several times without disturbing the pellet.
- Drain off the supernatant.
- Add 1 ml ice cold 70% ethanol to each tube.
- Invert each tube gently several times without disturbing the pellet.
- Drain off the supernatant.
- Centrifuge at ~9,200 × *g* for 30 sec. in a microcentrifuge.
- Remove any remaining liquid from the pellet with a 200-µl pipetter (or equivalent). Use a different tip for each tube.
- Place tubes at 37°C for 10 min.
- Resuspend dried pellet in 100 µl of TE buffer per tube. Store DNA at 4°, -20° or -70°C until used. If the pellet is small, then resuspending in 50 µl may make future work with the sample easier.

"Mini-prep" DNA extraction protocol (based on protocol adapted by Amgad Saleh and Kurt Zeller).

1. Prepare buffers as described in step one of the "Maxi-prep" procedure.
2. Approximately 1-2 hours prior to beginning a set of extractions, place the extraction buffer into a 65°C incubator to warm it. Alternately, the buffer can be heated to ~65°C in a microwave oven immediately prior to use.
3. Remove a frozen mycelial sample (~200 µg) from the freezer (-20 or -70°C) and grind to a fine powder in a mortar under liquid N_2.
   - Place enough of the ground mycelium into a labeled 1.5-ml microcentrifuge tube to fill the tube to approximately the 500-µl level. Work quickly to avoid having the sample thaw prematurely as this will allow DNAses to begin to act (and reduce the quality of the DNA extracted).
   - Place the partially filled tube into a regular freezer (-20°C) until all of the samples have been ground. Store any extra ground mycelia frozen (-20°C) in case it is necessary to repeat the extraction.
4. Remove microcentrifuge tubes containing ground mycelia from the freezer and add 700 µl of hot (65°C) 2% CTAB buffer with 2-mercaptoethanol to each sample. Again, work quickly to avoid prematurely thawing the samples. When the microcentrifuge tubes are open, work only in a fume hood to avoid releasing 2-mercaptoethanol fumes into the lab (it stinks!).
   - Briefly (2-3 sec.) vortex, or otherwise vigorously mix, each sample to disperse any clumps of mycelia.
   - Place the tubes in a 65°C incubator for 5-10 min.
   - Remove the extraction tubes from the incubator and vortex (2-3 sec.) each sample again to ensure that the contents are well mixed.
   - Return the tubes to the 65°C incubator and continue the incubation for another 20-25 min. (30 min. total incubation at 65°C).
5. Move the extraction tubes from the incubator to the fume hood.
   - Add 300-400 µl of chloroform:isoamyl alcohol (24:1, v:v) to each tube.
   - Vortex each sample briefly (a few seconds) to mix the aqueous and organic phases.
   - Continue to gently mix (by hand or on a rotary shaker) these samples for an additional 2-3 min.
6. Place the tubes in a table-top microcentrifuge and centrifuge at ~11,400 × *g* for 5 min. to separate the organic and aqueous phases.
   - Carefully remove the tubes at the end of the centrifuge run to avoid disturbing the layer of cellular debris that will form between the two liquid layers and move to the fume hood. If the middle layer of cell debris is disturbed, then repeat the centrifugation with the affected samples.
   - With a 1000-µl pipetter, transfer 600 µl (or as much of this layer as possible) of the aqueous (upper) phase of each sample to a fresh, sterile, microcentrifuge tube without disturbing the debris in the middle layer.
   - The debris layer and the organic phase should be discarded as appropriate, which in many cases is as a hazardous waste.
   - Add 600 µl of isopropanol (2-propanol) to the recovered aqueous phase of each extraction.
   - Invert each tube several times to mix well.
   - Allow the tubes to sit for ~5 min. at room temperature for the nucleic acids to precipitate. These tubes may be stored overnight at 4°C or -20°C, or you may continue immediately to the next step.
7. Centrifuge the tubes containing the samples for 5 min. at ~9200 × *g* to pellet the nucleic acids.
   - Remove the tubes from the microcentrifuge and decant and discard the aqueous/alcohol mixture from the pellets.

- Invert the tubes onto a clean paper towel and allow them to air-dry for 4-5 min.
- Add 600 µl of 1× TE buffer to the crude pellets and resuspend them by mixing. These tubes may be stored overnight at 4°C or -20°C, or you may continue immediately to the next step.

8. Once the pellets are completely resuspended, add 200-300 µl of phenol:chloroform:isoamyl alcohol (25:24:1 v:v:v) to each microcentrifuge tube.
   - Vortex each sample briefly (a few seconds) to mix the aqueous and organic phases.
   - Continue to gently mix (by hand or on a rotary shaker) these samples for an additional 1-2 min.
   - Place the tubes in a microcentrifuge and centrifuge them at ~11,400 × *g* for 5 min. to separate the organic and aqueous phases.
   - Carefully remove the tubes from the microcentrifuge at the end of the centrifuge run to avoid disturbing the layer of denatured proteins that will form between these two layers. If the middle layer is disturbed, repeat the centrifugation with the affected samples.
9. With a 1000-µl pipetter, transfer 500 µl (or as much of this volume as possible) of the aqueous (upper) phase of each sample to a fresh, sterile, microcentrifuge tube without disturbing the debris in the middle layer.
   - The debris layer and the organic phase should be discarded as appropriate, which in many cases is as a hazardous waste.
   - Add 1 µl of RNAse A (Sigma cat. no. R-4642, purchase as a solution in 50% glycerol and use without diluting) to each sample.
   - Vortex each sample briefly to disperse the RNAse.
   - Incubate at 37°C for 30 min. These tubes may be stored overnight at 4°C or -20°C, or you may continue immediately to the next step.
10. Remove the microcentrifuge tubes from the incubator.
    - Add 500 µl of isopropanol (2-propanol) to each tube.
    - Invert each microcentrifuge tube several times to mix.
    - Allow the microcentrifuge tubes to sit for ~5 min. at room temperature for the nucleic acids to precipitate. You may store these tubes overnight at 4°C or -20°C or continue to the next step.
11. Centrifuge the tubes containing the samples for 5 min. at ~11,400 × *g* to pellet the DNA.
    - Remove the tubes from the microcentrifuge.
    - Decant and discard the aqueous/alcohol mixture from the pellets.
    - Invert the tubes onto a clean paper towel and allow them to air-dry for 4-5 min.
12. Wash the DNA pellets with 1 ml of ice cold 70% (v/v) ethanol (may keep the 70% ethanol in a -20°C freezer) by adding the ethanol, and then decanting. Be careful not to wash out the pellet, as it can be easily dislodged and decanted along with the ethanol.
    - Allow the tubes to air-dry inverted on a clean paper towel for 2-3 minutes.
    - Repeat the wash one time.
    - After the second wash, briefly (4-5 minutes) dry the <u>open</u> tubes in a 65°C incubator. Most of the remaining ethanol will evaporate, but small amounts of water may remain in the tubes (don't worry about any small remaining droplets).
    - Resuspend each pellet in 50 µl of TE buffer and store at 4°C.

For both the "maxi-" and "mini-" DNA preparations, DNA should be quantified before storing. PCR applications require relatively little DNA, and the DNA suspensions may need to be significantly diluted before use.

"Micro-prep" DNA Extraction Protocol [after DuTeau and Leslie (530) and Lee *et al.* (1207)]. This protocol uses SDS, which can interfere with DNA restriction enzymes and other enzymes commonly used to process extracted DNA. If there are problems with DNA prepared using this method it usually is because not all of the SDS has been removed. Due to the relatively small quantity of DNA recovered, it usually is preferable to repeat the process for the affected sample(s), rather than to try to remove the excess SDS from the samples that have already been processed.

1. Grow a culture on a 75 mm slant of minimal (or other appropriate) medium for 2-5 days. The test tube in which the slant is made should be made of Pyrex glass (Corning cat. no. 9820-10), or equivalent, to help prevent shattering of the tube. Cheaper disposable tubes that may work well for other applications often do not work well for this DNA isolation protocol.
2. Prepare the lysis buffer [50 mM Tris (pH 8.0), 50 mM EDTA, 3% SDS, 1% β-mercaptoethanol, and 0.1 mg/ml Proteinase K]. The following recipe is for 500 ml of buffer.
   - To a 1-L beaker add 350 ml of double-distilled water.

- Add the Tris [25 ml of 1 M Tris-HCl (pH 8.0) stock or 3.03 g solid]. Allow the Tris to dissolve and mix before continuing.
- Add the EDTA ($Na_4EDTA \cdot 2H_2O$; 50 ml of 0.5 M EDTA stock), and continue mixing. The pH must be ≥ 8 for the EDTA to stay in solution.
- Add 150 g SDS (sodium dodecyl sulfate or sodium lauryl sulfate, Sigma cat. no. L-6026). Continue mixing until everything is dissolved.
- Transfer the solution to a graduated cylinder and adjust the final volume to 500 ml by adding distilled water.
- Adjust the final volume to 1 L with double-distilled H_2O and autoclave in a loosely capped bottle.
- Add 2-mercaptoethanol (β-mercaptoethanol) at 1% (v:v) and Proteinase K (Sigma cat. no. P-6556) at 0.1 mg/ml to the solution immediately before use.

3. Transfer a cube (5 mm^3) of mycelia-covered agar to a 1.5-ml microcentrifuge tube.
   - Fill the tube with liquid N_2.
   - Let the liquid N_2 evaporate.
   - Refill the tube with liquid N_2 and let it evaporate again. No grinding is necessary.
4. Add 500 μl of hot (65°C) lysis buffer to the microcentrifuge tube with a 1000-μl pipetter.
   - Vortex tube and incubate for 30 min. at 65°C.
   - Vortex and incubate for an additional 30 min (1 hr. total) at 65°C.
5. Add 500 μl of phenol:chloroform (1:1) with a 1000-μl pipetter.
   - Vortex each sample briefly (a few seconds) to mix the aqueous and organic phases.
   - Continue to gently mix (by hand or on a rotary shaker) these samples for an additional 2-3 min.
   - Place the tubes in a microcentrifuge and centrifuge at ~9200 × *g* for 5 min. to separate the organic and aqueous phases.
   - Carefully remove the tubes at the end of the centrifuge run to avoid disturbing the layer of cellular debris that will form between these two layers and move to the fume hood. If this middle layer is disturbed, then repeat the centrifugation with the affected samples.
   - With a 1000-μl pipetter, transfer 450 μl (or as much of this layer as possible) of the aqueous (upper) phase of each sample to a fresh, sterile, microcentrifuge tube without disturbing the debris in the middle layer.
   - The debris layer and the organic phase should be discarded as appropriate, which in many cases is as a hazardous waste.
6. Add 450 μl of phenol:chloroform (1:1) with a 1000-μl pipetter.
   - Vortex each sample briefly (a few seconds) to mix the aqueous and organic phases.
   - Continue to gently mix (by hand or on a rotary shaker) these samples for an additional 2-3 min.
   - Place the tubes in a microcentrifuge and centrifuge at ~9200 × *g* for 5 min. to separate the organic and aqueous phases.
   - Carefully remove the tubes at the end of the centrifuge run to avoid disturbing the layer of cellular debris that will form between these two layers and move to the fume hood. If this middle layer is disturbed, then repeat the centrifugation with the affected samples.
   - With a 1000-μl pipetter, transfer 400 μl (or as much of this layer as possible) of the aqueous (upper) phase of each sample to a fresh, sterile, microcentrifuge tube without disturbing the debris in the middle layer.
   - The debris layer and the organic phase should be discarded as appropriate, which in many cases is as a hazardous waste.
7. Add 400 μl of chloroform:isoamyl alcohol (24:1) with a 1000-μl pipetter.
   - Vortex each sample briefly (a few seconds) to mix the aqueous and organic phases.
   - Continue to gently mix (by hand or on a rotary shaker) these samples for an additional 2-3 min.
   - Place the tubes in a microcentrifuge and centrifuge at ~9200 × *g* for 5 min. to separate the organic and aqueous phases.
   - Carefully remove the tubes at the end of the centrifuge run to avoid disturbing the layer of cellular debris that will form between these two layers and move to the fume hood. If this middle layer is disturbed, then repeat the centrifugation with the affected samples.
   - With a 1000-μl pipetter, transfer 350 μl (or as much of this layer as possible) of the aqueous (upper) phase of each sample to a fresh, sterile, microcentrifuge tube without disturbing the debris in the middle layer.
   - The debris layer and the organic phase should be discarded as appropriate, which in many cases is as a hazardous waste.

8. Add 50 µl of 7.5 M NH_4 acetate to the microcentrifuge tube with a 100-µl or 200-µl pipetter.
   - Mix gently.
   - Add 880 µl of ice cold 95% ethanol to the microcentrifuge tube with a 1000-µl pipetter.
   - Invert to mix and/or vortex briefly.
   - Place tube in -20°C freezer for 30 min. to overnight to allow the DNA to precipitate.
9. Centrifuge the tubes containing the samples for 5 min. at ~11,400 × *g* to pellet the DNA.
   - Remove the tubes from the microcentrifuge.
   - Decant and discard the aqueous/alcohol mixture from the pellets.
   - Invert the tubes onto a clean paper towel and allow them to air-dry for 4-5 min.
10. Wash the DNA pellets with 1 ml of ice cold 70% (v/v) ethanol (may keep the 70% ethanol in a freezer at -20°C) by adding the ethanol, and then decanting. Be careful not to wash out the pellet, as it will be quite small and can be easily dislodged and decanted along with the ethanol.
    - Briefly dry (4-5 min.) the open tubes in a 65°C incubator. Most of the remaining ethanol will evaporate, but small amounts of water may remain in the tubes (don't worry about any small remaining droplets).
    - Resuspend each pellet in 10-20 µl of TE buffer and store at 4°C.

7.2 PCR – Mating-Type Alleles

Traditionally mating type is diagnosed through sexual crosses with an appropriate strain in the same species but of the opposite mating type (section 6.5). This process requires identification of female-fertile strains, an appropriate medium and crossing conditions, and considerable time (sometimes 6-7 weeks) and incubator space. Portions of the mating type genes are sufficiently conserved across species that PCR primers can be designed for their amplification. The primers developed by Kerényi *et al.* (1039) and Steenkamp *et al.* (2041) work well for strains belonging to *F. oxysporum*, or to the species closely related to it or contained within the *Gibberella fujikuroi* species complex. A less restrictive set of somewhat degenerate primers designed by Kerényi *et al.* (1037) can be used to amplify *MAT* genes from at least 22 different *Fusarium* species that have no known sexual stage or that have teleomorphs in genera such as *Gibberella*, *Nectria*, and *Calonectria*. The protocols for the Steenkamp *et al.* (2041) and the Kerényi *et al.* (1037) primers are similar. Differences between them are noted in the protocol below.

Specific PCR Amplification of Mating Type Alleles from *Fusarium*.

1. Isolate DNA from a pure culture of *Fusarium* (the microprep protocol usually suffices). After isolation, determine the approximate DNA concentration.
2. Dilute the DNA in sterile double-distilled water to a working concentration of ~20 ng/µl.
3. Most researchers run these reactions for multiple strains at the same time. Separate the commonly used strip tubes (200 µl capacity) into two separate groups, or label one set "*MAT-1*" and the other "*MAT-2*", with one tube in each set for each strain being tested (or use tubes of two different colors to distinguish the two reaction classes).
4. Transfer 1.0 µl of each DNA sample into two replicate PCR tubes, so that each 200-µl tube in a strip has a different sample. Change pipette tips between DNA transfers to avoid cross-contamination. Record the loading order of each sample in the strip of tubes using the molded letters/numbers on the tubes as a key. At the end of this step, there should be two identical sets of tubes containing DNA from the strains to be tested. Place all of these tubes on ice until used.
5. Prepare reaction "cocktails." A separate set of reaction cocktails is made for each allele or primer used. The recipes assume 10 reactions per set, and should be scaled up as necessary. The reaction cocktail contains the common reagents, often everything except the DNA from the specific strain, and permits the transfer of a single larger volume to the reaction tube. This process is more accurate and less error-prone than is separately adding several small volumes to many reactions. The order in which the ingredients are added does not matter; however, it usually is good practice to add the reagent with the smallest volume first and that

Table 7-1. Primers for use in diagnostic amplification of mating type (*MAT*) region.

	Steenkamp *et al.* (2041)	Kerényi *et al.* (1037)
MAT-1	5'-GTTCATCAAAGGGCAAGCG-3'	5'-CGCCCTCTKAAYGSCTTCATG-3'
	5'-TAAGCGCCCTCTTAACGCCTTC-3'	5'-GGARTARACYTTAGCAATYAGGGC-3'
MAT-2	5'-AGCGTCATTATTCGATCAAG-3'	5'-CGACCTCCCAAYGCYTACAT-3'
	5'-CTACGTTGAGAGCTGTACAG-3'	5'-TGGGCGGTACTGGTARTCRGG-3'

A – Adenine; C – Cytosine; G – Guanine; T – Thymine; K – G or T; R – A or G; S – C or G; Y – C or T.

with the largest volume last. By so doing, the addition of the larger volumes can be used to mix the other ingredients added in smaller volumes. Change pipette tips between all stocks of reagents to avoid cross-contamination.

- Steenkamp *et al.* (2041) reaction cocktails. Mix each reaction cocktail in a separate, labeled 1.5-ml microcentrifuge tube. To each tube add 10 µl 10× PCR buffer (usually supplied by supplier of *Taq* or other thermostable polymerase), 10 µl 25 mM Mg^{++}, 10 µl 2 mM dNTPs (0.5 mM of each dNTP), 10 µl of 1 µM stocks of both primers (Table 7-1) for either *MAT-1* or *MAT-2*, 0.5 µl *Taq* DNA polymerase (5 units/µl), and 39.5 µl sterile distilled water. Keep the reaction cocktails on ice until they have been dispensed into the final reaction tubes.
- Kerényi *et al.* (1037) reaction cocktails. Mix each reaction cocktail in a separate, labeled 1.5-ml microcentrifuge tube. To each tube add 10 µl 10× PCR buffer (usually supplied by supplier of *Taq* or other thermostable polymerase), 10 µl 25 mM Mg^{++}, 10 µl 2 mM dNTPs (0.5 mM of each dNTP), 25 µl of 1 µM stocks of both primers (Table 7-1) for either *MAT-1* or *MAT-2*, 0.2 µl *Taq* DNA polymerase (5 units/µl), and 9.8 µl sterile distilled water. Keep the reaction cocktails on ice until they have been dispensed into the final reaction tubes.

6. Prepare reaction mixtures by transferring 9.0 µl of the *MAT-1* reaction cocktail to each tube to be used for *MAT-1* diagnosis (giving a final reaction volume of 10.0 µl). Change pipette tips after each transfer. When finished with all of the tubes in a strip, loosely place the strip-caps across the tops of the strip-tubes and place the finished reaction mixtures back on ice. Repeat the process for the *MAT-2* reaction cocktail and the tubes to be used for *MAT-2* diagnosis.
7. Amplify the *MAT* alleles. Place the caps snugly onto the top of each strip of tubes, and place in the thermocycler.
   - Steenkamp *et al.* (2041) amplification program: 94°C for 60 sec., followed by 29 cycles of 94°C for 30 sec., 65°C for 45 sec., and 72°C for 30 sec., followed by 72°C for 5 min, and then 4°C indefinitely.
   - Kerényi *et al.* (1037) amplification program: 94°C for 2 min., followed by 35 cycles of 94°C for 30 sec., 60°C for 30 sec., and 72°C for 30 sec., followed by 72°C for 10 min., and then 4°C indefinitely.
8. Resolve the PCR amplification products on an agarose gel. The presence of an amplification product indicates the mating type of the strain.

7.3 Amplified Fragment Length Polymorphisms (AFLPs)

AFLPs (2266) are very useful in the construction of genetic maps and in the evaluation of genetic variation within populations. Both of the published genetic maps of *Fusarium* species rely heavily on AFLPs for their content (994, 995). AFLPs from these maps have also been used to define population genetic analyses in *G. zeae* (2409, 2410). Strains in the same species usually share 60-70% or more of their AFLP bands, while strains in clearly distinguishable species usually share 40% or less of their bands (1346, 2412). Strains in the intermediate range, *i.e.*, between 40 and 60% similarity, may be intermediate in terms of their cross fertility and taxonomic differentiation as well, *e.g.*, the *G. fujikuroi* mating populations C and D (1236).

In an AFLP reaction (Fig. 7-1), genomic DNA is digested to completion with a restriction enzyme with a six bp recognition site and another with a four bp recognition site; in the protocol below *Eco*RI and *Mse*I, respectively. If the DNA fragments from the double digests were separated on a gel, then there would be so many that the digest would appear as a smear. Further resolution is obtained by analyzing only a subset of the fragments selectively amplified by PCR. To this end all of the fragments are ligated to linkers. Once ligated to the fragments, the linkers destroy the existing restriction site (by removing the symmetry) and are long enough (~20 base pairs) to provide specificity for PCR amplification. A "preamplification" PCR reaction with sequences based on the linkers and the restriction site, termed the core primers, is used to amplify the fragments to which the linkers have been attached. The reduction in the number of fragments analyzed is obtained by adding additional bases to the core primers, with each base added reducing the number of bands amplified by 75%.

For *Fusarium* amplifications, four bases normally suffice, although for some especially complex patterns five may be needed. Usually two bases are added to the 3' end of each core primer, generically termed *Eco*++ and *Mse*++, but in theory the additional bases can be distributed in any manner. In practice, the number of bases added to the *Mse* core primer should be ≥ to the number of bases added to the *Eco* core primer, to reduce the number of amplified fragments with *Mse*I sites on both ends. The target of the analysis is fragments with an *Eco*RI site on at least one end. Thus, only the *Eco*++ primer is labeled in the final reaction. A second PCR reaction with the additional bases is then conducted and, when four bases are added, 1/256 of the original bands are now detected. These bands can be resolved on a polyacrylamide sequencing gel, and the labeled fragments detected via autoradiography. Considering the combined length of the primers (~40-45 bp), the smallest size fragments that should be included in an analysis should be > 100 bp (55-60 bp of genomic DNA), and probably > 200 bp (55-60 bp of genomic DNA).

Some common concerns regarding the suitability of AFLPs as markers also need to be addressed. First, our experience is that the size of the fragment does not affect the independence of the distribution of the location of the fragments in the genome, and that analyzing all of the fragments from 200-500 bp in size gives markers that are as randomly dispersed throughout the genome as would an analysis of those sized from 500-800 bp. Second, AFLP markers are of a band/no band nature, and bands of the same size are generally assumed to be homologous, although this assumption may require experimental proof under some conditions. (The larger the band the more likely this assumption is to be true.) In a diploid setting these markers would be dominant, but as all of the *Fusarium* species are vegetatively haploid, this potential complication is not a concern. Third, while bands of the same size generated from the same PCR reaction may be homologous, the equivalence of the "no band" alleles cannot be generally assumed, and the data analyses need to be adjusted accordingly.

The protocol given below is one that has been used successfully for several years in the Leslie lab. It assumes that the bands will be scored manually and that ^{33}P will be used as the label. The largest fragments clearly detected with this system usually are ~1000 bp, although with some changes in the gel running conditions fragments up to twice that size can be resolved. This protocol will require modifications if non-radioactive, usually fluorescent, labels are used or if the results are scanned and scored by computer. Up to the final labeling, this protocol also can be used to prepare AFLP reactions that will be analyzed on a DNA sequencer.

AFLP Protocol

1. Isolate DNA from culture (section 7.1).
2. Quantify the DNA using agarose gel electrophoresis or spectrophotometry.
3. Digest DNA to completion with restriction enzymes *Eco*RI and *Mse*I. The restriction digest reaction mixture contains 1 unit of *Eco*RI, 1 unit of *Mse*I, 1× OPA restriction digest buffer [50 mM K acetate, 10 mM Mg acetate, and 10 mM Tris-acetate (pH 7.4)] or the buffer recommended by the supplier of the restriction enzyme, and 100 ng of DNA in a 10 μl reaction volume. Incubate the sample at 37°C for at least 2 hours.

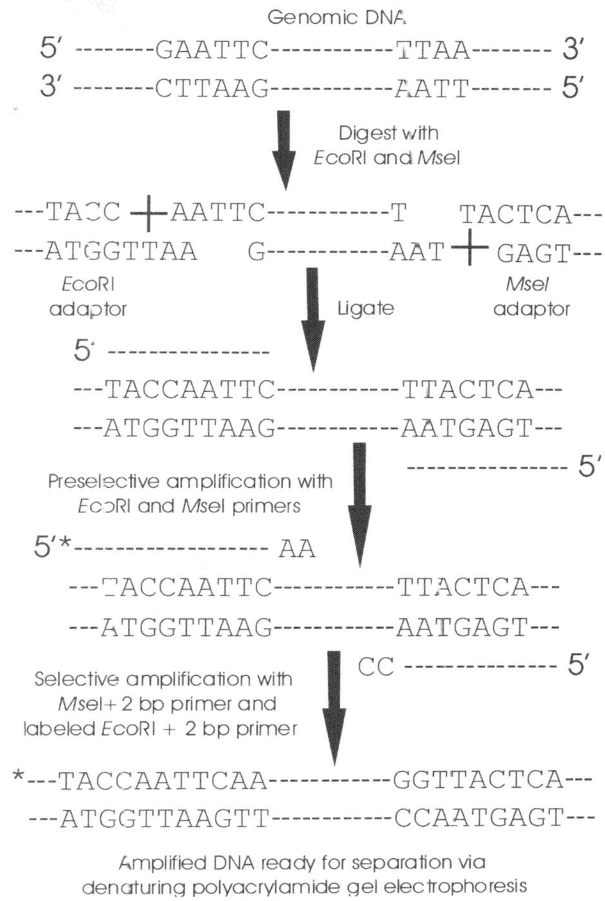

Figure 7-1. Cartoon of the amplified fragment length polymorphism (AFLP) protocol. Genomic DNA is digested to completion with two restriction enzymes, commonly *Eco*RI and *Mse*I, although other enzyme pairs also can be used. Digested DNA is mixed with adapters of known sequence that match the overhangs left on the genomic DNA by the restriction enzymes. The adapters are ligated in place and when ligated destroy the restriction enzyme recognition sites, which permits the restriction digestion and the ligation reactions to be performed at the same time. The entire mixture is amplified through PCR with primers that complement the sequence of the adapters (the "preamplification" step). A second PCR reaction (the AFLP amplification step) is conducted with primers that are two nucleotides longer than those used in the preamplification step. An addition of two bases to each primer works well for most *Fusarium* analyses. If a fifth base is needed to increase selectivity, a third base should be added to the *Mse* primer. One of the primers, usually the *Eco* primer, is radioactively or fluorescently labeled(*), so that amplified fragments can be detected via autoradiography following separation on a polyacrylamide DNA sequencing gel.

The restriction enzymes do not need to be inactivated before proceeding to the ligation step.

4. Ligate primer adapters to the digested DNA; *Eco*RI primer adapters – 5'-CTCGTAGACTGCGTACC-3' + 5'-AATTGGTACGCAGTC-3'; *Mse*I primer adapters – 5'- GACGATGAGTCCTGAG-3' + 5'-TACTCAGGACTCAT-3'.
   - Prepare the adapters (usually done just once, and then stored as a stock in the freezer at -20°C). Mix, in distilled water, the two *Eco*RI adapter sequences together to a final concentration of 5 pM/ml, and the two *Mse*I adapter sequences together to a final concentration of 50 pM/ml. Heat each mixture of single-stranded primers to 60°C, and then let it cool slowly to room temperature to allow the primers to anneal to one another.
   - Cool the digestion mixture (from step 3) to 20°C and then add 10 µl of ligation cocktail. The ligation cocktail contains 1× NEB ligation buffer [50 mM Tris-HCl, 10 mM $MgCl_2$, 10 mM dithiothreitol, 1 mM ATP, and 25 µg/ml BSA (pH 7.5)], 1× OPA restriction digest buffer, 0.4 µl of the 5 pM/µl stock of the *Eco*RI adapter, 0.4 µl of the 50 pM/µl stock of the *Mse*I adapter, and 0.17 µl (one Weiss unit = 7 NEB units) of NEB T4 DNA ligase at 400 units/µl.
   - Mix well and incubate for 2 hr. at 16°C, 4-6 hr. at 20°C, or overnight at room temperature.
5. Preamplify the digested/ ligated DNA with core primers.
   - Dilute the ligation mixture from step 4 with 9 volumes of 1× TE [10 mM Tris, 1 mM EDTA (pH 8.0)]. Add 180 µl of TE to the 20 µl of the ligation mixture from step 4, and store at -20°C until used.
   - Add 5 µl of this diluted mixture to the following cocktail:

     0.4 µl – 100 mM dNTPs (25 mM each; dATP, dCTP, dGTP, and dTTP).

     1.3 µl – 50 ng/µl stock of the *Eco*RI core primer (5'-CTCGTAGACTGCGTACCAATTC-3').

     1.3 µl – 50 ng/µl stock of the *Mse*I core primer (5'-GACGATGAGTCCTGAGTAA-3').

     5.1 µl – 10× AFLP buffer [500 mM KCl, 200 mM Tris-HCl (pH 9.0), 15-40 mM $MgCl_2$, 1% triton X-100].

     0.14 µl – *Taq* DNA polymerase (~7 units/µl).

     37.75 µl – distilled H_2O for a total of approximately 46.0 µl.

     Note: For multiple samples make a master mix containing the above proportions of each reagent. Because of pipeting inaccuracies make at least 5% more than you need.
   - PCR amplify the samples with an amplification program that denatures at 94°C for 5 min., followed by 20 cycles of 94°C for 30 sec., 56°C for 60 sec., and 72°C for 60 sec., and finishes with a 5 min. extension at 72°C, and then is held at 4°C indefinitely.
6. Dilute the preamplified DNA – this DNA is the template for AFLP analysis. The preamplification mixture from step 5 above is diluted with 49 volumes of 1× TE and the dilution is used as the template for the AFLP analysis. Usually 3 µl from the preamplification reaction is diluted with 147 µl of TE.
7. Prepare labeled AFLP primer.
   - Choose an *Eco*RI++ AFLP primer and end label with $^{33}\gamma$P using polynucleotide kinase. Prepare enough primer to label all of the reactions to be conducted during the day at one time. The primers are 5'-end labeled with $^{33}\gamma$P in the following reaction mixture:

Total volume - µl	50	25	12.5
*Eco*RI++ primer (27.8 ng/µl) - µl	18	9	4.5
10× kinase buffer - µl	5	2.5	1.25
ATP (10 µCi/µl) - µl	10	5	2.5
H_2O - µl	15	7.5	3.75
T_4 kinase (6 NEB units/µl) - µl	2	1	0.5

   The kinase buffer may be supplied by the manufacturer, or use as 10× buffer: 700 mM Tris-HCl, 100 mM $MgCl_2$, and 50 mM dithiothreitol (pH 7.6).
   - Incubate the reaction mixture for 1 hr. at 37°C then 15 min. at 70°C to inactivate the kinase. The resulting reagent is used as the labeled primer in the AFLP amplification reaction.
8. Mix labeled primer with unlabeled *Mse*I primer for use in the AFLP amplification reaction. The reaction is done in a 5-µl volume. Each reaction contains:

   1.3 µl – template DNA solution (from step 6).

   1.925 µl – H_2O.

   0.5 µl – 10× AFLP buffer (see step 5).

   1.125 µl – specific *Mse*I++ primer plus dNTPs (6.7 ng/µl *Mse*I primer and 0.89 mM dNTPs; 0.222 mM each).

   0.025 µl – *Taq* DNA polymerase (7 units/µl).

   0.125 µl – $^{33}\gamma$P labeled *Eco*RI++ primer (from step 7).

   5.0 µl – Total

   A master mix usually is made that contains all of the components except the template. Make at least 5% more than the measured need to allow for pipetting inaccuracies. Aliquot 3.7 µl into each reaction tube containing the DNA templates.

9. Run AFLP-PCR amplification program. Place the reaction tubes in the thermocycler and use the following amplification program:
   - One cycle: 94°C for 30 sec.; 65°C for 30 sec.; 72°C for 60 sec.
   - Lower the annealing temperature in each cycle by 0.7°C for each of the next 12 cycles for a total "touchdown" phase of 13 cycles.
   - Twenty-three cycles of 94°C for 30 sec.; 56°C for 30 sec.; 72°C for 60 sec.
   - One cycle of 72°C for 5 min.
   - Hold at 4°C indefinitely.
10. Prepare reagents for a 6% sequencing gel. The following recipes and descriptions are for a Model S2 sequencing gel electrophoresis apparatus (Life Technologies, Rockville, Maryland). When using other apparati adjustments may be required. Wear gloves when making the acrylamide gel and treat the unpolymerized acrylamide with caution since it is a neurotoxin that can be taken up through the skin.
    - In a 100-ml beaker, place 27 g urea, 7.2 ml 50% stock LongRanger acrylamide gel solution (BioWhittaker Molecular Applications, Rockland, Maine), 6 ml 10× TBE electrophoresis buffer [1 M Tris, 900 mM boric acid, 10 mM EDTA (pH 8.3)], and water to 55 ml.
    - Heat for 15 sec at full power in a microwave oven and stir at room temperature until the urea dissolves.
    - Add 30 µl of temed and bring the total volume to 60 ml with water.
    - Make a 10% ammonium persulfate solution by weighing 0.8-1.0 g of NH_4 persulfate into a 1.5 ml microcentrifuge tube and adding water to dissolve.
11. Prepare the gel apparatus.
    - Clean the glass plates by scrubbing with a plastic scouring sponge (non-abrasive) and soap.
    - Rinse the plates with warm tap water and dry them with paper towels.
    - Treat the shorter glass plate with Gel Slick (FMC Bioproducts, Rockland, Maine).
    - Wash the gel-side surfaces of the plates with 95% ethanol. Dry with a paper towel and then assemble the plates, spacers, and boot according to the manufacturer's instructions.
12. Pouring the sequencing gel (~ 31 × 38.5 cm).
    - Mix 300 µl of the 10% ammonium persulfate solution with the acrylamide solution (step 10).
    - Fill the gap between the glass plates through the hole in the bottom of the boot. Use a 60-ml syringe to inject the gel solution. Avoid air bubbles, as they will cause distortions in the separation.
    - Insert combs, flat surface down, so that the edge of the comb is below the edge of the short plate about halfway between the edge of the glass plate and the slot in the comb (~5 mm). Each gel uses two combs of either 24 or 48 teeth each, and results in gels with either 49 or 97 wells (one additional well is created where the combs meet in the center of the gel).
    - Clamp the plates on the combs with four binder clips (one on each end of each comb.)
    - Let the gel polymerize and age for at least 30 minutes, preferably for at least one hour, after the gel initially polymerizes. If gels are wrapped to prevent dehydration they can be used up to four days later.
    - Remove the gel from the boot. Remove the combs from the top of the gel. Mount the gel/glass plate sandwich on the gel apparatus, and clamp in place.
    - Fill the upper and lower chambers of the gel apparatus with 1× TBE buffer as the running buffer.
    - Rinse out the slot left by the combs. Insert the combs teeth down so that the teeth enter the surface of the gel to form wells between the teeth.
    - Run the gel for at least 20 minutes at 60 watts before loading any samples into the gel.
13. Prepare loading dye. This step need not be performed every time, as aliquots of loading dye may be stored at -70°C and thawed for use as needed. Mix together 10 ml deionized formamide (an "ultrapure" or molecular biology grade), 200 µl 0.5 M EDTA (pH 8.0), 10 mg bromophenol blue dye (Sigma cat. no. B0126), and 10 mg xylene cyanol FF dye (Sigma cat. no. X4126). Mix thoroughly. Conveniently stored as ~1-ml aliquots in 1.5-ml microcentrifuge tubes.
14. Running the gel.
    - Add an equal volume of loading dye to each of the samples, *i.e.*, samples are diluted by doubling their volume with formamide dye. Denature the samples by heating for 3 min at 90°C, and then placing them immediately on ice.
    - Load 1.4 µl of a sample into a well. A multichannel pipetter facilitates this process, but is not essential.
    - Run the gel at 60 watts (usually 1400-1500 volts) until the second blue dye (xylene cyanol blue) is ~3/4 of the way down the gel (~2 hr).

Table 7-2. Species of *Fusarium* with the number of entries in the GenBank database (July, 2005), and accession numbers for representative DNA sequences associated with genes commonly used in phylogenetic studies.

Species	Number of Available Sequences	*tub-2*	ITS	mtSSU	*tef-1*	*cmd*
F. acuminatum	20	U85567	U85533	U85550	AY337439	-
F. acutatum	11	U34431	AY569567	U34515	AF160276	AF158329
F. anthophilum	14	AY222292	-	U61593	AF160292	AF158345
F. avenaceum	148	AF405461	U26738	-	AY337423	-
F. begoniae	15	U61621	U61673	U61595	AF160293	AF158346
F. beomiforme	12	U61622	X94178	U61596	-	-
F. brevicatenulatum	8	U61623	U61675	U61597	AF160265	AF158318
F. bulbicola	11	U61624	U61676	U61598	AF160294	AF158347
F. camptoceras	6	-	-	-	-	-
F. chlamydosporum	8	-	AY213655	-	AY337448	-
F. circinatum	21,648	AF366555	-	AF366534	AF333930	AF366531
F. concentricum	18	AF333951	U61678	AF333943	AF333935	AF158335
F. crookwellense	85	AF212782	AF006340	U85551	AF212464	-
F. culmorum	2569	AF212779	AY260958	U85552	AF212463	-
F. decemcellulare	13	-	-	-	-	-
F. denticulatum	8	U61628	U61680	U61602	AF160269	AF158322
F. dimerum	3	-	-	-	-	-
F. dlaminii	9	U34485	X94177	U34514	AF160277	AF158330
F. equiseti	49	-	AY147368	-	AY337424	-
F. foetens	68	AY320125	-	AY320107	AY320089	-
F. fujikuroi	161	U34415	AY188916	AF366543	AY337451	AF366527
F. globosum	13	U61635	U61687	U61609	AY337440	AF158338
F. guttiforme	10	U34475	U34562	-	AF160297	AF158350
F. heterosporum	11	AF405459	AF111064	-	-	-
F. hostae	101	AF331812	-	AF324290	AF324327	-
F. konzum	10	AY222294	-	-	-	-
F. lactis	8	U61629	U61681	U61603	AF160272	AF158325
F. lateritium	55	-	AY188920	-	AY337435	-
F. mangiferae	2	-	AF430129	-	-	-
F. merismoides	6	-	-	-	-	-
F. miscanthi	17	AF060385	-	AF060372	AF324332	-
F. musarum	4	-	-	-	-	-
F. napiforme	10	U34483	AY303609	U34512	AF160266	AF158319
F. nelsonii	9	-	-	-	-	-
F. nisikadoi	10	U61633	U61685	U61607	AF324330	-
F. nygamai	22	U34481	X94174	U34510	AY337445	AF158326
F. phyllophilum	8	U34487	U34574	U34516	AF160274	AF158327
F. poae	215	AF404237	AY188915	U85555	AY337446	-
F. polyphialidicum	8	U34493	AJ538042	U34522	-	-
F. proliferatum	96	AF336910	AF291061	AF291056	AY337436	AJ560773
F. pseudoanthophilum	9	U61631	U61683	U61605	AF160264	AF158317
F. pseudocircinatum	8	U34482	U34569	U34511	AF160271	AF158324
F. pseudograminearum	107	AF107882	AJ491294	-	AF212471	-
F. pseudonygamai	8	U34476	U34563	U34505	AF160263	AF158316
F. ramigenum	8	U61632	U61684	U61606	AF160267	AF158320
F. redolens	100	AF324256	X94169	AF324282	AF456449	-
F. sacchari	20	U34469	AY533376	U34498	AY337447	AF158331
F. sambucinum	83	AF484166	AY188921	U85557	AY337422	-

Table 7-2. Continued

Species	Number of Available Sequences	*tub-2*	ITS	mtSSU	*tef-1*	*cmd*
F. semitectum[2]	35	-	AF162902	-	-	-
F. sporotrichioides	7872	AF404190	AY510069	-	AY337442	-
F. scirpi	1	-	-	-	-	-
F. sterilihyposum	2	AY222290	AF430130	-	-	-
F. subglutinans	246	AF374025	X94167	U34501	AY337444	AY374046
F. succisae	10	AY222289	U34561	U34503	AF160291	AF158344
F. thapsinum	12	U34473	U34550	U34502	Af160270	AF158323
F. torulosum	17	-	-	-	-	-
F. tricinctum	39	AF405457	AY188923	-	-	-
F. udum	8	U34488	U34575	U34517	AF160275	AF158328
F. venenatum	127	U85577	AY188922	U85560	-	-
F. verticillioides	90,972	U34468	X94166	AF273312	AY337450	AF158315

tub-2 – β-tubulin. **ITS** – nuclear internally transcribed spacer regions one and two. These sequences usually include the entire 5.8S rRNA coding region and portions of the 18S and 28S rRNA coding regions as well. **mtSSU** – mitochondrial small subunit rRNA coding region. ***tef-1*** – transcription elongation factor 1α. ***cmd*** – calmodulin. ***H3*** – histone H3.
[1]Listed in GenBank as *F. pallidoroseum* and *F. incarnatum*.

Blot the gel onto a single large piece of 3MM filter paper. Vacuum dry the gel for ~2 hr. at 80°C [temperature turned all the way up on the dryer (Gel Dryer GD2A, Labnet, Woodbridge, New Jersey)]. Note: Water-aspirator vacuum pumps (Boekel Scientific, Feasterville, Pennsylvania) are commonly used to accelerate the drying of these gels. These pumps must be at room temperature or cooler to pull an efficient vacuum (see manufacturer's instructions). During the gel-drying process, the water in the pump reservoir will warm up. Consequently, when drying two or more gels in succession, the water in the aspirator pump's reservoir must be cooled either by replacing it with chilled water or, more conveniently, by adding crushed ice to the water. Be sure to keep the level of the water between the red operating lines on the side of the pump's reservoir as the pump will not work efficiently if the water level is either too high or too low.

15. Place a sheet of 14" × 17" X-ray film on the gel in a cassette in the dark and expose the film at room temperature for 72 hours (or longer depending on the activity of the radioactivity). There are intensifying screens available for use with ^{33}P, but they often result in grainy images that are difficult to interpret if there are either faint bands on the gel, or if there are several bands relatively close together. Our experience has been that intensifying screens are not necessary to obtain useful AFLP results.

7.4 Sequence Analysis and Sequenced Loci

Developing DNA primer pairs that are specific for species diagnostics has begun for some *Fusarium* species. The list of references for such studies is relatively long, but none of the primers has been widely tested, and even fewer of these studies use a large number of species or develop more than one species specific primer for a species. Thus these primers should, at least at this time, be viewed as hypotheses in need of testing and verification rather than as carefully validated fail-safe protocols. Such diagnostic DNA sequences will become of greater importance if the number of recognized species continues to increase, and if the morphological differences between economically important species are difficult, or impossible, to detect or diagnose by anyone other than an expert in the field. There has been a great deal of work done with species such as *F. graminearum*, *F. pseudograminearum*, and *F. culmorum*, due to their ability to cause head blight and crown rot of wheat and barley. Studies in which species-specific PCR primers were developed for this group of fungi include: Akinsanmi *et al.* (30), Chelkowski *et al.* (345), Doohan *et al.* (517), Knoll *et al.* (1097), Mishra *et al.* (1455), Mulfinger *et al.* (1494), Schilling *et al.* (1910), Williams *et al.* (2334), and Yoder & Christianson (2392). Other studies have as their focus species in the *G. fujikuroi* species complex including: Asran *et al.* (89), Möller *et al.* (1511), or *forma speciales* within *F. oxysporum*, *e.g.*, Livens *et al.* (1257). Some include one or more *Fusarium* species amongst a group of,

Table 7-3. Partial or total gene sequences available in GenBank for various *Fusarium* species. The listing is not exhaustive. Sequences may be for portions of genes with or without introns and flanking regions, intact coding regions, or gene clusters.

Species	Gene	Accession number
F. avenaceum	*cbh-c*	AF490902
	esyn-1	AF351600
	mat-1	AJ535625
	mat-2	AJ535629
	o-methyltransferase	AY123622
	top-2	AF546129
F. circinatum	*H3*	AF374070
	mat-1	AY219876
	mat-2	AY219874
	*pg*A	AF207825
F. crookwellense	*H3*	AF314471
	Tri cluster	AY102569
	tub-3	AF212821
F. culmorum	*arp-2*	AY077706
	mat-1	AJ535626
	mat-2	AJ535630
	tri cluster	AY102602
	tub-3	AF212817
F. fujikuroi	*car*A	AJ426417
	*car*B	AJ426418
	*car*O	AJ566362
	cbh-c	AF546135
	*cre*A	Y16626
	GA_4 desaturase	AJ417493
	gdh-1	AJ310444
	gdh-2	AJ315471
	gs	AJ310443
	gst	AB071861
	H3	AF291063
	mat-1	AF100925
	mat-2	AF100926
	mk-1	AJ309177
	nit-1	AF403140
	nnu	Y11006
	*pac*C	AJ514259
	pks-1	AJ315370
	pks-2	AJ421889
	pks-3	AJ414140
	pks-4	AJ278141
	top-2	AF546131
F. guttiforme	*mat-1*	AY219875
F. lateritium	*cbh-c*	AY188914
	chy-1	M99046
	top-2	AY188913
F. nygamai	*H3*	AF150855
	mat-1	AF236763
	pks	AF327783
F. poae	*cbh-c*	AF546134
	mat-1	AJ535627
	mat-2	AJ535631
	top-2	AF546130
	tri-5	AY130294
F. proliferatum	*H3*	AF291602
	lac-1	AY026413
	lac-2	AY026414
	lac-3	AY026415
	mat-1	AF236760
	mco-4	AY026416
F. pseudograminearum	*H3*	AF314483
	tri cluster	AY102585
	tub-3	AF212824
F. sacchari	*mat-1*	AY219877
	mat-2	AF236766
F. sambucinum	*cbh-c*	AF490901
	esyn-1	AF351606
	top-2	AF546125
	tri-5	M64348
	tri-10	AF386074
F. semitectum [1]	*esyn-1*	AJ345016
F. sporotrichioides	*cbh-c*	AF546136
	top-2	AF546132
	tri cluster	AF359360
F. subglutinans	*H3*	AF374068
	mat-1	AY219878
	mat-2	AF025888
F. thapsinum	*mat-1*	AF236762
	mat-2	AF236770
F. tricinctum	*cbh-c*	AF490897
	top-2	AF546127
F. venenatum	*asp*	AF462062
	cbh-c	AF490895
	top-2	AF546126
F. verticillioides	*fum* cluster	AF155773
	gpd	AY393789
	hog-1	AY393790
	mat-1	AF236757
	mat-2	AF236765
	pac-1	AY216461
	pks-1 – 15	AY495591-605
	rnp-2	AY533830

Table 7-3. Legend.

arp-2 – 60s ribosomal protein P2. ***car*A** – carotene cyclase. ***car*B** – phytoene dehydrogenase. ***car*O** – opsin-like protein. ***cbh-c*** – cellobiohydrolase - c. ***chy-1*** – cyanide hydratase. ***cre*A** – carbon catabolite regulatory transcription factor. ***esyn-1*** – enniatin synthetase. ***fum*** – fumonisin biosynthesis. ***gdh-1*** – NAD-dependent glutamate dehydrogenase. ***gdh-2*** – NADP-dependent glutamate dehydrogenase. ***gpd*** – glyceraldehyde-3-phosphate dehydrogenase. ***gs*** – glutamine synthetase. ***gst*** – glutathione S transferase. ***H3*** – histone H3. ***lac*** – laccase; may be more than one gene. ***mat-1*** – mating type idiomorph 1. ***mat-2*** – mating type idiomorph 2. ***mco-4*** – multi-copper oxidase. ***mk-1*** – mitogen activated protein kinase. ***nit-1*** – nitrate reductase holoenzyme. ***nnu*** – nitrogen catabolite regulatory transcription factor; also termed *are*A. ***pac*C** – pH regulatory transcription factor. ***pg*A** – endopolygalacturonase. ***pks*** – polyketide synthetatse; may be more than one such gene per genome. ***rnp-2*** – RNA polymerase II. ***top-2*** – topoisomerase 2. ***tri*** – trichothecene biosynthesis. ***tub*** – tubulin (not β-tubulin).
[1]Also listed in GenBank as *F. pallidoroseum* and *F. incarnatum*.

usually, very diverse fungi, for purposes such as medical, soil, air or other environmental monitoring purposes, *e.g.*, Carbone & Kohn (313), Jaeger *et al.* (952), Wilson *et al.* (2341), Wu *et al.* (2369, 2370). Diagnostic primer pairs based on ribosomal ITS or IGS sequences appear unlikely to work on a broad scale as there is significant overlap of sequences between some species (155, 538, 1601). Many "phylogenetic species" are being distinguished on the basis of the nucleotide sequences in regions encoding the nuclear and mitochondria rRNA coding regions, β-tubulin, mating type, transcription elongation factor 1α, calmodulin, and histone H3.

Many phylogenetically useful sequences are available from GenBank (Table 7-2), but not all sequences should be viewed as having the same authority. Older sequences often were generated and scored by hand rather than with automated sequencers, and often are less accurate and shorter than are more recent sequences of the same gene. Sequences based on repeated PCR amplifications may have errors introduced by the *Taq* polymerase or the sequencing protocol itself. Sequences tied to a strain available from a publicly accessible culture collection generally are of greater value than those tied to a strain available only from an individual or those tied to no particular strain at all. A database specifically designed for the identification of *Fusarium* strains is available at www.fusarium.cbio.psu.edu. This database contains sequences that are associated with strains that have been characterized morphologically and are available through either the Fusarium Research Center at the Pennsylvania State University or from other public culture collections (688).

As seen in Tables 7-2 and 7-3, representation of the species by sequences in the databases is uneven. In some cases there are multiple sequences for the same region from several strains of the same species, while in other cases there is one or only a few strains that are not tied to any strain in particular. If either the species or the gene in question is not available in the database, then the necessary comparisons cannot be made. Adequate representation in the databases remains problematic for many *Fusarium* species. Not all sequences are equally informative for species in all portions of the genus either. In general nuclear ITS sequences are of limited use in strains from the *G. fujikuroi* species complex, and β-tubulin sequences are of limited utility in the *F. solani* species complex. The *tef-1* sequence has proven useful across a wide range of taxa, *e.g.*, Geiser *et al.* (688), Kristensen *et al.* (1139), and O'Donnell *et al.*, (1596, 1601). Cellobiohydrolase C and topoisomerase II also have been suggested as genes whose sequences might be phylogenetically informative across a broad range of *Fusarium* species (827). Finally, if the sequence is not from an *ex*-type strain, then the utility of the sequence also depends on the ability of the person who identified the strain from which the DNA sequence was taken to make a correct morphological species diagnosis.

Table 7-2 contains a listing of the species in this book, with the exception of *F. graminearum*, *F. oxysporum* and *F. solani*, the number of sequences reported for the species in GenBank as of July 2005, an accession number for a representative sequence for a number of sequences commonly used for phylogenetic studies. Species in Table 7-2, are listed by their *Fusarium* name, but sequences may be present in the database under either the *Fusarium* name or the name of a corresponding teleomorph (Table 10-1). *F. oxysporum* and *F. solani* were excluded from this listing because each contains numerous phylogenetic lineages, and sequences that typify these species could not be easily identified. *F. graminearum* was excluded since the entire genome of this organism has been sequenced (section 7-5).

GenBank also contains sequences associated with *Fusarium* species that are not included in this book. These species (followed by the number of sequences in GenBank) include: *F. ambrosium* (3), *F. anguioides* (2), *F. annulatum* (8), *F. arthrosporioides* (37), *F. bactridoides* (9), *F. buharicum* (5), *F. cavispermum* (2), *F. ciliatum* (1), *F. coccidicola* (1), *F. coeruleum* (5), *F. concolor* (6), *F. epistroma* (1), *F. falciforme* (3), *F. flocciferum* 11), *F. fractiflexum* (25), *F. inflexum* (10), *F. kyushuense* (23), *F. langsethiae* (178), *F. larvarum* (1), *F. lichenicola* (11), *F. lunulosporum* (18), *F. melanochlorum* (1), *F. pulverosum* (106), *F. reticulatum* (2), *F.*

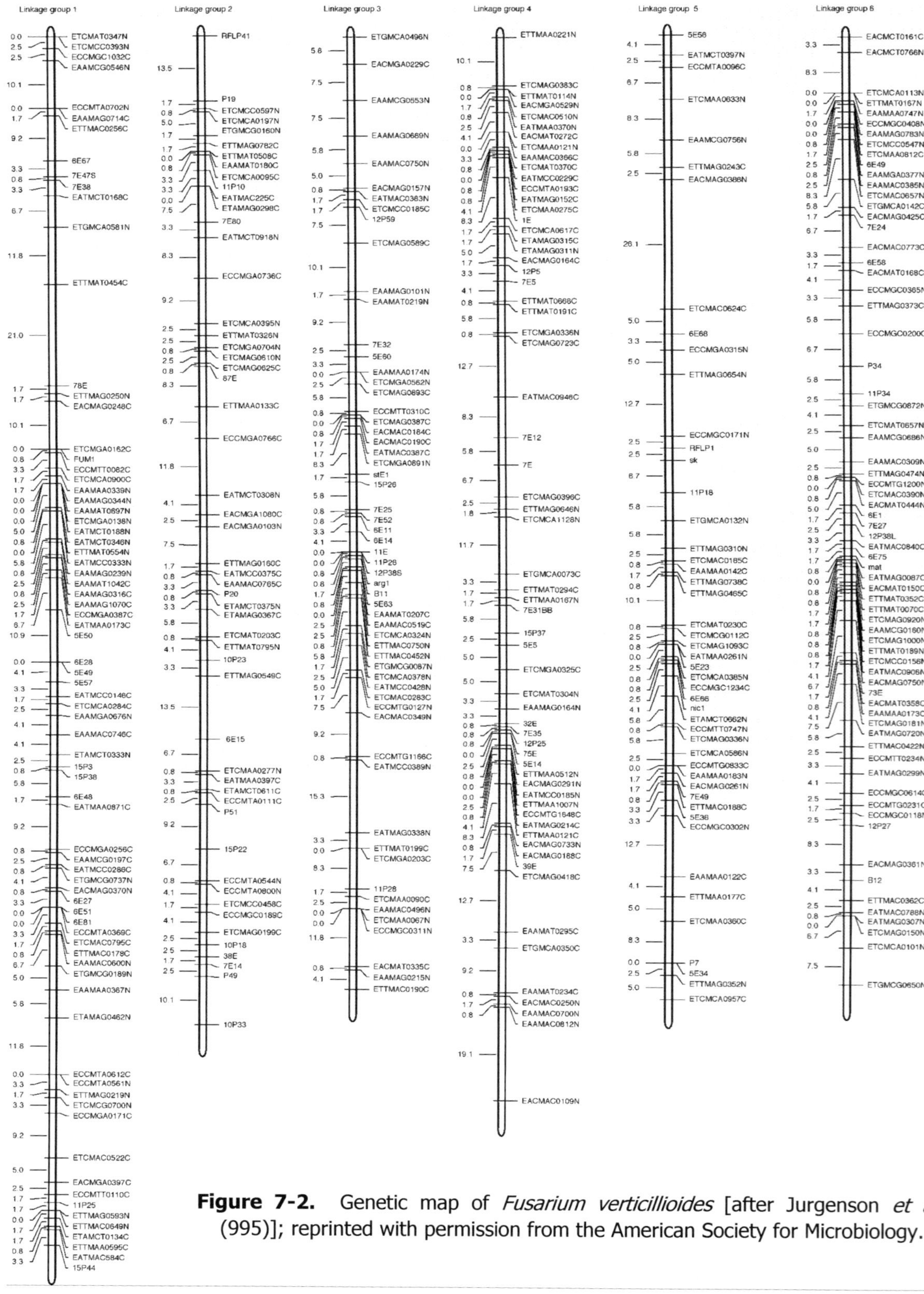

Figure 7-2. Genetic map of *Fusarium verticillioides* [after Jurgenson *et al.* (995)]; reprinted with permission from the American Society for Microbiology.

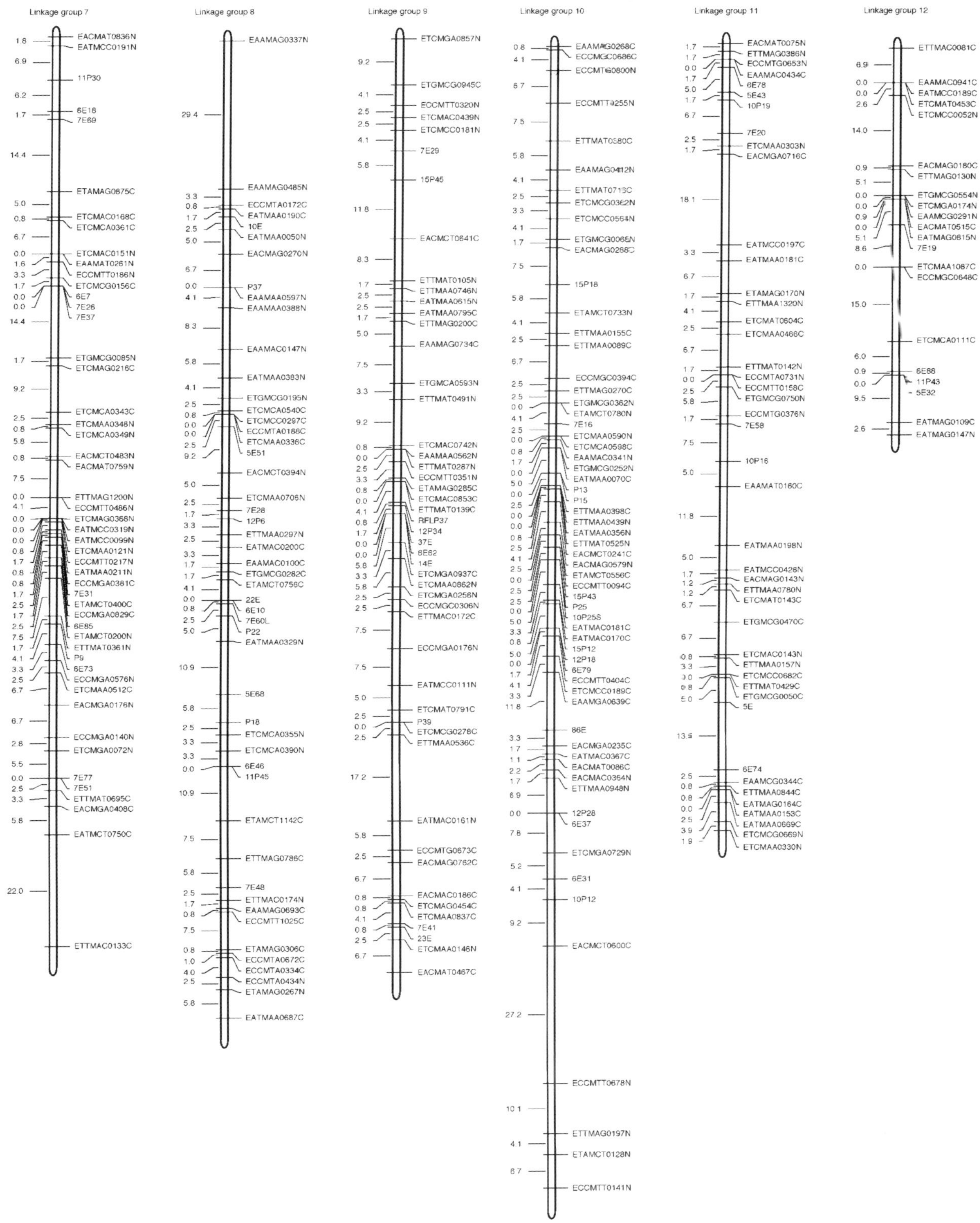
Linkage group 7
Linkage group 8
Linkage group 9
Linkage group 10
Linkage group 11
Linkage group 12

robustum (4), *F. setosum* (1), *F. sublunatum* (1), *F. sulphureum* (2), *F. tumidum* (8), *F. virguliforme* (4189) and *F. xylarioides* (98).

Finally there are some species for which there are no clearly identifiable sequences included in GenBank. Species represented in this book, but not represented in GenBank include: *F. andiyazi*, *F. armeniacum*, *F. aywerte*, *F. babinda*, *F. compactum*, *F. longipes*, and *F. nurragi*.

Loci sequenced in *Fusarium*. A relatively large number of loci, which encode diverse macromolecules, have been sequenced from one or more *Fusarium* species. Species with the most available sequences are *F. graminearum*, whose entire genome is available, *F. verticillioides* with ~91,000 sequences, *F. circinatum* with ~22,000 sequences, *F. sporotrichioides* with ~7900 sequences, and *F. culmorum* with ~2600 sequences (Table 7-2). For most species, the sequences available have potential phylogenetic utility (Table 7-2), but a number of other loci also have been analyzed (Table 7-3). The listing in Table 7-3 excludes ribosomal coding sequences and all of the specific sequences included in Table 7-2. The listing is not exhaustive, but instead should be viewed as indicative of the basic genetic and physiological studies being conducted with these organisms that go beyond studies of species definition and delimitation. As with Table 7-2, *F. graminearum*, *F. oxysporum*, and *F. solani* are not included in Table 7-3.

7.5 Genetic maps

Recombination-based maps. There are two detailed genetic maps of *Fusarium* species available – one for *Fusarium verticillioides* (*Gibberella moniliformis*) and the other for *Fusarium graminearum* (*Gibberella zeae*). Both maps contain primarily AFLP markers, but also include RFLPs, auxotrophic markers, morphological markers, and differences in mycotoxin production.

Genetic maps may have a physical or a recombinational base. Physically based maps include those based on electrophoresis of large DNA molecules, usually inferred to be whole chromosomes, those formed from overlapping fragments cloned in BAC or cosmid vectors, and complete genomic sequences. This area is an active one for research, and a careful literature search will be needed to remain current. A number of BAC and cosmid libraries are available, but no contigs covering the genomes of the species from which these libraries were generated have been constructed, except for *F. graminearum*.

Pulsed field gel electrophoresis studies have been conducted with *F. oxysporum* (211) and six of the species in the *G. fujikuroi* species complex (2374). Results from *F. oxysporum* suggest that its karyotype is subject to a great deal of variation and change. Such variation can presumably be tolerated in an asexual fungus in which meiotic pairing with a chromosomal homolog is not an important part of the life cycle. In the members of the *G. fujikuroi* species complex, sexual reproduction appears to be a more important portion of the life cycle, and the karyotypic variation observed in *F. oxysporum* is not observed. Instead, the six species examined all have the same number of chromosomes – 12 – of similar size with a relatively high degree of synteny.

The karyotypes of these six species, as well as *F. oxysporum* (211) and *Nectria haematococca* (2231) can contain dispensable "B" chromosomes. These chromosomes are known to carry expressed sequences, some of which are important for pathogenicity (1075, 1434), but are not essential for normal vegetative growth or sexual reproduction under laboratory conditions. These chromosomes do not pair and proceed through meiosis in an atypical manner and can be rearranged or lost during meiosis (2373, 2374). They also have reduced levels of recombination relative to the rest of the genome. The role these chromosomes and the genes they carry play in the evolution and speciation of *Fusarium* is unknown.

Genetic maps also may be based on the amount of recombination that occurs between different genes. Geneticists first constructed such maps in *Drosophila* in the early 1900s (2056), but until polymorphic DNA-based markers became available genetic maps required numerous crosses, and the induction and characterization of numerous morphological and nutritional mutations. The resulting maps, *e.g.*, for *Neurospora crassa* (1686), were based on hundreds of crosses, the scoring of thousands of progeny, and required years of work to assemble.

With DNA-based markers, most commonly RFLPs and AFLPs, the time needed to generate a genetic map has decreased significantly. A genetic map based on the analysis of a single cross with as many as 1000 polymorphic AFLP markers can be made in as little as 3-6 months. In general, 20-25% of the AFLP bands will be polymorphic between the parents in an effective mapping cross. If more than 30-40% of the AFLP bands differ, confirmation of conspecificity is important. The results of PCR amplification with 20-50 AFLP primer pairs usually suffices for the construction of a saturated genetic map. The results of these initial crosses provide detailed maps that can be used to correlate physical markers with recombination-based ones, and provide skeletons that can be used for orienting other crosses. Amongst *Fusarium* species, these types of maps are available for *F. verticillioides* (995, 2373) and *F. graminearum* (994), with addi-

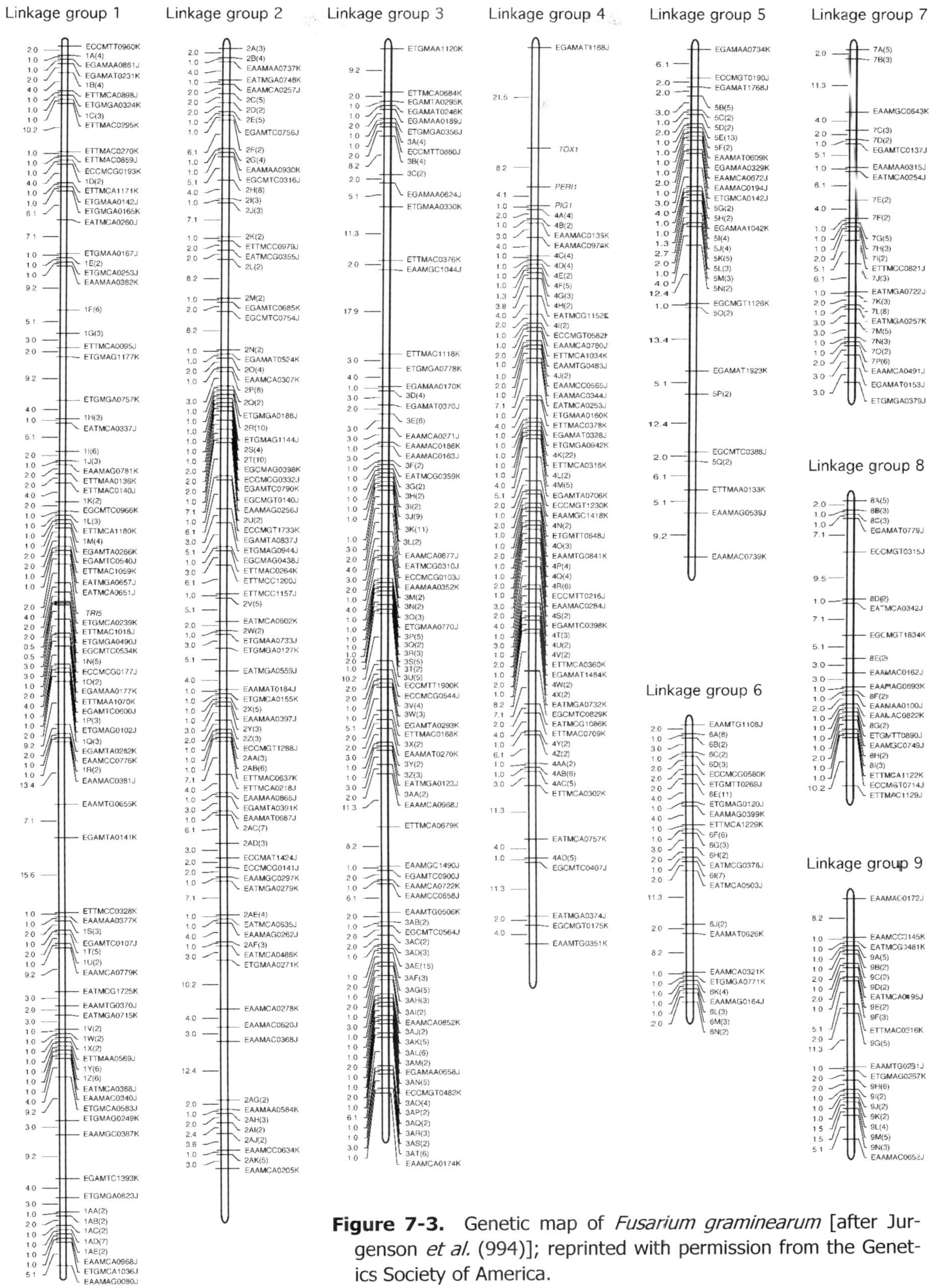

Figure 7-3. Genetic map of *Fusarium graminearum* [after Jurgenson *et al.* (994)]; reprinted with permission from the Genetics Society of America.

tional maps of these and other species being developed as of this writing. As more maps become available, it should be possible to use them to study synteny and karyotypic evolution within these species. The parents and the progeny of both the published mapping crosses described below are available from the Fungal Genetics Stock Center at the Department of Biological Sciences, University of Missouri – Kansas City, Kansas City, Missouri.

Fusarium verticillioides. For *F. verticillioides*, the existing genetic map (995, 2373) contains approximately 600 markers, the majority (~70%) of which are AFLPs. The linkage groups in this map have all been associated with a physical chromosome, and the numbers for both the linkage groups and the chromosomes are based on the size of the chromosomes as assessed in a CHEF gel separation (2374). The resulting map (Fig. 7-2) has an average marker interval of 3.9 cM, and averages ~21 kb/cM, with a total map length of approximately 2200 centiMorgans (map units) distributed across 12 chromosomes, one of which is a dispensable "B" chromosome. This map is close to saturated, as approximately 10% of the markers mapped to a locus at which more than one marker mapped. The maximum distance between any two adjacent markers is 29 cM, and only five intervals are greater than 20 cM in length. There are AFLP or RFLP markers less than 1 cM from the mating type (*MAT*) locus, or 2.5 cM from the spore killer (*SK*) locus, and eight AFLP markers within 8.5 cM of the *FUM1* (fumonisin biosynthetic) locus. The average crossover frequency for each chromosome varies from a low of 0.77 for chromosome 12 to a high of 2.5 for chromosome 1. The distribution of the number of crossovers per chromosome appears to be random. The physical size of the chromosome is not strongly correlated with the number of markers ($r = 0.72$), the number of loci ($r = 0.71$), or the length of a chromosome in map units ($r = 0.68$). The recombinational length of the chromosomes is strongly correlated with the number of markers ($r = 0.94$), and with the number of loci ($r = 0.95$), and the number of markers and loci per chromosome also are strongly correlated ($r = 0.98$).

Fusarium graminearum. The map of *F. graminearum* (994) has more markers than that of *F. verticillioides* (app. 600 vs. app. 1000), but has a similar number of loci and a somewhat shorter map length (Fig. 7-3). This map is based almost entirely on AFLP markers, and is composed of nine linkage groups, although the reported number of chromosomes presently is four (904, 2102). If the physical genomic sequence is used to order and reanalyze these groups (1205), then Linkage groups 7 and 8 go to the top of linkage group 1 (as drawn in Figure 7-3), and linkage group 9 goes to the bottom of the same linkage group. Similarly, linkage groups 5 and 6 go to the top of linkage group 4. The total map length is approximately 1300 cM with an average interval of 2.8 cM between loci. The linkage groups vary in total genetic length from 281 cM for linkage group I to 52 cM for linkage group IX.

Three of the nine linkage groups contain regions in which there are high levels of segregation distortion. Selection for nitrate-utilizing recombinant progeny (required to ensure that the progeny were of biparental origin) can explain two of the three skewed regions, but the explanation for the third remains unknown.

Two linkage groups (II and IV) have recombination patterns that are consistent with the presence of intercalary inversions. These inversions await confirmation in other crosses, but if present could represent post-zygotic isolating mechanisms that might play a part in the evolution of *F. graminearum* from one to more than one species. Loci governing trichothecene toxin amount and type (deoxynivalenol or nivalenol) map on linkage groups IV and I, respectively. The locus governing the type of trichothecene produced (nivalenol or deoxynivalenol) co-segregated with the *TRI5* gene (which encodes trichodiene synthase) and probably maps in the trichothecene biosynthesis gene cluster. The *TOX1* locus associated with the amount of toxin produced maps near genes for female fertility (*PER1*) and pigment production (*PIG1*) (994). QTLs for aggressiveness towards wheat assessed under field conditions also map near *TOX1* and *TRI5* (425). The seven primer pairs that include the EAA AFLP primer can be used to generate a map that contains approximately 70% of the linkages discerned in the entire map.

The distribution of crossovers across the linkage groups is not random. In particular, there is an excess of progeny with no crossovers within a linkage group and a reduction in the number of progeny with linkage groups in which recombination has occurred. For each of linkage groups VI, VII, VIII, and IX, at least half of the progeny have no detectable crossover, which could make estimates of linkage distances less accurate than expected given the number of progeny analyzed. The relatively high number of intact parental linkage groups observed in the progeny could be due to a lack of pairing and/or synapsis that could lead to crossing-over. In such cases chromosome distribution during meiosis presumably follows a distributive disjunction mechanisms such as has been proposed and tested in *Drosophila, e.g.*, Grell (769) and Hawley *et al.* (828). Alternatively, non-viable progeny could result if crossing-over occurs other than in a few specific patterns, and the relative lack of such recombinants could then skew the relative frequency amongst the viable progeny of the remaining progeny classes.

Additional genetic maps for *G. zeae* are in preparation (667) based on crosses between strains that are different from those used to synthesize the map described here. Once these maps become available, *G. zeae* will be one of the few fungi for which multiple detailed genetic maps are available.

Physical maps. The genome of one *Fusarium* species – *Fusarium graminearum* – is publicly available (www.broad.mit.edu/annotation/fungi/fusarium/) and preliminary correlations of both the Jurgenson *et al.* (994) map and a second map based on different strains with the physical sequence have been made (667, 1205). Genomes representing other *Fusarium* species have been sequenced by private companies (who have announced plans to make their sequences public during calendar year 2005) or are in the queue for a public sequencing project. Annotation of the physical sequence of *F. graminearum* is in progress, but is not complete. Applications to fund a genome sequence for *F. solani* as been funded, and one for sequencing *F. oxysporum* has been prepared, but as of this writing has not yet been funded. The completion of the sequences of several species from this genus should tell us much about the genetic relationships of these fungi and provide new insights into how they have evolved and/or are still evolving.

Taxonomy & Identification of *Fusarium*

8

A Brief History of *Fusarium* Taxonomy

The generic concept of *Fusarium* was first diagnosed by Link (1263) with the primary character being the presence of the distinctive canoe- or banana-shaped conidia well known to all that work with the genus. Yet Seifert (1935) also notes that the production of the distinctive macroconidia is not necessarily the sole realm of *Fusarium*, as a number of coelomycetes produce spores that could easily be identified as belonging to a *Fusarium* spp., if these spores were the only criterion that was used for identification. Many of the *Fusarium* species currently used in taxonomic systems are poorly defined and in many cases type specimens no longer exist. If the descriptions of many of these species were published today they would be unlikely to meet the current requirements of the International Code of Botanical Nomenclature (770) because the description is vague or poorly defined.

Following the diagnosis of Link, much of the early research on *Fusarium* focused on the diagnosis, identification and enumeration of taxa that caused plant diseases. A common view was that fungal isolates from individual plant species represented individual fungal species that were host specific. There were numerous species, numerous proposed characters, and multiple culture media that combined to make strain identification and species diagnosis very complicated. Within a relatively short period of time a huge number (> 1000) of *Fusarium* species were either described or attributed to various disease outbreaks. This state of confusion persisted until the mid-1930s when Wollenweber & Reinking's work (2359) reformulated the species concept within the genus, and provided guidelines for identification that were independent of plant host and based on the mycological characters of the strains involved.

Wollenweber and Reinking. The taxonomy of *Fusarium* has undergone a number of changes during the last 100 years. The concept of a species within the genus has varied greatly between broader and narrower concepts. The basis for all modern taxonomic systems in *Fusarium* is the work of Wollenweber & Reinking (2359). Prior to this landmark publication there were around a thousand *Fusarium* species described, often with a different species described for every host. Wollenweber and Reinking brought order to this near chaos and provided a subgeneric system based on 16 sections, 65 species, and 77 sub-specific varieties and forms. Many of these sections and species are still in common use even though it is well-recognized that the species within a section may not be monophyletic, *e.g.*, Kistler (1075) and O'Donnell *et al.* (1598).

The sections erected by Wollenweber and Reinking allowed the users of this taxonomic system to more readily identify the isolate in question. Each section contained species that were united by critical morphological characters, *e.g.*, macroconidial morphology and pigment. Each section contained only a relatively few species, so the task of assigning an isolate to a species was a relatively straight-forward two-step process. The first step associated an isolate with a section and the second step associated it with a species within that section. Although this two-step process simplified the identification process, it also was a weakness. Wollenweber and Reinking used a number of media (some quite complex) in their analyses. The multiple cultures combined with the identification first of section and then of species, meant that species identification could take some time. Although Wollenweber and Reinking made a tremendous reduction in the number of species and brought order and routine to the identification process, their species concept remained based on the differences between strains, rather than on the similarities that they shared.

Snyder and Hansen. In the 1940s and 1950s Snyder and Hansen (2102, 2013, 2015, 2018) reduced the number of species within the genus to nine, and demonstrated that only by using cultures derived from a single spore could species identifications be made reliably. The nine species recognized by Snyder and Hansen were: *F. oxysporum*, *F. solani*, *F. moniliforme*, *F. roseum*, *F. lateritium*, *F. tricinctum*, *F. nivale*, *F. rigidiuscula*, and *F. epi-*

Table 8-1. The relationship between the nine species of Snyder and Hansen, the sections of Wollenweber and Reinking, and the species used in this book.

Snyder and Hansen Species	Wollenweber and Reinking Sections	Species Described in Chapter 13 of this Book
F. episphaeria	*Eupionnotes* *Macroconia*	*F. merismoides, F. dimerum*
F. lateritium	*Lateritium*	*F. lateritium*
F. moniliforme	*Liseola*	*F. verticillioides, F. thapsinum, F. proliferatum, F. subglutinans, F. anthophilum, F. circinatum,* and other species in the *Gibberella fujikuroi* species complex
F. nivale	*Arachnites*	No longer considered *Fusarium*
F. oxysporum	*Elegans*	*F. oxysporum*
F. rigidiuscula	*Spicarioides*	*F. decemcellulare*
F. roseum	*Discolor* *Gibbosum* *Roseum* *Arthrosporiella*	*F. culmorum, F, crookwellense, F. graminearum, F. pseudograminearum, F. compactum, F. armeniacum, F. acuminatum, F. longipes, F. avenaceum, F. torulosum, F. heterosporum, F. equiseti, F. semitectum, F. polyphialidicum*
F. solani	*Martiella* *Ventricosum*	*F. solani*
F. tricinctum	*Sporotrichiella*	*F. poae, F. tricinctum, F. sporotrichioides. F. chlamydosporum*

Wollenweber and Reinking also included the sections Submicrocera (*F. ciliatum* and *F. cerasi*) and Pseudomicrocera (*F. juruanum* and *F. orthoconium*). None of the species in these sections currently are considered within the generic concept of *Fusarium.*

sphaeria (the relationship between these species and those currently used are outlined in Table 8-1). The Snyder and Hansen species concepts were popular with diagnosticians because they were easy to apply and virtually every isolate could be identified to species with relative ease. But the taxa circumscribed by these species were polyphyletic, led to huge losses in information, and render much of the data generated during this time difficult, if not impossible, to interpret or evaluate in any meaningful manner.

Two of Snyder and Hansen's species, *F. oxysporum* and *F. solani*, are still in wide use, but there is little doubt that these taxa contain more than a single species as defined with more current species concepts (105, 1593, 2065). As these newer species concepts are applied to strains currently identified as *F. oxysporum* and *F. solani*, a number of new species probably will be described. We do not expect these new species to appear all at once, but rather to result from efforts with groups of strains that are relatively coherent and sufficiently large for robust species descriptions to result.

Gordon. W. L. Gordon was based in Canada from the 1930s to the 1960s and published a number of papers (745-752) examining *Fusarium* species isolated from diverse substrates and environmenttal conditions. He developed a pragmatic approach to *Fusarium* taxonomy that used aspects of the Wollenweber and Reinking philosophy with some components of the Snyder and Hansen system. He also considered teleomorphs in his descriptions.

Messiaen and Cassini. These French researchers (1430, 1431) developed a system modeled on that of Snyder and Hansen but used botanical varieties rather than cultivars at the subspecific level.

Matuo. Matuo, based in Japan, also produced a system that was effectively identical to that of Snyder and Hansen, but included an additional species, *F. splendens* (1388). This species was reported as the anamorph of *Hypocrea splendens*, but later work indicated that the *Fusarium* probably was a hyperparasite.

Raillo. Raillo was a Russian researcher who in many ways pioneered the used of single spore culture methods. She also had an interest in variation in a number of species and published a taxonomic system in the mid-1930s (1763-1765) that was based on the shape of the macroconidia, and the presence of microconidia and chlamydospores. Unfortunately her major work (1765) was published posthumously in an incomplete form. This delayed publication along with her relative isolation from researchers from the western *Fusarium* community has limited the recognition of her work.

Bilai. Also based in Russia, Bilai was one of the few early workers who examined variability in cultural and physiological characters especially in response to temperature, moisture and media composition. She showed that the range of variability in some characters in several species was equivalent to that observed in whole sections of the genus. She produced her own revision of the taxonomy of the genus (182-185) and recommended combining some of the Wollenweber and Reinking sections, *e.g.*, *Liseola* with *Elegans* and *Gibbosum* with *Discolor*. These combinations were criticized by Nelson *et al.* (1551), but now appear more reasonable given some of the phylogenetic studies of *Fusarium*.

Booth. The next significant development in the taxonomy of *Fusarium* was made by Booth during the 1960s and 1970s and published as a monograph, *The Genus* Fusarium (222). Booth included keys to the sections and species of *Fusarium* in a taxonomic system that borrowed heavily from Wollenweber and Reinking's approach. Booth introduced the use of the morphology of the conidiogenous cells, especially those producing the microconidia, as a species-level diagnostic character. Conidiogenous cell morphology is now essential for distinguishing some of the species in sections *Liseola* and *Sporotrichiella*.

Gerlach and Nirenberg. Based in Wollenweber's laboratory in Germany these researchers continued the research initiated in Wollenweber & Reinking's *Die Fusarien*. They published a treatment of the genus in 1982 (697) that recognized a large number of species. At the time this treatment came in for significant criticism, particularly from Nelson *et al.* (1551). These criticisms focused, reasonably, on the small number of isolates examined for each species and the differentiation of species on the basis of seemingly minor morphological characters. Despite these criticisms, however, this work was a significant step forward in our understanding of *Fusarium* taxonomy, and many of the proposed species have withstood the test of time and now are generally accepted. The species of Gerlach & Nirenberg (697) sometimes provide important contrasts to those of Nelson *et al.* (1551). Collectively these two works clearly defined a number of hypotheses that are now being tested with genetic and molecular data.

Joffe. Abraham Joffe began studying *Fusarium* while living in Russia in the late 1940s and continued his studies after moving to Israel. He used isolates collected in both Russia and Israel in addition to those collected by others to evaluate taxonomic and mycotoxicological issues. His most substantial work, the monograph Fusarium *Species: Their Biology and Toxicology* was published in 1986 (983). His taxonomic approach followed that of Wollenweber & Reinking (2359) and Gerlach & Nirenberg (697), with macroconidial and microconidial shape and culture characters as its primary basis.

Nelson, Toussoun and Marasas. Paul E. Nelson and T. A. Toussoun of Pennsylvania State University, and Walter F. O. Marasas at the Medical Research Council in South Africa, produced an identification manual for *Fusarium*, in 1983 (1551) that had a somewhat simpler predecessor (2169). This very popular guide to the genus combined useful information on the isolation and culturing of *Fusarium*, color photographs to distinguish pigment variation, and excellent photomicrographs of many species. This volume continues to be heavily cited today. A companion volume (1341) covered mycotoxins and mycotoxin producing strains in more deta[l (and has recently been updated by Desjardins (468)]. The Nelson *et al.* taxonomic approach began a definitive shift towards a more complicated taxonomy and a larger set of recognized species and away from the nine species Snyder and Hansen system, especially in the United States. In essence the Nelson *et al.* (1551) volume was a formal acknowledgment that the Snyder and Hansen system was inadequate for distinguishing the species in the genus. A number of species that were recognized by Gerlach & Nirenberg (697), however, were not recognized by Nelson *et al.* (1551). A similar volume, but that emphasized species common in Australasia was published by Burgess *et al.* (279).

The great debates. Both Gerlach & Nirenberg (697) and Nelson *et al.* (1551) took threads developd by Wollenweber and Reinking's, Snyder and Hansen's, and Booth's work and wove them into taxonomic systems with reasonably robust species definitions based exclusively on morphological characters. Much has been written about the differences between these taxonomic systems; however, there is more in common between them than there are differences. Both use many of Wollenweber and Reinking's sections and many of the species definitions. There were several species with different names, but these differences generally were nomenclatural differences rather than differences in species definitions. The majority of *Fusarium* researchers currently use portions of both of these systems as the basis for identifying *Fusarium* species and describing new taxa *e.g.*, Burgess & Trimboli (281), Klittich *et al.* (1093), Marasas *et al.*, (1344, 1346), Nirenberg (1568), and Nirenberg & O'Donnell (1570)

***Formae speciales* and other sub-specific nomenclature.** Many of the taxonomic systems developed for *Fusarium* have made extensive use of subspecific taxa ranks. These ranks include varieties, cultivars, subspecies and formae speciales. Both Wollenweber & Reinking (2359) and Gerlach & Nirenberg (697) made extensive use of the variety rank to describe subspecific taxa that differed relatively little from each other in morphology. This taxa rank is clearly defined in the International Code of Botanical Nomenclature (770). The term "cultivar" was used by Snyder and Hansen to describe pathogenic taxa within their broad species definitions. These cultivar definitions were informal and no guidelines for their regular application were developed. Consequently a number of problems developed in the application of these names and they quickly lost favor.

Subspecific rankings have been used occasionally to separate strains that fit a common species description but that can be reliably distinguished, usually on the basis of morphological characters of untested taxonomic significance. In both cases where subspecies were used formally, *F. acuminatum* ssp. *acuminatum* and *F. acuminatum* ssp. *armeniacum* (270), and *F. avenaceum* ssp. *avenaceum*, *F. avenaceum* ssp. *aywerte* and *F. avenaceum* ssp. *nurragi* (1890), subsequent work has shown that the subspecies were sufficiently different to warrant promotion to species rank (169, 278).

Formae speciales are the most commonly used subspecific taxa ranking currently in use in *Fusarium*. These taxa distinguish pathogenic forms of a particular species that usually are not morphologically distinguishable. In some cases, further subdivision, *e.g.*, into races or VCGs occur within a form species. The *formae speciales* designation is not specific to *Fusarium* and also is used in other plant pathogenic fungi, *e.g.*, rusts. The usual form is the species followed by f. sp., for *formae speciales*, the name of the *formae specialis*, and finally the race or VCG name (or number), or both, if either of these designators exists. In *Fusarium*, *formae speciales* names are used most commonly in *F. oxysporum*. For example, *F. oxysporum* f. sp. *vasinfectum* can attack cotton and *F. oxysporum* f. sp. *cubense* can attack banana. Both of these groups can be subdivided further into VCGs and races.

Use of *formae speciales* combined with the very broad morphological definition of *F. oxysporum* has lead many to consider *formae speciales* to be monophyletic taxonomic/evolutionary units. These groups, however, were defined on the basis of plant pathogenicity, and are monophyletic only when this trait is a homologous character, which often it is not. When *formae speciales* were evaluated with molecular markers, strains in the same *formae speciales* often were found to be no more than distantly related to each other, *e.g.*, Baayen (101) and O'Donnell *et al.* (1598). With molecular markers it is not uncommon for strains in different *formae speciales*, especially if they are from the same geographic region, to be more closely related to one another than they are to other strains in the same *formae speciales*, *e.g.*, Australian strains of *F. oxysporum* f. sp. *vasinfectum* (2286). The realization that *formae speciales* need not be monophyletic has prompted reevaluation of genetic lineages within *F. oxysporum* for taxonomic distinctness. This reevaluation should result in major revisions of some taxa in the coming years. In general, we expect large poorly/broadly defined species such as *F. oxysporum* to dwindle in breadth as important groups of strains are redefined as species, *e.g.*, *F. hostae* (689), one group of strains at a time rather than from a single major overhaul of the entire group.

The *Gibberella fujikuroi* species complex. Wollenweber & Reinking (2359) established Section *Liseola* within *Fusarium* for species that produce microconidia in chains and/or false heads and macroconidia, but that do not produce chlamydospores. A chain-forming *Fusarium* sp. isolated from moldy maize in Nebraska described as *F. moniliforme* Sheldon (1954) was included in Section *Liseola* by Wollenweber & Reinking (2359) together with two other species and three varieties, *i.e.*, a total of six taxa. All six taxa were reduced to a single species, *F. moniliforme* Sheldon emend. Snyder & Hansen by Snyder & Hansen (2015).

Booth (222) initiated the reevaluation of the species in Section *Liseola*, which is still in progress, by separating the variety *F. moniliforme* var. *subglutinans* from *F. moniliforme* due to the absence of microconidial chains and the presence of polyphialidic conidiogenous cells. Gerlach & Nirenberg (697) expanded the number of taxa in Section *Liseola* to 10 and pointed out that, based on priority, the correct name for *F. moniliforme* is *F. verticillioides* (Sacc.) Nirenberg (1566), and that the name of the teleomorph is *Gibberella moniliformis* Wineland (2347); a position now generally accepted in the *Fusarium* research community (1936). The use of *F. verticillioides* to mean the asexual state of *G. moniliformis* is a much narrower species concept than that used by Snyder & Hansen (2015). As a result the identity of many of the isolates originally identified as "*F. moniliforme*" is now in question as these isolates are not *F. verticillioides* and it often is not clear to what other species within the *G. fujikuroi* species complex they belong.

Nelson *et al.* (1551) accepted four taxa in Section *Liseola*, including *F. moniliforme* and another chain-forming species, *F. proliferatum*, which was separated because of the presence of polyphialidic conidiogenous cells. They also returned *F. moniliforme* var. *subglutinans* to species rank as *F. subglutinans*, and recognized *F. anthophilum*, which produces napiforme microconidia. In the last 20 years, four additional chain-forming *Fusarium* species in Section *Liseola* have been described, *F. globosum* (1802), *F. thapsinum* (1093), *F. nisikadoi*

(1569), and *F. miscanthi* (672) as well as a range of related non-chain forming species, *e.g., F. dlamini* (1342).

The biological species concept also has been widely applied in the *Liseola* section and provided the genesis for the grouping of some of the species in the *Gibberella fujikuroi* species complex. The biological species were termed "mating populations" of a morphological species, in this case *Gibberella fujikuroi*. The ability to distinguish multiple mating populations within a single morphological species resulted in some of the first published data that the available morphological characters still were insufficient to distinguish all of the biologically meaningful groups. The first three biological species in section Liseola were described by Hsieh *et al.* (905). Kuhlman (1147) and Kathariou (1020) added a fourth mating population. *Fusarium thapsinum* was first recognized as a distinct mating population within *F. moniliforme* (1092), and then later elevated to species rank (1093). A biological species concept also resulted in a clear distinction of *F. subglutinans* from *F. sacchari* (1219), although a formal description of this distinction has only recently been published (1229). The biological species concept also was an important part of the descriptions of *F. circinatum* (242, 243, 1570). Klaasen & Nelson (1085) added a biological species from *F. nygamai* to this list, while Zeller *et al.* (2412) described the most recent mating population in the *G. fujikuroi* species complex as part of their description of *F. konzum*. Although members of different mating populations usually are completely infertile when they are crossed with one another, there are cases where strains that are at least somewhat cross fertile have been described (1236, 2039, 2042). These exceptions reinforce the need for a taxonomy that is based on multiple independent characters to ensure that all of the members of a species are correctly grouped and identified.

To further complicate both the taxonomy and the identification of strains in the *Gibberella fujikuroi* species complex, there are a number of *Fusarium* species that morphologically resemble those assigned to section *Liseola*, but that also produce chlamydospores, *e.g., F. nygamai* (281) and *F. napiforme* (1344), which along with *F. dlamini* (1342) were placed in the new Section *Dlaminia* by Kwasna *et al.* (1164). Establishing a new section was not the correct solution to this problem, however. Instead, sections *Liseola* and *Elegans* (672, 1569) have been functionally combined into a single entity, which now often is referred to as the "*Gibberella fujikuroi* species complex" or the *Gibberella fujikuroi* clade. DNA sequence data indicate that Section *Liseola* is paraphyletic and that *F. nygamai, F. napiforme*, and *F. dlamini* are "phylogenetically nested within Section *Liseola*" (1596, 1600). The "*Gibberella fujikuroi* species complex" now contains at least 46 phylogenetically distinct lineages (1596, 1600), including 10 new species described by Nirenberg & O'Donnell (1570), two species described by Nirenberg *et al.* (1571), and most recently, *F. andiyazi* (1346), *F. konzum* (2412), *F. mangiferae* (244) and *F. sterilihyposum* (244). The utility and distinctness of the many recently described species in the *Gibberella fujikuroi* species complex remain the subject of significant research.

Future prospects. Species descriptions, even very thorough ones, usually rely on a limited set of characters relative to all of those that could potentially be described. While most species still are based on morphological concepts, the application of biological and phylogenetic species concepts to species of *Fusarium* has begun and will undoubtedly continue. These approaches have been used most commonly for species whose status has been questioned, and their application has reignited many of the controversies that have swirled in past discussions of *Fusarium* taxonomy. These approaches have led to the clear split of *F. thapsinum* from *F. moniliforme* (1092, 1093), the redefinition of *F. moniliforme/verticillioides* to mean *Gibberella moniliformis* (1936), and the separation of *F. graminearum* Groups 1 and 2 into *F. graminearum* and *F. pseudograminearum* (71, 72). Novel morphological characters, *e.g.*, circinate hyphae and pseudochlamydospores, have been used, usually in combination with molecular characters such as AFLPs and DNA nucleotide sequences, in the description of other species, *e.g., F. andiyazi* (1346) and *F. circinatum* (242, 1570). The proposed splitting of *F. graminearum* into nine different species relies on a handful of single nucleotide polymorphisms (SNPs) in seven genes, but completely lacks clearly distinguishing morphological characters (1601). In many cases species groups were identified first on the basis of differences in sexual cross fertility or DNA sequences and the morphological distinctions consistent with the fertility or DNA sequence data followed and were used as the basis for the species descriptions.

Unfortunately many of the recently described species are based on relatively few, often < 10 and sometimes only two, strains. In such cases, the significance of the described morphological distinctions remains to be determined. Such small samples almost never suffice to represent the genetic variation present in a species, no matter what the character being assessed might happen to be. Increasing the number of verified strains of many of these newly described species, preferably from multiple hosts and/or geographic locations, is needed to validate these proposed species and their identifying characters, and to permit the evaluation of the risks, if any, that these species might pose to international agriculture.

As important as studies of agroecosystems are, however, we think that the most informative studies on systematics, population genetics, and evolutionary biology of *Fusarium* will be (i) those that incorporate isolates from natural ecosystems, (ii) those that focus on isolates from native host populations of sibling species to domesticated host species, or (iii) those that evaluate isolates from fields tended by indigenous, subsistence farmers who may have preserved ancient plant pathogen genotypes within correspondingly diverse host and pathogen populations. Such populations are more likely to include the full diversity of genetic variation found within the species and to provide the insights needed to properly understand the evolution, phylogenetic relationships, and genetic diversity within this large, complex, and continually evolving genus.

9
Species Concepts in *Fusarium*

"Why is the name of the fungus I work on changing?" is a common question raised by plant pathologists. Sometimes the answer is because a new character has been discovered, and what had been one species is being separated into several pieces based on our increased knowledge. At other times it is because two groups of strains have been found to be the same and the older, perhaps less-used, name takes precedence. Yet another reason why names change though, is because the concept of what the species is has changed. Must species share particular characters (a morphological species), be cross-fertile with other members of the same group (a biological species), or all originate from a single common ancestor (a phylogenetic species)? Different individuals place different values on the answers given to these questions.

A robust species usually remains the same no matter the approach that is used to describe it, but each of these concepts can fail for informative biological reasons. For example, morphological species descriptions can fail due to genetic mutations that alter critical morphological features; biological species definitions are not possible unless strains are sexually fertile; and phylogenetic species definitions can fail if hybridization between species occurs sufficiently frequently on an evolutionary time scale. Given these potential problems and the relatively large number of scientists who have studied *Fusarium* spp., and who often had different species concepts underlying their descriptions, the relative level of agreement on many of the major issues and important species is impressive. Yet, simultaneously, the potential level of confusion, especially for a novice working from both the current and the older literature, is quite high.

Species descriptions, even very thorough ones, usually rely on a limited set of characters relative to all of those that could potentially be described. Often the characters used are limited to those that can be easily seen and often quantified. To the extent that these characters are continuous in nature or are dependent on culture conditions, the easier it is for averages to be misleading and for ranges to overlap in some significant manner.

One approach to the study of a species is to use more in-depth analytical tools to identify previously undetected similarities, or to confirm the significance of distinctions that might initially have been considered marginal or of only minor importance. Alternatively, a different species concept can be employed and the hypothesis that both species concepts yield the same result (in terms of strains grouped together) tested. In either case the insights generated often lead to the reworking of species definitions, usually not only for the species on which the work was originally done, but also for sister taxa with similar characters.

In species that are studied primarily by mycologists and that are of limited economic importance, taxonomic problems can take years to resolve, due to the relatively few researchers studying them and the relatively small sets of strains available for study. The situation is very different for a fungal genus such as *Fusarium*, however, which contains species associated as endophytes or plant diseases with virtually every plant family (279, 1150), results in billions of dollars of damage due to direct losses at the farm gate (2342), produce toxic and carcinogenic secondary metabolites (468, 1341, 1482, 1550, 2083), and cause opportunistic infections of humans, especially the immunocompromised, and other animals (1783). As *Fusarium* spp. are relatively common in addition to being economically important, their names often acquire importance that transcends the scientific literature as they become targets for breeding and advertising efforts by commercial seed companies or as they are incorporated legislation, commercial trade agreements, or quarantine regulations. An effective way to generate a non-tariff trade barrier is to describe a new species (perhaps by splitting an old one) and then blocking the movement of the affected agricultural commodity until the presence of the new species has been established at the recipient's location, *e.g.*, the numerous proposed species within *Gibberella zeae* (1601). Thus, names of *Fusarium* species carry far more importance than do those for most fungi, and the unsettled taxonomy can have significant economic implications.

Representatives of *Fusarium* occur in all major geographic regions of the world (265). Individual species may have a cosmopolitan geographic distribution or be limited to one or a few crops, climates, or ecological zones (279, 1551). This type of distribution complicates identification and introduces uncertainty in assessing the significance of the results. For example, is an isolate that is somewhat different from the standard for a widely dispersed species an unusual member of that species? Or, is it instead a representative of a species that is less common, and perhaps restricted in distribution either by host or by geographic region? The variation in distribution pattern also means that most researchers work with only a fraction of the known species within the genus, and may be unfamiliar with closely related species or with the range of variation that can occur within a species. Consequently subtle differences and unusual features may be discounted in favor of more commonly used criteria.

Diseases caused by *Fusarium* occur in countries with very different levels of scientific infrastructure. In most western countries, access to molecular technologies usually is possible and often is expected, while in more poorly developed countries laboratories, if they exist, may have little more than a functional microscope. The diverse approaches used by and the differing needs of investigators results in many descriptions of "species". Merging these various observations into a single consistent picture is often far from trivial. A functional taxonomic system and species definitions must be robust enough for use in these very different settings and should enable researchers to collect useful information that is at least somewhat independent of their access to particularly expensive or highly specialized equipment.

9.1 Generic Problems in Speciation in *Fusarium*

The generic concept of *Fusarium* was first diagnosed in 1809 by Link (1263) with the primary character being the presence of the distinctive canoe- or banana-shaped conidia well known to all researchers with even a passing acquaintance with the genus. Many of the *Fusarium* species currently used in taxonomic systems are poorly defined and type specimens may not even exist. These descriptions probably could not be published today as they would not satisfy the requirements of the International Code of Botanical Nomenclature because the description is vague or poorly defined, but have been conserved due to priority and their wide usage.

Differences in the shape of the macroconidia are central to the identification of many *Fusarium* species, although other characters also are used, and in some cases are critical to distinguish sister species. Unfortunately with *Fusarium*, as with many microfungi, the morphological characters are limited in number, are probably subjected to selection, and their expression is sensitive to environmental conditions. To add further confusion these morphological characters usually are subtle, easily misinterpreted, and can vary in importance depending upon the species. As a final complication, Seifert (1935) has noted that the production of the distinctive macroconidia is not necessarily confined to members of the genus *Fusarium*, as a number of coelomycetes produce spores that could easily be identified as belonging to a *Fusarium* spp., if these spores were the only criterion that was used for identification.

With the introduction of molecular technologies *Fusarium* taxonomy is again a matter of great controversy. At the heart of the controversy is the species concept(s) employed and its practical implementation. In particular, the combination of the phylogenetic species concept with DNA-based data has resulted in a recent wave of new *Fusarium* species, including a number of sibling species that are all but impossible to distinguish morphologically, *e.g.*, Nirenberg & O'Donnell (1570), Nirenberg *et al.* (1571), and O'Donnell *et al.* (1601). In some cases these new species result from the naming of groups long suspected to be distinct, but for which clear morphological evidence of a species level difference was lacking. In other cases, however, the newly described species are being created by splitting species and using previously identified strains from culture (usually) or field (occasionally) collections. Consequently many researchers, especially those with limited experience with these fungi, now find the process of identifying taxa of *Fusarium* a daunting process. There is a significant risk that names will be misapplied and that further confusion, instead of increased clarity, will result. An understanding of the species concept underlying the delineation of the species, and an understanding of the history of *Fusarium* taxonomy (Chapter 8), helps in making decisions as to the types of techniques needed to identify a species, *e.g.*, Figure 11-1.

Morphological (or typological) species concepts, biological species concepts, and phylogenetic species concepts (819) all have a contribution to make toward a concept of what a species is in *Fusarium*. Traditionally morphological species concepts have dominated, *e.g.*, Nelson *et al.* (1551), but more recently biological (1219, 1221, 1222) and phylogenetic (1593, 1595, 1596, 1599-1601) concepts have become much more important, and have provided a different focus on the field. Each of these concepts is accompanied by techniques that are used to implement them in a real-world situation.

9.2 Morphological Species Concepts

Morphological (typological) species concepts are based on the idea that a morphological "type" or individual can represent the variation within an entire species. Morphological species under Linnaean definitions are delimited with two primary criteria. These are: (i) within-species (morphological) consistency, and (ii) sharp breaks in consistency between species (1394). For many purposes, morphologically-based species concepts and taxonomies are useful tools for (at least) initial classification of biodiversity. As noted by Taylor *et al.* (2121), the greatest strengths of the morphological species concept for fungi are its general applicability to any fungal taxon and its widespread and historical use. The Gerlach & Nirenberg (697) and Nelson *et al.* (1551) taxonomies are both morphological in nature and provide the base systems against which biological and phylogenetic species concepts are being tested and into which new species are being grafted.

Both physical and physiological characters have been used as morphological characters to distinguish *Fusarium* species, with the chief problem now being that the number of readily detectable characters is far smaller than the number of species that need to be distinguished. The shape of the macroconidia often is given the greatest weighting when defining species, but differences in macroconidial shape and size can be confusing, subjective, and dependent upon the environment in which they are produced. Other spores, *e.g.*, microconidia and chlamydospores, also are important in morphological species definitions, although the evolutionary significance of these characters is being reassessed in light of recent molecular data.

Both the Gerlach & Nirenberg (697) and the Nelson *et al.* (1551) systems require that cultures be grown on particular media under specific incubation conditions if strains are to be identified correctly. Both systems also divide the genus into a series of sections that follow those first proposed by Wollenweber & Reinking (2359), and which still serve as a straightforward means to reduce the number of species identities possible for any particular culture. Unfortunately, while the Sections provide a convenient means to subdivide the genus on the basis of morphology, they generally are not monophyletic, and a systematic revision of the genus will need either to replace them, redefine them, or abolish them. Some researchers have already abandoned the Sections, but the two step identification process they encourage – first to section and then to species – greatly simplifies the identification process for many others. Despite the need to use specialized culture conditions and media and the subtlety of some of the characters scored, the current widespread utilization of morphological criteria by many diagnosticians and the practical need to routinely identify many *Fusarium* cultures means that these morphological characters, and thus a morphological species concept, will remain important in *Fusarium* species concepts for the foreseeable future.

The value of physiological characters including growth rates, mycotoxin production, and secondary metabolite profiles produced, manifested on different media varies. Wollenweber & Reinking (2359), for example, placed heavy emphasis on the use of a variety of complex media and the resultant growth patterns and rates. At present, growth rates, most commonly 25° and 30°C, sometimes are used by some researchers to separate closely related species (Table 4-1), but this character is never the primary character for a species definition. The production of secondary metabolites, including mycotoxins, also may be used as an important character in *Fusarium* taxonomy, *e.g.*, Thrane (2142), but is technically difficult and requires equipment and chemical expertise that many mycologists and plant pathologists lack. It is unlikely that production of these compounds will be used to define species, even though the ability of a species to produce a particular secondary metabolite(s) often is a character of critical ecological and economic importance.

In this manual (Chapter 13) we place primary emphasis on the morphological species because of their widespread practical use, and because these characters remain sufficient to accurately and rapidly identify many *Fusarium* species. When the number of species exceeds the number of morphological characters identified for that group, however, biological and phylogenetic species concepts must be invoked for accurate and meaningful identifications to result.

9.3 Biological Species Concepts

The biological species concept as put forth by Mayr (1393, 1394) considers "... species as groups of populations that actually or potentially interbreed with each other." Dobzhansky (513) defined biological species as "the largest and most inclusive, ... reproductive community of sexual and cross-fertilizing individuals that share a common gene pool." Templeton (2125) described the Cohesion Species Concept, which is essentially that of Mayr and Dobzhansky, with the focus shifted to the cohesion within a gene pool, rather than the differences between gene pools. These species concepts shift the focus from individuals to populations and define species in terms of how the members of these populations interact and relate to one another rather than by comparisons with a static standard. Although a type specimen is still required for a valid description under the International Code of Botanical Nomenclature (770), thinking of a species as a population means characters may be continuous and that no individual isolate is likely to embody all of the variation present in a species.

The properties common to most biological species concepts are defined by Mayr (1394). First, shared characters within a species are more important than differences between species. Second, *populations*, as opposed to an *individual*, are required to clearly delimit and define the extent of variation within species. Finally, interfertility (or the lack thereof) between individuals in a population is not the definitive criterion for subdividing species. Rather, it is the reproductive isolation of the populations within a species from sympatric or allopatric populations of other species; for a more detailed discussion of this topic with particular reference to fungi see Perkins (1685). Biological species concepts treat species as categories defined by an actually or potentially shared gene pool, rather than as a taxon or type as is always done with a morphological species concept, and is commonly done with many phylogenetic species concepts (section 9.3).

There are practical difficulties with applying a biological species concept to *Fusarium*, including high levels of asexual reproduction, unequal relative frequencies of the mating type alleles (*MAT-1*/*MAT-2*), and often limited numbers of strains that are fertile as females, all of which are known to occur in field populations *e.g.*, Britz *et al.* (246), Chulze *et al.* (372), Leslie & Klein (1224), and Mansuetus *et al.* (1334). Defining a biological species for a homothallic organism, such as *Fusarium graminearum* (teleomorph *Gibberella zeae*), also is difficult even if the organism is known to outcross under laboratory conditions (230), and for the existing data to support significant outcrossing under field conditions (229, 668, 2409, 2410). In predominantly heterothallic species, it is necessary to develop female-fertile tester strains of both mating types, which may require backcrossing and selection (243, 1085, 1092, 1093, 1219, 1221, 1222, 1229, 2412). Even after several rounds of backcrossing, the level of female fertility in many of the progeny may not be very high, and careful selection and testing is required to ensure that the female fertility in the strain selected is stable and consistent. The genetic basis of female sterility in *Fusarium* has not been studied in any detail, although the crossing protocols we recommend (section 6.5) permit the identification of the male and female parents in crosses of the heterothallic species (section 6.4). Although the "legend" within the *Fusarium* research community is that geographically separated strains are more likely to be cross fertile than are those that were geographically proximal, our experience with the development of female-fertile tester strains is not consistent with this hypothesis.

With the identification of mating-type-allele-specific PCR primers (section 7.2) for most *Fusarium* spp. (414, 1037, 1039, 2041, 2402) crosses need be made only with the tester with which a fertile cross is expected. This process reduces the number of crosses required by a factor of two if the biological species is known, and by that amount or more if the goal is to identify a new biological species and produce its teleomorph (section 6.7). Biological species have been used most extensively within the *Gibberella fujikuroi* (*Fusarium* section *Liseola*) species complex *e.g.*, Leslie (1219, 1221, 1222), where nine biological species have been described and where others are known to await description. By utilizing molecular markers, such as DNA sequences or AFLP markers *e.g.*, Marasas *et al.* (1346), O'Donnell (1593), O'Donnell & Cigelnik (1594), O'Donnell *et al.* (1596, 1599, 1600), and Steenkamp *et al.* (2040), to identify likely conspecific isolates, the ease with which biological species within *Fusarium* can be identified should be greatly increased. As the development of the biological species concept in *Fusarium* progresses, the problems of interfertility between some of the biological species anticipated by Perkins (1685) have begun to be detected, *e.g.*, between mating populations C and D of the *Gibberella fujikuroi* species complex (1236).

9.4 Phylogenetic Species Concepts

Application of the phylogenetic species concept to *Fusarium* is a relatively new development that can in some situations clarify taxonomic difficulties but, if inappropriately applied or misinterpreted, can result in confusion and inherent contradictions. The goal of taxonomy is to categorize taxonomic diversity into logical units that allow descriptive sorting of diversity and species identification. Taxonomists try to develop order from a previously chaotic assemblage of organisms, to facilitate communication, and to develop methods that can be applied by others for utilitarian sorting and naming. Taxonomic classification systems may, but need not and often do not, represent evolutionary history or relatedness between taxa, which is the goal of phylogenetic analyses. In phylogenetic analyses, taxa are sorted on the basis of their (presumed) shared evolutionary history. Given these divergent goals, it is not surprising that taxonomic and phylogenetic analyses of an intensively studied genus such as *Fusarium* can come to different conclusions.

Harrington & Rizzo (819) defined a phylogenetic species in fungi in a manner very similar to that of an earlier definition by Nixon & Wheeler (1574). In both definitions, a species is differentiated from other species by being the smallest phylogenetic subgrouping (clade) of individuals or of populations that share a fixed suite of diagnostic characters. The significant change proposed by Harrington & Rizzo (819) was that phenotypic characters (presumably with biological importance or a shared evolutionary history) were needed to gauge the biological relevance of the observed phylogenetic subclustering. Although DNA sequences are now the most commonly used characters for identifying clades and delineating phylogenetic species, in practice any marker that is sufficiently informative, including morphological characters, can be used in the phylogenetic process.

Phylogenetic species concepts are just now being extensively applied in *Fusarium*, and may be especially useful for those species that do not regularly produce a sexual stage, or for those that are primarily or exclusively homothallic. Strictly asexual lineages, by definition, cannot be accommodated in a biological species concept. Homothallic fungi, *e.g.*, *Aspergillus nidulans*, *Neurospora tetrasperma*, *Podospora anserina*, and *F. graminearum*, much like automictic populations of plant or animal taxa, also may not regularly exchange genes with other members of the same "species", although significant exchange apparently does occur in field populations of *F. graminearum* (229, 668, 2409, 2410). Phylogenetic species concepts usually utilize molecular markers, most commonly DNA sequence data, that transcend mating and morphological character problems associated with the biological and morphological species concepts. In place of these problems, phylogenetic species concepts usually must deal with the problem of where to draw the line between "species", *i.e.*, "How different must two strains be to belong to different taxa?" Applications of phylogenetic species concepts vary in terms of whether a single isolate is used to characterize the species (as is common with the morphological species concept), or whether distinct populations and the variation that they may harbor is to be considered (as must be done with the biological species concept).

In practice, many fungal phylogenetic studies have relied on DNA sequences of one or two loci, with DNA isolated from one or a few representative or well-characterized isolates. The limitations and pitfalls of single gene phylogenies in describing species-level phylogenies have been discussed in detail elsewhere (521, 1312). For example within *Fusarium*, phylogenies based on ribosomal internally transcribed spacer (ITS) sequences can be misleading as two nonorthologous sequences types can be isolated from a single *Fusarium* biological species and even from single isolates (1594, 1598, 2273). In other cases, sequences from the β-tubulin gene can lead to erroneous conclusions because it is present in two copies per genome of some species rather than the usual one. Such problems are not unique to fungi. Avise & Ball (99) proposed a species concept that fuses molecular phylogenetic data with a biological species concept-based methodology. This method looks for concordance between multiple, independent, genetically based characteristics. The assumption is that as reproductive barriers (either pre- or post-zygotic) appear, gene-flow among subpopulations is reduced, and genetic differences among populations can become fixed (lineage sorting). This strategy has evolved into what is known as the Coalescent approach (98), which is sensitive to both the number of individuals and the number of genes analyzed (544). In general, this combination of phylogenetic and biological species concepts has the greatest potential for characterizing species-level groupings in the fungi.

Within *Fusarium*, molecular data have been used to help resolve groups that were later described as separate species (1346, 1570, 1571, 2412), usually on the basis of distinctive morphological characters. In the case of the mating populations within *Gibberella fujikuroi*, groups defined on the basis of the biological species concept and the phylogenetic species concept are the same (1222,

1596), even though the questions as to the "correct" names to be applied to these different groups is only now being laid to rest (1229, 1882). More species likely remain to be described in this group (1600), however, and the distinctness of some of these groups, *i.e.*, *F. fujikuroi* and *F. proliferatum*, is being reevaluated given the description of naturally occurring hybrids between these two species (Leslie *et al.*, 2004*b*, *c*).

In contrast to the general consistency of biological and phylogenetic species definition within the members of the *G. fujikuroi* species complex are the recent proposals (1599, 1601, 2291) to subdivide *Fusarium graminearum* into at least nine phylogenetic species. Members of at least some of these phylogenetic species however, are cross fertile, at least in the laboratory (230, 994). Thus for *F. graminearum* the applied phylogenetic and biological species concepts do not presently yield the same result. We suggest that in such cases the species defined by the biological species concept should be viewed as "correct" and that the putative phylogenetic species should be treated as isolated populations and taxonomically termed varieties, if they are given formal names at all. Strains in species such as *F. graminearum*, and perhaps *F. fujikuroi* and *F. proliferatum* as well, essentially represent intermediates in the process of fungal species evolution in that separation has begun, as indicated by the available molecular data, but is not yet complete, as is indicated by the existing cross-fertility. We think that cases in which the different species concepts appear to give different answers may provide very important groups for studies of fungal evolution and differentiation. These cases provide the opportunity to observe these processes and their intermediates in action rather than needing to infer them from studies of the putative starting and ending points, *i.e.*, current well-resolved species.

9.5 How Many Strains Make a Species?

Names assigned under the International Code of Botanical Nomenclature are tied to a dried type. This material may contain one or several strains, although the current practice by those working with *Fusarium* is to make the dried types from a culture that originated from a single conidium. The living cultures that correspond to these dried cultures are *ex*-type cultures. When a morphological species concept is applied, this single culture is all that is required. Biological species concepts are necessarily based on populations, and, if heterothallic, at least two strains of opposite mating type need to be identified to enable the formation of the sexual stage. Morphological and phylogenetic species descriptions also are significantly strengthened if more than a relatively small number of strains are available to support the description.

A significant number of newly described *Fusarium* species have been based on the analysis of isolates from culture collections *e.g.*, Nirenberg & O'Donnell (1570) and O'Donnell *et al.* (1601), which virtually always are small in number. Thus, the number of isolates examined frequently is too small to reflect the diversity present in a field population. In such cases the implementation of a biological species concept-based approach is difficult, if not impossible, since it often is not possible to distinguish the point at which differences between species begin from that at which variability within a population ends. Despite this limitation, many of the recently described *Fusarium* species probably are "true species," primarily because the morphological species descriptions in wide use often have been too broad. It is likely that advances from population genetic analyses and the determination of biogeographical boundaries in *Fusarium* will result from studies in which relatively large numbers of isolates, particularly from diverse natural ecosystems, are analyzed with a variety of techniques – morphological, biological and molecular – to provide an integrated species concept.

The fewer the number of isolates examined, the greater the risk that a species description might fail as more isolates are examined and the extent of variation within a species is better described. A binomial distribution can be used to estimate the statistical power for detecting variation within a sample as: $\log(\alpha) = \log(1\text{-}x)^n$, where α is the desired statistical significance level (typically < 0.05 or 0.01), x is the expected frequency of a character within a sample, and n is the sample size. In practice, to be 95% certain of detecting all markers present in a population at $\geq 5\%$ requires at least 59 independent isolates, and for the $\geq 1\%$ level requires at least 299 isolates. Conversely, beyond these sample sizes there is little additional information to be gained for the extra work. The small samples used in some studies however, should raise cautionary flags. For example, with a sample size of 10, only markers present in the population with a frequency of 26% or greater can be detected with 95% certainty. If the sample size drops to five, then the frequency of the markers detectable with 95% confidence increases to 45%.

Another consequence of insufficient sampling has been referred to as the Iceberg Bias (1235, 2145). The iceberg bias occurs when most, or all, of the attention is focused on a limited number of characters or individuals, *i.e.*, the portion of the iceberg that is above the water. Pieces of the iceberg may seem unconnected based on their above water appearance even if below the water's surface they are solidly frozen together. Analogously, fungal strains may appear in different species if character differences are drawn too finely even if a more in depth examination with more characters or strains might reveal connections that were not initially discerned.

If a microbial species inhabits a diverse range of hosts or geography, and if there are differences in allelic frequencies between these populations, sampling only a limited subset of the ecological or geographic range of a given species can bias the resulting estimates of phylogenetic, inter- or intra-population divergence. As many *Fusarium* species have been described from only one or a few locations, *e.g.*, *F. beomiforme* (1549), *F. brevicatenulatum* (1571), *F. konzum* (2412), and *F. nurragi* (1890), or from samples of one or a few agricultural crops or hosts, *e.g.*, *F. andiyazi* from sorghum (1346), *F. guttiforme* from pineapples (1570), *F. begoniae* from begonias (1570), and *F. brevicatenulatum* from *Striga parasitica* (1571), this bias may apply.

The iceberg bias also may be pronounced in species with a reportedly broad geographic distribution or host range if only a small portion of those ranges is included in a given sample. Thus, simply analyzing even a relatively large but non-systematic sampling of well characterized isolates drawn from existing culture collections or limited by host or geographic distribution, may not be sufficient to describe extant genetic diversity across a species. An ideal description of intra-specific variation, which will strengthen any species description, will incorporate intensive local, as well as ecologically and geographically broad sampling *e.g.*, Klittich & Leslie (1092), Klittich *et al.* (1093), McDonald *et al.* (1402), and Zeller *et al.* (2409, 2410).

9.6 Species Names

The philosophy and practicality of assigning species names can become important issues. In fungal groups in which there is relatively little research activity or economic significance, assigning a name to one or a limited number of strains may be the only method to guarantee their preservation in a herbarium or public culture collection. Changes in nomenclature of such groups may be of importance for basic science, but these changes often have relatively few consequences outside the mycological community.

For economically-important fungi, *e.g.*, most plant pathogens, species terminology carries additional weight. In most cases those working on these fungi are not just mycologists and plant pathologists, but also scientists from allied disciplines such as plant breeding and food science. These names also can acquire a meaning outside the scientific realm alone as they may be written into quarantine, and other government, regulations and trade contracts, and used in legal proceedings, especially if members of the species are involved in toxin production and other safety issues. Unstable species terminology also discourages scientists who are not primarily mycologists from working with these fungi to study aspects of their biology that may have little bearing on their taxonomic status or evolutionary relationships. These scientists often are looking for a research community within which to work that has a focus on a subject other than nomenclature.

It also is important to maintain the utility and information content of a name, as the name should have both descriptive and predictive properties. For example, the change of the name of the fungal strain (A3/5) that is used commercially to produce Quorn from *F. graminearum* to *F. venenatum* based on molecular evidence (1595, 2392) has resulted in monitoring of the fermentation product for a somewhat different set of possible toxin contaminants than was monitored for when this strain was identified as *F. graminearum*. In contrast, the name *Fusarium roseum* subsumed so many different taxa that it is presently of little value and no longer used because the information content of the name is so low (Table 8-1).

Recently, the name *Fusarium verticillioides* has been accepted as a replacement for the name *Fusarium moniliforme* (1936). Reasons for the replacement were primarily nomenclatural and were concerned with maintaining the use of an older name. Yet *F. verticillioides* was accepted by many because the name *F. moniliforme*, as used by Snyder & Hansen (2012, 2018), could now refer to any one of at least 25 different species, so that the *F. moniliforme* name had become both meaningless and confusing. To keep *F. verticillioides* a meaningful name it is important that it be used for only a single entity, in this case *Gibberella moniliformis* or *Gibberella fujikuroi* mating population A. *F. verticillioides* should not simply be substituted mechanically wherever *F. moniliforme* was used in the past. Should the mechanical substitution become the rule, then there will soon be two meaningless and confusing names instead of just one. As other species are split, the need to maintain the integrity of the old name will be a challenge. Finally, teleomorph species epithets should agree with their anamorph counterparts wherever possible. In some cases older names already exist and have precedence, *e.g.*, *F. graminearum* and *Gibberella zeae*. As new teleomorphs are recognized, however, we strongly encourage the use of the same root for their name as was used for the anamorph from which they were derived, *e.g.*, *F. thapsinum* and *G. thapsina* or *F. nygamai* and *G. nygamai*.

9.7 Subspecific Terminology

One approach to some of the problems associated with the need to recognize groups that do not yet warrant distinction at the species level is to use subspecific nomenclature either until the problems are resolved or as an indication that clearly defined differences exist, but are insufficient to warrant resolution at the species level. The International Code of Botanical Nomenclature (770) recognizes varieties as an infraspecific taxon for use in such cases, and a type specimen is required to support the designation. Variety designations have been used in the past, but generally not widely, to describe intraspecific variation within *Fusarium*.

The most common infraspecific designation currently used within *Fusarium* is form species, or *formae speciales* (*f. sp.*). This infraspecific subdivision is recognized in the International Code of Botanical Nomenclature, but the rules of the code do not govern *formae speciales*, and no type specimen is required to validate the name.

Within *Fusarium*, *formae speciales* are most commonly encountered in *F. oxysporum*. Physiological differences *per se* may be used to distinguish *formae speciales*, but in *Fusarium formae speciales* are virtually always defined in terms of plant pathogenicity towards a host or a group of, usually related, hosts. *Formae speciales* may be further subdivided into races, again usually on the basis of pathogenicity, but this time towards a particular set of host lines, varieties or cultivars. In practice strains often are assigned to a single *formae speciales*, usually on the basis of a pathogenicity test against the particular host. Once assigned to a particular *formae speciales*, strains are rarely, if ever, checked for pathogenicity towards other hosts. Thus while a strain could belong simultaneously to more than one *formae speciales*, this possibility is rarely tested and the degree of overlap between various *formaes speciales* generally is unknown.

The phylogenetic significance of *formae speciales* and races depends on the origin of the pathogenicity determinants. If pathogenicity is not an ancestral character, *e.g.*, if it can be acquired independently via mutation or the acquisition of a "B" chromosome, then strains within the race or *formae speciales* need not be monophyletic in origin. Treating such groups as monophyletic may lead to errors in studies of the mechanism of pathogenicity, to difficulties in breeding for resistance, and to inappropriate quarantine measures. We do not suggest that the *formae speciales* terminology be discarded, as it is well-established in the literature and generally conveys useful information regarding pathogenicity quickly and efficiently. It is important, however, to realize that the basis for these names need not be grounded in traits that are monophyletic in origin *e.g.* Baayen *et al.* (105) and O'Donnell *et al.* (1598), and that strains in the same race or *formae speciales* need not even be very closely related to one another.

There is a need to actively expand subspecific nomenclature in *Fusarium* to reflect differences now being detected that are not accurately documented by the *formae speciales* concepts. In particular there is a need for nomenclature to deal with situations when different species concepts yield conflicting answers. For example the C and D mating populations of *Gibberella fujikuroi* are known to exchange genetic information through reasonably fertile intercrosses of some strains (1236). These two groups have been given separate species names even though molecular studies, *e.g.*, O'Donnell & Cigelnik (1594) and O'Donnell *et al.* (1596), show that these two groups are very difficult to distinguish. Perkins (1685) discussed such problems in fungi, especially in *Neurospora* spp., and concluded that some genetic exchange between different species could occur and that the exchanging groups could still be recognized as distinct species. Yet recent work with *Neurospora* has redefined the species concept for these fungi and concluded that no inter-specific hybrids exist (488, 489).

The question, as always, is where to draw a line within a gray area? A cross that produces an occasional perithecium with a handful of viable ascospores is very different from one producing ~1% of the spores produced by a cross of two tester strains selected for high fertility under laboratory conditions, yet both are probably in the gray area termed interfertility. We have found that within the *Gibberella fujikuroi* species complex, strains in the same biological species usually share at least 60-70% of their bands on an AFLP gel, while those in separate species usually share no more than 40% of these bands, and often significantly less (1346, 2412). The C and D mating population isolates that we have examined share 50-55% of such bands and are clearly intermediates by this criterion. The proposed phylogenetic species into which *G. zeae* recently has been split (1599, 1601, 2291), whose members are known to be cross-fertile with one another (230, 994), have a comparable or higher level of genetic similarity based on AFLPs.

In such cases we suggest that the recognized infraspecific taxon rank, variety, be used to designate the different groups if names are required. If the variety names are chosen with care, then they could be easily elevated to species level should data that warrant such a change become available. These names could be permanent if there is evidence of the potential for genetic exchange, *e.g.*, laboratory crosses, between the groups, but the groups are distinct enough, *e.g.*, based on morphological or molecular characters, that there is reason to distinguish them. Variety names also could be dropped when there is a property of the larger group that is of primary concern, or if data that blur the distinctions, *e.g.*, a significant number of naturally occurring hybrid strains, between the different groups, become available. For example, are differences in gene frequencies that are observed between geographically distinct populations indicative of incipient speciation as identified through a phylogenetic analysis, or are they instead indicative of populations that are still fully interfertile but have simply been isolated and accumulated differences through mechanisms such as genetic drift.

Subspecific terminology is needed to match the different levels of relatedness known to exist between different groups of *Fusarium*. The already observed differences are strong evidence that evolution is continuing within this group of organisms, and that evolutionary studies of the relatedness of these strains may lead to new insights into how both fungi and plant pathogens evolve. As an ever-larger number of strains and species is examined with powerful molecular tools, the apparently distinct lines used to separate some groups of strains into species are likely to become fuzzier, perhaps much fuzzier, than they are at present.

9.8 A Species Concept for *Fusarium*

The species concept used in the species definition essentially defines the criteria through which species can be recognized and the bases upon which they can be differentiated. The broad bases for species concepts in *Fusarium* generally have been stable since the publication in the earlier 1980s of the works by Gerlach & Nirenberg (697) and Nelson *et al.* (1551). The species concept developed in these works has served well, in the sense that everything identified as different has been proven to be so. This species concept, although morphological in origin, has been robust enough to support additions developed through biological and phylogenetic approaches. Yet there are major shortcomings in that many of the well-established species are not monophyletic, and that the use of a common name, *e.g.*, *F. oxysporum* or *F. solani*, for multiple biological entities probably is hindering the identification of important plant pathological, mycotoxicological, and ecological characters.

We advocate the continued use of this morphological base, even while recognizing its limitations and its need for refinement and amendment. As new taxa are described it is important that the descriptions meet the requirements laid down by the International Code of Botanical Nomenclature (770), so that unnecessary changes in nomenclature do not further complicate the already complex taxonomic issues. In some cases variety may be a more appropriate taxon rank than species. Obviously, unambiguous and clearly distinguishable morphological characters are preferable in the description of new species, as many scientists will continue to rely on such characters for at least initial identifications. If taxa descriptions that rely solely on DNA sequence data are proposed, then a practical means that can be used to routinely distinguish members of the various taxa also should be provided to ensure that the entities described can be accurately diagnosed on a regular basis by most of the practitioners in the field. The current problems with this group of fungi suggest that clear, distinguishing morphological features are relatively few and that characters such as sexual cross-fertility and variations in DNA sequence will be important for accurate identifications. Thus, the integration of morphological, biological and phylogenetic species should become the rule rather than the exception.

All species concepts are subject to errors that can result from the "Iceberg effect" (2145), wherein a relatively small number of cultures are analyzed (*i.e.*, the tips of the iceberg) and differences/similarities discernable only when large numbers of strains are analyzed are discounted or ignored. Thus, species descriptions based on a relatively large set of strains collected at multiple times or locations usually are more reliable than those based on a relatively small collection (< 20-30) from a single field sample or from culture collections. Although sample size is not a crucial variable in making diagnostic strain identifications, the soundness and reliability of the species description can be affected by this variable.

A species description should not rise or fall on any single character or character type. It is far too easy for a mutation to occur that results in modification of such a character leading to errant identification. Similarly the description needs to be based on a reference population rather than being restricted to any single isolate. Within the species, gene sequences, or at least combined gene trees, should be monophyletic and AFLP similarities for strains within the population should be > 60-70% (1346, 2412). There are relatively few instances in which all three types of information exist for a *Fusarium* species, but in these cases the data have been used to develop a "polyphasic" species description, *e.g.*, *Gibberella circinata* (*Fusarium circinatum*) (242, 243, 1570) and *Gibberella konza* (*Fusarium konzum*) (2412). In these cases the species definitions are extremely robust.

Each type of species definition is subject to a characteristic set of errors. The most common error resulting from a morphological species description in *Fusarium* is the grouping of isolates that should be separated into distinct species into a common species, *i.e.*, "lumping". This problem is the classic one of "running out of characters before running out of species." With biological species, the most common error is no diagnosis, because the appropriate tester strains may not be available or suitable laboratory crossing conditions have not been identified. With phylogenetic species, the most common error is to resolve isolates into more groups than is biologically meaningful, *i.e.*, "splitting".

Needs to address and their implementation. Identification and description of *Fusarium* species is a process that should be expected to continue, with the splitting of older species into two or more new ones the most likely event. The most active area of recent research has been within section *Liseola*, where the number of species has increased by four-fold in the last 10-15 years, and the clear morphological demarcation between sections *Liseola* and *Elegans* has been blurred and found to have little, if any, evolutionary meaning. We expect there to be an equally massive, if not larger, subdivision of the two remaining Snyder and Hansen species, *F. oxysporum* and *F. solani*, *e.g.*, Baayen *et al.* (105), O'Donnell (1593), and Suga *et al.* (2065).

With respect to many of the other species and sections within the genus the situation is less clear. At the moment many of the morphological species appear to hold reasonably well, *e.g.*, Benyon *et al.* (169). Only a few of the existing species descriptions, however, have been rigorously challenged with tools used to delimit either biological or phylogenetic species, and the question of the integrity of these morphological species remains open. Lack of living ex-type strains for many of these species will make it essential for them to be retypified as the existing morphological species are tested for congruence with biological and/or phylogenetic species definitions. The most fruitful challenges to the existing morphological species concepts probably will come from studies of strains that do not seem to reside "comfortably" within any of the known species. In some cases strains already in culture collections may provide a place from which such studies can begin, and indicate the need for more in-depth studies and additional field collections for confirmation. Collections from tropical regions, both native and agricultural, as well as from temperate native regions also should yield new strains that could challenge existing species concepts and aid in understanding the evolution that has occurred and is still occurring within the genus.

As new groups of strains are identified, we recommend that the biological species concept, *i.e.*, sexual cross fertility, be the dominant concept applied, wherever it is applicable. If sexual cross-fertility tests are not available then molecular markers almost certainly will dominate. For these studies to be useful, however, the samples analyzed must be large enough and diverse enough to be representative of the population. Within species differentiation should be treated with care. Populations separated by host or geography could be named, probably at the variety level, but we think this temptation should generally be resisted unless the population is of peculiar economic or ecological importance.

In general, we expect many investigators to use molecular markers to sort like strains together and then to test for cross-fertility and to identify unique morphological characters. These studies also could proceed in the opposite direction (morphology and/or cross-fertility first followed by molecular analyses), but we think most such studies would be focused on testing the integrity of an existing species or on determining the range of variation within an already known species. No matter the order in which the studies are conducted, however, investigators must ensure (i) that representative strains are available for examination by others, preferably by depositing strains in public and personal collections to increase their accessibility, (ii) that collectors have been properly acknowledged, and (iii) that geographic and host/substrate origins are accurately documented, *cf.*, Nirenberg & O'Donnell (1570) and Leslie *et al.* (1233).

10

Teleomorphs

Studies of the mechanistic development of sexual structures in *Fusarium* has been limited. Most investigators have focused on whether perithecia are produced for purposes of identifying biological species or for recovering ascospores for use in genetic analyses. In this chapter we summarize some of what is known about the development of sexual structures, describe an unusual feature of ascus development and ascospore maturation, and provide information on links known between *Fusarium* anamorphs and their teleomorphs in *Gibberella*, *Albonectria* and *Haematonectria*. Recently described teleomorphs often have not been compared with type material of species described in the late 19th and early 20th centuries, and types of many of these species are no longer available. Thus neotypification may be important to stabilize nomenclature for many of these species.

10.1 Taxonomy of Teleomorphs

A number of teleomorph genera have been associated with species of *Fusarium*. Most of the teleomorphs are members of the Hypocreales in the Ascomycetes although some were in other orders. Gradually most of the more divergent species have been removed from *Fusarium*, *e.g.*, *Microdochium nivale* and *Plectosporium tabucinum*, leaving associations with three teleomorph genera, *Gibberella, Haematonectria* and *Albonectria* (1837, 1838, 1881, 1882). Of these *Gibberella* is the most common, being linked with the majority of *Fusarium* species (1882) and includes most of the important pathogens including *Gibberella zeae* (*Fusarium graminearum*), *Gibberella moniliformis* (*Fusarium verticillioides*), and the rest of the *Gibberella fujikuroi* species complex (Fig. 10-1). More details on a dozen of the *Gibberella* teleomorphs can be found elsewhere (467). *Haematonectria* includes *H. haematococca*, which was formally known as *Nectria haematococca* (1838), the teleomorph of *Fusarium solani* (Fig. 10-1). The genus is differentiated from *Nectria* on the basis of morphological and phylogenetic characters. *Albonectria,* which was formerly included in *Calonectria* (1838), is associated with a small number of species of *Fusarium* the most important of which is *Fusarium decemcellulare* (*Albonectria rigidiuscula).*

10.2 General Teleomorph Characters

Morphological differences within genera of the teleomorphs often are difficult to discern, and of limited use for diagnostic purposes, although there are exceptions such as the large ascospores produced by *G. sacchari* (1229). Teleomorphs are not commonly observed for most species other than *G. zeae* under field conditions, *e.g.*, Afanide *et al.* (22), Broadhurst & Johnston (248), Porcelli & Frisullo (1720), Summerell *et al.* (2078, 2079), and Watanabe & Umehara (1977). It is more common to generate the teleomorph under laboratory conditions for further study. Culture conditions can be major influences on the size, and to a lesser extent shape, of perithecia, asci, and ascospores, *e.g.*, Headrick *et al.* (835). Thus, comparisons should be made only between

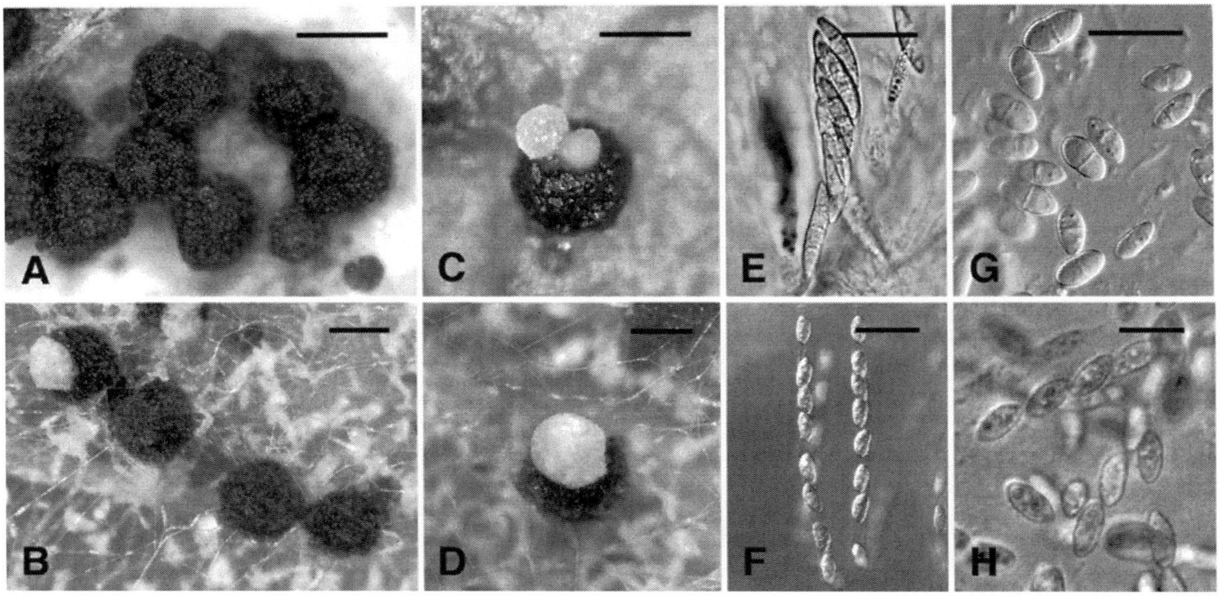

Figure 10-1. Perithecia, asci and ascospores of *Haematonectria haematococca, Gibberella zeae* and *Gibberella moniliformis*. **A.** Cluster of perithecia of *G. zeae* on wheat straw, bar = 200 μm. **B.** Perithecia of *H. haematonectria* on carnation leaf pieces from CLA, bar = 200 μm. **C.** Perithecium of *G. zeae* oozing ascospores in a cirrhus, bar = 200 μm. **D.** Perithecium of *H. haematococca* oozing ascospores in a cirrhus, bar = 200 μm. **E.** Asci and ascospores of *G. zeae*, note 3-septate ascospores, bar = 25 μm. **F.** Asci and ascospores of *H. haematococca*, bar = 25 μm. **G.** Ascospores of *G. moniliformis*, note 1-septate ascospores, bar = 10 μm. **H.** Ascospores of *H. haematococca*, bar = 10 μm.

structures that originate from cultures growing on similar types of material. Naturally occurring mutants that are abnormally pigmented also are known, *e.g.*, Babai (109) and Chaisrisook & Leslie (333).

Gibberella perithecia are dark purple-black, and become red in 3% KOH with an obovoid or subglobose shape and a warty exterior that makes them appear to be rough. Asci are relatively narrow and clavate, and usually lack an apical discharge mechanism. Ascospores are fusoid shaped, 1-3 septate, straight or slightly curved. Ascospores are initially hyaline, but become a light brown when discharged, often as part of a cirrhus of spores that ooze from the perithecium. Strains are not sexually dimorphic, although female-fertile strains can be rare in field populations of some species (1224).

Haematonectria perithecia are yellow to red with globose to pyriform shape and are superficial on the substrate. They darken slightly in KOH. The asci are clavate, and contain striated, ellipsoid, 1-septate ascospores.

Albonectria perithecia are white to pale, subglobose, globose to ellipsoid and do not react with KOH. Asci contain 4-8 spores, and ascospores are ellipsoid to long-ellipsoid and 3-septate.

10.3 Sexual Development & Differentiation

Many of the events that occur during *Neurospora*'s mating process are thought to have parallels in other ascomycetes including *Fusarium*. For a successful cross to occur, a conidium or hyphal fragment of a strain of one mating type must donate a nucleus to a protoperithecium of a strain of the opposite mating-type. These nuclei usually are retrieved by a specialized structure called a trichogyne. A similar structure with similar function is presumed to occur in *Fusarium*, but has not been described. The trichogyne of each mating type is apparently sensitive to a sex pheromone produced, perhaps constitutively, by cells of the opposite mating type (194). The chemical nature of these pheromones, or even if they exist, in *Fusarium* is not known. Early reports that zearalenone, *e.g.*, Wolf & Mirocha (2353, 2354) and Wolf *et al.* (2352), can serve as stimulant or phermone for sexual reproduction have been disproven (2345). Once fertilization has occurred, perithecial development begins leading to meiosis and the formation of asci and ascospores.

Perithecial development in *Fusarium* teleomorphs has not been studied in detail other than studies of *G. zeae* (781, 2171) and *G. pulicaris* (1649). Ascospores of

G. zeae are ejected with great force and high acceleration, 870,000×*g* (2172), the highest acceleration reported for any biological system. There are some mapped/cloned genes that are known to affect perithecial development in *G. zeae*, *e.g.*, *MAP1* (2204), *MGV1* (902), and *PER1* (994), and a number of ESTs specific for developing perithecia have been identified and are available for further study (2175).

Development in *G. zeae* probably is typical of what occurs in the other members of this genus, even though *G. zeae* is homothallic and most of the other members of this genus are heterothallic. In *G. zeae* perithecial initials are formed following treatment with aqueous 0.25% Tween 60. Hyphae forming these initials are short, curved, and stain darkly with toluidene blue and hematoxylin. These hyphae are found in areas of dense mycelial growth. Cells in these hyphae are small and uninucleate. As the perithecial initial enlarges two cell types are observed – central cells, which presumably are the origin of the ascogenous cells, and outer cells, which presumably are the origin of the perithecial walls. The central cells are thinner, lighter-staining, and generally uninucleate, unless dividing.

The outer cell walls differentiate into three identifiable layers that intergrade from one to the next. The outer layer is two-three cells thick and composed of large, pigmented, spherical, thick-walled, highly vacuolated cells. Cells in the middle layer also are thick-walled, but become more flattened as the perithecium develops and contain prominent dark-staining granules in the cytoplasm. Cells in the three-to-five-cell thick inner layer are thin-walled and round in two dimensions, but flat in the third. Cells in both the middle and inner layer are multinucleate, and are compressed as the perithecium develops. A clear ostiole is visible in mature perithecia.

The central cells differentiate to form apical paraphyses that elongate to fill the central cavity of the perithecium. The ascogenous cells are formed at the base of the perithecium and displace the apical paraphyses as the asci grow and differentiate. Mature asci normally contain eight spores. It is assumed, but not proven, that all of the asci within a perithecium originate from the same two haploid progenitor nuclei. The ascus tips are simple, and ascospores are released through a small tear or slit at the apex. Ascospores presumably are discharged forcefully due to changes in turgor pressure, perhaps attributable to mannitol dehydrogenase activity (2173). Ascospore release is generally higher in the light than in the dark and when relative humidity is >92% (1662, 2174).

In heterothallic *Gibberella* species (1768), the asci are normally eight-spored and the ascospores two-celled and hyaline. Meiosis and ascospore delimitation are presumed to be typical of many ascomycetes; see Raju (1767) for details in *Neurospora*. Crozier cells are formed in which haploid nuclei undergo karyogamy, followed by the formation of a large diploid nucleus that divides only meiotically. The two meiotic divisions are followed by a single mitotic division resulting in eight nuclei that are each enclosed within a single ascospore. Further mitotic divisions occur after ascospore delimitation resulting in multinucleate ascospores, although all of the nuclei in any given ascospore are genetically identical. The tetrads in *Gibberella* are unordered, because of spindle overlap and/or irregular positioning of the nuclei prior to ascospore delimitation. Thus although the spores are arranged in the ascus in a linear manner, their position does not reflect segregation of genetic markers during the first and second meiotic divisions. For genetic analysis, this process means that *Gibberella* tetrads should be treated as unordered.

10.4 Spore killer

Spore killer is a genetic trait described in *G. moniliformis* (1021), *G. intermedia* (1768), and *G. subglutinans* (1978), which also may occur in other *Fusarium* species. There are three basic types of spore killer alleles – killer (sk^k), sensitive (sk^s), and resistant (sk^r). In homozygous crosses of $sk^k \times sk^k$, $sk^s \times sk^s$, and $sk^r \times sk^r$, all eight of the ascospores in an ascus potentially are viable. The heterozygous crosses of $sk^k \times sk^r$ and $sk^s \times sk^r$ also are capable of producing eight viable ascospores per ascus. In a heterozygous cross of $sk^k \times sk^s$, however, only four viable spores are produced per ascus and all of them carry the sk^k allele. The non-viable spores do not mature, and asci in such heterozygous crosses generally appear to have four spores per ascus rather than eight. Often these spores do not appear to completely fill the ascus.

Spore killer is an example of a segregation distortion phenomenon that is known in other fungi (808, 1768), *e.g.*, *Neurospora crassa*, *Podospora anserina*, and *Cochliobolus heterostrophus*, and in many other organisms such as mosquitoes, mice, and *Drosophila* (674). The segregation distortion results from a relative increase in the number of sk^k alleles and a relative decrease in the number of sk^s alleles. In a population, the effect is one of meiotic drive, with the sk^k allele being driven towards fixation and the sk^s allele being driven towards extinction. The meiotic drive effect is the most pronounced when sk^k and sk^s are equally frequent and the number of heterozygous crosses is maximized, and is of least importance when heterozygous crosses are rare and one of the alleles is relatively rare. The mechanistic basis for spore killer is known in some higher organisms (674). In fungi

Table 10-1. Teleomorph names for species with a *Fusarium* anamorph (222, 1229, 1882, 2412).

Teleomorph	*Fusarium* species
Albonectria species	
A. rigidisciula	*F. decemcellulare*
Gibberella species	
G. acuminata	*F. acuminatum*
G. avenacea	*F. avenaceum*
G. baccata	*F. lateritium*
G. buxi	*F. lateritium* var. *buxi*
G. circinata	*F. circinatum*
G. coronicola	*F. pseudograminearum*
G. cyanea (= *G. gordonii*?)	*F. reticulatum*[a]
G. fujikuroi	*F. fujikuroi*
G. heterochroma	*F. flocciferum*[a]
G. indica	*F. udum*
G. intermedia	*F. proliferatum*
G. intricans	*F. bullatum*
G. konza	*F. konzum*
G. moniliformis	*F. verticillioides*
G. nygamai	*F. nygamai*
G. pseudopulicaris	*F. sarcochroum*[a]
G. pulicaris	*F. sambucinum*
G. pulicaris var. *minor*	*F. torulosum*
G. sacchari	*F. sacchari*
G. stilboides	*F. stilboides*[a]
G. subglutinans	*F. subglutinans*
G. thapsina	*F. thapsinum*
G. tricincta	*F. tricinctum*
G. tumida	*F. tumidum*
G. xylarioides	*F. xylarioides*
G. zeae	*F. graminearum*
Haematonectria species	
H. haematococca	*F. solani*

[a]Relationship between anamorph and teleomorph presumed but not proven.

it is known only that the killing process occurs after ascospore delimitation in *Neurospora* when the nuclei carrying the two alleles have segregated from one another (1768). This segregation pattern has lead to the hypothesis that *sk* loci and the alleles present in isolated populations may have played an important role in the evolution of homothallic and pseudohomothallic fungal reproductive strategies (1770).

In *Neurospora* and *Cochliobolus*, the sk^k and sk^r alleles are associated with chromosome rearrangements. These rearrangements greatly reduce the amount of recombination near the *sk* region, and may serve to prevent recombination between various portions of the *sk* alleles. In *Neurospora* the region affected by the recombination block is quite large, approximately 30 map units (302). No such recombination block is known in *G. moniliformis* (2373), where the *sk* locus maps to chromosome 5 (Fig. 7-2). In *Neurospora*, multiple *sk* loci occur that are closely linked to one another. The effects of at least two of these loci, *sk-2* and *sk-3*, are independent of each other in the sense that a $sk\text{-}2^s$ strain may be simultaneously $sk\text{-}3^k$ and vice versa. In *Fusarium*, only a single *sk* locus has been identified in the species examined. This locus could be amenable to map-based cloning in *G. moniliformis*.

In most fungi, the percentage of viable sk^s ascospores from a heterozygous $sk^k \times sk^s$ cross is low, usually < 1%. A number of *Fusarium* alleles are known in which a far greater percentage of the sk^s ascospores survive. The meiotic drive effect on these alleles is much less intense than it is on alleles in which fewer spores carrying the sk^s alleles survive, which could help explain the polymorphism for *sk* known in existing *Fusarium* field populations. The existence of this polymorphism also could be dependent on selection against the sk^k allele (or for the sk^s allele) during the vegetative portion of the life cycle, as a means to counter the selection occurring for the sk^k allele during the sexual portion of the life cycle.

10.5 Anamorph-Teleomorph connections

Many *Fusarium* species have both an anamorph (asexual state) and teleomorph (sexual state name). The International Code of Botanical Nomenclature (Article 59.1) indicates that the correct name in such settings is the teleomorph name, although reference to an anamorphic name alone is permissible under some conditions (Article 59.5). For most *Fusarium* species the *Fusarium* name is the more commonly applied, although there are exceptions, *e.g.*, *G. zeae* is widely used for the fungus that has *F. graminearum* as the anamorphic state. As noted above, sexual stages associated with *Fusarium* species are not commonly observed under field conditions, which means that most scientists are referring solely to the anamorph when they describe experimental materials and observations. The absence of an observed teleomorph under field conditions, even if the teleomorph can be readily synthesized in the lab, is sufficient reason to continue using an anamorphic name (Article 59, example 1). We recommend that both the *Fusarium* name and the teleomorph name, if one has been assigned, be included in publications, but that after the equivalence is established that only one of the names is used throughout the rest of the paper.

Sexual states for many of the well-known *Fusarium* species are listed in Table 10-1. This table is arranged by teleomorph name. To search for teleomorphs by the *Fusarium* name, refer to the individual species descriptions in chapter 13.

11

Practical Approaches to Identification

Most plant pathologists, at some time in their career, must identify at least one culture of a *Fusarium* species. The complexity of the problem varies, depending on the host from which the culture originated and the degree of resolution required in the identification. As noted in earlier chapters, *Fusarium* species cause a huge number of diseases on an extraordinary range of host plants. These fungi can be soilborne, airborne, or carried in plant residue, and can be recovered from any part of a plant from the deepest root to the highest flower. In addition, *Fusarium* taxonomy has been plagued by changing species concepts, with as few as nine or well over 1000 species being recognized by various taxonomists during the past 100 years depending on the species concept employed (Chapter 8). This volume (Chapter 13) contains summary descriptions of the best described of these species, but most plant pathologists will not work with all of these species.

Our objective in this chapter is to outline a practical approach to identifying species of *Fusarium*, with special consideration given to those that cause plant diseases. This approach will allow sufficient identification of most cultures for routine work in a plant disease diagnostic laboratory, and indicate when assistance from other scientists with more specialized equipment or skills should be sought. In such identifications, spore type and morphology usually are the important pieces of data, but the diagnosis process can be simplified if additional information, *e.g.*, the host plant, the disease symptoms, and observations made in the isolation and recovery process, are included as part of the diagnosis process. As most who study *Fusarium* are plant pathologists, we have focused on problems associated with the diagnosis and identification of plant diseases caused by *Fusarium*; however, the principles outlined can be used when identifying *Fusarium* cultures that originate from other sources.

11.1 Overall Identification Strategy

The critical question for all who identify *Fusarium* cultures is, "How much information is needed?" Part of the problem in identifying *Fusarium* is that the answer to this question changes depending on the species being identified, and the use to which the answer will be put. The species concept employed in the species description (Chapter 9) also can play a critical role in the type and amount of information needed. Often the most difficult part of the diagnostic identification of a *Fusarium* culture is to clearly define the question being asked, to determine the amount of information needed to answer that question, and to know if the information obtained suffices to answer the question.

An identification strategy that we usually employ is diagrammed in Fig. 11-1. The first step in the identification process is to clearly describe the plant disease and the symptoms observed on the diseased plant, and to note the weather conditions under which the disease occurred. Next, isolation and recovery protocols are evaluated to determine the spectrum of species that are likely to be recovered. Finally, the strains of interest are purified and then evaluated by using morphological, molecular, and cross-fertility criteria.

11.2 The Diseased Plant and its Geographic Origin

Strains of *Fusarium* species can cause a very wide range of plant diseases. The nature of the disease provides important clues as to the species that will be recovered, and often limits the range of species that need to be distinguished. Thus, the possible outcomes for most routine diagnoses usually can be identified relatively quickly, as the number of species associated with particular hosts or disease symptoms often is relatively limited. If symptoms are novel or unusual for the diseased host, however, then the information provided by this approach may not be particularly valuable.

Fusarium species recovered from both natural and agricultural ecosystems have distinct climatic preferences (114, 1894). The climate, and even local variations in weather, can limit the range of species observed even if several are present, and influence their relative frequency of recovery (457, 1480, 1835, 2048, 2251). In some cases these differences have been used to establish disease forecasting systems for *Fusarium* diseases, *e.g.*, chickpea (1535), Douglas Fir (206, 207), maize (2048, 2049), mango (1646), potato (1300) and wheat (457, 891, 1663, 2219). In broad terms there are species that prefer tropical climates, hot arid climates or temperate climates, and a fourth group that has a cosmopolitan range. A concern with many of the species in the fourth group is that these may be sets of sibling species, *i.e.*, species that are morphologically indistinguishable but genetically distinct, that have yet to be adequately resolved. For example, maize stalk rot from plants in relatively warm dry areas, *e.g.*, Kansas, is most likely caused by *F. verticillioides*, while the same disease in cooler moister areas, *e.g.*, Minnesota or the highlands of Mexico, is more likely to be caused by *F. subglutinans*. In wheat, *F. culmorum* can cause crown disease in cooler temperate areas while *F. pseudograminearum* is more common in warmer subtropical and arid regions.

From a diagnostic perspective these general rules can be used to guide the identification process. Both climate and host are potentially informative. A species causing a winter wheat disease in Siberia, for example, is not a likely candidate for the causal agent of a tree disease in the tropics. The Disease Compendia series published by the American Phytopathological Society provide good initial guidance in many cases, especially for temperate crops. Some bioclimatic modeling of *Fusarium* diseases has been done as well, *e.g.*, Backhouse *et al.* (114), but generally these studies are not as comprehensive as are studies of host specificity. Finally, simply because a species is detected in a particular setting it does not necessarily follow that that species is the causal agent of the disease symptoms observed. In summary, routine diagnoses should focus first on known causal agents of diseases on a crop and on species known from that particular geographic region. Only after these "likely" candidates have all been excluded as possibilities should the process be broadened and species not previously described from the area and/or the host being evaluated be considered for the possible identity of an unknown strain.

11.3 Native and Agricultural Populations

Fusarium spp. are ecological members of both agricultural and native communities. These communities and the species and populations within them can behave quite differently. Pathogens of agricultural crops are much more prone to boom-and-bust type cycles that can lead to selection for clonality of particularly successful haplotypes. These populations also can be subjected to severe, but irregular, selection pressures as the crop and cropping system and practices in the field change. Pathogen populations of agricultural crops usually are larger, in terms of biomass, and the niches in which they are found are much more uniform than are the populations found in native or non-agricultural niches. Computer models (2139) suggest that *Fusarium oxysporum* persists in both native and agricultural soils whether or not a host is present, but the differences in parameters that could result in loss of the pathogen, extinction of the host, or stable coexistence of the host and the pathogen were relatively small, which makes predicting the results for any particular situation difficult.

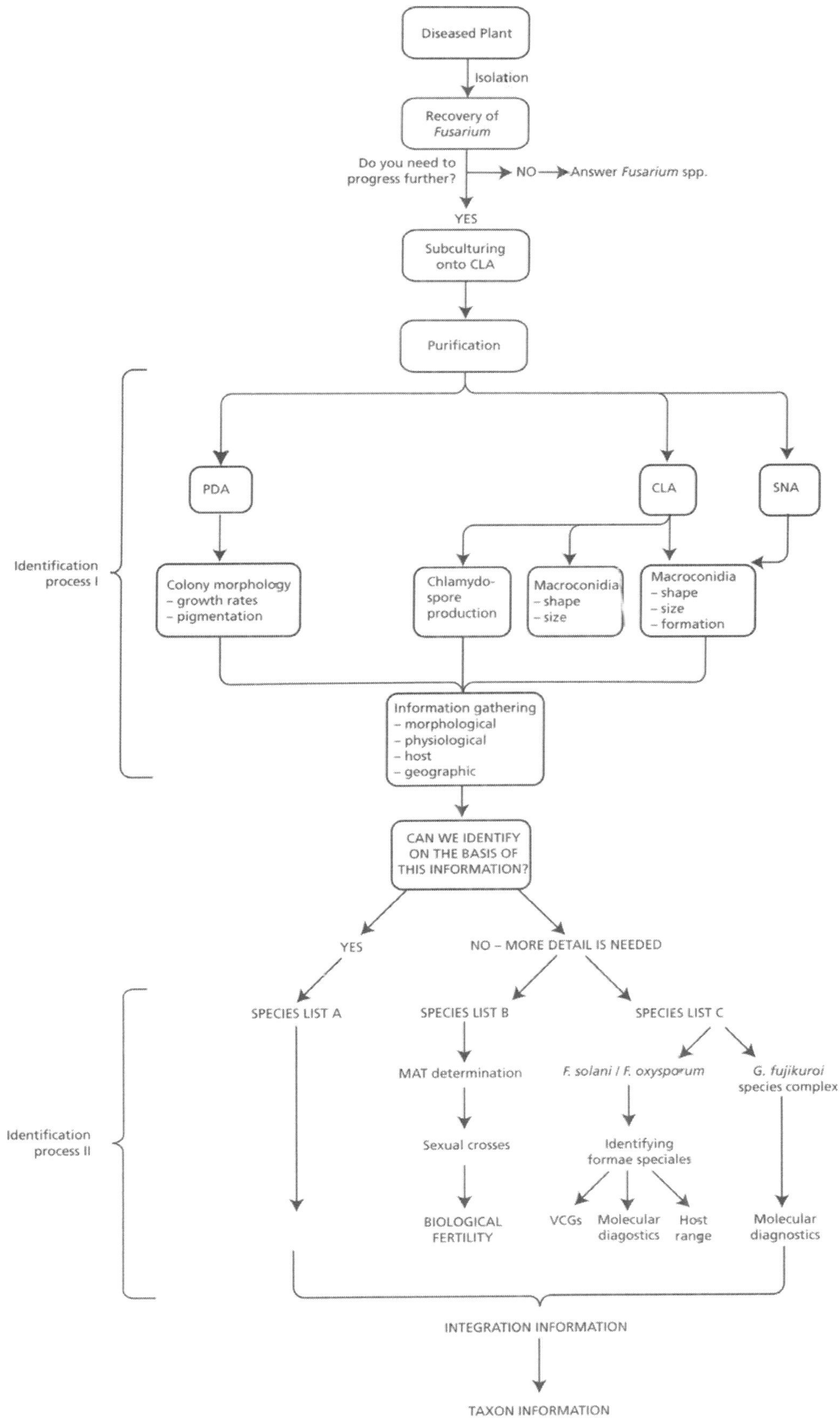

Fig. 11-1. Flow chart of identification protocol used for identifying *Fusarium* species.

Native populations should be capable of sustaining more genetic variation as they will not be subjected to the uniform, and often intense, selection pressure associated with a monoculture agricultural plant population. Populations from native areas, however, are likely to be found in fewer locations than are agricultural pathogens of major crops and probably are not dispersed as often since the environment will not be disturbed in a regular manner such as is found in production agriculture fields with routine tillage and crop rotation practices. The lower levels of mixing expected for the native populations and their smaller physical sizes should make genetic drift more important in these populations than in their agricultural counterparts. An agricultural analog of a native plant pathogen population might be found in subsistence agriculture settings where the field cultivated is small, and the planting of (usually) a mixture of land race crop varieties that may even be planted as intercrops could provide small, diversified and relatively isolated niches in which genetic drift also could be of considerable importance. Samples from multiple locations are likely to be important for studies of both agricultural pathogen populations and native, non-agricultural populations to ensure that all of the variation present has been sampled and is represented in the final study population.

Questions regarding the nature and distribution of genetic variation within a pathogen population often overlap with questions related to species-level differences. In both native and agricultural populations, it is important to identify potential causes of population bottlenecks such as competitors, the ability to persist between seasons, the presence of preferred or alternate hosts, and physical aspects of the environment, *e.g.*, climate, soil type, and tillage regime. The time of sampling and the size of the samples can be important factors in both native and agricultural populations as well. Sample sizes considerably larger than those needed to detect uncommon alleles may be required to estimate parameters such as effective population number (1224) or community richness (916), and to detect gametic disequilibrium or migration (2410). Finally, it is important to remember that if variation in pathogenicity is the central question regarding a species, then scoring the frequency and distribution of anonymous, presumably neutral markers need not provide useful information. In populations where reproduction is primarily or exclusively asexual, correlated markers may be quite useful, *e.g.*, the correlation between vegetative compatibility groups (VCGs) and pathogenic form species in *F. oxysporum* as first proposed by Puhalla (1745) has held up remarkably well (1087, 1220). In some cases, VCGs have been used to track *F. oxysporum* pathogens and infer their geographic origins, *e.g.*, Appel & Gordon (75), Cai *et al.* (295), Gale *et al.* (669), Koenig *et al.* (1104), Lori *et al.*, (1295), and Rosewich *et al.* (1834). In a population where even irregular sexual reproduction occurs, however, searches for correlated markers may be fruitless unless one (or more) of the relevant pathogenicity genes is identified. VCGs in such populations may be composed of multiple genetically diverse individuals united only by their common VCG (372).

Given their broad host range and geographic distributions, *Fusarium* spp. might be expected to have played a large role in our understanding of the host-pathogen evolutionary process. Certainly they have contributed to our understanding in this area, with the identification of dispensable "B" chromosomes (1434) and the determination that horizontal transfer of pathogenicity genes can occur between some species (1075, 2231). Recent studies of populations of *G. zeae* in Asia (668) and in the United States (2409, 2410) have found that little or no linkage disequilibrium exists in these populations. In the United States the genetic distance and geographic distance are correlated. This correlation could reflect the time it takes for the different genotypes to diffuse from one population to the next (2410). Similarly, the use of molecular markers to track the spread of some *F. oxysporum* form species and to study their evolution also has been informative, *e.g.*, *F. oxysporum* f. sp. *cubense* (1470). Beyond these types of studies the contributions made with *Fusarium* to the basic study of host-pathogen interactions has been quite modest.

Some of these limitations are due to problems with the nomenclature leading to expected, or unexpected results. As an example, consider *F. oxysporum*. Evolution of new form species or races informs us about host-pathogen evolution only if the new race or form species is arising from the same genetic material as yielded the original pathogen. If the new form species or race emerges in a cryptic species that shares only superficial morphological characters with the other pathogens, then an understanding of the evolution of pathogenicity will be difficult to extract from the comparisons of the new and the old strains. The *formae speciales* terminology imposes expectations of genetic similarity and relatedness that may not be supported by more detailed genetic analyses *e.g.*, Baayen *et al.* (105), Clark *et al.* (378), Koenig *et al.* (1104), and Marlatt *et al.* (1360). On the other hand, comparisons of groups of very closely related and cross-fertile, but isolated strains, *e.g.*, the different phylogenetic lineages of *G. zeae* (1599, 1601, 2291) or unusual populations of *F. verticillioides* (1476), could provide significant new insights. We think that the most valuable insights into host-*Fusarium* evolution to be provided by isolates of *Fusarium* are likely to come from the examination of groups of strains that are known to be closely related to one another.

11.4 Culture Preparation

Whether it originated from a native or an agricultural population, if a culture of *Fusarium* warrants further work, then it must be subcultured and purified before proceeding further. A common mistake is to try to identify the culture directly from the isolation medium. The most common medium for recovering *Fusarium* species is PPA, a peptone-PCNB based medium (Section 2.3), that is semi-selective for *Fusarium*. Colonies growing on this medium for more than 2-3 weeks usually die due to the accumulation of toxic metabolites that are byproducts of the fungal growth on this medium. Similarly, cultures on many rich media, *e.g.*, PDA, do not produce the uniform macroconidia that are necessary for an accurate morphological identification. PDA cultures can give rise to spores that are characteristic of many different species even if the culture producing them was purified by a single spore subculture and is genetically homogeneous. PDA culture often results in cultures that are "laboratory horrors," *i.e.*, morphologically degenerate cultures that also may have lost sexual cross fertility and pathogenicity capabilities. Such cultures must be identified on the basis of DNA sequence similarity if they can be identified at all. Finally, cultures that have not been purified through a single-spore subculture (section 4.3) often are mixed, and the resulting irregular mixture of morphological characters leads, at best, to inaccurate diagnoses and, at worst, to both nonsense and frustration. The recent resolution of cultures of *Fusarium xylarioides* into two different species (690) demonstrates clearly how important this step is and why it can not be omitted!

In some cases conidia from colonies on PPA can be used for starting single spore cultures from which identifications can be made. Some species yield only abnormal cultures if started from single spores from PPA-grown colonies, however. Unless there is previous reason to expect colony stability, colonies should be transferred from the isolation medium to a nutritionally weak medium, *e.g.*, water agar or SNA, with a bit of sterilized plant tissue and spores from the resulting colonies used for the single-spore subculturing process. Macroconidia, microconidia, and ascospores all can be used to initiate these subcultures. Microconidia are the most common spore type produced by some species and are easily separated by micromanipulation or dilution plating. Macroconidia often are preferred because their larger size makes them easier to handle, but in cultures with both micro- and macroconidia being certain that a macroconidium is not accompanied by a microconidium riding "piggyback" is important. Single spores separated by either dilution plating or micromanipulation on water agar are incubated overnight for germination. Germlings are subcultured from the water agar to a medium for growth, identification, and/or preservation. In some instances it may be necessary to purify cultures by isolating a hyphal tip instead of subculturing a single spore as some species, *e.g.*, *F. graminearum,* occasionally produce relatively few conidia and others, *e.g.*, *F. longipes*, mutate readily when subcultured as single spores.

Accurate identification of a culture requires growing it on at least two media – CLA and PDA (Section 2.1). Many species of *Fusarium* readily form sporodochia with robust, uniform macroconidia on the carnation leaf-pieces that are particularly useful for identification purposes (or for use in making single spore subcultures). PDA cultures are used primarily to assess pigmentation and gross colony morphology. These PDA cultures should generally be regarded as "dead ends" and should not be subcultured further. We usually use cultures growing in 6-cm diameter plastic Petri dishes on CLA and PDA to make initial evaluations, and then make decisions as to whether the fungus should be grown on SNA (section 2.1) or a chemically defined liquid medium. Cultures grown on SNA usually are evaluated for microconidia, which may be more abundant and diverse on this medium, and for chlamydospores, which often are more common and may be produced more rapidly on this medium. Cultures grown in a liquid medium usually are used as a source of DNA (section 7.1) for DNA sequencing or for studies of molecular genetic diversity.

Incubation conditions for *Fusarium* vary depending on the laboratory and the facilities available. Critical parameters are temperatures of 20-25°C and the presence of light, particularly some exposure to UV (black) light. A diurnal (day/night) cycle often is used and probably is beneficial, although some species, *e.g.*, some members of the *Gibberella fujikuroi* clade (1570) are reported to form critical diagnostic characters only in total darkness or under continuous UV or white light. For PDA cultures, light usually is not essential but does increase the total amount of pigment and the speed with which the pigments are produced. Light is more important for CLA and SNA cultures as it increases sporodochial production.

11.5 The Essence of Morphological Identifications

After a purified colony is subcultured to CLA, SNA and/or PDA, it usually is incubated for 7-10 days, depending on its rate of growth before any attempts are made to identify it. Consistency and completeness, *i.e.*, being sure to collect all of the data that are needed, are critical characteristics of this process. A data sheet similar to that in Table 12-1 is a common means to collate the resulting observations. A typical morphological identification strategy follows.

1. Record gross colony morphology features.
   - What is the color of the PDA culture? Check Nelson *et al.* (1551) and Summerell *et al.* (2087) for color pictures of typical pigments.
   - What pigments are produced in the agar?
   - Are sporodochia obvious in the center of the PDA or SNA plates? or on the carnation leaf pieces? If so, what color are they?
   - How thick and high is the hyphal growth?
2. Examine the CLA plate *in situ* under the 10× or 20× objective (100-200× total magnification) of a compound microscope. Metallurgical objectives have a longer working distance and a greater depth of field than do the comparable lenses commonly used for making biological observations. If numerous *Fusarium* identifications are to be made, then the purchase of a metallurgical lens usually is worth its extra cost. For micromanipulation a metallurgical objective usually is essential.
   - Place the plate containing the CLA or SNA culture on the stage of the microscope and examine directly the features of the fungus. Increasing magnification from 100× to 200× (by switching from the 10× to the 20× objective) may be necessary for a detailed examination. Determine if microconidia are present, and if so note their shape, their size, and the manner in which they are formed (Fig. 12-1).
   - If microconidia are present, determine the nature of the conidiogenous cells or phialides, *i.e.*, the cells of the hyphae on which the microconidia are borne. It is not unusual for it to be difficult to make this determination at this stage, and this feature may need to be re-evaluated following observations of a microscope slide preparation.
   - Chlamydospores may be observed in the hyphae, on the surface of the agar, or embedded in the agar. Chlamydospores are more common in older cultures than younger ones, and may occur more abundantly on media such as mud agar (Section 2.2) or SNA than on CLA. Chlamydospores normally form in older cultures, and it may be necessary to re-examine cultures for up to two months to determine definitively if chlamydospores are present.

   It is easy to confuse true chlamydospores, pseudochlamydospores, and swollen cells as they often appear similar. True chlamydospores have a thick wall, a warty appearance (verrucose), and light coloration, usually a yellow-brown. Chlamydospores can be formed in chains, clumps or singly and can be found in the hyphae above, on, or below the agar surface, or within the carnation leaf pieces. Pseudochlamydospores have been described, so far, for only a single species, *F. andiyazi* (1346), and their value as a diagnostic character has not yet been widely evaluated. Pseudochlamydospores are thin and smooth walled and may be found either singly or as short chains in the hyphae. Swollen cells are found in a number of species, especially in section *Liseola*. By definition they are found in the hyphae, and usually are less round in shape than are either pseudochlamydospores or chlamydospores.
3. Make microscope slide preparations for detailed examinations of macroconidia, microconidia (if present), and the conidiogenous cells producing the microconidia (if present). Mounts in plain water will suffice, but some researchers use a stain to highlight the features. If water mount preparations are used, then the microscope must be adjusted such that all of the features of the hyaline structures can be clearly observed. The use of differential interference microscopy (Nomarski) can be useful in discerning the hyaline features of these structures. Photographs of spores in chapter 13 were all taken with differential interference microscopy.
   - **Macroconidia.** These spores usually should be the first structure examined. Preferably they should be from sporodochia on the CLA plate as these spores will be the most consistent in terms of shape and size. Key macroconidial characters are the shape of the spore, its size, the number of septa, and the shape of the apical and basal cells. Develop an overall impression and do not focus on rare or unusual shapes or sizes, or on small nuances in size or shape. First impressions are very important, often are correct, and generally increase in importance and accuracy with experience.
   - **Microconidia.** Variation in microconidial spore morphology often is easier to see in a microscope water mount (Fig. 12-1) than in the *in situ* observations of the producing culture. Results from the *in situ* observations should be confirmed at this time.
   - **Conidiogenous cells.** As with the microconidia, the difference between monophialides and polyphialides often is clearer in a microscope water mount

Table 11-1. Commonly recovered *Fusarium* species that can be identified based on different types of data collected as part of the evaluation process outlined in Fig. 11-1.

Morphology alone (Species list A)	+ Sexual Cross-Fertility (Species list B)	+ DNA-based data (Species list C)
F. acuminatum, F. avenaceum	*F. circinatum*	Remainder of the *Gibberella fujikuroi* species complex
F. chlamydosporum, F. compactum	*F. fujikuroi*	Many formae speciales of *F. oxysporum* and *F. solani*
F. crookwellense, F. culmorum	*F. konzum*	
F. decemcellulare, F. dimerum	*F. nygamai*	
F. equiseti, F. graminearum	*F. proliferatum*	
F. longipes, F. merismoides	*F. sacchari*	
F. oxysporum, F. poae	*F. subglutinans*	
F. pseudograminearum, F. scirpi	*F. thapsinum*	
F. semitectum, F. solani	*F. verticillioides*	
F. sporotrichioides, F. torulosum		
F. tricinctum		

(Fig. 12-1) than in the *in situ* observation of the producing culture. Again, results from *in situ* observations should be confirmed at this time.

4. Secondary characters. The most commonly used secondary characters are odor and growth rate. Usually these are used to confirm a diagnosis based on other characters rather than to make an initial identification. Growth rates may be based on linear growth in a race tube (1852) or on radial growth in a Petri dish, usually at 25 or 30°C (Table 4-1). The Petri dish and race tube growth rates are not directly comparable, and if growth rate is used as a character, then a comparison with the correct reference data set or to *ex*-type cultures is essential. Other useful secondary characters can be physiological data on toxins and other metabolites produced (2142), but such data are not commonly available for routine diagnoses.

5. Organizing the data. The data collected on an isolate at this stage often is qualitatively diverse and quantitatively large. We usually use data recording sheets (Table 12-1) that have separate fields for each of the important characters, to organize results in a consistent manner and to provide a place for making systematic notes of both routine and unusual properties of a culture.

6. Making an identification. At this stage there is enough information to identify some species accurately but not others. Traditionally the genus *Fusarium* has been separated into sections, with the species within a section sharing important morphological characters. Strains are identified in a two-step process, first to section and then to a species within the section. Although the sections are not monophyletic on the basis of DNA sequence characters, they are a useful means for sorting cultures based on morphological identification characters, *e.g.*, presence/absence of chlamydospores, microconidia, or various types of conidiogenous cells. There are computer based diagnostics available that may help with this process as well (757).

With the data collected thus far, at least 21 species can be identified (Table 11-1, column A), but differentiating the remaining species requires additional information. The species that can be identified at this stage are a mixture of well-defined species with strong morphological characters that we think will withstand further scrutiny, *e.g.*, *F. poae* and *F. chlamydosporum*, as well as species that are relatively poorly defined and likely to be polyphyletic, *e.g.*, *F. compactum*, *F. semitectum*, and *F. lateritium*, but have not yet warranted intensive research. Molecular studies of these poorly defined species could alter the placement of these species on this list.

The utility of species identifications based solely on morphological criteria will depend on the question the identification results are needed to help answer. For example, separating *F. verticillioides*, *F. thapsinum* and *F. andiyazi*, based solely on morphological criteria alone is difficult, and the best name probably is *Fusarium moniliforme sensu lato*, *i.e.*, *F. moniliforme* in the broad sense. Another alternative is the "*Gibberella fujikuroi* species complex." For a basic diagnosis this result might suffice, and would be more meaningful and useful than "*Fusarium* sp.", which is the most likely alternative, but in other cases, *e.g.*, analysis of mycotoxin production, quarantine or many journal publications, such a limited diagnosis would be inadequate, if not confusing. Separating species clusters such as these requires either tests of sexual cross-fertility (section 6.7), which could be used to positively identify *F. verticillioides* or *F. thapsinum*, or analysis of molecular markers, which could be used to definitively identify *F. andiyazi* as well (Table 11-1). Until enough cultures of many of the relatively rare species have been examined for easily recognized morphological characters to be discerned and accepted, however, data other than traditional morphology will be needed to make some species-specific diagnoses both accurately and with confidence.

11.6 Beyond Morphology – Sexual Cross Fertility

Although morphological observations may not suffice for a complete identification, a great deal of information usually has been obtained on a culture by this stage of the process, and the possible species identities usually have been narrowed to a relative few. Again the relative cost and time for the additional work must be weighed against the need for the additional information. There are two general techniques that can be used to extend the range of questions being asked. One is studies of sexual cross fertility, and the second is the analysis of DNA sequences. Cross-fertility studies are relatively inexpensive in terms of reagents and technical sophistication, but relatively expensive in terms of the waiting time required for an answer and the need for adequate incubator space.

The ability to cross and to produce the teleomorph (sexual or perfect state, Chapters 6 and 10) with standard testers of defined species groups is the ultimate assurance of a correct species identification. Most analyses of cross fertility for species identification presently are done with strains that are members of the *Gibberella fujikuroi* species complex (1219, 1221, 1222). These crosses are made by crossing an unknown strain as the male parent with a standard tester strain of a known species and mating type, which serves as the female parent (section 6.7). Tester strains for the (currently nine) known mating populations in the *G. fujikuroi* species complex are available at a minimal cost from the Fungal Genetics Stock Center (Department of Biological Sciences, University of Missouri-Kansas City, Kansas City, Missouri).

The number of crosses to be made with each unknown culture can be reduced by half if existing molecular protocols (section 7.2, 1037, 1039, 2041) are used to determine whether the strain in question carries the *MAT-1* or the *MAT-2* allele at the mating type locus. Positive crosses are definitive, but negative crosses (which must have proper controls to be reliably interpreted!) while indicating that a strain is not a particular species, usually are not informative beyond this result.

11.7 Beyond Morphology – Molecular Diagnostics

When critical morphological characters are not available, researchers increasingly are turning to DNA-sequenced based characters to assist in identifications. DNA studies (sections 7.2 and 7.3) are more demanding of equipment, reagents and technical expertise than are studies of sexual cross fertility, but answers can be obtained more rapidly and for more species than presently can be obtained with studies of sexual cross-fertility. The future of many diagnostic protocols for *Fusarium* will undoubtedly involve a DNA study, probably through PCR amplification with a diagnostic primer (or more likely a set of primers) being the tools most commonly used by diagnosticians (section 7.4), and AFLPs (section 7.5) or other fingerprinting techniques, and DNA sequencing the tools more commonly used by more specialized researchers.

Many techniques ranging from restriction enzyme digests of genomic DNA to the sequencing of appropriate genes can be used to identify and distinguish cultures that are morphologically inseparable. We usually use AFLPs to generate multigenic fingerprints as the first step in this second sorting process (section 7.3). AFLPs can be used to group like strains together as well as to associate unidentified strains either with each other or with reference strains of individual species. Based on our experience, strains that share > 65-70% of the bands in an AFLP profile are in the same species and those that share < 40% of the bands are in different species. Strains that share 40-65% of the bands in the profile often represent taxa in which speciation presumably is in progress. Thus, by running an unknown isolate alongside a set of reference strains it often is relatively easy to identify a strain to species with a high degree of certainty.

Direct sequencing of one or more genes also can be used to make species assignments. In this process one, and preferably more, genes is amplified and sequenced. Some of the more commonly sequenced genes for *Fusarium* include β-tubulin, histone, and translocation elongation factor-1α (Table 7-2). The resulting sequence is then compared with similar sequences from strains in related species. The limitations to this analysis have been cost, which is decreasing rapidly, and the availability of sequence data from reference strains in an accessible data bank, which is increasing rapidly, to enable the correct comparison to be made. In most cases, DNA for such analyses is amplified from a genomic DNA mini- or micro-preparation (section 7.1), which can be recovered by using a commercially available kit. The PCR-amplified DNA is then submitted to a sequencing lab, and the sequence returned as a computer file that can be manipulated and compared with other sequences from standard data bases, *e.g.*, GenBank.

The diagnostic value of a single sequence often is inadequate, but once several genes have been sequenced their diagnostic value can be very high. In many cases only one, or at most a handful, of strains has been sequenced for any single species, so the possibility of variation within a species usually has not been addressed (Table 7-2). Similarly, the number of differences that are required for two strains to be said to belong to different species requires calibration and validation with data independent of the sequences themselves to have biological meaning. For example, the phylogenetic lineages in *G. zeae* described by O'Donnell *et al.* (1599) and Ward *et al.* (2291) have been described as separate species (1601), based on the DNA sequences of eight genes. The DNA sequences of other genes, however, are not consistent with this conclusion (2291) and at least some strains from different lineages are cross-fertile (230). Thus the validity and utility of these nine proposed phylogenetic species remains open to question. In general, exact matches in sequence analysis, especially if several genes are used, generally are definitive for species identification, but if there are mismatches, then the significance of the mismatches must be determined before a diagnosis is made.

11.8 The Special Case of *Fusarium oxysporum*

Fusarium oxysporum is responsible for an enormous range of plant diseases, usually involving a vascular wilt syndrome. The majority of the isolates causing vascular wilts are specific strains that infect only a small number of host plants and that are differentiated on the basis of pathogenicity as *formae speciales*. For example the strains commonly attacking banana are *F. oxysporum* f. sp. *cubense,* cotton – *F. oxysporum* f. sp. *vasinfectum*, and tomato – *F. oxysporum* f. sp. *lycopersici*. Morphologically these strains are very similar, or identical, and cannot be differentiated morphologically from non-pathogenic or saprophytic strains, of which there is a huge diversity especially in soil. From a diagnostic point of view the subdivision of this species into *formae speciales* has important diagnostic and quarantine implications. Identification of these strains has traditionally involved pathogenicity testing with sets of host differentials appropriate for the *formae speciales* in question. Pathogenicity tests, however, are time consuming to set up under the best of conditions, and can require long periods of time, *e.g.*, 4-6 months for *F. oxysporum* f. sp. *canariensis* (1736), before they can be scored definitively.

The vegetative compatibility group (VCG) technique (1220, 1745) works well for some formae speciales, *e.g.*, *F. oxysporum* f. sp. *cubense* (1470), but not for others, *F. oxysporum* f. sp. *asparagi* (568). When the number of VCGs in the *formae speciales* of interest is relatively small, then VCGs can be a very effective diagnostic technique. In such cases, the tester strains needed for the *formae speciales* usually are relatively easy to obtain. Even in such cases, however, such diagnostics may not be definitive as some VCGs contain both pathogenic and non-pathogenic strains towards a common host (378, 2365). If there are a large number of VCGs, however, or if the *formae speciales* of interest has not already been evaluated for the VCG diversity that it contains, then VCG analyses usually are more suitable as a research tool than as a diagnostic protocol.

Species- and form species-specific PCR-primer pairs for making identifications of strains of *F. oxysporum* are of commercial and regulatory importance. Again, the research into the identification of such primers is well underway, but the reliability and validation of these PCR primers in large scale tests remains unproven, and a subject for significant future research. We expect significant developments in this field before the next edition of this book appears.

11.9 Differences Between Temperate and Tropical Regions

Similar to their counterparts in temperate regions, *Fusarium* species in the tropics are quite diverse in terms of numbers of species, distribution, host range and virulence (752). However, there are some notable differences between species found in tropical and temperate regions. Plant diseases caused by *Fusarium* spp. in the tropics are becoming more significant with the introduction of intensive, high-yielding production systems and genetically uniform cultivars (1875, 2284). *F. oxysporum* and *F. solani* are the most common plant pathogens in this region. They also are widely dispersed saprophytic soil inhabitants throughout the tropics and the vascular wilts they cause are the most economically important diseases caused by *Fusarium* in the region. The polyphyletic nature of both of these species, means that strains from the tropics are likely to be quite different from similar strains from temperate regions, even though they share a common name. The ambiguous relationship between strains between strains from *Gibberella fujikuroi* mating populations C and D, *i.e.*, *F. fujikuroi* and *F. proliferatum* (1236), and *F. verticillioides* strains from banana to those from maize (1476) are probably indicative of differences that remain to be identified and resolved. Members of the

Gibberella fujikuroi clade are frequently associated with agricultural produce or plants, particularly those in the family Poaceae and with newly introduced crops from outside the tropical region, *e.g.*, asparagus. Species common on some crops of tropical origin, *e.g.*, mango, millets and sorghum, are morphologically similar to those from temperate regions but belong to different species.

Due to deficiencies in resources, plant pathologists in the tropics commonly have problems resolving morphologically similar *Fusarium* species. In the past there was a tendency to name pathogenic *Fusarium* species on the basis of their host, *e.g.*, *F. batatas*, *F. coffeanum*, *F. cubense*, *F. javanicum* and *F. sacchari*, which has added significantly to the problems associated with identification (1875). The name, *per* Snyder and Hansen, *F. roseum* was widely used to identify many tropical *Fusarium* isolates that produced bright red pigments in the culture media, *e.g.*, Benigno & Quebral (165), Semangun (1940), and Singh (1990). Consequently much of the recorded information on plant diseases caused by *Fusarium* in the tropics lacks modern specificity, and many records and disease lists should be viewed with skepticism. In addition, there are many instances in which a tropical strain is identified as a member of a temperate species for which it is the "best fit" or "most like," but which on further study turns out to represent a new species. This problem points to the need to retain cultures, and/or deposit them in an appropriate international collection, so that unusual strains can be re-examined with more sophisticated techniques or in the light of refined species concepts. Certainly a great deal of additional research on *Fusarium* diseases, and the identity of the species that cause them, in the tropics is still needed.

There also are difficulties associated with recovering *Fusarium* species in the tropics. Usually these difficulties have at their core the inability to recover clean isolates from diseased material. Temperature and humidity conditions in the tropics, for example, often result in more rapid decay of diseased plants, leading to samples that are more heavily colonized by secondary fungal and bacterial invaders. This secondary contamination probably is one reason for the numerous records of *F. oxysporum* and *F. solani* as plant pathogens of diverse crops in many tropical countries. Thus, it is important to take samples at the earliest stages of colonization or when disease symptoms first appear. Stringent pathogenicity tests also should be performed with recovered isolates that have unusual or unexpected morphological profiles or that do not conform to previously described and tested disease profiles.

11.10 Conclusions

The primary objective of this chapter was to illustrate the rationale and complexity that underlie the accurate identification of *Fusarium* species and the necessity to follow an identification process, such as that in Fig. 11-1, systematically through all of the steps. The history of *Fusarium* biology is littered with publications and reports that are now difficult to interpret because such a process was not followed. Consequently many species reports are ambiguous or wrong. Unfortunately, such publications are still occurring. The identification process is simplified if a standard protocol is applied, the steps outlined carefully followed, and the limitations of the different types of information that can be acquired clearly understood and identified. The key to success is to collect enough of the information necessary to make an acceptable identification. Unfortunately the amount of information needed for such a goal varies by species, and the use to which the diagnosis will be put. There is not yet a substitute for the judgment of a skilled diagnostician in determining the amount of information that is necessary to make an acceptable diagnosis. Neither is there any single DNA sequence, or set of DNA sequences, that will invariably lead to the correct conclusion. Clearly, "one size" in terms of *Fusarium* diagnostics does not fit all.

We encourage the deposition of cultures in recognized, publicly accessible culture collections and disease herbaria. If cultures are available for analysis, then the names for critical cultures can shift as the nomenclature evolves while cultures remain available for testing new hypotheses regarding pathogenicity, evolutionary speciation, and genetic variability. The *Fusarium* research community is one of the largest and most diverse for any filamentous fungus or plant pathogen, with regular meetings dedicated solely to work on these fungi, *e.g.*, US Wheat and Barley Head Blight forums, European *Fusarium* Seminars, and satellite meetings associated with the International Congresses of Plant Pathology. There also are opportunities for training in identification practices available through the *Fusarium* Laboratory Workshops that are held regularly in the United States and elsewhere in the world. Through these collective efforts, our knowledge of this complex fungal genus should continue to increase and to be utilized more effectively to control the many plant diseases caused by these fungi.

Species Descriptions

12

Morphological Characters

Morphological characters are by far and away the most commonly used criteria for identifying *Fusarium* species. When strains are grown under proper conditions on appropriate media, the identifications usually can be made with reasonable accuracy after some practice. Sets of "standard" cultures are available (for a fee) from the organizers of the *Fusarium* Laboratory workshops and can be useful for comparison purposes when working with an unusual or unfamiliar set of cultures. As many researchers see only a limited number of species in their day-to-day routine, the amount of variation observed in some traits also is likely to be limited.

A necessary prelude to good morphological characterizations are healthy cultures grown from single spores on appropriate media under favorable growing conditions, as outlined elsewhere in this manual. Bypassing these steps may save time and/or effort, but is likely to result in cultures that either are misidentified or unidentifiable. Finally a good identification guide, such as this one, with pictures of critical features to which the examined culture can be compared is essential to confirm the identity of the species.

In general it is important to focus on the general characters of the trait being examined rather than trying to pin down every nuance in form and structure. Focusing on the details can lead to confusion, frustration and errant identifications. First general impressions of morphological features often are correct, and more detailed observations can lead to more confusion than clarity. Recording observations in a systematic manner in a form, such as that provided in Table 12-1, helps insure that all of the critical characters have been examined and considered in making the species diagnosis.

12.1 Macroconidia

Macroconidia (Fig. 12-1A) are the single most important cultural character in the identification of *Fusarium* species. In many cases the morphology of this spore alone is sufficient to identify a culture to species. We recommend the use of CLA rather than SNA for identifying macroconidia as macroconidia formed on the carnation leaf pieces and on the agar surface may differ from one another in an important manner for some species. Avoid sugar-rich media such as minimal, complete and PDA for recovering macroconidia for identification, as macroconidia from cultures grown on these media often are highly variable and quite different from those observed on CLA or other nutrient-poor media. Macroconidia from the carnation leaf pieces also may differ significantly from those produced on the agar, and those from sporodochia tend to be more representative and uniform than those produced elsewhere in the culture. The macroconidial characters described below can all be observed from cultures grown on CLA.

The first macroconidial character is size. Look at length in terms of long or short, without regard to width or curvature. Look at the entire length of the spore including the apical and foot cells, and not just the main

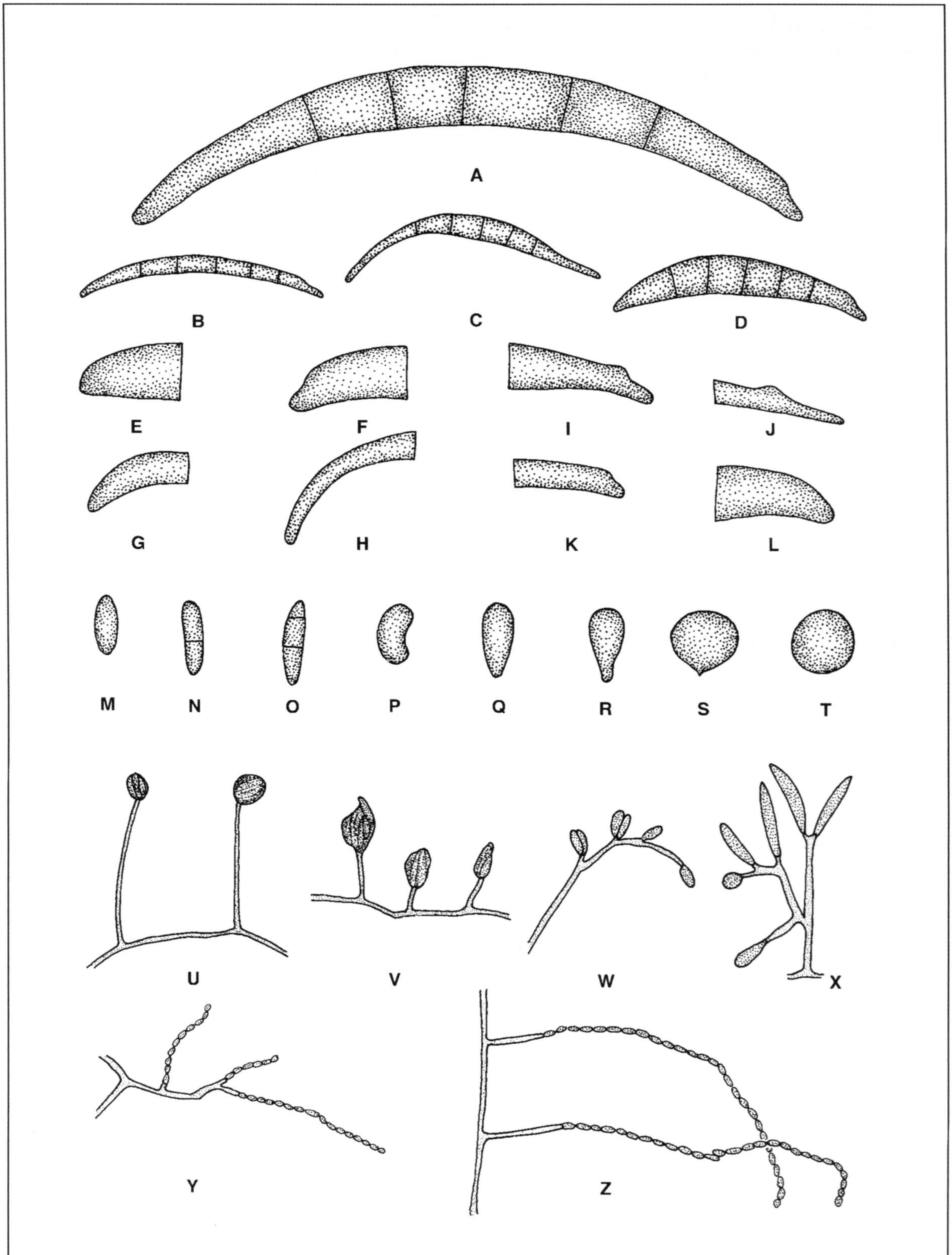
A
B
C
D
E
F
I
J
G
H
K
L
M
N
O
P
Q
R
S
T
U
V
W
X
Y
Z

body of the spore. Look only for gross differences and focus on the general length of the entire group of spores in the slide mount, not on individual spores that may be significantly longer or shorter. In most cases, spore size is a relatively constant character, and large variations may mean that there are problems with the culture conditions used. Count the number of septa present in the macroconidia and determine both the range and the average number of septa per spore. When looking at size, determine whether the spore seems to be elongate or squat. As a comparison look at the macroconidia of *F. armeniacum* (Fig. 13-5) and *F. culmorum* (Fig. 13-19). The macroconidia of *F. armeniacum* are relatively long and elongate, while those of *F. culmorum* are relatively short and squat.

With respect to shape there are three basic shapes (Fig. 12-1B-D). First, there are straight macroconidia (12-1B), which can appear almost needlelike if they are thin. Next are macroconidia with dorsiventral curvature along all or a portion of the spore (Fig. 12-1C). These spores usually, but not always, are approximately the same width along their entire length. Finally, there are macroconidia in which the dorsal (upper) side is distinctly more curved than the ventral (lower) side (Fig. 12-1D). In these spores, the cells in the middle of the spore usually are wider than the cells at the ends.

The ends of the macroconidia often are important in morphological determinations as well. As drawn in Fig. 12-1A, the more curved portion of the cell is to the top, the apical cell is to the left, and the basal cell is to the right. The shape of both the apical and the foot cell can be important species delimiters as well.

There are four general forms of apical cells: blunt (Fig. 12-1E), papillate (Fig. 12-1F), hooked (Fig. 12-1G), and tapering (Fig. 12-1H). The drawings show the general form only and are not necessarily indicative of the length of the cells. Again cell length can vary widely between species, but is usually constant within a species. Cells that are particularly squat or elongate also may have distorted versions of these models for their apical cell. The most important diagnostic characters are the degree of curvature, relative length, and general form.

The basal, or foot, cell also has four general forms: foot-shaped (Fig. 12-1I), elongated foot shape (Fig. 12-1J), distinctly notched (Fig. 12-1K), and barely notched (Fig. 12-1L). Again the drawings show idealized general forms and are not necessarily indicative of cell length or the distortions that squat or elongate cell length may induce. The basal cell usually is the least curved of the cells in the spore, if it is curved at all. The most important characters are the relative length of the cell and its general form.

12.2 Microconidia

Microconidia are not produced by all *Fusarium* species, so their presence alone is an important character. The microconidia themselves, the conidiogenous cell on which they are borne, and the arrangement of the microconidia on and around the conidiogenous cell all are important and potentially diagnostic characters.

The first microconidial character is size. As with the macroconidia, the goal is to determine the general average and not to closely inspect and characterize spores at the edges of the range. Again the number of septations should be counted. The number of septations for the microconidia usually is either 0 or 1, but some species may produce conidia with two septa. Both the average and the range for length and number of septa present should be determined.

Microconidial shape also is an important character. These shapes apply to the entire microconidium and not just single cells or portions of the microconidium. The most common shapes are: oval (Figs. 12-1M-O), reniform, *i.e.*, kidney shaped (Fig. 12-1P), obovoid, *i.e.*, almost oval, with a truncate base (Fig. 12-1Q), pyriform, *i.e.*, pear-shaped (Fig. 12-1R), napiform, *i.e.*, turnip-like (Fig. 12-1S), globose, *i.e.*, spherical (Fig. 12-1T), and fusiform, *i.e.*, spindle-like or cigar-shaped. A single cul-

Figure 12-1. Spore morphology characters used in the identification of *Fusarium* species. Drawings are idealized and not necessarily to the same scale. **A-D: Macroconidial shapes. A.** Typical *Fusarium* macroconidium. Apical cell on left, basal cell on right. **B.** Slender, straight, almost needle-like macroconidium, *e.g., F. avenaceum.* **C.** Macroconidium with dorsoventral curvature, *e.g., F. equiseti.* **D.** Macroconidium with the dorsal side more curved than the ventral, *e.g., F. crookwellense.* **E-H: Macroconidial apical cell shapes. E.** Blunt, *e.g., F. culmorum.* **F.** Papillate, *e.g., F. sambucinum.* **G.** Hooked, *e.g., F. lateritium.* **H.** Tapering, *e.g., F. equiseti.* **I-L: Macroconidial basal cell shapes. I.** Foot-shaped, *e.g., F. crookwellense.* **J.** Elongated foot-shape, *e.g., F. longipes.* **K.** Distinctly notched, *e.g., F. avenaceum.* **L.** Barely notched, *e.g., F. solani.* **M-T: Microconidial spore shapes. M.** Oval. **N.** Two-celled oval. **O.** Three-celled oval. **P.** Reniform. **Q.** Obovoid with a truncate base. **R.** Pyriform. **S.** Napiform. **T.** Globose. **U-X: Phialide morphology. U.** Monophialides, *e.g., F. solani.* **V.** Monophialides, *e.g., F. oxysporum.* **W.** Polyphialides, *e.g., F. polyphialidicum.* **X.** Polyphialdes, *e.g., F. semitectum.* **Y-Z: Microconidial chains. Y.** Short chains, *e.g., F. nygamai.* **Z.** Long chains, *e.g., F. verticillioides.*

ture may produce microconidia with more than one shape. In some cases the relative frequency of spores with the different shapes may vary as well.

The conidiogenous cells are of two basic types: monophialides (Figs. 12-1U-V) and polyphialides (Figs. 12-1W-X). The difference is due to the number of openings per cell through which conidia are produced. Monophialides have only a single opening per cell, while polyphialides have more than one. Many of the species that produce polyphialides, *e.g.*, *F. proliferatum*, may produce more monophialides than they do polyphialides. Thus it may take some time to determine conclusively that no polyphialides are present. The length of the conidiogenous cells also is important, with "long" (Fig. 12-1U) and "short" (Figs. 12-1V-W) the two relevant categories.

Microconidia may be arranged on the phialides singly (Fig. 12-1X), in chains (Figs. 12-1Y-Z), or in false heads (Figs. 12-1U-V). The length of the chains observed may vary from just a few spores to very long "roller coaster"-like patterns. Chain formation may depend on the medium used. For example, some species will form chains on KCl agar, but not on CLA. Chain length can be an important species character and the general length of the chains in the aerial mycelia should be noted. *Fusarium* species do not produce true heads of conidia such as those found in some other fungi, *e.g.*, *Aspergillus*, but can produce clumps of spores at the end of the phialides that superficially resemble spore heads. The false heads could result from the collapse of spore chains back onto the phialides. The false heads may be the only form in which the microconidia are present, or the microconidia may be present in both chains and false heads.

12.3 Chlamydospores

Chlamydospores (Fig. 12-2) are an important character in many *Fusarium* species descriptions. They are not well-conserved evolutionarily, however, and species that produce chlamydospores may be very closely related to those that do not. Chlamydospores may be formed singly (Figs. 12-2A, B, G), doubly (Fig. 12-2F, H), in clumps (Fig. 12-2C, D, I) and in chains (Fig. 12-2E, J-L). They often take a long time (6+ weeks) to produce, and may not be produced in large numbers. Chlamydospores often form more readily on SNA than on CLA, but in some cases an even more specialized medium (soil agar) may be required for their observation. Chlamydospores may be found in the aerial mycelia or embedded in the agar, and the location often is important in species identification.

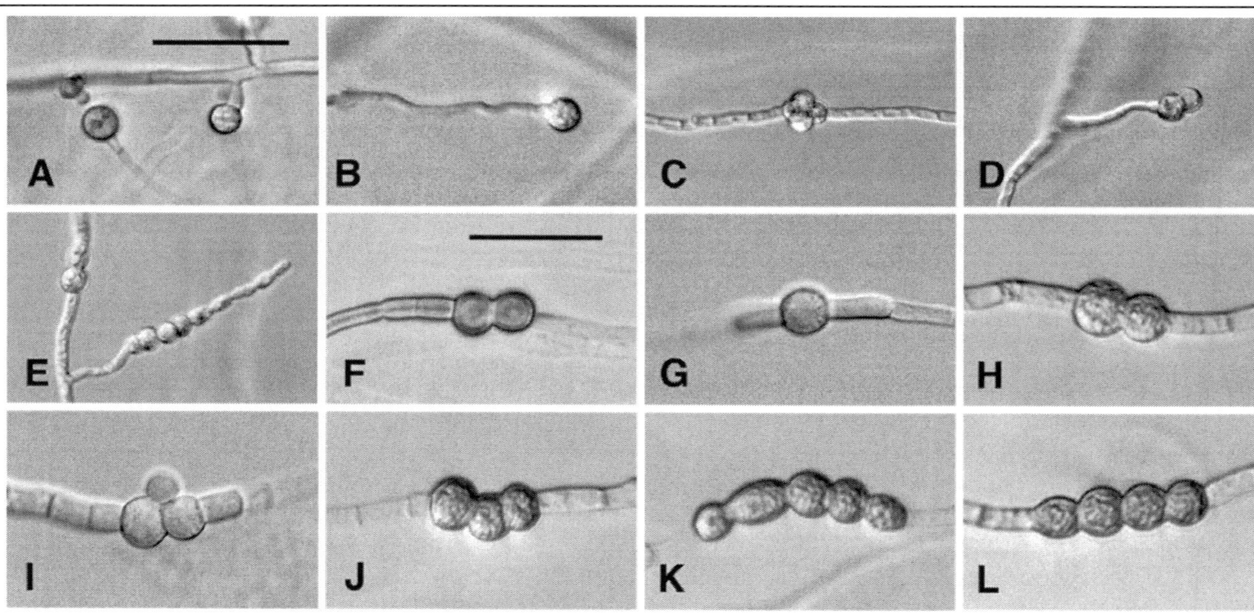

Figure 12-2. Chlamydospores of *Fusarium* species. **A-B:** Single, verrucose chlamydospores of *F. solani.* **C-D:** Clustered chlamydospores of *F. compactum.* **E:** Chain of verrucose chlamydospores of *F. compactum.* **F:** Paired, smooth-walled chlamydospores of *F. solani.* **G:** Single, verrucose chlamydospore of *F. scirpi.* **H:** Paired, verrucose chlamydospores of *F. compactum.* **I:** Clustered, smooth-walled chlamydospores of *F. scirpi.* **J** and **L:** Chains of verrucose chlamydospores of *F. compactum.* **K:** Chain of verrucose chlamydospores of *F. scirpi.* A-E: Scale bar = 50 um; F-L: Scale bar 25 um.

There are a number of structures that are related to, but distinct from, chlamydospores such as pseudochlamydospores and swellings in the hyphae. These structures may be the primary distinguishing feature for some species, *e.g.*, *F. andiyazi*, and usually are more common in the carnation leaf pieces than on the agar surface. These structures may take as long to form as do chlamydospores. In general only cultures from CLA have been evaluated for these structures, and it is not known if using other media would speed, retard, or otherwise alter their formation. As these structures have been used only in the description of a relatively few species, their general utility as species delimiters remains to be critically evaluated.

12.4 Other Characters

In addition to macroconidia, microconidia, and chlamydospores, and related structures, a few other characters also have been used for species delimitation. Unless working with one of the species for which these traits are diagnostic characters, however, most researchers will not observe these characters. Circinate (coiled) hyphae are important for some species, particularly *F. circinatum*, within the *G. fujikuroi* species complex. Some species also produce crystals in the media, but while this character often is indicative it usually is not diagnostic. Similarly, the ability to produce sclerotial-like structures is considered to be an important diagnostic character by some researchers. Our experience with sclerotial-like structures is that they are of primary importance as an indicator of which strains are likely to be relatively highly female-fertile strains, and that the production of such structures can vary widely within a species. We do not think this character is proven to be a good one for use in species delimitations.

Mesoconidia is a term proposed by Pascoe (1653) to describe the fusoid conidia that are typically produced by some species of *Fusarium* from polyphialides in the aerial mycelium on CLA or SNA (*e.g.*, Figs. 13-13E-F, 13-56D-E, and 13-59C-F). These spores are not produced in sporodochia. This term was used because these conidia were considered to be neither macro- nor microconidia. They are significantly larger than microconidia and normally have a fusoid shape with up to 3-4 septa. They are, however, not normal macroconidia, as they often are somewhat smaller than the macroconidia typical for their species of origin and the mesoconidia often lack a foot-shaped or notched basal cell. Mesoconidia generally stick up in the aerial mycelium and often have a "rabbit ears" appearance when viewed through the microscope. Species that produce mesoconidia include *F. camptoceras*, *F. chlamydosporum*, *F. semitectum*, *F. sporotrichioides*, *F. polyphialidicum* and some isolates of *F. avenaceum* and *F. subglutinans*. The term sporodochial conidia has been proposed for macroconidia, and the term aerial conidia for meso- and microconidia, but neither of these terms has gained wide acceptance or is in general use.

Teleomorphs (sexual stages – Chapter 10) can be a definitive species character. Normally these sexual stages are observed under controlled laboratory conditions rather than in the field. In such cases the joint production of the sexual stage by a known laboratory tester strain and the field strain crossed with it is interpreted to mean that the field strain belongs to the same biological species as the laboratory tester strain. Teleomorph morphology usually is not sufficiently different to be useful as a morphological marker for identification purposes, but the formation of the teleomorph itself can be very important.

12.5 Secondary characters

A number of secondary characters also are used in the identification of *Fusarium* species. The most prominent of these secondary characters is pigmentation. Culture conditions and media are critical for the production of comparable pigments, some of which are light sensitive and others of which are sensitive to the pH of the agar medium into which they have been excreted. PDA usually is the medium of choice, and a 12:12 hr light:dark cycle is preferred, unless otherwise specified in the species description. To make comparable comparisons of colors, a standard color guide such as that of Kornerup & Wanscher (1125) usually is required to differentiate shades of particular colors. In the present manual we have not provided information of this precision, and have used only more general terms such as "red", "brown", "violet", "orange", "yellow", "gray" and "white" as color indicators. Colors can be observed either on plates or slants [see Nelson *et al.* (1551) and Summerell *et al.* (2087) for pictures of representative cultures], and usually can be reliably evaluated after approximately a week of incubation. Pigments usually are not diagnostic for a species, with the yellow pigment produced by cultures of *F. thapsinum* a prominent exception.

Growth rate is another commonly used secondary character. There can be some variation in this trait (see Table 4-1). Traditionally measurements are made on PDA plates inoculated with single spores and grown for three days at either 25° or 30°C. The colony diameter is measured, and it is important to make several replicate plates to insure an accurate measurement has been made. In this manual, we have not placed great reliance on growth rates. In most cases the differences are subtle and the differences generated are not always clear cut. There are some species that grow considerably slower (or faster) than others, however, and such slow- (fast-) growing species and the conditions resulting in their slow (fast) growth are noted in the individual species descriptions.

Secondary metabolites and mycotoxins also may be used as secondary identification characters, even though a strain's potential ability to form a mycotoxin often is a reason for seeking to identify it. Some strains produce volatile compounds that result in an odor that is diagnostic if present. In other cases complex chemical profiles of all of the secondary metabolites present can be used to group strains and then these groups associated with particular species (2142). In general, we have presented only limited information on the mycotoxins and other secondary metabolites due to their chemical diversity, their generally complex biosynthetic pathways, and the plethora of analytical techniques used to characterize, identify and quantify them.

Table 12-1. *Fusarium* culture identification checklist.

ID	Macroconidia				Microconidia			Chlamydo-spores	Colony Color	Comments	Species
no.	Size	Shape	Apical	Foot	Size	Shape	Phialide				

13

Species Descriptions

Fusarium acuminatum Ellis & Everhart

Sexual Stage. *Gibberella acuminata* Wollenweber.

Common Synonyms. *Fusarium acuminatum* ssp. *acuminatum*, *Fusarium scirpi* var. *acuminatum*, *Fusarium scirpi* ssp. *acuminatum*, *Fusarium gibbosum* var. *acuminatum*.

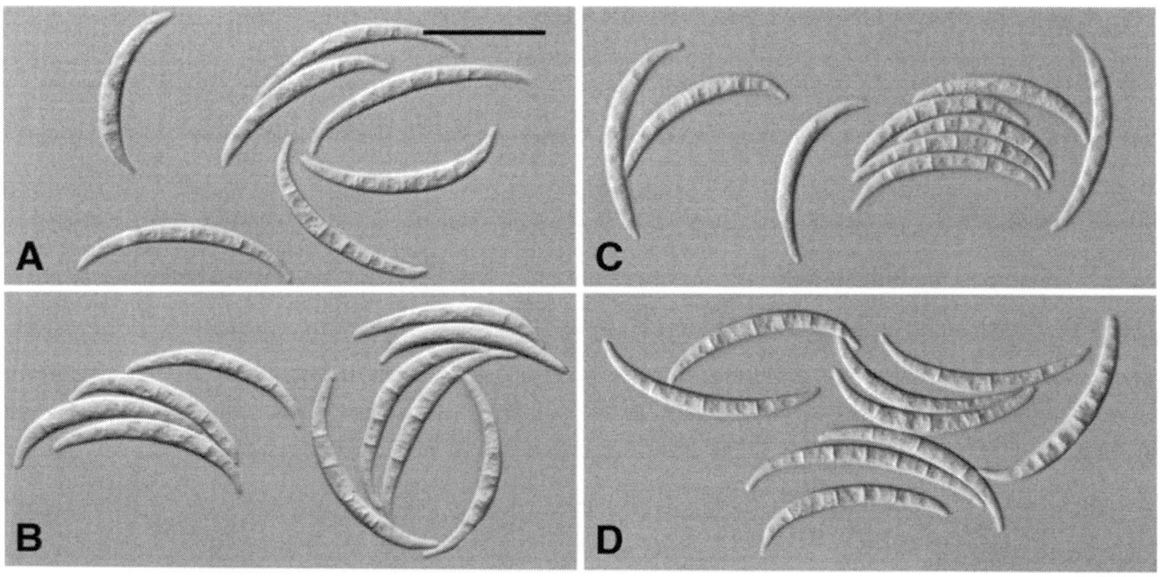

Figure 13-1. *Fusarium acuminatum.*
A – D: Macroconidia; scale bar 25 μm.

Reported Geographic and Host Distribution

Found in temperate regions usually as a soil saprophyte or associated with roots and crowns of plants. It can occasionally be associated with root and crown diseases of a variety of hosts, especially legumes, and has been recovered from cereal grains in Canada, Europe and the former Soviet Union.

Media to Examine and Key Characters

Characters on CLA. Macroconidia are formed in pale orange sporodochia. Microconidia usually are absent, but may be formed sparsely by some isolates. When formed, the microconidia may be 0- or 1-septate.

Chlamydospore formation is very slow in most isolates. They may be formed in chains or in clusters. Their presence is not a reliable taxonomic criterion.

Characters on PDA. *Fusarium acuminatum* is a relatively slow-growing species that produces white mycelium which is abundant in some isolates. The mycelium is floccose with rose to burgundy pigmentation that can be grayish-rose at the periphery. Sporodochia form in the center of the colony in a small central spore mass and are pale orange to dark brown. Red pigments (sometimes brown) are formed in the agar.

Macroconidia

- Sporodochia: Pale orange color. Found on carnation leaf pieces in CLA.
- General morphology: Moderate equal curvature. Thick walls.
- Basal cell morphology: Distinct foot shape to the cell, but not as prominent as in species such as *F. equiseti* or *F. armeniacum.*

- Apical cell morphology: The apical cell tapers and can be relatively elongate. There often is a distinct curve to the latter half of the apical cell.
- Number of septa: Typically 5-septate, but 3- and 4-septate macroconidia are not uncommon.
- Abundance: Very abundant in sporodochia.

Microconidia. Microconidia are not a reliable taxonomic indicator for this species. The isolates that produce microconidia have not been compared molecularly with those that do not to confirm taxonomic identity or relatedness.

- Shape/septation: Fusiform or reniform. Usually 1-septate, but occasionally 0-septate.
- Aerial mycelium presentation: Singly or in pairs.
- Conidiogenous cells: Monophialides.
- Abundance: Sparse.

Chlamydospores

- Abundance/Speed of formation: Varies by isolate, but usually sparse. Chlamydospores usually form slowly on CLA and may require more than six weeks.
- Location: Formed in hyphae on and in the agar.
- Appearance: In chains and in clusters.

Taxonomy, Pathology and Ecology

Fusarium acuminatum was first described by Ellis & Everhart (564) from potato stems in Geneva, New York. Wollenweber & Reinking (2359) considered *F. acuminatum* to be a variety of *F. scirpi*. Gordon (746) made *F. scirpi* a synonym of *F. equiseti* and restored *F. acuminatum* to species rank. In 1993, Burgess *et al.* (270), split *F. acuminatum* into two subspecies, *F. acuminatum* subsp. *armeniacum* and *F. acuminatum* subsp. *acuminatum* because the shape of the macroconidium did not differ significantly between the two taxa and the two subspecies "intergraded with respect to morphological and ecological criteria and some physiological characteristics". Nagy & Hornok (1519), used pulse-field electrophoresis to show that *F. armeniacum* and *F. acuminatum* differ in chromosome size and number.

Considerable differences exist in the toxin producing ability of cultures of the two species (1279) as well. Rabie *et al.* (1759) had also noted that some atypical isolates (= *F. armeniacum*) of *F. acuminatum* produced high levels of T-2 toxin. *F. armeniacum* produces A-type trichothecenes, but these compounds are not produced by *F. acuminatum* (2348, 2350). These data collectively resulted in the decision to raise both taxa back to species rank (278). *F. acuminatum* may contain more than one phylogenetic lineage (2389).

Fusarium acuminatum is most often confused with isolates of *F. avenaceum* or *F. armeniacum*. *Fusarium acuminatum* can be differentiated from *F. avenaceum* by the shape of the macroconidia which in *F. acuminatum* has some dorsiventral curvature. *F. acuminatum* also has thick walls and a distinct foot-shape to the base of the basal cell. *F. avenaceum* macroconidia have parallel sides, thin walls and a notched base to the basal cell. Unfortunately there are some isolates that are very difficult to differentiate because the macroconidia are intermediate in shape and are not typical of either species. This problem can be minimized by examining only macroconidia from sporodochia on CLA. The growth rates of the two species on PDA can be similar. *F. avenaceum* does not form chlamydospores, but because chlamydospores form slowly in cultures of *F. acuminatum* this difference is not always useful and should not be relied on. PCR primers that are reported to be species specific for *F. acuminatum* have been developed, but have not been widely tested (2334).

Fusarium acuminatum is more common in temperate areas, often in grassland and cultivated soils; it is less common in tropical areas (112, 1181, 1622). Sangalang *et al.* (1888, 1889) have shown that *F. acuminatum* survives poorly at higher temperatures and presumably this is one reason why the fungus is not found in regions where temperatures are generally higher. Note that *F. acuminatum* also was a poor competitor with representative strains of *F. compactum*, *F. equiseti*, *F. sambucinum*, and *F. solani* at temperatures ranging from 18-30°C (1893). Some isolates have ice-nucleation activity (1727) and others have been examined for antioxidant enzyme activity (1028).

Although *F. acuminatum* is generally regarded as a saprophyte and a secondary colonist of necrotic or senescent tissues, some isolates can cause severe root rot in some legume species (810, 1179, 1181, 1413, 1564). This species also been reported to be involved in diseases of pumpkin (567) and wheat (615, 1424).

Fusarium acuminatum may produce trace levels of trichothecenes (2348, 2350), but earlier reports of high level toxin production by *F. acuminatum* (3, 1759) probably are best attributed to *F. armeniacum*, from which *F. acuminatum* was formally separated in 2000 (278). The many toxic strains characterized by Marasas *et al.* (1341) as *F. acuminatum* also need to be re-evaluated and their species identity confirmed. *F. acuminatum* can produce enniatin B (1286) and steroids (284). Strains producing moniliformin that are toxigenic to sheep (3, 347, 1181) also have been identified.

Current Species Description: Booth (222).

Fusarium acutatum Nirenberg & O'Donnell

Sexual Stage. None known.

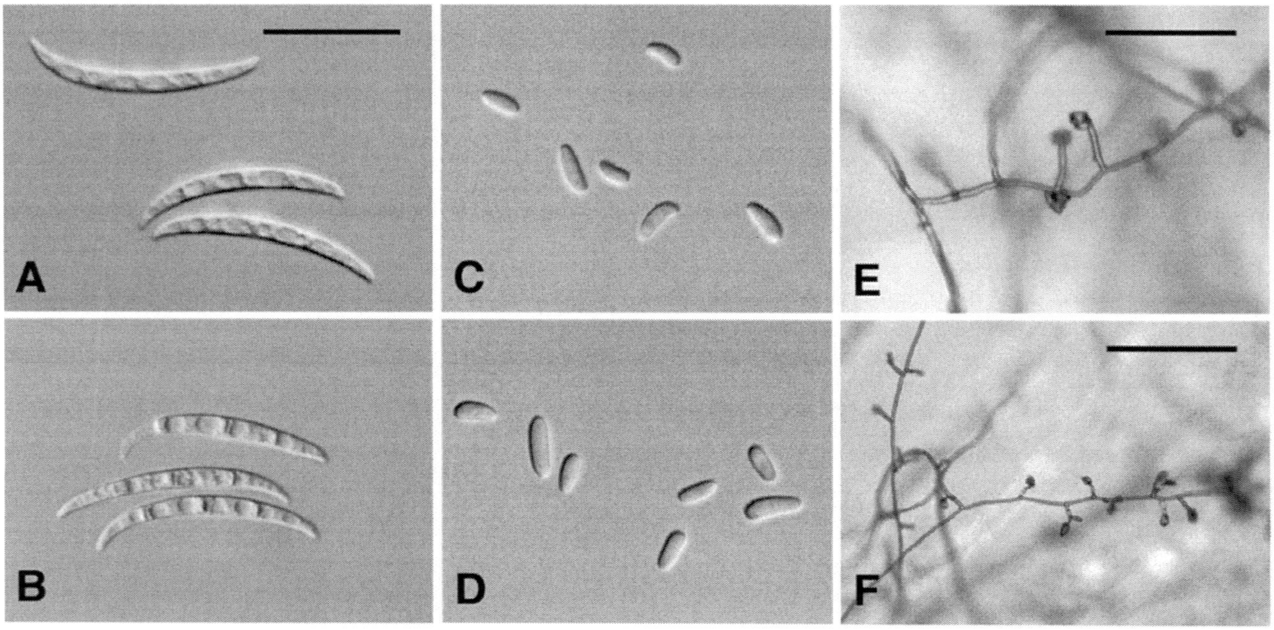

Figure 13-2. *Fusarium acutatum.*
A – B: Macroconidia; **C – D:** Microconidia; **E – F:** Microconidia *in situ* on CLA. A – D, scale bar = 25 μm; E – F, scale bar = 50 μm.

Reported Geographic and Host Distribution

Recovered from *Cajanus* sp. and from aphids from wheat in India and Pakistan.

Media to Examine and Key Characters

Characters on CLA. Macroconidia are falcate and thin walled, but produced only sparsely. Microconidia are abundant in false heads primarily on monophialides. Microconidia may be more distinctive on SNA than on CLA.

Characters on PDA. Mycelium is white to pinkish-white with light orange pigments in the agar.

Macroconidia

- Sporodochia: Uncommon and slow to form, usually requiring 10-21 days on CLA.
- General morphology: Typical of species in the *Gibberella fujikuroi* species complex. Falcate and thin-walled.
- Apical Cell morphology: Bent.
- Basal Cell morphology: Foot-shaped.
- Number of septa: Usually 3-septate.
- Abundance: Sparse, due to the relatively limited number of sporodochia.

Microconidia

- Shape/septation: Oval, allantoid and fusoid. Usually 0-septate.
- Aerial mycelium presentation: Microconidia are found only in false heads.
- Conidiogenous cells: Primarily monophialides, but occasionally from polyphialides with two openings.

- Abundance: Abundant in the aerial mycelium.

Chlamydospores

- Abundance/Speed of formation: Present, but generally formed slowly, *i.e.*, in 4-6 weeks on CLA.
- Location: In hyphae in the aerial mycelia and on the agar surface.
- Appearance: In chains and in clusters.

Taxonomy, Pathology and Ecology

The description by Nirenberg & O'Donnell (1570) is based on only five cultures and the current species description is considered to be invalid (*Index of Fungi* **6:** 435, 1999). The species needs to be confirmed and validated with a larger number of cultures from diverse sources. Reportedly it is easy to confuse *F. acutatum* with *F. udum* (1570), but the apical cell of a *F. acutatum* macroconidium is acute, *i.e.*, bent (Fig. 13-2B), while the apical cell of a *F. udum* macroconidium is uncinate, *i.e.*, distinctly hooked (Fig. 13-68A).

Isolates of *F. acutatum* produce trace levels of beauvericin and fumonisins and no detectable fusaproliferin or moniliformin (638).

Current Species Description: Nirenberg & O'Donnell (1570).

Fusarium andiyazi Marasas, Rheeder, Lamprecht, Zeller & Leslie

Sexual stage. None known.
Common Synonyms. *Fusarium moniliforme*, *Fusarium verticillioides*.

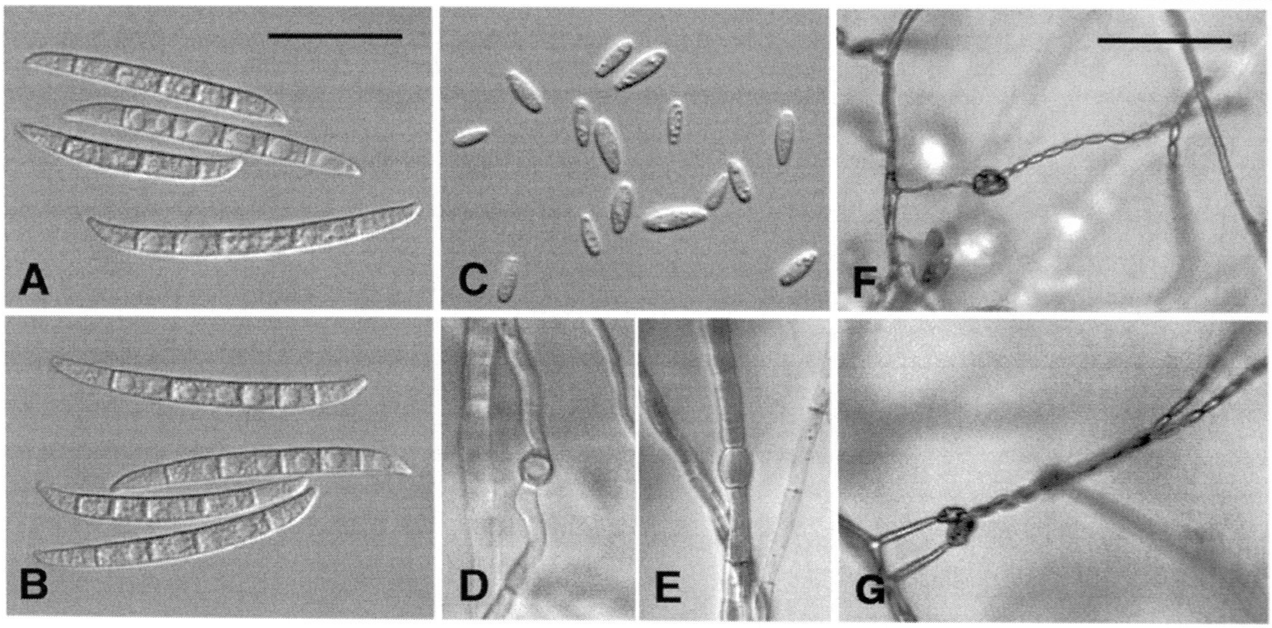

Figure 13-3. *Fusarium andiyazi.*
A – B: Macroconidia; **C:** Microconidia; **D – E:** Pseudochlamydospores; **F – G:** Microconidia *in situ* on CLA.
A – E, scale bar = 25 µm; F – G, scale bar = 50 µm.

Reported Geographic and Host Distribution

From sorghum in Australia, Ethiopia, Nigeria, South Africa, and the United States.

Media to Examine and Key Characters

Characters on CLA. Macroconidia are produced in orange sporodochia and are straight to slightly curved. Microconidia are produced in abundance in chains from monophialides. Pseudochlamydospores are formed on this medium, mostly near, on and in the carnation leaf pieces.

Characters on PDA. Floccose to powdery mycelium that initially is white but may become violet. Violet pigmentation in the agar may vary from pale to dark purple.

Macroconidia

- Sporodochia: Orange sporodochia on CLA. Sporodochia usually begin to form when microconidial chains begin to collapse.
- General morphology: Hyaline, thin walled, straight to slightly curved.
- Apical Cell morphology: Slightly curved.
- Basal Cell morphology: Pedicillate.
- Number of septa: 3- to 6-septate, mostly 3-septate.
- Abundance: Found in sporodochia on monophialides or branched conidiophores.

Microconidia

- Shape/septation: Clavate to ovoid with a flattened base. Usually 0-septate.

- Aerial mycelium presentation: False heads, or more commonly long chains that often collapse to short chains as the colony ages, particularly near the center of the colony.
- Conidiogenous cells: Long monophialides.
- Abundance: Common in the aerial mycelia.

Chlamydospores. Absent.

Pseudochlamydospores. Strains of *F. andiyazi* form pseudochlamydospores which, as the name suggests, appear similar to chlamydospores but have a smooth, thinner wall. Pseudochlamydospores are found in hyphae in carnation pieces on CLA or in nearby aerial mycelial, but never in submerged mycelia. They may be at the end of or intercalary in the hyphae. Pseudochlamydospores usually are solitary, but may occasionally be found in short chains.

Taxonomy, Pathology and Ecology

This species was described by Marasas *et al.* (1346) using morphological and molecular characteristics. It is likely that this species was included within *F. moniliforme* or *F. verticillioides* in the past as the morphology is very similar.

Fusarium andiyazi is pathogenic to sorghum in *in vitro* seedling pathogenicity tests, but does not produce either the fumonisin or the moniliformin mycotoxins (1233). When multiple strains were tested for growth at both 25 and 30°C, growth was consistently, but not statistically better at 30°C than it was at 25°C.

At some locations *F. andiyazi* may be the dominant *Fusarium* species on sorghum. It has not been searched for extensively, but probably has a much broader distribution with sorghum than has been reported thus far.

Current Species Description: Marasas *et al.* (1346).

Fusarium anthophilum (A. Braun) Wollenweber

Sexual stage. None known.
Common Synonyms. *Fusarium moniliforme* var. *anthophilum*.

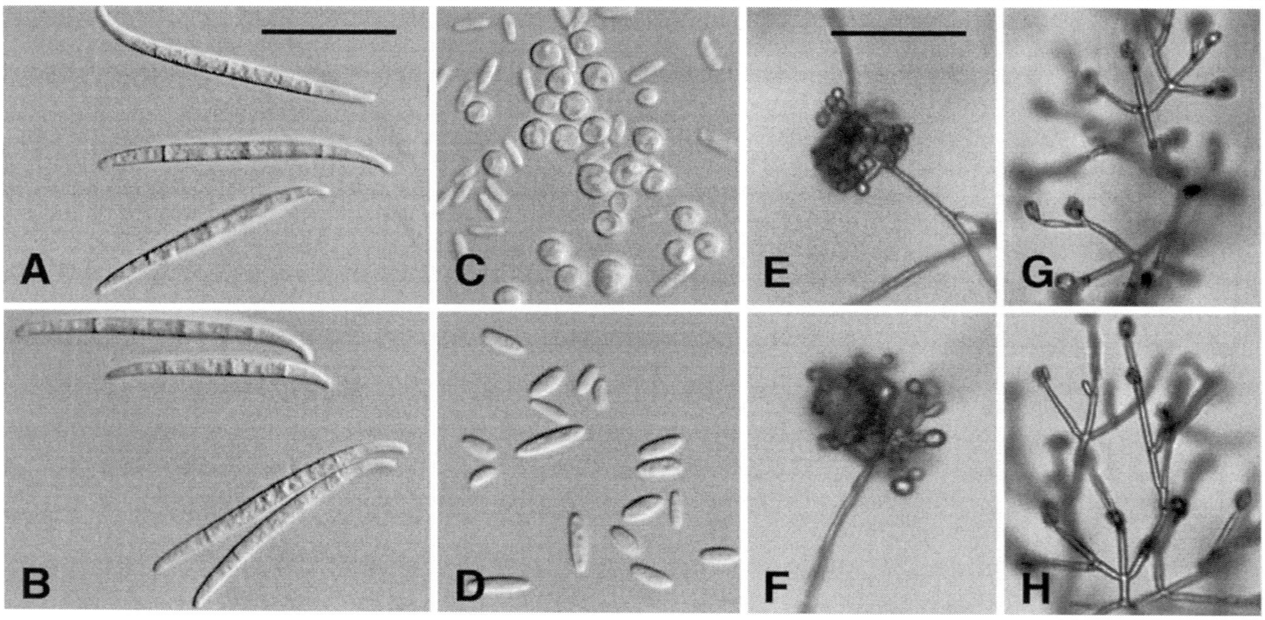

Figure 13-4. *Fusarium anthophilum.*
A – B: Macroconidia; **C:** Globose microconidia; **D:** Ovoid microconidia; **E – F:** Globose microconidia *in situ* on CLA; **G – H:** Ovoid microconidia *in situ* on CLA. A – D, scale bar = 25 μm; E – H, scale bar = 50 μm.

Reported Geographic and Host Distribution

Cosmopolitan. Found on various plant species in temperate regions.

Media to Examine and Key Characters

Characters on CLA. The microconidia are the most important character, as distinctive globose or pyriform microconidia are formed on mono- and polyphialides.

Characters on PDA. Aerial mycelia are white and grow rapidly. Sporodochia, if present, are tan to orange in color and may be discrete or confluent. The agar may be colorless or contain violet pigments.

Macroconidia

- Sporodochia: Orange to tan color, when present.
- General morphology: Typical of species in the *Gibberella fujikuroi* species complex. Relatively slender with no significant curvature. Thin walled.
- Apical Cell morphology: Curved.
- Basal Cell morphology: Usually relatively poorly developed, but foot shaped when clearly developed.
- Number of septa: Usually 3-, but occasionally 4-septate.
- Abundance: Varies by strain, but often difficult to find.

Microconidia

- Shape/septation: Pyriform, globose and ovoid. Usually 0-septate, but occasionally 1-septate.
- Aerial mycelium presentation: False heads. No chains.

- Conidiogenous cells: Monophialides and branched and unbranched polyphialides.
- Abundance: Common in the aerial mycelia.

Chlamydospores. Absent.

Taxonomy, Pathology and Ecology

This species was recognized by Wollenweber & Reinking (2359), Gerlach & Nirenberg (697), and Nelson *et al.* (1551). It can be easily confused with *F. subglutinans* and related species if the microconidia are not carefully examined on the CLA or SNA plates. Production of the different microconidial types is likely to be enhanced on SNA. This species is very difficult to differentiate from *F. konzum* and it may be necessary to use mating tests or molecular analysis to distinguish these species (2412). O'Donnell *et al.* (1596) found that *F. anthophilum* was 10/1067 steps from *F. succisae*. This distance is less than the 15/1067 steps that separate *F. proliferatum* and *F. fujikuroi*, two species between which fertile interspecific hybrids are known to occur under both laboratory and field conditions (1236). Maximum linear growth is reported to occur at 25°C and an osmotic potential of -1.0 MPa (1541).

Fusarium anthophilum has been reported to be recovered from a number of countries and host substrates, *e.g.*, Engels & Kramer (580), Nirenberg (1566), Nyvall *et al.* (1588) and Sangalang *et al.* (1889), although it is not thought to be pathogenic to plants. This fungus is not associated with any human or animal diseases, although extracts from two cultures have been reported to be mildly dermotoxic to rabbits (1341). Some strains of *F. anthophilum* have been reported to moniliformin (347, 1349). Some strains also have been reported to produce fumonisins and to carry the necessary biosynthetic genes to encode fumonisin biosynthesis (1548, 1740).

Current Species Description: Wollenweber & Reinking (2359).

Fusarium armeniacum (Forbes, Windels & Burgess) Burgess & Summerell

Sexual Stage. None known.
Common Synonyms. *Fusarium acuminatum* ssp. *armeniacum*.

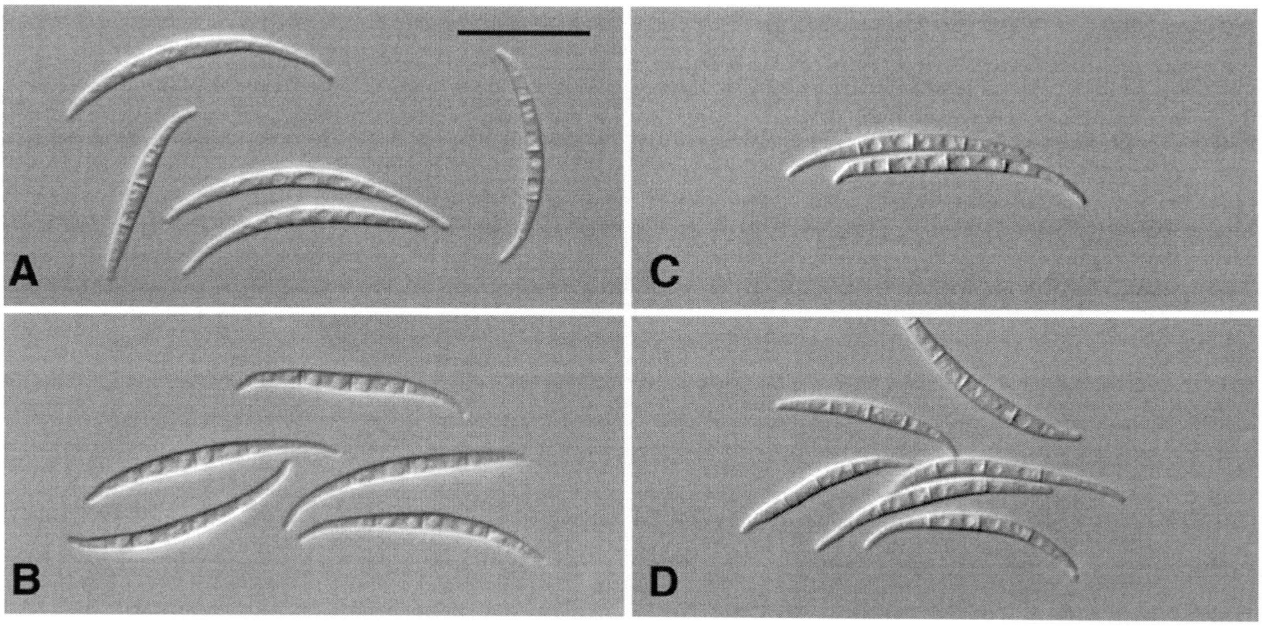

Figure 13-5. *Fusarium armeniacum.*
A – D: Macroconidia; scale bar = 25 µm.

Reported Geographic and Host Distribution

Subtropical and tropical regions as a soil saprophyte.

Media to Examine and Key Characters

Characters on CLA. Macroconidia are formed in bright orange sporodochia on carnation leaves. They are distinctly curved and have a prominent foot-shaped basal cell. Always use macroconidia from sporodochia on carnation leaves when identifying this species. Microconidia are not produced. Chlamydospore formation is rapid in most isolates and they are found in chains and in clusters.

Characters on PDA. This species is relatively fast-growing and produces abundant white mycelium. Apricot to reddish-orange sporodochia develop in the center of the culture in a conspicuous central spore mass. These sporodochia may become brown with age. *F. armeniacum* produces various pigments ranging from bright orange to red and reddish-brown in color in the agar. This species mutates readily to the pionnotal form or a slow-growing mycelial form on this medium.

Macroconidia

- Sporodochia: Common on the surface of carnation leaf pieces and the agar surface.
- General morphology: Prominently curved.
- Apical Cell morphology: Long and tapering.
- Basal Cell morphology: Distinct foot shape.
- Number of septa: Usually 5-septate, but 3- and 4-septate macroconidia also are produced.
- Abundance: Abundant in sporodochia on carnation leaf pieces.

- Unusual feature: A germ tube may be produced by one of the cells of a macroconidium that never appears to develop past a single cell stage.

Microconidia. Absent.

Chlamydospores

- Abundance/Speed of formation: Formed abundantly and rapidly (within 2 weeks) on CLA.
- Location: With hyphae both on and in the agar.
- Appearance: In chains and in clusters.

Taxonomy, Pathology and Ecology

This species was initially referred to as *F. acuminatum* and then described as a subspecies of *F. acuminatum* (270) due to similarities in macroconidial morphology. However, subsequent studies, *e.g.*, Altomare *et al.* (49), Benyon & Burgess (168) and Benyon *et al.* (169), showed that this taxon was a distinct species and it was raised to species rank by Burgess & Summerell (278). Based on molecular chromosomal karyotypes, *F. armeniacum* is more closely related to *F. sporotrichioides* than it is to *F. acuminatum* (1519). *F. acuminatum* and *F. armeniacum* can be differentiated on the basis of pigmentation, speed of production of chlamydospores and growth rates.

Fusarium armeniacum has been found in Minnesota (1118) and was later recovered in Australia (270, 275, 636) and South Africa (1336, 1759, 2223). It appears to be more abundant in high rainfall areas or in clay soils on flood-plain country in drier areas. *F. armeniacum* appears to be a saprophyte.

Strains of *F. armeniacum* degenerate relatively easily in culture, even if single germinated macroconidia are used for subculturing. Toxin, pigment, and chlamydospore production, however, may be unchanged in the degenerate strains (2349).

This species is highly toxigenic in a chick assay and can produce trichothecenes (2348, 2350). Strains of *F. armeniacum* contain the *tox5* gene that is the first unique step in trichothecene biosynthesis, but *F. acuminatum* strains do not (1519). Thus, earlier reports of the production of high levels of trichothecenes (specifically T-2 toxin, HT-2 toxin and neosolaniol) by *F. acuminatum*, *e.g.*, Rabie *et al.* (1759) and Logrieco *et al.* (1279), probably are best attributed to *F. armeniacum*.

Current Species Description: Burgess *et al.* (270).

Fusarium avenaceum (Fries) Saccardo

Sexual Stage. *Gibberella avenacea* Cook.
Common Synonyms. *Fusarium avenaceum* ssp. *avenaceum*.

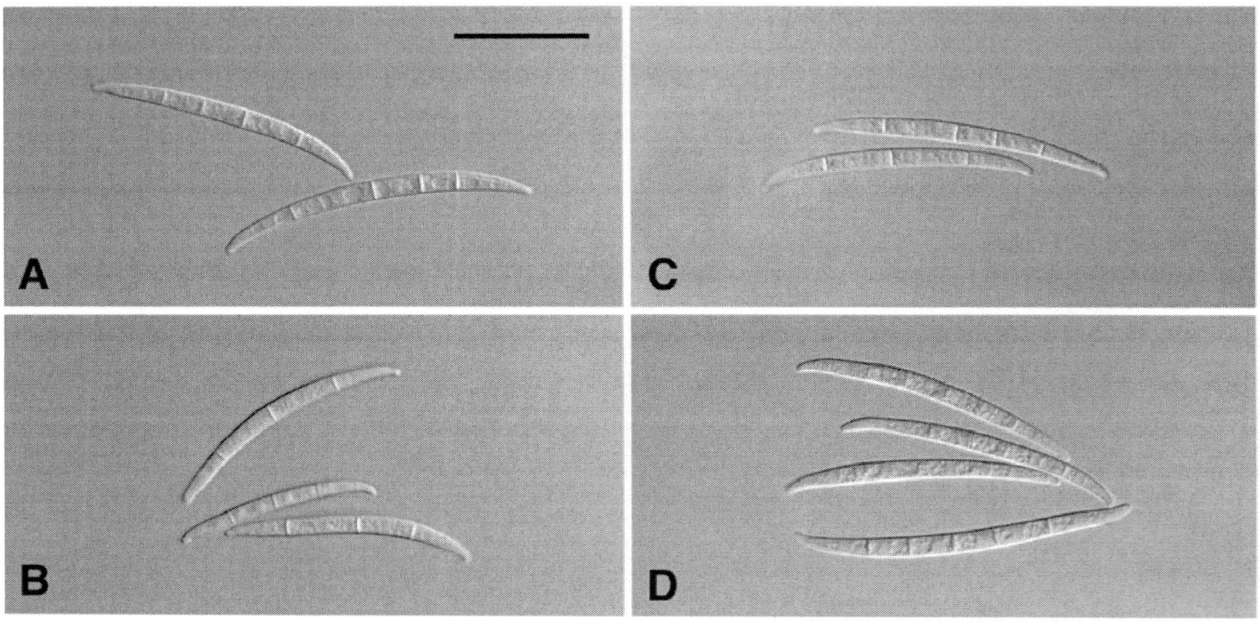

Figure 13-6. *Fusarium avenaceum.*
A – D: Macroconidia; scale bar 25 = µm.

Reported Geographic and Host Distribution

Found in temperate regions as a soil saprophyte and as a pathogen of legumes, carnations, and various perennial plant species. May be common on some cereal grains, *e.g.*, wheat and barley.

Media to Examine and Key Characters

Characters on CLA. Straight to slightly curved macroconidia are formed in pale orange sporodochia on carnation leaf pieces and the agar surface. Always use macroconidia from sporodochia on carnation leaves when identifying this species. Microconidia are produced sparsely by some isolates, but are variable in shape and septation. No chlamydospores are produced.

Characters on PDA. Growth rate varies from slow to relatively rapid. *Fusarium avenaceum* forms abundant mycelium which varies from white to light yellow to grayish rose. Abundant pale-orange to brown sporodochia form in a central spore mass. The pigment formed in the agar is grayish rose to burgundy but may appear brownish because of the light reflected from the central spore mass. Colony morphology is highly variable. Cultures on PDA often mutate to a pionnotal form and less often to a white mycelial form.

Macroconidia

- Sporodochia: Pale orange color. Found on carnation leaf pieces and directly on the agar surface of CLA.
- General morphology: Long and slender. Thin walled, and straight to slightly curved.
- Apical Cell morphology: Long and tapering to a point, may be bent.
- Basal Cell morphology: Usually notched, although some isolates may have foot-shaped basal cells.

- Number of septa: Usually 5-septate, but 3- and 4-septate macroconidia may be observed.
- Abundance: Moderately abundant in sporodochia.

Microconidia/Mesoconidia

- Shape/septation: Fusoid. 1- to 2-septate. May vary in size.
- Aerial mycelium presentation: Usually singly.
- Conidiogenous cells: Monophialides and polyphialides.
- Abundance: Produced only by some isolates. When produced they usually are rare.

Chlamydospores. Absent.

Taxonomy, Pathology and Ecology

Fusarium avenaceum was split into three subspecies following the discovery of two taxa (now known as *Fusarium aywerte* and *Fusarium nurragi*) in unique environments in Australia (1889, 1890). *F. avenaceum* was differentiated from the other two subspecies by shorter macroconidia, a broader geographic distribution, and molecular markers (169). *F. avenaceum* may contain more than one phylogenetic lineage, and shares alleles at some loci used widely for phylogenetic purposes with strains identified as *F. arthrosporioides* (2389). Primers and protocols for detecting *F. avenaceum* through real-time PCR assays are available (2275). Although both the *MAT-1* and *MAT-2* mating types occur (1036) and are transcribed in *F. avenaceum* (1037), no sexual stage for this species has been identified.

Fusarium avenaceum may be confused with *F. acuminatum*. *F. acuminatum* produces chlamydospores, but their formation often is quite slow and their numbers may be few. Thus the presence of chlamydospores is a reliable character for distinguishing these species, but their absence is not. Both colony morphology and rate of growth can overlap and so are not reliable criteria. Thus, the shape of the macroconidia is the principle character to use for distinguishing these species. Generally this character is consistent and easily applied if the cultures are carefully prepared and macroconidia from sporodochia on carnation leaves on CLA are used for identification. Macroconidia with parallel sides, thin walls, and a notched basal cell are expected in cultures of *F. avenaceum* (Figs. 13-6A-D), while those of *F. acuminatum* should have significant dorsiventral curvature, thick walls, and usually a distinctly foot-shaped base to the basal cell (Figs. 13-1A-D).

Fusarium avenaceum is predominantly soilborne and common in temperate regions throughout the world. It is particularly common in subterranean clover-dominant pasture soils and can cause stem and root rot diseases in various pasture legumes (1034, 1179, 1181, 1403, 1413, 1898), may be a seedborne pathogen of some legume species (1034, 1308), and can cause head blight of wheat (1005).

F. avenaceum has been recovered from soils from temperate forests (2085) and is sensitive to SO_2 pollution (1557). It also has been recovered from desert soil in Israel (984). *F. avenaceum* is frequently found in the grain, stems and stubble of a number of cereals including wheat and barley, *e.g.*, Golinski *et al.* (720) Hall & Sutton (807), Smith & Olien (2007), and Turkington *et al.* (2193), where it probably is splash-dispersed (968). It also can cause an economically important stem and stub rot of carnations (131, 247) and *Eustoma grandiflorum* (813, 1666), and economically important diseases of broccoli (1423), Douglas Fir (1478), lentils (921,1261), linseed (1422), raspberry (815), sour cherries (1621), and peaches and nectarines (821). *F. avenaceum* has ice-nucleating activity (1727).

Fusarium avenaceum is not associated with any human or animal toxicoses, but ground cultures are toxic to chicks and mice and are dermotoxic to rabbits (1341). It is resistant to most clinical antifungals, with amphotericin B reported as the most effective (1748).

Fusarium avenaceum also may produce toxins such as beauvericin (1289), fusarin C (3, 600, 2140), and moniliformin (3, 347, 1339, 1341, 1758) under field conditions (15, 975, 1289). Some isolates were toxic both to ducklings and to sheep (1181). Enniatin cyclic peptides, primarily enniatins A, B and B_1, are synthesized by some strains of *F. avenaceum* under laboratory (1289) and field conditions (975), and may have a role in plant pathogenicity (852). Early reports of the production of butenolide, diacetoxyscirpenol, neosolaniol, T-2, and zearalenone have not been confirmed (1341). *F. avenaceum* is reported not to carry the *tri5* gene, which is essential for the production of trichothecenes (2112).

Current Species Description: Wollenweber & Reinking (2359).

Fusarium aywerte (Sangalang & Burgess) Benyon & Burgess

Sexual Stage. None known.
Synonym. *Fusarium avenaceum* ssp. *aywerte.*

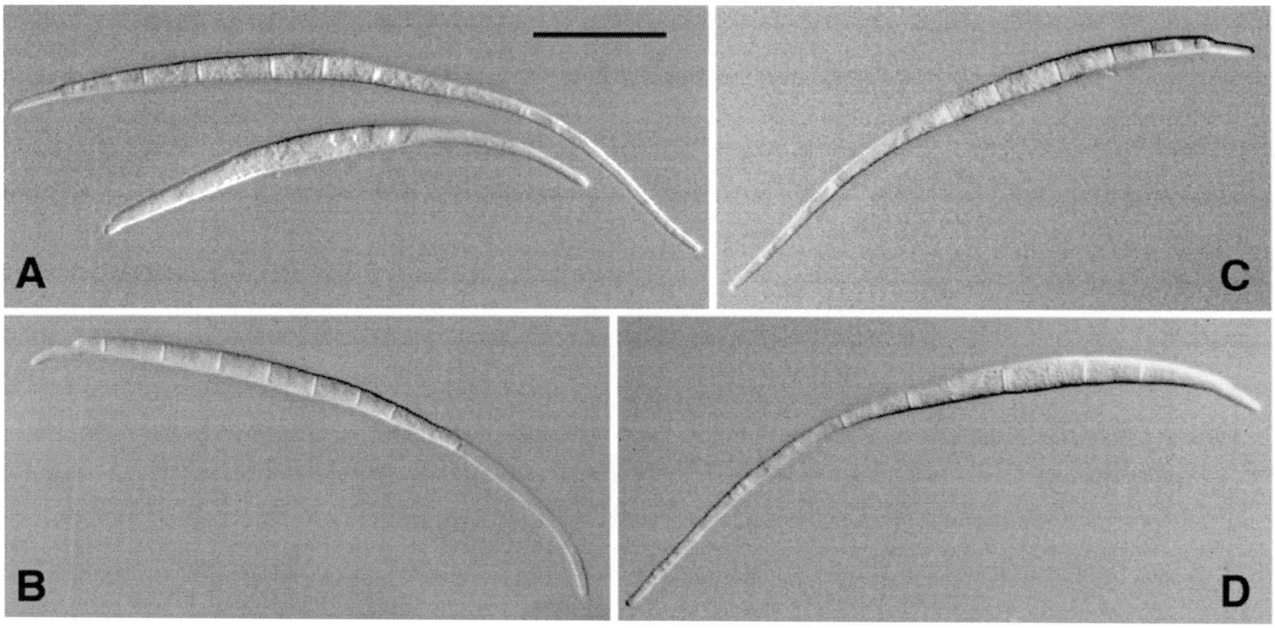

Figure 13-7. *Fusarium aywerte.*
A – D: Macroconidia; scale bar = 25 μm.

Reported Geographic and Host Distribution

Found in arid central Australia in association with tussock-forming grasses (spinifex) belonging to the genus *Triodia.*

Media to Examine and Key Characters

Characters on CLA. Abundant characteristic macroconidia in sporodochia on carnation leaf pieces in CLA.

Characters on PDA. White to grayish rose mycelium. Produces red pigment in the agar.

Macroconidia

- Sporodochia: Abundant orange sporodochia on carnation leaf pieces.
- General morphology: Long and thin with a whip-like appearance.
- Apical Cell morphology: Long and tapering.
- Basal Cell morphology: Elongate foot shape.
- Number of septa: 6- to 8-septate.
- Abundance: Abundant in sporodochia, but also can be found on the agar surface.

Microconidia. Absent.

Chlamydospores. Absent.

Taxonomy, Pathology and Ecology

This species was first described by Sangalang *et al.* (1890) as a subspecies of *F. avenaceum* but was recognized as a species when it was shown that the two could be clearly differentiated with molecular markers (169). Morphologically *F. aywerte* is quite distinctive, although it could potentially be confused with *F. nurragi. F. aywerte* has been recovered only from stems and from soil around the roots of the spinifex grasses in central Australia. The distribution of *F. nurragi* is different and includes shrubby coastal heath species in Victoria and Western Australia. Thus the geographic distribution and host species from which the isolate is recovered normally should suffice to distinguish *F. aywerte* from *F. nurragi*.

Toxin production by *F. aywerte* has not been reported.

Current Species Description: Sangalang *et al.* (1890).

Fusarium babinda Summerell, Rugg & Burgess

Sexual Stage. None known.

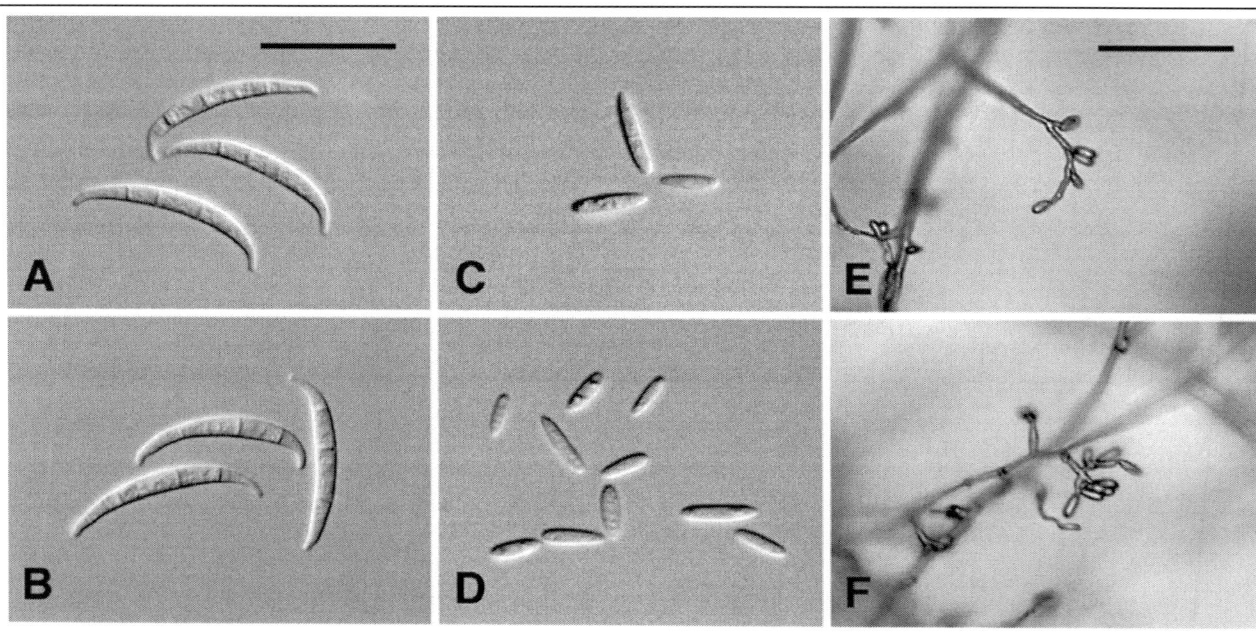

Figure 13-8. *Fusarium babinda.*
A – B: Macroconidia; **C – D:** Microconidia; **E – F:** Microconidia *in situ* on CLA. A – D, scale bar = 25 µm; E – F, scale bar = 50 µm.

Reported Geographic and Host Distribution

Recovered from soil and plant debris in soil from rain forests and wet schlerophyll forests in Australia.

Media to Examine and Key Characters

Characters on CLA. Sporodochia are orange and produce macroconidia that are 3-5 septate, falcate with a pedicellate basal cell, and a slightly curved to hooked apical cell. Fusiform microconidia are found on mono- and polyphialides. The polyphialides may be elaborate and proliferating. Chlamydospore production requires 4-6 weeks.

Characters on PDA. Produces abundant mycelia that initially are white but become pale orange with a violet center or are violet across the entire plate. Violet pigment may be excreted into the agar. The reverse of the colony usually is pale orange to pink with a violet center.

Macroconidia

- Sporodochia: Orange sporodochia found on carnation leaf pieces in CLA.
- General morphology: Relatively slender and falcate.
- Apical Cell morphology: Slightly curved to distinctly hooked.
- Basal Cell morphology: Foot shaped.
- Number of septa: Predominantly 3-septate, but 4- and 5-septate macroconidia may be observed.
- Abundance: Moderately abundant in sporodochia.

Microconidia

- Shape/septation: Fusiform and hyaline. Usually 0-septate, but occasionally 1-septate.

- Aerial mycelium presentation: Always in false heads and never in chains, even on KCl medium.
- Conidiogenous cells: Monophialides and polyphialides. Polyphialides may be elaborately and distinctly proliferating.
- Abundance: Abundant in aerial mycelial of colonies on CLA and SNA.

Chlamydospores

- Abundance/Speed of formation: Sparse and slow to form. Usually present on both CLA and SA after 4-6 weeks.
- Location: In aerial hyphae and on the agar surface.
- Appearance: Singly, and in pairs, clusters and chains.

Taxonomy, Pathology and Ecology

According to Nirenberg & O'Donnell (1570) and O'Donnell *et al.* (1596), the type of *F. babinda* (DAR 70287) is inconsistent with the published description. The most distinctive characteristic of cultures of this species is the production of fusiform microconidia from elaborate proliferating polyphialides and the production of chlamydospores.

The fungus also has been recovered from soil samples collected from other highland sites in eastern Australia. All of the sites from which this fungus was isolated were characterized by high rainfall.

No information is available regarding the pathogenicity or mycotoxicology of *F. babinda*.

Current Species Description: Summerell *et al.* (2086).

Fusarium begoniae Nirenberg & O'Donnell

Sexual Stage. None known.

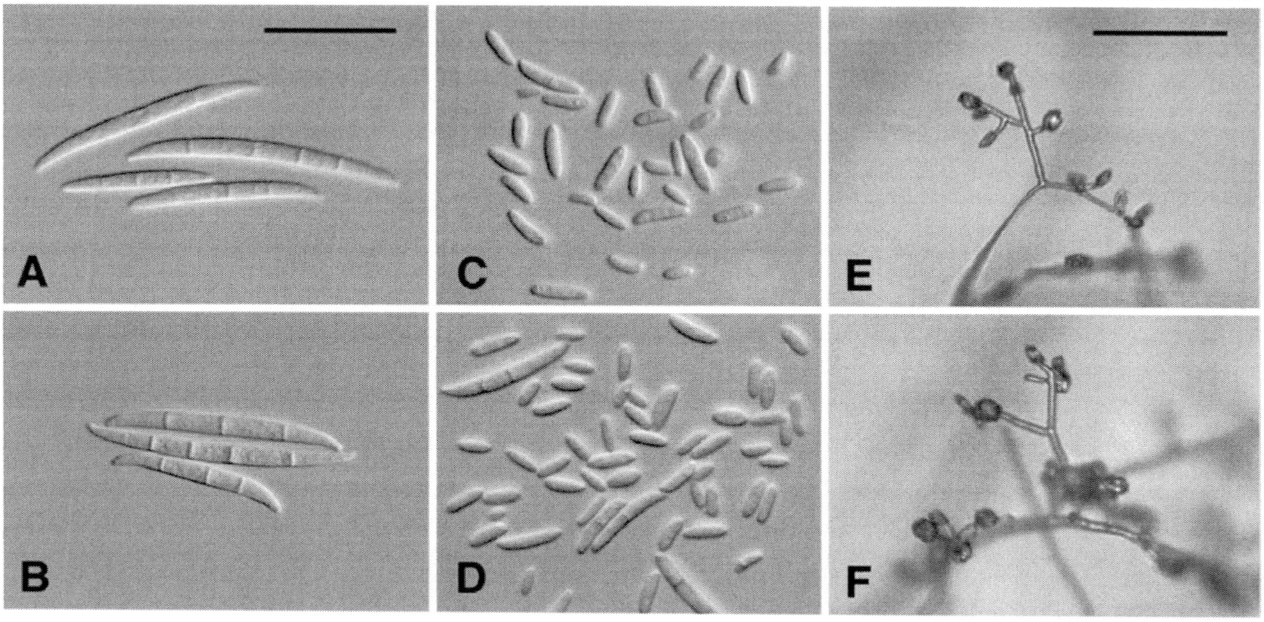

Figure 13-9. *Fusarium begoniae.*
A – B: Macroconidia; **C – D:** Microconidia; **E – F:** Microconidia *in situ* on CLA. A – D, scale bar = 25 µm; E – F, scale bar = 50 µm.

Reported Geographic and Host Distribution

Recovered from *Begonia* hybrids in Germany although potentially of wider distribution.

Media to Examine and Key Characters

Characters on CLA. Orange sporodochia with macroconidia on carnation leaf pieces. Abundant microconidia on both monophialides and polyphialides in the aerial mycelia. Prostrate form and limited branching of the phialides are important characters.

Characters on PDA. White mycelium with grayish-yellow pigmentation in the agar.

Macroconidia

- Sporodochia: Orange. Abundant after a week on carnation leaf pieces in CLA.
- General morphology: Typical of species in the *Gibberella fujikuroi* species complex. Slender, thin-walled and falcate.
- Apical Cell morphology: Beak-like.
- Basal Cell morphology: Foot shaped.
- Number of septa: Usually 3-septate, but occasionally 4-septate.
- Abundance: Common in sporodochia.

Microconidia

- Shape/septation: Oval, allantoid and obovoid. Usually 0-septate, but occasionally 1- to 2- septate.
- Aerial mycelium presentation: False heads.
- Conidiogenous cells: Monophialides and polyphialides.
- Abundance: Abundant in aerial mycelia.

Chlamydospores. Absent.

Taxonomy, Pathology and Ecology

This species was described by Nirenberg & O'Donnell (1570) from two isolates recovered from *Begonia elatior* hybrids where it apparently causes a disease. It is difficult to distinguish *F. begoniae* from *F. bulbicola*. *F. begoniae* has conidiophores that generally are prostrate in the aerial mycelia and are rarely branched, while the aerial conidiophores of *F. bulbicola* are mostly erect and frequently branched. Both *F. begoniae* and *F. foetens* are pathogens of *Begonia*, but can be easily distinguished by the presence of chlamydospores and short monophialides in *F. foetens*. Molecular characters are useful for distinguishing the two described strains of this species, but more strains are needed to validate the species description and its uniqueness.

F. begoniae produces high levels of moniliformin, trace levels of fumonisin B_1, and no detectable beauvericin or fusaproliferin (638). Little else is known of the ecology and pathology of this fungus.

Current Species Description: Nirenberg & O'Donnell (1570).

Fusarium beomiforme Nelson, Toussoun & Burgess

Sexual Stage. None known.

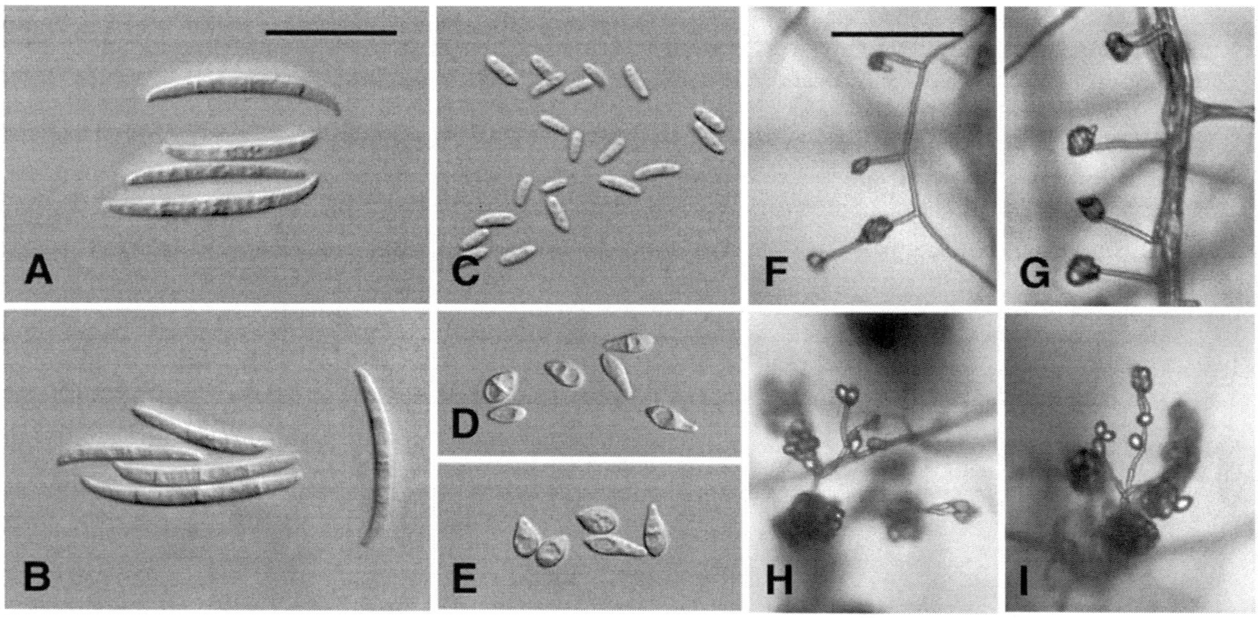

Figure 13-10. *Fusarium beomiforme.*
A – B: Macroconidia; **C:** Oval microconidia; **D – E:** Napiform microconidia; **F – G:** Oval microconidia *in situ* on CLA; **H – I:** Napiform microconidia *in situ* on CLA. A – E, scale bar = 25 µm; F – I, scale bar = 50 µm.

Reported Geographic and Host Distribution

Recovered from soil and soil debris in warm, moist areas, *e.g.*, the Rockhampton area of Australia, the Markham Valley of Papua-New Guinea, and Natal in South Africa.

Media to Examine and Key Characters

Characters on CLA. Sporodochia with macroconidia. Distinctive napiforme and globose microconidia in false heads on monophialides.

Characters on PDA. The mycelium is pale orange to white, with orange-red to reddish-brown pigmentation in the agar.

Macroconidia

- Sporodochia: Orange. Usually on carnation leaf pieces in CLA.
- General morphology: Long, falcate, and thick walled.
- Apical Cell morphology: Slightly curved.
- Basal Cell morphology: Notched or foot shaped.
- Number of septa: Predominantly 3- to 5-septate.
- Abundance: Common in sporodochia.

Microconidia. Two types are produced.

- Shape/septation: The most diagnostic microconidia are large, globose to napiform to lemon shaped and papillate. These microconidia usually are 0-septate, but may occasionally be 1-septate. Small oval to fusiform microconidia that are usually 0-, but sometimes 1-septate, also may form, but are not usually useful for diagnostic purposes.
- Aerial mycelium presentation: False heads.
- Conidiogenous cells: Monophialides.
- Abundance: Abundant overall, but the globose and napiforme microconidia usually are much less common than those with an ovoid-fusiform shape.

Chlamydospores

- Abundance/Speed of formation: Sparse and slow to form, usually requiring 4-6 weeks of incubation on CLA before observation.
- Location: In aerial mycelia and in hyphae on the agar surface.
- Appearance: Singles, chains and clumps.

Taxonomy, Pathology and Ecology

This species was described by Nelson *et al.* (1549) from a number of isolates from wetter tropical regions of Australia, Papua New Guinea and South Africa. Maximum linear growth occurs at a relatively high 30°C and an osmotic potential of -1.0 MPa (1541). O'Donnell *et al.* (1596) excluded *F. beomiforme* from the *Gibberella fujikuroi* species complex and the "Dlaminia" clade on the basis of its phylogenetic placement.

This species can produce moniliformin (1350) and beauvericin (1283), but does not produce fumonisin (1548). *F. beomiforme* has not been reported to be plant pathogenic and probably is a saprophyte.

Current Species Description: Nelson *et al.* (1549).

Fusarium brevicatenulatum Nirenberg, O'Donnell, Kroschel & Andrianaivo

Sexual Stage. None known.

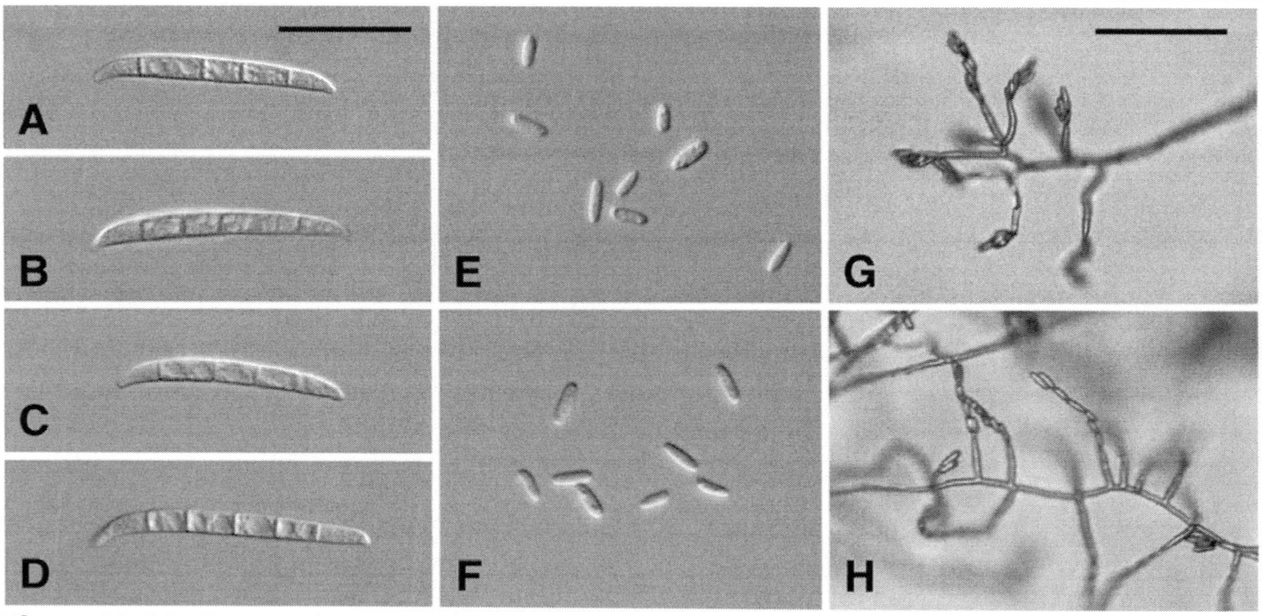

Figure 13-11. *Fusarium brevicatenulatum.*
A – D: Macroconidia; **E – F:** Microconidia; **G – H:** Microconidia *in situ* on CLA. A – F, scale bar = 25 μm; G – H, scale bar = 50 μm.

Reported Geographic and Host Distribution

Recovered from *Striga asiatica* from Madagascar and from millets in east Africa.

Media to Examine and Key Characters

Characters on CLA. Microconidia in the aerial mycelia produced primarily on monophialides. Very short conidial chains are formed when incubated under continuous black light

Characters on PDA. White mycelium with orange to dark blue-gray pigmentation in the agar.

Macroconidia

- Sporodochia: Pale orange sporodochia are relatively slow to form and usually are rare.
- General morphology: Typical of species in the *Gibberella fujikuroi* species complex. Slender, thin-walled and falcate.
- Apical Cell morphology: Bent.
- Basal Cell morphology: Foot shaped.
- Number of septa: Predominantly 5-septate, but 3- and 4-spetate macroconidia also may be observed.
- Abundance: Sparse and usually limited to the sporodochia.

Microconidia

- Shape/septation: Long-oval to obovoid.
- Aerial mycelium presentation: False heads and short chains. Spores in chains may appear side-by-side rather than end-to-end. The chains are more common

when the cultures are incubated under continuous black light.

- Conidiogenous cells: Usually monophialides with occasional polyphialides.
- Abundance: Abundant in the aerial mycelia.

Chlamydospores. Absent.

Taxonomy, Pathology and Ecology

This species was described by Nirenberg *et al.* (1571) from isolates recovered from the parasitic witchweed, *Striga asiatica*, and was associated with symptoms of disease in this plant. Pathogenicity tests, however, have not been carried out to confirm Koch's postulates, nor have studies been made to determine if *F. brevicatenulatum* is pathogenic either to maize or to sorghum, the most common hosts for *Striga* spp. We have preliminary information that indicates that this species is conspecific with *F. pseudoanthophilum*, a species to which it is reported to be phylogenetically very closely related (1596).

Isolates of *F. brevicatenulatum* can produce trace levels of fumonisin B_1, but are not known to produce detectable levels of beauvericin, fusaproliferin or moniliformin (638).

Current Species Description: Nirenberg *et al.* (1571).

Fusarium bulbicola Nirenberg & O'Donnell

Sexual Stage. None known.
Common Synonym. *F. sacchari* var. *elongatum*.

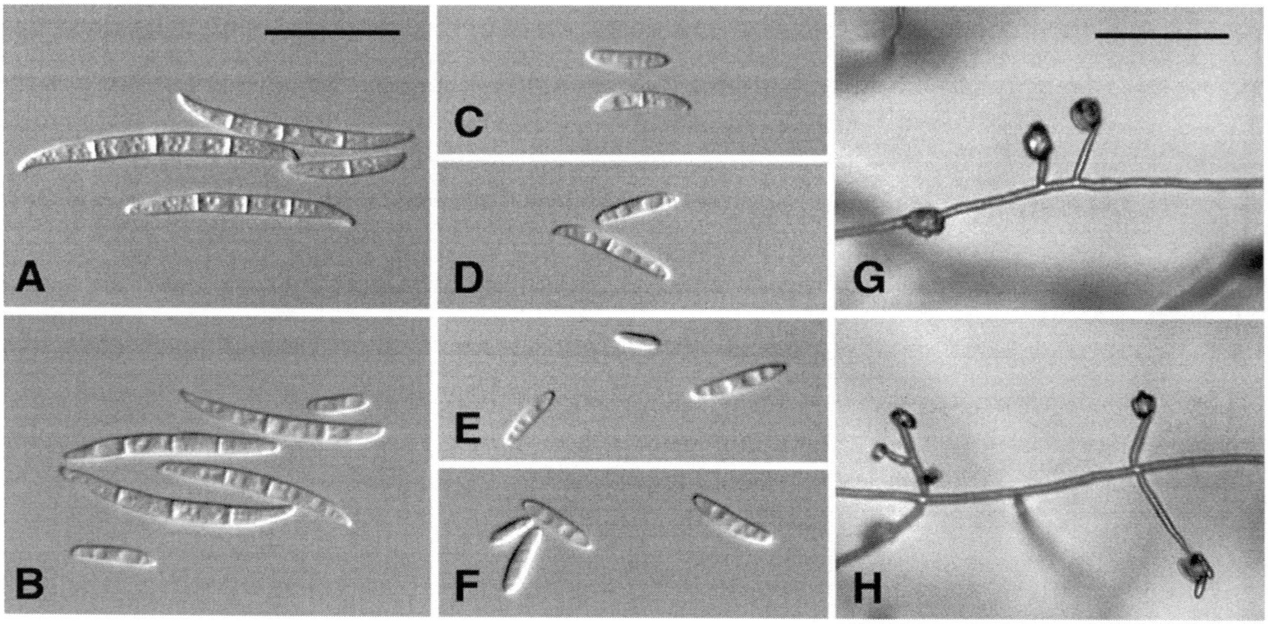

Figure 13-12. *Fusarium bulbicola.*
A – B: Macroconidia; **C – F:** Microconidia; **G – H:** False heads of microconidia *in situ* on CLA. A – F, scale bar = 25 μm; G – H, scale bar = 50 μm.

Reported Geographic and Host Distribution

Various bulb species, *e.g.*, *Haemanthus*, *Narcissus* and *Nerine*, in Europe.

Media to Examine and Key Characters

Characters on CLA. Long, slender macroconidia from sporodochia, which are slow to form and require black light illumination. Abundant microconidia borne on branched, elongated phialides in the aerial mycelium. Erect, branched phialides are an important diagnostic character.

Characters on SNA. Elongate and branched phialides.

Characters on PDA. White mycelium with red pigmentation in the agar.

Macroconidia

- Sporodochia: Form after 14 days of incubation with continuous black light illumination.
- General morphology: Typical of species in the *Gibberella fujikuroi* species complex. Falcate and thin walled.
- Apical Cell morphology: Elongate and tapering relative to others in the *G. fujikuroi* species complex.
- Basal Cell morphology: Usually notched, but sometimes with a distinct foot shape.
- Number of septa: Predominantly 3-septate.
- Abundance: Moderately abundant in sporodochia when incubated under continuous black light.

Microconidia

- Shape/septation: Long-oval to obovoid. Usually 0-septate, but occasionally 1-septate.
- Aerial mycelium presentation: False heads.
- Conidiogenous cells: Monophialides and polyphialides.
- Abundance: Abundant in the aerial mycelia.

Chlamydospores. Absent.

Taxonomy, Pathology and Ecology

This taxon was raised to species status by Nirenberg & O'Donnell (1570) and was previously described as a variety of *F. sacchari* (*F. sacchari* var. *elongatum*) by Nirenberg (1566). It is difficult to differentiate this species from isolates of *F. sacchari* and *F. subglutinans* morphologically and it is likely that differentiation will require molecular techniques. It also is difficult to distinguish *F. bulbicola* from *F. begoniae*. *F. bulbicola* has conidiophores that generally are erect the aerial mycelia and are frequently branched, while the aerial conidiophores of *F. begoniae* are mostly prostrate and only rarely branched.

This species has been associated with bulb rots of several species including *Nerine* and *Narcissus* (1566). Two isolates of *F. bulbicola* were mildly dermotoxigenic to rabbits, but the nature of the mycotoxin(s) produced by these strains has not been determined (985, 1341). *F. bulbicola* does not produce detectable levels of beauvericin, fumonisins, fusaproliferin or moniliformin (638).

Current Species Description: Nirenberg & O'Donnell (1570).

Fusarium camptoceras Wollenweber & Reinking emend Marasas & Logrieco

Sexual Stage. None known.

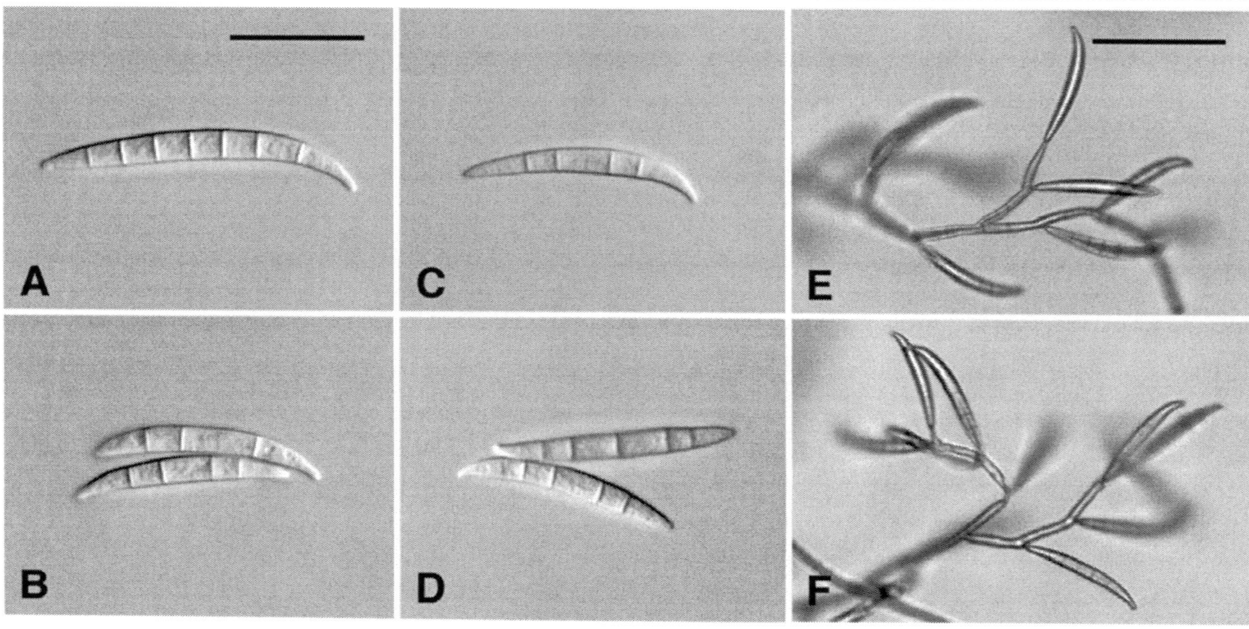

Figure 13-13. *Fusarium camptoceras.*
A: Macroconidia; **B – D:** Mesoconidia; **E – F:** Mesoconidia *in situ* on CLA. A – D, scale bar = 25 µm; E – F, scale bar = 50 µm.

Reported Geographic and Host Distribution

Restricted to subtropical and tropical regions, recovered from decaying bananas and cacao.

Media to Examine and Key Characters

Characters on CLA. Macroconidia are produced in cream to orange colored sporodochia. Mesoconidia, very similar in appearance to macroconidia, are produced on polyphialides in the aerial mycelia.

Characters on PDA. White to beige mycelium with brown pigmentation in the agar.

Macroconidia

- Sporodochia: Cream to pale orange color. Contain loosely branched conidiophores with polyphialides that usually have 2-3 openings and sometimes 4. May be slow to form.
- General morphology: Somewhat squat, but falcate.
- Apical Cell morphology: Pointed.
- Basal Cell morphology: Conical to pointed and notched.
- Number of septa: Usually 3- to 5-septate, but some may be 6- to 7-septate.
- Abundance: Abundant once the sporodochia form.

Microconidia/Mesoconidia

- Shape/septation: These conidia are not clearly distinguishable in shape from the macroconidia. They may be up to 6-septate.
- Aerial mycelium presentation: Mesoconidia are produced singly from each phialide.

- Conidiogenous cells: Polyphialides.
- Abundance: Common in the aerial mycelia.

Chlamydospores

- Abundance/Speed of formation: Formed slowly and sparsely.
- Location: In aerial hyphae and in hyphae submerged in the agar.
- Appearance: Pairs, chains or clusters.

Taxonomy, Pathology and Ecology

This taxon was described by Wollenweber & Reinking (2358) and was recognized by Gerlach & Nirenberg (697) and Nelson *et al.* (1551). It is similar to *F. semitectum*, but the macroconidia of *F. camptoceras* are more squat and curved. The emended description (1347) differs from the original in recording pedicellate macroconidia from sporodochia and the description of mesoconidia on polyphialides in the aerial mycelia. The sexual stage of *F. camptoceras* probably is heterothallic (1036).

Two other species were separated from *F. camptoceras* by Marasas *et al.* (1347) – *F. musarum* and *F. nelsonii.* Reports of *F. camptoceras* prior to 1998 need to be re-evaluated to determine to which of the three currently described species the isolates examined belong, *e.g.*, Fekete *et al.* (611), Jimenez *et al.* (977, 978), Marasas *et al.* (1336), and Mulè *et al.* (1491). *F. camptoceras* can be distinguished from *F. nelsonii* and *F. musarum* on the basis of pigmentation in PDA, the size and septation of the mesoconidia, the presence/absence of sporodochia, and the pattern in which chlamydospores are formed.

Fusarium camptoceras has been recovered from decayed bananas and cacao in various tropical regions, but does not appear to have a role as a plant pathogen.

Current Species Description: Marasas *et al.* (1998).

Fusarium chlamydosporum Wollenweber & Reinking

Sexual Stage. None known.
Common Synonyms. *Fusarium sporotrichioides* var. *chlamydosporum*, *Fusarium fusarioides*.

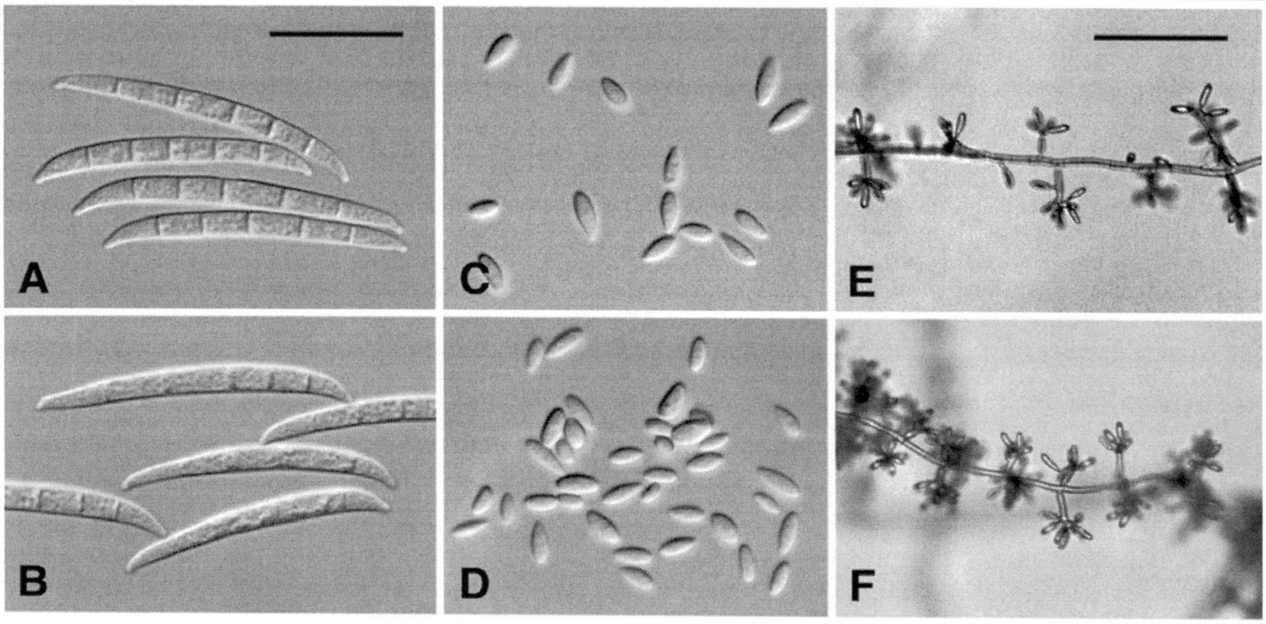

Figure 13-14. *Fusarium chlamydosporum.*
A – B: Macroconidia; **C – D:** Microconidia; **E – F:** Microconidia *in situ* on CLA. A-D, scale bar = 25 µm; E – F, scale bar = 50 µm.

Reported Geographic and Host Distribution

Found commonly in arid and semi-arid regions in the soil and as a saprophyte on a variety of substrates.

Media to Examine and Key Characters

Characters on CLA. Macroconidia are not usually used to identify this species due to their rarity. When macroconidia are formed they are found in orange sporodochia that may be obscured by the mycelium. Microconidia are abundant and produced on both mono- and polyphialides. Conidiophores in the aerial mycelia may be elaborately branched. Chlamydospores are abundant and form rapidly.

Characters on PDA. White mycelium usually with grayish rose to burgundy pigment in the agar, although yellow to pale-brown pigmentation also has been reported (1551). Sporodochia are rarely formed on PDA. Chlamydospore production by older cultures may result in a mycelium with a pale brown coloration.

Macroconidia

- Sporodochia: Not always produced. They usually are found on carnation leaf pieces and may be hidden under an overgrowing mycelium
- General morphology: Thick walled and moderately curved. Unequal dorsiventral curvature, *i.e.*, the upper wall is curved and the lower wall is almost straight.
- Apical Cell morphology: Short, curved and pointed.
- Basal Cell morphology: Usually notched, but occasionally foot shaped.
- Number of septa: 3- to 5-septate.

- Abundance: May be difficult to find if sporodochium production is rare.

Microconidia

- Shape/septation: Straight to comma-shaped. Usually 0-septate, but 1- and 2-septate microconidia may be observed.
- Conidiogenous cells: Elaborate polyphialides that result in branching conidiophores with a tree-like appearance when viewed *in situ*. Microconidia may be produced singly, or sometimes in pairs from each phialidic opening. The mechanism resulting in these structures has been evaluated microscopically (2229).
- Abundance: Abundant in the aerial mycelia.

Chlamydospores

- Abundance/Speed of formation: Produced abundantly and rapidly. Usually requires 2-4 weeks of incubation on CLA.
- Location: In aerial mycelia, on the agar surface, or in submerged hyphae.
- Appearance: May occur singly, but are more common in chains and in clusters. Usually are verrucose with pale brown coloration.

Taxonomy, Pathology and Ecology

Morphologically this species is well defined, but the morphological integrity has not been verified through sexual crosses or with molecular markers.

This species is common in soil in warmer areas of the world and can be abundant in semi-arid and arid grassland soils. Burgess & Summerell (277) and Sangalang *et al.* (1889) reported that it was common in semi-arid parts of Australia, but infrequently isolated from subtropical moist grasslands at the same latitude. This species is commonly isolated from grain in drier areas particularly in the Middle East, southern Europe, central Asia, and Australia, *e.g.*, Kanaan & Bahkali (1002).

F. chlamydosporum also has been isolated occasionally from the tissues of various plants that it presumably colonized as a secondary invader, *e.g.*, Tinline *et al.* (2150). *F. chlamydosporum* has been associated with tobacco where it may reduce seed quality and germination (595) and serve as an antagonist to the more serious tobacco pathogen *Phytophthora nicotianae* (1561). It also is reported to cause a blight on kangaroo paw (*Anigozanthos* spp.) plants (1897), damping off of rooibos tea plants (578), and a stem canker of okra (653).

Fusarium chlamydosporum has been implicated in some human mycoses (1373), particularly in immunocompromised patients (1049, 1760, 1933). It is resistant to most clinical antifungals, with amphotericin B reported as the most effective (1748). Some strains of this species make moniliformin (1757, 1758) and other secondary metabolites such as acuminatopyrone (2020, 2262), chlamydosporol (1905, 2020), and steroids (284).

Current Species Description: Wollenweber & Reinking (2359).

Fusarium circinatum Nirenberg & O'Donnell emend. Britz, Coutinho, Wingfield & Marasas

Sexual Stage. *Gibberella circinata* Nirenberg & O'Donnell emend. Britz, Coutinho, Wingfield & Marasas.
Common Synonyms. *Fusarium subglutinans* f. sp. *pini*, *Fusarium lateritium* f. sp. *pini*, *Gibberella fujikuroi* Mating population H.

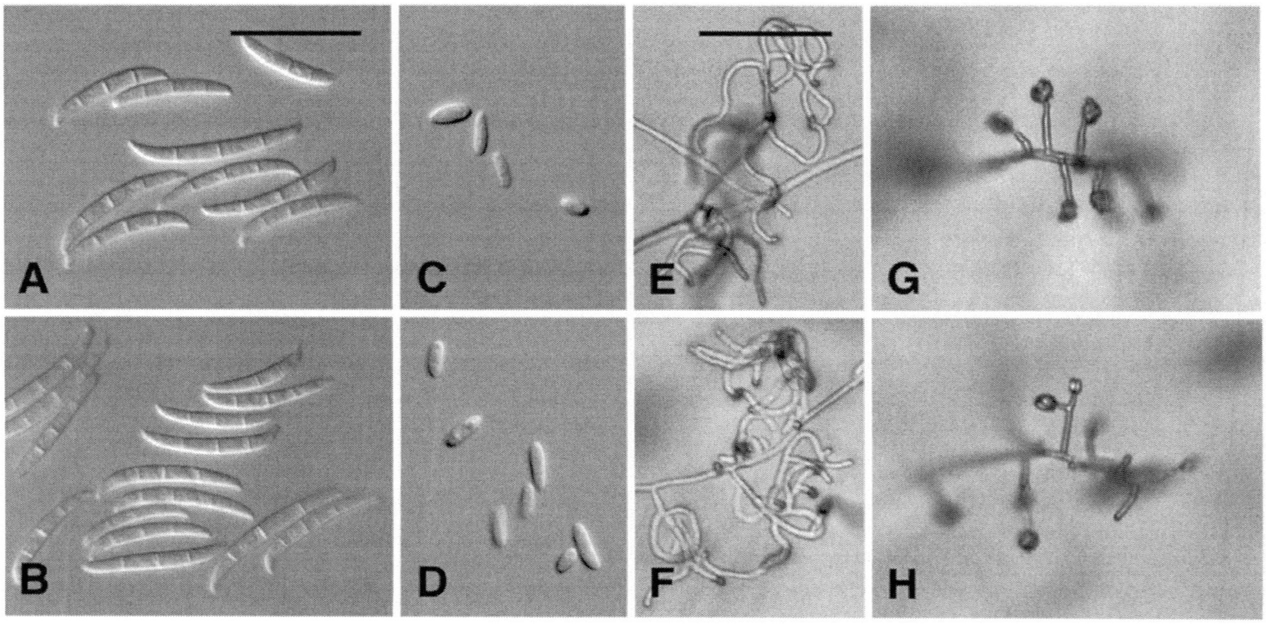

Figure 13-15. *Fusarium circinatum.*
A – B: Macroconidia; **C – D:** Microconidia; **E – F:** Hyphal coils; **G – H:** Microconidia *in situ* on CLA. A – D, scale bar = 25 μm; E – H, scale bar = 50 μm.

Reported Geographic and Host Distribution

A pathogen on *Pinus* species and some other conifers. Found in the United States, South Africa, Mexico, Spain, Japan, and Chile.

Media to Examine and Key Characters

Characters on CLA. Sporodochia with macroconidia are sparse and may be difficult to find. Macroconidia are typical of those produced by most species in the *Gibberella fujikuroi* species complex being slender and thin walled. Obovoid, oval or allantoid microconidia are borne on mono- and polyphialides that are similar in structure and shape to those of *F. subglutinans*. The coiled hyphae from which the species epithet is derived may be found on CLA, but these hyphae usually are more easily and more reliably observed on SNA.

Characters on SNA. The coiled, sterile hyphae from which the specific epithet is derived are distinctive for this species and can be found reliably only on SNA. The degree of coiling usually is greater on SNA than it is on CLA.

Characters on PDA. White mycelium, but may produce violet pigment in the agar. Grows relatively rapidly.

Macroconidia

Sporodochia: Sparse and may be difficult to find. When they are found they usually are on carnation leaf pieces and are pale orange in color.

- General morphology: Relatively slender with no significant curvature. Typical of species in the *Gibberella fujikuroi* species complex.
- Apical Cell morphology: Curved.
- Basal Cell morphology: Relatively poorly developed.
- Number of septa: Usually 3-septate.
- Abundance: Often difficult to find.

Microconidia

- Shape/septation: 0-septate. Obovoid, oval or allantoid shape.
- Aerial mycelium presentation: Produced singly or in aggregations of a few microconidia.
- Conidiogenous cells: Polyphialides and monophialides. Polyphialides may proliferate extensively.
- Abundance: Abundant in the aerial mycelia of CLA and SNA cultures.

Chlamydospores. Absent. Some isolates produce swollen hyphae that may superficially resemble chlamydospores or pseudochlamydospores.

Other Features. Coiled, sterile hyphae are diagnostic for this species. They may be found reliably on SNA, but their production on CLA varies. The number of coiled hyphae and the degree of coiling usually is higher on SNA as well.

Taxonomy, Pathology and Ecology

The original species description (1570) was based on only four isolates and lacked critical information for the name to be valid under the International Code of Botanical Nomenclature (770). The emended description (242) differs little in the morphological description, but provides the remaining information required to validate the name and identifies numerous additional strains that have been characterized genetically. The teleomorph, *G. circinata*, has been produced only in culture by crossing compatible mating strains of the fungus (242, 243, 246, 414, 1570). *G. circinata* has not been observed to date in nature, although there is reason to believe that sexual recombination could occur under field conditions (241, 246, 2327). Formation of the teleomorph is sensitive to temperature, with ~17°C being optimal (242, 414).

Fusarium circinatum is easily confused with a number of other species in the *Gibberella fujikuroi* clade, particularly *F. subglutinans* and related species, and for a number of years this species was defined as a *forma specialis* of *F. subglutinans* (401), and before that as a *forma specialis* of *F. lateritium* (1148). Morphologically, *F. circinatum*, *F. pseudocircinatum* and *F. sterilihyphosum* are the only species that produce sterile coiled hyphae. *F. pseudocircinatum* produces short chains of microconidia that can be used to differentiate it from *F. circinatum*. These species also can be distinguished by the production of erect conidiophores by *F. circinatum* and prostrate conidiophores by *F. pseudocircinatum*. *F. circinatum* and *F. sterilihyphosum* can be distinguished on the basis of macroconidial morphology with the macroconidia of *F. sterilihyphosum* being longer and more slender than those of *F. circinatum*.

Differences in the sequences of any of several conserved genes (1596, 1600, 1929, 2040) are the most reliable method for the identification of this species, as the coiled hyphae may be difficult to observe on any medium other than SNA. A fertile sexual cross with one of the available tester strains (Table 6-1) also is a reliable species diagnostic. Relatively large and diverse collections of *F. circinatum* are available for study. Microsatellite markers for diversity studies have been developed specifically for this fungus (245).

The pitch canker disease caused by this fungus, is a serious disease of pines in the United States (532), Mexico, Japan, South Africa and many other locations. Genetic structure of populations of this fungus varies by location, with some populations appearing to be nearly clonal while others are much more outbred, *e.g.*, Correll *et al.*, (400), Gordon *et al.* (744) and Viljoen *et al.* (2254). This species is of quarantine importance in New Zealand and Australia (504, 894). *Pinus radiata* is the best known host for this pathogen (1635, 2328), but numerous other pine species are sensitive as well, *e.g.*, Enebak & Stanosz (577), Gordon *et al.* (743), Schmale & Gordon (1913). Endopolygalacturonases are important for the penetration of the host by this pathogen (350). Disease severity can be reduced by fungicide treatment for up to a year after seedlings have been transplanted from the nursery to the field (1458). The disease may be transmitted by cone- and twig-infesting beetles (892, 2051).

Isolates of *F. circinatum* can produce beauvericin, but are not known to produce fumonisins, fusaproliferin or moniliformin (638).

Current Species Description: Britz *et al.* (242).

Fusarium compactum (Wollenweber) Gordon

Sexual Stage. None known.
Common Synonym. *Fusarium equiseti* var. *compactum*.

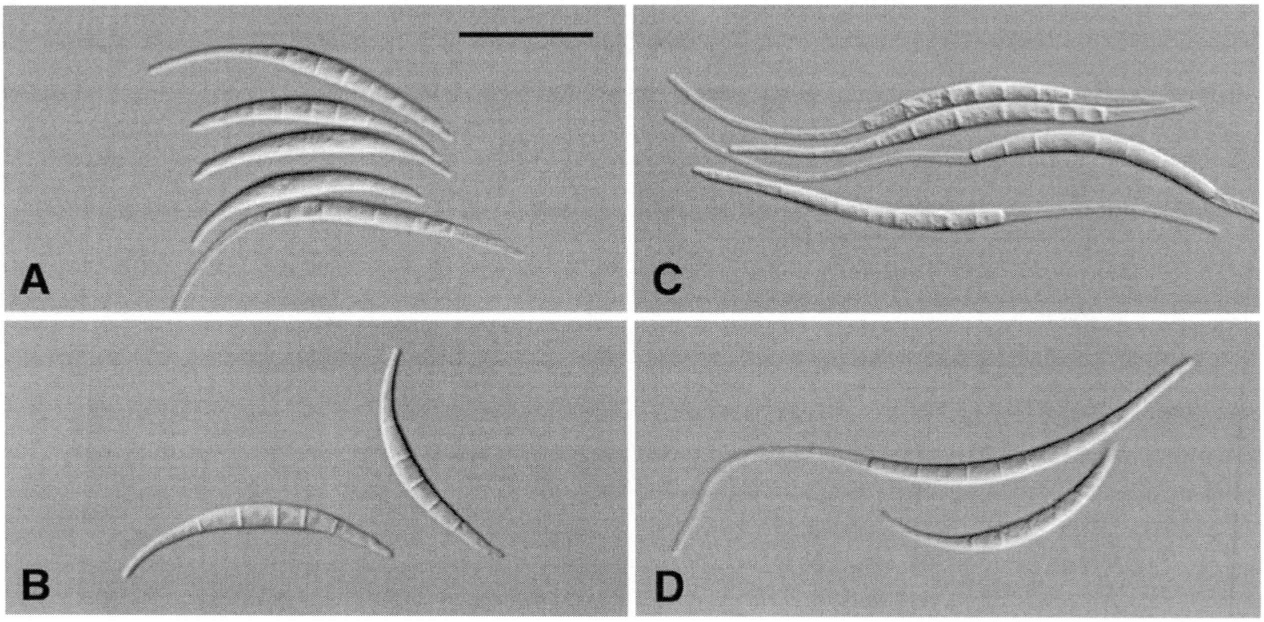

Figure 13-16. *Fusarium compactum.*
A – B: Short macroconidia; **C – D:** Long macroconidia. Scale bar = 25 µm.

Reported Geographic and Host Distribution

Generally found in hot arid to semi-arid climates as a soil saprophyte.

Media to Examine and Key Characters

Characters on CLA. Macroconidia are formed in orange sporodochia on the carnation leaf pieces and on the agar. The macroconidia have strong dorsi-ventral curvature with an exaggerated foot cell and tapering elongate apical cell that is often needle like in appearance. The macroconidia are produced from monophialides on branched conidiophores in sporodochia and to a minor extent from monophialides formed directly on the hyphae.

Characters on PDA. Colonies produce white mycelia that may become pigmented with age. In some isolates orange sporodochia may form at the center of the culture, while for others the macroconidia are found in orange or pale orange spore masses in the aerial mycelium. Pigment formation in the agar varies. Some isolates produce no pigment, while others form brown or red pigments. Brown pigment flecks also may form in the agar of cultures. *F. compactum* mutates readily when cultured on PDA, usually to a pionnotal form. Growth at 30°C is relatively rapid.

Macroconidia

- Sporodochia: Orange color. Found on the agar and on carnation leaf pieces.
- General morphology: Robust, thick walled with strong dorsi-ventral curvature, and quite wide at the midpoint of the macroconidium. Length can vary with some isolates producing long macroconidia and others producing spores that are quite short.
- Apical Cell morphology: Elongate and tapering. May taper sufficiently to give an almost needle-like appearance.

- Basal Cell morphology: Pronounced and distinctive foot shape.
- Number of septa: Usually 5-septate.
- Abundance: Usually abundant in sporodochia. Macroconidia produced in the sporodochia usually are more uniform in shape and size than are those produced from hyphae growing directly on the agar surface.

Microconidia. Absent.

Chlamydospores

- Abundance/Speed of formation: Abundant. Usually formed rapidly (2-4 weeks) on CLA. Chlamydospores may be produced by pionnotal cultures (2349).
- Location: In aerial hyphae, and hyphae on the agar surface or in submerged hyphae.
- Appearance: In chains and in clusters. Usually become verrucose and pigmented with age.

Taxonomy, Pathology and Ecology

Macroconidia of *F. compactum*, especially those producing brown pigment, may be similar to those of *F. equiseti*, and the two species can be difficult to differentiate solely on spore morphology. The apical cells provide the clearest difference since apical cells of macroconidia of *F. compactum* have a needle-like appearance while the apical cells of macroconidia of *F. equiseti* usually are more rounded. Cultural characters also may be useful in distinguishing these species. Strains of *F. compactum* normally form a sparse mycelium on the agar of CLA while most isolates of *F. equiseti* form relatively abundant floccose mycelia on the agar between the carnation leaf pieces. Isolates of *F. compactum* also grow faster than do isolates of *F. equiseti* at 30°C.

Fusarium compactum commonly is recovered from grassland and desert soils in dry, hot climates, *e.g.*, Australia, the Middle East, and Africa (277, 1327, 1336). Some climactic models suggest that it will occur rarely, if at all, in cooler areas (112). This species often is regarded as a saprophyte, but has been implicated in a corm and root rot of banana (648), a fatal canker on Italian cypress (1311), and a wilt and root rot of peanuts (1870). *F. compactum* has been associated with dermatophytic diseases in humans (1320).

F. compactum can be toxigenic and groups phylogenetically with other species that produce type A trichothecenes (1493). Isolates of *F. compactum* first reported as *F. acuminatum*, and others, can produce neosolaniol (68, 2259) and acuminatin (2259). *F. compactum* has been implicated in the death of large numbers of sandhill cranes in New Mexico following ingestion of peanuts that were heavily contaminated with *F. compactum* (1542). Significant quantities of trichothecenes were present in the peanut waste (391, 2346). Strains of *F. compactum* also have been reported to produce enniatins (2257) as well as uncharacterized secondary metabolites, some with antimicrobial activity (240, 1329, 2111).

Current Species Description: Wollenweber & Reinking (2359).

Fusarium concentricum Nirenberg & O'Donnell

Sexual Stage. None known.

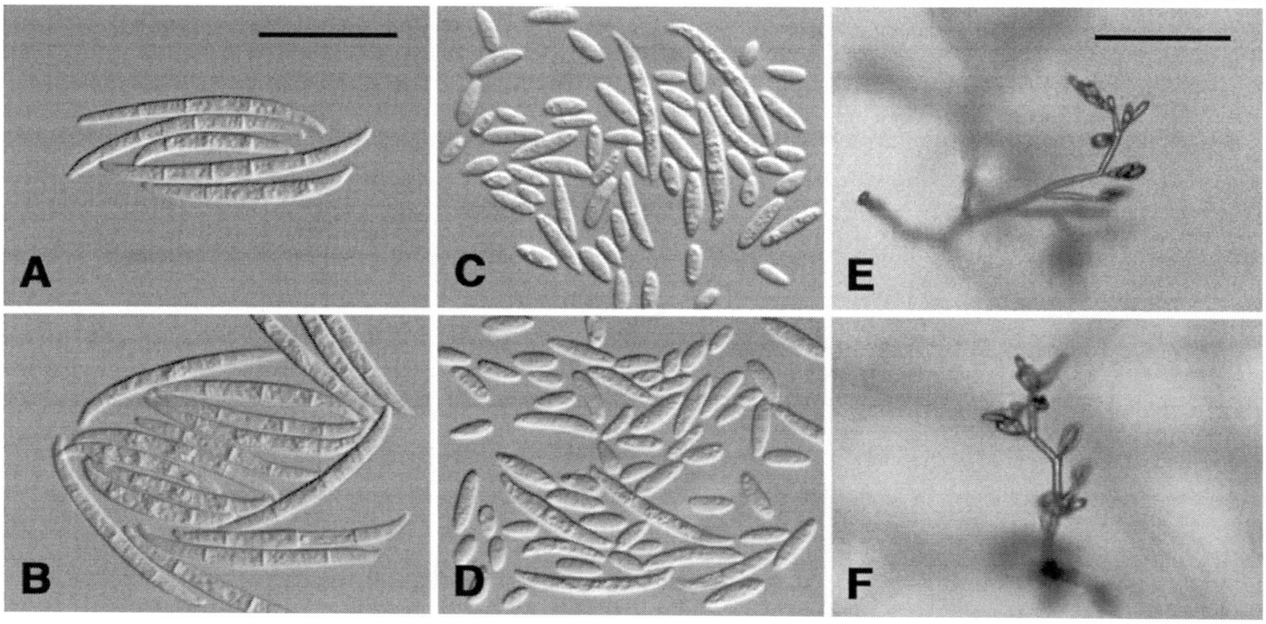

Figure 13-17. *Fusarium concentricum.*
A – B: Macroconidia; **C – D:** Microconidia; **E – F:** Microconidia *in situ* on CLA. A – D, scale bar = 25 μm; E – F, scale bar = 50 μm.

Reported Geographic and Host Distribution

Mainly from *Musa* sp. in Central America; one isolate recorded from an aphid in Korea.

Media to Examine and Key Characters

Characters on CLA. Macroconidia found in orange sporodochia; 3-5 septate, slender with a "beaked" apical cell and a foot-shaped basal cell. Obovoid to oval or allantoid, 0-septate microconidia borne primarily on monophialides but also on polyphialides in the aerial mycelia.

Characters on PDA. Reddish-white aerial mycelium. Pigments in the agar are formed in alternating red and orange concentric rings. These concentric rings are the morphological feature on which the species epithet is based.

Macroconidia

- Sporodochia: Sporodochia are orange in color and are most common on the carnation leaf pieces in CLA. They form after microconidia have formed in the aerial mycelia.
- General morphology: Relatively slender with no significant curvature. Typical of species in the *Gibberella fujikuroi* species complex.
- Apical Cell morphology: Beaked.
- Basal Cell morphology: Foot shaped.
- Number of septa: 3- to 5-septate.
- Abundance: Sparse to moderately abundant.

Microconidia

- Shape/septation: Oval, obovoid to allantoid. Usually 0-septate, but occasionally 1-septate.

- Aerial mycelium presentation: False heads.
- Conidiogenous cells: Primarily monophialides with some polyphialides. The conidiophores may be very elaborate with aggregations of long phialides.
- Abundance: Abundant in the aerial mycelium.

Chlamydospores. Absent.

Taxonomy, Pathology and Ecology

This species was described by Nirenberg & O'Donnell (1570) predominantly from strains isolated from bananas in Central America and another isolate from an aphid in Korea. Aerial conidia produced by *F. circinatum*, *F. concentricum* and *F. guttiforme* are similar. *F. circinatum* strains can be distinguished by the formation of coiled hyphae on SNA. The aerial conidiophores of *F. guttiforme* are more strongly branched and contain more polyphialides than do those of *F. concentricum*. Aoki *et al.* (74) identified additional strains of *F. concentricum* from Japan and also described the production of microconidia with a pointed apex and a wedge-shaped base when the strains were cultured under black light.

Strains initially identified as *F. subglutinans* (2247) and *F. semitectum* (796) and later reclassified as *F. concentricum* (1570) were reported to produce moniliformin and beauvericin, respectively. Fotso *et al.* (638) confirmed the production of beauvericin, but not of moniliformin, and also found that the *ex*-type strain could produce neither fumonisins nor fusaproliferin when cultured on maize grits. Little else is known of this species' ecology or biology, and no plant diseases or animal toxicoses are associated with *F. concentricum*.

Current Species Description: Nirenberg & O'Donnell (1570).

Fusarium crookwellense Burgess, Nelson & Toussoun

Sexual Stage. None known.
Common Synonym. *Fusarium cerealis*.

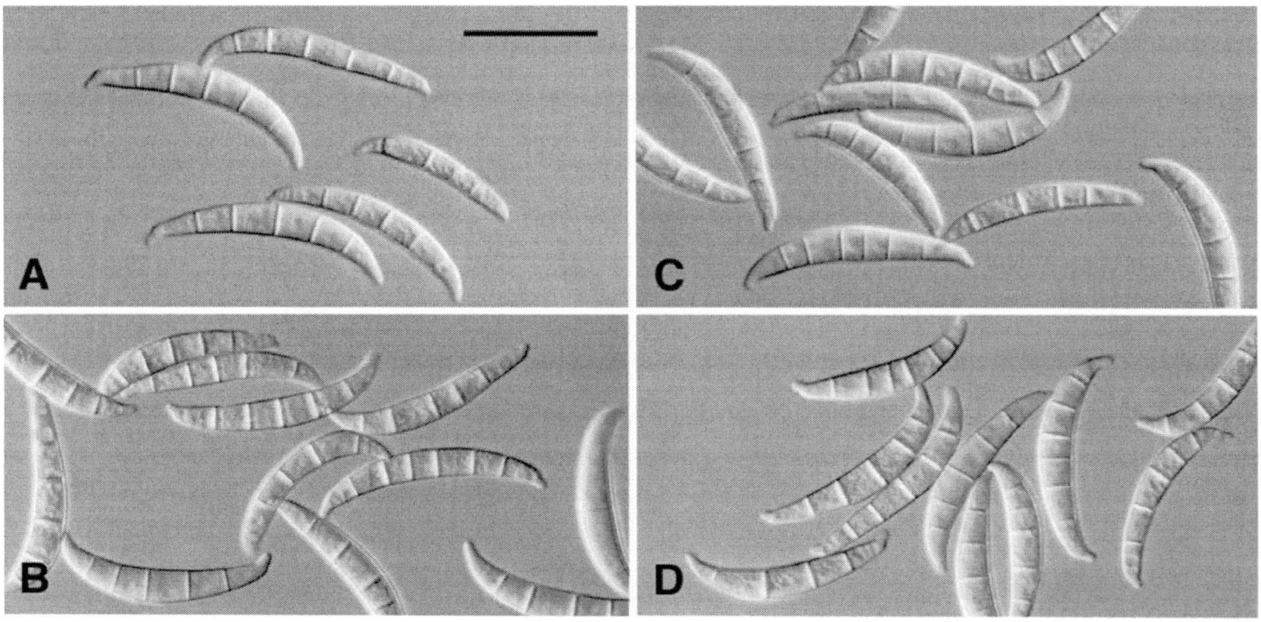

Figure 13-18. *Fusarium crookwellense.*
A – D: Macroconidia. Scale bar = 25 μm.

Reported Geographic and Host Distribution

Generally more frequent in temperate regions. Associated with plant debris in soil, potatoes, crowns of cereal plants, and other plant species.

Media to Examine and Key Characters

Characters on CLA. Abundant macroconidia are formed on monophialides in pale orange to dark brown sporodochia. They are medium length, thick walled, falcate with the dorsal side more curved than the ventral, usually 5-septate, and normally are widest at the mid-point. No microconidia are produced. Chlamydospores are moderately abundant but form slowly.

Characters on PDA. *F. crookwellense* usually grows quite rapidly with a dense aerial mycelium that covers a standard 9-cm Petri plate in ~4 days. Most isolates produce abundant sporodochia, that aggregate together in a conspicuous spore mass in the center of the culture. This spore mass becomes brown with age, and may have annular zonations if the culture is incubated under alternating conditions of light and dark. Many isolates that produce abundant sporodochia have sparse mycelia. Most isolates will produce red pigments in the agar.

Macroconidia

- Sporodochia: Pale orange to dark brown color. Abundant on carnation leaf pieces, but may also form directly on the agar.
- General morphology: Robust, thick-walled macroconidia are widest at the mid-point of the spore. Generally, the macroconidia are intermediate in length with pronounced curvature to the dorsal side while the ventral side is almost straight. The delimitation of macroconidia of *F. crookwellense* has been studied in some detail with light microscopy and scanning and transmission electron microscopy (2226-2228).

- Apical Cell morphology: Distinctly curved and tapering to a point at the tip.
- Basal Cell morphology: Pronounced foot shape.
- Number of septa: Usually 5-septate.
- Abundance: Macroconidia usually are abundant in sporodochia. Macroconidia produced in sporodochia usually are uniform in shape and size while those produced from hyphae growing directly on the agar surface often are highly variable.

Microconidia. Absent.

Chlamydospores

- Abundance/Speed of formation: Moderately abundant. Form slowly (4-6 weeks) on CLA.
- Location: In aerial hyphae, hyphae on the agar surface or in submerged hyphae. Also may form in cells of the macroconidia.
- Appearance: In chains and clusters. More smooth-walled than verrucose.

Other characters. Plectenchymatous bodies (aggregations of hyphae) may form in some isolates of this species.

Taxonomy, Pathology and Ecology

F. crookwellense is similar to the description of *F. cerealis* and Nirenberg (1567) suggested that *F. cerealis* is the preferred name. There is no type material available for *F. cerealis*, however, so the name *F. crookwellense* is preferable for isolates that fit this description.

Fusarium crookwellense often is confused with *F. culmorum*, as cultures on PDA are similar and their geographic ranges overlap. The macroconidia of *F. crookwellense* are longer and narrower than those of *F. culmorum*, and the basal cell of *F. crookwellense* is conspicuously foot shaped, while the foot shape of the *F. culmorum* basal cell is much less obvious. The apical cell of *F. crookwellense* also is distinctly curved and tapers gradually to a narrow point, while the apical cell of *F. culmorum* usually is rounded or is occasionally slightly papillate. Usually strains of *F. crookwellense* are stable in culture, but pionnotal mutants may occur, especially when cultured on PDA, in which the aerial mycelium is completely absent and the colony surface is a sheet of macroconidia. The sexual stage of *F. crookwellense* probably is heterothallic (1036).

Fusarium crookwellense was first isolated from dimple-like lesions on potato tubers from the rural area of Crookwell in New South Wales, Australia in 1971 (273). Since then representative isolates have been recovered from a variety of plant species (mainly crown or root tissues) and from debris from grassland and cultivated soils. The fungus usually is more common in temperate areas. Isolates of *F. crookwellense* have been found in the northern areas of the U.S.A., Canada, South Africa, Japan, temperate parts of Australia, New Zealand, many parts of Europe, Columbia, and China.

Fusarium crookwellense can cause mild root rot and foot rot of wheat in glasshouse trials with pasteurized field soil (1253). It also can cause necrosis of hops (1691). *F. crookwellense* often is associated, usually as a minor component of the fungal community, with wheat head blight epidemics in various parts of the world. Scanning electron micrographs suggest that strains of *F. crookwellense* can effectively penetrate wheat heads (225). *F. crookwellense* is associated with seedling death (poor stand establishment) under field conditions and is pathogenic to oat seedlings under laboratory conditions (933). It also may be recovered from maize stems and leaves, but it not closely associated with disease symptoms in this host.

Isolates of *F. crookwellense* are reported to produce nivalenol and zearalenone, but not deoxynivalenol (492, 1446, 2094), but deoxynivalenol-producing strains should occur if mutations in the *Tri13* trichothecene biosynthetic pathway gene similar to those found in *F. graminearum* occur under field conditions. The trichothecene gene cluster of *F. crookwellense* is similar to that of *F. graminearum* and *F. culmorum* and it is possible that inter-specific exchange amongst these species has occurred with respect to this region (341, 2291). Isolates of *F. crookwellense* also can produce fusaric acid (119) and fusarin C (2140).

Current Species Description: Burgess *et al.* (273).

Fusarium culmorum (W.G. Smith) Saccardo

Sexual Stage. None known.

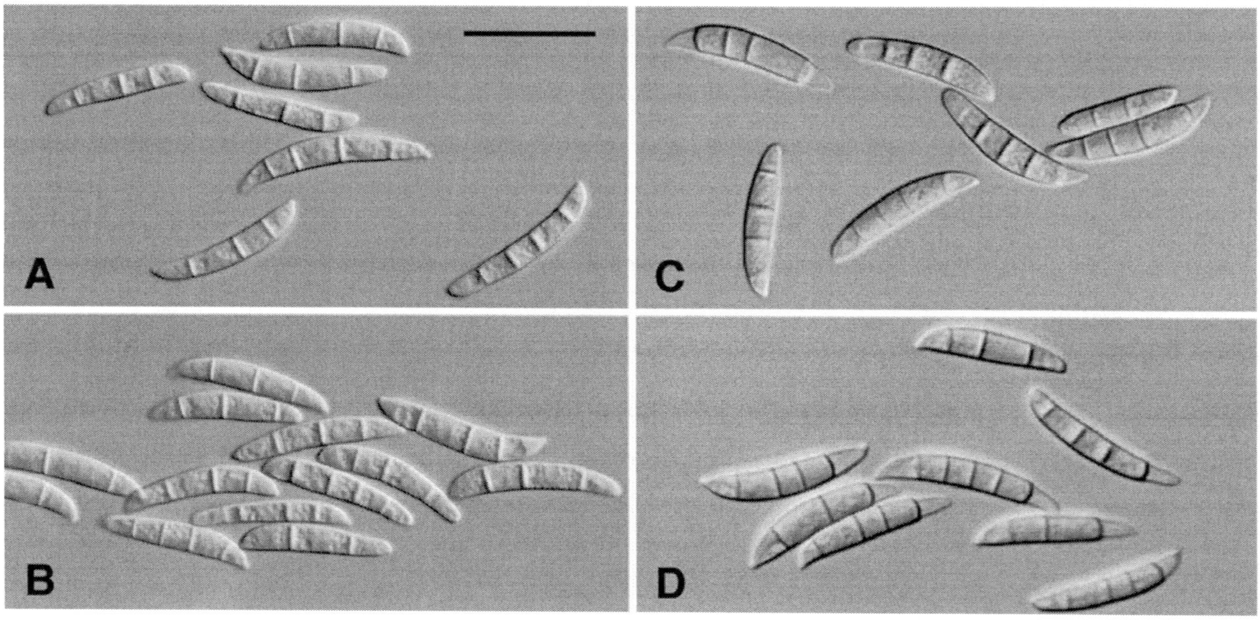

Figure 13-19. *Fusarium culmorum.*
A – D: Macroconidia. Scale bar = 25 μm.

Reported Geographic and Host Distribution

Commonly found in temperate regions. Is associated with cereal crowns and grain, and plant debris in soil.

Media to Examine and Key Characters

Characters on CLA. Macroconidia are formed abundantly in orange sporodochia. The macroconidia are short and stout with a poorly developed foot cell and rounded apical cell. Macroconidia are formed from monophialides on branched conidiophores in sporodochia and to a minor extent from monophialides on the hyphae.

Characters on PDA. *Fusarium culmorum* grows rapidly producing abundant sporodochia in a large central spore mass (1 to 2 cm diameter), that is initially pale orange but becomes brown to dark brown with age. Rings of spore masses may be formed by some isolates under alternating conditions of light and temperature. Most strains form red pigments in the agar, but some may have olive brown mycelium and olive brown pigment in the agar.

Macroconidia

- Sporodochia: Orange to brown color and relatively common.
- General morphology: Robust, relatively short, and thick walled. Widest at the midpoint of the macroconidium. The dorsal side is somewhat curved, but the ventral side is almost straight. Quite wide relative to length.
- Apical Cell morphology: Rounded and blunt.
- Basal Cell morphology: Notched and without a distinct foot shape.
- Number of septa: Usually 3- or 4-septate.
- Abundance: Usually abundant in sporodochia. These macroconidia usually are uniform in shape and size.

Microconidia. Absent.

Chlamydospores

- Abundance/Speed of formation: Usually abundant and form relatively quickly, requiring 3-5 weeks on CLA.

The absence of chlamydospores is not a reliable character for identification.

- Location: In hyphae and macroconidia. Those found in the macroconidia persist longer than those found in the hyphae under field conditions (929).
- Appearance: Found singly, in clumps or chains.

Taxonomy, Pathology and Ecology

Fusarium culmorum may be confused with *F. sambucinum* or *F. crookwellense*, as all can be isolated from similar hosts and climatic regions. The apical cell of *F. culmorum* usually is blunt (rounded) but in some isolates it is slightly papillate which can lead to confusion with those of *F. sambucinum*. The relatively rapid growth rate of *F. culmorum* distinguishes it from *F. sambucinum*, which grows slowly. *F. culmorum* is distinguished from *F. crookwellense* by the shape of the macroconidium. The macroconidia of *F. crookwellense* are longer, have a distinct foot-shaped base to the basal cell, and have a tapering apical cell while *F. culmorum* macroconidia are short, stout and lack a distinctive foot-shaped basal cell. Phylogenetically, *F. culmorum* is most closely related to *F. graminearum* and *F. crookwellense* (1493). The sexual stage of *F. culmorum* probably is heterothallic (1036). In addition to DNA-based differences, these species also may be distinguished by their secondary metabolite profiles (2141).

Fusarium culmorum is most common in temperate regions, and often is associated with cereal foot rots and head blights, particularly in cool parts of Europe. Interactions between *F. culmorum* and the Barley Yellow Dwarf Virus may intensify the disease symptoms observed (1102). Plants with head blights have shrunken "tombstone" kernels, but these diseased kernels are not responsible for disease transmission under field conditions (701). The importance of this fungus as a cause of head blight of wheat may be decreasing, however, with *F. graminearum* becoming the more commonly observed cause of this disease (2274). An ABC transporter protein is required for pathogenesis of *F. culmorum* on wheat, suggesting that wheat synthesizes an antifungal compound that this protein is capable of exporting from the cell (1997).

Although a soil fungus, *F. culmorum* can be splash-dispersed to the heads of some cereals, *e.g.*, wheat and barley (968), where it can cause head blight symptoms and colonize the grain extensively (941). Splash dispersal is important as *F. culmorum* does not appear to be a systemic pathogen of wheat (2009). *F. culmorum* may be pathogenic towards the model plant species *Arabidopsis thaliana* that may enable more rapid studies of host-pathogen interactions for this fungus (2203).

Fusarium culmorum has been isolated at a low frequency from subterranean clover – rye grass pasture soils (598, 719, 887). There are preliminary reports that some strains may be useful in an integrated management program of the aquatic weed *Hydrilla verticillata* (1949, 1950). *F. culmorum* also is a causal agent of diseases of carnations (247), hops (1857), leeks (1106), Norway spruce (997), and strawberries (374), and of reduced seed germination in sorghum (1658).

Although no sexual stage is known for *F. culmorum*, population analyses suggest that recombination is an important part of this fungus's life cycle (1456, 2167). Although both the *MAT-1* and *MAT-2* mating types occur and are transcribed in *F. culmorum* (1037), no sexual stage for this species has been identified. Microsatellite markers are available for use in population analyses of this fungus (706) as are primers and protocols for detecting *F. culmorum* through real-time PCR assays (2275).

A genetic transformation protocol for *F. culmorum* is available (430, 519). Protoplasts of *F. culmorum* that take up DNA of *F. graminearum* may have a significantly altered morphological phenotype (1313). A physical map of the relatively large (116 kb) mitochondrial DNA of *F. culmorum* has been made (1103, 1526). Detailed observations of the process by which the hypal tip grows and the Spitzenkorper have been made (1293, 1294). Macroconidial germination also has been studied in some detail in this fungus (352). *F. culmorum* can reversibly sequester significant amounts of cadmium (1159).

Fusarium culmorum produces steroids (284) and mycotoxins such as moniliformin (1930), deoxynivalenol and related trichothecenes (1341, 1753), fusarin C (600, 2140) and zearalenone (1341). Reports of the production of T-2 toxin and neosolaniol by *F. culmorum* have not been substantiated. Trichothecene production is an important factor in Fusarium head blight of small grains (1439, 2010), and is correlated with the expression of the *tri5* and *tri6* genes (412). Sublethal levels of fungicides may increase mycotoxin production (511), but higher levels of fungicides can stop trichothecene production (412). *F. culmorum* has been associated with dermatitis in humans (2011). Both domesticated animals, *e.g.*, chickens and pigs, and experimental animals, *e.g.*, mice, rabbits, and rats are sensitive to toxins produced by strains of *F. culmorum*, but not all of these toxins have been identified (1341). Mycotoxin biosynthesis appears to be more sensitive to temperature than it is to water availability (1274, 1275).

Current Species Description: Wollenweber & Reinking (2359).

Fusarium decemcellulare Brick

Sexual Stage. *Albonectria rigidiuscula* (Berkeley & Brick) Rossman & Samuels.

Common Synonym. *Fusarium rigidiuscula.*

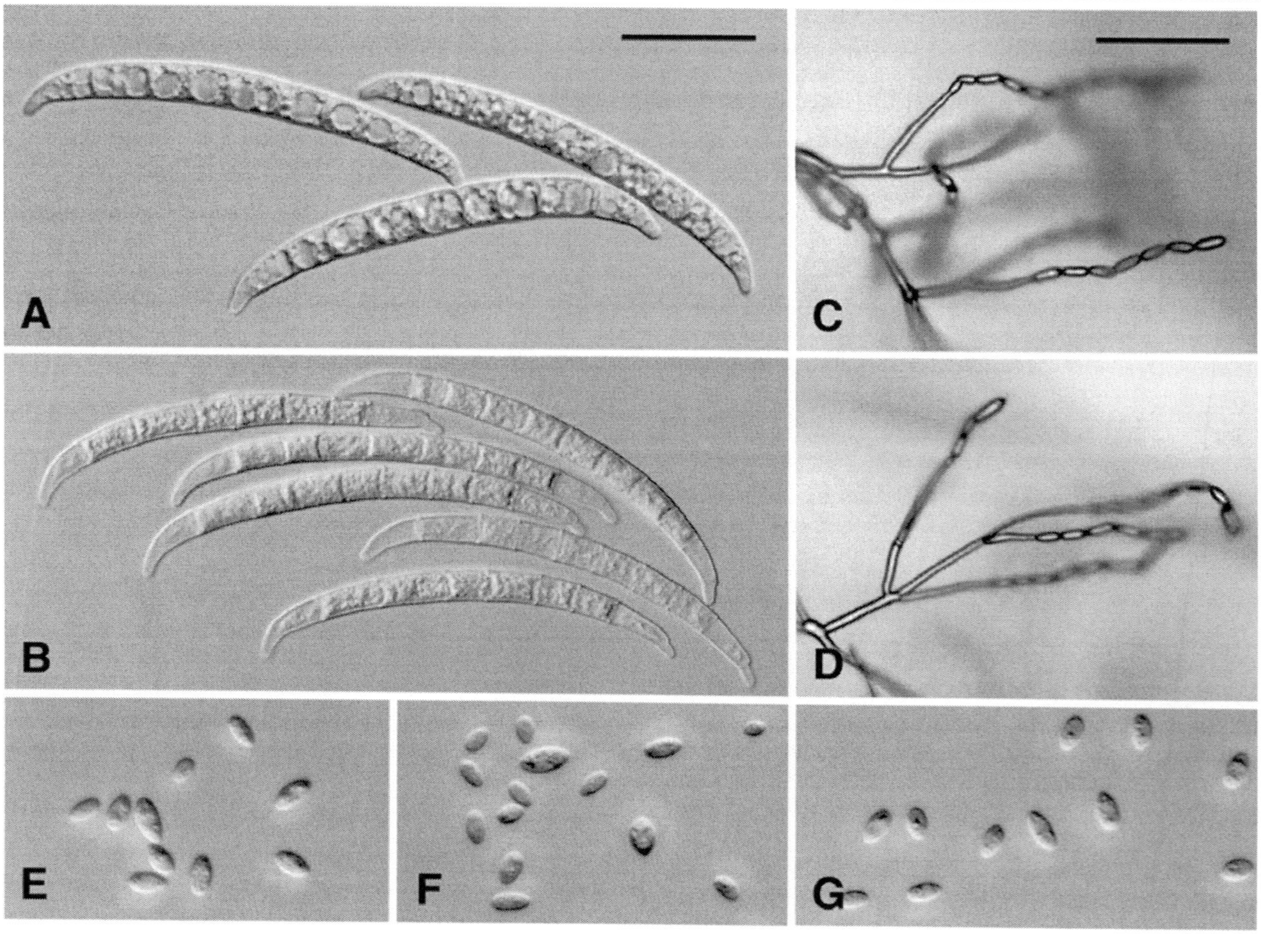

Figure 13-20. *Fusarium decemcellulare.*
A – B: Macroconidia; **C – D:** Microconidia *in situ* on CLA; **E – G:** Microconidia. A – B, E – G, scale bar = 25 µm; C – D, scale bar = 50 µm.

Reported Geographic and Host Distribution

Found more frequently in tropical regions. Is associated with canker diseases of various tree species.

Media to Examine and Key Characters

Characters on CLA. Macroconidia are formed on monophialides found in yellow sporodochia and on hyphae. The macroconidia are rarely confused with those of any other species being very large, both long and wide, either straight or slightly curved, and robust with thick walls. They are 5- to 9-septate with a hooked apical cell and a foot-shaped basal cell. Macroconidia are produced on monophialides on branched conidiophores in sporodochia and on the hyphae. Microconidia are found in long chains produced from monophialides in branched conidiophores or directly from the hyphae. The microconidia are oval in shape and usually 0-septate. Chlamydospores are absent.

Characters on PDA. *Fusarium decemcellulare* is a slow growing fungus on PDA. The mycelium is white to cream colored, but may darken with age. Most characteristically these cultures produce yellow sporodochia from which droplets of exudate may form to give the colony a moist appearance. Red pigments usually are produced in the agar.

Macroconidia

- Sporodochia: Yellow, sometimes with a liquid exudate. Usually common.
- General morphology: Very long, robust and thick walled. Quite wide relative to many other species with even curvature on both sides of the macroconidia.
- Apical Cell morphology: Rounded and blunt.
- Basal Cell morphology: Foot shaped.
- Number of septa: 5- to 9- septate, usually 7- to 9-septate.
- Abundance: Usually abundant in sporodochia. Macroconidia from sporodochia usually are uniform in shape and size.

Microconidia

- Shape/septation: Oval shaped and 0-septate.
- Aerial mycelium presentation: Usually long chains, but occasionally false heads.
- Conidiogenous cells: Monophialides.
- Abundance: Abundant in the aerial mycelia.

Chlamydospores. Absent.

Taxonomy, Pathology and Ecology

Fusarium decemcellulare is rarely confused with any other *Fusarium* species because of the large size of the macroconidia and unique morphological features. The sexual stage, *Albonectria rigidiuscula*, is homothallic (2063).

Fusarium decemcellulare usually is found only in tropical and sub-tropical areas, and has been consistently associated with branch canker and die-back of a range of tropical fruit trees, *e.g.*, avocado (443, 444), cacao (1341, 1777), guarana (21), and mango (1710). Other reports of hosts on which this fungus can cause a disease include *Averrhoa bilimbi* – cucumber tree (154), *Annona squamosa* × *A. cherimola* – atemoya (2155), *Bixa orellana* – annatto (1830), *Cordia alliodora* – Spanish Elm (34), *Melicocca bijuga* – Spanish lime (1830), *Toona ciliata* – Australian red cedar (1296), and *Zizyphus mauritiana* – Indian jujube (2278).

Fusarium decemcellulare can grow on wood treated with borax and sodium pentachlorophenoxide (134) and on ceiling tiles in buildings in humid areas (1259). Fenpropimorph is known to inhibit ergosterol biosynthesis in *F. decemcellulare* (458).

There is a report of *F. decemcellulare* being toxigenic to mice (936), but this report has not been confirmed (1341). Other reports of animal toxigenicity attributed to *F. decemcellulare* have been found to be caused by other *Fusarium* species (1341). The napthoquinone pigments of *F. decemcellulare* are phytotoxic (1414). *F. decemcellulare* carries a gene that encodes a functional trichothecene 3-O-acetyltransferase gene, which can be used as a defense against trichothecene mycotoxins (2156).

Current Species Description: Wollenweber & Reinking (2359).

Fusarium denticulatum Nirenberg & O'Donnell

Sexual stage. None known.

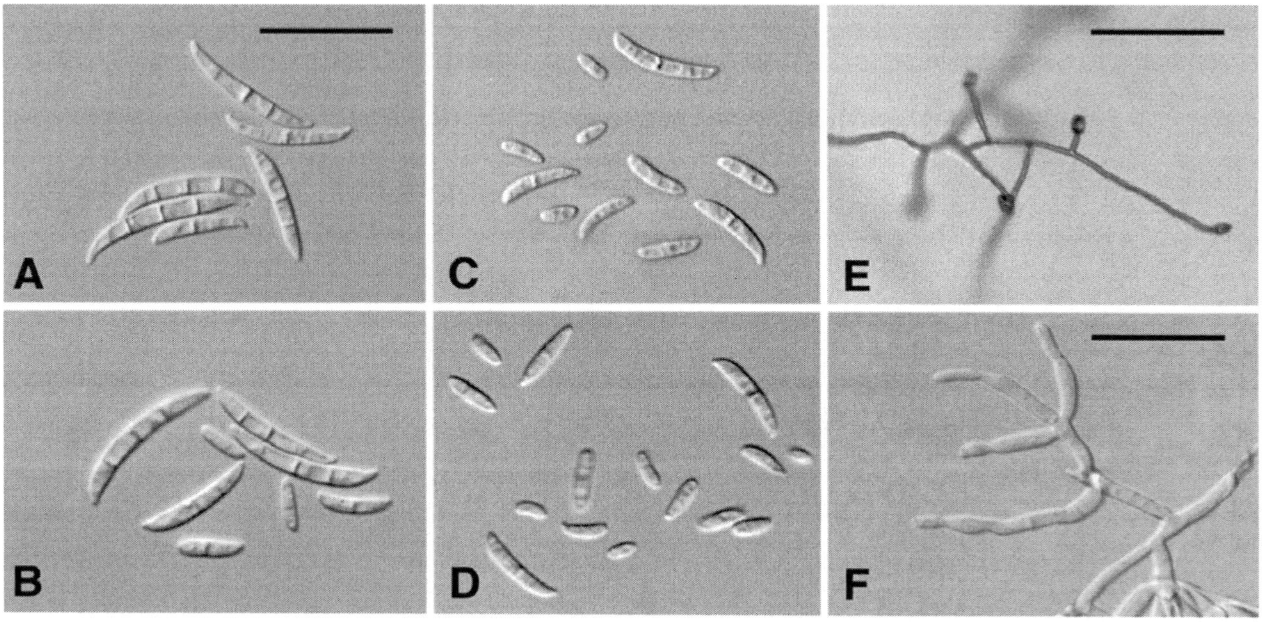

Figure 13-21. *Fusarium denticulatum.*
A – B: Macroconidia; **C – D:** Microconidia; **E – F:** Microconidia *in situ* on CLA. A – D, scale bar = 25 μm; E – F, scale bar = 50 μm.

Reported Geographic and Host Distribution

Recovered from sweet potato (*Ipomoea batatas*) in many parts of the world, *e.g.*, Cuba, USA, Brazil, Zambia, and Indonesia.

Media to Examine and Key Characters

Characters on CLA. Macroconidia typical of those seen in other species of the *Gibberella fujikuroi* species complex are formed on small orange sporodochia on the carnation leaf pieces. The key character is the formation of microconidia and mesoconidia from denticulate polyphialides in the aerial mycelia.

Characters on PDA. Relatively slow growing with white mycelium. Orange and brown pigments are excreted into the medium. The center of the colony often is darkly pigmented.

Macroconidia

- Sporodochia: Small orange sporodochia, usually on the carnation leaf pieces.
- General morphology: Relatively slender and slightly curved. Typical of species in the *Gibberella fujikuroi* species complex.
- Apical Cell morphology: Beaked.
- Basal Cell morphology: Foot shaped.
- Number of septa: 3- to 5-septate.
- Abundance: Relatively abundant in sporodochia, but sporodochium formation may be sparse in some isolates.

Microconidia/Mesoconidia

- Shape/septation: Long, oval to obovoid. 0- to 1-septate. Fusiform mesoconidia that are up to 3-septate also may be observed.

- Aerial mycelium presentation: False heads.
- Conidiogenous cells: Monophialides and polyphialides with characteristic denticulate (finger-like) phialides that are diagnostic and are the basis for the species epithet.
- Abundance: Abundant in the aerial mycelia

Chlamydospores. Absent.

Taxonomy, Pathology and Ecology

Fusarium denticulatum causes chronic leaf distortion of sweet potato (377, 730, 1545). These isolates were initially identified as *F. lateritium* (1545). *F. denticulatum* and *F. lateritium* are similar on PDA in terms of pigmentation and lobed colony margins. *F. lateritium* never produces conidia from polyphialides in the aerial mycelium. *F. denticulatum* produces distinctive microconidia on polyphialides that have consistent rectangular denticle-like openings to the polyphialides. These denticle-like conidiogenous openings are unique to this species.

Isolates of *F. denticulatum* can produce moniliformin, but are not known to produce detectable levels of beauvericin, fumonisins, or fusaproliferin (638).

Current Species Description: Nirenberg & O'Donnell (1570).

Fusarium dimerum Penzig

Sexual Stage. None known.

Common Synonyms. *Fusarium episphaeria, Microdochium dimerum.*

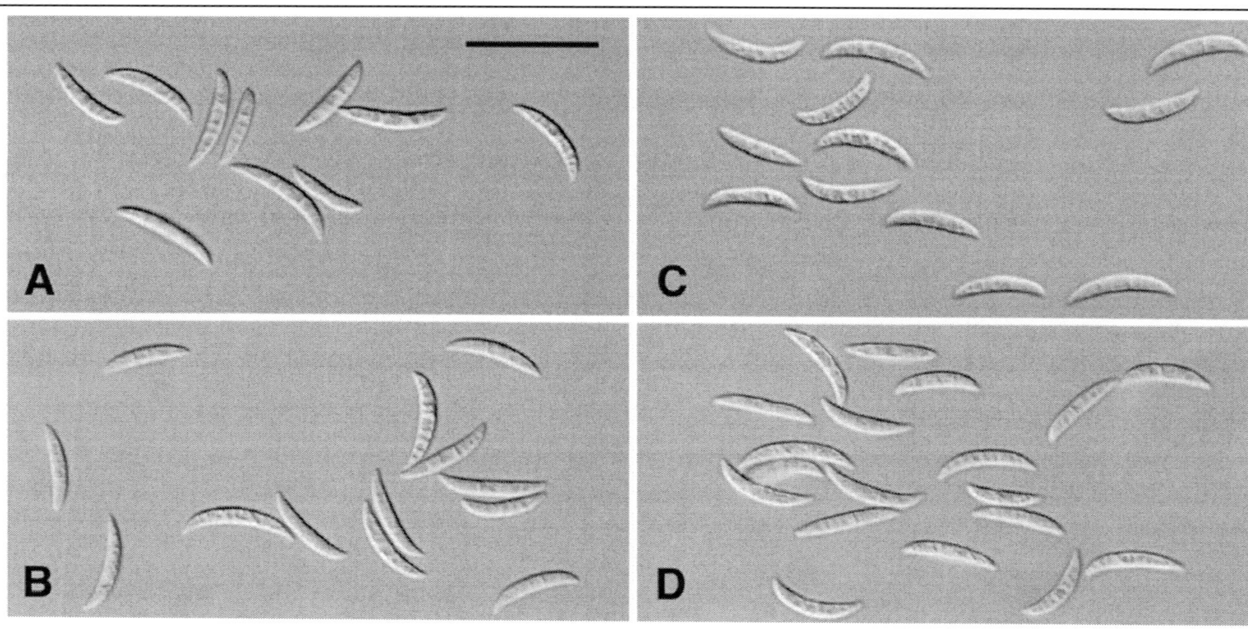

Figure 13-22. *Fusarium dimerum.*
A – D: Macroconidia. Scale bar = 25 µm.

Reported Geographic and Host Distribution

Commonly recovered from soil in various geographic regions. Important as a human pathogen.

Media to Examine and Key Characters

Characters on CLA. Sporodochia are absent and macroconidia are found in the mycelium across the colony. Macroconidia are formed on monophialides on branched conidiophores. The macroconidia are small, distinctly curved, and 1- to 2-septate. No microconidia. Chlamydospores are rare.

Characters on PDA. Very distinctive, slow-growing, slimy, almost yeast-like colonies that are white to orange in color. There is no obvious aerial mycelium. Little or no pigments are excreted into the media.

Macroconidia

- Sporodochia: No distinct sporodochia.
- General morphology: Very short with even curvature on both sides of the spore. Quite wide relative to length, and widest in the center of the macroconidium
- Apical Cell morphology: Rounded and often hooked.
- Basal Cell morphology: Blunt (rounded) or may be slightly notched.
- Number of septa: Usually 0- or 1-septate. Occasionally 2-septate.
- Abundance: Usually abundant across the colony.

Microconidia. Absent.

Chlamydospores

- Abundance/Speed of formation: May be rare and difficult to find. Usually present after 6-8 weeks of incubation on CLA. The absence of chlamydospores is not a diagnostic character.
- Location: In mycelia on the agar surface and in hyphae submerged in the agar.

- Appearance: Singly, in pairs, and in short chains.

Taxonomy, Pathology and Ecology

This species is very distinctive and there are few other *Fusarium* species that it will be confused with, although the macroconidia resemble those of *Microdochium nivale*. The morphology of the colonies of these two species on PDA, however, is quite different. Von Arx (2265) proposed that *F. dimerum* be reclassified as *M. dimerum* because it appeared congeneric with *M. nivale*, but this proposal has not received significant support. *F. dimerum* has relatively little in common with most of the *Fusarium* species that are regularly isolated.

Fusarium dimerum is in general of minor importance for production agriculture, but has been reported to be present at high levels in wheat (657).

Fusarium dimerum has been associated with several human diseases. In particular it is associated with keratomycoses and other eye infections (659, 2263, 2404), endocarditis (301), and disseminated infections in immunocompromised patients (95, 181, 210, 1135, 1237). Fluconazole and flucytosine are not effective against *F. dimerum*, and the effectiveness of amphotericin B is isolate dependent (1748). There is a report of the successful treatment of a *F. dimerum* infection with voriconazole (181).

Little is known of the biology of this species. The fungus has been recovered, usually from soil, from a number of geographic locations, *e.g.*, Australia (1889), Bahrain (1329), Brazil (657), California (2059), Great Britain (1408), and North Dakota (1410), and thus has a broad climatic range.

Current Species Description: Wollenweber & Reinking (2359).

Fusarium dlamini Marasas, Nelson & Toussoun

Sexual Stage. None known.

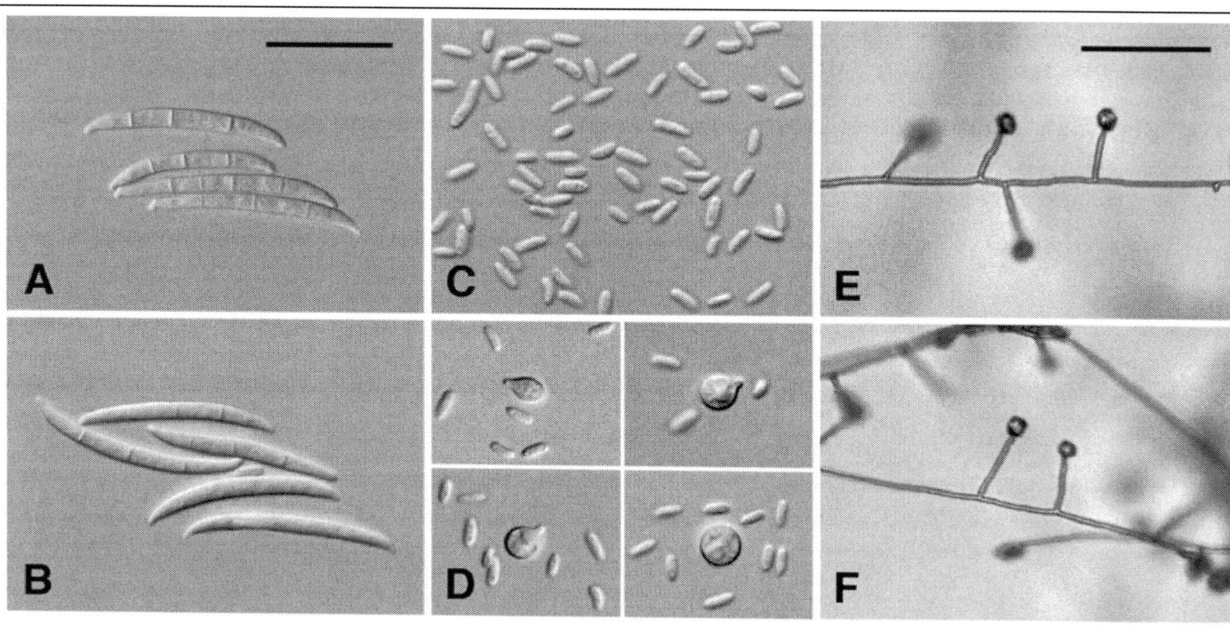

Figure 13-23. *Fusarium dlamini.*
A – B: Macroconidia; **C:** Oval microconidia; **D:** Napiform microconidia; **E – F:** Microconidia *in situ* on CLA. A – D, scale bar = 25 μm; E – F, scale bar = 50 μm.

Reported Geographic and Host Distribution

Reported from plant debris in soil in southern Africa.

Media to Examine and Key Characters

Characters on CLA. Abundant orange sporodochia on carnation leaf pieces produce macroconidia typical of the *Gibberella fujikuroi* species complex. Two types of microconidia, fusiform and napiform, are produced in false heads from monophialides in the aerial mycelia.

Characters on PDA. The mycelium is floccose to powdery. It is white in color and may become light to dark purple with age. Purple pigments are produced in the agar.

Macroconidia

- Sporodochia: Orange color. Found on carnation leaf pieces.
- General morphology: Moderately long. Hyaline and thin walled. Falcate to straight.
- Apical Cell morphology: Curved and tapering.
- Basal Cell morphology: Foot shaped.
- Number of septa: 3- to mostly 5-septate.
- Abundance: Abundant in sporodochia.

Microconidia

- Shape/septation: Produces two types of microconidia – allantoid to fusiform (0-septate) and napiforme (0- to 1-septate). Most of the microconidia are fusiform. Close observation of the culture may be necessary to observe the napiform microconidia.
- Aerial mycelium presentation: False heads.
- Conidiogenous cells: Monophialides.
- Abundance: Abundant in the aerial mycelia.

Chlamydospores

- Abundance/Speed of formation: Produced relatively abundantly after 4-6 weeks on CLA.

- Location: In aerial or submerged hyphae. May be terminal or intercalary.
- Appearance: Solitary, or in pairs, chains or clumps. Hyaline to pale brown. Smooth or rough walled.

Taxonomy, Pathology and Ecology

Fusarium dlamini was first described (1342) from cultures isolated from plant debris present in soil from maize fields in South Africa, a finding confirmed by Jeschke *et al.* (974). Since then *F. dlamini* also has been reported from maize in Argentina (1319). *F. dlamini* may cause a pre-emergence disease problem for Norway Spruce (1483).

This species has some similarity to a number of species in the *Gibberella fujikuroi* complex and to *F. oxysporum*. *F. dlamini* can be distinguished from *F. oxysporum* because *F. dlamini* produces napiforme microconidia. *F. dlamini* resembles *F. anthophilum* in the production of napiforme microconidia, but *F. anthophilum*, does not produce chlamydospores. Phylogenetically, the species most closely related to *F. dlamini* is *F. acutatum* (1596).

This species does not produce moniliformin and was not toxic in duckling assays (1350). Some isolates can produce beauvericin (1283), moniliformin (347) and fumonisins (1548).

Current Species description: Marasas *et al.* (1342).

Fusarium equiseti (Corda) Saccardo

Sexual Stage. *Gibberella intricans* Wollenweber.

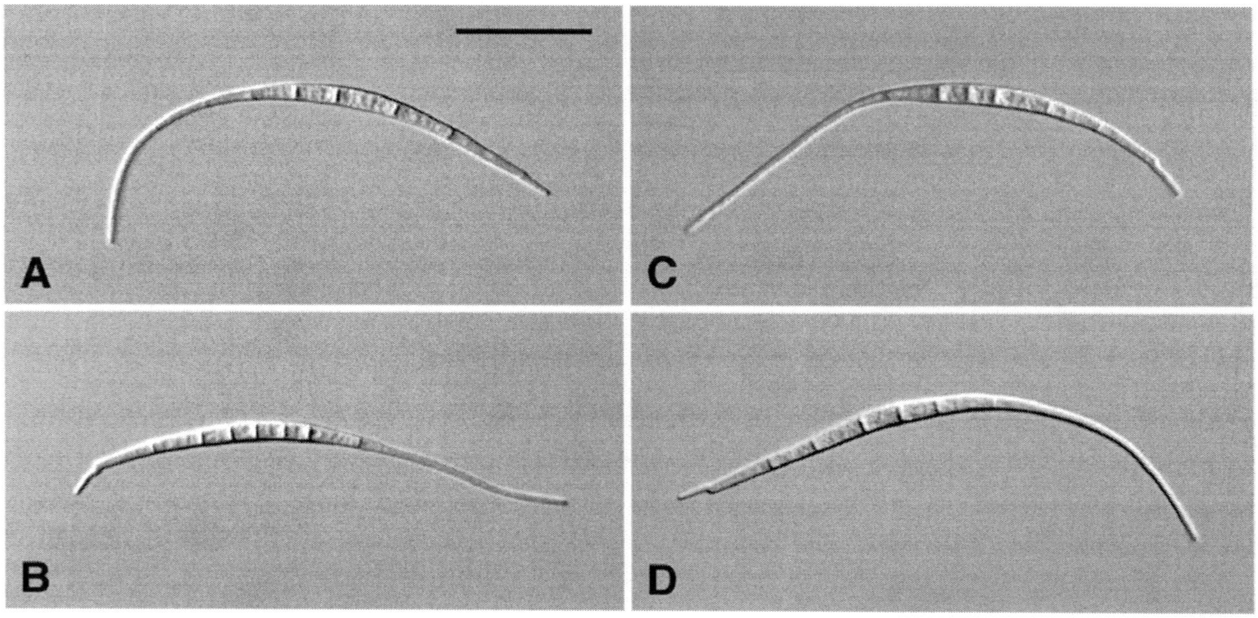

Figure 13-24. *Fusarium equiseti.*
A – D: Macroconidia. Scale bar = 25 µm.

Reported Geographic and Host Distribution

Cosmopolitan in areas ranging from cool and temperate to hot and arid, primarily as a saprophyte or secondary invader.

Media to Examine and Key Characters

Characters on CLA. Abundant macroconidia form in orange sporodochia. Use only macroconidia from the sporodochia for identification purposes. The thick-walled macroconidia have strong dorsiventral curvature and usually are 5- to 7-septate, with a distinctly foot-shaped basal cell and a tapering apical cell that may be filamentous or whip-like. Macroconidial length varies from 25 to > 120 µm. Macroconidia are produced from monophialides on branched conidiophores in sporodochia. No microconidia are produced. Chlamydospores are formed relatively rapidly on CLA. The chlamydospores become verrucose and pale brown with age and may give a CLA culture a golden or brown appearance.

Characters on PDA. Abundant mycelium that initially is white, but becomes brown with age. Sporodochia may form in a central spore mass but may not be obvious since they can be obscured by the mycelium. The spore mass may be pale orange to dark brown and annular zonations may develop in response to a light-dark cycle. This species forms a pale brown to dark brown pigment where the colony contacts the agar. Dark brown spots or flecks of pigment usually are formed in the agar.

Macroconidia

- Sporodochia: Orange color. Found on carnation leaf pieces in CLA. The macroconidia from the sporodochia usually are uniform in shape and size and should be used for identification purposes for this fungus.
- General morphology: Long to very long and slender. The macroconidia have pronounced dorsiventral curvature typical of those from species in section *Gibbosum*.
- Apical Cell morphology: Tapered and elongate or even whip-like.
- Basal Cell morphology: Prominent foot shape that may be elongated in appearance.

- Number of septa: Usually 5- to 7-septate.
- Abundance: Usually abundant in sporodochia

Microconidia. Absent. Nelson *et al.* (1551) report that some isolates of *F. equiseti* produce microconidia but these authors also included *F. scirpi* in their definition of *F. equiseti*. Gerlach & Nirenberg (697) accepted both *F. equiseti* and *F. scirpi* and in their description of *F. equiseti* state that only macroconidia are formed.

Chlamydospores

- Abundance/Speed of formation: Usually abundant after 2-6 weeks on CLA.
- Location: In aerial hyphae and hyphae in the agar.
- Appearance: May be found singly or in chains or in clumps. A brown pigment and thick verrucose walls appear with time.

Taxonomy, Pathology and Ecology

Fusarium equiseti can be confused with a number of other *Fusarium* species. Nelson *et al.* (1551) included *F. scirpi* in their description of *F. equiseti*, with Burgess *et al.* (276) formally separating the two when the description of *F. scirpi* was emended. *F. equiseti* is most easily confused with *F. compactum* (brown pigmented isolates), *F. semitectum*, and *F. scirpi*. *F. compactum* and *F. equiseti* are distinguished on the basis of growth rate on PDA. *Fusarium scirpi* produces abundant microconidia on polyphialides whereas *F. equiseti* produces no microconidia. *F. semitectum* produces abundant spindle-shaped macroconidia from polyphialides that are never seen in cultures of *F. equiseti*. Some isolates also produce very long macroconidia with a filamentous or whip-like apical cell and may resemble the macroconidia formed by *F. longipes*. PDA cultures of *F. longipes* produce red pigment in the agar, however, while PDA cultures of *F. equiseti* produce brown pigment. *F. equiseti* may be a species complex, but molecular data are needed to rigorously test this possibility.

Fusarium equiseti is a cosmopolitan soil inhabitant that has been recovered from many parts of the world and is equally effective when colonizing plant roots at either 13°C or 30°C (1894). As a root colonizer it may reduce or prevent the association of a mycorrhizal fungus with the root (1395). *F. equiseti* is especially common in drier areas, *e.g.*, Bahrain (1327) or Israel (984), although it also can be recovered from soil near the Arctic Circle (1116). It is a common colonizer of senescent or damaged plant tissue. It is not unusual to recover *F. equiseti* from diseased plants, however records of *F. equiseti* as a pathogen should be treated cautiously. The fungus has been associated with a few diseases such as cankers of sour cherry trees (1621), rots of pumpkin (567) or cucurbit fruits in contact with soil (19, 987), and diseases of date palm (6), but in many cases disease reports are based only on the recovery of the fungus from the diseased tissue, and not on the completion of Koch's postulates. *F. equiseti* may produce enough linamarase enzyme (926) or ursodeoxycholic acid (1152) to be considered as a commercial biosynthetic source for these compounds.

With respect to humans, *F. equiseti* is known to be allergenic (2240, 2244), and may be recovered as a human pathogen (718). It is resistant to fluconazole and flucytosine (1748), although it is sensitive to a combination of flurocytosine and amphotericin B (1938). The cyclomunine cylic peptide produced by *F. equiseti* has anti-lymphoblastic activity, which could increase the potential of this fungus to cause disease in humans (1558).

Fusarium equiseti is associated with toxicities to cattle, chickens, ducklings, guinea pigs, mice, pigeons, pigs, rabbits and rats (1341). In numerous cases, the toxins responsible for animal disease symptoms have not been identified, but the toxins known to be produced by at least some strains of *F. equiseti* include: butenolide (1341), beauvericin (1283), trichothecenes such as diacetoxyscirpenol, nivalenol and T-2 toxin (1341), equisetin (854, 1341), fusarochromanone (2371), and zearalenone (854, 1341, 1539). Equisetin (1694), a poison of rat mitochondria (1105), may be important as a plant pathogenicity factor (2308). Some strains also produce steroids of unknown significance (284).

Current species description: Wollenweber & Reinking (2359).

Fusarium foetens Schroers, O'Donnell, Baayen & Hooftman

Sexual Stage. None known.

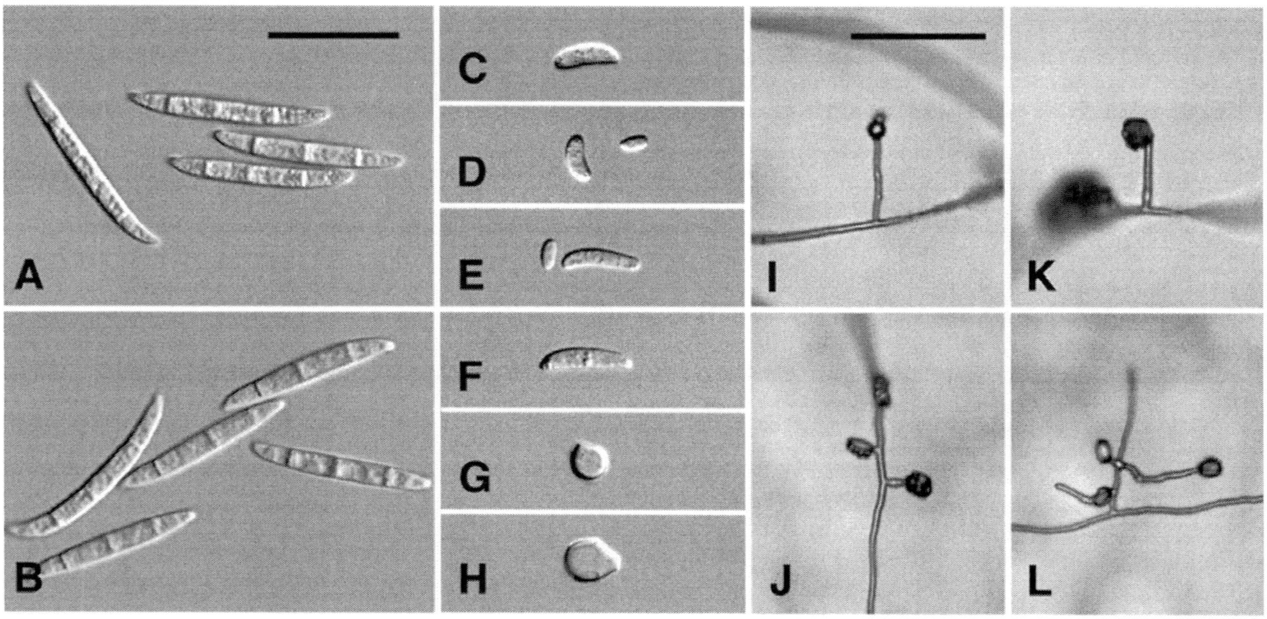

Figure 13-25. *Fusarium foetens.*
A – B: Macroconidia; **C – H:** Microconidia; **I – L:** Microconidia *in situ* on CLA. A – H, scale bar = 25 µm; I – L, scale bar = 50 µm.

Reported Geographic and Host Distribution

Recovered from *Begonia* sp. in nurseries in the Netherlands and Germany.

Media to Examine and Key Characters

Characters on CLA. Mycelia are most abundant on and adjacent to the carnation pieces. Mycelia may grow as "ropes" of intertwined hyphae. Microconidia are abundant in the aerial mycelia. Pale orange sporodochia contain macroconidia that are nearly straight, and usually 3-septate with a curved apical cell and an indistinct basal cell.

Characters on PDA. Abundant white mycelia may be present as tufts or as an irregular pattern over the surface of the plate and produce red to brown pigments in the agar. These colonies may have a strong odor (stench), which results in the species epithet – *foetens* – for stinking.

Macroconidia

- Sporodochia: Pale to light orange on the agar surface. Color develops best if colonies are exposed to near-UV light during growth. They may be unpigmented if incubated in the dark.
- General morphology: Falcate. The central two cells frequently are almost straight and are widest in the middle.
- Apical Cell morphology: Curved.
- Basal Cell morphology: Rounded to barely foot shaped.
- Number of septa: Usually 3-septate, but may be 3- to 5-septate. Macroconidia with 4 or 5 septa are not noticeably longer than those with only 3 septa.
- Abundance: Abundant in sporodochia.

Microconidia/Mesoconidia

- Shape/septation: Microconidia are ovoid to ellipsoid and usually 0-septate. Mesoconidia are fusiform, but without a distinct foot cell, and may be up to 3-septate.

- Aerial mycelium presentation: False heads.
- Conidiogenous cells: Usually monophialides, but occasionally polyphialides.
- Abundance: Microconidia are abundant in the aerial mycelia. Mesoconidia are rare.

Chlamydospores

- Abundance/Speed of formation: Frequency varies depending on strain. Absence of chlamydospores is not a diagnostic character.
- Location: Usually at the end of hyphae.
- Appearance: Usually singly. May be smooth or verrucose.

Taxonomy, Pathology and Ecology

This species was described from isolates recovered from *Begonia* × *hiemalis* showing symptoms of tracheomycosis in nurseries in The Netherlands and Germany although it may also have spread to some nurseries in the Southern hemisphere (1926). There are no reports of this disease in Europe before 2000, although the host has been cultured there for at least 100 years. Schroers *et al.* (1926) completed Koch's postulates with established isolates of this species, confirming its pathogenicity. The susceptibility of other *Begonia* species to *F. foetens* is unknown, but may of the species also cultured in commercial greenhouses with the host do not appear to be susceptible to this pathogen. It is possible that some infections are latent or endophytic, which may aid the spread of the pathogen.

Morphologically *F. foetens* is most similar to strains of *F. oxysporum*. *F. foetens*, however, produces polyphialides (although not abundantly) and may produce long monophialides, characters that are never observed in strains of *F. oxysporum*. *F. foetens* also produces a distinctive odor on PDA (but not CLA or SNA). This odor is even stronger if the fungus is cultured on Oatmeal Agar. Both *F. begoniae* and *F. foetens* are pathogens of *Begonia*, but these species can be easily distinguished by the presence of chlamydospores, the short monophialides, and distinctive odor of *F. foetens*.

Phylogenetically, *F. foetens* also is closely related to *Fusarium oxysporum* (1926). The two species can be readily separated based on DNA sequences, *e.g.*, β-tubulin, translation elongation factor-1α, and the mitochondrial small subunit ribosomal DNA. Many of the strains identified by Schroers *et al.* (1926) have the same RAPD banding pattern, and the possibility that these strains are clones and not genetically independent individuals has not been rigorously excluded.

Current Species Description: Schroers *et al.* (1926).

Fusarium fujikuroi Nirenberg

Sexual Stage. *Gibberella fujikuroi* (Sawada) Ito in Ito & K. Kimura.

Common Synonyms. *Fusarium proliferatum*, *Gibberella fujikuroi* Mating Population C.

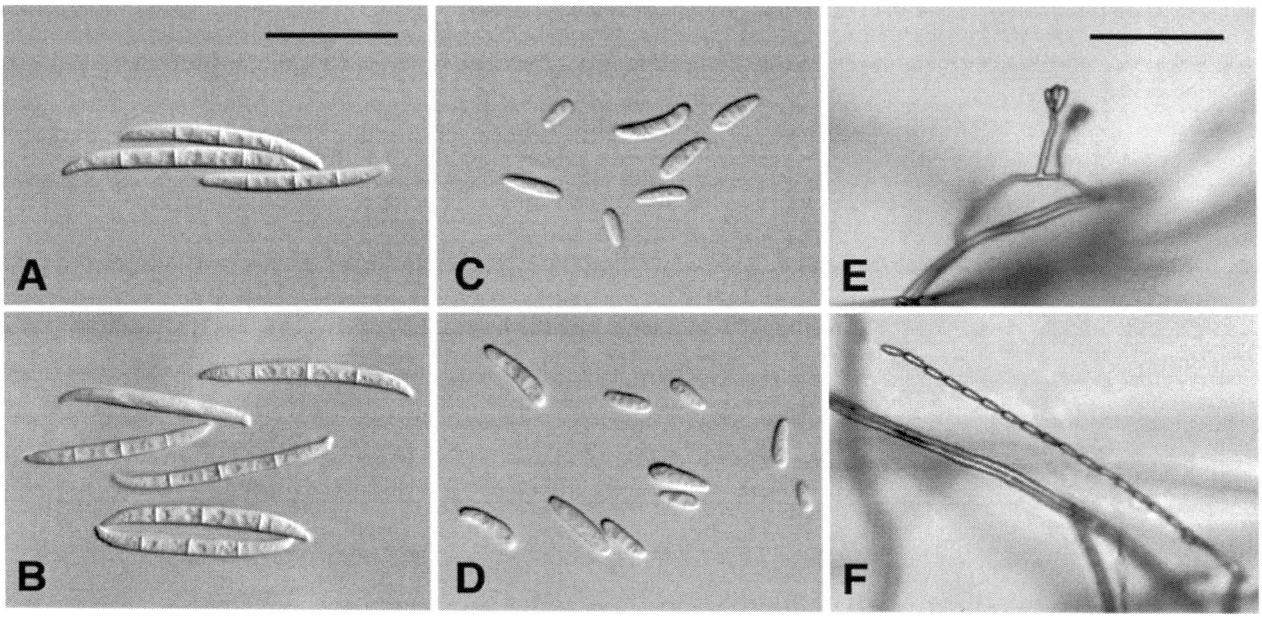

Figure 13-26. *Fusarium fujikuroi.*
A – B: Macroconidia; **C – D:** Microconidia; **E – F:** Microconidia *in situ* on CLA. A – D, scale bar = 25 µm; E – F, scale bar = 50 µm.

Reported Geographic and Host Distribution

Primarily with rice as the cause of bakane disease.

Media to Examine and Key Characters

Characters on CLA. Pale orange sporodochia may be produced that often are obscured by mycelium and chains of microconidia. Macroconidia are long, slender, usually 3- to 5-septate, and thin walled. Macroconidia are produced on monophialides of branched conidiophores in the sporodochia and only rarely from monophialides on hyphae. Microconidia are abundant in false heads or chains from polyphialides, which may proliferate, and less often from monophialides. The microconidia are oval to club shaped, usually 0- to 1-septate, and have a flattened base. Microconidial chains usually are shorter than those of *F. verticillioides* or *F. proliferatum* and often are formed in pairs from polyphialides to give a characteristic 'V' shape when observed *in situ*.

Characters on PDA. *Fusarium fujikuroi* is very similar in colony morphology to *F. verticillioides* and almost indistinguishable from *F. proliferatum*. It forms white mycelia that become gray, violet or magenta with age. Sporodochia normally are absent but, if present, they are pale orange. Dark colored sclerotia may develop in some isolates, but these are more likely indications of potential female fertility than they are of a species diagnosis. Pigmentation in the agar varies and ranges from no pigmentation or grayish orange in some isolates to violet grey, dark violet or dark magenta (almost black) in others.

Macroconidia

- Sporodochia: May produce orange sporodochia. When sporodochia are present, the macroconidia they contain usually are uniform in shape and size. Many isolates form very few or no sporodochia or lose this ability following repeated subculturing.

- General morphology: Relatively slender, medium length with no significant curvature. Typical of species in the *Gibberella fujikuroi* species complex.
- Apical Cell morphology: Tapered.
- Basal Cell morphology: Poorly developed.
- Number of septa: 3- to 5-septate.
- Abundance: Abundant in sporodochia.

Microconidia

- Shape/septation: Oval or club shaped with a flattened base and 0- to 1-septate. A few pyriform microconidia may be present in some isolates.
- Aerial mycelium presentation: False heads and chains of short to medium length.
- Conidiogenous cells: Polyphialides (commonly) and monophialides.
- Abundance: Abundant in the aerial mycelia.

Chlamydospores. Chlamydospores are absent. Note that some swollen cells may develop in the hyphae of this species and appear, superficially, like chlamydospores or pseudochlamydospores.

Taxonomy, Pathology and Ecology

Fusarium fujikuroi is effectively morphologically indistinguishable from *F. proliferatum* and can be accurately distinguished only by making tests of sexual cross fertility or through DNA sequencing. The sexual stage, *G. fujikuroi*, is known to occur under field conditions (2297) and numerous vegetative compatibility groups, which could result from sexual recombination, are found under field conditions (2088). Phylogenetically, *F. proliferatum* and *F. fujikuroi* are very closely related as well (1596, 1690), although their karyotypes are different (2374). The separation between *F. fujikuroi* and *F. proliferatum* with respect to mating is not complete (1236), and naturally occurring hybrids between these two species have been recovered (1234). Electron microscope studies of conidium formation have also been made to identify characters that can be used to distinguish *F. fujikuroi* from other species in the *G. fujikuroi* species complex (2147).

Studies of the genetics and molecular biology of *Fusarium fujikuroi* are relatively numerous. Many genes have been cloned and characterized, with the most notable including those related to carotenoid (1265, 1730) and gibberellic acid biosynthesis (331, 2184, 2187), and the regulation of nitrogen (1445, 2123, 2185) and carbon (2186) metabolism. *F. fujikuroi* requires transforming plasmids to contain a piece of *F. fujikuroi* DNA, and transformants that result from recombination occur primarily, if not exclusively, at a chromosomal site that contains the *F. fujikuroi* DNA on the transforming plasmid (617). Adding the *AMA1* sequence from *Aspergillus nidulans* to a vector used to transform *F. fujikuroi* enables autonomous replication of the vector in the transformed strain and approximately doubles the rates at which transformants are recovered (258).

This fungus causes "bakanae", or foolish seedling, disease of rice in a number of rice growing areas. The disease is characterized by the formation of elongated stems by the rice plant in response to gibberellins (2088). These gibberellins also appear to stimulate fungal spore germination and the elongation of young hyphae (1527). The gibberellic acid pathways in plants and *F. fujikuroi* have many common steps, but also differ significantly, suggesting that the biosynthetic pathways in plants and fungi have evolved separately and not simply by horizontal gene transfer (842, 2183). Gibberellic acid biosynthesis is triggered by light (988) and has been studied intensively from both physiological *e.g.*, Candau *et al.* (303), Duran-Paramo *et al.* (529), Fernandez-Martin *et al.* (618), Giordano *et al.* (705), Krasnapol'skaya *et al.* (1133), Machado *et al.* (1306), Oller-Lopez *et al.* (1620), Rojas *et al.* (1825), Tudzynski *et al.* (2188) and Urrutia *et al.* (2205), and molecular biological perspectives, *e.g.*, Linnemanstons *et al.* (1266), Malonek *et al.* (1324) Sanchez-Fernandez *et al.* (1884) and Tudzynski *et al.* (2184, 2187). *F. fujikuroi* can be fermented commercially for gibberellic acid production (1154, 1213, 1752, 1975, 1976) and is a potential commercial source of carotenoids (675).

Fusarium fujikuroi produces moniliformin, and is widely used as a source of this toxin in animal feeding studies, *e.g.*, Broomhead *et al.* (251), Harvey *et al.* (824) and Morgan *et al.* (1477). It also can produce high levels of beauvericin (1474) and fusaric acid (119). *F. fujikuroi* carries a gene that encodes a functional trichothecene 3-O-acetyltransferase gene, which can be used as a defense against trichothecene mycotoxins (2156). Some strains carry the genes required for fumonisin biosynthesis and can produce low levels of this toxin (1740).

Current Species Description: Nirenberg (1566).

Fusarium globosum Rheeder, Marasas & Nelson

Sexual Stage. None known.

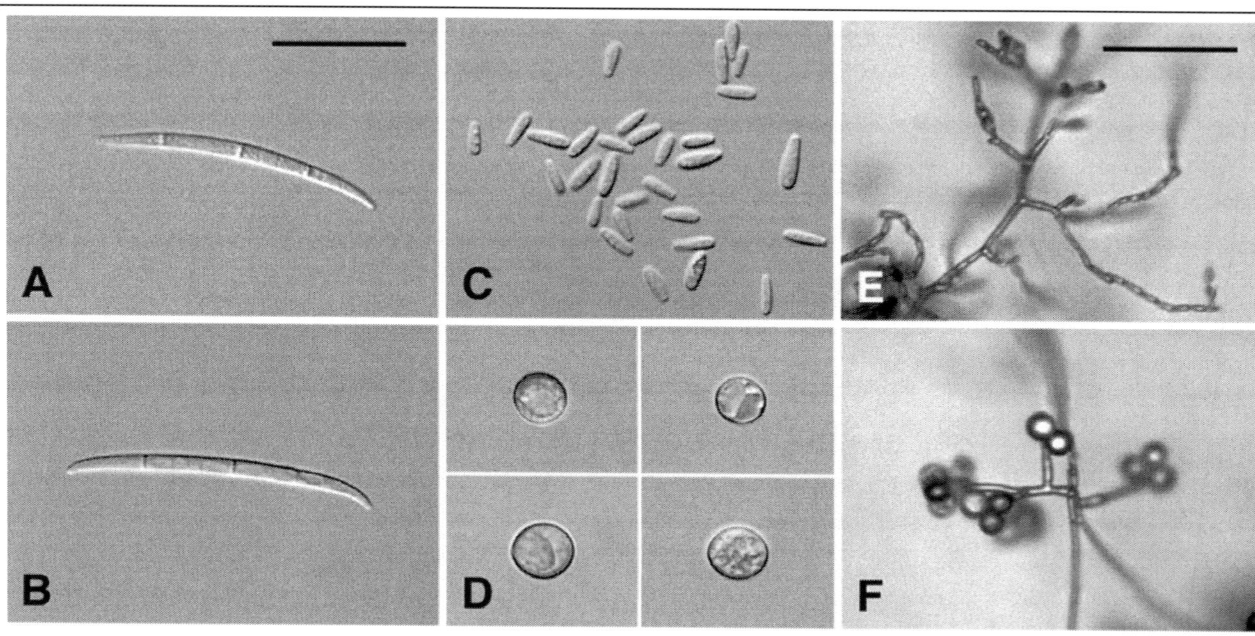

Figure 13-27. *Fusarium globosum.*
A – B: Macroconidia; **C:** Oval microconidia; **D:** Globose microconidia; **E:** Oval microconidia *in situ* on CLA; **F:** Globose microconidia *in situ* on CLA. A – D, scale bar = 25 μm; E – F, scale bar = 50 μm.

Reported Geographic and Host Distribution

Maize in southern Africa and wheat in Japan.

Media to Examine and Key Characters

Characters on CLA. Produces orange sporodochia, with speed of production dependent upon culture conditions. Macroconidia may be found in sporodochia or be produced by hyphae growing on the agar surface. Macroconidia are thin walled, slender, and straight to slightly curved with a foot-shaped basal cell and a slightly curved apical cell. Produces three types of microconidia of which the oval-clavate form may be found in short and lengthy chains arising from monophialides and polyphialides. Globose microconidia are usually diagnostic and are most common either singly or in small clumps. Chlamydospores are absent.

Characters on PDA. Mycelia are floccose and initially white, but may become violet with age. Violet pigments are formed in the agar and may darken with age.

Macroconidia

- Sporodochia: Medium to bright orange color. Produced slowly (4-6 weeks) on carnation leaf pieces in CLA plates incubated under a mixture of white and black light. Produced more abundantly and more rapidly (2-4 weeks) on CLA slants incubated under black light only.
- General morphology: Thin walled, slender and straight to slightly curved. Typical of species in the *Gibberella fujikuroi* species complex.
- Apical Cell morphology: Slightly curved.
- Basal Cell morphology: Foot shaped.
- Number of septa: 3- to 5-septate, primarily 3-septate.

- Abundance: Produced in the aerial mycelia, but more common in sporodochia.

Microconidia

- Shape/septation: Produces three types of microconidia – clavate with a truncate base (0- to 3-septate), napiforme/pyriform (0-septate), and globose (0- to 1-septate) which often have a distinct papilla.
- Aerial mycelium presentation: Clavate microconidia are produced in chains which may be very long (> 50 microconidia) or quite short (< 10 microconidia). The napiforme/pyriform and globose microconidia are borne singly or in clumps.
- Conidiogenous cells: Monophialides and polyphialides.
- Abundance: Oval microconidia in chains are abundant in the aerial mycelia. The napiform and globose microconidia usually are much rarer.

Chlamydospores. Absent.

Taxonomy, Pathology and Ecology

This species was described by Rheeder *et al.* (1802) following the recovery of isolates from maize kernels in South Africa in 1992 and also is known from wheat in southern Japan (70).

Fusarium globosum bears some similarity to *F. verticillioides* and *F. proliferatum*, especially if the microconidial chains are relatively long. Even though all three species can be recovered from maize, the presence of globose microconidia clearly distinguishes *F. globosum* from the other two species. A preliminary differentiation of these species can be made by culturing contaminated maize kernels on malt extract agar with *F. globosum* identified based on denser aerial mycelia and fewer and shorter microconidial chains. The globose microconidia may resemble those found in *F. beomiforme*, but *F. beomiforme* produces chlamydospores and lacks polyphialides and microconidial chains.

Spore morphology is sensitive to light conditions with clavate conidia elongated, aerial globose conidium production suppressed, and falcate sporodochial conidia produced under continuous black light (70). The napiform/pyriform microconidia are not diagnostic and are thought to be a precursor to the globose microconidia (1802).

Fusarium globosum produces fumonisins (1956, 2095) and is known to carry the *FUM* gene cluster required for fumonisin biosynthesis (1740). Strains of *F. globosum* also produce low levels of beauvericin (1956), and some strains also produce fusaproliferin (1956).

Current Species Description: Rheeder *et al.* (1802).

Fusarium graminearum Schwabe

Sexual Stage. *Gibberella zeae* Schwein (Petch).

Common Synonym. *Fusarium graminearum* Group 2.

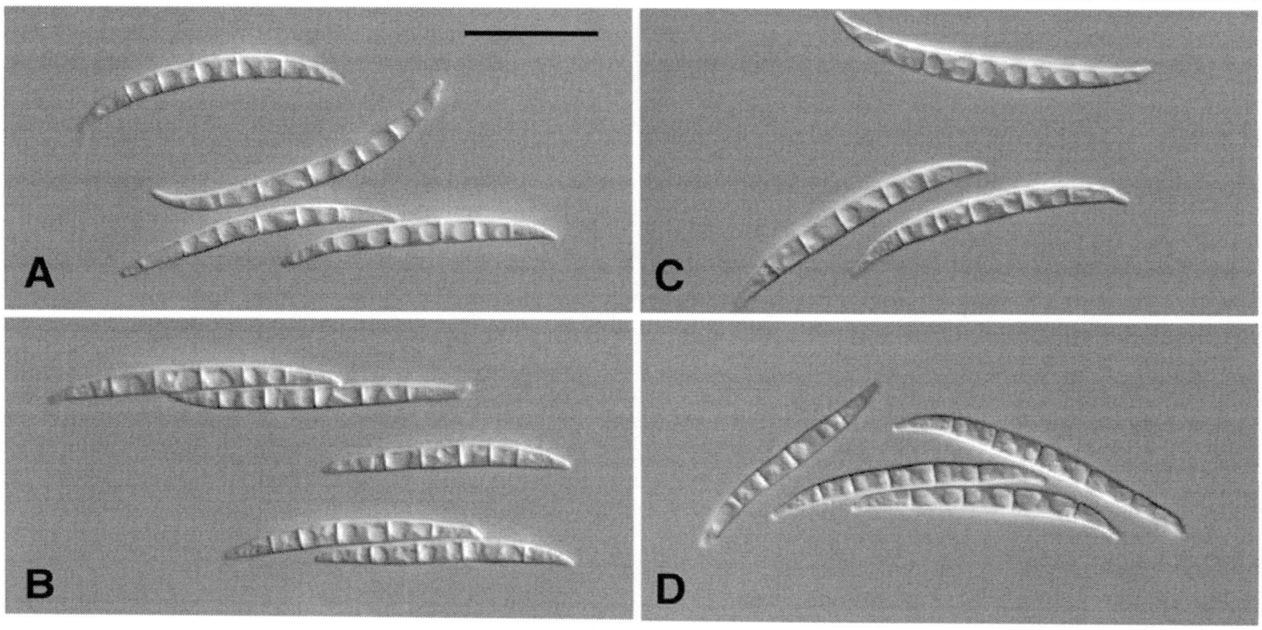

Figure 13-28. *Fusarium graminearum.*
A – D: Macroconidia. Scale bar = 25 µm.

Reported Geographic and Host Distribution

Cosmopolitan. Found primarily with maize, wheat and barley, but also known from other annual and perennial plants.

Media to Examine and Key Characters

Characters on CLA. Sporodochia often are sparse, but when present are pale orange and may be hidden under the mycelia. Macroconidia are relatively slender, sickle-shaped to almost straight, thick walled, 5- to 6-septate, with a tapered apical cell and a distinctly foot-shaped basal cell. Chlamydospore formation varies, and they often are formed in the macroconidia. Microconidia are absent. The teleomorph stage, *Gibberella zeae*, usually forms on carnation leaf pieces and on the agar. The fungus is homothallic and thus perithecia can be formed without the need to cross isolates.

Characteristics on PDA. Colonies grow rapidly and produce relatively large amounts of dense mycelia that vary from white to pale orange to yellow in color. Produces red-brown to orange sporodochia slowly (> 30 days). Cultures form red pigments in the agar. The pigment is pH sensitive and may change from red to yellow as the pH drops.

Macroconidia

- Sporodochia: Pale orange, but often rare or difficult to find. Macroconidia in the sporodochia usually are uniform in shape and size.
- General morphology: Slender, thick-walled, and of medium length. Moderately curved to straight with the

ventral surface straight and the dorsal side smoothly arched.

- Apical Cell morphology: Tapered and occasionally constricted to a snout-like shape.
- Basal Cell morphology: Well developed foot shape.
- Number of septa: 5- to 6-septate. Septa usually are quite distinct.
- Abundance: Relative to other species macroconidia are relatively rare in *F. graminearum* cultures. Macroconidia are most abundant in sporodochia.

Microconidia. Absent.

Chlamydospores

- Abundance/Speed of formation: Variable, but often very slow to form. Lack of chlamydospore production is not diagnostic.
- Location: Most commonly in the macroconidia, but may also form in the mycelia.
- Appearance: Produced singly, in clusters, and in chains. Usually are globose with a finely roughened, but not verrucose, appearance.

Taxonomy, Pathology and Ecology

For many years *Fusarium graminearum* was split into two taxa known as *F. graminearum* Group 1 and *F. graminearum* Group 2. Morphologically it is very difficult to differentiate the two groups, but there are important ecological and pathological differences. Aoki and O'Donnell (71) described the Group 1 population as *F. pseudograminearum* and the Group 2 population retained the original *F. graminearum* name and the associated *Gibberella zeae* teleomorph. From a practical point of view the easiest way to distinguish the two species is the formation of perithecia by single-spore purified isolates on CLA or carrot agar. *F. graminearum* is homothallic and most strains will produce perithecia on either of these media, although the perithecia usually are produced more quickly (in some cases in as little as four days) and in greater number on carrot agar than on CLA. A type specimen for *G. zeae* is available (1934). *F. pseudograminearum* is heterothallic and single-spore purified cultures cannot form perithecia alone on either medium. *Fusarium graminearum* is one of the most difficult species in the genus to preserve in long-term storage. Stored cultures should to be checked with some regularity to ensure their continued viability.

Further subdivisions of *F. graminearum* also have been proposed. Cullen *et al.* (423) proposed two types, termed 'A' and 'B', that differed in pathogenicity, cultural morphology, and zearalenone production, but this terminology has not been widely used and the genetic basis for these differences has not been examined. Carter *et al.* (319) identified three different groups, termed "A", "B" and "C" from Europe, Nepal and the United States, and Desjardins *et al.* (476) identified three groups from Nepal, but how these groups relate to the lineages described by O'Donnell and coworkers is not yet known. O'Donnell and co-workers, *i.e.*, O'Donnell *et al.* (1599, 1601) and Ward *et al.* (2291), identified nine phylogenetic lineages in *F. graminearum* that can be resolved by sequencing selected genes. The eight described lineages and a ninth that had not been given a lineage designation have been described as separate species by O'Donnell *et al.* (1601). These species are: *Fusarium brasilicum* Aoki, Kistler, Geiser & O'Donnell (no previous lineage designation); *Fusarium austroamericanum* Aoki, Kistler, Geiser & O'Donnell (previous lineage designation – 1); *Fusarium meridionale* Aoki, Kistler, Geiser & O'Donnell (previous lineage designation – 2); *Fusarium boothii* O'Donnell, Aoki, Kistler & Geiser (previous lineage designation – 3); *Fusarium mesoamericanum* Aoki, Kistler, Geiser & O'Donnell (previous lineage designation – 4); *Fusarium acaciae-mearnsii* O'Donnell, Aoki, Kistler & Geiser (previous lineage designation – 5); *Fusarium asiaticum* O'Donnell, Aoki, Kistler & Geiser (previous lineage designation – 6); *F. graminearum* (previous lineage designation – 7); and *Fusarium cortaderiae* O'Donnell, Aoki, Kistler & Geiser (previous lineage designation – 8). Four additional lineages/species also have been proposed but have yet to be formally described (2037). Based on the cross-fertility of strains from different lineages there is no indication that these lineages belong to more than a single biological species. Thus, we recommend that the single name, *F. graminearum*, be used for all of the phylogenetic lineages/species associated with this group of fungi.

Cultures of *F. graminearum* can be confused easily with *F. pseudograminearum*, *F. crookwellense* and *F. culmorum* and with species in section *Sporotrichiella*, e.g., *F. sporotrichioides*, *F. tricinctum*, *F. poae* and *F. chlamydosporum*. Differences in the morphology of the macroconidia allow the differentiation of *F. graminearum* from *F. culmorum* and *F. crookwellense*. The absence of microconidia distinguishes isolates of *F. graminearum* from isolates of some members of the Section *Sporotrichiella* which form colonies on PDA similar to those of *F. graminearum*.

Most isolates form perithecia abundantly in nature and release ascospores that cause diseases of aerial plant parts such as head blight (head scab) of wheat (702, 755, 2325), barley (1378), oats (2124), and wild rice (1588), stalk and cob rot of maize (2311), sorghum grain mold

(1421), branch dieback of *Acacia* and *Eucalyptus* (1840), and stub dieback of carnations (247, 1547, 2034). The perithecia release ascospores in response to rain or high humidity (1793, 2172, 2174), although the timing of this response under field conditions is not completely settled (1662).

Distribution of ascospores onto grain heads under field conditions appears to be random and nearly continuous during the growing season (1912, 1914). Perithecia are reported to retain viability for up to 16 months on maize kernels (1792) or 23 months on wheat straw residue (1677) under field conditions, with viability usually lasting longer than the ability to sporulate (1047). The optimal reported temperature for perithecium formation (29°C) is somewhat higher than the maximum temperature (26°C) at which ascospores are released (2182). Ascospores and macroconidia can both effectively initiate plant infections (2033), but ascospore and macroconidial germination are both dependent on humidity with ascospores capable of germinating at a lower relative humidity than macroconidia (180). By using a transformed strain carrying a green fluorescent protein marker the details of the infection process of wheat and barley by *F. graminearum* have been worked out (1449, 1995).

Fusarium graminearum can be a pathogen on the model plant species *Arabidopsis thaliana*, which may enable more rapid studies of host-pathogen interactions for this fungus (1995, 2203).

Maize lines vary widely in their sensitivity to *F. graminearum* (820) with lines resistant to *F. graminearum* usually resistant to *Diplodia maydis* as well (2310). *F. graminearum* can degrade the antimicrobial benzoxazinoids 6-methoxy-2-benzoxazolinone and 2-benzoxazolinone produced by maize (712). Increased ferulic acid content in maize kernels is correlated with resistance to the ear rot induced by *F. graminearum* (90). *F. graminearum* also is sensitive to the plant defense chemical apigeninidin (1928). Maize ear infection through silk channels depends on the host genotype (1791) and is most efficient when the silks are newly emergent (1790). Maize kernels infected with *F. graminearum* are significantly less likely to be infected with *F. verticillioides* than are kernels that were not infected with *F. graminearum* (Rheeder *et al.*, 1990*b*). "BT" hybrid maize, which is less susceptible to the European corn borer *Ostrinia nubialis*, may have a higher level of infection with *F. graminearum* than do comparable hybrids without the transgene (685), but may also have lower levels of deoxynivalenol (1648).

Wheat and barley lines also vary in their sensitivity to *F. graminearum*, but fungal strains cannot always be differentiated based on their aggressiveness towards wheat (259, 1432, 2375). Trichothecene production is an important character for virulence (1183, 1738) and for spread of the fungus within an infected wheat head (124, 1183). The MAP kinase encoded by *Gpmk1* regulates the expression of secreted cell-wall degrading enzymes required for pathogenicity (966). Head blighted wheat and barley plants have shrunken "tombstone" kernels, but these diseased kernels are not responsible for disease transmission under field conditions (701). Instead splash dispersal of spores on plant residue appears to suffice for dispersal of both macroconidia and ascospores within a field (1661). Regular meetings on Fusarium Head Blight of Wheat and Barley are held annually in the United States and coordinated by the US Wheat and Barley Scab Initiative (www.scabusa.org). References to *F. graminearum* as the causal agent of crown rot of wheat are virtually always to the group of strains that now compose *F. pseudograminearum*.

Field populations of *G. zeae* are genetically quite diverse as determined from studies that use a number of different genetic markers (229, 665, 668, 1396, 1457, 1466, 2064, 2409, 2410). In the northern United States these populations contain only lineage 7 and are essentially randomly mating. The amount of genetic variation limited to a single population is ~2-2.5% and the genetic distance between populations is correlated with their physical distance (2410). Primers and protocols for detecting *F. graminearum* through real-time PCR assays are available (208, 1794, 2275), and have been used to show that *F. graminearum* dominates in Fusarium Head Blight epidemics in Europe.

Significant genetic work has been done with *F. graminearum* and auxotrophic, morphological and regulatory mutants are available for study (482, 505, 809, 902, 1216, 1218, 1374, 2204). Studies of vegetative incompatibility can be made using both *nit* mutants (229, Chapter 5) or the barrage interactions that result when some field strains are paired (1397). Meiotic progeny from heterozygous crosses can be recovered either by selecting for recombinant prototrophic progeny from crosses between complementary *nit* mutants (230) or by forcing a cross in which a wild-type strain is the male parent and a strain with a partially inactivated mating type is the female parent (1206, 2402). The entire genome has been sequenced (www.broad.mit.edu/annotation/fungi/fusarium) and ESTs based on cDNA libraries of strains cultured under different conditions are available for study (2175). A physical map of the mitochondrial genome also is available (1526).

A detailed genetic map is available for *G. zeae* (994). Preliminary correlations between the Jurgenson *et al.* (994) genetic map and a second genetic map based on different strains with the physical sequence also have been made (667, 1205). Aggressiveness towards wheat maps as two QTLs, both in regions known to carry genes that affect the production of the trichothecene mycotoxins (425, 426) that are associated with increased disease severity (2289).

A number of physiological studies have been made with *F. graminearum* as well, but studies of strain A 3/5, which is used for Quorn mycoprotein production, are included in the section on *F. venenatum*, the species to which this strain now has been assigned (1595, 2392). *F. graminearum* can utilize a broad range of compounds as the sole nitrogen source including, nitrate, ammonium, urea, and most purines and amino acids that do not contain sulfur (1217) and has a fairly typical fungal cell wall composed of chitin and cross-linked galactose, mannose, arabinose and glucuronic acid (147). Macroconidia germinate best at higher relative humidity (> 80%) at ~20°C in the dark (179). Hyphal extension decreases and the intracellular sugar alcohol content increases as soil water potential decreases (1772). Freeze-thaw cycles do not affect percent germination, but may reduce the amount of hyphal growth that is produced by the resulting colonies (179). dsRNAs that may severely alter the morphology and physiology of the strains that carry them have been isolated from *F. graminearum* (370, 371). The number of enzymes examined from *F. graminearum*, other than those associated with toxin biosynthesis, is not large and includes xylanases (161).

F. graminearum is not usually associated with humans as a direct pathogen. It is resistant to most clinical antifungals, with amphotericin B reported as the most effective (1748).

Isolates of *F. graminearum* may produce three important mycotoxins, zearalenone, nivalenol and deoxynivalenol (1341) as well as aurofusarin (88, 766), culmorins (1668), fusarin C (600, 694, 2140), fusarochromanone (1657), and steroids (284). *F. graminearum* is not reported to produce moniliformin (602). The genes involved in the aurofusarin biosynthetic pathway have been identified and cloned, but their regulation and interrelationships are not yet well understood (1061, 1325). The molecular genetics of trichothecene production is generally well understood (253) and is regulated by the *tri6* gene product (1386). Production of deoxynivalenol or nivalenol usually is determined by the functioning of a single gene, *tri13* (1056, 1209, 1210), and PCR primers for identifying the common functional and non-functional alleles are available (341, 1210). These primers have been used to analyze populations of *F. graminearum* in the United Kingdom (971) and Australia (2113). Primers that distinguish strains producing DON and NIV also have been developed based on differences in the *tri5-tri6* intergenic region (1250).

Marasas *et al.* (1341) has a lengthy discussion of the animal and human toxicoses associated with *F. graminearum*, which include hyperestrogenism and feed refusal and emetic syndromes of domesticated animals, primarily swine, and scabby grain intoxication of humans. In general these diseases can be explained by one of the three primary mycotoxins produced by *F. graminearum*, but there are some cases that suggest that additional toxigenic compounds remain to be identified. In chickens, culture material was toxic if a trichothecene was being produced, but not if zearalenone alone was being produced (1341). Cattle also were relatively insensitive to zearalenone, although this compound can be transmitted into the milk of dairy cattle (1453) if zearalenone is present at a sufficiently high level in the diet (1735).

Some strains of *F. graminearum* can produce up to 60 g/kg zearalenone (1341), and are used for the commercial production of zearalenone, which is then chemically modified and sold as a cattle growth promotant (858). Zearalenone was initially reported to serve as a stimulant or phermone for sexual reproduction for *F. graminearum*, *e.g.*, Wolf & Mirocha (2353, 2354) and Wolf *et al.* (2352), but these initial claims have not withstood more detailed examination (2345). Cu^{++} and Fe^{++} both stimulate zearalenone production (421).

Mycotoxin biosynthesis in *F. graminearum* in general appears to be more sensitive to temperature than it is to water availability (893, 1274, 1275) and is subject to global genetic regulation (260). Under field conditions, timing of sample collection affects the amount of mycotoxinn recovered (340). Under laboratory conditions the production of deoxynivalenol increases with relative humidity (180), an observation consistent with anecdotal field observations. The use of fungal mycelia, perhaps detected through a species or toxin-specific gene, as a surrogate for detection of toxins has been proposed but has not been widely accepted (1908).

Current Species Description: Booth (222).

Fusarium guttiforme Nirenberg & O'Donnell

Sexual Stage. None known.

Common Synonym. *Fusarium subglutinans* f. sp. *ananas*.

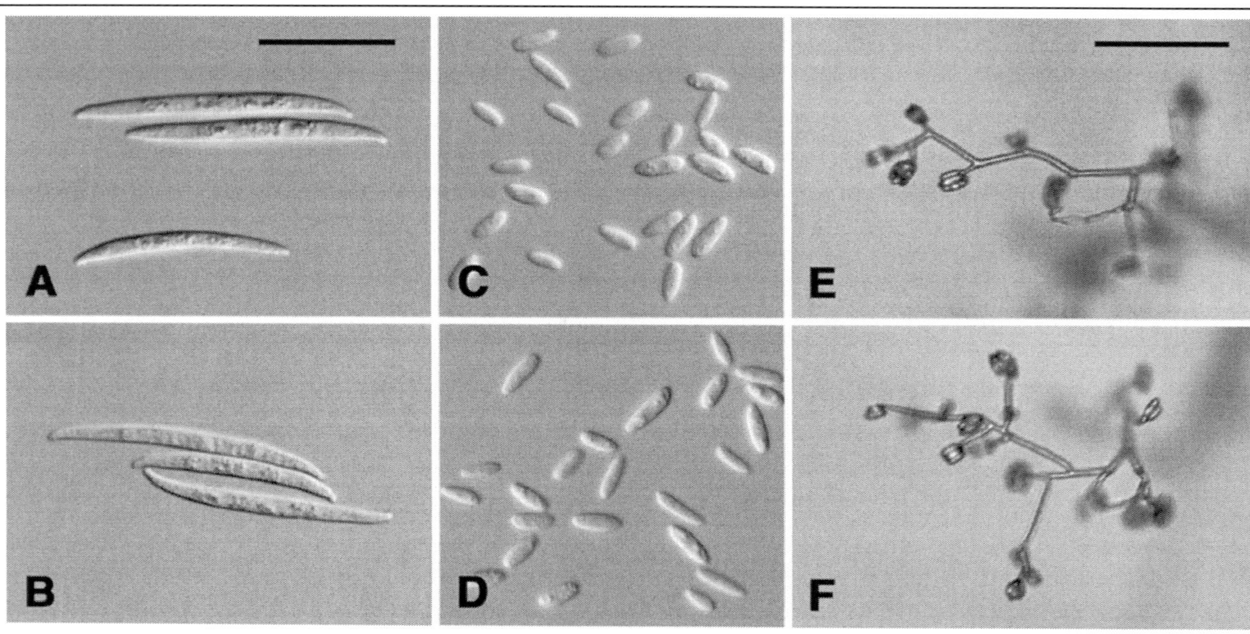

Figure 13-29. *Fusarium guttiforme.*
A – B: Macroconidia; **C – D:** Microconidia; **E – F:** Microconidia *in situ* on CLA. A – D, scale bar = 25 μm; E – F, scale bar = 50 μm.

Reported Geographic and Host Distribution

Recovered from pineapple (*Ananus comosus*) primarily in Brazil, but also from Cuba and elsewhere in South America.

Media to Examine and Key Characters

Characters on CLA. Produces few or no sporodochia. When present the sporodochia are orange and contain thin-walled macroconidia that are slender and straight to slightly curved. Oval microconidia are found in the aerial mycelia in false heads from either monophialides or polyphialides. No chlamydospores

Characters on PDA. White mycelia that may become gray or purple with age. Orange and/or violet pigments in the agar.

Macroconidia

- Sporodochia: Orange in color when formed, but not commonly produced.
- General morphology: Thin walled, slender and straight to slightly curved. Typical of species in the *Gibberella fujikuroi* species complex.
- Apical Cell morphology: Tapering.
- Basal Cell morphology: Foot shaped.
- Number of septa: Usually 3-septate.
- Abundance: Sparse. Rarely found in sporodochia.

Microconidia

- Shape/septation: Obovoid. Primarily 0-, but occasionally 1-septate.
- Aerial mycelium presentation: False heads.

- Conidiogenous cells: Primarily polyphialides, but some monophialides.
- Abundance: Abundant in the aerial mycelia.

Chlamydospores. Absent.

Taxonomy, Pathology and Ecology

This species was described by Nirenberg & O'Donnell (1570) from nine isolates associated with fruitlet core rot or basal rot of pineapples in Brazil. This species probably has been referred to by several names, the most likely of which are *F. subglutinans* and *F. sacchari*, but not all isolates of *F. sacchari* or *F. subglutinans* from pineapple are necessarily *F. graminearum*. Aerial conidia produced by *F. circinatum*, *F. concentricum* and *F. guttiforme* are similar. *F. circinatum* strains can be distinguished by the formation of coiled hyphae on SNA. The aerial conidiophores of *F. guttiforme* are more strongly branched and contain more polyphialides than do those of *F. concentricum*. The microconidia of *F. sacchari* tend to be somewhat smaller than those of *F. guttiforme* and are produced only from prostrate conidiophores. Conidiophores of *F. subglutinans* have fewer conidiogenous openings than do those of *F. guttiforme*. Sexual stages are known for both *F. sacchari* and *F. subglutin*ans, and crosses with the standard mating type tester strrains of these species can be used to distinguish *F. guttiforme* from *F. sacchari* and *F. subglutinans*.

The optimal temperature for vegetative growth and conidiation is 25°C, with microconidia produced on most carbon sources, but macroconidia never were produced when starch, glucose or fructose was the sole carbon source (216). Light is not required for either growth or conidiation. Addition of 0.1 M or higher KCl to the medium increases the production of microconidia but may result in the complete loss of macroconidial formation (216).

Much of the literature on the pineapple disease attributed to *F. guttiforme* is based on work done in Brazil and published in Portugese, although results of some studies are available in English, *e.g.*, Bolkan *et al.* (215), Dianese *et al.* (503), and Rohrbach & Pfeiffer (1823). The disease is favored by daytime temperatures of 21-27°C (1824). The fungus may persist as conidia in the soil for up to 13 weeks or on colonized plant debris for up to a year (214). Resistance to the disease can be screened *in vitro* by using culture filtrate produced by the fungus (856, 857). This filtrate contains fusaric acid, but its presence does not correlate with disease resistance (224).

In Brazil, the fungus causes the major disease of pineapple in the country, although some pineapple lines resistant to the pathogen are now available. Reports of "*F. subglutinans*" and "*F. moniliforme*" from pineapple have been included in our discussion of *F. guttiforme*, but it is not clear that these reports are of either *F. guttiforme* or the disease that it causes. Strains recovered from pineapple prior to the description of *F. guttiforme* need to have their identity verified as *F. guttiforme* before this species name is applied to them.

Isolates of *F. guttiforme* can produce beauvericin and fusaproliferin, but not fumonisins or moniliformin (638).

Current Species Description: Nirenberg & O'Donnell (1570).

Fusarium heterosporum Nees *ex* Fries

Sexual Stage. *Gibberella cyanea* (Sollm.) Wollenweber.

Common Synonyms. *Fusarium graminum*, *Fusarium reticulatum*.

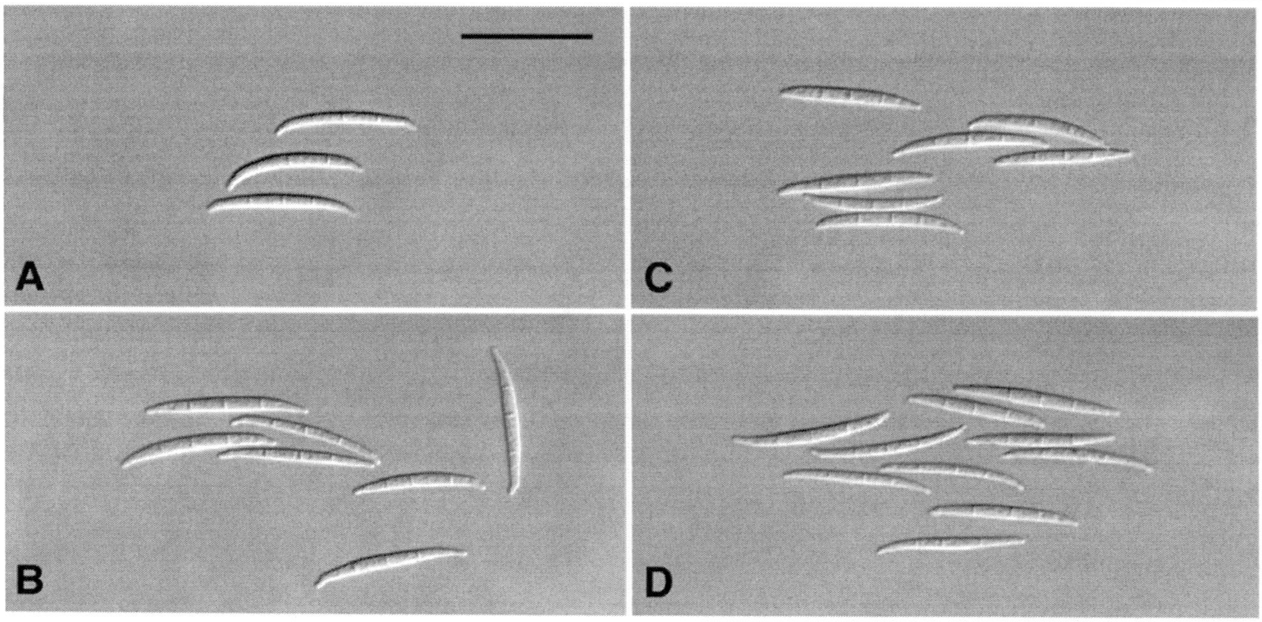

Figure 13-30. *Fusarium heterosporum.*
A – D: Macroconidia. Scale bar = 25 µm.

Reported Geographic and Host Distribution

Cosmopolitan. May be associated with an ergot caused by *Claviceps* spp. or with head blight of millets and other grasses, especially in Africa.

Media to Examine and Key Characters

Characters on CLA. Macroconidia form abundantly in bright orange sporodochia. The macroconidia are of medium length, slender, falcate to almost straight, thin-walled, and 3- to 5-septate. The basal cell may be either distinctly foot-shaped or notched. Macroconidia usually form on monophialides on branched conidiophores in the sporodochia and to a lesser extent from monophialides on hyphae. Microconidia and chlamydospores are absent.

Characters on PDA. Abundant bright orange sporodochia form in the center of the colony. There may be rings of spores formed if the cultures are incubated under alternating temperature or light/dark conditions. A watery exudate forms on the sporodochia of most isolates. The mycelium is quite dense and is white to pinkish-white in color. No obvious pigment is excreted into the agar, but the undersurface of the colony may appear brownish-orange to caramel in color because of the reflected light from the sporodochia. Strains of *F. heterosporum* mutate readily to a white, sterile mycelial form when cultured on this medium. In some cases the mutated forms may sporulate again after being exposed to black light (222).

Macroconidia

- Sporodochia: Bright orange color. Produced abundantly on carnation leaves and on the agar surface near the carnation leaves in CLA.
- General morphology: Medium length, thin walled, slender, and straight or gently curved.

- Apical Cell morphology: Tapered.
- Basal Cell morphology: May be either foot shaped or notched.
- Number of septa: 3- to 5-septate.
- Abundance: Very abundant in sporodochia.

Microconidia. Absent.

Chlamydospores. Absent.

Taxonomy, Pathology and Ecology

There has been some confusion as to the most appropriate name for this fungus. The binomial *Fusarium graminum* has been used by some researchers, but as there is no difference between *F. graminum* and *F. heterosporum*, the *F. heterosporum* name is preferred since it has nomenclatural priority. The macroconidia are similar to, but not as long as, the macroconidia of *F. avenaceum*.

Some studies have been made of the teleomorph of this species. Gordon (753) described it as *Gibberella cyanea*, but also associated the *G. cyanea* perfect state with a form of *F. sambucinum*. Booth (222) renamed the perfect stage of *F. heterosporum* as *G. gordonia*; however, Gerlach & Nirenberg (697) suggested that *G. gordonia* was the perfect stage for *F. reticulatum*, which also may be conspecific with *F. heterosporum* (222). Gordon (753) included *F. heterosporum* on his list of heterothallic species, but noted that mating type and male/female fertility were not segregating from one another – a segregation pattern that could result if female-fertile and female-sterile strains from a homothallic species were crossed with one another.

Fusarium heterosporum has one of the widest documented geographic ranges for any *Fusarium* spp., being recovered from subantarctic soils (1805) and Malaysian oil palms (869), and both relatively dry, *e.g.*, North Dakota grasslands (1410, 1412), and wet, *e.g.*, *Spartina anglica* marsh grasses in the United Kingdom (1734, 1782). *F. heterosporum* is commonly associated with grass inflorescences infected by *Claviceps* spp. Bright orange sporodochia form on the ergot produced by the *Claviceps* that are clearly visible to the naked eye. *F. heterosporum* may be used as a food source by some insects (318). It also may function as a biological control agent of white mold (*Sclerotinia sclerotiorum*) on bean (213, 930) and dollar spot of creeping bentgrass (734).

Some strains of *F. heterosporum* produce a thermostable (50°C) lipase that has been cloned, well characterized, and may have a commercial application (1513-1517).

Fusarium heterosporum usually is regarded as nontoxic, but some strains can produce fusaric acid (119). One strain of *F. heterosporum* that was reported to produce trichothecenes (390) has since been reidentified as *F. acuminatum* (1341).

Current Species Description: Booth (222).

Fusarium hostae Geiser & Juba

Sexual Stage. *Gibberella hostae* Geiser & Juba.

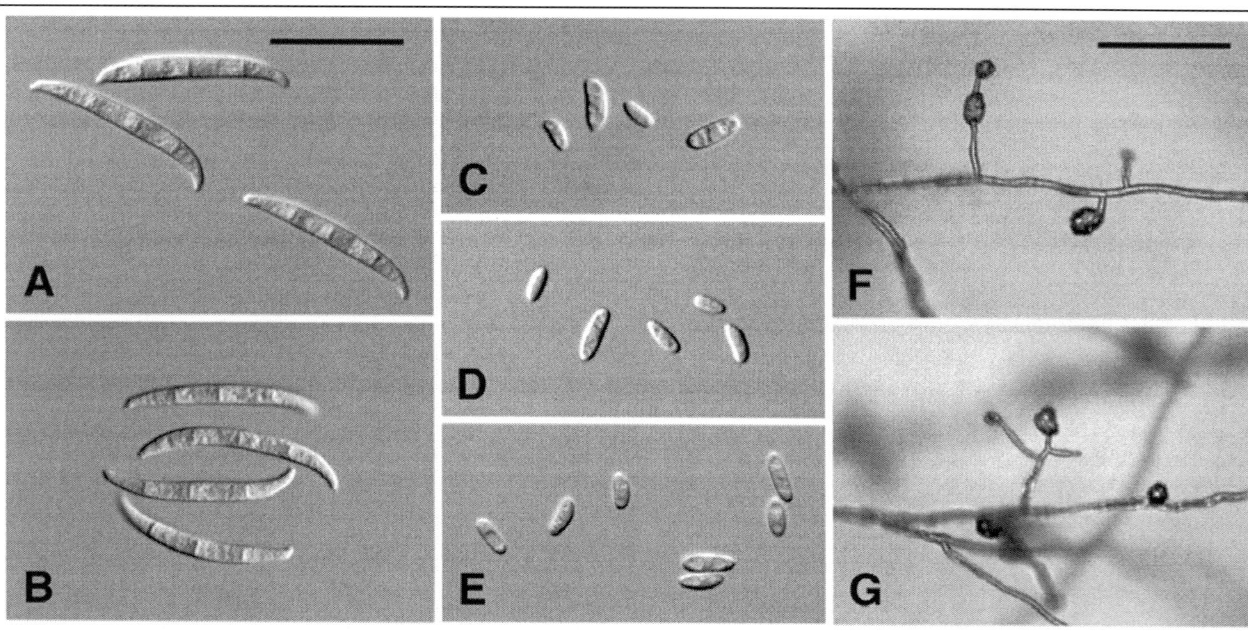

Figure 13-31. *Fusarium hostae.*
A – B: Macroconidia; **C – E:** Microconidia; **F – G:** Microconidia *in situ* on CLA. A – E, scale bar = 25 µm; F – G, scale bar = 50 µm.

Reported Geographic and Host Distribution

Found on *Hosta* spp. (Liliaceae) in South Carolina, USA causing a root and crown rot of container grown plants.

Media to Examine and Key Characters

Characters on CLA. Limited aerial mycelia. Purplish-orange sporodochia produced directly on the agar surface and containing macroconidia borne on monophialides. Fusiform, usually 0-septate, microconidia borne predominantly on monophialides in the aerial mycelia. Abundant chlamydospores found singly, in chains and small clumps. May produce perithecia on CLA.

Characters on PDA. Limited aerial mycelia with light orange, yellow or violet pigmentation. Pigmentation in the agar is similar to that in the mycelia.

Macroconidia

- Sporodochia: Initially slimy purple-yellow-pink that becomes a purple-orange-pink at maturity. Produced on the agar surface.
- General morphology: Usually falcate, but occasionally fusiform.
- Apical Cell morphology: Curved or hooked.
- Basal Cell morphology: Foot shaped.
- Number of septa: May be 2- to 4-septate, but usually 3-septate.
- Abundance: May be difficult to find if sporodochia are sparse.

Microconidia

- Shape/septation: Fusiform. Usually 0-septate, but may be 1- or 2-septate.

- Aerial mycelium presentation: False heads.
- Conidiogenous cells: Predominantly monophialides with occasional polyphialides.
- Abundance: Abundant in the aerial mycelia.

Chlamydospores

- Abundance/Speed of formation: Abundant, but form slowly, *i.e.*, in 4-6 weeks on CLA.
- Location: Within and alongside hyphae. May be interstitial or terminal. Singly or in chains or clumps.
- Appearance: Various globose, subglobose and doliiform shapes. Finely roughened, not verrucose.

Taxonomy, Pathology and Ecology

This species was described on the basis of twelve isolates from container-grown hostas from South Carolina. Disease severity can be reduced by cultural practices such as container mix, temperature and watering regime (2287). Phylogenetically, *F. hostae* is most closely related to *F. redolens* from which it can be separated by differences in the sequence of the β-tubulin and translation elongation factor 1-α genes. Together with *F. redolens*, *F. hostae* forms a clade that is distinct and intermediate between those for the *F. oxysporum* and *G. fujikuroi* species complexes (106).

Morphologically *F. hostae* also is closely related to *F. redolens* from which it can be distinguished by the slower radial growth of *F. hostae* on PDA, and by the lack of production of polyphialides by *F. redolens*. Other species with morphological similarities to *F. hostae* include *F. polyphialidicum* and *F. nygamai*. *F. hostae* differs from *F. polyphialidicum* as the macroconidia of *F. hostae* are stouter, its microconidia are more clearly fusiform, and its polyphialides have no more than two conidiogenous openings. *F. nygamai* can be distinguished from *F. hostae* since *F. nygamai* produces microconidia in chains and *F. hostae* does not.

Production of the *Gibberella* stage on CLA by an isolate purified through a single-spore subculture is unusual and suggests that this species may be homothallic. Maintaining this capability through subcultures of single ascospores strengthens this conclusion. If homothallism is confirmed, then this species will be only the second in *Gibberella* that is homothallic (the other is *Gibberella zeae*/*Fusarium graminearum*). Only 1/12 isolates examined produced perithecia, however, so the formation of perithecia in culture by field strains should not be relied on for identification purposes. If this low rate of perithecium formation is typical for the species, then it suggests that sexual reproduction is relatively unimportant in the life cycle of *F. hostae*.

Fusarium hostae has not been evaluated for mycotoxin production.

Current Species Description: Geiser *et al.* (2001).

Fusarium konzum Zeller, Summerell & Leslie

Sexual Stage. *Gibberella konza* Zeller, Summerell & Leslie.

Common Synonym. *Gibberella fujikuroi* Mating Population I.

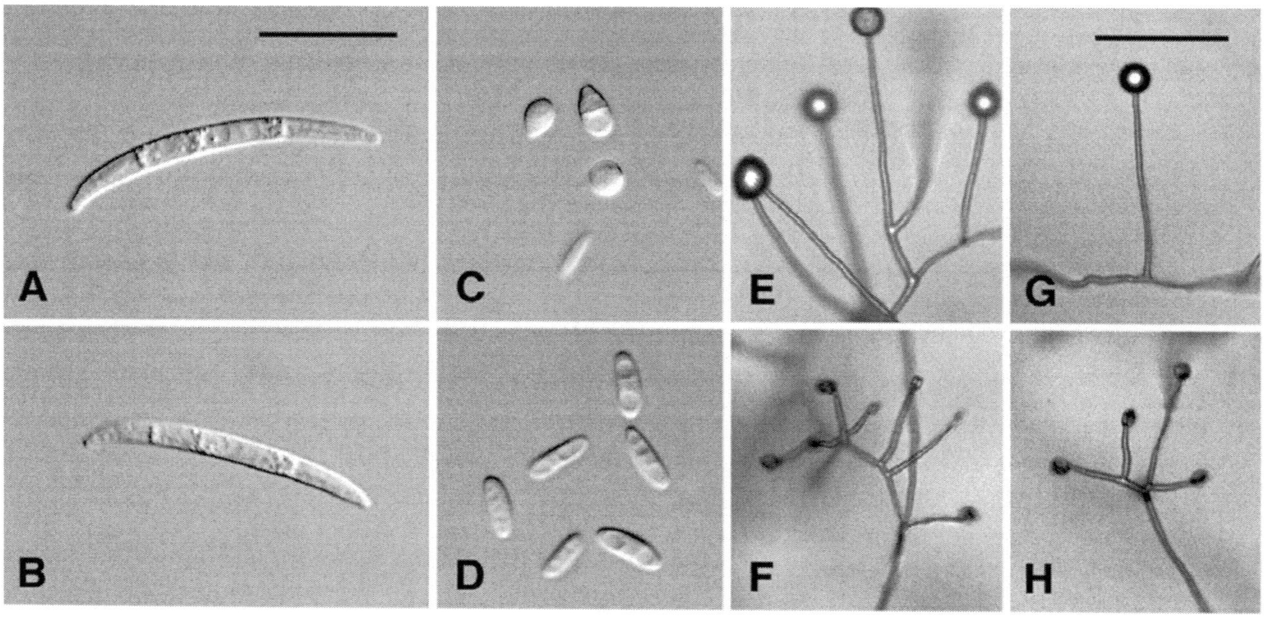

Figure 13-32. *Fusarium konzum.*
A – B: Macroconidia; **C:** Pyriform and napiform microconidia; **D:** Oval to obovoid microconidia;
E & G: Napiform microconidia *in situ* on CLA; **F & H:** Oval microconidia *in situ* on CLA. A – D, scale bar = 25 µm; E – H, scale bar = 50 µm.

Reported Geographic and Host Distribution

Recovered from prairie grasses in Kansas, USA.

Media to Examine and Key Characters

Characters on CLA. Sporodochia and macroconidia are rare. When present sporodochia are pale orange and found on carnation leaf pieces. Macroconidia are falcate, 3- to 5-septate with a foot-shaped basal cell and a barely curved apical cell. Oval (common) and pyriform or napiform (less common) microconidia borne singly or in small false heads from monophialides or polyphialides. No chlamydospores.

Characters on PDA. Colonies have floccose mycelia that is initially white but may become violet with age. Agar may be pigmentless or become violet or red.

Macroconidia

- Sporodochia: Rare. When present they are pale orange and found on carnation leaf pieces in the CLA.
- General morphology: Thin walled, slender and straight to slightly curved. Typical of species in the *Gibberella fujikuroi* species complex.
- Apical Cell morphology: Slightly curved.
- Basal Cell morphology: Foot shaped.
- Number of septa: Usually 3-septate, but may be up to 5-septate.
- Abundance: Sparse

Microconidia

- Shape/septation: Produces three types of microconidia. Oval, hyaline and 0- to 1-septate. Pyriform, and

0- or, occasionally, 1-septate. Larger napiform to globose, and 0-septate.

- Aerial mycelium presentation: Singly or small false heads. May see occasional chains of 2-3 conidia that collapse to a false head when they dry.
- Conidiogenous cells: Monophialides and polyphialides.
- Abundance: Abundant in the aerial mycelia. Pyriform and napiform microconidia are more common on SNA than they are on CLA.

Chlamydospores. Absent.

Taxonomy, Pathology and Ecology

This species was described by Zeller *et al.* (2412) from isolates recovered from prairie grasses (*Andropogon* spp., *Bouteloua* spp., *Sorghastrum* spp., *Sporobolus* spp.) in a tallgrass prairie preserve in Kansas, USA. No *F. konzum* isolates were found in adjacent maize and sorghum fields suggesting that this species may have a relatively limited host range and not be an economically important pathogen. Only 2/52 field isolates were female fertile, suggesting that sexual reproduction is not common in this fungus, but multi-locus tests for association were not significantly different from those expected for a randomly mating population.

For positive identification, isolates of this species can be crossed to produce the teleomorph, *G. konza*. It is preferable to use positive sexual crosses with standard tester strains to identify this species. The ascospores of *G. konza* are smaller than those of *G. moniliformis*, *G. intermedia*, *G. sacchari* and *G. thapsina*, but are similar in size to those of many other *Gibberella* species, *e.g.*, *G. fujikuroi*. *F. konzum* is morphologically similar to *F. anthophilum*, with the size and shape of the macroconidia and the oval and pyriform microconidia falling within the range previously reported for *F. anthophilum*. The monophialides of *F. konzum* usually are longer than those of *F. anthophilum* and the polyphialides somewhat more swollen. *F. anthophilum* also does not produce globose microconidia (it may be necessary to observe SNA cultures of *F. konzum* to observe these spores). Phylogenetically *F. konzum* is closely related to, but clearly distinguishable from, *F. subglutinans* and its sibling species.

Isolates of *F. konzum* may produce low levels of fumonisins, or may produce none of this toxin at all (1234). They also may produce low to moderate levels of both beauvericin and fusaproliferin (1234).

Current species description: Zeller *et al.* (2412).

Fusarium lactis Pirotta & Riboni

Sexual stage. None known.

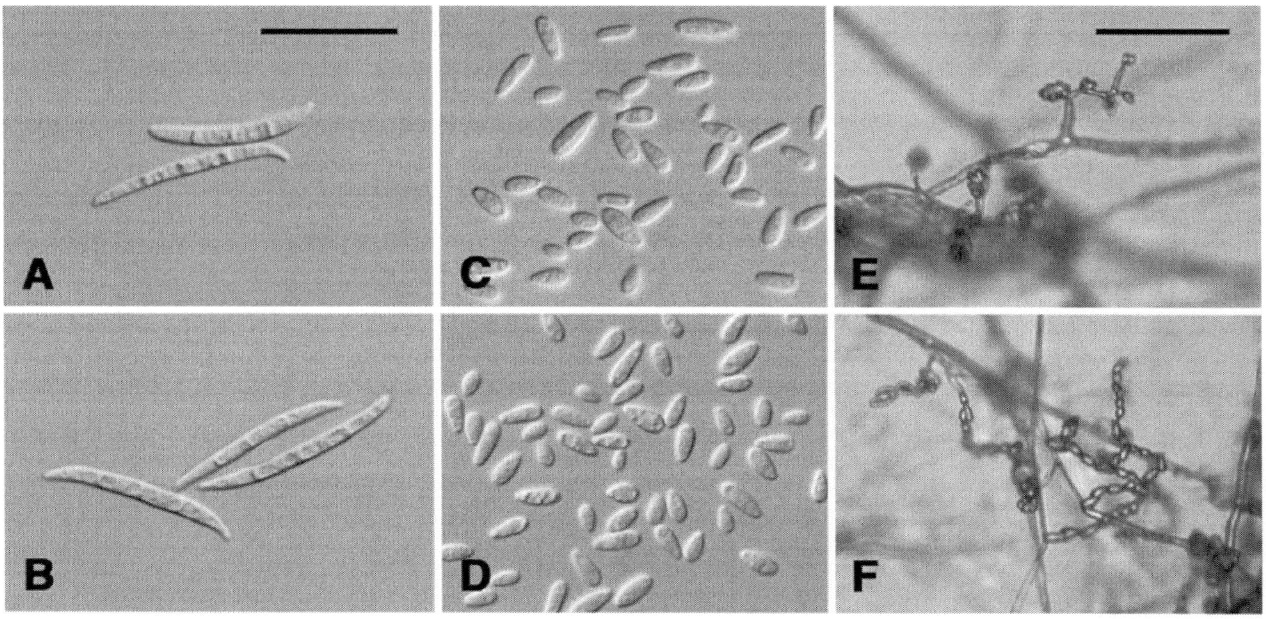

Figure 13-33. *Fusarium lactis.*
A – B: Macroconidia; **C – D:** Microconidia; **E – F:** Microconidia *in situ* on CLA. A – D, scale bar = 25 µm; E – F, scale bar = 50 µm.

Reported Geographic and Host Distribution

From fig (*Ficus carica*) in California, USA.

Media to Examine and Key Characters

Characters on CLA. Sporodochia and macroconidia are rare. Obovoid microconidia are produced in chains and false heads from monophialides and polyphialides. The chains may be distinctively zigzagged in appearance.

Characters on PDA. White mycelia with some violet pigments in the agar.

Macroconidia

- Sporodochia: Rare. When present they are light orange in color and found on carnation leaf pieces in CLA.
- General morphology: Thin walled, slender and straight to slightly curved. Typical of species in the *Gibberella fujikuroi* species complex.
- Apical Cell morphology: Bent.
- Basal Cell morphology: Notched.
- Number of septa: Usually 3-septate.
- Abundance: Sparse and often difficult to find.

Microconidia

- Shape/septation: Obovoid, often with a flattened base. Usually 0-, but occasionally 1-septate.
- Aerial mycelium presentation: False heads and short- to medium-length chains. Chains may have a distinctive zigzag appearance.
- Conidiogenous cells: Monophialides and polyphialides.
- Abundance: Abundant in the aerial mycelia.

Chlamydospores. Absent.

Taxonomy, Pathology and Ecology

This species was described in 1879 by Pirotta & Riboni (1700) and was recognized by Wollenweber & Reinking (2359). However it was not considered by Nelson *et al.* (1551) and was thought to be a *nomen confusum* by Gerlach & Nirenberg (697). The variety name *lactis* has been used with several different *Fusarium* species, and it is not clear that all of these references are to the same biological entity. Nirenberg & O'Donnell (1570) examined further cultures and confirmed that the original description was valid and retypified the species.

Fusarium lactis was originally identified as a milk contaminant in association with other fungi. The six isolates used by Nirenberg & O'Donnell for their redescription all originate from Calimyrna figs in California. This species is known to cause endosepsis of cultivated figs in California (1437, 1438, 2059, 2060). *F. lactis* almost certainly occurs outside California given its description by Wollenweber & Reinking (2359), but it is not universally found in fig growing areas, even when careful taxonomic observations are made (1280). Some isolates of *F. lactis* were previously described as *F. moniliforme* or *F. moniliforme* var. *fici* (1435, 1436, 2058, 2061).

Isolates of *F. lactis* can produce moniliformin, but not beauvericin, fumonisins or fusaproliferin (638).

Current Species Description: Nirenberg & O'Donnell (1570).

Fusarium lateritium Nees

Sexual Stage. *Gibberella baccata* (Wallroth) Saccardo.

Common Synonym. *Fusarium stilboides*.

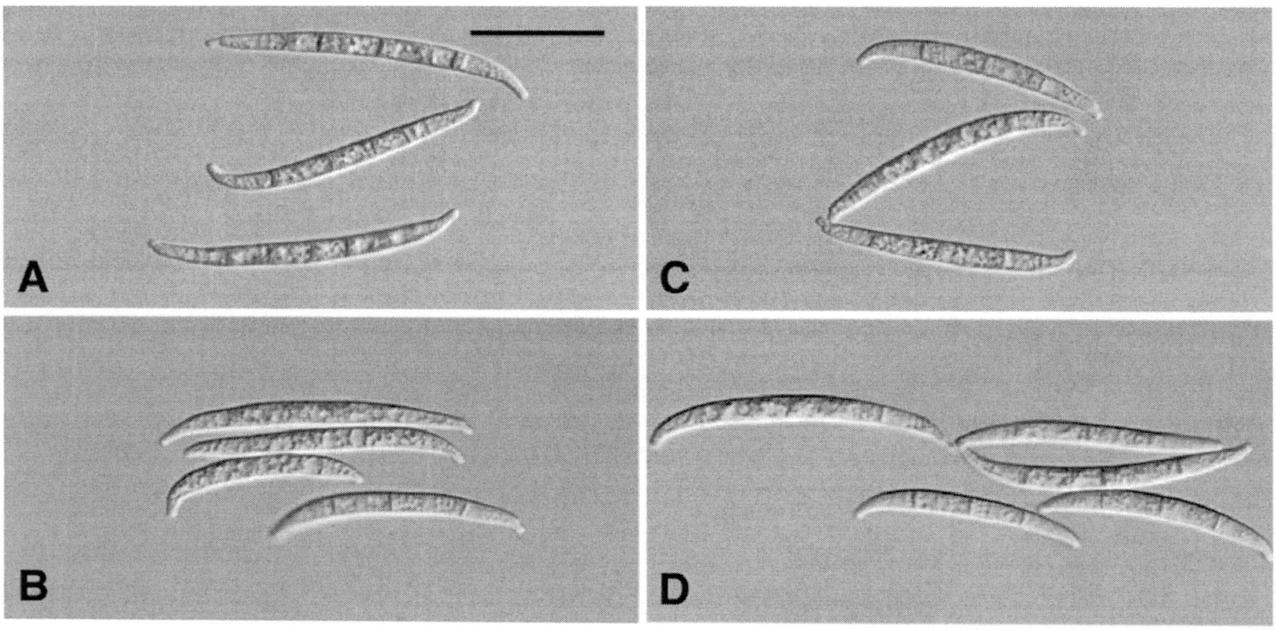

Figure 13-34. *Fusarium lateritium*.
A – D: Macroconidia. Scale bar = 25 μm.

Reported Geographic and Host Distribution

Worldwide in soil and from woody plants, particularly trees.

Media to Examine and Key Characters

Characters on CLA. *Fusarium lateritium* forms little or no aerial mycelium on CLA. Macroconidia are formed in pale orange sporodochia. They are long, thin, falcate to almost straight, with parallel walls, and variable septation. The apical cell has a distinct hook or beak and the basal cell is foot-shaped or notched. Microconidia usually are absent but are abundant when present. Microconidia are elliptical, oval, spindle- or club-shaped. Chlamydospores may be produced by some isolates but the presence/absence of these spores is not taxonomically useful.

Characters on PDA. *Fusarium lateritium* is relatively slow growing with a broad range of morphologies on PDA. Mycelia are sparse and may be white, pale orange or pale pink. Abundant macroconidia are produced in orange or pink sporodochia that may spread over the entire surface of the colony. Many isolates produce a reddish orange pigment in the agar. This species commonly mutates to mycelial and pionnotal forms on this medium. These changes usually result from changes at a single gene (1197).

Macroconidia

- Sporodochia: Abundant pale orange sporodochia are found on or near the carnation leaf pieces in CLA. In some cultures the sporodochia may be nearly confluent across the agar surface giving the culture a slimy appearance.

- General morphology: Medium-long, thin, and relatively straight to gently curved with parallel walls for most of the spore length.
- Apical Cell morphology: Characteristic hook or beak.
- Basal Cell morphology: Foot shaped or notched.
- Number of septa: Varies from 4- to 7-septate, but predominantly 5-septate.
- Abundance: Abundant in sporodochia.

Microconidia

- Shape/septation: Ellipsoid, spindle or club shaped. 0- to 3-septate.
- Aerial mycelium presentation: Singly.
- Conidiogenous cells: Monophialides.
- Abundance: Generally absent, but in some cases may be present in abundance.

Chlamydospores

- Abundance/Speed of formation: Usually present, but not a taxonomically useful trait. The speed with which chlamydospores varies by strain; some may be quite rapid (2-3 weeks) and others may be quite slow (6+ weeks) on CLA.
- Location: Intercalary in submerged and aerial hyphae.
- Appearance: Singly or in chains.

Taxonomy, Pathology and Ecology

It is well accepted that *Fusarium lateritium* is a species complex and that it probably contains several taxa that need to be resolved and characterized. *F. denticulatum* and *F. circinatum* were both at one time classified as *F. lateritium*. A number of varieties and *formae speciales* have been described by various investigators, but the nomenclature remains unsettled and the phylogenetic relationships generally not well-explored.

F. lateritium has been isolated from numerous woody trees and shrubs where it may cause wilt, tip or branch dieback, or cankers. Hosts include citrus, hardwood, coffee, plum, apple and mulberry trees (222, 1341). *F. lateritium* has been reported to symbiotic with a beetle in canker formation of black walnut (1042).

Abundant blue-back perithecia, representative of the sexual stage, *Gibberella baccata*, are produced on diseased mulberry branches in cool wet weather. *F. lateritium* causes stem rot of celosia in New South Wales, Australia (2176) and isolates have been recovered from coffee from Papua New Guinea which correspond to descriptions of *F. stilboides* (222) which is, here, included within *F. lateritium*. Work by Saito & Matuo (1864) and Lawrence *et al.* (1195, 1196) have shown that *F. lateritium* is heterothallic and that there may be multiple mating populations that are not interfertile, *i.e.*, there may be several biological species in this species complex. Strains in this species require further research to clarify their taxonomic relationships.

Fusarium lateritium is being developed as a potential biological control of *Sclerotinia sclerotiorum* on lettuce (1994), and *Eutypa* spp. on grapes (1409, 1500) and apricots (320).

Fusarium lateritium has a cyanide hydratase enzyme that may be useful for bioremediation of sites contaminated with cyanide or nitriles (252, 386, 1580). It also has dimethylsulfoniopropionate lyase activity that could enable it to degrade leaf litter in salt water marshes (110) and sufficient pectinase activity to be potentially useful in the retting of flax (850).

Fusarium lateritium has not been associated with any human or animal toxicoses (1341). Strains identified as *F. lateritium* and associated with alimentary toxic aleukia in humans and fescue foot of cattle have been reidentified as *F. sporotrichioides* (1341). Fenpropimorph inhibits ergosterol biosynthesis in *F. lateritium* (458).

F. lateritium is reported to produce enniatins (192, 1698) and lateropyrone (288). None of the reports of production of the trichothecenes, butenolide or zearalenone mycotoxins by strains correctly identified as *F. lateritium* were confirmed by Marasas *et al.* (1341)

Current Species Description: Wollenweber & Reinking (1935).

Fusarium longipes Wollenweber & Reinking

Sexual Stage. None known.

Common Synonym. *Fusarium equiseti.*

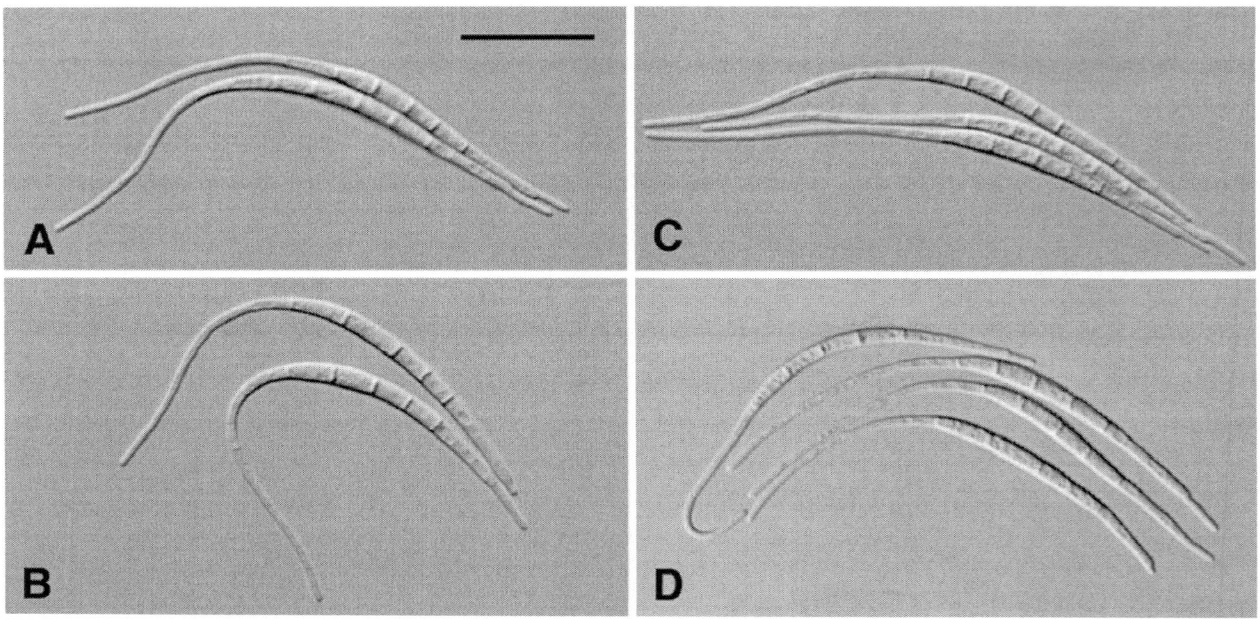

Figure 13-35. *Fusarium longipes.*
A – D: Macroconidia. Scale bar = 25 μm.

Reported Geographic and Host Distribution

Pan tropical. Usually recovered from soil or as a saprophyte.

Media to Examine and Key Characters

Characters on CLA. Abundant macroconidia produced in bright orange sporodochia that can ooze into a columella 4-6 mm in height. The macroconidia are extremely long and slender, with thick walls, usually 5- to 7-septate with a distinct dorsi-ventral curvature with a distinctly foot-shaped and elongate basal cell and an apical cell that is whip-like and may twist around on itself to form a coil or a ring. These macroconidia are not usually confused with those of any other species. Microconidia are absent. Chlamydospores are formed in chains or clusters and become pale brown and verrucose with age.

Characters on PDA. Grows rapidly and produces a dense white to grayish-rose mycelium with red pigments in the agar. Tan to orange sporodochia usually develop in the center of the colony as the colony ages with the abundant macroconidia often forming a columella which may be 4-6 mm in height. Strains of *F. longipes* cultured on PDA can mutate quickly to become pionnotal or mycelial.

Macroconidia

- Sporodochia: Large bright orange sporodochia that may form columella.
- General morphology: Relatively thin, extremely long, with thick walls and pronounced dorsi-ventral curvature.
- Apical Cell morphology: Long, tapering and whip-like. May twist around upon itself.
- Basal Cell morphology: Elongated foot shape.

- Number of septa: 5- to 7-septate.
- Abundance: Abundant in sporodochia.

Microconidia. Absent.

Chlamydospores

- Abundance/Speed of formation: Form relatively abundantly and rapidly (2 weeks on CLA).
- Location: Intercalary in hyphae on the agar surface and submerged in the agar. Produced singly, in chains and in clusters.
- Appearance: Become pale brown and verrucose with age.

Taxonomy, Pathology and Ecology

Generally the unique combination of characters for *F. longipes* means that this species is unlikely to be confused with any other species of *Fusarium*. Some isolates of *F. equiseti* can produce macroconidia that resemble those of *F. longipes*, but *F. longipes* produces red pigments in the agar while *F. equiseti* produces brown pigments.

Fusarium longipes is a tropical species (112) and has been recovered from many locations throughout the tropics (275, 277, 490, 1627, 1876, 1889, 2085). *F. longipes* has been associated with a disease of guava in India (533, 534) and with crown rot of wheat in southern Africa (962, 2222, 2223), but is generally regarded as a saprophytic fungus.

Some strains of *F. longipes* are reported to produce beauvericin (1283), but this species is generally thought to be nontoxigenic (2348).

Current Species Description: Wollenweber & Reinking (2359).

Fusarium mangiferae Britz, Wingfield & Marasas

Sexual Stage. Not known.

Common Synonym. *Fusarium subglutinans*.

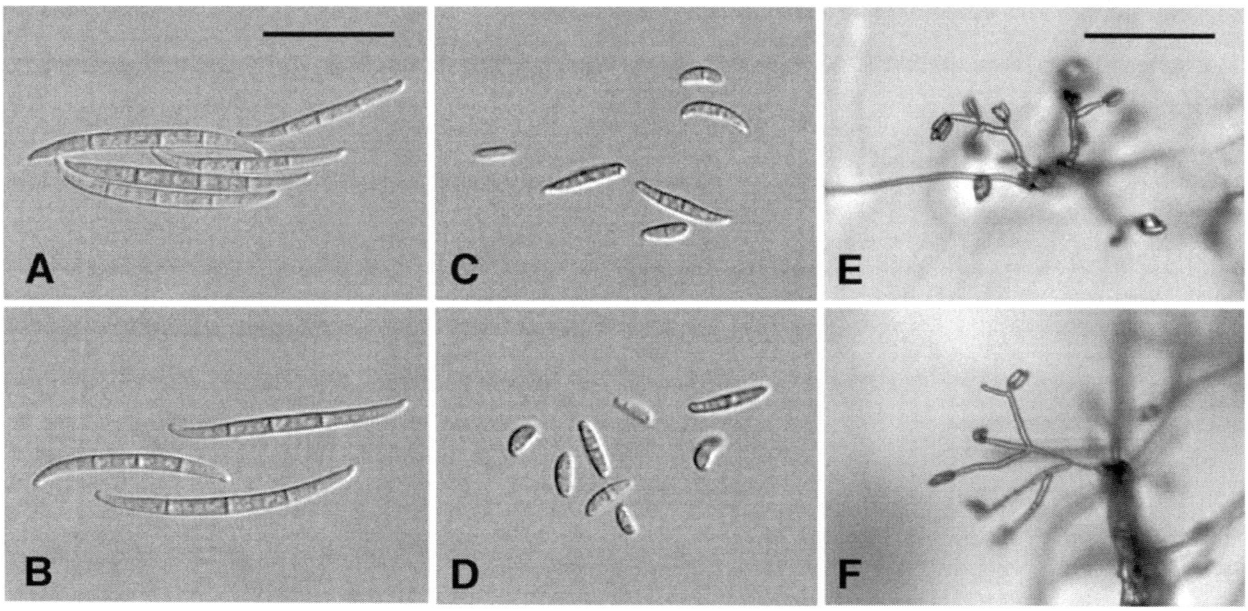

Figure 13-36. *Fusarium mangiferae*.
A – B: Macroconidia; **C – D:** Microconidia; **E – F:** Microconidia *in situ* on CLA. A – D, scale bar = 25 μm; E – F, scale bar = 50 μm.

Reported Geographic and Host Distribution

Causes vegetative and inflorescence malformation on mango (*Mangifera indica*) in Brazil, Egypt, India, Israel, Malaysia, Pakistan, South Africa, and the USA.

Media to Examine and Key Characters

Characters on CLA. Cream colored sporodochia produced on carnation leaves. Macroconidia typical of the *G. fujikuroi* species complex. Obovoid microconidia are produced in small false heads from mono- and polyphialides.

Characters on PDA. White floccose mycelium with light to dark purple pigments in the agar.

Macroconidia

- Sporodochia: Cream-orange color. Relatively abundant on and around carnation leaf pieces in CLA.
- General morphology: Thin walled, long, slender and straight to slightly curved. Typical of species in the *Gibberella fujikuroi* species complex.
- Apical Cell morphology: Curved to almost bent.
- Basal Cell morphology: Foot shaped.
- Number of septa: 3- to 5-septate.
- Abundance: Relatively abundant in sporodochia.

Microconidia

- Shape/septation: Obovoid microconidia are the most common, although oval to allantoid conidia also may be observed. Usually 0-septate, but 1-septate microconidia may occur.
- Aerial mycelium presentation: Small false heads (usually < 5 spores).
- Conidiogenous cells: Monophialides and polyphialides. Polyphialides may have 2-5 conidiogenous openings.

Conidiophores are sympodially branched giving a zig-zag appearance.

- Abundance: Abundant in the aerial mycelia.

Chlamydospores. Absent.

Taxonomy, Pathology and Ecology

This species was described by Britz *et al.* (244) from isolates causing mango malformation disease in Israel (644) and elsewhere that were previously described as *F. subglutinans*. The morphological characters used to describe this species are similar to those of *F. subglutinans* and related species. *F. mangiferae* can be distinguished from *F. sterilihyphosum* as *F. mangiferae* produces longer macroconidia, grows more slowly at 25°C, and does not produce sterile coiled hyphae. Morphologically, *F. mangiferae* is most similar to *F. concentricum* and *F. guttiforme*. *F. mangiferae* can be distinguished from *F. concentricum* by the sympodial branching of the conidiophores that occurs in *F. mangiferae* but not in *F. concentricum*. *F. mangiferae* can be distinguished from *F. guttiforme* because *F. guttiforme* produces only obovoid microconidia and 3-septate macroconidia that are somewhat shorter than the macroconidia produced by *F. mangiferae*. Phylogenetically, *F. mangiferae* groups with the clade of the *G. fujikuroi* species complex that includes *F. proliferatum*, and not with the one that includes *F. subglutinans*, the name with which it is probably most often misidentified.

Molecular sequences have been defined that can be used to differentiate this species from others to which it is closely related (2038). Attempts to generate a sexual stage under laboratory conditions have not yet been successful, even though isolates of opposite mating type have been identified from field populations (244). The isolates examined were not fertile with tester strains of the known biological species in the *G. fujikuroi* species complex. Molecular biological techniques probably should be used to confirm morphological identifications of this species. Field strains are diverse with respect to vegetative compatibility and several types of DNA polymorphisms (1711, 2417).

Fusarium mangiferae is one of three species reported from mango, the others are *F. sterilihyphosum* and an undescribed taxon (244), and is the only species proven to cause the mango malformation disease (644). This disease is economically important in mango cultivation in many parts of the world since infected flowers produce no fruit and result in severe economic losses. In some cases vegetative shoots may be affected as well (1709). Thus while references to either *F. subglutinans* or *F. moniliforme* from mango are discussed below, the certainty of the identity of the referenced isolates as *F. mangiferae* in general has not been confirmed.

Mango malformation was first reported in India over 100 years ago (1153) and has probably spread from there to other areas in which mangos are cultivated. The disease may be soilborne (1731), but it is not clear whether this means of transmission is important for generating new infections. Seedborne transmission has been suggested (1731), but not verified (643). Disease severity can be reduced by pruning of infected branches and the use of disease free seedlings (1583, 1711, 1731). Plant age also is significant in symptom expression (1646). The disease can be vectored by insects and the use of insecticides can reduce disease spread and severity (520). The fungus attacks both the anthers and pistils of infected plants, and may be associated with the young buds very early during their development (1130), which is consistent with the conclusion of Ploetz (1708) that malformation symptoms develop in floral tissue only after extensive colonization by the fungus. The ability to use nitrogen sources present within the flower may be important in determining the severity of infection (343, 344) as may the presence of a dsRNA (795).

There are no reports of mycotoxin production by strains of *F. mangiferae*, although strains in species to which it is phylogenetically closely related are capable of synthesizing gibberellic acid and a variety of mycotoxins including, fumonisins, beauvericin, fusaproliferin and moniliformin. *F. mangiferae* may synthesize cytokinins that are responsible for at least some of the abnormal plant development associated with the mango malformation disease (2221).

Current Species Description: Britz *et al.* (244).

Fusarium merismoides Corda

Sexual Stage. None known.

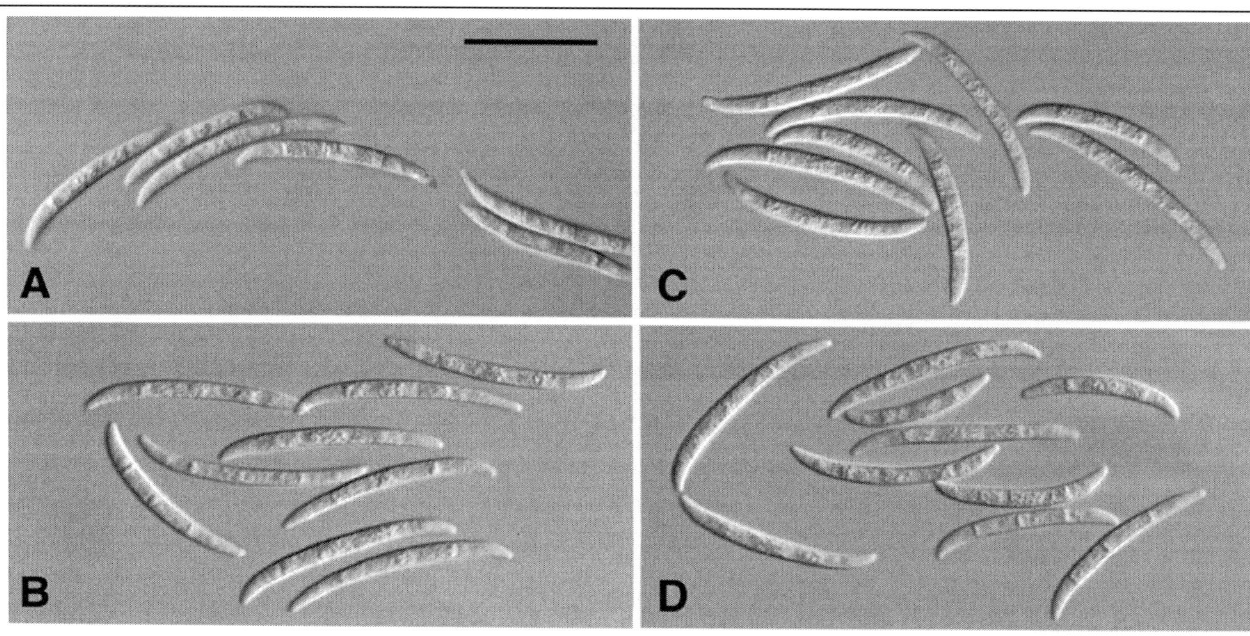

Figure 13-37. *Fusarium merismoides.*
A – D: Macroconidia. Scale bar = 25 µm.

Reported Geographic and Host Distribution

Cosmopolitan from soil.

Media to Examine and Key Characters

Characters on CLA. Abundant sporodochia may be white, yellow, orange or burgundy. Typical macroconidia are small, curved, have a hooked apical cell and a poorly developed basal cell. No microconidia. Chlamydospore formation varies and is not useful as a diagnostic character.

Characters on PDA. Colonies grow very slowly and usually appear to be slimy or yeast-like with little if any aerial mycelia. The colonies may be white, yellow, orange or burgundy, with orange the most common, but there usually is no pigmentation in the agar. In some cases colonies are pigmented in the center and have a white fringe.

Macroconidia

- Sporodochia: Abundant sporodochia may be white, yellow, orange or burgundy and are produced on carnation leaf pieces and the agar surface.
- General morphology: Small and slightly curved.
- Apical Cell morphology: Usually blunt and occasionally hooked.
- Basal Cell morphology: Poorly developed; neither foot shaped nor notched.
- Number of septa: Usually 3- to 4-septate when mature, but immature macroconidia may appear to be 0-septate.
- Abundance: Abundant in the sporodochia.

Microconidia. Absent.

Chlamydospores

- Abundance/Speed of formation: Slow to form and often not observed. The absence of chlamydospores is not a diagnostic character for this species.
- Location: Intercalary within the hyphae.
- Appearance: Singly, in pairs or in chains with smooth walls.

Taxonomy, Pathology and Ecology

Fusarium merismoides is a distinctive species that is not easilyconfused with other species of *Fusarium*. Gerlach & Nirenberg (697) described five varieties, several of which also were recognized by Wollenweber & Reinking (2359). These varieties are differentiated primarily on the basis of differences in the shape of the macroconidia, but in some cases colony pigmentation is an important criterion as well. Neither Booth (222) nor Nelson *et al.* (1551) recognized these varieties. Putative plasmids have been identified in *F. merismoides* (1847) that may be associated with senescence (687).

Isolates of *F. merismoides* have been recovered from many ecosystems, but this species seems to be more abundant in cool temperate regions, *e.g.*, from sub-Antarctic Islands (1805) and inside the Arctic circle in Norway (1116).

There are relatively few studies in which *F. merismoides* is the focus of the research rather than an incidental participant. There are records of this fungus as a plant pathogen on a number of different species including grapes (525), ironwood (*Ostrya virginiana*; 525), and tomato (631). Booth (222) noted that some diseases previously attributed to *F. merismoides* probably should be attributed to *F. oxysporum* instead. Hoch & Abawi (870) noted that *F. merismoides* is a parasite of *Pythium ultimum* oospores and might be useful as a biocontrol for *pythium*. We think that *F. merismoides* is probably best regarded as a saprophyte, but that it has the potential to cause some plant diseases if the environmental conditions are appropriate.

Marasas *et al.* (1341) identified one strain of *F. merismoides* from potato tubers in northern Germany that was slightly toxigenic to brine shrimp, but the compound responsible for the toxigenicity was not identified. *F. merismoides* has been reported to produce azepinostatin, a protein kinase C inhibitor (1607).

Current Species Description: Gerlach & Nirenberg (697).

Fusarium miscanthi W. Gams, Klamer & O'Donnell

Sexual Stage. None known.

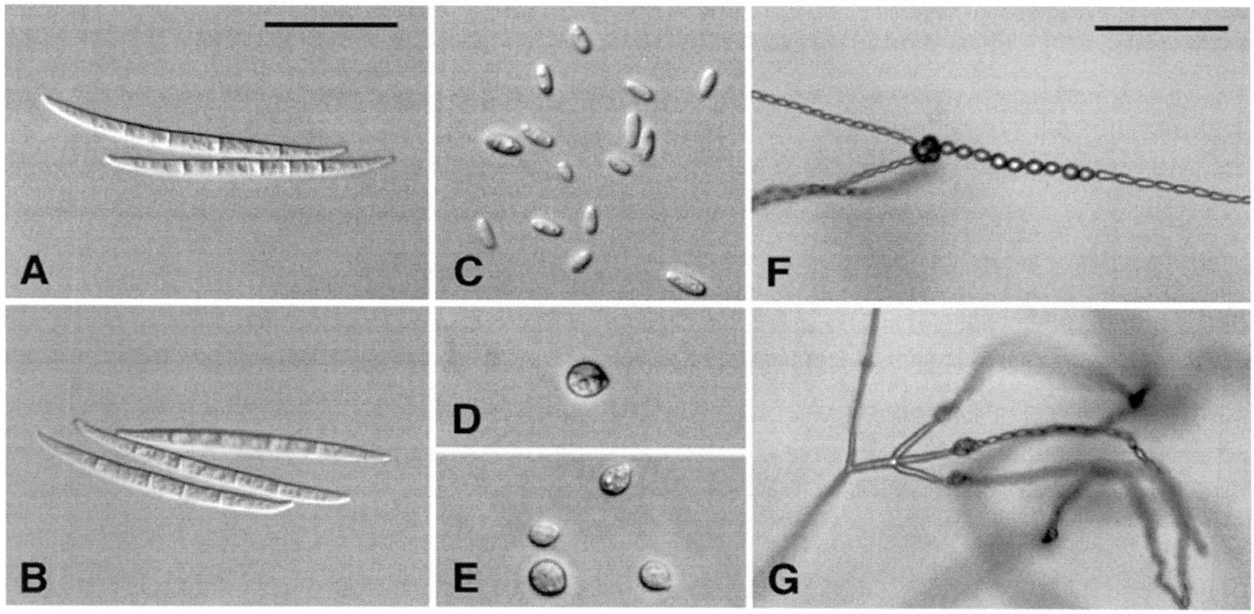

Figure 13-38. *Fusarium miscanthi.*
A – B: Macroconidia; **C:** Oval microconidia; **D – E:** Napiform microconidia; **F – G:** Microconidia *in situ* on CLA. A – E, scale bar = 25 μm; F – G, scale bar = 50 μm.

Reported Geographic and Host Distribution

Isolated from the litter of *Miscanthus sinensis* buried in agricultural soil in Denmark.

Media to Examine and Key Characters

Characters on CLA. Pale orange sporodochia are produced sparsely on carnation leaf pieces, particularly in the older parts of the colony. Macroconidia are typical of species in the *G. fujikuroi* species complex. Two types of microconida are produced. One is fusiform to clavate and the other pyriform. Micrconidia usually are found in chains from monophialides. No chlamydospores are produced.

Characters on PDA. White, floccose but scanty aerial mycelium. Purple pigments are produced in the agar.

Macroconidia

- Sporodochia: Pale orange sporodochia appear after ~3 weeks, usually in the center of the colony on PDA plates and on carnation leaf pieces in older parts of the colony on CLA plates.
- General morphology: Thin walled and straight to falcate. Typical of those found in species of the *G. fujikuroi* species complex.
- Apical Cell morphology: Curved and tapering.
- Basal Cell morphology: Foot shaped.
- Number of septa: 3- to 5-septate.
- Abundance: Abundant in sporodochia once the sporodochia form.

Microconidia

- Shape/septation: Two types of 0-septate microconidia are formed – pyriform and clavate to fusiform. The clavate to fusiform microconidia are more common, but the pyriform microconidia are of diagnostic importance.
- Aerial mycelium presentation: Long chains and false heads. Cultures grown on SNA may be required to ob-

serve the long chains of pyriform microconidia, as pyriform microconidial production may not be common on CLA.

- Conidiogenous cells: Primarily monophialides with occasional polyphialides.
- Abundance: Abundant in the aerial mycelia.

Chlamydospores. Absent.

Taxonomy, Pathology and Ecology

This species was described by Gams *et al.* (672) from isolates recovered from litter of *Miscanthus sinensis* (Poaceae) that was used in composting experiments in Denmark. The grass is an ornamental species introduced from Japan.

Morphologically *F. miscanthi* is most closely related to *F. proliferatum*, *F. globosum* and *F. nisikadoi*. *F. miscanthi* is distinguished from *F. proliferatum* by the production of pyriform microconidia in chains. *F. proliferatum* may occasionally produce pyriform microconidia, but these are not present in the long chains found in this species. *F. globosum* is readily distinguished from *F. micanthi* by the formation of globose microconidia by *F. globosum*. *F. nisikadoi* and *F. miscanthi* are more difficult to distinguish. The fusiform microconidia of these species are quite different in size. Although both species are reported to produce pyriform microconidia, the strains of *F. nisikadoi* examined by Gams *et al.* (672) did not produce these spores, which could provide a second trait that can be used to distinguish these closely related species.

Phylogenetically *F. miscanthi* is most closely related to *F. nisikadoi* (672), and together these species are most closely related to *F. oxysporum* (106). The distance from *F. miscanthi* to *F. nisikadoi* is four fixed nucleotide changes in the 28 mtSSU gene and seven fixed nucleotide changes in the β-tubulin gene.

Attempts to form the sexual stage of *F. miscanthi* under laboratory conditions have not been successful, although some strains form sclerotia-like structures that often are indicative of the ability to serve as a female parent in a cross. Crosses between *F. miscanthi* and *F. nisikadoi* have not been attempted to determine if the phylogenetic and biological species concepts are coincident for these two species.

Little else is known of the biology or ecology of this species.

Current Species Description: Gams *et al.* (672).

Fusarium musarum Logrieco & Marasas

Sexual Stage. None known.

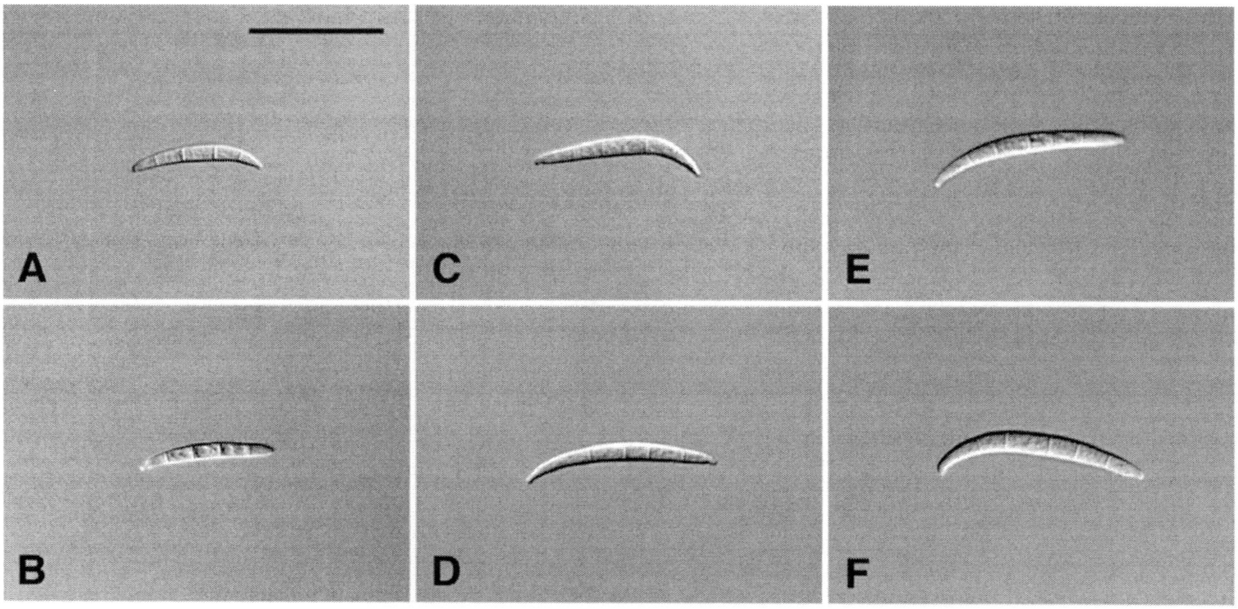

Figure 13-39. *Fusarium musarum.*
A – F: Mesoconidia. Scale bar = 25 µm.

Reported Geographic and Host Distribution

From banana in Panama.

Media to Examine and Key Characters

Characters on CLA. Sporodochia and macroconidia are absent. Large mesoconidia are produced in the aerial mycelia from mono- and polyphialides. Chlamydospores usually are present but are slow to form.

Characters on PDA. Relatively fast growing colony with floccose to powdery, red mycelia. Produces red pigments in the agar.

Macroconidia. Absent.

Microconidia/Mesoconidia

- Shape/septation: Mesoconidia are 2- to 9-septate, but are primarily 5- to 6-septate. These spores are hyaline, falcate and thick walled with a pointed apical cell and a conical basal cell. The basal cell is not foot shaped.
- Aerial mycelium presentation: Produced singly from monophialides and polyphialides.
- Conidiogenous cells: Polyphialides with 2-4 openings.
- Abundance: Abundant from aerial hyphae in the aerial mycelium. No sporodochia are produced.

Chlamydospores

- Abundance/Speed of formation: Usually present, but produced slowly and sparsely. Their absence is not a reliable taxonomic character.
- Location: Intercalary in aerial or submerged hyphae.
- Appearance: Hyaline and globose or subglobose with smooth walls. May be present singly, in clusters, or in short chains.

Taxonomy, Pathology and Ecology

This species was described by Marasas *et al.* (1347) from cultures isolated in Panama from banana fruit. It is unusual in that it does not produce macroconidia but all of its other characters align it closely with species in the section *Arthrosporiella* such as *F. semitectum* and *F. nelsonii.* The absence of sporodochia with macroconidia clearly differentiates *F. musarum* from *F. nelsonii* and *F. camptoceras*, both of which produce both sporodochia with macroconidia and mesoconidia. In practice the term mesoconidium is used here for any spore formed in the aerial mycelium and the term macroconidium is reserved for spores formed in sporodochia. The mesoconidia produced by *F. musarum* are generally longer and wider than the mesoconidia produced by either *F. camptoceras* or *F. nelsonii* or the macroconidia produced by *F. nelsonii*. *F. musarum* also grows more rapidly than do either *F. nelsonii* or *F. camptoceras* on PDA incubated at both 25° and 30 °C. *F. nelsonii* produces chlamydospores rapidly, while both *F. camptoceras* and *F. musarum* produce chlamydospores slowly and sparsely if they are produced at all. Little is known of the ecology or biology of *F. musarum* other than that it occurs on banana fruits in Central America.

Isolates of *F. musarum* can produce trichothecenes such T-2 and HT-2 toxins (1491). This species is one of the few known from tropical regions that can produce trichothecene toxins.

Current Species Description: Marasas *et al.* (1347).

Fusarium napiforme Marasas, Nelson & Rabie

Sexual Stage. None known.

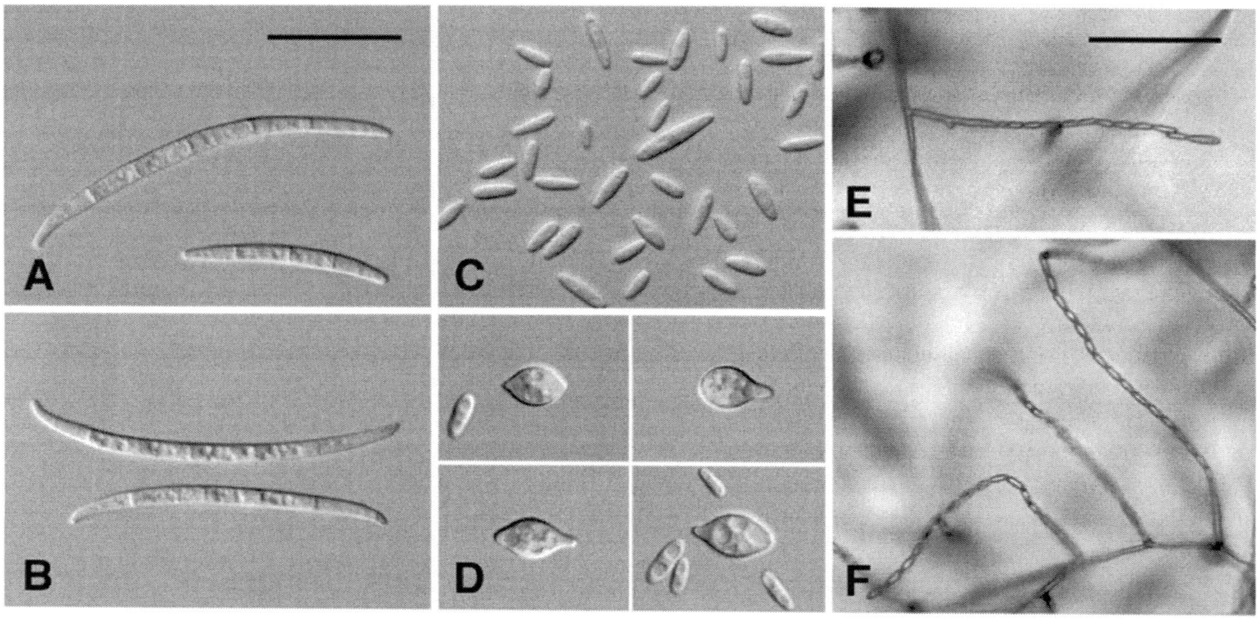

Figure 13-40. *Fusarium napiforme.*
A – B: Macroconidia; **C:** Ovoid to obovoid microconidia; **D:** Napiform microconidia; **E – F:** Microconidia *in situ* on CLA. A – D, scale bar = 25 μm; E – F, scale bar = 50 μm.

Reported Geographic and Host Distribution

Recovered from soil, sorghum and millet grains and poultry feed in Africa, Argentina and Australia.

Media to Examine and Key Characters

Characters on CLA. Orange sporodochia are produced in the aerial mycelia. Macroconidia are moderately long, falcate to straight, hyaline and thin walled. Three types of conidia are formed in the aerial mycelia. Obovoid microconidia are abundant in chains and false heads from monophialides. Napiform microconidia are diagnostic and are produced slowly either singly or in small false heads (<5 spores). Fusiform mesoconidia are produced sparsely and singly. Chlamydospores are produced relatively slowly.

Characters on PDA. Floccose white colonies with purple pigments in the agar.

Macroconidia

- Sporodochia: Medium to bright orange sporodochia are found in the aerial mycelia.
- General morphology: Moderately long, hyaline, and falcate to straight.
- Apical Cell morphology: Tapering and slightly curved.
- Basal Cell morphology: Foot shaped.
- Number of septa: 3- to 5-, mostly 5-septate.
- Abundance: Abundant in sporodochia and found occasionally on the agar surface.

Microconidia/Mesoconidia

- Shape/septation: Hyaline, 0- to 1-septate, obovoid, lemon-shaped, and napiform microconidia are produced by all strains. The obovoid form is most abundant. Napiform microconidia are sparse and may take up to three weeks to appear. 1- to 3-septate fusiform mesoconidia are produced rarely by some strains.

- Aerial mycelium presentation: False heads and short chains ($\leq$ 10 spores) on CLA, with longer chains (up to 25 spores) on KCl media.
- Conidiogenous cells: Monophialides.
- Abundance: Abundant in the aerial mycelia.

Chlamydospores

- Abundance/Speed of formation: Produced relatively slowly, requires at least 3 weeks, on CLA. The chlamydospores often occur sparsely and are difficult to observe.
- Location: Terminal or intercalary in both submerged and aerial hyphae.
- Appearance: Hyaline to pale brown and smooth walled. Smooth walled, usually in chains or clumps but occasionally singly.

Taxonomy, Pathology and Ecology

Fusarium napiforme was first described by Marasas *et al.* (1344) from cultures isolated from millet and sorghum grain in South Africa and Namibia. This species bears some similarity to a number of species in the *Gibberella fujikuroi* species complex, most notably *F. dlamini* and *F. nygamai*, both of which also can produce chlamydospores. Both the chlamydospores and napiform conidia are slow to form, and young cultures may be mistakenly identified as *F. verticillioides*. *F. napiforme* can be easily distinguished from *F. dlamini* since *F. napiforme* produces microconidia in chains and *F. dlamini* does not. The production of napiform and lemon-shaped microconidia by *F. napiforme* and the production of polyphialides by *F. nygamai* can be used to morphologically distinguish these species. Maximum linear growth is reported to occur at 25°C and an osmotic potential of -1.0 MPa (1541).

Although first identified from southern Africa, *F. napiforme* also has been identified from sorghum grain (729) and poultry feed in Argentina (1319), from sorghum grain (1626), millet grain (1628) and soil in various parts of Africa (974), and soil in various locations in Australia (277, 1889). *F. napiforme* has been associated with human cases of hypersensitivity pneumonitis (1208) and disseminated hyalohyphomycosis (1418).

Some strains of *F. napiforme* can produce moniliformin (1350), fusaric acid (119), and fumonisins (1548).

Current Species Description: Marasas *et al.* (1344).

Fusarium nelsonii Marasas & Logrieco

Sexual Stage. None known.

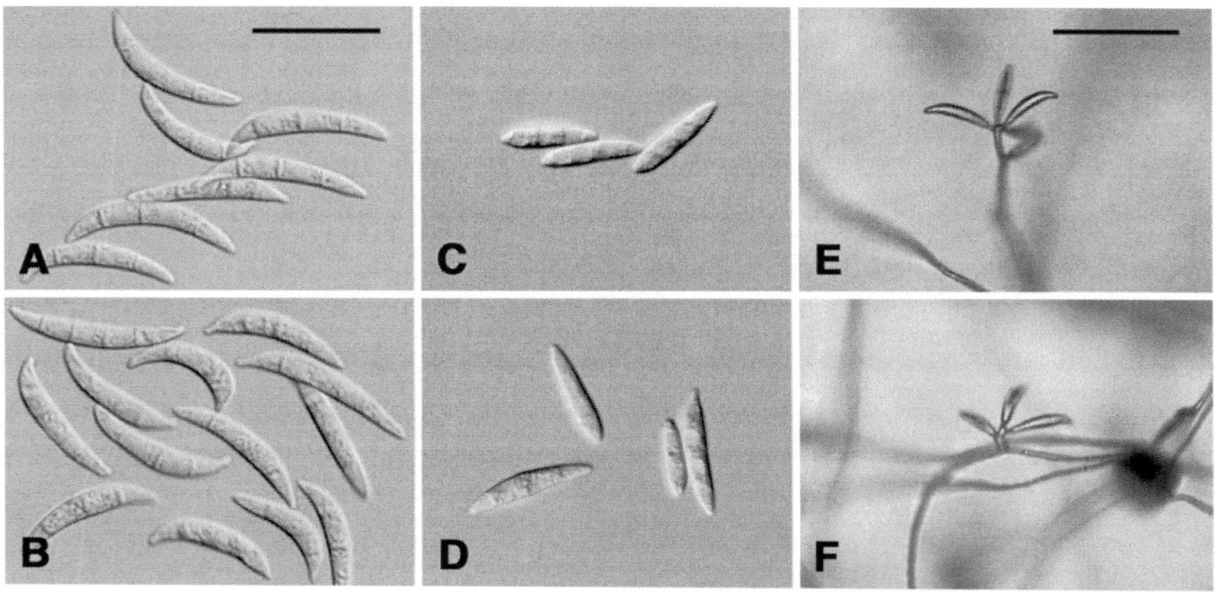

Figure 13-41. *Fusarium nelsonii.*
A – B: Macroconidia; **C – D:** Mesoconidia; **E – F:** Mesoconidia *in situ* on CLA. A – D, scale bar = 25 μm; E – F, scale bar = 50 μm.

Reported Geographic and Host Distribution

Found in South Africa and isolated from plant debris in soil, *Medicago* roots, and sorghum malt.

Media to Examine and Key Characters

Characters on CLA. Sporodochia appear as pale cream-colored droplets after 2-6 weeks, and may become confluent on the plate and orange in color. Macroconidia are predominantly curved with a distinctive and strongly curved and beak-shaped apical cell. Fusiform mesoconidia are produced in the aerial mycelium from mono- and polyphialides. Cross-shaped polyphialides are commonly observed. Chlamydospores are abundant and form rapidly.

Characters on PDA. Floccose to powdery mycelia with white to red pigmentation. Produces a distinctive carmine red pigment in the agar.

Macroconidia

- Sporodochia: Pale cream-colored droplets are common but produced slowly in the aerial mycelia. May become orange with age and confluent on the plate.
- General morphology: Straight or curved and falcate.
- Apical Cell morphology: Curved and beak shaped.
- Basal Cell morphology: Foot shaped.
- Number of septa: Usually 3-, but occasionally 5-septate.
- Abundance: common in sporodochia.

Microconidia/Mesoconidia

- Shape/septation: Mesoconidia may be fusiform or lanceolate and are straight to slightly curved. They are 0- to 3-septate, primarily 3-septate.
- Aerial mycelium presentation: Singly.
- Conidiogenous cells: Polyphialides.
- Abundance: Abundant in the aerial mycelia.

Chlamydospores

- Abundance/Speed of formation: Produced rapidly and abundantly.
- Location: Intercalary or terminal in submerged or aerial hyphae.
- Appearance: Initially smooth walled and hyaline becoming verrucose and pale brown with time. Most commonly found in terminal pairs, but may also be found singly, in clumps, and in short chains.

Taxonomy, Pathology and Ecology

This species was first described by Marasas *et al.* (1347) and was placed in section *Arthrosporiella*. This section includes *F. semitectum* and *F. camptoceras* which share some similarities to *F. nelsonii,* most notably the presence of mesoconidia in the aerial hyphae (particularly on CLA). In practice the term mesoconidium is used here for any spore formed in the aerial mycelium and the term macroconidium is reserved for spores formed in sporodochia. *F. nelsonii* produces both macroconidia and mesoconidia which distinguishes it from *F. musarum* which produces only mesoconidia. *F. nelsonii* mesoconidia and macroconidia are shorter than those of *F. camptoceras. F. camptoceras* also produces chlamydospores more slowly than does *F. nelsonii* and produces brown pigments on PDA in the dark while *F. nelsonii* produces red pigments. Little else is known of the ecology of this species and there are no reports of its ability to produce mycotoxins.

Current Species Description: Marasas *et al.* (1347).

Fusarium nisikadoi T. Aoki & Nirenberg

Sexual Stage. None known.

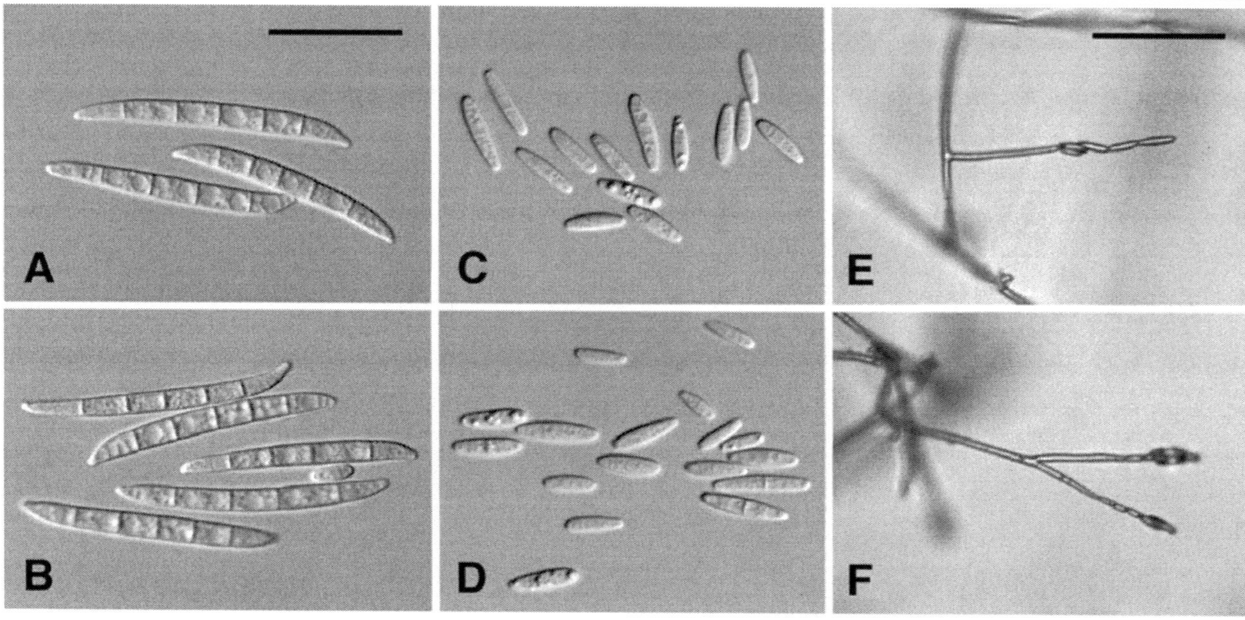

Figure 13-42. *Fusarium nisikadoi.*
A – B: Macroconidia; **C – D:** Microconidia; **E – F:** Microconidia *in situ* on CLA. A – D, scale bar = 25 μm; E – F, scale bar = 50 μm.

Reported Geographic and Host Distribution

Recovered from nuts, wheat and bamboo in Japan.

Media to Examine and Key Characters

Characters on CLA. Sporodochia are rare and long macroconidia are produced in the hyphae. Chains of clavate microconidia are produced in the aerial mycelium, normally from long monophialides. The chains often have a distinctive zig-zag appearance.

Characters on PDA. White floccose mycelia with gray to orange pigment in the agar.

Macroconidia

- Sporodochia: Are produced rarely, but when produced are orange and found on the carnation leaves.
- General morphology: Long and straight to falcate.
- Apical Cell morphology: Acute.
- Basal Cell morphology: Foot shaped.
- Number of septa: 1- to 9-septate with 3- to 7-septate the most common.
- Abundance: Macroconidia usually are produced abundantly by hyphae in the agar.

Microconidia

- Shape/septation: Both 0-septate pyriform microconidia and 0- to 3-septate clavate microconidia are produced. The clavate microconidia are common while the pyriform microconidia are produced only occasionally.
- Aerial mycelium presentation: Long chains and false heads. The chains often have a zig-zag appearance.
- Conidiogenous cells: Predominantly monophialides with occasional polyphialides
- Abundance: Very abundant in the aerial mycelia.

Chlamydospores. Absent.

Taxonomy, Pathology and Ecology

This species was first described by Nirenberg & Aoki (1569) from isolates from bamboo and wheat in Japan. It is superficially similar to other species that form chains of microconidia, *e.g.*, *F. verticillioides, F. thapsinum* and *F. proliferatum*. However, the presence of microconidia in these chains that are up to 3-septate allows easy differentiation of this species. *F. nisikadoi* and *F. miscanthi* are more difficult to distinguish. The fusiform microconidia of these species are quite different in size. Although both species are reported to produce pyriform microconidia, the strains of *F. nisikadoi* examined by Gams *et al.* (1569) did not produce these spores. Thus, the relative frequency of pyriform microconidia may be a secondary character that can be used to distinguish these closely related species.

Phylogenetically *F. nisikadoi* is most closely related to *F. miscanthi* (672), and together these species are most closely related to *F. oxysporum* (106). The distance from *F. miscanthi* to *F. nisikadoi* is four fixed nucleotide changes in the 28 mtSSU gene and seven fixed nucleotide changes in the β-tubulin gene.

Isolates of *F. nisikadoi* produce trace amounts of moniliformin but no detectable amounts of beauvericin, fumonisins, or fusaproliferin (638). Little else is known of the biology or ecology of this species.

Current Species Description: Nirenberg & Aoki (1569).

Fusarium nurragi (Summerell & Burgess) Benyon, Summerell & Burgess

Sexual Stage. None known.

Common Synonyms. *Fusarium avenaceum* ssp. *nurragi.*

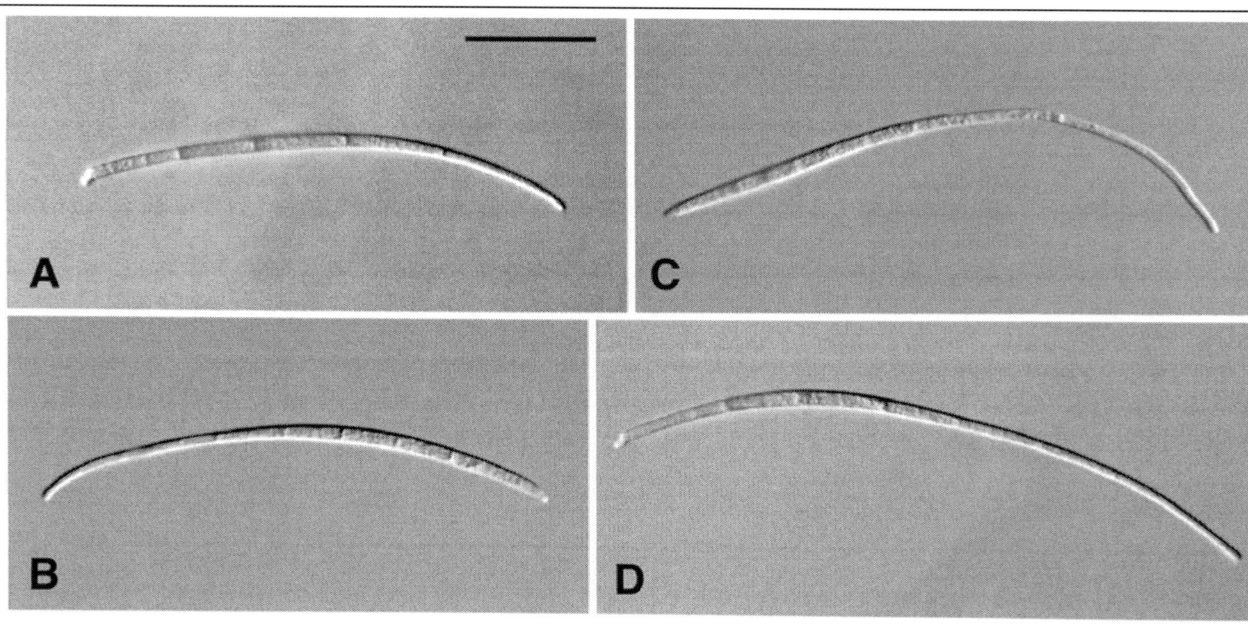

Figure 13-43. *Fusarium nurragi.*
A – D: Macroconidia; scale bar = 25 μm.

Reported Geographic and Host Distribution

Found in southern Australia in soil associated with coastal heathland vegetation.

Media to Examine and Key Characters

Characters on CLA. Extremely long macroconidia are produced in orange sporodochia on and around the carnation leaf pieces. The macroconidia are long and thin and usually gently curved with a tapering apical cell and an elongate foot-shaped basal cell. Clumps of macroconidia also are found in the aerial mycelia. Neither microconidia nor chlamydospores are produced.

Characters on PDA. Produces dense aerial mycelia that is white to orange and pink to orange pigments in the agar.

Macroconidia

- Sporodochia: Orange sporodochia are produced on the carnation leaves and on the agar surrounding the carnation leaves. In some isolates the sporodochia are found in annular zonations (rings) in response to diurnal lighting.
- General morphology: Long, thin and whip-like in appearance.
- Apical Cell morphology: Long tapering apical cell.
- Basal Cell morphology: Elongate foot shape.
- Number of septa: Usually 5-septate.
- Abundance: Very abundant in both the sporodochia and the aerial mycelium.

Microconidia. Absent.

Chlamydospores. Absent.

Taxonomy, Pathology and Ecology

This species was first described by Sangalang *et al.* (1890) as a subspecies of *F. avenaceum* but was recognized as a species in its own right following molecular analysis showed that it could be clearly differentiated from *F. avenaceum* and *F. aywerte* (169). Morphologically *F. nurragi* is quite distinctive, although it could potentially be confused with *F. aywerte*. *F. aywerte* has been recovered only from stems and from soil around the roots of the spinifex grasses in central Australia. The known distribution of *F. nurragi* is quite different and includes shrubby coastal heath species in Victoria and Western Australia. Thus the geographic distribution and the host species from which the isolate is recovered normally should suffice to distinguish *F. aywerte* from *F. nurragi*. Toxin production by *F. nurragi* has not been reported.

Current Species Description: Sangalang *et al.* (1890).

Fusarium nygamai Burgess & Trimboli

Sexual Stage. *Gibberella nygamai* Klaasen & Nelson.

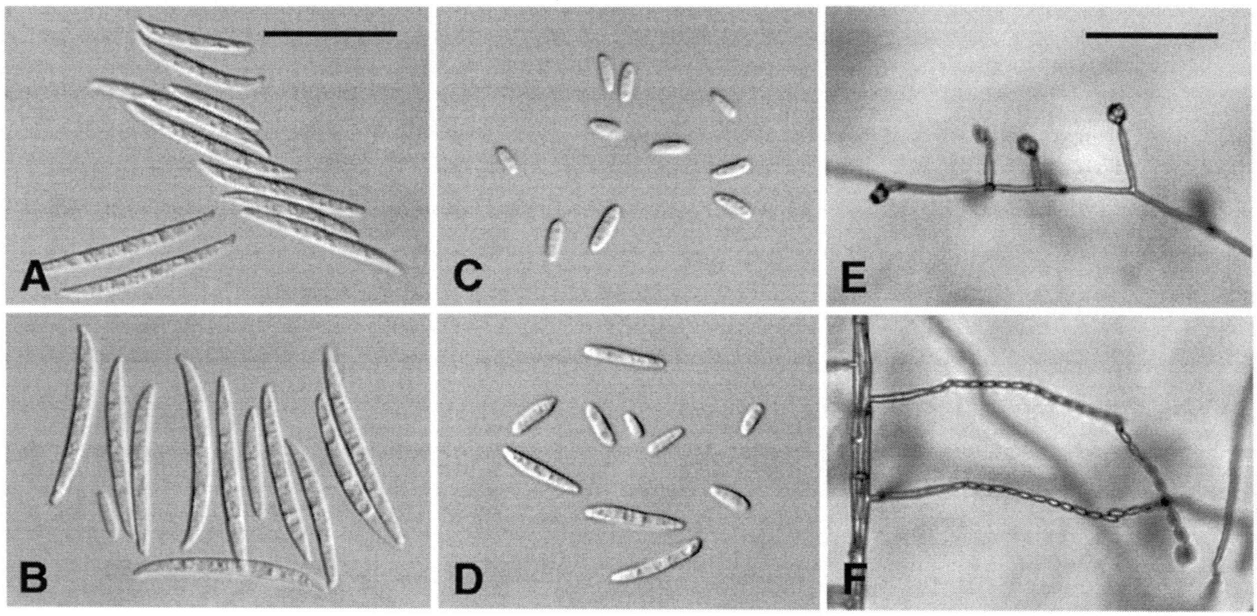

Figure 13-44. *Fusarium nygamai.*
A – B: Macroconidia; **C – D:** Microconidia; **E – F:** Microconidia *in situ* on CLA. A – D, scale bar = 25 µm; E – F, scale bar = 50 µm.

Reported Geographic and Host Distribution

Hot, often arid locations. Frequently associated with sorghum, but also may be recovered from cotton, maize, native grasses, rice and soil.

Media to Examine and Key Characters

Characters on CLA. Macroconidia are abundant in orange sporodochia and are short to medium in length, slender, falcate to almost straight, usually 3-septate, thin walled with a short tapered apical cell and a notched or foot-shaped base to the basal cell. Microconidia are small, 0-septate, oval to elliptical in shape and formed predominantly from monophialides in false heads. Polyphialides and short chains ($\leq$ 10 spores) also can be observed, primarily in older cultures or at the edge of the colony. Chlamydospores are formed, but their frequency varies by strain from rare to very abundant.

Characters on PDA. The mycelium initially is white becoming violet with age, with violet pigments produced in the agar. A grayish-orange to dark violet spore mass is present in the center of most colonies.

Macroconidia

- Sporodochia: Abundant orange sporodochia found on carnation leaf pieces in CLA.
- General morphology: Slender, thin walled, hyaline, straight to slightly curved and medium length.
- Apical Cell morphology: Short and tapered.
- Basal Cell morphology: Notched or foot shaped.
- Number of septa: 3- to 5-septate, but primarily 3-septate.
- Abundance: Abundant in sporodochia.

Microconidia

- Shape/septation: Small, oval or club shaped. Usually 0-septate, but occasionally 1-septate.

- Aerial mycelium presentation: False heads and chains. False heads are common. Chains are usually short (≤ 10 spores), but may be up to 20 microconidia in length. Chains are more common at the edges of the culture plate and in older cultures on CLA. Chain length is a more constant character on SNA than it is on CLA.
- Conidiogenous cells: Monophialides and occasional polyphialides. Polyphialides are more common at the edges of the culture plate and in older cultures.
- Abundance: Abundant in the aerial mycelia.

Chlamydospores

- Abundance/Speed of formation: Varies by strain from uncommon to very abundant. On CLA, chlamydospores usually can be observed after 2-4 weeks, but some strains must be cultured on soil agar for chlamydospores to appear. Chlamydospore formation is more consistent on SNA after 4-6 weeks.
- Location: In hyphae on the agar surface, but more commonly in the aerial mycelium. Formed singly, in chains and in clusters in intercalary positions in the hyphae.
- Appearance: Smooth- or rough-walled. Hyaline to pale yellow color.

Taxonomy, Pathology and Ecology

With respect to morphological characteristics, *F. nygamai* is intermediate between *F. verticillioides*, *F. thapsinum* and *F. oxysporum*. All of these species include representatives that produce violet pigmentation in colonies on PDA. The macroconidia of *F. nygamai* could be mistaken for those of any of these species but usually are more like the macroconidia of *F. verticillioides* and *F. thapsinum*. *F. nygamai*, however, forms abundant sporodochia on CLA, a characteristic of *F. oxysporum*. The microconidia of *F. nygamai* resemble those of *F. verticillioides*. *F. nygamai* cannot be included in section *Liseola* as defined by Wollenweber & Reinking (2359) because it forms chlamydospores. It cannot be included in section *Elegans* because it forms microconidia in chains. Phylogenetically, *F. nygamai* is most closely related to *F. thapsinum* (1596), but *F. nygamai* also may be a species complex that is in need of further resolution (1233) which could make it difficult to position accurately on a phylogenetic tree. The stability of the morphological features of *F. nygamai* was studied by Burgess *et al.* (274) who showed the key characters of the fungus to be stable in culture and following long-term storage. Polyphialide production and culture pigmentation varied and were not considered reliable characters for species identification. Maximum linear growth is reported to occur at 25°C and an osmotic potential of -1.0 MPa (1541).

The biological species defined by *G. nygamai* probably does not include all of the strains identified as *F. nygamai* (1233). The two mating type tester strains (FGSC 8933 and FGSC 8934) described by Klassen & Nelson (1085, 1086) are not highly fertile as female parents, and obtaining the sexual stage is more difficult for *F. nygamai* than it is for other species in the *G. fujikuroi* species complex.

Fusarium nygamai was first observed during surveys of fungi associated with basal stalk and root rot of grain sorghum in New South Wales in 1977 and 1978 (2178) when it was identified as *F. moniliforme* because no chlamydospores were observed. Later, some of these isolates produced chlamydospores when inoculated on sorghum roots and grown under greenhouse conditions. *F. nygamai* is abundant in hot dry areas, *e.g.*, soils in the Alice Springs region of central Australia (1889), although it also can be found in wet tropical regions, *e.g.*, the Malaysian peninsula (1876) and the Indonesian archipelago. It can dominate in the soil of some native grasslands (277, 1630). *F. nygamai* can cause a root rot of broad beans (*Vicia faba*) in the Sudan (1160) and is associated with asparagus (570), cotton (2414), maize (590), millet (136), rice (1628), and sorghum (1626). Some strains of this species are pathogenic towards sorghum in an *in vitro* seedling pathogenicity assay (1233).

Strains of *F. nygamai* are toxic to ducklings (1233) and may cause systemic infections in humans (1145). This species can produce beauvericin (1283) and a number of variants of fusaric acid (52, 309). Some strains of *F. nygamai* produce high levels of fumonisins (1233, 1507, 1548, 2132) and carry the *FUM* biosynthetic gene cluster (1740) The *F. nygamai* strains from millet described by Marasas *et al.* (1350) that produce moniliformin and that were toxic to ducklings probably are *F. pseudonygamai*. No other *F. nygamai* strains have been reported to produce moniliformin (1233).

Current Species Description: Burgess & Trimboli (281).

Fusarium oxysporum Schlechtendahl emend. Snyder & Hansen

Sexual Stage. None known.

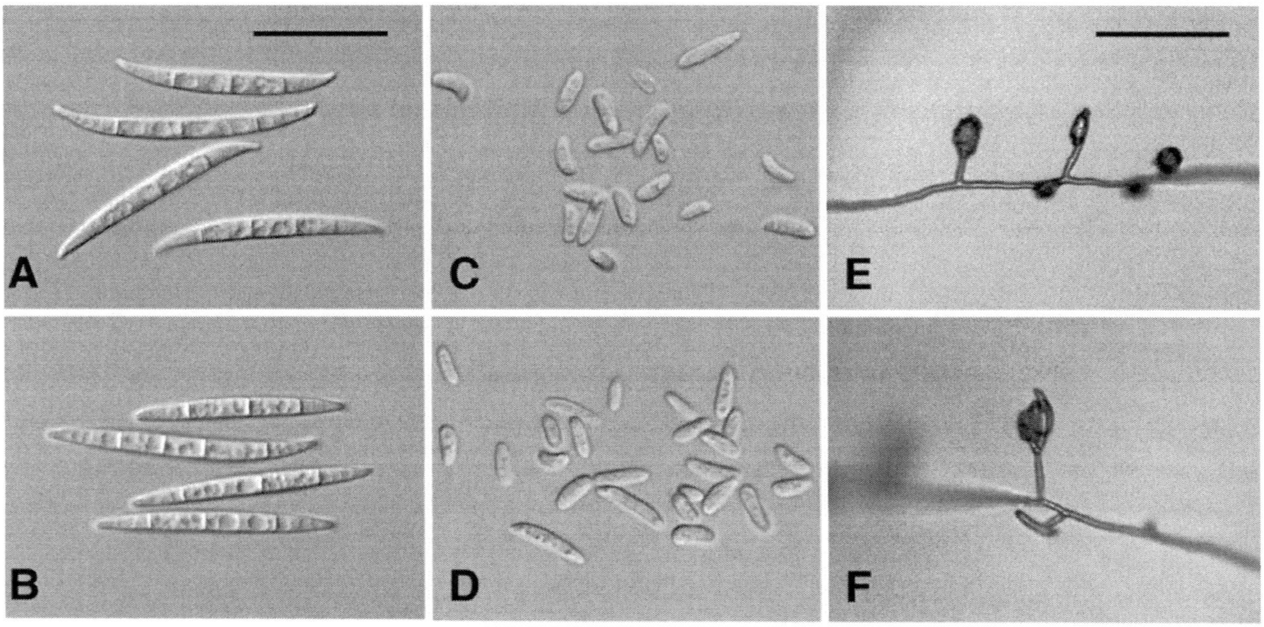

Figure 13-45. *Fusarium oxysporum.*
A – B: Macroconidia; **C – D:** Microconidia; **E – F:** Microconidia *in situ* on CLA. A – D, scale bar = 25 μm; E – F, scale bar = 50 μm.

Reported Geographic and Host Distribution

Cosmopolitan. An important vascular wilt pathogen on many plant species worldwide. A common soil saprophyte.

Media to Examine and Key Characters

Characters on CLA. Macroconidia are formed in pale orange, usually abundant, sporodochia. The macroconidia are short to medium in length, falcate to almost straight, thin walled and usually 3-septate. The apical cell is short and is slightly hooked in some isolates. The basal cell is notched or foot-shaped. Macroconidia are formed from monophialides on branched conidiophores in sporodochia and to a lesser extent from monophialides on hyphae. Microconidia usually are 0-septate, may be oval, elliptical or reniform (kidney-shaped), and are formed abundantly in false heads on short monophialides. Chlamydospores are formed abundantly in hyphae on the agar surface by most isolates, especially saprophytic clones from soil, but may be slow (4-6 weeks) to form in some isolates.

Characters on PDA. Colony morphology on PDA varies widely. Mycelia may be floccose, sparse or abundant and range in color from white to pale violet. Abundant pale orange or pale violet macroconidia are produced in a central spore mass in some isolates. Small pale brown, blue to blue-black or violet sclerotia may be produced abundantly by some isolates. *F. oxysporum* usually produces a pale to dark violet or dark magenta pigment in the agar but some isolates produce no pigment at all. Some isolates of *F. oxysporum* mutate readily to the pionnotal form or to a flat "wet" mycelial colony with a yellow to orange appearance when cultured on PDA.

Macroconidia

- Sporodochia: Most isolates produce abundant pale orange sporodochia; however, in some isolates sporodochia may be sparse or non-existent.
- General morphology: Short to medium length, straight to slightly curved, relatively slender and thin walled.
- Apical Cell morphology: Tapered and curved, sometimes with a slight hook.
- Basal Cell morphology: Foot shaped to pointed.
- Number of septa: Usually 3-septate.
- Abundance: Sparse in some strains, but usually abundant in sporodochia and occasionally from hyphae growing on the agar surface.

Microconidia

- Shape/septation: Oval, elliptical or kidney shaped and usually 0-septate.
- Aerial mycelium presentation: False heads.
- Conidiogenous cells: Short monophialides.
- Abundance: Abundant in the aerial mycelia.

Chlamydospores

- Abundance/Speed of formation: Formed abundantly and quickly (2-4 weeks on CLA) by most isolates, but some isolates form chlamydospores slowly if they form them at all.
- Location: Usually formed singly or in pairs, but also may be found in clusters or on short chains. May be either terminal or intercalary in aerial, submerged, or surface hyphae.
- Appearance: Smooth or rough walled.

Taxonomy, Pathology and Ecology

Fusarium oxysporum is the most widely dispersed of the *Fusarium* species and can be recovered from most soils – Arctic (1116), tropical or desert (984, 1328), and cultivated or not (1410, 1412). It also can be dispersed by insects (704) and recovered from marine algae (764). *F. oxysporum* also is without a doubt the most economically important species in the *Fusarium* genus given its numerous hosts and the level of loss that can result when it infects a plant. The *F. oxysporum* species is one of two of the nine species descriptions of Snyder and Hansen that remains in general use. Yet *F. oxysporum* is clearly heterogeneous and composed of at least dozens of species that need to be clearly defined and separated in a usable manner. Both its dispersion and its economic importance have made this the *Fusarium* species on which the most scientific research has been conducted, with ~6000 articles available through a computerized search of *Biological Abstracts* as of this writing. Clearly the comments below can identify only general trends and provide a few examples.

The critical morphological features of *F. oxysporum* include the production of microconidia in false-heads on short phialides formed on the hyphae, the production of chlamydospores and the shape of the macroconidia and the microconidia. The number and morphology of the macro- and microconidia formed can be drastically altered by apparently single-gene mutations (1605, 1606). Isolates of *F. oxysporum* are most difficult to distinguish from those of *F. solani* and *F. subglutinans*. *F. solani* forms microconidia in false-heads on very long monophialides formed on the hyphae. *F. subglutinans* is distinguished from *F. oxysporum* by the formation of microconidia from polyphialides and the absence of chlamydospores. Polyphialides are, however, difficult to find in some isolates of *F. subglutinans*, and chlamydospores may be formed slowly by some isolates of *F. oxysporum*, so careful and patient observation may be required to distinguish these species morphologically. Strains of *F. oxysporum* that carry dsRNAs have an apparently normal morphological phenotype and do not confer hypovirulence on infected strains (1050). Maximum linear growth is reported to occur at 25°C and an osmotic potential of -1.0 MPa (1541).

Fusarium oxysporum includes many representatives that are pathogenic to plants often causing vascular wilt diseases, *e.g.*, Beckman (159), Nelson *et al.* (1540) and Summerell & Rugg (2084), damping-off problems, *e.g.*, Nelson *et al.* (1544), and crown and root rots, *e.g.*, Jarvis & Shoemaker (960). A strain originally isolated as a tomato pathogen also can cause disseminated infection of immunodepressed mice (1632). The vascular wilt diseases of plants often result when xylem vessels are blocked, with the blockage in at least some cases being due to gels composed of neutral sugars commonly found in the host plant's cell wall (1960, 2212). Recommended disease reduction measures include host plant resistance and numerous biological and chemical control methods. Economically important plant pathogenic strains can be recovered from non-host, usually native, plants (845, 2286). *F. oxysporum* can be dispersed by many different means including wind and in soil, seeds, or infected planting material (683).

Host specificity and *formae speciales*. Many *F. oxysporum* isolates appear to be host specific, which has resulted in the subdivision of the species into *formae speciales* and races that reflect the apparent plant pathogenic specialization (see section 11.8). Over 100 *formae speciales* and races of *F. oxysporum* have been described, *e.g.*, Booth (222). Some times these groups are distinguishable using some molecular characters, *e.g.*, isozymes

(1907), DNA fingerprinting probes (166, 1331, 1484, 1530, 2312), DNA reassociation kinetics (1157, 2097, 2098), mitochondrial DNA polymorphisms (238), and specific antisera (923, 998), but in general they are not monophyletic (84, 105, 1598, 1998) and should be used only as phenotypic characterizations and not as indications of genetic or evolutionary relatedness. *F. oxysporum* f. sp. *ciceris* may be an exception to this general rule (980).

Differences in pathogenicity may arise within the same VCG or other genetically very closely related strains (45, 140, 561, 1998), and the same, or very similar, pathogenic capabilities can develop in clearly distinct genetic backgrounds (452, 1104, 1360). Some pathogenic isolates are clearly related to fungi associated with native flora, *e.g.*, Wang *et al.* (2286). That convergent evolution could result in the independent and recurrent selection for pathogenicity is a hypothesis with very different expectations than one in which the mutation(s) responsible for pathogenicity occurred just once and then was transmitted as a homologous trait (981, 2082).

Physiological studies of host-pathogen interactions involving *F. oxysporum* are not common, but suggest that there several different mechanisms may be involved (79, 428, 429, 461, 865, 931, 937, 939, 954-956, 969, 970, 1633, 1784, 2074, 2090). Some strains of *F. oxysporum* are pathogenic towards *Arabidopsis* (1391) and some of the proteins and other molecules involved in this host-pathogen interaction have been identified (67, 178, 582, 583, 2252). MAP kinase activity is essential for some strains to be pathogenic (495). Fusaric acid also has been implicated as a partial determinant of pathogenicity in bulb rot of *Gladiolus* spp. (1795).

The number of genetic differences separating pathogenic and nonpathogenic strains often is small, but enumerating these differences can be difficult given the lack of a functional sexual cycle and classical genetic analyses (2128, 2130). Some REMI-induced mutants with altered pathogenicity are available, but in general such mutants have not been characterized in great detail (932, 1026, 1529).

There are numerous studies that involve microscopic analyses of the interaction between strains of *F. oxysporum* and both host and non-host plants (1142, 1617, 1871, 1872). In cotton, *F. oxysporum* can penetrate and grow within root tips without inducing damage such as plasmolysis, cell degeneration or host necrosis (1818). Tomato roots can alter growth and branching patterns of mycelia of *F. oxysporum* that develop near them (2044). Strains transformed with markers that permit easy visualization and quantification of the fungal strain within host tissue, *e.g.*, GUS (142, 143, 411, 1616) and fluorescent proteins (388, 1172, 1521, 1581, 1582), also will increase our understanding of host-pathogen interactions. Vital fluorochrome dyes (2047) also may be used to follow early stages of infection, as the dyes can be incorporated into conidia prior to germination and the hyphae from the germinating conidia followed microscopically.

Cereals and grasses are not generally affected by *F. oxysporum* although they may be infected with it (1118, 1119, 1226, 1630). Saprophytic members of *F. oxysporum* commonly colonize necrotic roots, as secondary invaders, and are readily isolated, and therefore often mistakenly assumed to be the primary cause of necrosis. Isolates of *F. oxysporum* need to be tested for pathogenicity before conclusions are drawn about their role in a disease.

Biological control. Isolates of *F. oxysporum* have often been proposed for use in biological control programs. In some cases the *F. oxysporum* strains are mycoparasites (164, 794, 1887, 2301) or are parasites of a weed or other undesirable plants (56, 57, 209, 293, 375, 388, 389, 573, 604, 838, 860, 1361-1363, 2134, 2135, 2393). Strains that are better pathogens of target weeds have been constructed following transformation with genes that produce plant hormones (389). There also is a report that *F. oxysporum* may be pathogenic towards some insects (2233) and sea turtles (1696).

In other cases apparently non-pathogenic strains are an important component of a suppressive soil in which a disease is either sharply reduced or completely eliminated (54, 205, 640, 650, 1187, 1188, 1191, 1214, 1330, 1450, 1636, 1725, 1726, 2131, 2378), perhaps through the induction of systemic resistance (164, 335, 524, 925, 1188, 2200, 2377). Nonpathogenic mutants of known pathogenic strains also have been shown to be effective in controlling diseases of cucurbits (645). Competition between pathogenic and biocontrol strains for nutrients or space on the roots of the hosts may be an important factor in determining the efficacy of some biocontrol strains (141, 524, 1188). The heterogeneity within *F. oxysporum* makes interpreting the literature in this area particularly difficult, and some strains originally identified as *F. oxysporum* have been assigned to other species as identification techniques improved and species boundaries became better defined.

Economically feasible and physiologically efficient conditions for the production (396, 574-576, 838-841, 1488, 1754) and dispersal (125, 126, 640, 761-763, 1489, 1951) of inoculum to be used for biocontrol purposes can be difficult to identify. In some cases a level that is effective in one setting, *e.g.*, a greenhouse, may not be effective in another, *e.g.*, the field (604, 651, 1189).

Vegetative Compatibility Groups (VCGs). Numerous studies have been made of vegetative compatibility and the associated vegetative compatibility groups (VCGs) in *F. oxysporum* (1087, 1220, 1745, Chapter 5). Isolates in the same VCG can form a stable heterokaryon, while strains in different VCGs form a heterokaryon that is unbalanced and/or transitory. There are some studies of heterokaryosis that precede the genetic studies of VCGs in *Fusarium* that began in the mid-1980s (1883, 1980). Differences in VCG can be overcome by using protoplast fusion and stringently selectable markers (1315).

If two strains belong to the same VCG it is not a guarantee of genetic uniformity, as strains in the same VCG may carry mitochondrial DNA molecules with the same (76, 612, 741) or clearly different sequences (740). One mitochondrial DNA restriction pattern also may be associated with strains that belong to multiple VCGs (612, 742, 949). Growth characters and carbon source utilization patterns generally are conserved within a VCG and vary at least somewhat between VCGs (2043). In some cases, diagnostic molecular protocols have been developed to identify VCGs of pathogenic importance (2272, 2364).

Individual VCGs usually are given a number that correlates with the *forma specialis* with which their members are associated (1008, 1010, 1076). Members of the same *forma specialis* may be in one or a few VCGs, *e.g.*, *F. o. adzukicola* (1121), *F. o. albedinis* (613, 2115), *F. o. amaranthi* (349), *F. o. basilicum/basilici* (572, 1011), *F. o. canariensis* (1713), *F. o. cubense* (1705), *F. o. cucumerinum* (24), *F. o. cyclaminis* (2365), *F. o. dianthi* (45, 104, 108, 556, 1012, 1295, 1332, 2366), *F. o. erythroxyli* (1887), *F. o. gladioli* (454, 501, 1428), *F. o. lactucae* (1604, 1655, 2380), *F. o. lentis* (160), *F. o. lilii* (102), *F. o. lycopersici* (140), *F. o. melonis* (75, 947, 948, 1923, 2425), *F. o. niveum* (1190), *F. o. pisi* (2312), *F. o. radicis-cucumerinum* (2208), *F. o. radicis-lycopersici* (1014, 1015, 1834), *F. o. spinaciae* (623), *F. o. tuberosi* (2238), *F. o. tulipae* (102) and *F. o. vasinfectum* (239, 452, 612, 1013, 1999), or in numerous VCGs, *e.g.*, *F. o. asparagi* (204, 569, 1178), *F. o. betae* (823), *F. o. cepae* (2093), *F. o. cucumerinum* (2209), *F. o. lini* (1485), *F. o. lycopersici* (558, 559), *F. o. phaseoli* (50, 555, 2362) or *F. o. pisi* (1718).

The number of VCGs identified may depend on the population sampled, and published results should be confirmed before larger diagnostic surveys of a new population are begun. When the number of VCGs in a *formae specialis* is limited, then the VCG groupings can potentially be used to identify pathogenic strains without conducting plant pathogenicity assays. However, there are examples of pathogenic and non-pathogenic strains that belong to the same VCG, (76, 2364). In studies of *F. oxysporum* from wild species of carnations, the *F. oxysporum* strains recovered were in numerous VCGs and were not pathogenic to cultivated carnations, so that these fungal strains are more similar to those recovered from soil than they are to those recovered from cultivated carnations (1009). Strains from soil (75, 404, 560, 739, 742, 1295, 2008) and orchids (2008) usually are quite diverse with multiple members of the same VCG relatively rare occurrences.

The *nit* mutants (section 5.3) generated to work with VCGs generally retain their pathogenicity under field conditions, which means they can be used in "release and recapture" type experiments (799, 2108, 2109). Other auxotrophic mutants and transformants apparently can be used in a similar manner (1980, 2170).

Physiology and biochemistry. Studies of the biochemical capabilities of *F. oxysporum* also are common, with studies focusing on cell wall degrading enzymes, *e.g.*, cellulases, glucanases, glucosidases, pectinases, polygalacturonidases and xylanases, amongst the most common (35, 36, 79, 316, 358, 361, 366., 367, 455, 494, 497, 500, 616, 677, 707, 723, 724, 782, 915, 1146, 1323, 1370, 1371, 1416, 1640, 1660, 1679, 1680, 1724, 1848, 1849, 1985, 2024, 2025, 2053, 2075, 2234, 2235, 2396). Some of these enzymes may be part of the means used by this fungus to pathogenize its host (498). Other examined enzymes include acetyl esterase (365), α-L-fucosidase (2379), α-amylases (342), α-L-arabinofuranosidase (332, 359, 1643), β-galactosidase (234), calmodulin (899), catalase (1122), chitinase (1587), cytochrome P450s (1974), D-amino acid oxidase (658), feruloyl esterase (2159-2162), fructosyl lysine oxidase (1866), fructosyl transferase (791), hydroxymethyl-glutaryl coenzyme A reductase (1316, 1317), inulinases (792), invertase (791, 1375), lactonohydrolase (1100), lipases (300, 315, 1779), lipoxygenase (900, 1383, 1384), nitric oxide reductase (436, 1523, 1524, 1652); nitroalkane oxidase (448, 627, 660-664, 1048, 1518), pantothenate synthetase (1681), pectate lyase (499), proteases (146, 496, 1854, 1855), sterol ester hydrolase (1318, 1610), trehalase (53), and xylitol dehydrogenase (1642).

Various strains of *F. oxysporum* can metabolize, and in some cases catabolize, numerous sugars (2071, 2073) and a variety of other compounds including aromatics (220), butyl benzyl phthalate (1064), glycerol (329), nitriles (717), penicillin V (1299), sodium monofluoroacetate (2195), and xylose (1987, 1988, 2073, 2307). They also can accumulate significant levels of arsenic (764) and reduce asbestos toxicity by chelating iron from the asbestos fibers (1372). *F. oxysporum* can serve as model for mammalian degradation of protriptyline, an extensively used antidepressant (523). Strains of *F. oxysporum* also can be used to solubilize lignite and other low grade coals (872, 873), and have ligninolytic capabilities (596).

Heat shock proteins have been characterized from strains of *F. oxysporum* (642) as have some stress-responsive mRNAs (354). Some strains have ice-nucleating activity (1806). Superoxide dismutase activity differs in vegetative hyphae and chlamydospores (1122), although the role of this enzyme in the differentiation process remains unknown.

The biochemistry of the host-pathogen interaction has usually been evaluated primarily in terms of plant reactions to fungal infection. There are reports of strains of *F. oxysporum* that can degrade various anti-fungal defense compounds synthesized by some host plants (937-939, 1173-1175, 1826). Both *F. oxysporum* and *F. solani* can degrade α-tomatine, a phytoalexin synthesized by tomato, but a polyclonal antibody that recognizes the enzyme from *F. oxysporum* does not recognize the *F. solani* enzyme (1176). Chitin synthase activity also is required for full expression of some pathogenic phenotypes (1367). Signals transmitted by G proteins within the fungus are known to be an important part of the pathogenicity process (461, 954-956). The ability of some strains of *F. oxysporum* to synthesize ethylene (901, 969) and some small peptides (1798) could have important implications for host-pathogen interactions as well.

Some strains of *F. oxysporum* are relatively insensitive to cyanide (150) and have enzymes such as cyanide hydratase (1672-1674, 2382), also termed formamide hydrolyase, that can be used in the degradation process. These strains also have a cyanide insensitive alternative to the conventional cytochrome respiratory chain for the uptake of O_2 that can be used when high levels of cyanide are present. The presence of at least two mitochondrial denitrification systems that can serve as an alternative to oxidative respiration (1099, 1101, 1163, 1972, 2103-2107, 2197, 2415, 2418) could generate the ATP necessary to sustain a strain growing under high cyanide conditions. Unlike similar bacterial systems, this process in *F. oxysporum* occurs under microaerobic conditions rather than under conditions of strict anaerobiosis (2419).

Commercially, *F. oxysporum* is used to resolve chemically synthesized stereoisomers of pantoyl lactone (889, 890, 907, 1016, 1017, 1968, 1969). It also produces thermostable endoglucanase and β-glucosidase with high temperature optima in submerged culture at high levels (360). A phosophlipase produced by *F. oxysporum* is available commercially from Novo under the trade name Lecitase and is used for an oil-degumming step in the refining of vegetable oils (381). Finally, *F. oxysporum* can be used to reduce the phenol content of dry olive residue sufficiently to enable it to be used as an organic fertilizer (1880).

The ability of *F. oxysporum* to convert pentose sugars to ethanol also has been evaluated in a commercial context (1986, 2072). A liquid fermentation has been described that uses *F. oxysporum* to convert wood hydrolyzate to ethanol (992, 1641, 1644). Solid-state fermentations of sorghum (1243, 1244, 1326) and straw (362, 363) by *F. oxysporum* to produce ethanol have been described as has a fermentation of potato pulp to acetic acid (1155). These fermentations are not particularly sensitive to levels of cellobiohydrolase or carboxymethylcellulase, but were quite sensitive to the level of β-glucosidase present (364).

Genetics and molecular biology. Genetic work with *F. oxysporum* has been rather limited, probably due to the lack of a known sexual stage, although functional mating type genes have been identified in at least some isolates (80, 2402). No meiotically based genetic map is available for *F. oxysporum*, but a mitotic map based on parasexual segregation patterns of AFLPs associated with a transposable is available (2129). Parasexual recombination and reassortment can result in extensive chromosome rearrangements if the heterokaryon in which the parasexual event occurs resulted from protoplast fusion (2130). No genomic sequence is available for this organism, although there are some mitochondrial maps and sequences available (740, 741, 1046, 1053, 1077, 1366, 1678). A proposal to sequence the genome of a *F. oxysporum* strain has been submitted, but as of this writing this proposal has not been funded. This sequence probably will contain a significant amount of repetitive DNA, unlike the situation in *F. graminearum* and *F. verticillioides*.

Transformation protocols for this fungus have been developed (508, 1185, 1495, 1728), including one for a self-replicating linear plasmid (679) and another for particle bombardment (12). Gene regulation mechanisms appear to be generally similar in *F. oxysporum* and in model filamentous fungi such as *Neurospora crassa* and *Aspergillus nidulans* (37, 314, 435, 1605). A protocol for making and regenerating protoplasts is available (1302). Some transformants are sufficiently stable that they can be released and recovered in biological control studies without noticeable changes in the foreign DNA sequences (1442). Genes for host specificity are associated with the nuclear and not the mitochondrial DNA (1463). DNA methylation has been associated with the loss of pathogenicity that occurs when some strains are maintained long term under laboratory conditions (1052).

Neither the number of chromosomes, ranging from 5 to 14, nor the genome size, ranging from 16 to 59 Mb, were constant in all strains of *F. oxysporum*, although chromosome numbers in strains in the same VCG usually were the same (50, 211, 1054, 1440, 1441, 1463, 2153). Some of these differences may be due to duplications of material that normally is present only as single copy sequences (1078, 1082).

Plasmid-like DNA, retrotranspons and other mobile genetic elements are known to occur in strains of *F. oxysporum* (61, 339, 431, 433, 434, 493, 722, 909-913, 993, 1184, 1427, 1443, 1615, 1797, 1984). There is evidence for horizontal transfer between species within the *Fusarium* genus for some of these elements (431). Some of these elements can move in other fungi once they have been introduced, even if the species into which they were introduced has no known instances of similar type of movement by related endogenous elements (913, 1562, 2255).

The distribution of transposable elements within the genome of *F. oxysporum* is not random, and although these elements do not appear responsible for chromosome breaks *per se*, recombination between homologous elements on non-homologous chromosomes can account for much of the variation in karyotype observed in this fungus (449). In *F. oxysporum*, the movement of transposable elements can result in mutations in genes that are important for pathogenicity or changes in expression patterns that have not been obtainable when conventional mutagenesis techniques were used (465, 1444, 2181). No phenotype has been associated with the known plasmid-like DNA elements (1080). Some of these plasmids are similar to mitochondrial plasmids from other fungi, can also be found as full length RNA molecules (1079), and may be novel telomere-like retroelements (2285). In at least some cases, exposure of *F. oxysporum* strains to media containing chlorate, as is done to generate the *nit* mutants commonly used for VCG analyses (section 5.3), may result in the movement of retrotransposons within the *F. oxysporum* genome (62). Mutants generated by γ-irradiation may undergo chromosome rearrangements and incur multiple changes that are attributed to differences in multiple copy DNA sequences (1429).

Various DNA sequences have been used to evaluate genetic diversity, usually of relatively small strain sets. Some of the sequences evaluated include DNA fingerprinting probes (537, 696, 1276, 2008), microsatellites (153), mitochondrial DNA (76, 91, 612, 740-742, 949, 1713), multiple copy repeat sequences (537, 540, 1531), ribosomal IGS spacer sequences (50, 334, 537, 540, 1055, 2364), ribosomal ITS sequences (516, 1982), and single copy DNA sequences (516, 537, 1712). In most cases the DNA sequences examined were polymorphic in the set of strains evaluated.

There are numerous reports of DNA fragments that are specific either for *F. oxysporum* as a whole or for a particular *forma specialis*, race or VCG. Many, but not all, of these studies are based on relatively small strain sets and none of these diagnostics has yet been accepted for use on a broad scale (51, 454, 539, 614, 678, 848, 914, 979, 1035, 1257, 1455, 1492, 1654, 1656, 1712).

Human infections. *Fusarium oxysporum* has been associated with various human infections that include corneal infections (847, 1831), various types of dermatitis (777, 1565, 2198), burn wounds (2309), and localized (145, 699, 716, 718, 1676, 1812, 1827, 2210, 2332, 2397) and systemic internal infections (33, 210, 777, 849, 1136, 1460, 1886).

Fusarium oxysporum has been found contaminating hospital water systems, and the water supply in a hospital could serve as a reservoir for repeated infections over a period of years (60). The fungus also has been introduced into patients via contaminated catheters (562, 1755). This fungus may pose a serious problem for some patients with HIV (562) or that receive transplants of major organs (2294). *F. oxysporum* was not pathogenic in an experimental test with immunocompetent mice (1392).

Amphotericin B, hamycin, and amphotericin B plus 5-fluorocytosine were the most effective antimycotics against *F. oxysporum* infections (1241, 1748, 1799, 1938, 2029). A combination of terbinafine and amphotericin B was effective even in an immunocompromised patient (1839). Most strains were resistant to fluconazole and 5-fluorocytosine alone (1748, 1799, 1938, 2029). Reports of sensitivity to itraconazole and ketoconazole are inconsistent (83, 1241, 1748, 1799, 1938, 2029). Voriconazole is reported to have good *in vitro* activity, but reports of *in vivo* effectiveness are lacking (376, 587, 588, 1241, 1404). Rapamycin (2361), nanoemulsions (1508) and pentamidene (1267) also have been suggested as a potentially effective clinical antibiotics against *F. oxysporum* but have not been widely tested. Fenpropimorph inhibits ergosterol biosynthesis in *F. oxysporum* (458).

Antifungal compounds that inhibit vegetative growth do not always effectively inhibit spore germination and *vice versa* (2196). This differential inhibition may affect the choice of antifungal compound used to treat a *Fusarium* infection. *F. oxysporum* does not elicit as large an oxidative burst by human phagocytes as do some other fungi, which may explain its association with diverse human infections (2351).

Toxins. Strains of *F. oxysporum* are generally viewed as non-toxigenic, although they often can synthesize various polyketide secondary metabolites with unknown functions and toxicities (162). Some strains of *F. oxysporum* have been associated with moldy sweet potato toxicosis of cattle (1341), which results in a severe pulmonary edema and which may also be attributable to *F. solani*, and tibial dyschondroplasia in chickens (369).

There are reports of a few strains of *F. oxysporum* that can produce zearalenone and one or more of the trichothecene mycotoxins (1341), but most strains of *F. oxysporum* do not produce either of these types of mycotoxins and lack the *tri5* gene required for trichothecene biosynthesis (2112). Some strains of *F. oxysporum* carry functional alleles of some of the genes that encode enzymes in the trichothecene biosynthetic pathway (1067). *F. oxysporum* carries a gene that encodes a functional trichothecene 3-O-acetyltransferase gene, which can be used as a defense against trichothecene mycotoxins (2156).

Fusarium oxysporum also has been reported to produce beauvericin (1283, 1473), bikaverin (1407), enniatins (1473, 2422, 2423), fusaric acid (52, 119, 388, 429, 546, 1341), fusarin C (3, 308), isoverrucarol (1063), moniliformin (3, 347, 1339), naphthoquinone pigments (132), sambutoxin (1057, 1059) and wortmannin (2, 3, 5). Some strains of *F. oxysporum* can produce fumonisins and related compounds (388, 1941, 1942, 1946) and are known to carry the *FUM* biosynthetic gene cluster (1740). A relatively large (> 6-8 kDa) uncharacterized compound produced by *F. oxysporum* is known to be responsibility for the toxicity of this fungus to penaeid shrimp (2028).

Current Species Description: Gerlach & Nirenberg (697).

Fusarium phyllophilum Nirenberg & O'Donnell

Sexual Stage. None known.

Common Synonyms. *Fusarium proliferatum* var. *minus*.

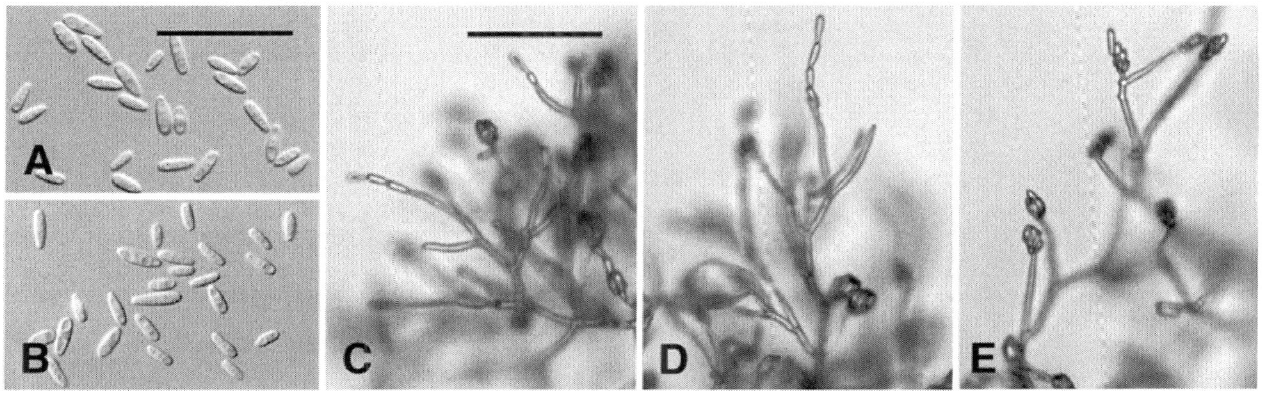

Figure 13-46. *Fusarium phyllophilum.*
A –B: Microconidia; **C – E:** Microconidia *in situ* on CLA. A – B, scale bar = 25 µm; C – E, scale bar = 50 µm.

Reported Geographic and Host Distribution

In Europe and Japan from the foliage of some species in the Dracaenaceae (especially *Dracaena* and *Sansevieria*) and Asphodelaceae.

Media to Examine and Key Characters

Characters on CLA. This species is unusual in that macroconidia and sporodochia are rarely, if ever, produced. The key feature is the presence of clavate microconidia in short chains and false heads from monophialides and polyphialides. No chlamydospores are produced.

Characters on PDA. White mycelia that may become grayish-violet with time. Pigments in the agar usually have a gray tint to them, but otherwise can be white, orange, violet, or blue.

Macroconidia. Macroconidia are extremely sparse and difficult to find. Nirenberg & O'Donnell (1570) note that strains may need to be cultured for years to produce thin, falcate macroconidia that are up to 5-septate. We have never observed macroconidium production by confirmed cultures of *F. phyllophilum*.

Microconidia

- Shape/septation: Usually clavate and 0-septate, although occasionally 1- to 2-septate.
- Aerial mycelium presentation: False heads and short chains.
- Conidiogenous cells: Monophialides and polyphialides.
- Abundance: Abundant in the aerial mycelia.

Microconidia. Clavate and borne in false heads and in short chains on monophialides and polyphialides.

Chlamydospores. Absent.

Taxonomy, Pathology and Ecology

This species was described by Nirenberg & O'Donnell (1570) but was previously included within the definition of *F. proliferatum* var. *minus* (1566). Note that *F. phyllophilum* is not the same as *F. moniliforme* var. *minus* Wollenweber *sensu stricto*, since this variety contains all members of the *Liseola* section that produce conidia in chains but do not produce sporodochia (2359). *F. phyllophilum* is morphologically similar to *F. brevicatenulatum*, but these two species can be distinguished by the shape of the microconidia – clavate in *F. phyllophilum* and oval to obovoid in *F. brevicatenulatum* – and the conditions under which the microconidia are produced – in complete darkness for *F. phyllophilum* and under continuous black light for *F. brevicatenulatum*.

The fungus can cause a leaf spot canker on *Sansevieria* and related hosts with thick leaves and was first reported from these hosts in Europe (1570). It also has been identified as the cause of a leaf spot on candelabra aloe (*Aloe arborescens*) in Japan (1074). *F. phyllophilum* can produce low levels of fumonisins and high levels of moniliformin, but does not produce significant levels of beauvericin or fusaproliferin (638).

Current Species Description: Nirenberg & O'Donnell (1570).

Fusarium poae (Peck) Wollenweber

Sexual Stage. None known.

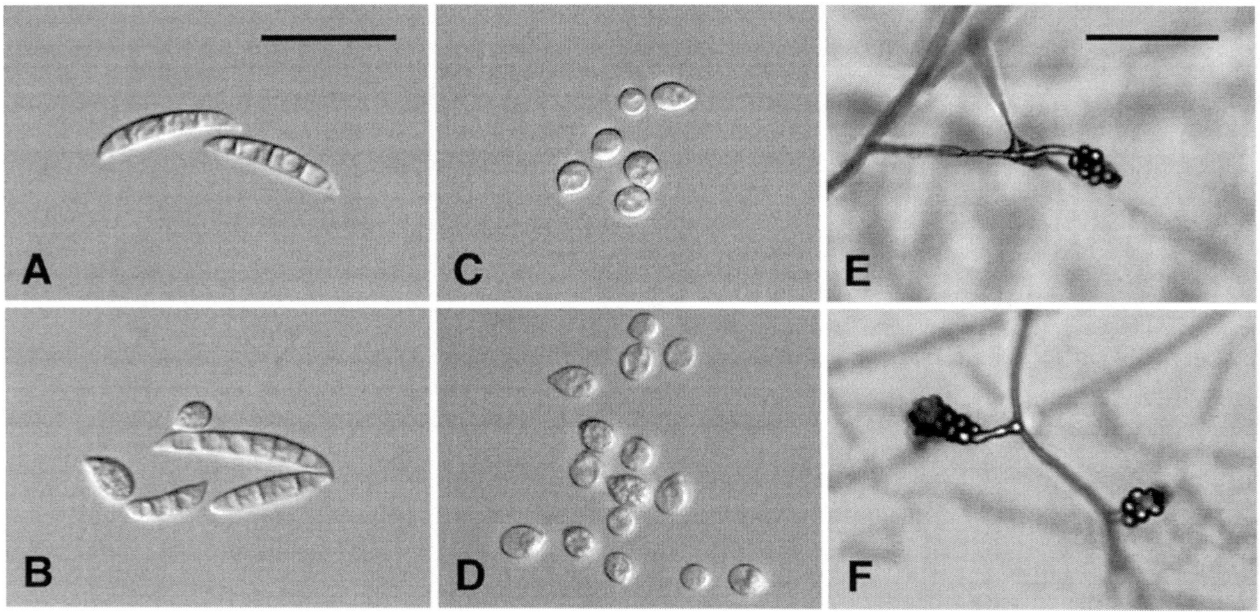

Figure 13-47. *Fusarium poae.*
A – B: Macroconidia; **C – D:** Microconidia; **E – F:** Microconidia *in situ* on CLA. A – D, scale bar = 25 μm; E – F, scale bar = 50 μm.

Reported Geographic and Host Distribution

Widespread, but most common in temperate regions where it usually is isolated from seed and grain heads or woody seedlings.

Media to Examine and Key Characters

Characters on CLA. Macroconidia are rare unless the culture is exposed to near-UV light. They are relatively short, falcate and 3-septate with a curved and tapering apical cell and a foot shaped basal cell. Globose to napiform, 0-septate microconidia are produced abundantly on urn-shaped monophialides with a distinct collarette. The short, branched conidiophores give the appearance of a "bunch of grapes." Chlamydospores are rare, if they are found at all.

Characters on PDA. The aerial mycelium is abundant appearing hairy or felted and changes to powdery as microconidia are formed. The mycelium is initially pale in color and darkens to a reddish-brown with age. Red (most common) to yellow pigments are produced in the agar. The cultures may have a distinctive sweet odor.

Macroconidia

- Sporodochia: Sporodochia are not formed by all strains, but when they do form they may be found on the carnation leaves and the agar surface of both CLA and PDA cultures.
- General morphology: Slender, relatively short, and falcate to almost lunate.
- Apical Cell morphology: Curved and tapering.
- Basal Cell morphology: Well-developed foot cell.
- Number of septa: 3- to 5-septate, but mainly 3-septate.
- Abundance: Generally rare and not formed by all cultures even following stimulation with UV light.

Microconidia

- Shape/septation: Globose or napiform, sometimes with a distinct papilla, and usually 0- but occasionally 1-septate.

- Aerial mycelium presentation: Found in clusters that look like bunches of grapes.
- Conidiogenous cells: Urn-shaped monophialides with a distinctive collarette.
- Abundance: Abundant in the aerial mycelia.

Chlamydospores

- Abundance/Speed of formation: Formed very infrequently and not a reliable taxonomic character.
- Location: Clumps or chains in mycelia of older cultures.
- Appearance: Gerlach & Nirenberg (697) state that true chlamydospores are not formed by *F. poae*. They describe inflated cells with thickened walls and granulated plasma that occur in older cultures, and these structures may be what were seen by Nelson *et al.* (1551) and used in their description of this species.

Taxonomy, Pathology and Ecology

This species was first described by Wollenweber (in 1240) and was recognized by Wollenweber & Reinking (2359), Booth (222), Gerlach & Nirenberg (697), and Nelson *et al.* (1551). Although both the *MAT-1* and *MAT-2* mating types occur and are transcribed in *F. poae* (Kerényi *et al.*, 2004), no sexual stage for this species is known. A group of cultures that are morphologically similar to *F. poae*, sometimes termed "powdery" *F. poae*, but produce a toxin spectrum similar to that of *F. sporotrichioides* have recently been characterized exensively and identified as a separate species – *F. langsethiae* (1098, 1123, 1305, 1563, 1915, 1916, 2143, 2163, 2164, 2335, 2388). Field isolates of *F. poae* belong to a relatively large number of VCGs, suggesting that sexual recombination occurs at least infrequently in this species (1038, 1270, 1271).

On PDA, cultures of *F. poae* may resemble *F. sporotrichioides, F. chlamydosporum*, *F. tricinctum*, and *F. graminearum*. The lack of polyphialides differentiates *F. poae* from *F. sporotrichioides* and *F. chlamydosporum*, and the distinctive phialides of *F. poae* differentiates it from *F. tricinctum*. These species also can be differentiated with various PCR assays (517, 2275).

Some strains of *F. poae* may be infected with dsRNA mycoviruses, but the presence of these mycoviruses is not known to alter the morphological phenotype of these strains (395, 608, 896). dsRNAs appear to be very widely dispersed in *F. poae* (608).

Based on pulse-field electrophoresis, *F. poae* has five chromosomes, one or two additional dispensable chromosomes, and at least one mitochondrial plasmid (611, 896). A moderately repetitive DNA element may be responsible for the karyotype polymorphism observed in some strains of *F. poae* (609).

Fusarium poae is regularly recovered from a variety of grains and the seeds of numerous native and domesticated plant species (928, 1127, 1128, 1280, 1621, 2124). It may be spread by a mite (*Siteroptes graminum*) that feeds on the fungus (1200) or by airborne conidia (1359). *F. poae* is usually considered a weak pathogen and although associated with head blight of wheat it is not usually as important as are the main causal agents of this disease, *i.e.*, *F. graminearum* and *F. culmorum* (518, 1868, 2166). Isolates of *F. poae* that cause a central bud rot in carnation can cross-infect wheat (398).

Some isolates of *F. poae* have significant lipase activity (315). Strains of *F. poae* have been used to convert octanoic acid to 2-heptanone, a dairy flavor component (2213).

Many isolates of *F. poae* are weakly toxic or non-toxic, and cases of human and animal toxigenicity due to trichothecene mycotoxins and attributed to *F. poae* may be due to *F. sporotrichioides*, which often is present along with *F. poae*. Isolates of *F. poae* from the Urov valley of eastern Siberia are associated with an osteoarthritic disease, Kashin-Beck disease. The mycotoxin and/or nutritional deficiency responsible for this disease has not been identified, but the disease symptoms have been replicated in dogs and rats fed grain contaminated with *F. poae* (1341). *F. poae* has been recovered from human tonsils (1321).

When inoculated onto maize *F. poae* can produce fusarenone-X and nivalenol (760) and when inoculated onto barley, nivalenol is produced (1868). Isolates of *F. poae* can produce beauvericin (1283), fusarin C (308, 600, 2140), and the trichothecenes diacetoxyscirpenol, nivalenol and T-2 toxin (1341, 1539, 1693, 2069, 2096, 2258).

Current Species Description: Gerlach & Nirenberg (697).

Fusarium polyphialidicum Marasas, Nelson, Toussoun & van Wyk

Sexual Stage. None known.

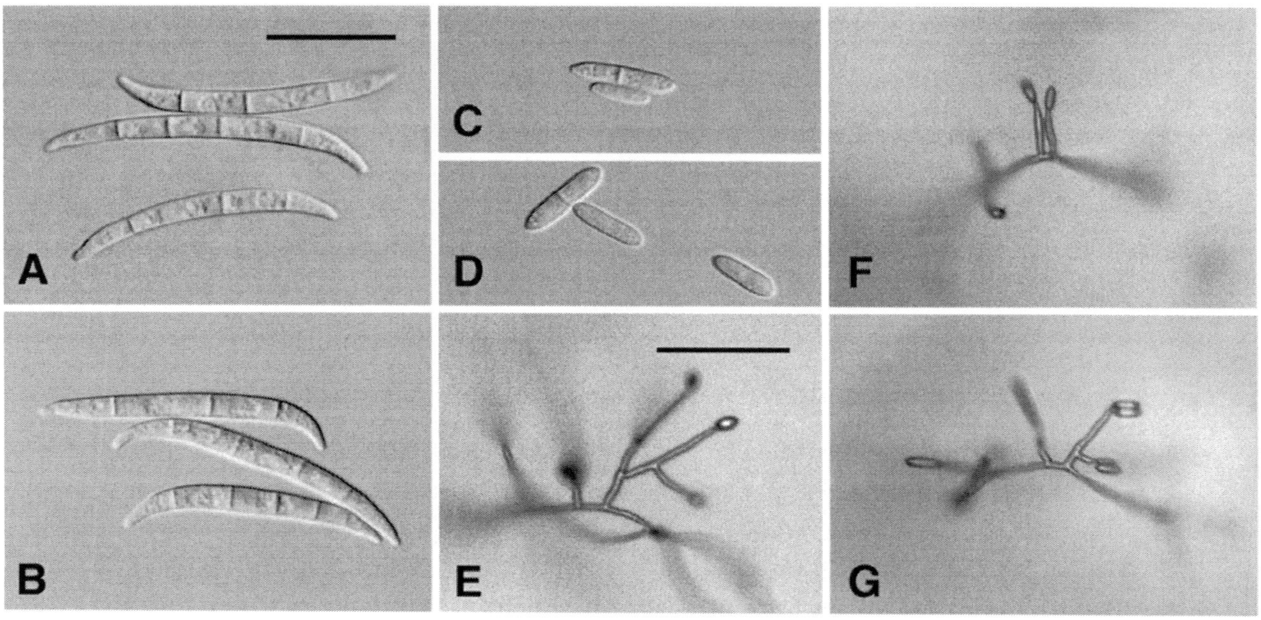

Figure 13-48. *Fusarium polyphialidicum.*
A – B: Macroconidia; **C – D:** Microconidia; **E – G:** Microconidia *in situ* on CLA. A-D, scale bar = 25 μm; E – G, scale bar = 50 μm.

Reported Geographic and Host Distribution

Recovered from plant debris in soil and sorghum grain in Italy, South Africa and Australia.

Media to Examine and Key Characters

Characters on CLA. White to pale orange sporodochia are produced in the aerial mycelium and contain thick walled, predominantly 5-septate macroconidia with a curved apical cell and a foot-shaped basal cell. Fusiform to subclavate microconidia are produced in pairs or false heads from polyphialides that often have numerous openings. Chlamydospores are formed moderately quickly in pairs, clumps or chains.

Characters on PDA. Abundant white to pale orange mycelium on PDA with white to yellow pigments in the agar.

Macroconidia

- Sporodochia: Found in the aerial mycelia and are white to pale orange.
- General morphology: Robust and quite large, slightly falcate, and thick walled.
- Apical Cell morphology: Curved and tapering.
- Basal Cell morphology: Foot shaped.
- Number of septa: 3- to 7-septate, but mainly 5-septate.
- Abundance: Often difficult to find.

Microconidia

- Shape/septation: Hyaline microconidia may be fusiform or subclavate. They usually are 0-septate, but may be up to 3-septate.
- Aerial mycelium presentation: Pairs or false heads, but never in chains.
- Conidiogenous cells: Both monophialides and polyphialides may be observed. The polyphialides are quite distinctive and may be very complex.
- Abundance: Abundant in the aerial mycelia.

Chlamydospores

- Abundance/Speed of formation: Form moderately quickly (2-4 weeks on CLA) and are relatively abundant.
- Location: Chlamydospores may be terminal or intercalary in either aerial or submerged hyphae. They are found singly, in pairs, clumps and chains.
- Appearance: Hyaline to pale brown with smooth or rough walls.

Taxonomy, Pathology and Ecology

Fusarium polyphialidicum is most likely to be confused with *F. semitectum*, *F. subglutinans* and related species, and *F. chlamydosporum*. *F. polyphialidicum* can be differentiated from *F. semitectum* by the brown pigment produced in PDA by *F. semitectum* and the white to pale yellow pigment produced by *F. polyphialidicum*. The macroconidia of *F. polyphialidicum* also are longer, wider and more robust in appearance than are those of *F. semitectum*. *F. polyphialidicum* can be distinguished from *F. subglutinans* using these criteria as strains of *F. subglutinans* usually produce violet pigments when cultured on PDA and have macroconidia that are smaller and narrower than those produced *F. polyphialidicum*. *F. polyphialidicum* produces macroconidia that are wider, longer and have thicker walls than do the macroconidia produced by *F. chlamydosporum*.

The geographic and host range of *F. polyphialidicum* is not well defined. *F. polyphialidicum* has been recovered from plant debris and soil from a number of locations in southern (1343, 1627) and western (1627) Africa, Italy (135), and arid portions of central Australia (756, 1889). It also has been isolated from moldy sorghum grain (42) and pine seedlings (1589). *F. polyphialidicum* has been tested for use as an insect control agent (806). Reports (7) that *F. polyphialidicum* can produce fumonisins have not been confirmed.

Current Species Description: Marasas *et al.* (1343).

Fusarium proliferatum (Matsushima) Nirenberg

Sexual Stage. *Gibberella intermedia* (Kuhlman) Samuels, Nirenberg & Seifert.

Common Synonyms. *Gibberella fujikuroi* mating population D, *Gibberella fujikuroi* var. *intermedia.*

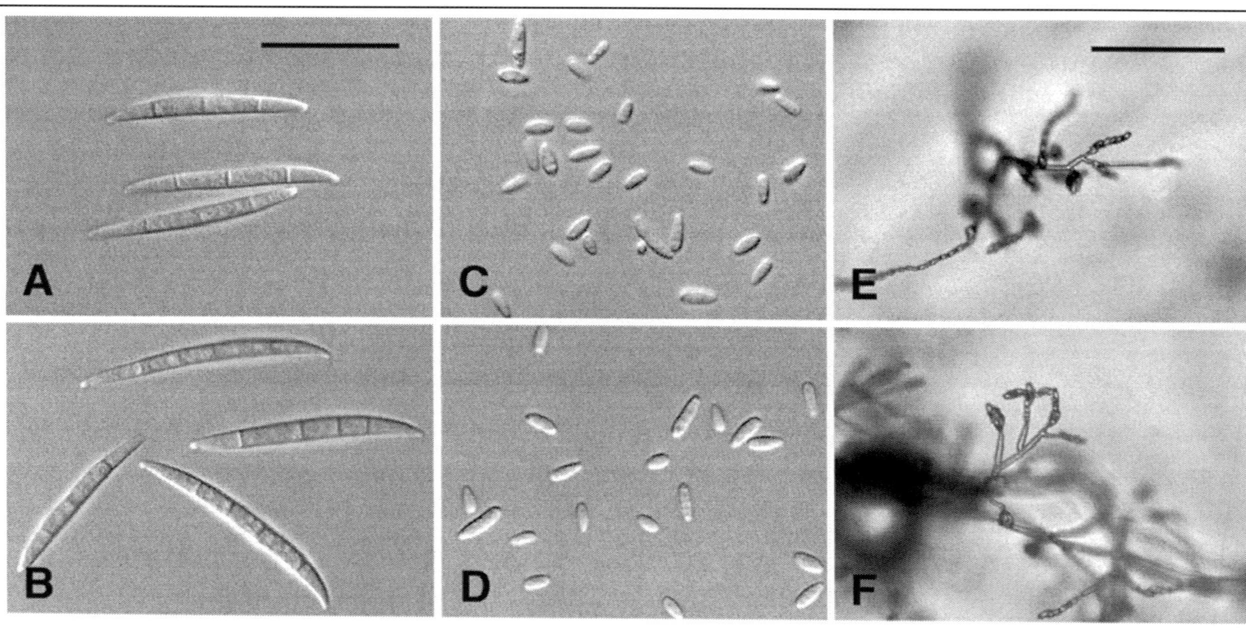

Figure 13-49. *Fusarium proliferatum.*
A – B: Macroconidia; **C – D:** Microconidia; **E – F:** Microconidia *in situ* on CLA. A – D, scale bar = 25 μm; E – F, scale bar = 50 μm.

Reported Geographic and Host Distribution

Worldwide distribution on a variety of agricultural and non-agricultural substrates. Causes diseases on maize, sorghum, mango and asparagus.

Media to Examine and Key Characters

Characters on CLA. Macroconidia are typical of those formed by species in the *G. fujikuroi* species complex and are found in pale orange sporodochia that may be produced infrequently or difficult to find. The macroconidia are slender, almost straight, and usually 3- to 5-septate. Microconidia form in chains and, less commonly, in false heads from monophialides and polyphialides. The polyphialides may proliferate extensively. The microconidial chains are of moderate length and usually are shorter than those formed by *F. verticillioides*. The microconidia are club shaped with a flattened base and pyriform microconidia may be formed by some strains (1566). Chlamydospores are absent.

Characters on PDA. The abundant aerial mycelium initially is white but may become purple-violet with age. Sporodochia may be present as discrete entities or nearly confluent over portions of the colony. Violet pigments usually are produced in the agar, but with overall pigmentation varying in intensity from nearly colorless to almost black. Blue-black sclerotia may develop in some isolates, but are not diagnostic although they may be indicative of a high level of sexual female fertility.

Macroconidia

- Sporodochia: Tan to pale orange sporodochia often are produced infrequently or are difficult to find.
- General morphology: Slender, thin-walled, relatively straight and typical of those produced by species in the *G. fujikuroi* species complex.

- Apical Cell morphology: Curved.
- Basal Cell morphology: Poorly developed.
- Number of septa: Usually 3- to 5-septate.
- Abundance: The abundance of macroconidia varies since this character can be lost in this species following repeated subcultures. Fresh cultures usually produce large numbers of macroconidia in sporodochia.

Microconidia

- Shape/septation: Club shaped with a flattened base and 0-septate. Pyriform microconidia also may occur but generally are rare.
- Aerial mycelium presentation: May be found in chains of varying, but usually moderate, length, false heads, or aggregates of a few microconidia.
- Conidiogenous cells: Monophialides and polyphialides.
- Abundance: Abundant in the aerial mycelia.

Chlamydospores. Absent.

Taxonomy, Pathology and Ecology

Fusarium proliferatum was first described as a *Cephalosporium* species by Matsushima (1387) and described as a *Fusarium* species by Nirenberg (1566). It was recognized as a species by Gerlach & Nirenberg (697) and Nelson *et al.* (1551). Prior to the Nirenberg (1566) description, most isolates of *F. proliferatum* probably would have been identified as *F. moniliforme*, and much of the research on *F. proliferatum* is probably imbedded in and inseparable from that on *F. moniliforme*.

Fusarium proliferatum is most likely to be confused with *F. fujikuroi*, *F. oxysporum*, *F. thapsinum* and *F. verticillioides*. Morphologically *F. proliferatum* and *F. fujikuroi* can not be effectively differentiated solely on morphology and are sibling species. They usually, but not always, can be differentiated by using mating tests or by DNA sequences. The separation between them as biological species is not complete (1236), and naturally occurring hybrids between these two species have been recovered (1234). PDA cultures of *F. proliferatum* and *F. oxysporum* often appear similar, but these species are easily distinguished by the presence of microconidia in chains for *F. proliferatum* and the presence of chlamydospores and microconidia in false heads for *F. oxysporum*. *F. proliferatum* can be differentiated from *F. thapsinum* and *F. verticillioides* by the presence of polyphialides and shorter chains of microconidia. Morphological differences between *F. proliferatum*, *F. subglutinans* and *F. verticillioides* are reported to be enhanced by culturing on minimal medium (section 2.5) with 20% saccharose instead of 3% sucrose (382). Carnation leaf agar containing 4-8 g/l KCl (section 2.2) also can be useful in the identification of *F. proliferatum* (626).

Fusarium proliferatum has not often been studied in depth from a population genetics perspective. Strains pathogenic to asparagus belong to numerous VCGs, with some of them very widely dispersed and some limited to a single strain from a single location (565, 571). Sexual recombination probably occurs often enough to ensure that most populations usually will appear to be randomly mating. The effective population size is affected much more severely by a decrease in the number of female-fertile strains than it is by any known inequalities in the relative numbers of strains carrying the different mating type idiomorphs (1224).

There are reports of DNA fragments that are specific for *F. proliferatum*. Many, but not all, of these studies are based on relatively small strain sets and none of these diagnostics has yet been accepted for use on a broad scale (516, 1492, 1493, 1690, 2040, 2274). Polymorphism in mtDNA sequences has been reported and could be used to show that multiple strains of the fungus can simultaneously attack a single host plant (1170). AFLP analysis of cDNAs can be used to differeniate cell growth and mycotoxin producing stages (967).

Some strains of *F. proliferatum* carry one or more dsRNA molecules, but these dsRNA molecules do not have any known effects on the morphological phenotype (837).

Maximum linear growth for *F. proliferatum* is reported to occur at 25°C and an osmotic potential of -1.0 MPa (1541). Commercially available antioxidants can be used to treat grain and regard growth by *F. proliferatum* and reduce the amount of fumonisins produced *in situ* (591, 1801). Conidia germinate optimally at 30°C (1358). *F. proliferatum* can persist in maize stalk debris either on the soil surface or buried in a field for at least 21 months (410). Injection of a spore suspension through maize husk leaves at blister stage (R2) can be used to differentiate maize lines that are resistant/susceptible to Fusarium ear rot from one another (384).

The region around the mating type locus in *F. proliferatum* has significant synteny with that of *F. verticillioides* and a number of other ascomycetes (2276). The fumonisin biosynthetic gene clusters in *F. proliferatum* and *F. verticillioides* have a high degree of synteny, but the flanking regions are completely different (2276). These data may be interpreted to mean that more than one event resulted in the acquisition of these toxin genes by these fungi and that toxin production is not a homologous character that can be used for phylogenetic studies.

Fusarium proliferatum has been recovered from numerous environments worldwide. It is a cause of root rot of pine seedlings (1589), Fusarium crown and root rot of asparagus (566, 568, 570, 571, 1924) and date palm decline (10). As a cause of stalk and cob rot of maize (1285, 2311), multiple genetically distinct isolates of *F. proliferatum* may be recovered from a single maize plant (1029). "BT" hybrid maize, which is less susceptible to the European corn borer *Ostrinia nubialis*, had a lower level of infection with *F. proliferatum* than did comparable hybrids without the transgene (685, 1499). Isolates have been associated with the rhizosphere of *Livistona* palms in arid regions of Australia (1556), and recovered from several grass species in native North American tallgrass prairies (1234). *F. proliferatum* may be endophytic in wheat (190), and alters the defense response in those plants (191). Other reported hosts for *F. proliferatum* include banana (978), citrus fruits (922), orchids (170, 924, 1201), rice (477), and sorghum (1226).

Fusarium proliferatum may be mycoparasitic and is being evaluated as a biological control of *Plasmopara viticola*, the causal agent of grape downy mildew (Bakshi *et al.*, 2001; Falk *et al.*, 1996). It also is being evaluated as a biological control of leafy spurge (*Euphorbia* spp.; Caesar, 1996; Caesar *et al.*, 1998).

Fusarium proliferatum may produce both proteases and β-glucosidases under solid-state fermentation conditions that might suffice for simple commercial enzyme production from waste plant material (606). This species also can produce a β-xylosidase capable of degrading corn fiber xylan (1862, 1863). Some isolates of *F. proliferatum* produce laccases and related enzymes that may be involved in biodegradation of lignin-containing compounds (1165, 1167, 1168, 1788, 1789). Other strains can break down ethynylestradiol (Shi *et al.*, 2002), an estrogenic compound that can persist in both water and agricultural soils. Other enzymes characterized from *F. proliferatum* include catalase (1166), lipoxygenase (188, 919, 920) and superoxide dismutase (1166).

Fusarium proliferatum was not pathogenic in an experimental test with immunocompetent mice (1392). There is at least one reported case of *F. proliferatum* being responsible for the death of an immunocompromised human patient (2076), and of other infections of both healthy (1565) and immunocompromised patients (210, 846, 851). *F. proliferatum* is resistant to most antifungal drugs, being the least resistant to amphotericin B (1748) and posaconazole (851).

In tests with contaminated feed, the toxicity to ducklings is correlated with the amount of moniliformin produced and not with the amount of fumonisins produced (2248). Contaminated grain also causes death, hemorrhage and diarrhea in experimental rats (4).

Isolates of *F. proliferatum* can produce gibberellic acid (1341) and a wide variety of mycotoxins often at high levels including beauvericin (1234, 1285, 1704, 1800, 1956), fusaproliferin (1234, 1774, 1800, 1813, 1956), fusaric acid (119), fusarins (1448) and moniliformin (347, 1285, 1349, 1448).

Fumonisins are produced by *F. proliferatum*, often at high levels (118, 440, 1234, 1285, 1340, 1448, 1507, 1546, 1548, 1800, 1956, 2132), and the *FUM* gene cluster encoding the genes required for fumonisin biosynthesis has been sequenced and characterized in some detail (1740, 2276). The wide host range of this fungus can result in fumonisins being recovered from seemingly unlikely sources such as asparagus and garlic (1282, 1932). Some strains of *F. proliferatum* can rapidly degrade fumonisins (1032).

Current Species Description: Gerlach & Nirenberg (697).

Fusarium pseudoanthophilum Nirenberg, O'Donnell & Mubatanhema

Sexual Stage. None known.

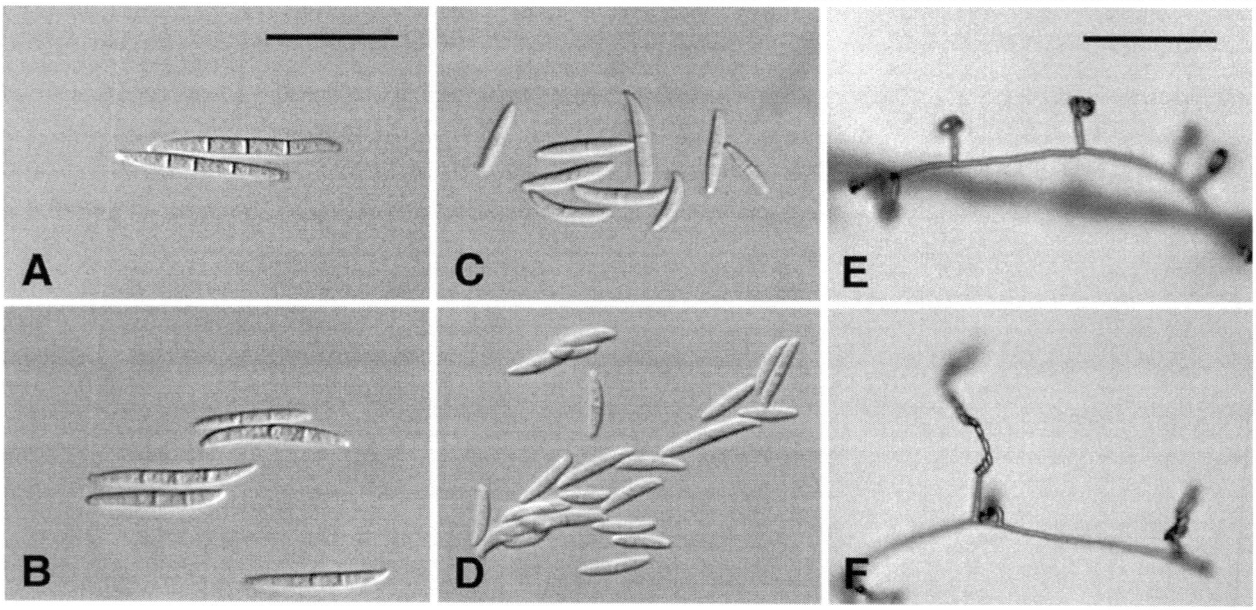

Figure 13-50. *Fusarium pseudoanthophilum.*
A – B: Macroconidia; **C – D:** Microconidia; **E – F:** Microconidia *in situ* on CLA. A – D, scale bar = 25 µm; E – F, scale bar = 50 µm.

Reported Geographic and Host Distribution

Maize in southern Africa.

Media to Examine and Key Characters

Characters on CLA. Pale orange sporodochia are produced sparsely and contain slender 3-5 septate macroconidia. The most characteristic feature is the production of aerial microconidia in short chains on monophialides. The microconidia may be oval, obovoid, clavate or pyriform. Chlamydospores may be formed, but are not a useful diagnostic character.

Characters on PDA. White to pale orange aerial mycelium with orange to dark blue-gray pigments in the agar.

Macroconidia

- Sporodochia: Produced sparsely after 2-4 weeks and often difficult to find.
- General morphology: Slender, falcate and thin-walled. Typical of those produced by species in the *Gibberella fujikuroi* species complex.
- Apical Cell morphology: Bent or pointed.
- Basal Cell morphology: Foot shaped.
- Number of septa: Predominantly 3- to 5-septate.
- Abundance: Abundant in sporodochia.

Microconidia

- Shape/septation: Long-oval to obovoid, clavate or pyriform, and usually 0- but occasionally 1-septate.
- Aerial mycelium presentation: False heads and short chains.
- Conidiogenous cells: Usually monophialides with occasional polyphialides.
- Abundance: Abundant in the aerial mycelia.

Chlamydospores

- Abundance/Speed of formation: Found occasionally but usually not abundant. It may require up to two months of incubation for chlamydospores to form on CLA, but these structures are formed more rapidly by cultures growing on SNA.
- Location: Formed in aerial mycelium and in the agar.
- Appearance: Produced in chains

Taxonomy, Pathology and Ecology

This species was described by Nirenberg *et al.* (1571) from isolates recovered from maize; however it is not known if it is a pathogen or a saprophyte. We have preliminary information that indicates that this species is conspecific with *F. brevicatenulatum*.

Morphologically *F. pseudoanthophilum* resembles *F. anthophilum* in the formation of pyriform conidia on branched conidiophores from both monophialides and polyphialides. The two species can be distinguished since *F. pseudoanthophilum* produces very short chains of microconidia and chlamydospores in chains, characters that are never observed for *F. anthophilum*. Phylogenetically, *F. pseudoanthophilum* and *F. anthophilum* are not closely related, with *F. pseudoanthophilum* being most closely related to *F. brevicatenulatum*, *F. pseudonygamai* and *F. verticillioides*, and *F. anthophilum* being most closely related to *F. bulbicola* and *F. succisae* (1596).

Isolates of *F. pseudoanthophilum* can produce low levels of beauvericin, but are not reported to produce detectable levels of fumonisins, fusaproliferin or moniliformin (638).

Current Species Description: Nirenberg *et al.* (1571).

Fusarium pseudocircinatum O'Donnell & Nirenberg

Sexual Stage. None known.

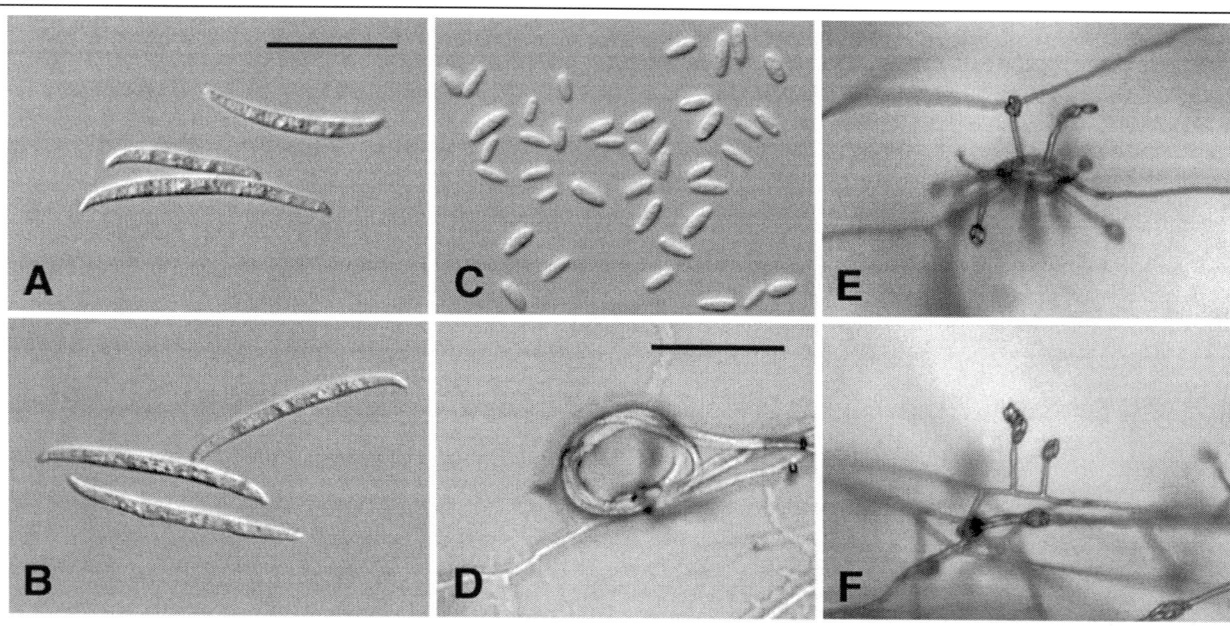

Figure 13-51. *Fusarium pseudocircinatum.*
A – B: Macroconidia; **C:** Microconidia; **D:** Coiled hyphae; **E – F:** Microconidia *in situ* on CLA. A – D, scale bar = 25 μm; E – F, scale bar = 50 μm.

Reported Geographic and Host Distribution

From a variety of substrates from tropical regions, *e.g.*, Ghana, Papua New Guinea, Panama and the Philippines.

Media to Examine and Key Characters

Characters on CLA. Sporodochia are sparse and consequently macroconidia are rare. When present, the macroconidia are slightly falcate. Microconidia are produced abundantly from monophialides in the aerial mycelium in false heads and short chains. Coiled sterile hyphae are produced and are a very important character. Chlamydospores are absent.

Characters on PDA. White feathery mycelium. Produces orange to violet pigments in the agar, especially in the center of the culture.

Macroconidia

- Sporodochia: Sparse and hard to find. Cultures may need to be incubated for up to two weeks under black light before sporodochia form on the agar surface.
- General morphology: Slender, slightly falcate and thin walled. Typical of those produced by species in the *Gibberella fujikuroi* species complex.
- Apical Cell morphology: Beaked.
- Basal Cell morphology: Foot shaped.
- Number of septa: Predominantly 3-septate.
- Abundance: Abundant in sporodochia, but rare, if present at all, when sporodochia are absent.

Microconidia

- Shape/septation: Oval to obovoid and 0- to 1-, but usually 0-septate.
- Aerial mycelium presentation: False heads and short (< 10 spores) chains.
- Conidiogenous cells: Usually monophialides, but occasionally polyphialides.
- Abundance: Abundant in the aerial mycelia.

Chlamydospores. Absent.

Other Features. Coiled, sterile hyphae are diagnostic for this species. They may be found reliably on SNA, but their production on CLA may vary. The number of coiled hyphae and the degree of coiling usually is higher on SNA as well.

Taxonomy, Pathology and Ecology

This species was described by Nirenberg & O'Donnell (1570) from cultures that had been received under a variety of names. Little is known of the ecology or biology of this species at present other than it appears to be pantropical in distribution and apparently not involved in plant disease. Morphologically it has similarities with *F. circinatum* and *F. sterilihyphosum* as these three species are the only ones with coiled hyphae. *F. circinatum* and *F. pseudocircinatum* can be differentiated by the formation of short chains of microconidia by *F. pseudocircinatum*, whereas *F. circinatum* produces microconidia only in false heads. As the short chains can be difficult to find, molecular techniques may be necessary for a definitive answer. *F. pseudocircinatum* and *F. sterilihyphosum* can be distinguished on the basis of macroconidial morphology with those of *F. sterilihyphosum* being longer and more slender than those of *F. pseudocircinatum*.

Phylogenetically, *F. circinatum* and *F. sterilihyphosum* are not closely related to *F. pseudocircinatum*, with *F. pseudocircinatum* being most closely related to *F. lactis* and *F. denticulatum*, and *F. circinatum* and *F. sterilihyphosum* being most closely related to *F. anthophilum*, *F. bulbicola*, and *F. succisae* (244, 1596). Thus the unusual hyphal coiling pattern probably is a trait that has developed convergently rather than being evolutionarily conserved.

Isolates of *F. pseudocircinatum* do not produce detectable levels of beauvericin, but can produce moniliformin, low levels of fusaproliferin, and trace levels of fumonisins (638).

Current Species Description: Nirenberg & O'Donnell (1570).

Fusarium pseudograminearum Aoki & O'Donnell

Sexual Stage. *Gibberella coronicola* Aoki & O'Donnell.

Common Synonym. *Fusarium graminearum* Group 1.

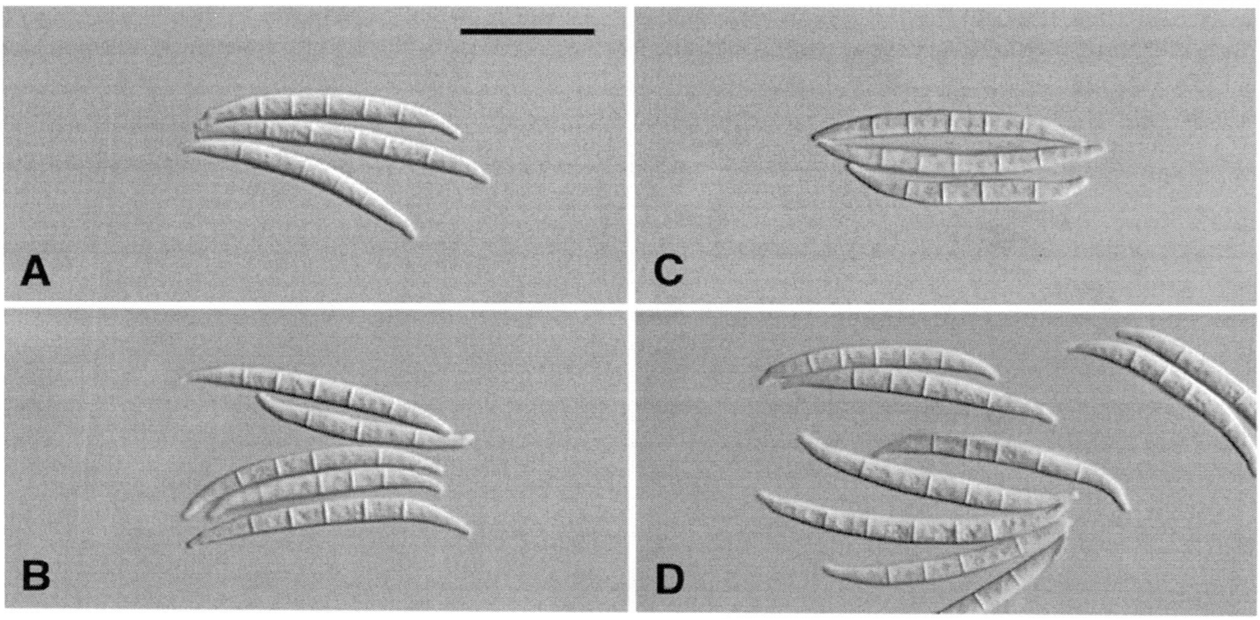

Figure 13-52. *Fusarium pseudograminearum.*
A – D: Macroconidia; scale bar = 25 µm.

Reported Geographic and Host Distribution

A pathogen on wheat, triticale and barley. Found in drier areas where these grains are cultivated.

Media to Examine and Key Characters

Characters on CLA. Macroconidia are produced in pale orange sporodochia. The macroconidia are relatively slender, curved to almost straight, usually 5- to 6-septate, with a curved apical cell and a foot-shaped basal cell. Microconidia are absent. Chlamydospore formation varies and is not a useful taxonomic criterion. Chlamydospores may be found in the macroconidia and in the hyphae.

Characters on PDA. Abundant white to yellow to red mycelia can completely fill the Petri dish. Red pigments are formed in the agar.

Macroconidia

- Sporodochia: Abundant pale orange sporodochia are found on the carnation leaf pieces in CLA and on the water agar adjacent to the carnation leaf pieces.
- General morphology: Relatively slender and almost straight to moderately curved.
- Apical Cell morphology: Curved.
- Basal Cell morphology: Foot shaped.
- Number of septa: 1- to 11-septate, with 5- to 6-septate the most common.
- Abundance: Relatively abundant.

Microconidia. Absent.

Chlamydospores

- Abundance/Speed of formation: Presence varies, but usually formed within 4 weeks when present. Lack of chlamydospores is not a reliable diagnostic character.

- Location: May be found in macroconidia or in terminal or intercalary positions in hyphae growing aerially or submerged in the agar. Chlamydospores are found singly, in chains, and in clumps.
- Appearance: Hyaline to pale yellow or gray in color. May be smooth or rough walled.

Taxonomy, Pathology and Ecology

This species was formally described in 1999 (71) and prior to that was known as *F. graminearum* Group I. Purss (1749, 1750) was the first to recognize that two definable forms existed within *F. graminearum*: one that usually causes aerial diseases of wheat and maize, *i.e.*, *F. graminearum*, and can homothallically produce the sexual stage (*Gibberella zeae*), and the other that usually causes a crown disease of wheat, *i.e.*, *F. pseudograminearum*, is heterothallic, and apparently poorly fertile. Francis & Burgess (639) defined these as Group 1 (*F. pseudograminearum*) and Group 2 (*F. graminearum*). The teleomorph, *G. coronicola*, was described by Aoki & O'Donnell (72) and is heterothallic. *G. coronicola* may be found under field conditions but is rare, although the widespread occurrence of *G. coronicola* may occur in some years as described by Summerell *et al.* (2078).

Fusarium pseudograminearum is easily confused with *F. graminearum* as they are morphologically identical. Most isolates of *F. graminearum* will produce the *G. zeae* teleomorph on CLA or Carrot Agar. *F. pseudograminearum* also has some similarities to *F. crookwellense* and *F. culmorum*, but careful examination of the macroconidia usually suffices to resolve these species. The macroconidia of *F. pseudograminearum* are more elongate than those of *F. crookwellense* and *F. culmorum*. The PDA cultures also usually are quite distinct; the pigments in cultures of *F. pseudograminearum* are red whilst those found in *F. crookwellense* and *F. culmorum* tend to be browner. Based on molecular markers *F. pseudograminearum* is as closely related to *F. culmorum*, *F. lunulosporum*, and *F. crookwellense* as it is to *F. graminearum* (71, 169, 1169, 1171, 2112). Some PCR primers are available for the specific identification of *F. graminearum*, but these primers have not been extensively tested (2334).

Fusarium pseudograminearum is a debris and soilborne pathogen that causes crown rot on cereals including wheat, barley, rye and triticale and grassy weeds (30, 111, 267, 269, 397, 639, 1550). This disease is particularly important in Australia, where it is found throughout the entire grain-growing region (112, 113). The disease is favored by farming practices in which infested wheat stubble is retained on the soil surface to prevent soil erosion (266, 2077, 2080, 2091). The fungus can, on rare occasions, cause outbreaks of head blight of wheat but only in exceptionally wet seasons (271). References to *F. graminearum* as the causal agent of crown rot of wheat are virtually always to *F. pseudograminearum*.

Studies of *F. pseudograminearum* for toxin production are limited. Strains of this species are known to produce deoxynivalenol, 3-acetyldeoxynivalenol, and zearalenone (197).

Current Species Description: Aoki & O'Donnell (71).

Fusarium pseudonygamai O'Donnell & Nirenberg

Sexual Stage. None known.

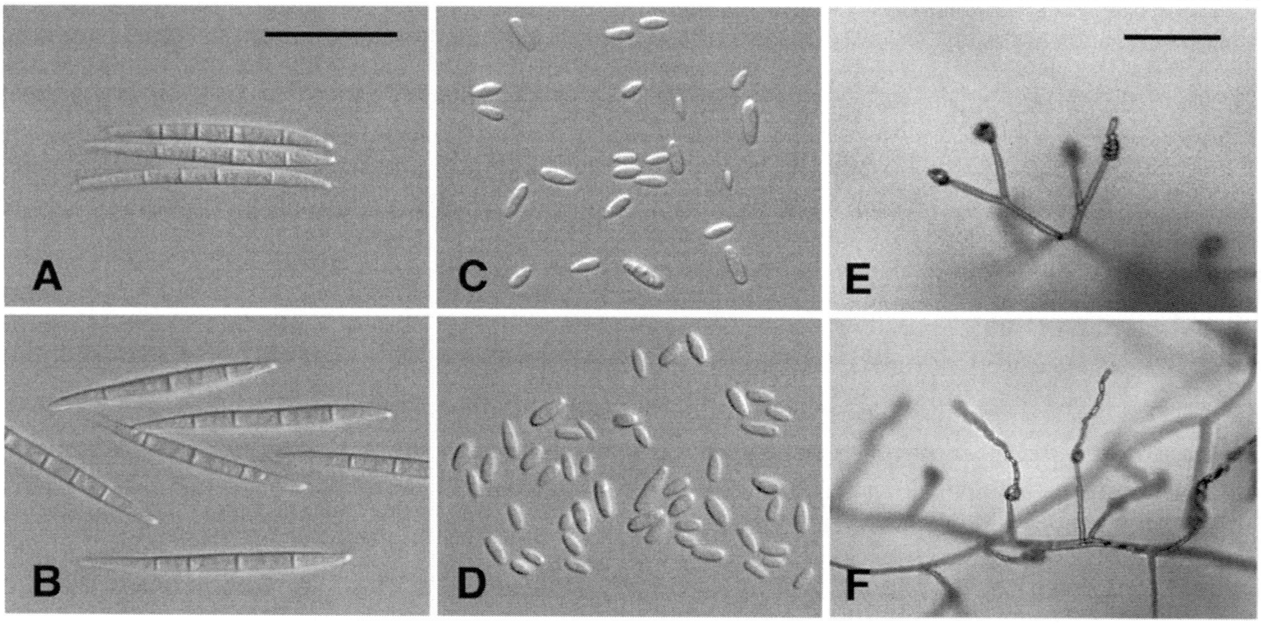

Figure 13-53. *Fusarium pseudonygamai.*
A – B: Macroconidia; **C – D:** Microconidia; **E – F:** Microconidia *in situ* on CLA. A – D, scale bar = 25 µm; E – F, scale bar = 50 µm.

Reported Geographic and Host Distribution

Recovered from *Pennisetum* spp. (pearl millet) in Africa and the United States.

Media to Examine and Key Characters

Characters on CLA. Sporodochia are rarely observed and it may be difficult to find macroconidia. When present the macroconidia are almost straight and 3-4 septate. Obovoid to clavate microconidia are produced abundantly in medium length chains mainly from monophialides. No chlamydospores are produced.

Characters on PDA. Abundant off-white colored aerial mycelia. Produces orange to violet pigments in the agar, and often is very heavily pigmented in the center of the colony.

Macroconidia

- Sporodochia: Not usually observed.
- General morphology: Almost fusoid.
- Apical Cell morphology: Tapering.
- Basal Cell morphology: Poorly developed foot shape.
- Number of septa: 3- to 5-septate, but usually 3- to 4-septate.
- Abundance: Very rarely formed.

Microconidia

- Shape/septation: Obovoid to clavate and 0-septate.
- Aerial mycelium presentation: False heads and chains of short to medium length.
- Conidiogenous cells: Usually monophialides, but polyphialides may be observed.
- Abundance: Abundant in the aerial mycelium.

Chlamydospores. True chlamydospores are absent, but hyphal swellings that may be misidentified as chlamydospores may be observed.

Taxonomy, Pathology and Ecology

This species was described by Nirenberg & O'Donnell (1570) from cultures previously recorded as *F. nygamai* (1336, 1345). These strains were collected from pearl millet growing in Namibia (1345), not Nigeria, as reported in the original species description (1233, 1570). We have recovered this species from a number of locations in Africa in surveys of *Fusarium* spp. associated with pearl millet.

Fusarium pseudonygamai is morphologically very similar to *F. nygamai* but can be differentiated from it by the absence of chlamydospores in *F. pseudonygamai* and the presence of slightly longer microconidial chains. Phylogenetically, *F. pseudonygamai* is most closely related to *F. verticillioides*, *F. pseudoanthophilum* and *F. brevicatenulatum* (1596).

Strains of *F. pseudonygamai* are toxic to ducklings and can produce moniliformin, but produce little or no fumonisins (638, 1233). The unusual *F. nygamai* strains from millet described by Marasas *et al.* (1350) that produced moniliformin and were toxic to ducklings probably are *F. pseudonygamai* as well. Strains of this species also may produce fusaproliferin, but do not produce detectable levels of beauvericin (638). Some strains of this species are pathogenic towards sorghum in an *in vitro* seedling pathogenicity assay (1233).

Current Species Description: Nirenberg & O'Donnell (1570).

Fusarium ramigenum O'Donnell & Nirenberg

Sexual stage. None known.

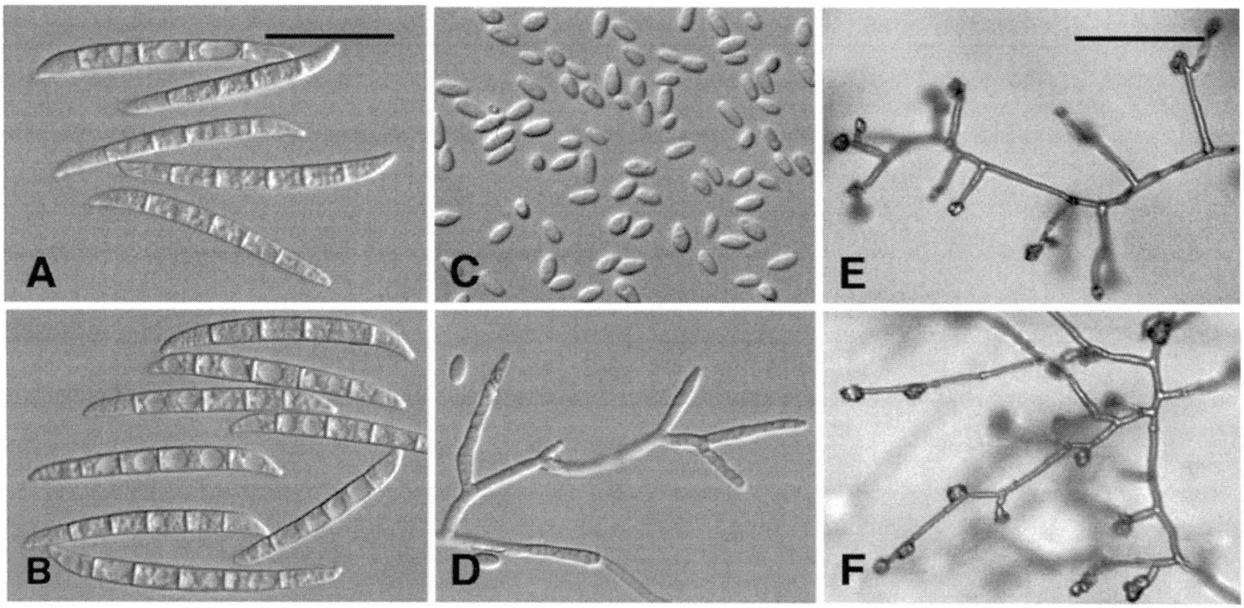

Figure 13-54. *Fusarium ramigenum.*
A – B: Macroconidia; **C:** Microconidia; **D:** Polyphialide; **E – F:** Microconidia *in situ* on CLA. A – D, scale bar = 25 μm; E – F, scale bar = 50 μm.

Reported Geographic and Host Distribution

Recovered from *Ficus carica* in California, USA.

Media to Examine and Key Characters

Characters on CLA. Macroconidia usually are found in sporodochia and are rare unless sporodochia, which also are uncommon, are present. Obovoid, generally 0-septate microconidia are common in small false heads borne on long phialides that branch extensively. Chlamydospores are absent.

Characters on PDA. The aerial mycelium is white to pink with orange to reddish-brown pigmentation in the agar.

Macroconidia

- Sporodochia: Usually rare, but when present are orange and found on the carnation leaf pieces.
- General morphology: Robust and falcate.
- Apical Cell morphology: Bent.
- Basal Cell morphology: Notched.
- Number of septa: Predominantly 5-septate.
- Abundance: Usually found in sporodochia, so generally rare unless sporodochia are present.

Microconidia

- Shape/septation: Obovoid and 0-1 septate.
- Aerial mycelium presentation: Usually seen as small false heads, but short chains of microconidia may be found following incubation under a 12 hour alternating period of light and dark.
- Conidiogenous cells: Monophialides and polyphialides. Phialides may be quite distinctive as they are prostrate, branch extensively, proliferate sympodially, and may be very long.
- Abundance: Abundant in the aerial mycelia.

Chlamydospores. Absent.

Taxonomy, Pathology and Ecology

This species was described from isolates from various figs in California USA (1438, 1570). A preliminary report (1496) has been made of a sexual stage that would be a member of the *Gibberella fujikuroi* species complex, *Gibberella ramigena*. *F. ramigenum* is quite similar to *F. lactis* and the two species are differentiated by the production of zig-zag chains of microconidia by *F. lactis* but not by *F. ramigenum*.

Isolates of *F. ramigenum* can produce moniliformin, but do not produce detectable levels of beauvericin, fumonisins or fusaproliferin (638). Little else is known of the ecology or biology of this fungus or of its ability to act as a plant pathogen.

Current Species Description: Nirenberg & O'Donnell (1570).

Fusarium redolens Wollenweber

Sexual stage. None known.

Common Synonym. *Fusarium oxysporum* var. *redolens*.

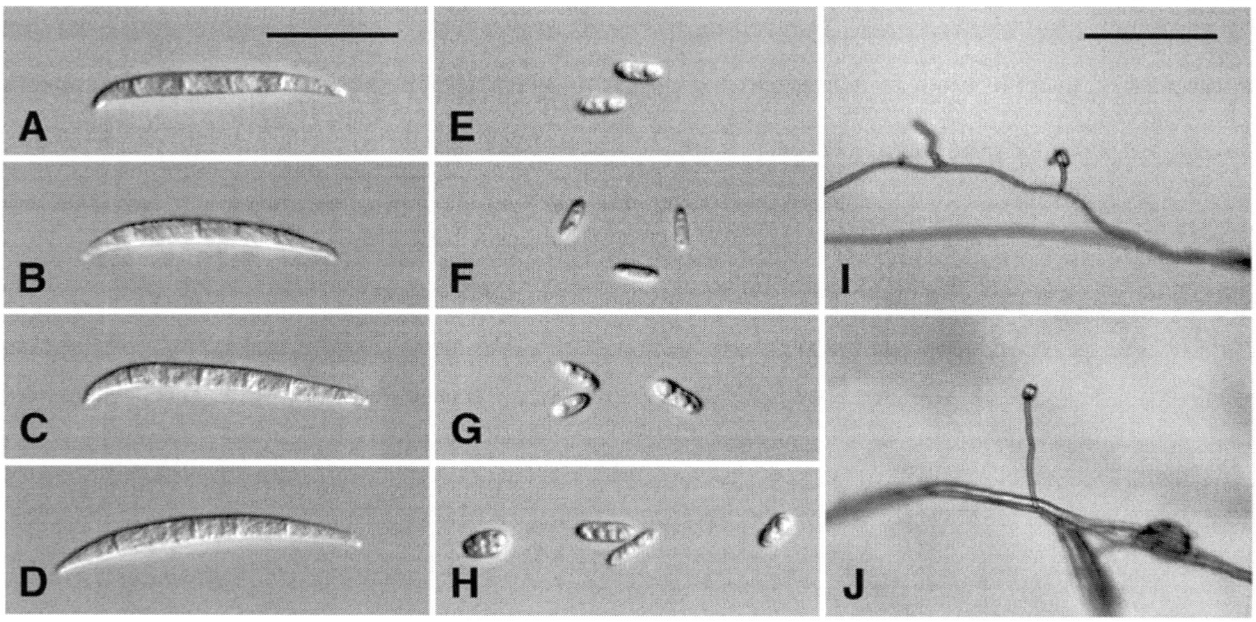

Figure 13-55. *Fusarium redolens.*
A – D: Macroconidia; **E – H:** Microconidia; **I – J:** Microconidia *in situ* on CLA. A – H, scale bar = 25 µm; I – J, scale bar = 50 µm.

Reported Geographic and Host Distribution

Fusarium redolens is a relatively common soilborne fungus found in temperate areas. It has been recorded on a variety of hosts associated with root rot diseases.

Media to Examine and Key Characters

Characters on CLA. Cream to pale brown sporodochia are formed sparsely and contain macroconidia that often resemble those formed by *F. solani*. Macroconidia are robust, thick-walled, widest towards the apical end of the spore, and 3-5 septate with a hooked apical cell and a foot-shaped basal cell. Oval to cylindrical microconidia are borne on monophialides in false heads and are 0- to 1-septate and often pointed on one end.

Characters on PDA. Colonies have a relatively flat aerial mycelium that is white to pink with some brown pigmentation in the agar.

Macroconidia

- Sporodochia: Sporodochia are small and cream to pale brown in color, but macroconidia also may be found in a slimy pionnotal layer that forms in some cultures.
- General morphology: Robust and thick walled. The upper third of the conidia are the widest.
- Apical Cell morphology: Hooked.
- Basal Cell morphology: Foot shaped.
- Number of septa: 3-5 septate.
- Abundance: Abundant, especially in pionnotal cultures.

Microconidia

- Shape/septation: Oval to cylindrical with 0 or 1 septa and often pointed on one end.
- Aerial mycelium presentation: False heads.
- Conidiogenous cells: Monophialides.
- Abundance: Varies, but usually common in the aerial mycelia.

Chlamydospores

- Abundance/Speed of formation: Usually quite abundant, especially as the culture ages. Commonly found after 4-6 weeks of incubation on CLA.
- Location: Usually found in chains that are terminal or intercalary in hyphae in the air, on the agar surface, or submerged in the agar.
- Appearance: Spherical to oval shape, rough walled, and slightly pigmented.

Taxonomy, Pathology and Ecology

This species was first described by Wollenweber (2355). The taxonomic status of this species has changed with the authorities who treated the genus. Snyder & Hansen (2012) included *F. redolens* in their concept of *F. oxysporum*. Booth (222) reduced *F. redolens* to a variety of *F. oxysporum*. Nelson *et al.* (1551), following Snyder & Hansen (2012), retained *F. redolens* within *F. oxysporum*. Gerlach & Nirenberg (697) recognized *F. redolens* as a separate species. Certainly there are similarities between *F. oxysporum* and *F. redolens* – intermediate forms have been recognized in both (103), strains from the two species are reported to be capable of forming heterokaryons with one another (1780), and serological and protein electrophoretic banding patterns may be similar (1781). In some characters, especially macroconidia, *F. redolens* resembles *F. solani*, and it has occasionally been grouped with that species instead of *F. oxysporum*. Baayen *et al.* (106) used molecular characters to show that *F. redolens* was quite distinct from *F. oxysporum* and that it should be recognized as a distinct species. Phylogenetically, *F. hostae* is most closely related to *F. redolens* from which it can be separated by single nucleotide polymorphisms (SNPs) in the sequence of the β-tubulin and translation elongation factor 1-α genes.

Morphologically *F. redolens* is closely related to *F. hostae* from which it can be distinguished by the slower radial growth of *F. hostae* on PDA, and by the lack of production of polyphialides by *F. redolens*. Together with *F. redolens*, *F. hostae* forms a clade that is distinct and intermediate between those for the *F. oxysporum* and *G. fujikuroi* species complexes (106).

Fusarium redolens has been associated with root rot disease of a large number of plant species in temperate regions but most particularly with asparagus, beans, carnation, peas, roses and spinach (222, 380, 754, 1193, 2400). It also has been recovered from native grasslands (McMullen & Stack, 1983*a*, 1984).

Formae speciales names for the isolates causing diseases on various hosts have been recognized (103, 105, 107, 108, 697). Some of these *formae speciales* have been subdivided into VCGs (104, 108, 1331). As with *F. oxysporum*, the *formae speciales* within *F. redolens* are not monophyletic and should be used only as phenotypic descriptions and not as indications of evolutionary or genetic relatedness.

Toxin studies of *F. redolens* often are difficult to locate since for many years these strains were identified simply as *F. oxysporum*. *F. redolens* has been associated with human disease (1876). There is one report of fumonisin production by *F. redolens* (8). *F. redolens* is known to produce fusaric acid, although it is not clear if this compound has a role in the determining the pathogenicity phenotype of the strains that produce it (1385).

Current Species Description: Gerlach & Nirenberg, (697).

Fusarium sacchari (E. J. Butler) W. Gams

Sexual Stage. *Gibberella sacchari* Summerell & Leslie.

Common Synonyms. *Fusarium subglutinans*, *Fusarium neoceras*; *Gibberella fujikuroi* Mating Population B.

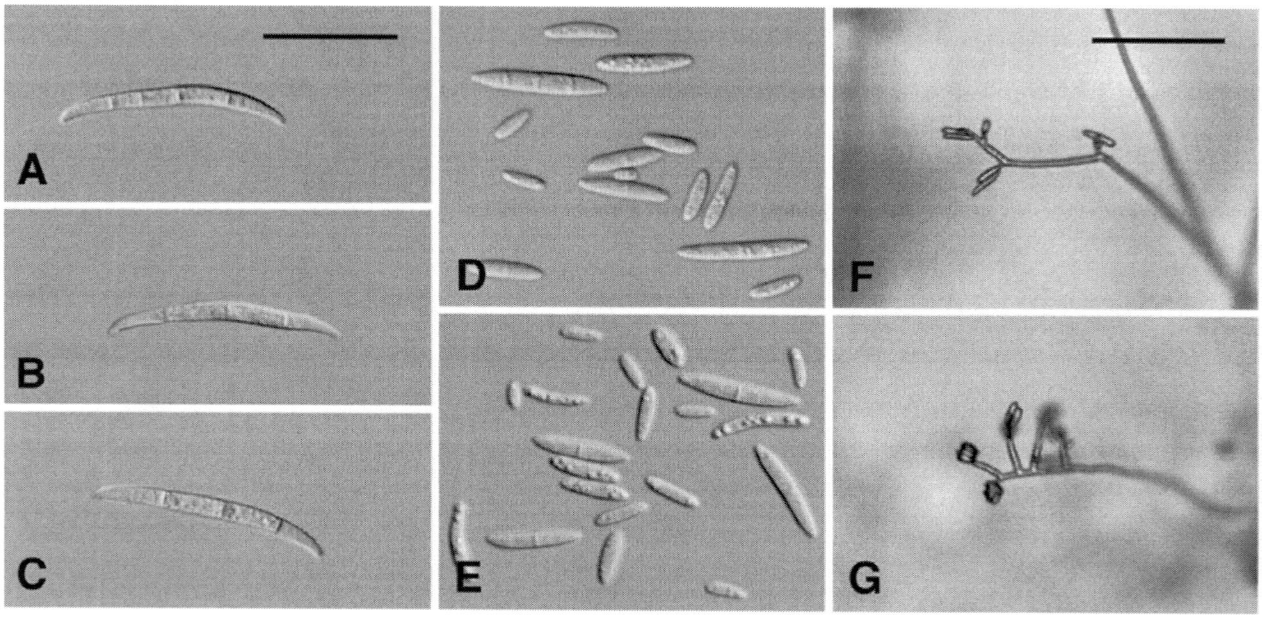

Figure 13-56. *Fusarium sacchari.*
A – C: Macroconidia; **D – E:** Microconidia and mesoconidia; **F – G:** Microconidia *in situ* on CLA. A – E, scale bar = 25 µm; F – G, scale bar = 50 µm.

Reported Geographic and Host Distribution

Found in Asia on sugar cane as the cause of pokkah boeng disease and from sorghum in the Philippines and Mexico. Also recovered occasionally from maize and from orchids.

Media to Examine and Key Characters

Characters on CLA. Macroconidia are formed sparsely if at all. Exposure of cultures to UV light may help to stimulate the production of more macroconidia in some isolates. Abundant microconidia are produced in false heads from mono- and polyphialides and are oval shaped and 0-septate. The polyphialides may proliferate extensively.

Characters on PDA. Abundant mycelial growth, initially pale but becoming violet with age. Violet pigments may be produced in the agar. Blue-black sclerotia may develop in some isolates, but are not diagnostic although they may be indicative of a high level of sexual female fertility.

Macroconidia

- Sporodochia: Are produced rarely, but when produced are orange and found on carnation leaf pieces in CLA. Exposure to UV light may stimulate the production of sporodochia.
- General morphology: Relatively slender, slightly falcate and thin walled. Typical of those produced by species in the *Gibberella fujikuroi* species complex.
- Apical Cell morphology: Curved.
- Basal Cell morphology: Poorly developed.
- Number of septa: Usually 3-septate.
- Abundance: Often very difficult to find; exposure to UV light may stimulate greater production.

Microconidia/Mesoconidia

- Shape/septation: Oval, slender and primarily 0-septate, although 1- or 2-septate elongated mesoconidia also may be formed.
- Aerial mycelium presentation: False heads.
- Conidiogenous cells: Mono- and polyphialides.
- Abundance: Abundant in the aerial mycelia.

Chlamydospores. Absent.

Taxonomy, Pathology and Ecology

Fusarium sacchari was first described as *Cephalosporium sacchari* Butler & Hafiz Kahn from sugar cane in India in 1913 (291). The original description did not include a description of macroconidia, which is not surprising given the scarcity of macroconidia formed by this species in *in vitro* culture. Cultures with macroconidia were described by Wollenweber & Reinking (2358) as *Fusarium neoceras* Wollenweber & Reinking. Gams (671) synonymized the two names to the present form. As the original type specimen was no longer available Leslie *et al.* (1229) neotypified it. *F. sacchari* was recognized by Gerlach & Nirenberg (697) but not by Nelson *et al.* (1551) who synonymized it with *F. subglutinans*.

The heterothallic teleomorph and fertile tester strains were described by Leslie *et al.* (1229). The morphology of the teleomorph can be used to help identify this species morphologically as the ascospores it produces are larger than those of most of the other *Gibberella* species. Female fertility is lost in cultures that degenerate to a thin wispy aerial mycelium instead of a thicker and denser form. The current mating-type tester strains have been selected for cultural stability and are the progeny of field strains originally used for this purpose. On rare occasions a few perithecia may be produced by single-spored cultures of the tester strains and a few other strains of *F. sacchari* (243). Any crosses in which a few perithecia are produced on a plate should be viewed as suspect until the progeny are shown to have a biparental origin. Chen (348) described the occurrence of pyriform microconidia in some isolates of *F. sacchari* (that were confirmed in crosses with tester strains) and segregated 1:1 in crosses with globose microconidia, a trait previously described by Booth (222). We have not observed this feature, which is important for distinguishing this species from *F. anthophilum*.

Fusarium sacchari is most likely to be confused with *F. subglutinans* and a range of other similar species, *e.g.*, *F. circinatum*. It cannot be differentiated from these species reliably other than through the use of mating tests or molecular markers.

Fusarium sacchari is the cause of an important disease of sugar cane, pokkah boeng (545), and can reduce the quality of the harvested crop (515) especially amongst varieties with high sugar yields (531). The fungus may be spread by contaminated seed (1532) and pupae and adults of sugarcane stem borers (228). Host resistance to this disease is available (1303). *F. sacchari* also has been recovered from orchids (924), sorghum (1229) and beach sand (1895). It may function as a biological control, probably through the activation of a systemic acquired resistance response, in pigeon pea (335, 336).

Fusarium sacchari caused disseminated fungemia in an immunocompromised patient, but the infection was cured following treatment with amphotericin B (779). Little is known about the toxicity of *F. sacchari per se* as this species and *F. subglutinans* were grouped together for many years. Presumably the toxin profile for *F. sacchari* is quite similar to that of the species to which it is most closely related such as *F. circinatum*, *F. mangiferae*, *F. sterilihyphosum*, and *F. subgluinans*.

Current Species Description: Leslie *et al.* (1229).

Fusarium sambucinum Fückel *sensu stricto*

Sexual Stage. *Gibberella pulicaris* (Fries) Saccardo var. *pulicaris*.

Common Synonyms. *Fusarium torulosum*, *Fusarium venentaum*.

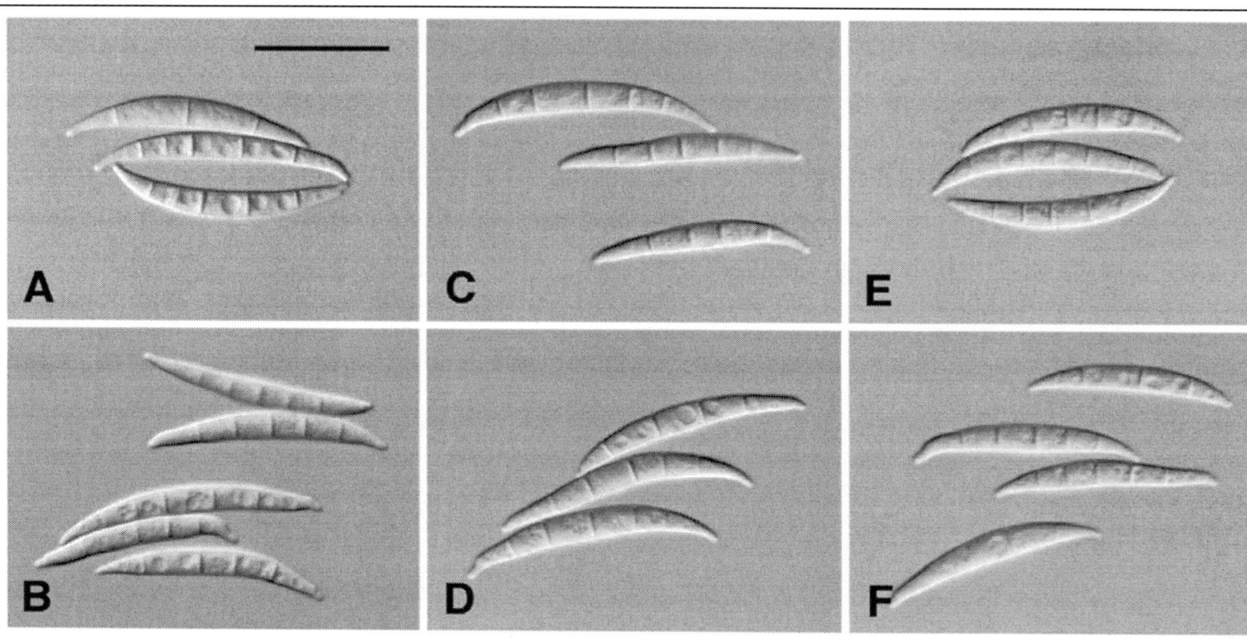

Figure 13-57. *Fusarium sambucinum*.
A – F: Macroconidia. Scale bar = 25 μm.

Reported Geographic and Host Distribution

Found on a diverse range of substrates in temperate parts of the world.

Media to Examine and Key Characters

Characters on CLA. Abundant macroconidia are formed in orange sporodochia. Macroconidia are comparatively short, slender, 5-septate, and falcate with a pointed apical cell and a foot-shaped basal cell. Very occasionally microconidia, 0- to 1- septate, may be present in some isolates. Chlamydospores are formed in chains or clusters.

Characters on PDA. Relatively rapidly growing, often with a series of apparent concentric rings. The colony margins often are lobed, and the absence of lobes may indicate that the strain has begun to degenerate. Aerial mycelia are generally abundant, floccose to felt-like and white to yellow to gray-orange in color. A mass of macroconidia may be formed at the center of the culture. Pigments in the agar usually are yellow, orange or ruby red, sometimes with brown dots. Orange to brown sclerotia are produced by some isolates.

Macroconidia

- Sporodochia: Orange sporodochia are common and found on carnation leaf pieces and on the agar surrounding the leaf pieces on CLA.
- General morphology: Falcate, slender, comparatively short, and usually rather uniform in size.
- Apical Cell morphology: Pointed.
- Basal Cell morphology: Foot-shaped.
- Number of septa: 3- to usually 5-septate.
- Abundance: Usually abundant in sporodochia.

Microconidia

- Shape/septation: Oval and 0- to 1-septate.
- Aerial mycelium presentation: Either singly or in small false heads.

- Conidiogenous cells: Monophialides.
- Abundance: Very rare, but found in the aerial mycelia when present.

Chlamydospores

- Abundance/Speed of formation: Present in small numbers that may take six weeks or longer to form on CLA. Their absence is not a reliable diagnostic character.
- Location: Found in chains and clusters in hyphae either on the surface of or submerged in the agar.
- Appearance: Smooth and not usually pigmented.

Taxonomy, Pathology and Ecology

Fusarium sambucinum is the type species for the genus *Fusarium* and thus is of particular importance in *Fusarium* taxonomy (673). The treatment of this species differs depending upon the authority (222, 697, 1551, 2359), with the currently most widely used species concept, and the one described here, that of Nirenberg (1568). She accepted *F. sambucinum* which is now used as *F. sambucinum sensu stricto*, and included *F. sulphureum* and *F. trichothecioides* within *F. sambucinum*. She made a new combination as *F. torulosum*, which includes the strains in *F. sambucinum* var. *coeruleum* of Booth (222) and Gerlach & Nirenberg (697). She created a new species, *F. venenatum*, and left unresolved the status of two other species, *F. bactridioides* and *F. sarcochroum*, that had been included in *F. sambucinum* by one or more authorities (222, 1551). Her subdivision of *F. sambucinum sensu lato* into these three species was supported by a morphometric analysis of macroconidial size and shape (805), but not by sequences of the ITS regions of the ribosomal DNA (1592). A sexual stage is known for *F. sambucinum sensu stricto*, but its formation under laboratory conditions is sufficiently problematic that only strains that produce positive crosses and form perithecia and ascospores are diagnostic (479).

Papers published prior to Nirenberg's (1568) subdivision of *F. sambucinum* should be reevaluated to determine to which of the three recognized species the strains originally described as *F. sambucinum* belong. Such reevaluations are particularly important for reports of toxin production (see below) but also are important for other studies, *e.g.*, Schneider & Seaman (1918-1920) in which the ontogeny of chlamydospore production by *F. sulphureum* is described, even though the strains used probably would now be called either *F. torulosum* or *F. venenatum*.

Fusarium sambucinum is most likely to be confused with *F. torulosum* and *F. venenatum* because of the very similar morphology. *F. sambucinum* can be distinguished from *F. torulosum* by growth rate on PDA, with *F. sambucinum* cultures growing much more rapidly than those of *F. torulosum*. *F. sambucinum* is the only one of these three species that does not produce chlamydospores. *F. sambucinum* also can be confused with *F. culmorum* due to similarities in PDA cultures and the morphology of the macroconidia. The macroconidia of *F. culmorum* are shorter and wider than those of *F. sambucinum*. Molecular markers also can be used distinguish *F. sambucinum* from these other species (1288, 1455, 1690, 2112, 2392).

Fusarium sambucinum is most frequently recovered from potato where it causes a dry rot of tubers (222, 466, 2046). An ATP-binding cassette (ABC) transporter protein is important for virulence as this protein provides *F. sambucinum* with resistance to potato phytoalexins such as rishitin and lubimin (629). Polyubiquitin production also is induced upon exposure to phytoalexins or following a heat shock (1297). The degradation of rishitin by *F. sambucinum* is an important virulence determinant (471, 472, 680, 2304), but this degradation is more efficient when the fungus is growing on an agar surface than when the fungus is growing submerged in a liquid culture (2303). Trichothecene production plays an important role in determining the pathogenicity of strains of *F. sambucinum* (174).

Populations of *F. sambucinum* from Europe are more diverse than those from the United States (466). Both mating types are common in European populations of the fungus but one of them is rare (~1%) in North American populations. By 1990, most of the strains of *F. sambucinum* causing dry rot of potatoes in the United States were resistant to thiabendazole, although prior to that time many of the strains examined were sensitive to this fungicide (470). Thiabendazole resistance in *F. sambucinum* need not result from mutations at the locus encoding β-tubulin (1027).

Fusarium sambucinum is more common in temperate parts of the world, and may be recovered from areas as far north Iceland and northern Norway near the Arctic Circle (1116, 1117). It also has been recovered from the seeds and roots of a number of plants including alfalfa, cereals, cowpea, beetroot, cabbage, *Medicago* spp. and soybeans (1140, 1181, 1568, 1622). Many diseases were attributed to *F. sambucinum* before it was split into *F. sambucinum sensu stricto*, *F. torulosum*, and *F. venenatum*. In most cases these diseases have not been reevaluated to determine which of these three species was responsible for the disease symptoms observed. Amongst these diseases are deterioration of avocado fruits (445, 2406), fruit blotch of strawberry (918), and stem rot of pigweed (202, 203).

Fusarium sambucinum can degrade the antimicrobial benzoxazinoids 2-hydroxy-1,4-benzoxazin-3-one and 2-

benzoxazolinone produced by maize and *Aphelandra tetragona* (2420). It also can degrade the saponins α-chaconine and α-tomatine, with the enzyme α-chaconinase (158, 2305, 2306), as well as lubimin (473, 681), sesquiterpene phytoalexins (474) and fusanocoumarins (487, 2030).

Amylases from *F. sambucinum* can be used to convert various raw plant starches to glucose (1364, 1365). Other enzymes studied from *F. sambucinum* include ATP citrate lyase (1586), cutinase (1260), and peptidyl-prolyl-*cis*/*trans*-isomerases (895).

Trichothecenes of several types are synthesized by many strains of *F. sambucinum* (46, 480, 1341, 1461, 1687, 1807-1809, 1836, 1891, 2144), and various culture conditions for the production of these toxins have been tested (483, 1467). Trichothecene biosynthesis has been particularly well studied in one strain of *F. sambucinum*, R-6380. Mutations blocking trichothecene biosynthesis have been induced in this strain, and transformation protocols that enable the dissection of the biosynthetic pathway developed (172, 173, 469, 1869). Loci that encode proteins with both regulatory and biosynthetic functions have been identified (879, 2099). Trichodiene synthase, the first enzyme in the trichothecene biosynthetic pathway that is specific for trichothecenes has been the subject of numerous molecular and biochemical studies (304, 876, 878). Disruption of trichodiene synthase lowers pathogenicity of transformed strains towards parsnip roots but not towards potato tubers (475). Trichothecene biosynthesis also has been studied in other *Fusarium* species, *e.g.*, *F. graminearum* and *F. sporotrichioides*.

Enniatin production by *F. sambucinum* also has been examined in some detail (46, 92-94, 285). Various enniatin analogues, some with antihelmentic activity, can be produced by *F. sambucinum* depending on the precursors in the media (1134). The polyketide synthetase responsible for enniatin biosynthesis has been identified and characterized (1697). D-hydroxyisovalerate dehydrogenase (1202-1204), the enzyme that catalyzes the synthesis of a key precursor for enniatin biosynthesis, also has been isolated and characterized.

Strains of *F. sambucinum* also produce beauvericin (1283), fusarin C (3, 600), fusaric acid (119), sambutoxin (1057-1060, 1917), steroids (284), wortmannin (3), and numerous sesquiterpenes (964). The antifungal fusacandins (papulacandin class) produced by *F. sambucinum* inhibit β-glucan biosynthesis by *Candida albicans in vitro*, but not in the presence of serum (871, 943). Strains of *F. sambucinum* also can degrade this class of antifungal compounds (2030).

Current Species Description: Nirenberg (1568).

Fusarium scirpi Lambotte & Fautrey

Sexual Stage. None known.

Common Synonyms. *Fusarium equiseti* var. *bullatum*, *Fusarium chenopodium*.

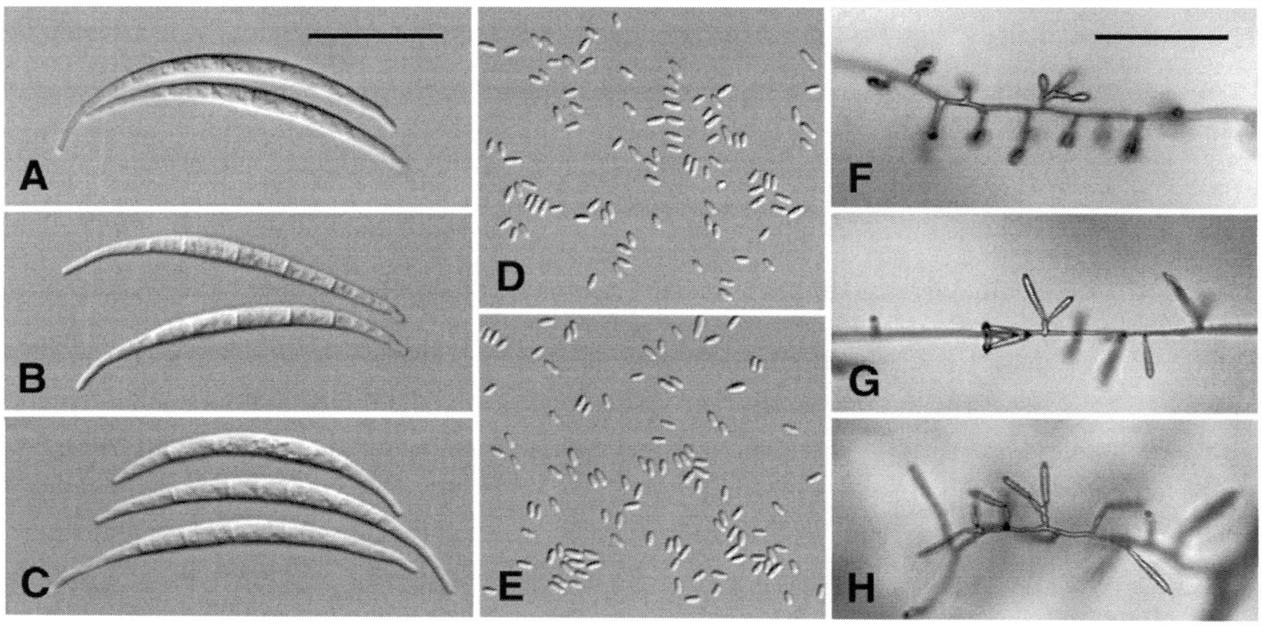

Figure 13-58. *Fusarium scirpi.*
A – C: Macroconidia; **D – E:** Microconidia; **F – H:** Microconidia *in situ* on CLA. A – E, scale bar = 25 µm; F – H, scale bar = 50 µm.

Reported Geographic and Host Distribution

Found in arid and semi-arid regions, particularly in soil.

Media to Examine and Key Characters

Characters on CLA. Orange sporodochia are common but may be slow to develop. When present they are found on and adjacent to the carnation leaves. The macroconidia they contain are long, slender, thin-walled and distinctly curved with six to seven septa, a tapering apical cell and a pronounced foot shape to the basal cell. Club shaped microconidia form on short phialides. Polyphialides are common and often have three openings arranged in a cross-shaped manner. Chlamydospores are abundant in clumps and chains.

Characters on PDA. *Fusarium scirpi* grows rapidly and produces abundant white mycelium that turns tan or brown with age. It is not uncommon for an orange or brown spore mass to be formed. Brown pigments develop in the agar and dark brown flecks often occur. Violet pigment also may be produced in the agar.

Macroconidia

- Sporodochia: Orange in color and often slow to develop. Usually located on or adjacent to the carnation leaves in CLA. Sporodochia also may form on PDA, but spores from the sporodochia formed on PDA should not be used for diagnostic purposes.
- General morphology: Relatively slender, but widest at the center. Strong, pronounced dorsi-ventral curvature.
- Apical Cell morphology: Tapering and elongate.
- Basal Cell morphology: Well-developed foot shape.
- Number of septa: Usually 6- to 7-septate. Septa in the macroconidia may be closer together in the middle of the conidium than at the end.

- Abundance: Abundant in sporodochia.

Microconidia

- Shape/septation: Ellipsoidal to club shaped and 0- to 3-septate.
- Aerial mycelium presentation: Singly and small false heads.
- Conidiogenous cells: Both mono- and polyphialides are present, but cross-shaped polyphialides with three openings are the most common and are diagnostic.
- Abundance: Moderately abundant in the aerial mycelia.

Chlamydospores

- Abundance/Speed of formation: Abundant in older cultures. Chlamydospores require 4-6 weeks to form on CLA.
- Location: Chlamydospores are found in clumps and chains in hyphae.
- Appearance: Usually smooth walled. Initially hyaline, but may become yellow or brown pigmented with age.

Taxonomy, Pathology and Ecology

Isolates conforming to the description of *F. scirpi* were illustrated and described by Wollenweber (2356) but were named as either *F. chenopodium* or *F. equiseti* var. *bullatum*. Wollenweber & Reinking (2359) included *F. chenopodinum* as a synonym of *F. scirpi*, recognizing the species. Gordon (746) and Booth (222) both synonymized *F. scirpi* with *F. equiseti*. Nelson *et al.* (1551) and Gerlach & Nirenberg (697) recognized *F. scirpi* as a separate species but Gerlach & Nirenberg (697) differentiated it into two distinct varieties. Burgess *et al.* (276) published an account of the nomenclatural history of *F. scirpi* and an emended description.

Fusarium scirpi is most likely to be confused with *F. equiseti* as the PDA cultures are similar and the macroconidia are of a similar size and shape. CLA cultures of *F. scirpi* have abundant microconidia and usually have diagnostic, short, cross-shaped polyphialides. The diagnostic microconidia and polyphialides are found in the aerial mycelia, and an accurate diagnosis cannot be made solely from the observation of macroconidia from sporodochia.

Isolates of *F. scirpi* have been recovered from a number of countries including Australia (277, 570, 756), India (793), Iran (2405), South Africa (2223) and Turkey (1124), all of which are characterized by a semi-arid to arid climate. Backhouse & Burgess (112) used a computer climate modeling program to determine that *F. scirpi* would be likely to be found in most of temperate and arid Australia, but would not be found in areas that were too cold, where winter rainfall was more than twice the summer rainfall, or in the hot summer rainfall areas of the tropics.

Fusarium scirpi probably is a saprophyte, but there are records of it association with decay of fruit of cucurbits (1785) and papaya (793). A protocol for preparing protoplasts of *F. scirpi* is available (186). α-amylases have been characterized from this fungus as well (342).

Fusarium scirpi usually is regarded as having little or no toxicity, especially relative to other species in section *Gibbosum* that can produce trichothecenes (2348). Various enniatin analogues, some with antihelmentic activity, can be produced by *F. scirpi* if the precursors in the media are varied (1134). Enniatin synthetase, which may be a constitutively produced enzyme (187B), has been cloned and characterized *in vitro* (285, 715, 798, 803, 804, 1697).

Current Species Description: Burgess *et al.* (276).

Fusarium semitectum Berkeley & Ravenel

Sexual Stage. None known.

Common Synonym. *Fusarium pallidoroseum, Fusarium incarnatum.*

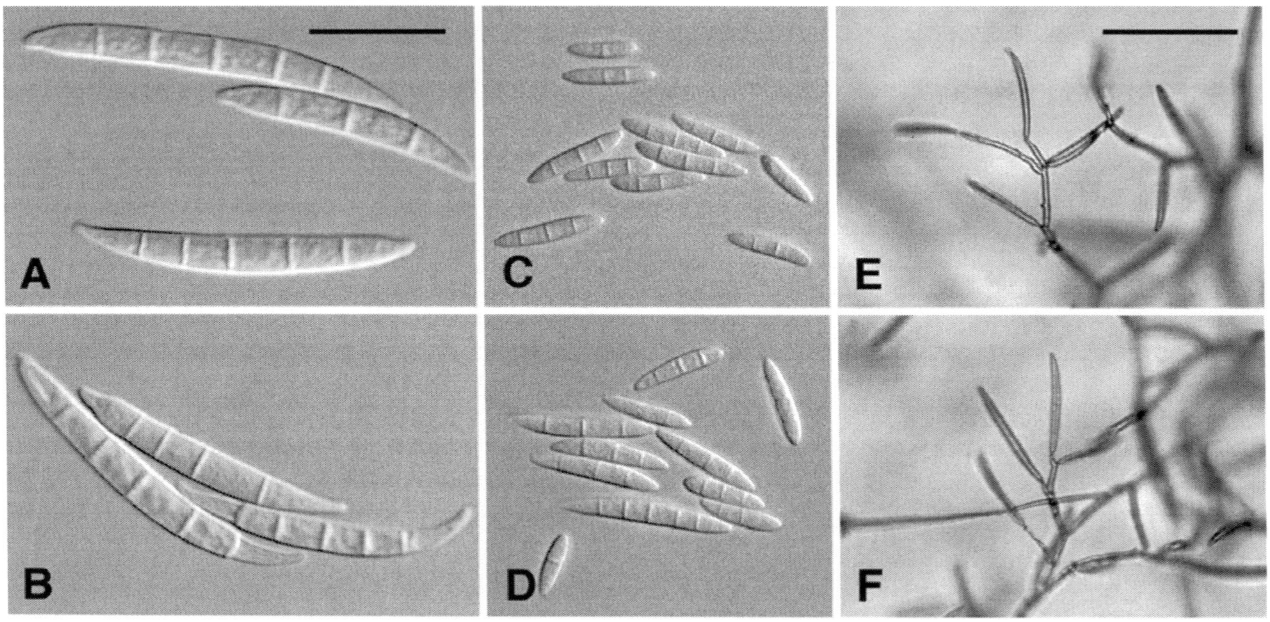

Figure 13-59. *Fusarium semitectum.*
A – B: Macroconidia; **C – D:** Mesoconidia; **E – F:** Mesoconidia *in situ* on CLA. A – D, scale bar = 25 μm; E – F, scale bar = 50 μm.

Reported Geographic and Host Distribution

Regularly found on aerial plant parts in subtropical and tropical regions. Often associated with storage rots in bananas.

Media to Examine and Key Characters

Characters on CLA. Macroconidia are not common, but when observed are slightly curved with a foot-shaped basal cell and may be formed in orange sporodochia. The most distinctive character is the abundant production of straight, spindle-shaped mesoconidia from polyphialides in the aerial mycelia. These mesoconidia are easily observed microscopically *in situ* and often have the appearance of "rabbit ears". Chlamydospore formation varies by strain and chlamydospores may be difficult to find in some cultures.

Characters on PDA. Cultures usually grow rapidly and produce abundant dense aerial mycelia that initially is off white and becomes beige or brown with age. Brown pigments also may be produced in the agar. Light orange sporodochia may be produced by some strains.

Macroconidia

- Sporodochia: Orange sporodochia produced on carnation leaves in CLA by some strains.
- General morphology: Relatively slender with a curved dorsal surface and a straighter ventral surface.
- Apical Cell morphology: Curved and tapering to a point.
- Basal Cell morphology: Foot shaped.
- Number of septa: 3- to 5- septate.
- Abundance: May be difficult to find in some cultures.

Microconidia/Mesoconidia

- Shape/septation: Microconidia are pyriform to obovate and usually 1-septate and most common in older cultures. Mesoconidia are fusoid and 3- to 5- septate.

- Aerial mycelium presentation: Individual spores per phialide, but often two spores per polyphialide to give a "rabbit ear" appearance.
- Conidiogenous cells: Monophialides and polyphialides.
- Abundance: Mesoconidia are abundant in the aerial mycelia. Microconidia are scarce and often are difficult to find.

Chlamydospores

- Abundance/Speed of formation: Present, but not common. The absence of chlamydospores is not a reliable diagnostic character.
- Location: Found in the hyphae both singly and in chains, and singly within conidia.
- Appearance: Globose and smooth. Initially hyaline, but may become a light yellow color with age.

Taxonomy, Pathology and Ecology

The taxonomic status of *F. semitectum* has varied over time. It was first described in 1875 (177) and was recognized by Wollenweber & Reinking (2359), Booth (222), Gerlach & Nirenberg (697), and Nelson *et al.* (1551). Gerlach & Nirenberg (697) described three varieties, *F. semitectum* var. *semitectum, F. semitectum* var. *majus*, and *F. semitectum* var. *violaceum*. Preliminary work with molecular techniques indicates that *F. semitectum* probably is a species complex, but at this time we have restricted the description to a single nomenclatural name. Booth & Sutton (223) proposed the name *F. pallidoroseum* for this species, but as pointed out by Subramanian (2062) and Nirenberg (1567) the species epithet *incarnatum* predates *pallidoroseum* by ~ 30 years and if the epithet were to change then it should be to *F. incarnatum*. Given this nomenclatural difficulty, we prefer to use the more widely recognized name of *F. semitectum* until more work is done on this species (or species complex).

This species is most likely to be confused with *F. subglutinans sensu lato*, *F. polyphialidicum* and *F. sporotrichioides*, which all can produce spindle-shaped microconidia from polyphialides on CLA. These microconidia often are produced in pairs and appear like "rabbit ears" thus resembling the formation of the spindle-shaped macroconidia of *F. semitectum* on CLA. There are a number of characters, however, that distinguish this species including the production of brown pigments on PDA, the size and shape of mesoconidia, and the absence of smaller microconidia. The sporodochial macroconidia resemble those found in *F. equiseti* and *F. heterosporum*, but these species do not produce either the microconidia or the spindle-shaped mesoconidia that are commonly produced by *F. semitectum*. *F. semitectum* cultures that degenerate to pionnotal types still produce the spindle-shaped mesoconidia. Although both the *MAT-1* and *MAT-2* mating types occur (1036) and are transcribed in *F. semitectum* (1037), no sexual stage for this species has been identified.

Fusarium semitectum commonly is isolated from soil (275, 1226) and from diverse aerial plant parts in tropical and sub-tropical areas, *e.g.*, banana fruits (1964, 2282) and palm fronds, but also can be recovered from soils in the Arctic (1116) and deserts (984). Although there are many reports of *F. semitectum* being implicated in various diseases it often is not regarded as an important plant pathogen. It has been reported to cause a canker of walnut (1944), a blight of the kangaroo paw ornamental plant (1897), pod and seed rot of beans (491), reduced seed germination and seedling growth of sorghum (735), corky dry rot of melons (321, 1406), and storage rot problems of mushrooms (1945), bananas and other fruits (771, 772, 1096, 1301, 1353, 2283), and is one of the dominant fungi on the grain of pearl millet (2337-2339). Pollen of sorghum and pearl millet promote germination of conidia of *F. semitectum* (736) and could explain the frequent association of this species with head mold and grain of these hosts (1380, 1381).

The use of *F. semitectum* as a biological control as a pathogen for water hyacinth (1533) and *Mimosa invisa* in rice (1301) has been proposed. *F. semitectum* may be a biological control of ergot on pearl millet by reducing sclerotia formation and development (1778) and by parasitism of *Rhizoctonia solani* (82) and *Sclerospora graminicola* (1776).

Fusarium semitectum is resistant to most clinical antifungals, *e.g.*, itraconazole, miconazole and flucytosine (1799), with amphotericin B (1748, 1799) and natamycin (1799) reported as the most effective. It can cause endocarditis (1405) and skin and disseminated infections in immunocompromised (210) and burn patients (927).

Fusarium semitectum has been associated with bovine pulmonary emphysema (Linnabary & Tarrier, 1988), a disease normally associated with *F. solani* (there are more details on this disease in the section on *F. solani*). It also can cause an important shell softening disease in tortoises (Rose *et al.*, 2001) and can parasitize the eggs of the roundworm *Toxocara canis* (Ciarmela *et al.*, 2002).

Fusarium semitectum is known to produce apicidins (1992), beauvericin (796, 1283), equisetin (2308), fusapyrone (47, 48, 592, 593), moniliformin (1758), sambutoxin (1059), trichothecenes (1341), and zearalenone (1341). Some extracts of *F. semitectum* cultures have insecticidal activity (198), but the compounds responsible for the insecticidal activity have not been identified. *F. semitectum* also has insecticidal activity that may not be attributable to toxin activity (814).

Current Species Description: Gerlach & Nirenberg (697) and Nelson *et al.* (1551).

Fusarium solani (Martius) Appel & Wollenweber emend. Snyder & Hansen

Sexual Stage. *Haemanectria haematococca* (Berkeley & Broome) Samuels & Nirenberg.

Common Synonym. *Nectria haematococca.*

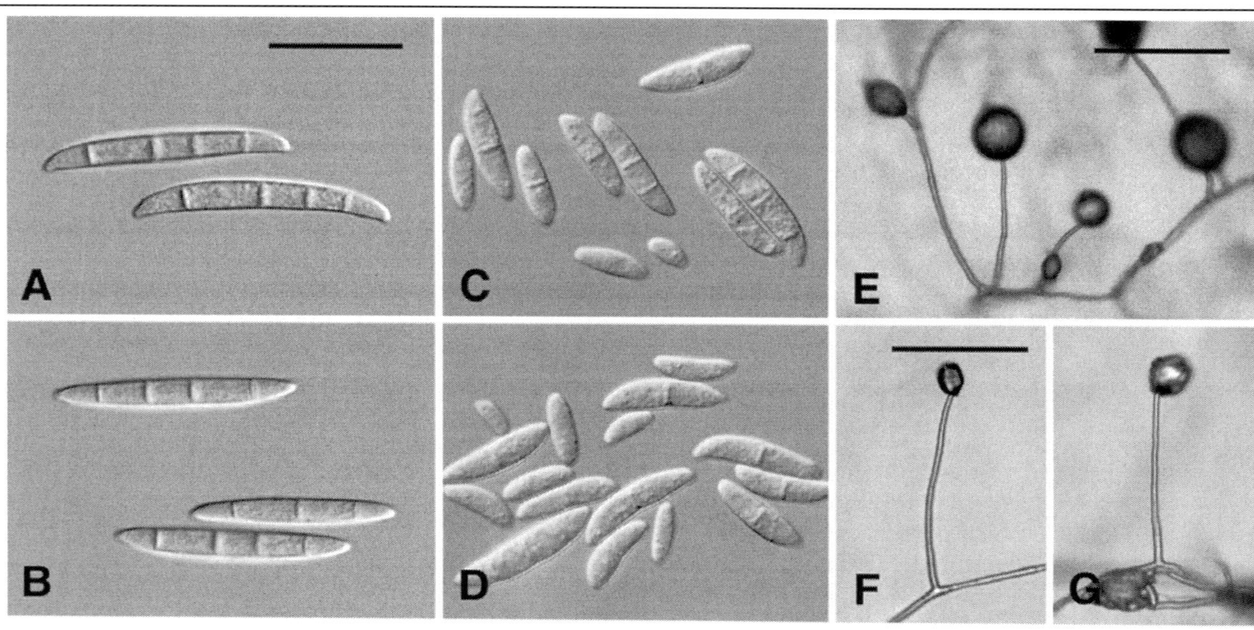

Figure 13-60. *Fusarium solani.*
A – B: Macroconidia; **C – D**: Microconidia; **E – G:** Microconidia *in situ* on CLA. A – D, scale bar = 25 μm; E, scale bar = 100 μm; F – G, scale bar = 50 μm.

Reported Geographic and Host Distribution

Cosmopolitan on a range of substrates. Regularly isolated from soils in a variety of environments. A pathogen of a large number of plant species, especially trees.

Media to Examine and Key Characters

Characters on CLA. Macroconidia are relatively wide, straight to slightly curved, 3- to 7-septate with rounded ends and found abundantly in cream and less frequently in blue or green sporodochia. Oval, ellipsoidal or reniform, 0- or 1-septate microconidia are formed in round false heads on relatively long monophialides. Some isolates are homothallic and may produce red or orange perithecia. Chlamydospores often are produced abundantly in pairs in hyphae and in the agar.

Characters on PDA. Cultures of *Fusarium solani* usually are white to cream with sparse mycelium. Sporodochia often are produced in abundance and may be cream, blue or green. Many isolates do not produce pigments in the agar although some violet or brown pigments may be observed.

Macroconidia

- Sporodochia: Cream, blue or green sporodochia are common on carnation leaf pieces and in the agar and contain numerous macroconidia (cream sporodochia usually contain more macroconidia than do the blue or green ones).
- General morphology: Relatively wide, straight, stout and robust.
- Apical Cell morphology: Blunt and rounded.

- Basal Cell morphology: May have a distinct foot shape or be poorly developed, straight to almost cylindrical, usually with a notched or a rounded end.
- Number of septa: 5- to 7-septate.
- Abundance: Usually abundant in sporodochia.

Microconidia

- Shape/septation: Oval, ellipsoid, reniform and fusiform with 0 or 1 to occasionally 2 septa.
- Aerial mycelium presentation: False heads.
- Conidiogenous cells: Monophialides, often quite long.
- Abundance: Abundant in the aerial mycelia.

Chlamydospores

- Abundance/Speed of formation: Commonly formed abundantly and rapidly, usually within 2-4 weeks on CLA. Chlamydospore formation by *F. solani* has been evaluated with electron microscopy (2215, 2217).
- Location: May be intercalary in the hyphae or formed terminally on short lateral branches usually singly or in pairs, but occasionally in short chains.
- Appearance: May be globose to oval in shape and smooth or rough walled.

Taxonomy, Pathology and Ecology

The taxonomic approach to *Fusarium solani* adopted here is based on that developed by Snyder & Hansen (2012) and used by Nelson *et al.* (1551). This species concept is a gross over simplification, but until the phylogenetic analyses currently in progress are complete, we believe this approach to be the best. A number of mating populations, *i.e.*, biological species, have long been recognized (1389, 1390, 1838) within *F. solani*, and both heterothallic and homothallic strains have been described. The described biological species are concordant with a phylogenetic species concept (1211, 1597, 1837, 2065), and these entities, at the least, need to be described as separate species. Strains with high levels of female fertility are available for some of these mating populations (655).

Many researchers confuse *F. solani* and *F. oxysporum*. This confusion reflects overlaps in some aspects of morphology and ecological niches. *F. solani* and *F. oxysporum* can be distinguished by examination of the phialides bearing the microconidia and the microconidia themselves. The long monophialides bearing microconidia found in *F. solani* are quite different from the relatively short monophialides bearing microconidia in *F. oxysporum*. Microconidia of *F. solani* tend to be somewhat wider, more oval in shape, and to have thicker walls than do the microconidia of *F. oxysporum*.

Distribution and occurrence. *Fusarium solani* has a cosmopolitan distribution. It can be found in numerous native soils (277, 379, 552, 563, 974, 984, 1116, 1226, 1336) and is one of the few species of *Fusarium* that can be found at high frequency from soils in rain forest habitats (2085). Perithecia are commonly observed in the field in the wet tropics, but are much rarer in temperate areas. *F. solani* is recorded as a pathogen on a vast and diverse range of host plants but the utility of many of these records is questionable given the taxonomic diversity likely contained within this species complex.

F. solani is, however, well documented as a pathogen of a number of legumes and other tropical plants where it often is associated with cankers and die-back problems of trees (1550). Some economically important plants with significant diseases caused by *F. solani* include avocado (445), beans (1981), citrus (1552, 2220), cocoyam (1602), cowpea (1865), orchids (170), passion fruit (1669, 1707, 1729), peas (776, 1131, 1132), peppers (630), potato (1931) and squash (990, 1878). The strains of *F. solani*, sometimes termed *F. solani* f. sp. *glycine*, that cause sudden death syndrome in soybeans (16, 353, 1252, 1575, 1841, 1850, 1851) have been described as two separate species – *F. virguliforme* in North America and *F. tucumaniae* from South America (73, 85). The fungus may serve as a food source for some tree-boring insects (812). The mating population to which a strain belongs need not be indicative of its phytopathogenic capabilities (831).

Fusarium solani has been proposed as a possible biological control for leafy spurge (*Euphorbia* spp.; 293, 294), morning glory (1), striga (1144), Texas Gourd (231, 2299, 2300) and water hyacinth (957). It also may function as a mycoparasite for *Phomopsis sclerotioides* (2211) and *Mucor spinosus* (522). Competitively, nonpathogenic isolates of *F. solani* can function as a biological control of Fusarium wilt of tomato and can be more effective than other microbial biocontrol agents such as *Burkholderia*, *Gliocladium*, and *Pseudomonas* and *Trichoderma* (1187, 1189).

Physiology. Germination of spores of *F. solani* is stimulated by aldehydes of some fatty acids such as heptanal, octonal, nonanal, decanal and undecanal (817) and flavanoids (122, 1844). Spore germination is disrupted when strains are exposed to inhibitors of cAMP-dependent protein kinases and cAMP phosphodiesterase (122, 1844). Chlamydospore viability declines if they are maintained in nonsterile soil (1465) under moist conditions at room temperature (1464) and increases if they are maintained at cool temperatures under dry conditions (1464). Macroconidia may require ethanol or serine to germinate under some conditions (1664) and may be inhibited by ammonia and pH under other conditions (1278). Lipid content of the macroconidia may affect chlamydospore formation, but the persistence of the chlamydospores under field conditions is not affected by their lipid content (2216). A double mutant in which microconidia germinate and immediately produce a phialide bearing microconidia, *i.e.*, microcycle conidiation, has been isolated and produces numerous conidia and can be used for mutagenesis experiments requiring replica plating (1114).

Colony morphology is sensitive to K^+ levels and changes from a zonation pattern to a feathery pattern when the level of K^+ drops below 3 mM (446). Zonation also occurs in response to a light:dark cycle (447). The fungus can grow under microaerobic conditions (700) and in the presence of creosote (1262).

Mechanisms of plant pathogenicity have been widely studied with *F. solani*. The cutinase enzyme(s) and modes of spore attachment and penetration of the host's surface (584, 801, 990, 991, 1878, 2421) are both important characters that help determine if a strain is an effective pathogen and the degree of disease that the strain is capable of causing. The genes responsible for pathogenicity against pea are carried on a dispensible chromosome and have a G+C content that differs significantly from that of other portions of the genome. This difference has been interpreted to mean that the genes for pea pathogenicity were acquired from another organism via a horizontal gene transfer mechanism (2126). Pathogenicity towards cucurbits is under polygenic control with three to fifteen effective factors segregating depending on the parents of the cross (829).

The genes for pea pathogenicity, including the one for pisatin demethylase, are subject to complex regulation at the transcriptional level both *in vitro* and *in planta* (833, 867, 1045, 1846, 2054, 2055). The gene encoding pisatin demethylase (sometimes abbreviated PDA) is located on a dispensible chromosome, but genes necessary for pathogenicity towards tomato and carrots are located elsewhere in the genome (656). Based on hybridization, there is a gene similar to that from *F. solani* in *F. oxysporum* f. sp. *pisi*, but not in other pathogenic fungi that are capable of rapidly detoxifying pisatin (462). *F. solani* also can degrade other phytoalexins such as kievitone (1245) that are produced as part of plant responses to some fungal attacks. Some fungal-produced proteins trigger specific and systemic acquired resistance responses by the host (800).

Mutants of *F. solani* that are insensitive to growth inhibition by tomatine can rot green tomatoes, which wild-type cannot, and are similar to their wild-type progenitors in pathogenicity towards peas and red tomatoes (459). These mutants also are insensitive to digitonin, have a low sterol content and have phenotypes that are inherited as single Mendelian alleles (460).

Naphthaquinones *e.g.*, dihydrofusarubin and isomarticin, are associated with chlorosis in citrus (1553-1555) and can be found in the xylem of diseased plants (2220). Pigment production is generally inherited in a Mendelian manner and probably is independent of plant pathogenicity, which appears to be inherited quantitatively (885).

Biochemistry. A number of different enzymes produced by *F. solani* have been studied biochemically. Some of these enzymes may play a role in the pathogenicity process including cutinase (see below) and the enzymes involved in the degradation of various phytoalexins (355, 385, 464, 1083, 1149, 1176, 1821, 2002-2004, 2190, 2191, 2326, 2330, 2331, 2416). Both *F. oxysporum* and *F. solani* can degrade α-tomatine, a phytoalexin synthesized by tomato, but a polyclonal antibody that recognizes the enzyme from *F. oxysporum* does not recognize the *F. solani* enzyme (1176). Other enzymes examined include D-amino acid oxidase (935), cellulases (189), chitosanases (1970, 1971), α-dialkyl amino acid transferase (586), esterases (619), levoglucosan kinase (1084), lipase (507, 536, 1322, 1504), pectate lyase (415, 731, 788-790), polygalacturonase (2413), serine protease (1619) and tannin acyl hydrolyase (128).

Cutinase in particular has been intensively studied, with these studies amongst the first to detail host-pathogen interactions at a molecular level (127, 130, 419, 1001, 1044, 1109, 1110, 1112, 1246, 1247, 1249, 1715, 1716, 1845, 2021, 2035, 2360). This enzyme has been extensively characterized biochemically in both free and immobilized forms and has been crystallized (11, 129, 392, 441, 547, 633, 726-728, 775, 965, 1065, 1111, 1113, 1177, 1260, 1290, 1291, 1333, 1368, 1369, 1420, 1503, 1560, 1689, 1699, 1714, 1743, 1953, 2022, 2023, 2218). Cutinase activity is regulated differently in MP I and MP V of *N. haematococca* than it is in MP VI (832), suggesting that this enzyme does not necessarily have a common role in the various diseases induced by these fungi. Cutinase also may be of use as an additive to laundry detergent where it helps remove fat-based stains (416, 417, 632), and is being evaluated for industrial production in a *S. cerevisiae* system (296-298, 323, 407, 427, 621, 622).

Fusarium solani has a cyanide hydratase (also termed formamide hydrolyase) enzyme that may be useful for bioremediation of sites contaminated with cyanide (149, 150, 526). This enzyme enables *F. solani* to utilize formamide as a sole nitrogen source, but enzyme production is not induced by the presence of formamide, only by the presence of cyanide (151). Note that this enzyme also has been well-studied in *F. oxysporum*.

Fusarium solani can metabolize steroids from various sources (55, 695, 1120, 1873, 1874, 1900, 1901, 2407, 2408), and various lignin components and their breakdown products (940, 999, 1000, 1018, 1019, 1528, 1556, 1585, 1892, 2089). It also can release sulfur from complex organic pollutants (1051) and degrade DDT (1459), bronopol – a pharmaceutical preservative (2136), the herbicides butachlor (337), oxadiazon (338) and pendimethalin (152), the antibiotic cephalosporin C (935), glaucine – an antitussive alkaloid (451, 1040, 1041), nitriles (87, 818, 1162), polyaromatic hydrocarbons (2239), phloroglucinol – a stain that is an important precursor in the synthesis of pharmaceuticals and explosives (2281), and some plastics (1025, 1504, 1829).

Genetics and molecular biology. Mitosis in *F. solani* has been studied in considerable detail with extensive fundamental studies of the role of the spindle and astral microtubules in the cell division process (26-28, 972). Mitotic recombination and chromosomal segregation in hybrids resulting from protoplast fusion can be used to assign markers to linkage groups (432). Based on CHEF gel karyotypes and light microscopic observations, the various mating populations known in *F. solani* have 5-13 chromosomes and a total genome size of ~ 40 Mb (1537, 2066, 2101), a result consistent with the suggestion that these entities should be described as species.

Genetically, *F. solani* carries a number of dispensable "B" chromosomes that do not segregate in a 1:1 manner following meiosis (1434). These "B" chromosomes may carry one or more copies of some transposable elements (581, 1963) and are implicated in pathogenicity towards pea (1081, 2296). Some strains of *F. solani* carry dsRNA mycoviruses (1577-1579) and linear mitochondrial plasmids (1877) that do not alter the morphological phenotypes of the strains.

Some stress-related mRNAs have been identified and characterized (354). A number of genes have been cloned from *F. solani* including a mitogen-activated protein kinase (1248).

Vegetative compatibility has been studied in *F. solani* (830). Recovering *nit* mutants can be difficult and *sul* (sulfate non-reducing) mutants often had to be used instead. A relatively high proportion of the strains examined were heterokaryon self-incompatible.

Transformation vectors and protocols have been developed for *F. solani* (420, 1352, 1512, 1811), including linear self-replicating plasmids (1728, 1879). Protocols that restrict colony growth and make replica plating of strains easier also have been developed (2230).

There also have been numerous reports of DNA fragments that are specific either for *F. solani* as a whole or for a particular subgroup within the species. An intron in the 3' end of the small subunit RNA coding region is polymorphic in length with at least seven different alleles identified (2067). Many, but not all, of these studies are based on relatively small strain sets and none of these diagnostics has yet been accepted for use on a broad scale (41, 313, 624, 827, 2065). These studies also include a PCR-based test developed specifically to detect *F. solani* from material in the eye (952). Species-specific antibodies also are available (1937).

Human pathogenicity. With respect to human pathogenicity, *F. solani* has been recovered from eyes, nails and skin (422, 548, 637, 646, 682, 716, 864, 934, 1256, 1298, 1487, 1565, 1608, 1637, 1783, 1828, 1906, 2198, 2279, 2302, 2329, 2333), bone (121), nasal cavities (69, 226, 1292, 1576), infected wounds (859, 1238, 2158, 2214), and systemically infected cancer (287, 599, 849, 1135, 1136, 1382, 1426, 1614, 1760, 2152, 2256) and HIV (778) patients. Patients with keratitis resulting from infection with *F. solani* also were more likely to be HIV positive (1486). *F. solani* also may cause endocarditis (783, 797) and lung disease (81), and has been shown to be allergenic (1623-1625, 2240-2245).

Culture of the fungus from infected tissue was more accurate than was direct observation of the fungal hyphae in the infected tissue (934). Fluorescein-conjugated lectins (1814), fluorescent-labeled antibodies (1024) and confocal microscopy (634), however, can improve the accuracy of direct observation identification.

Fusarium solani is resistant to most clinical antifungals, *e.g.*, clotrimazole (1487), flucytosine (780, 1799), itraconazole (986, 1241, 1799, 2138) and micoconazole (1799), with amphotericin B (986, 1241, 1748, 1799, 2280) and natamycin (1799) reported as the most effective, but in some systemic infections even these drugs may not suffice (59, 780). Amidoamine myristamidopropyl dimethylamine (917), amorolfine (1251), pentamidene (1267) and voriconazole (587, 986, 1062, 1241, 1717) have been suggested as potentially effective clinical antibiotics against *F. solani* but have not been widely tested and may require additional monitoring to prevent side effects, some of which may be severe (23, 44). Pentamidene is more active against other *Fusarium* species than it is against *F. solani* (1267). Fenpropimorph is known to inhibit ergosterol biosynthesis in *F. solani* (458).

Fusarium solani is more resistant to attack by human phagocytes than are most other fungi, which could account for its relatively frequent association with infection, especially in immunosuppressed patients (210, 2351). *F. solani* has been found contaminating hospital water systems, and the water supply in a hospital could serve as a reservoir for repeated infections over a period of years (60).

Animal Pathogenicity. *Fusarium solani* was pathogenic in an experimental test with immunocompetent mice (1392) and was the only *Fusarium* species in this test that was pathogenic. Immunocompromised mice had disseminated infections rather than the localized infections found in immunocompetent mice (1212). When soil strains of *F. solani* were passaged through mice, the recovered strains were more virulent than were the strains that were initially recovered from the soil (2070). Whether these results are due to selection for virulent strains from a mixed inoculum or to physiological changes in the passaged strains is not clear, but these results do suggest that strains of *F. solani* recovered from animals need to be treated with more caution than are those recovered from soil or diseased plants. Strains of *F. solani* that successfully colonize corneal material *in vivo* develop unique defensive structures to protect themselves from attack by neutrophils (1073).

Rabbit models for keratitis caused by *F. solani* are available (676, 737) and suggest that corneal infections by *F. solani* may be masked if the patient is taking corticosteroids (1591). Mouse models are available for studying corneal infections by *F. solani* (2367) and of the sensitivity of *F. solani* to antifungal agents such as voriconazole (767) and amphotericin (780). A rat model for the study of *F. solani* as a cause of endophthalmitis also is available (1590).

Fusarium solani is toxic and/or pathogenic to a number of animals other than humans. The pathogenicity may be direct, as it is for crayfish (351), dogs (1007), some insects (822), leeches (1896), penaeid shrimp (163, 394, 418, 897, 898), sharks (2001), snakes (1156) and turtles (292, 1695, 1696). Toxicity often is a presumed mode of action, but the compounds produced by these pathogenic cultures usually have not been well characterized.

Toxins. *F. solani* is associated with field outbreaks of moldy sweet potato toxicosis in cattle and with an androgenic syndrome in chickens (1341). The sweet potato associated toxins, furanoterpenoids, cause pneumonia. These compounds are not mycotoxins in the strict sense of the word as the fungus modifies phytoalexins synthesized by the plant in response to non-specific stresses to produce the toxic compounds, rather than synthesizing the compounds internally from simple starting materials. The toxins produced in moldy sweet potatoes also are active against a number of other animals, including rats (1766) and humans, and are not destroyed by normal cooking or food preparation processes. The androgen associated with the androgenic syndrome in chickens results from fungal metabolism of a steroidal alkaloid produced by a host tree.

Previous reports of the production of trichothecene mycotoxins and zearalenone by *F. solani* have not been confirmed (1341), and *F. solani* lacks the *tri5* gene required for biosynthesis of the trichothecenes (2112). *F. solani*, however, does carry a gene that encodes a functional trichothecene 3-O-acetyltransferase gene, which can be used as a defense against trichothecene mycotoxins (2156).

Some strains of *F. solani* produce the immunosuppressive compound cyclosporin A (2068), which could increase the pathogenic potential of this fungus towards animals, especially in a direct infection. *F. solani* synthesizes numerous naphthaquinone-type pigments (17, 55, 132, 695, 1341, 2117-2120) that often are a component of the plant pathogenicity process and may act as uncouplers of the mitochondrial electron transport pathway (843, 844, 1822). Other compounds known to be synthesized by *F. solani* include fusalanipyrone (13, 14), fusaric acid (119, 1341) and moniliformin (347). In addition to these compounds, a number of chemically unidentified toxic compounds also are synthesized by *F. solani* (1341).

Current Species Description: Gerlach & Nirenberg (697) and Nelson *et al.* (1551).

Fusarium sporotrichioides Sherbakoff

Sexual Stage. None known.

Common Synonyms. *Fusarium tricinctum*, *Fusarium sporotrichiella* var. *sporotrichioides*.

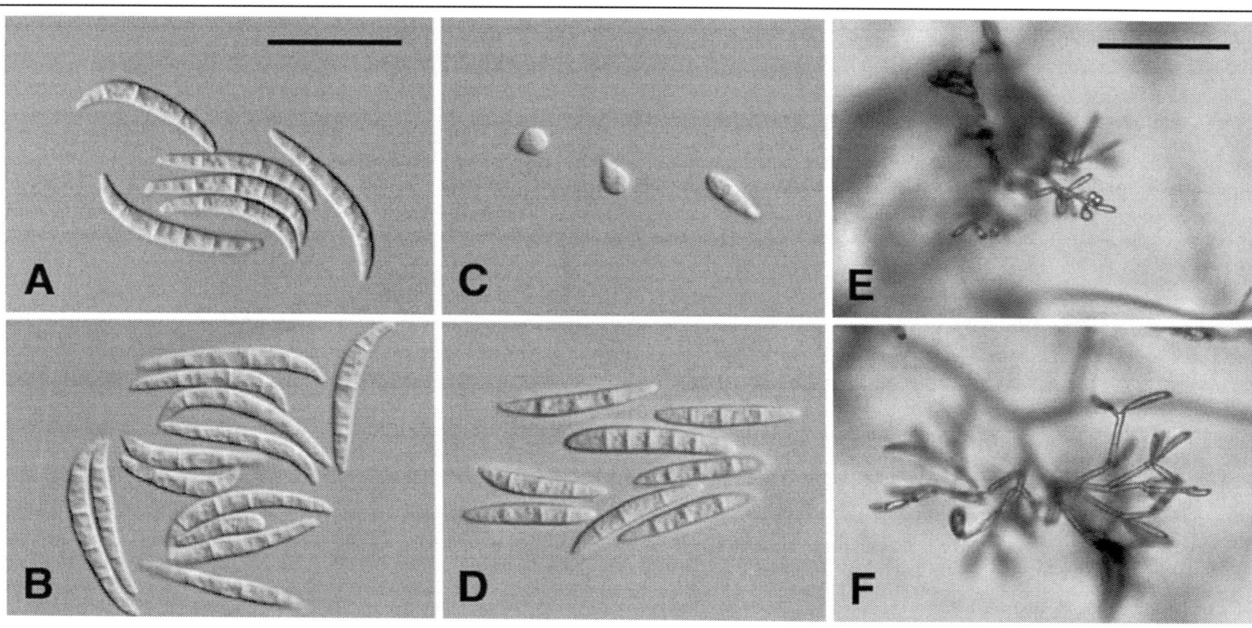

Figure 13-61. *Fusarium sporotrichioides.*
A – B: Macroconidia; **C:** Pyriform and napiform microconidia; **D:** Mesoconidia; **E – F:** Microconidia *in situ* on CLA. A – D, scale bar = 25 μm; E – F, scale bar = 50 μm.

Reported Geographic and Host Distribution

Widespread. Generally found in temperate parts of the world on a variety of substrates.

Media to Examine and Key Characters

Characters on CLA. Macroconidia are abundant in orange sporodochia. They are small to medium in length, falcate, 3- to 5-septate with a tapering apical cell and a foot-shaped basal cell. Microconidia are produced from monophialides and polyphialides and may be pyriform and 0-septate or fusoid and 0- to 1-septate. Fusoid mesoconidia may be formed and be up to 5-septate. Chlamydospores are abundant in older cultures, and darken with age.

Characters on PDA. Mycelia are profuse and grow densely and rapidly. Initially the culture is white or pale red, but with age the pigment darkens and orange sporodochia may appear. Red pigments are produced in the agar.

Macroconidia

- Sporodochia: Sporodochia are orange. Production of sporodochia can vary, although most cultures eventually will produce them. Exposure to near UV light can stimulate production. Older PDA cultures (4+ weeks) may produce sporodochia underneath the mycelium. Macroconidia from sporodochia produced on PDA should not be used for diagnostic purposes.
- General morphology: Falcate to almost lunate.
- Apical Cell morphology: Curved and tapering.
- Basal Cell morphology: Neither distinctly foot shaped nor notched.
- Number of septa: 3-5 septate, but predominantly 3- septate.
- Abundance: Relatively abundant.

Microconidia/Mesoconidia

- Shape/septation: 0-septate pyriform microconidia, 1-septate ellipsoid to fusoid microconidia, and fusoid mesoconidia that may be up to 5-septate. These spores may have a papilla at the base.

- Aerial mycelium presentation: Singly, in small clumps or in false heads.
- Conidiogenous cells: Mono- and polyphialides. Fusoid micro- and mesoconidia can form from polyphialides to give a "rabbit ear" appearance *in situ*.
- Abundance: Abundant in the aerial mycelia.

Chlamydospores

- Abundance/Speed of formation: Often formed rapidly, 2-4 weeks on CLA, and abundantly.
- Location: Found singly and in chains in the hyphae.
- Appearance: Globose, smooth-walled and initially hyaline becoming a light brown with age.

Taxonomy, Pathology and Ecology

This species, first described by Sherbakoff (1958), was recognized by Wollenweber & Reinking (2359), Gerlach & Nirenberg (697) and Nelson *et al.* (1551). *F. sporotrichioides* can grow at very low temperatures and can be isolated from grain that has overwintered under snow in Canada, the USA, and the former Soviet Union. It can be recovered from diverse substrates including grasses, small grains and alfalfa (450, 928, 1622, 1688, 1868, 2390), but generally is regarded as a weak pathogen.

Cultures on PDA can be confused with the related species *F. poae*, *F. chlamydosporum*, and *F. tricinctum*. The presence of polyphialides distinguishes *F. sporotrichioides* from *F. poae* and *F. tricinctum*, while the presence of two types of microconidia distinguishes it from *F. chlamydosporum*. The fusoid shaped microconidia are similar to the spindle-shaped conidia of *F. semitectum*. Conidium formation has been studied at the electron microscopic level (2229). The sexual stage of *F. sporotrichioides* probably is heterothallic (1036). Auxotrophic mutants of *F. sporotrichioides* have been made and used to force heterokaryons (424), but there are no formal studies of vegetative compatibility.

A PCR diagnostic for *F. sporotrichioides* based on the *ITS2* region of the rRNA-coding region has been developed, but it has been neither widely tested nor adopted (1151). mtDNA polymorphism can be used to distinguish *F. sporotrichioides* from most other *Fusarium* species, but not from *F. tricinctum* (1103). The karyotypes of *F. sporotrichioides* and *F. armeniacum* are similar enough that they can not be used to distinguish these species (1519). The *F. sporotrichioides* karyotype contains minichromosomes that are composites of uniques and dispersed sequences (1520).

This species is extremely important as a producer of the T-2 and diacetoxyscirpenol mycotoxins associated with alimentary toxic aleukia in humans and moldy corn toxicosis in animals (1341, 1539, 2096, 2258). The grain being colonized and the strain used both may influence the type of trichothecene produced (1351, 1377).

Trichothecene biosynthesis has been studied extensively in *F. sporotrichioides*, and many of the genes encoding enzymes in or regulatory molecules for this biosynthetic pathway in this fungus are clustered and have been cloned and sequenced (38, 39, 171, 253-255, 880, 882, 1398, 1400, 1401, 1415, 1650, 1651, 1670). Leucine is a precursor for trichothecene biosynthesis (175). The *tri5* gene encodes the enzyme responsible for the first step in the trichothecene biosynthetic pathway, trichodiene synthase, an enzyme that has been crystallized and extensively characterized (305, 306, 876, 877, 883, 884, 1853). The *tri5* sequence is the same in both *F. sporotrichioides* and *F. langsethiae* (610, 1563, 2335) and in some cases is used as a DNA marker for trichothecene production (543, 610). The *tri4* locus encodes a cytochrome P450 monooxygenase that is responsible for the first oxygenation step in the trichothecene biosynthetic pathway (880). *tri10* and *tri6* encode positive regulatory elements that are required for trichothecene biosynthesis and for other as yet unknown functions (881, 1739, 2099). Of the 26 transcripts regulated by *tri6*, 22 also are regulated by *tri10* (1671). The *tri15* locus appears to encode a negative regulatory protein and is coexpressed with other *tri* loci, but when it is disrupted, there is no effect on T-2 biosynthesis (40). *tri15* is highly expressed when exogenous T-2 is added to the culture medium.

Trichothecene biosynthesis is blocked by the plant growth regulator ancymidol and strains growing in the presence of this compound produce trichodiene instead (483). Trichothecene production may be enhanced by low levels of the fungicides azoxystrobin (512) or tridemorph (1481). G-protein mediated signals can alter trichothecene production by *F. sporotrichioides* (2100). The *tri101* gene from *F. sporotrichioides*, which encodes an acetyl transferase, is responsible for the "self-resistance" of *F. sporotrichioides* to trichothecenes (1066, 1068). If this gene is introduced into wheat then the severity of wheat head blight is reduced (1613). It also reduces the toxicity associated with trichothecenes towards tobacco (1490) and yeast (1399).

Fusarium sporotrichioides can produce butenolide (1341), fusarin C (690, 2140), moniliformin (1930), scirpentriol (1688), steroids (284), and zearalenone (1341). It also synthesizes the enzyme cyclophilin, which can degrade cyclosporin A (357, 654).

Current Species Description: Gerlach & Nirenberg (697) and Nelson *et al.* (1551).

Fusarium sterilihyphosum Britz, Marasas & Wingfield

Sexual Stage. None known.

Common Synonym. *Fusarium subglutinans*.

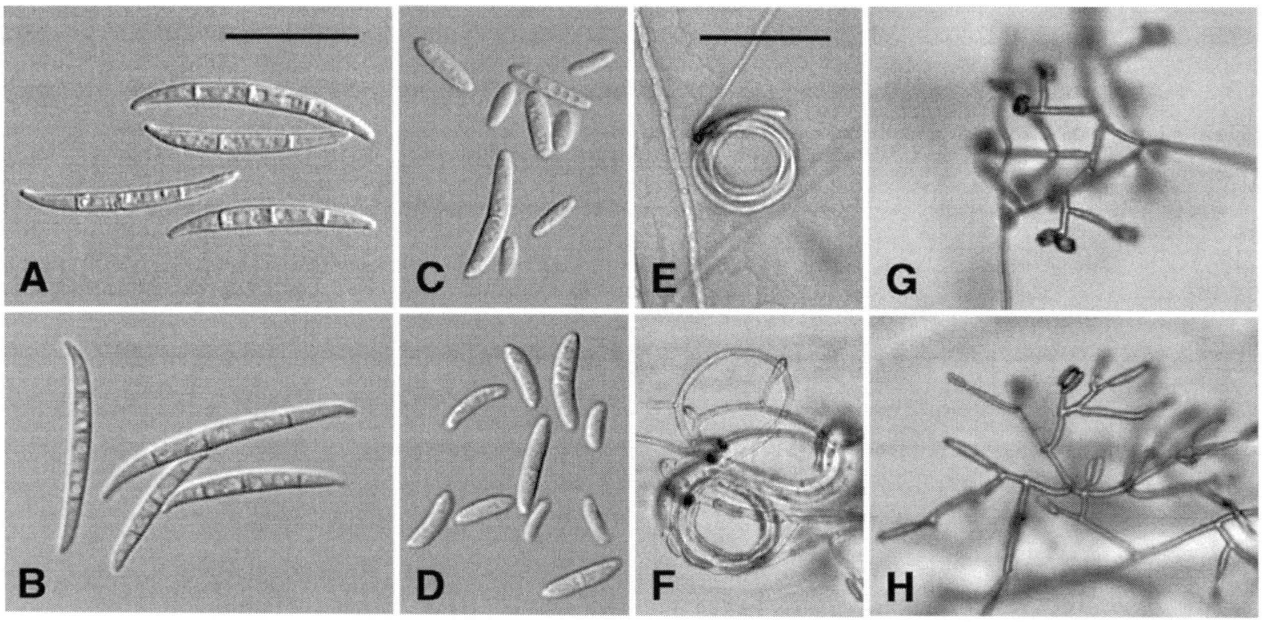

Figure 13-62. *Fusarium sterilihyphosum*.
A – B: Macroconidia; **C – D:** Microconidia; **E – F:** Coiled hyphae; **G – H:** Microconidia *in situ* on CLA.
A – D, scale bar = 25 μm; E – H, scale bar = 50 μm.

Reported Geographic and Host Distribution

Associated with inflorescence malformation on mango (*Mangifera indica*) in South Africa.

Media to Examine and Key Characters

Characters on CLA. Cream- to orange colored sporodochia with macroconidia are rare. When present, the macroconidia are long, slender and 3- to 5-septate. Microconidia are produced from mono- and polyphialides. These are obovoid, oval to allantoid in shape and usually are 0-septate but occasionally are 1-septate. Distinctive sterile coiled hyphae may be observed on the agar.

Characters on PDA. Colonies have a white, floccose mycelium with rose to light purple pigmentation in the agar.

Macroconidia

- Sporodochia: Cream- to orange-colored, but rare.
- General morphology: Relatively slender, slightly falcate and thin walled. Typical of those produced by species in the *Gibberella fujikuroi* species complex.
- Apical Cell morphology: Slight beak.
- Basal Cell morphology: Foot shaped.
- Number of septa: 3- to 5- septate.
- Abundance: Rare, and found almost exclusively in sporodochia, which also are rare.

Microconidia

- Shape/septation: Obovoid, ovoid and allantoid shape with 0 or 1 septa.
- Aerial mycelium presentation: False heads.

- Conidiogenous cells: Mono- and polyphialides. The conidiophores are sympodially branching and may have quite long phialides.
- Abundance: Abundant in the aerial mycelia.

Chlamydospores. Absent.

Other features. Sterile hyphae, similar to those found in *F. circinatum* and *F. pseudocircinatum* are present, and seen most clearly on SNA, but are not as tightly coiled.

Taxonomy, Pathology and Ecology

This species was described by Britz *et al.* (244) from isolates associated with mango malformation disease in South Africa; it has not been definitively shown to cause the disease. Based on the sequence of histone H3, *F. sterilihyphosum* is most closely related to *F. guttiforme*. Morphologically, *F. sterilihyphosum* most closely resembles *F. mangiferae* from which it can be distinguished by the presence of sterile coiled hyphae. *F. circinatum* and *F. pseudocircinatum* also produce sterile coiled hyphae, but these species can be distinguished from *F. sterilihyphosum* on the basis of macroconidial morphology. The macroconidia of *F. sterilihyphosum* are long, slender and 3- to 5-septate, while those of *F. circinatum* and *F. pseudocircinatum* are shorter and usually only 3-septate. Attempts to produce a sexual stage for *F. sterilihyphosum* have not yet been successful, although strains of both mating type have been described. Little else is known about the biology or ecology of *F. sterilihyphosum*.

Current Species Description: Britz *et al.* (244).

Fusarium subglutinans (Wollenweber & Reinking) Nelson, Toussoun & Marasas

Sexual Stage. *Gibberella subglutinans* Nelson, Toussoun & Marasas.

Common Synonyms. *Fusarium moniliforme* var. *subglutinans*, *Fusarium sacchari* var. *subglutinans*; *Gibberella fujikuroi* Mating Population E.

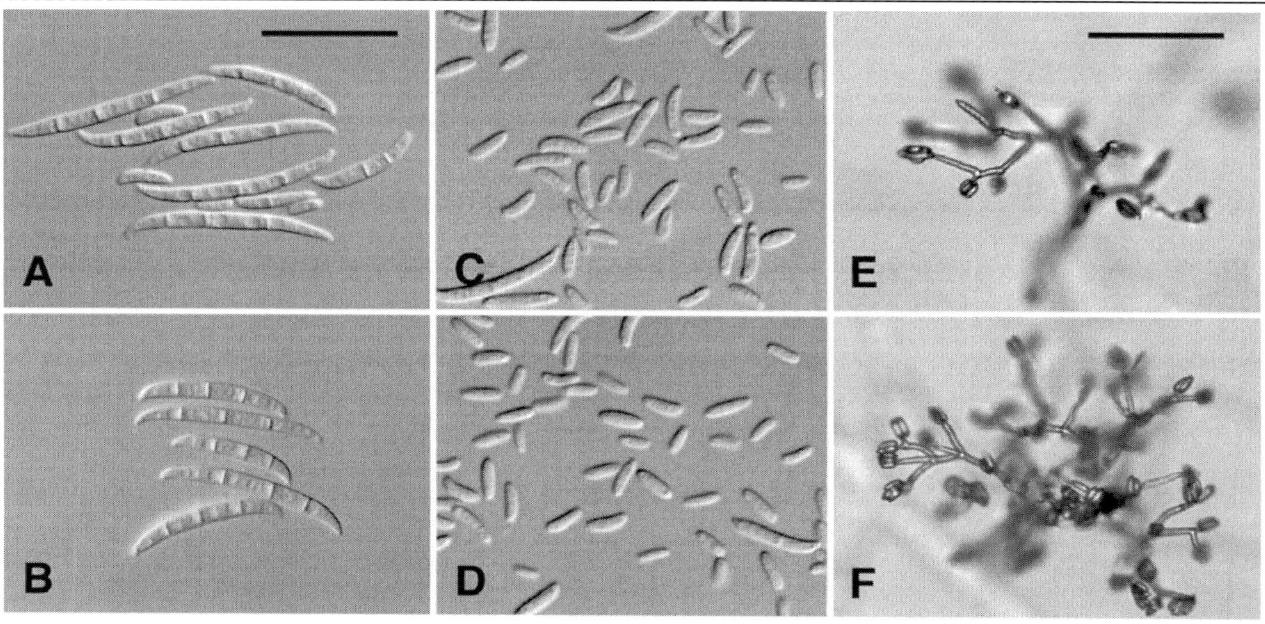

Figure 13-63. *Fusarium subglutinans.*
A – B: Macroconidia; **C – D:** Microconidia; **E – F:** Microconidia *in situ* on CLA. A – D, scale bar = 25 µm; E – H, scale bar = 50 µm.

Reported Geographic and Host Distribution

A pathogen of maize. Found primarily in cooler regions where maize is cultivated.

Media to Examine and Key Characters

Characters on CLA. Macroconidia are formed sparsely, if at all, and are typical of those found in the *G. fujikuroi* species complex. Microconidia are produced in false heads from mono- and polyphialides and are oval shaped and 0-septate. Longer 2- to 3-septate fusiform microconidia also may be produced and appear to be an almost smooth transition to the macroconidia. Chlamydospores are absent.

Characters on PDA. Abundant mycelial growth is initially white but becomes violet as the culture ages. Agar pigmentation ranges from colorless to a dark purple that is nearly black, with most cultures producing some shade of purple pigment. Blue-black sclerotia may develop in some isolates, but are not diagnostic although they may be indicative of a high level of sexual female fertility.

Macroconidia

- Sporodochia: Tan to orange color, found on carnation leaf pieces and the agar of CLA and occasionally on the surface of PDA.
- General morphology: Relatively slender, slightly falcate and thin walled. Typical of those produced by species in the *Gibberella fujikuroi* species complex.
- Apical Cell morphology: Curved.
- Basal Cell morphology: Relatively poorly developed.

- Number of septa: Usually 3-septate.
- Abundance: Usually abundant.

Microconidia

- Shape/septation: Oval and 0-septate.
- Aerial mycelium presentation: False heads. Spores are retained in the false heads by mucilage, and not by a direction connection with the other spores in the false head (Tiedt *et al.*, 1986).
- Conidiogenous cells: Mono- and polyphialides. The polyphialides may proliferate extensively.
- Abundance: Abundant in the aerial mycelia.

Chlamydospores. Absent.

Taxonomy, Pathology and Ecology

Fusarium subglutinans was first recognized as a separate species by Nelson *et al.* (1551). It is equivalent to *F. sacchari* var. *subglutinans* as described by Wollenweber & Reinking (2359) and Gerlach & Nirenberg (697). Occasionally the teleomorph, *G. subglutinans,* will form on residue of maize plants (542).

Fusarium subglutinans is relatively easily separated from other major species in the *G. fujikuroi* species complex. *F. subglutinans* produces microconidia only in false heads, which distinguishes it from both *F. proliferatum* and *F. verticillioides*, which both produce microconidia in chains. A mutant of *F. subglutinans* that produces apparent short chains instead of false heads has been reported (2146). The presence of polyphialides also serves to distinguish this species from *F. verticillioides*. The lack of chlamydospore production by *F. subglutinans* is the major morphological character distinguishing it from *F. oxysporum*. Maximum linear growth occurs at 25°C and an osmotic potential of -1.0 MPa (1541).

A number of species that are morphologically similar to *F. subglutinans* have been described, including *F. bulbicola*, *F. circinatum*, *F. guttiforme*, *F. mangiferae*, *F. pseudocircinatum*, *F. sacchari* and *F. sterilihyphosum* (242, 244, 1570). Many of these species are quite difficult to differentiate from one another or from *F. subglutinans* unless molecular markers or sexual crosses are used. Additional species within this group probably remain to be identified, *e.g.*, Desjardins *et al.* (484) and Steenkamp *et al.* (2042). The wide host and geographic ranges attributed to this species by Booth (222) and Marasas *et al.* (1341) may be due to the use of a single name for a number of discrete entities.

Fusarium subglutinans is more common in cooler areas where maize is grown and is associated with stalk rot and cob rot of maize (542, 2311) and may be seed transmitted (996). *F. subglutinans* can persist in maize debris either on the soil surface or buried in a field for at least 21 months (410). Multiple genetically distinct isolates of *F. subglutinans* may be recovered from a single maize plant (1029). "BT" hybrid maize, which is less susceptible to the European corn borer *Ostrinia nubialis*, had a lower level of infection with *F. subglutinans* than did comparable hybrids without the transgene (685, 1499).

Fusarium subglutinans also has been recovered from grasses in a native North American tallgrass prairie (1234, 1410, 1412). Other hosts for *F. subglutinans* include banana (978, 1095), cowpea (1140), millet (1628), orchids (170, 924), peppers (Mathur & Utkhede, 2004), sorghum (1626, 1756, 1899), soybean (1911), teosinte (484, 2039) and wild rice (1588). Isolates of "*Fusarium subglutinans*" from sugarcane probably are *Fusarium sacchari.*

Enzymes produced by *F. subglutinans* and at least partially characterized include chitinase (1587), galactose oxidase (148) and a milk-clotting enzyme (698). *F. subglutinans* can degrade anthracene (1141). It also can degrade the antimicrobial benzoxazinoids 6-methoxy-2-benzoxazolinone and 2-benzoxazolinone produced by maize (712, 2253). This ability may be one of the reasons why this fungus is a successful maize pathogen.

There are reports of DNA fragments that are specific for *F. subglutinans*. Many, but not all, of these studies are based on relatively small strain sets and none of these diagnostics has yet been accepted for use on a broad scale (1493, 1511).

Fusarium subglutinans can cause endophthalmitis (2031) and is resistant to most clinical antifungals, with amphotericin B reported as the most effective (1748).

Feed contaminated with *F. subglutinans* was toxic to rats, although the causal agent of the toxicity was not identified (4). Cultures of *F. subglutinans* often are toxic to ducklings due to the large amount of moniliformin that can be produced in these cultures (347, 579, 1137). One strain, MRC 115, is known to produce up to 11.3 mg/g of moniliformin in pure culture and ducklings that consumed culture material contaminated with strain were dead within 90-140 minutes after consuming some of the contaminated feed (1137). Two isolates of "*F. subglutinans*" reported as dermotoxic to rabbits (985, 1341) probably are *F. bulbicola*.

Fusarium subglutinans produces little or no fumonisins (1548, 1800), but can produce moniliformin (602, 1239, 1349, 1948, 1956, 2133, 2246), beauvericin (796, 1234, 1283, 1475, 1800, 1947, 1956), fusaric acid (119), and high levels of fusaproliferin (327, 1234, 1284, 1800, 1947, 1956, 2368). Both temperature and growth medium affect the amount of mycotoxin produced (1129). Previous reports of trichothecene production by *F. subglutinans* have not been confirmed (1341).

Current Species Description: Nelson *et al.* (1551).

Fusarium succisae (Schröter) Saccardo

Sexual stage. None known.

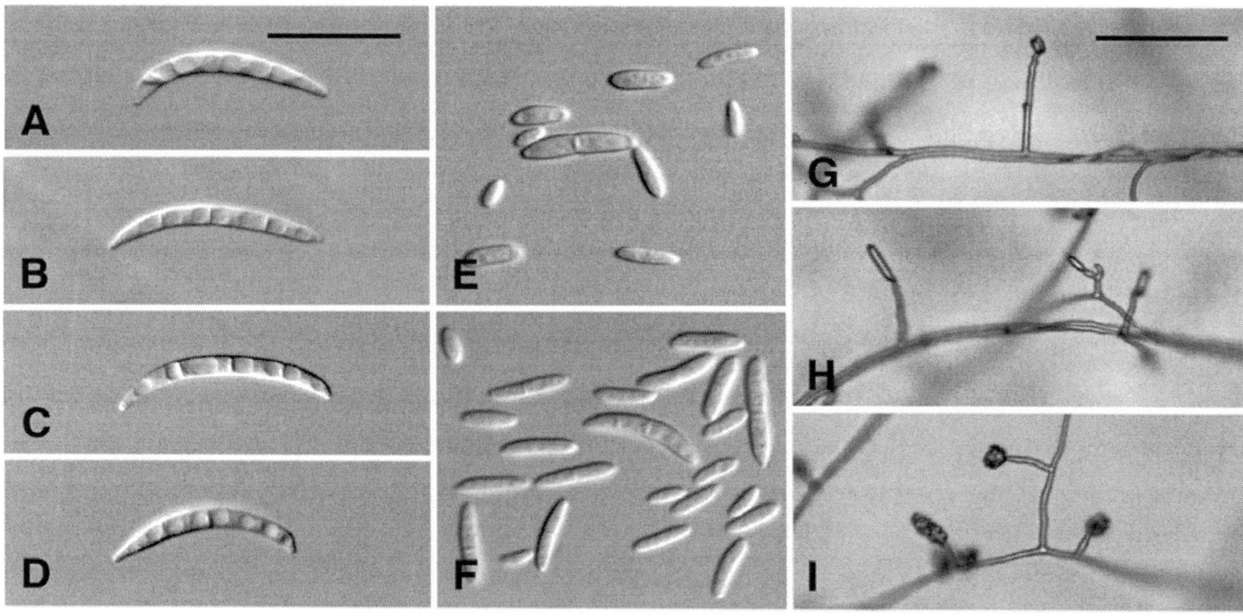

Figure 13-64. *Fusarium succisae.*
A – D: Macroconidia; **E – F:** Microconidia; **G – I:** Microconidia *in situ* on CLA. A – F, scale bar = 25 μm; G – I, scale bar = 50 μm.

Reported Geographic and Host Distribution

Recovered from *Succisa pratensis* (Dipsacaceae) in Europe.

Media to Examine and Key Characters

Characters on CLA. Orange sporodochia are sparse and slow to form. Macroconidia are falcate to distinctly curved and predominantly 3-septate. Microconidia are produced abundantly in false heads from monophialides and polyphialides. Microconidia are oval to allantoid, usually 0-septate but occasionally 1-2 septate.

Characters on PDA. Rapidly growing white colony that turns grey-violet and red with age and produces pink pigments in the agar.

Macroconidia

- Sporodochia: Bright orange in color, but very slow to form in culture (2-3 months).
- General morphology: Falcate to distinctly curved, sometime with almost a "U" shape. The single culture illustrated by Nelson *et al.* (1551) produced macroconidia with an almost horseshoe shape.
- Apical Cell morphology: Curved and tapering.
- Basal Cell morphology: Foot shaped.
- Number of septa: Predominantly 3-septate.
- Abundance: Often difficult to find since sporodochia form slowly.

Microconidia

- Shape/septation: Oval or allantoid and usually 0-, but occasionally 1- to 2-septate. Some may have papillae.
- Aerial mycelium presentation: False heads.
- Conidiogenous cells: Mono- or polyphialides.
- Abundance: Form rapidly (3-5 days) and abundantly in the aerial mycelia.

Chlamydospores. Absent.

Taxonomy, Pathology and Ecology

This species was first described as a *Fusisporium* (1927) and then as a species of *Fusarium* in 1892 (1858). Wollenweber (2357) did not recognize it as a separate species but rather treated it as con-specific with *F. moniliforme* var. *anthophilum*. Nirenberg (1566) recognized it as a separate species. O'Donnell *et al.* (1596) found that *F. succisae* was 10/1067 steps from *F. anthophilum*. This distance is less than the 15/1067 steps that separate *F. proliferatum* and *F. fujikuroi*, two species between which fertile interspecific hybrids are known to occur under both laboratory and field conditions (1236).

Fusarium succisae is not known to produce fumonisins (1548) or moniliformin (1349). There is one study of conidium formation by species in section *Liseola* that included *F. succisae* (2147). A report of *F. succisae* from date palm in Iran has not been verified (1472). Little else is known about the biology or ecology of this species.

Current Species Description: Gerlach & Nirenberg (697).

Fusarium thapsinum Klittich, Leslie, Nelson & Marasas

Sexual Stage. *Gibberella thapsina* Klittich, Leslie, Nelson & Marasas.

Common Synonyms. *Fusarium moniliforme*, *Gibberella fujikuroi* Mating Population F.

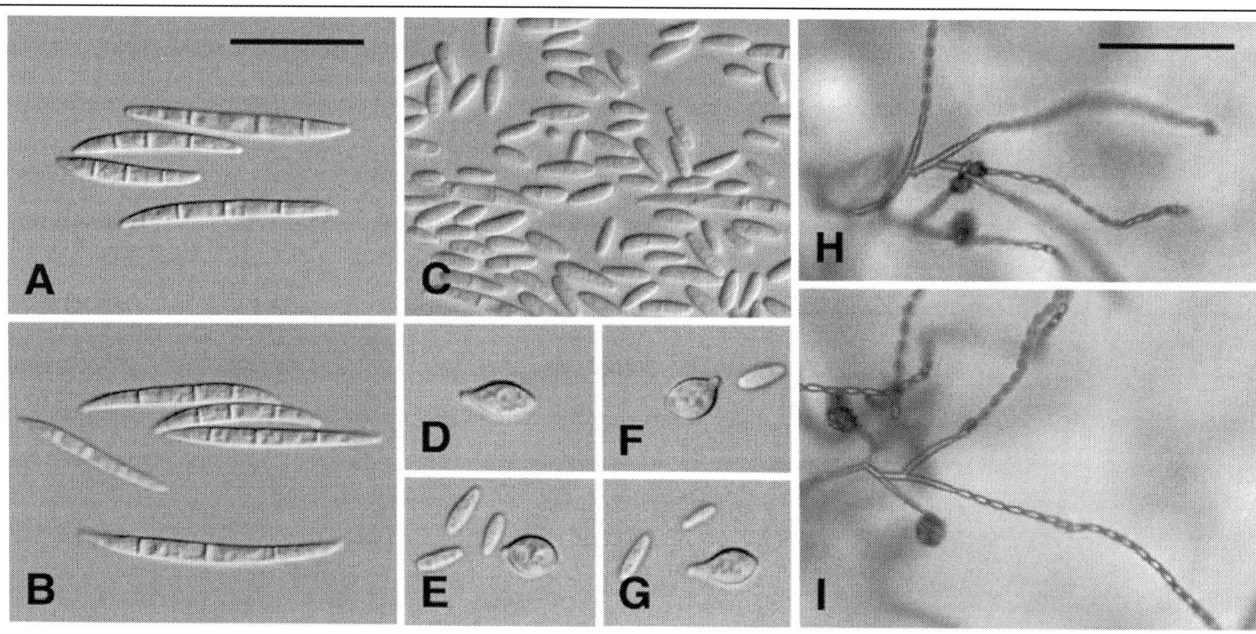

Figure 13-65. *Fusarium thapsinum.*
A – B: Macroconidia; **C:** Oval microconidia; **D - G:** Napiform microconidia; **H – I:** Microconidia *in situ* on CLA. A – G, scale bar = 25 µm; H - I, scale bar = 50 µm.

Reported Geographic and Host Distribution

A pathogen of sorghum that causes stalk rot and is found wherever sorghum is cultivated.

Media to Examine and Key Characters

Characters on CLA. Macroconidia are typical of those produced by other species in the *G. fujikuroi* species complex form in pale orange sporodochia. Sporodochia may be rare or difficult to observe as they can be obscured by thick mycelium and numerous chains of microconidia. Many isolates produce few or no sporodochia and macroconidia. 0-septate microconidia usually are club-shaped with a flattened base, but may occasionally be napiform. Microconidia are produced from monophialides usually in chains that can be quite long, but occasionally also may be found in false heads. Chlamydospores are absent.

Characters on PDA. *Fusarium thapsinum* forms abundant white mycelium which may darken (violet pigments) with age. Sporodochia usually are absent, but if present are pale orange. Dark colored sclerotia may be produced by some isolates. Pigmentation in the agar is quite variable. Most strains produce a distinctive yellow pigment that is diagnostic and is the basis of the species epithet. Other strains may produce either no pigment or violet pigments in the agar. Violet pigment may be produced in the agar.

Macroconidia

- Sporodochia: Often rare, but pale orange and found on carnation leaf pieces in CLA when produced.
- General morphology: Relatively slender, slightly falcate or straight, and thin walled. Typical of macroconidia produced by species in the *Gibberella fujikuroi* species complex.

- Apical Cell morphology: Curved and tapering.
- Basal Cell morphology: Relatively poorly developed foot shape.
- Number of septa: Usually 3- to 5-septate.
- Abundance: Often difficult to find.

Microconidia

- Shape/septation: Usually club-shaped with a flattened base, but occasionally napiform. Usually 0-septate.
- Aerial mycelium presentation: Long chains are common, but small aggregates of a few spores and false heads occur occasionally.
- Conidiogenous cells: Monophialides, which are occasionally produced in pairs to give a "rabbit ear" appearance.
- Abundance: Abundant in the aerial mycelia.

Chlamydospores. Absent.

Taxonomy, Pathology and Ecology

Fusarium thapsinum was first described by Klittich *et al.* (1093). Prior to that it was recognized as *F. moniliforme*. The splitting of *F. thapsinum* was made because of the recognition of the teleomorph (*G. thapsina*) as a separate mating population (1092) and a number of important physiological characteristics including host preference and toxin production.

Fusarium thapsinum isolates that do not produce the diagnostic yellow pigment are morphologically identical to *F. verticillioides*. The two species can be differentiated on the basis of mating tests and DNA sequencing. *F. thapsinum* also is similar to *F. andiyazi* but can be differentiated by the presence of pseudochlamydospores in *F. andiyazi* (1346). It also is similar to *F. proliferatum*, but *F. thapsinum* produces only monophialides and *F. proliferatum* produces both monophialides and polyphialides.

Fusarium thapsinum causes stalk rot and grain mold of sorghum (233, 641, 958, 1468, 1741, 1787, 2177) and is very pathogenic towards sorghum seedlings in an *in vitro* assay system (1233). The degree of pathogenicity is dependent on growth stage (2116) and genetic background of the host (86, 2127). *F. thapsinum* also is associated with grain mold of sorghum (137, 328, 1819), can colonize the peduncle of the plant, and can induce the production of 3'-deoxyanthocyanins (a stress response) in sorghum plants that produce those compounds (1269). Even if the level of grain mold is relatively low, seed from heads molded with *F. thapsinum* can show significantly reduced germination (1742). Although sorghum is a cold-sensitive plant, *F. thapsinum* can survive winters in Kansas, USA as spores and hyphae in above ground pieces of stalk without the formation of specialized protective structures (1335). Conidia of *F. thapsinum* can induce the synthesis of defense response genes in sorghum (1268). Susceptibility of sorghum to *F. thapsinum* is reduced in transgenic sorghum plants that carry a rice chitinase gene (1138, 2290).

In addition to sorghum, *F. thapsinum* also can be recovered from bananas, maize and peanuts (1093, 1227), and from native grasses in a North American native tallgrass prairie (1234). *F. thapsinum* can cause stalk rot of maize under greenhouse conditions (959), but is not commonly recovered from maize under field conditions. This species is geographically widely dispersed and has been reported from Africa, Asia and North America.

Fusarium thapsinum has 12 chromosomes and a total genome size of ~46 Mb, which is typical of other *Fusarium* species in this group (2374). Limited genetic studies are available, primarily of fertility and mating type in populations (1224), and of the heritability of pigment production and female fertility (1093).

Fusarium thapsinum is toxic to ducklings (1225, 1233) and can produce high levels of moniliformin, but little more than trace amounts of fumonisins (257, 1225, 1233, 1234, 1340). It also can produce fusaric acid (1721, 1723). Based on duckling assays it is possible that strains of *F. thapsinum* are producing additional compounds that are toxic to ducklings but that have not yet been described (1225).

Current Species Description: Klittich *et al.* (1093).

Fusarium torulosum (Berkeley & Curtis) Nirenberg

Sexual Stage. *Gibberella pulicaris* var. *minor* Wollenweber.

Common Synonyms. *Fusarium sambucinum, Fusarium sambucinum* var. *coeruleum, Fusarium venenatum.*

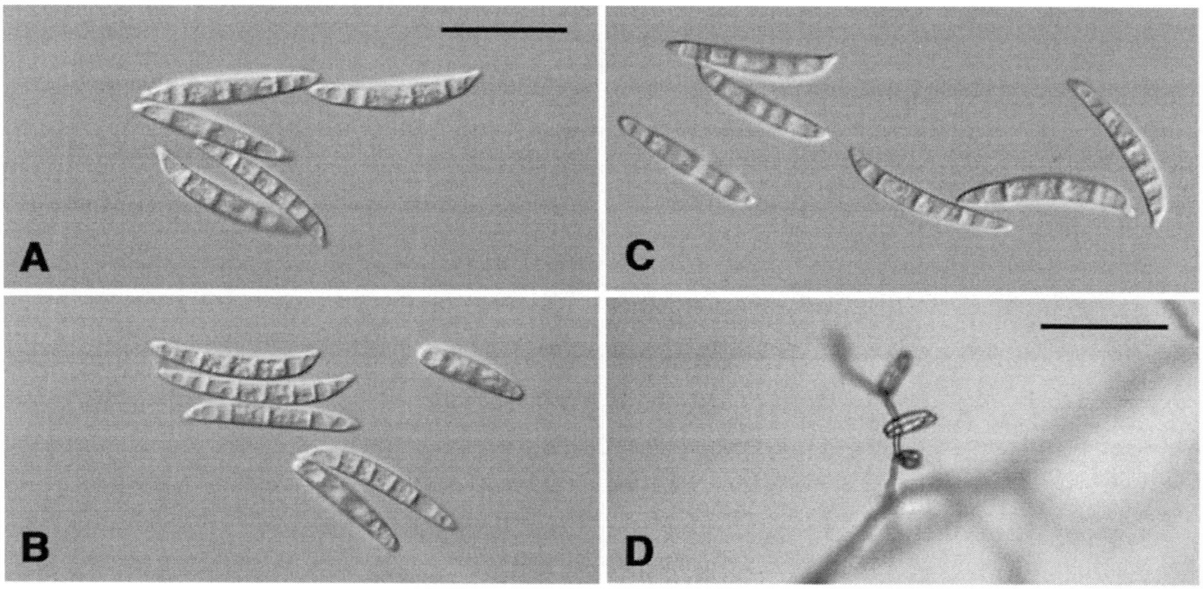

Figure 13-66. *Fusarium torulosum.*
A – C: Macroconidia; **D:** Microconidia *in situ* on CLA. A – C, scale bar = 25 μm; D, scale bar = 50 μm.

Reported Geographic and Host Distribution

Found in soil and on a variety of plant species in temperate regions.

Media to Examine and Key Characters

Characters on CLA. Abundant macroconidia are found in orange sporodochia, which may be slow to form. The macroconidia are comparatively short, falcate, and usually 5-septate with a pointed apical cell and a foot-shaped basal cell. Oval 0- to 1- septate microconidia may be produced occasionally by some isolates. Chlamydospores are found in chains or clusters.

Characters on PDA. Cultures are slow-growing with an initially white lannose to felty mycelium that becomes pigmented with age and may appear in narrow concentric rings. Red pigments commonly are produced in the agar, but production of yellow, white or brown pigments also may occur.

Macroconidia

- Sporodochia: Orange sporodochia are found on carnation leaf pieces in CLA, and may be slow to form.
- General morphology: Comparatively short and falcate.
- Apical Cell morphology: Pointed.
- Basal Cell morphology: Foot shaped.
- Number of septa: Usually 5-septate.
- Abundance: Abundant in sporodochia.

Microconidia

- Shape/septation: 0- to 1-septate and oval shape.
- Aerial mycelium presentation: Singly or in small false heads.
- Conidiogenous cells: Monophialides.
- Abundance: Very rare, but when found are seen in the aerial mycelia.

Chlamydospores

- Abundance/Speed of formation: Abundant and formed relatively quickly, 2-4 weeks, on SNA and CLA.

- Location: Found in long chains and in clusters, intercalary in the hyphae.
- Appearance: Smooth and usually unpigmented.

Taxonomy, Pathology and Ecology

Fusarium torulosum has been known by a number of names, the most common of which was *F. sambucinum* var. *coeruleum*. Nirenberg (1568) raised this group to species status combining the species name from the original description, *Fusidium torulosum*, with the *Fusarium* genus name. The link to the teleomorph, *Gibberella pulicaris* var. *minor* is presumed but not proven. *F. torulosum* is more common in temperate parts of the world.

Fusarium torulosum is most likely to be confused with *F. sambucinum* and *F. venenatum* because of the very similar morphology. Differentiation of these species can be made by comparing the rate of growth on PDA, as *F. torulosum* grows much more slowly than does either *F. sambucinum* or *F. venenatum*. *F. torulosum* also forms chlamydospores and *F. sambucinum* does not. Otherwise, differentiation of these species can be difficult and molecular markers may be necessary for a conclusive diagnosis (1288). *F. torulosum* also can be confused with *F. culmorum* due to similarities in PDA cultures and the morphology of the macroconidia. The macroconidia of *F. culmorum* are shorter and wider than those of *F. torulosum*. Germination of the macroconidia of *F. torulosum* has been studied in some detail via electron microscopy (2036).

Fusarium torulosum has been recovered from the roots of a number of plants including cereals, tomatoes, beet root and trees (1568). It can cause a root rot of various *Medicago* spp. (1413). *F. torulosum* is primarily found in temperate areas and has been recovered from the Himalayan foothills in Nepal (477).

Fusarium torulosum does not produce trichothecenes (46) and lacks the *tri5* gene that is essential for trichothecene production (2112). It can produce antibiotic Y (2144), enniatin B (46) and wortmannin (2144). It also produces a number of volatile organic compounds that may be used in the development of biosensors for determining if dry rot of potatoes is occurring in storage (408).

Current Species Description: Nirenberg (1568).

Fusarium tricinctum (Corda) Saccardo

Sexual Stage. *Gibberella tricincta* El-Gholl, McRitchie, Schoulties & Ridings.

Common Synonyms. *Fusarium sporotrichioides, Fusarium sporotrichioides* var. *tricinctum*.

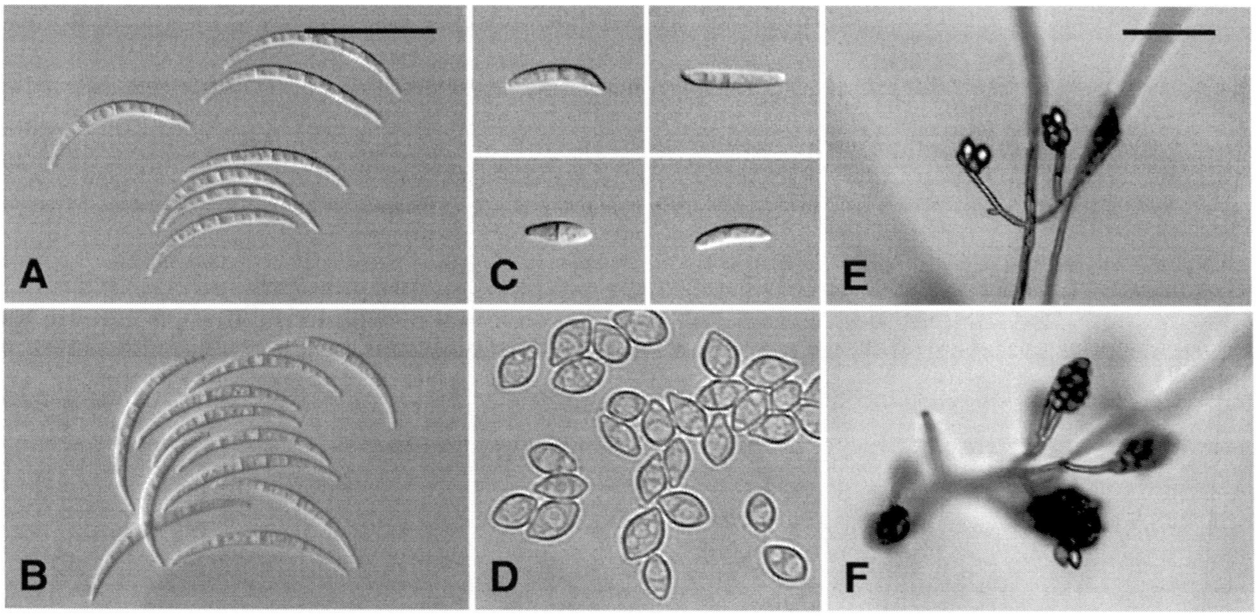

Figure 13-67. *Fusarium tricinctum.*
A – B: Macroconidia; **C:** Oval microconidia; **D:** Napiform microconidia; **E – F:** Microconidia *in situ* on CLA.
A – D, scale bar = 25 μm; E – F, scale bar = 50 μm.

Reported Geographic and Host Distribution

Has been isolated from various substrates in many parts of the world. Appears to be more common in temperate parts of the world.

Media to Examine and Key Characters

Characters on CLA. Macroconidia are relatively common in pale orange sporodochia found on carnation leaves. Macroconidia are short to medium in length, 3- to 5-septate, distinctly curved to lunate, and relatively slender with a curved tapering apical cell and an obviously foot-shaped basal cell. The macroconidia may appear to be pointed at both ends. Abundant microconidia are napiform, pyriform and citriform in shape and 0- to 1-septate. Microconidia may be clustered in false heads that look like a bunch of grapes. Chlamydospores may be formed but are not abundant, and when present are formed singly or in chains.

Characters on PDA. *Fusarium tricinctum* grows rapidly and forms abundant dense mycelia that are initially white, but that become pink, red or purple with age. It forms red pigments in the agar.

Macroconidia

- Sporodochia: Pale orange color and found on carnation leaves in CLA.
- General morphology: Relatively slender and falcate to almost lunate.
- Apical Cell morphology: Curved and tapering.
- Basal Cell morphology: Well developed foot shape.
- Number of septa: Predominantly 3- but occasionally 4- to 5-septate.

- Abundance: Relatively abundant.

Microconidia

- Shape/septation: Napiform, pyriform and occasionally citriform with usually 0 but occasionally 1 septa. Some microconidia may have papilla at the base of the microconidium.
- Aerial mycelium presentation: Small false heads with a few microconidia that may be clustered and appear like a bunch of grapes.
- Conidiogenous cells: Monophialides.
- Abundance: Abundant in the aerial mycelia.

Chlamydospores

- Abundance/Speed of formation: Formed only occasionally by some isolates, and when formed not useful as a diagnostic character. Strains that do form chlamydospores often do so very slowly, usually requiring six or more weeks of incubation on CLA.
- Location: May be intercalary, terminal or on short lateral branches. Usually in hyphae either singly or in chains.
- Appearance: Globose with a smooth outer wall that becomes brown with age.

Taxonomy, Pathology and Ecology

Fusarium tricinctum was neotypified by Neish (1538) and was recognized as a distinct species by Booth (222), Gerlach & Nirenberg (697), and Nelson *et al.* (1551). An epitype for this species was designated by Holubova-Jechova *et al.* (888). Many of the published references to *F. tricinctum* are not accurate due to the confusion of this species with other species in section *Sporotrichiella*, although the recognized species in this section can be clearly distinguished with both morphological and molecular criteria (1287, 2229). The *G. tricincta* sexual stage (550) can be formed in culture, requires light, and occurs more readily at 27°C than at 20°C (551).

Fusarium tricinctum is usually occurs as a saprophyte or a weak parasite in temperate regions such as Europe, North America and the former Soviet Union (64, 346, 721, 1128, 1182). It also may be found in warmer climates, but its distribution in these regions is not as well documented.

The PDA culture of *F. tricinctum* can be confused with that of other related species such as *F. poae*, *F. sporotrichioides* and *F. chlamydosporum*. There are a number of characters that differentiate *F. tricinctum* from these species, especially the microconidial shape and the monophialidic conidiogenous cell. *F. tricinctum* does not produce polyphialides, which differentiates it from *F. sporotrichioides* and *F. chlamydosporum*. *Fusarium poae* produces microconidia from urn-shaped phialides that are easily distinguished from the monophialides produced by *F. tricinctum*.

Other than mycotoxin synthesis the physiology of *F. tricinctum* has not been intensively examined. Some strains of *F. tricinctum* have ice nucleation activity (1806). Asparaginase (1909) and pectolytic and cellulolytic enzyme activity (2313) have been evaluated as well.

Auxotrophic mutants of *F. tricinctum* have been made and used to force heterokaryons, although vegetative compatibility *per se* has not been studied (424). These auxotrophs segregate in a Mendelian manner following sexual crosses.

This species generally is not regarded as a plant pathogen and has not been associated with any human or animal toxicoses (1341). It produces the toxins fusarin C (2140) enniatins (283) and moniliformin (347), and the secondary metabolites acuminatopyrone (2262), fosfonochlorin (2110) and visoltricin (2260, 2261). *F. tricinctum* is not reported to produce trichothecenes (1281) and lacks the *tri5* gene, which is essential for trichothecene production (2112), although weak hybridization to *tri5* probes may occur under some conditions (610).

Current Species Description: Neish (1538).

Fusarium udum Butler

Sexual Stage. *Gibberella indica* B. Rai & R. S. Upadhyay.

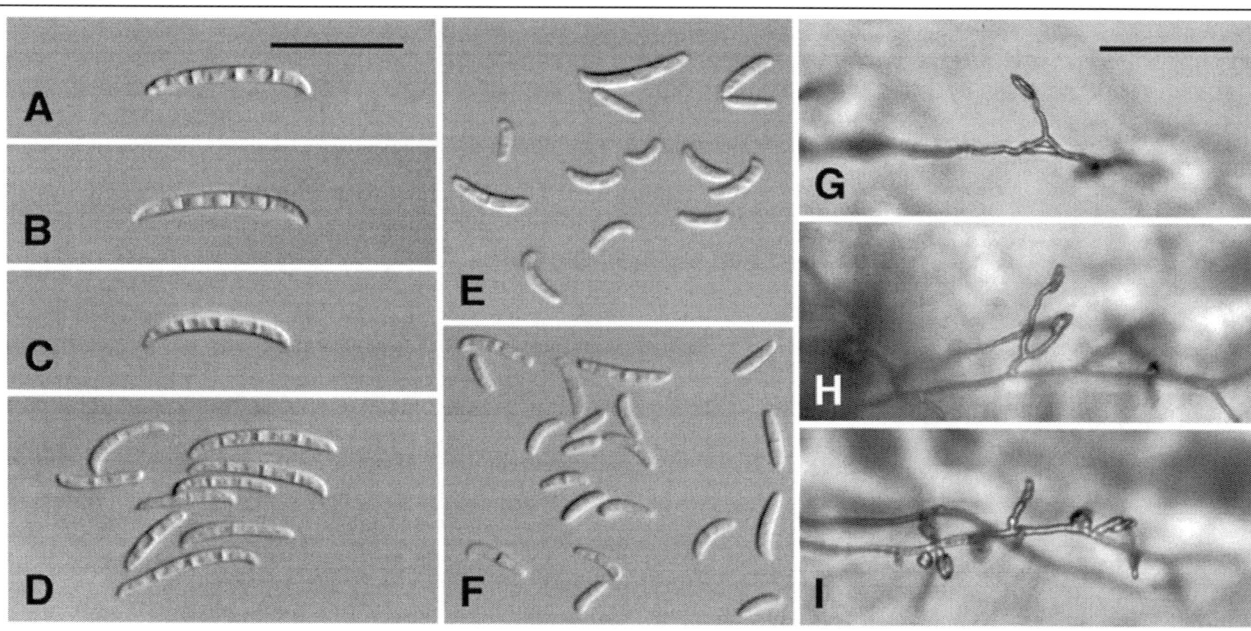

Figure 13-68. *Fusarium udum.*
A – D: Macroconidia; **E – F:** Microconidia; **G – I:** Microconidia *in situ* on CLA. A – F, scale bar = 25 μm. G – I, scale bar = 50 μm.

Reported Geographic and Host Distribution

Causes a serious wilt disease of pigeon pea (*Canjanus cajan*) in most regions where the host is grown (southern Asia and sub-Saharan Africa), but especially India.

Media to Examine and Key Characters

Characters on CLA. Pink to salmon colored sporodochia with abundant macroconidia. Macroconidia are straight to slightly curved with a distinctly curved apical cell and a foot shaped basal cell. Hyaline 0-septate oval to fusiform microconidia can be found in the aerial mycelia borne on false heads by monophialides. Chlamydospores are present in the hyphae either singly or in clumps.

Characters on PDA. White, felt-like mycelium with pink to purple pigments in the agar. Some cultures may be pionnotal with little aerial mycelium and be nearly covered by a sheet of conidia.

Macroconidia

- Sporodochia: Pink to salmon-colored. May be discrete and found on or adjacent to carnation leaf pieces in CLA or nearly pionnotal across the entire surface of the culture.
- General morphology: Straight to falcate and thin walled.
- Apical Cell morphology: Curved to almost hooked.
- Basal Cell morphology: Foot shaped.
- Number of septa: 1-5 septate, but predominantly 3-septate.
- Abundance: Abundant in sporodochia.

Microconidia

- Shape/septation: Fusiform to reniform and 0- to 1-septate.
- Aerial mycelium presentation: False heads.
- Conidiogenous cells: Monophialides.
- Abundance: Sparse in the portion of the colony growing on the agar surface of a CLA plate. Usually much

more abundant in mycelia that are on or near the carnation leaf pieces of a CLA plate.

Chlamydospores

- Abundance/Speed of formation: Usually abundant and formed in 4-6 weeks on CLA plates.
- Location: Commonly intercalary in the hyphae where they are found in pairs, chains and clusters. May occasionally be found singly within the macroconidia.
- Appearance: Globose, nonpigmented and smooth walled.

Taxonomy, Pathology and Ecology

This species was originally described by Butler and was recognized by Wollenweber & Reinking (2359), Booth (222), and Gerlach & Nirenberg (697). Nelson *et al.* (1551) presented it as an insufficiently documented species. *F. udum* bears some similarities to *F. oxysporum* and consequently some authorities placed it in section *Elegans*, *e.g.*, Gerlach & Nirenberg (697), while others place it in section *Lateritium*, *e.g.*, Booth (222). The sexual stage, *Gibberella indica*, is heterothallic, but highly fertile female tester strains have not been identified (1761). Studies of vegetative compatibility found only one or perhaps two VCGs amongst a group of 79 strains (1072).

Fusarium udum causes a serious wilt disease of pigeon pea (861, 1006) that was initially reported from India but has been recorded in a number of other countries where this crop is grown. Strains can vary significantly in their aggressiveness, but no race structure is recognized (686, 1612). Colonization of the host root by the fungus is most effective at soil temperatures of 22-30°C, pH 7-9, and 5-30% soil moisture (1762). Aggressiveness, growth rate and AFLP fingerprint are not correlated with one another (1071). Strains pathogenic to cotton also have been reported. Disease symptoms are those of a vascular wilt in which infected plants suddenly wilt and dry followed by yellowing of the leaves and then their loss. Control is almost exclusively through the use of resistant germplasm (953, 1611, 1647, 1786, 1989), although there has been extensive work to identify suitable biocontrol agents as well (144, 963, 1645, 1733, 1977, 2201, 2207, 2250).

Fusarium udum has been reported to produce fusaric acid and some derivatives (52, 222, 1732). It also is pathogenic to some freshwater fish (193) and has some mycoparasitic activity (2202).

Current Species Description: Gerlach & Nirenberg (697).

Fusarium venenatum Nirenberg

Sexual Stage. None known.

Common Synonyms. *Fusarium sambucinum*, *Fusarium sambucinum* var. *coeruleum*, *Fusarium torulosum*.

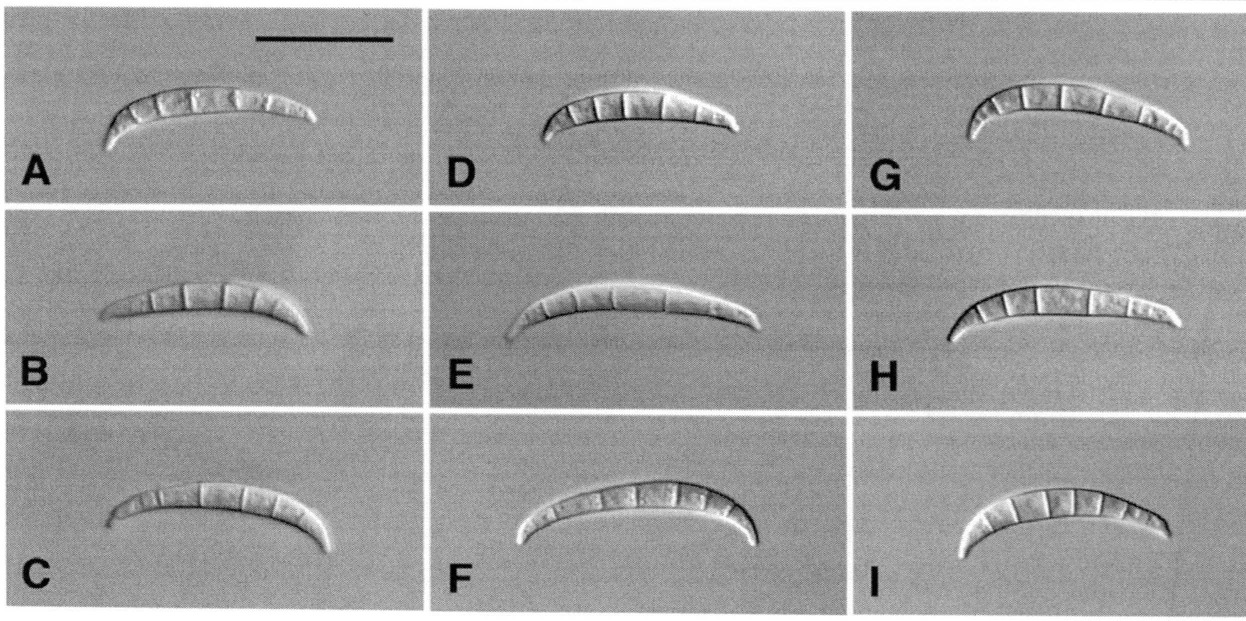

Figure 13-69. *Fusarium venenatum.*
A – I: Macroconidia. Scale bar = 50 µm.

Reported Geographic and Host Distribution

Found on a variety of plant species and soil from Europe.

Media to Examine and Key Characters

Characters on CLA. Abundant macroconidia are found in reddish-brown to brown sporodochia. Macroconidia are comparatively short, slender, falcate and mostly 5-septate with a pointed apical cell and a foot-shaped basal cell. Chlamydospores are formed in short hooked chains.

Characters on PDA. Cultures grow moderately rapidly with long, dense cottony mycelia that initially are white but that darken to orange, gray or red with age. Red pigments, sometimes very intensely red, are produced in the agar.

Macroconidia

- Sporodochia: Abundant reddish-brown to brown sporodochia found on the carnation leaf pieces and on the agar surrounding the leaf pieces.
- General morphology: Relatively short, slender and falcate.
- Apical Cell morphology: Pointed.
- Basal Cell morphology: Foot shaped.
- Number of septa: Mostly 5-septate.
- Abundance: Abundant in sporodochia.

Microconidia – absent.

Chlamydospores

- Abundance/Speed of formation: Relatively abundant. Formed after four weeks on CLA.
- Location: Found singly and in short hooked chains in the hyphae or from the middle cells of macroconidia.
- Appearance: Smooth and pale to light gray in color.

Taxonomy, Pathology and Ecology

Nirenberg created this species in 1995 (1568) as part of her re-evaluation of *F. sambucinum sensu lato*. Wollenweber & Reinking (2359) probably included this species as *F. culmorum* var. *cerealis* and Nelson *et al.* (1551) probably included it in *F. culmorum*. The best known strain of this species, A 3/5, was identified as *F. graminearum* on the basis of morphological characters. *F. venenatum* is most likely to be confused with *F. sambucinum* and *F. torulosum* because of the very similar morphology. *F. venenatum* grows more rapidly than *F. torulosum* on PDA, and can be differentiated from *F. sambucinum* by the presence of chlamydospores in cultures of *F. venenatum* and their absence in cultures of *F. sambucinum*. Molecular markers are available (1288) and may be required to make definitive identifications, especially if the strains have begun to degenerate.

The fungus has been recovered from several plant species and soil in Europe (1568).

A strain, A 3/5, originally identified as *F. graminearum* has been used in the United Kingdom to produce Quorn, a mycoprotein for human consumption that is sold in several European countries (541, 2179, 2317, 2318). This strain is grown in continuous flow culture (2180) and a number of mutants with unusual growth phenotypes have been derived from it (2319, 2324). This strain has been used to show that Ca^{++} (1815, 1816) and choline (2323) are involved in the regulation of hyphal branching and extension in fungi. Mutants that increase growth capabilities under some conditions, but not others, are recovered relatively commonly during chemostat culture of this strain (1983). Iron uptake is important for growth, and siderophore production is increased in the presence of an insoluble source of iron (2316).

The identity of the Quorn fungus as *F. venenatum* was confirmed with DNA sequencing (1595) and species-specific primers (2392). Quorn may induce allergenic responses in people who have allergies to other fungi (874, 1022, 2122) as well as those with no previous history of such allergies (951). People who used Quorn as a carbohydrate substitute had decreased LDL (low density lipoprotein), increased HDL (high density lipoprotein), and unchanged cholesterol (2194).

Strain A 3/5 has been used as an industrial host for producing transgenic proteins of economic importance (176, 199, 200, 738, 1667, 1843, 2322, 2372). This strain can be transformed efficiently with an *Agrobacterium tumefaciens*-mediated transformation protocol (453). A 3/5 produces heterologous proteins more efficiently in glucose-limited chemostat culture than in a fed-batch fermentation (2321). Protein production is sensitive to the pH of the medium, but not to the nitrogen source supplied (2320) and may occur during stationary phase (605).

Strain A 3/5 produces neither trichothecene (1595) nor zearalenone (261) mycotoxins. Other strains of *F. venenatum* produce trichothecenes such as diacetoxyscirpenol (46, 1917, 2144) and sambucinol (1917). Vectors have been developed to specifically delete the *tri5* gene in *F. venenatum* (1842). Deleting *tri5*, the gene encoding trichodiene synthase which is the first enzyme specific for the trichothecene biosynthesis pathway, prevents the biosynthesis of trichothecenes and also reduces the levels of other secondary metabolites (1477).

Current Species Description: Nirenberg (1568).

Fusarium verticillioides (Saccardo) Nirenberg

Sexual Stage. *Gibberella moniliformis* Wineland.

Common Synonyms. *Fusarium moniliforme*, *Gibberella fujikuroi* Mating Population A.

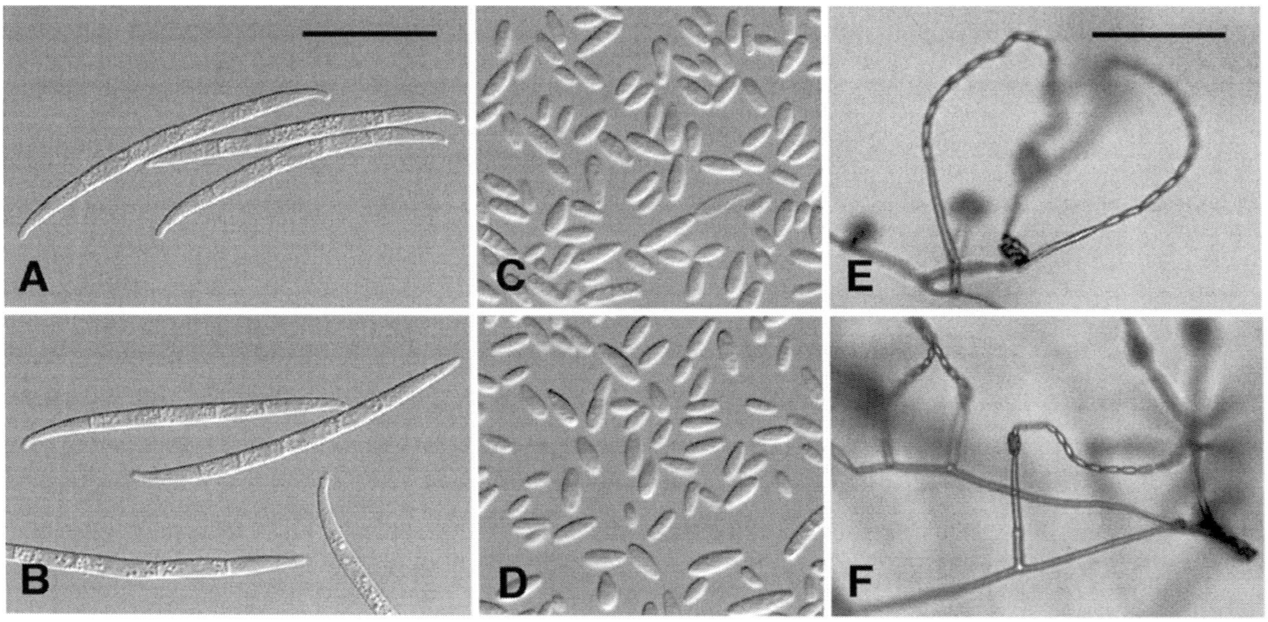

Figure 13-70. *Fusarium verticillioides.*
A – B: Macroconidia; **C – D:** Microconidia; **E – F:** Microconidia *in situ* on CLA. A – D, scale bar = 25 µm; E – F, scale bar = 50 µm.

Reported Geographic and Host Distribution

A pathogen on maize and found throughout the world wherever maize is cultivated.

Media to Examine and Key Characters

Characters on CLA. Macroconidia typical of those of other species in the *G. fujikuroi* species complex are formed in pale orange sporodochia that may be rare or obscured by thick mycelium or numerous microconidial chains. Macroconidia may be rare in some isolates. Microconidia are oval to club-shaped with a flattened base and 0-septate. They are always formed from monophialides, which may occur in V-shaped pairs to give a "rabbit ear" appearance, and are found in chains that may be quite long. Chlamydospores are not produced, but swollen cells in the hyphae may be mistaken for chlamydospores or pseudochlamydospores.

Characters on PDA. Initially cultures have white mycelia but may develop violet pigments with age. Pigmentation in the agar varies, ranging from no pigmentation or grayish orange to violet grey, dark violet or dark magenta (almost black) in others. Blue-black sclerotia may develop in some isolates, but are not diagnostic although they may be indicative of a high level of sexual female fertility.

Macroconidia

- Sporodochia: May be tan or orange in color and present as discrete entities or as a pseudo-pionnotal mass.
- General morphology: Relatively long and slender, slightly falcate or straight, and thin walled. Typical of macroconidia produced by species in the *Gibberella fujikuroi* species complex.
- Apical Cell morphology: Curved and often tapered to a point.

- Basal Cell morphology: Notched or foot shaped.
- Number of septa: 3- to 5-septate.
- Abundance: Varies by strain, but macroconidia may be difficult to find.

Microconidia

- Shape/septation: Oval to club-shaped with a flattened base and usually 0-septate.
- Aerial mycelium presentation: Long chains are common, but small aggregates of a few spores occur occasionally.
- Conidiogenous cells: Monophialides, which are occasionally produced in pairs to give a "rabbit ear" appearance.
- Abundance: Abundant in the aerial mycelia.

Chlamydospores. Chlamydospores are not produced, although some isolates may produce swollen cells in the hyphae that can easily be mistaken for chlamydospores or pseudochlamydospores.

Taxonomy, Pathology and Ecology

There has been significant controversy regarding the name of this taxon with some taxonomists calling it *F. moniliforme* and others *F. verticillioides*. The name *F. verticillioides* has priority and it is now generally accepted that this is the name that should be used for this species (1936). Note that *F. moniliforme sensu* Snyder & Hansen encompassed stains in species other than *F. verticillioides*. The name *F. verticillioides* should be used only for strains that have the *G. moniliformis* (*Gibberella fujikuroi* mating population A) teleomorph and not simply as a replacement for *F. moniliforme sensu* Snyder & Hansen. Information on strains of "*F. moniliforme*" that now likely would be called other species, *e.g.*, *F. thapsinum* from sorghum, *F. sacchari* from sugar cane, *F. mangiferae* from mango, or *F. fujikuroi* from rice, is summarized under those species, unless strains verified as *F. verticillioides* by more than morphological characters were reported in the study. The sexual stage of *F. verticillioides* can be readily generated under laboratory conditions (1090) and has been recovered under field conditions (2079). Based on the frequency of female fertile strains in field populations 1-3% of the population participates in sexual reproduction annually (372, 1224).

Fusarium verticillioides is morphologically identical to strains of *F. thapsinum* that do not produce the diagnostic yellow pigment. *F. verticillioides* also is similar to *F. proliferatum*, but the latter species is distinguished by its ability to form of chains of microconidia from polyphialides. The microconidial chains produced by *F. proliferatum* usually are shorter than those of *F. verticillioides* and *F. thapsinum*. *F. verticillioides* and *F. thapsinum* can be reliably differentiated only by using mating tests or molecular markers, although many isolates of *F. thapsinum* produce yellow pigments that are unique to *F. thapsinum*. *F. verticillioides* also is very similar to *F. andiyazi* but does not form pseudochlamydospores. *F. verticillioides* can produce swollen cells in hyphae that may be difficult to differentiate from pseudochlamydospores.

Fusarium verticillioides is similar in some respects to *F. nygamai* which forms microconidia in short chains or false heads from monophialides (281). *F. nygamai*, however, forms abundant macroconidia in sporodochia and chlamydospores in the aerial hyphae in older cultures. Strains that are morphologically similar to and phylogenetically closely related to *F. verticillioides* have been isolated from bananas, but these strains do not form fumonisins and have an apparently clonal population structure (1451, 1476). Whether these strains represent an isolated subpopulation of *F. verticillioides* or are instead a new species remains to be determined (863).

Conidia are produced on the phialides through an enteroblastic process in which the inner wall of the conidium and the phialide are continuous and to which the middle and outer layers are subsequently added (2148). This process reduces the amount of material available to block the phialidic openings and may be one of the reasons why the chains of *F. verticillioides* are so long. The shape and thickness of the colarette formed by the phialide during conidium production may be useful for taxonomic purposes (2147), but this trait is difficult to use and has been neither widely examined nor evaluated.

Host and geographic distribution. *Fusarium verticillioides* is widely distributed throughout the world, and is particularly associated with maize where it can cause stalk rot and cob rot that result in significant yield losses and reductions of grain quality. Disease severity is affected by the inoculation technique and by the time of both inoculation and scoring of the disease symptoms (784, 2049). A seedling test for pathogenicity also is available (117) as this fungus may also cause seed rot of infested kernels when they are planted (2383). *F. verticillioides* can enter a maize plant systemically from the seed (1631), through wounds in the plant or though infections of the silks. Of these different routes, kernel infection occurs most efficiently from strains that are inoculated onto the silks (1501). Disrupting husk integrity increases ear rot severity (2292).

Drought stress increases the amount of stalk rot and can be relieved, at least in part, by irrigation (356, 549, 554, 1921). High lysine corn (785, 2199, 2295), brown midrib maize (*bm3*; 1559) and sweet corn lines with the *sh-2* genotype (2057, 2403) are particularly susceptible to infection by *F. verticillioides*, although the host genomic background can mitigate disease severity (237, 836, 1277). *F. verticillioides* can cause root rot and decreased root growth in maize seedlings (2026). Maize lines developed to be resistant to infection by *Aspergillus* also may be more resistant to infection by *F. verticillioides* (256).

The fungus usually appears in the kernels as they near physiological maturity and continues to increase until the end of the growing season (201, 886, 1194, 2399) when it may represent 90%, or more, of the *Fusarium* isolates from the grain (139, 1119, 1996). Based on electron microscopic observations, the fungus is found at the tip cap of both symptomatic and asymptomatic maize kernels (115). In symptommatic kernels the embryo and endosperm also were extensively colonized and in some cases microconidia were produced inside the infected kernel.

Amongst harvested grain, *F. verticillioides* is more commonly recovered from the fines, bran and germ than it is from whole kernels or coarse grit fractions (1023). Particular strains for experimental purposes can be introduced to seeds either by spraying the silks with a spore suspension or by soaking seeds in a spore suspension (1497). The effect of seedborne *F. verticillioides* on germination and yield is a subject of debate (63, 1522, 1631, 2137), but appears heavily dependent on the conditions under which germination and subsequent growth occur.

If *F. verticillioides* is present in a maize kernel, then that kernel is less likely to be infected by *Aspergillus* spp., *F. graminearum*, *F. subglutinans*, and *Diplodia maydis* (1356, 1357, 1804, 2225, 2315). Kernels infected with *F. verticillioides* also may contain less of mycotoxins produced by other fungi, *e.g.*, aflatoxins (1354).

Multiple genetically distinct isolates of *F. verticillioides* may be recovered from a single maize plant (1029). Injection of a spore suspension through maize husk leaves at the blister stage (R2) can be used to differentiate maize lines that are resistant/susceptible to Fusarium ear rot from one another (384). "BT" hybrid maize, which is less susceptible to the European corn borer *Ostrinia nubialis*, had a lower level of infection with *F. verticillioides* than did comparable hybrids without the transgene (383, 685, 1499) and accumulates less fumonisins (1498, 1648).

Fusarium verticillioides has been recovered from teosinte (484), suggesting that the association between this fungus and maize is of long standing and possibly of evolutionary importance. *F. verticillioides* also can degrade the antimicrobial benzoxazinoids 6-methoxy-2-benzoxazolinone and 2-benzoxazolinone produced by maize (711-714, 1810, 2401). This ability may be one of the reasons why this fungus is a successful maize pathogen.

Fusarium verticillioides also may be recovered from finger millet, and native North American tallgrass prairie grasses (1234, 1412) and desert soils (984). It may be recovered from (1227) and cause disease on sorghum (758, 1233). This fungus has been associated with a broad range of diseases but because of the nomenclatural confusion and the lack of understanding that there were more than one species in the older species definitions it is difficult to determine the true causal agent in many cases. *F. verticillioides* has been reported to cause top rot of sugar cane, foot rot of rice and crown rot of asparagus (2045) and to infect as many as 11,000 plant species (119), but these pathogenic associations need to be re-evaluated to confirm that they are caused by *F. verticillioides* and not by another member of the *G. fujikuroi* species complex that used to be included in "*Fusarium moniliforme*".

The fungus commonly occurs as an endophyte within hosts such as maize (116, 2384), where it may protect against infection by other fungi (66, 1804, 2225, 2314, 2315). The proportion of asymptomatic infection in maize kernels may depend on the genotype of the host (1069). *F. verticillioides* also is well adapted to air and wind dispersal (1629) and can infect the grain of maize via the silk of the cob or through wounds created by *Lepidopterum* larvae (703, 2019). *F. verticillioides* persists in host residues on the soil surface or in the soil following mechanical incorporation and can survive for up to 630 days under Iowa field conditions (410) and up to 900 days in cool, dry conditions (1254). Reducing tillage increases the amount of *F. verticillioides* inoculum available for the following season (1996). Both the fungus *Trichoderma* and the bacterium *Pseudomonas* have potential to serve as biocontrol agents of *F. verticillioides* (120).

Genetics and genomics. Single gene mutants of *F. verticillioides* that do not produce the characteristic long chains of microconidia (or any microconidia at all) have been identified (714). Hydrophobins are found on the outside of the microconidia, and two of them must both be present if the microconidia are to be found in chains rather than in false heads (652).

The genetics of vegetative compatibility in the entire *Fusarium* genus is modeled on the basic results obtained primarily with *F. verticillioides* (Chapter 5). Critical studies include those determining the number of segregating *vic* loci (1747), determining the genetic basis for the *nit* mutants used to force heterokaryons (1090), and identification of the heterokaryon self-incompatible phenotypic class (403). Vegetative compatibility as a tool for population genetic analysis in *F. verticillioides* has not proven particularly useful as most of the strains in a population are in different VCGs and thus the information obtained primarily is that no two strains are identical, *i.e.*, clones in these populations are rare (1979).

The spore killer trait is polymorphic in field populations of *F. verticillioides* (1021). This trait has been placed on the genetic map of *F. verticillioides* (995, 2373). *F. verticillioides* has 12 chromosomes, one of which is dispensable, and a total genome size of ~46 Mb, which is typical of other *Fusarium* species in this group (1441, 2374). There is no evidence for chromosome rearrangements surrounding the spore-killer region, which could limit the amount of recombination that occurs near the spore killer locus. A number of other mutants have been identified either from field isolates or from laboratory mutagenesis experiments including benomyl tolerance (2381), cyclins (1967), fumonisin biosynthesis (289, 290, 1703, 1737), nitrate catabolism (1090, 2232), nutritional auxotrophs (1746, 1747), perithecial pigmentation (333), and pH response (628).

Protoplasts can be formed from *F. verticillioides* at relatively high efficiency (2236). *F. verticillioides* has been transformed and a number of common vectors and genes can be used as selectable markers, *e.g.*, β-glucuronidase (2385), green and other fluorescent proteins (227, 1631) and resistance to hygromycin B (1223, 2385). Duplicated DNA introduced to a strain by transformation may MIP if the transformant goes through meiosis (1223).

The region around mating type in *F. verticillioides* has significant synteny with that of *F. proliferatum* and a number of other ascomycetes (2276). The *impala* transposable element can move when introduced into the *F. verticillioides* genome and could provide an alternative method for creating tagged mutants in this fungus (913).

There are reports of DNA fragments that are specific either for *F. verticillioides* or for fumonisin biosynthetic genes. Many, but not all, of these studies are based on relatively small strain sets and none of these diagnostics has yet been accepted for use on a broad scale (208, 732, 862, 1493, 1502, 1511, 1659, 1690).

Physiology and biochemistry. Maximum linear growth is reported to occur at 25°C and an osmotic potential of -1.0 MPa (1541). Optimal spore germination is sensitive to a_w, with a 30°C optimum for an a_w between 0.90 and 0.94, and a broad range of 25-37°C as the optimum with an a_w between 0.96 and 0.98 (1358). The type (vegetative or reproductive), maturity and physical condition of the maize tissue can determine how well it can be colonized by *F. verticillioides* (2386), with the best growth occurring on wounded, relatively immature reproductive tissues. Fumonisin B_1 production also is highest under these conditions (2293) in the laboratory, but high levels of fungal growth, especially under field conditions, are not necessarily correlated with high levels of toxin formation (485, 1030, 1419, 1959).

Many strains of *F. verticillioides* produce relatively few macroconidia. The number of macroconidia produced increases if strains are grown on media with maltose or soluble starch as a carbon source (906). Addition of zinc to culture media increases fungal dry weight and ammonia assimilation but reduces the production of some secondary metabolites (945). Commercially available antioxidants can be used to treat grain and regard growth by *F. verticillioides* and reduce the amount of fumonisins produced *in situ* (591, 1801). The fungus also can grow under microaerobic conditions (700).

A number of enzymes from *F. verticillioides* have been examined biochemically including: β-D-galactosidase (1309), dextranase (553); D-lactonohydrolase (1273), pectate lyase (510, 1775), peptidases (1817), phosphatases (1003, 1004, 2395), polygalacturonase (217-219, 310-312, 442, 456, 607, 1572, 1573, 1939), oxygenase (2206), proteases (368, 1107, 1108), ribonuclease (1525, 2394), and β-xylosidase (1860, 1861). Some strains of *F. verticillioides* can degrade polylactic acid and other lactic acid containing polymers (2165) or transform steroids (409). Strains of *F. verticillioides* are used commercially to resolve mixtures of DL-pantolactone (908, 2114).

Human and animal pathogenicity. *Fusarium verticillioides* is known to be allergenic to humans (2244) and to be capable of systemically infecting cancer (210, 325, 603, 646, 849, 1510, 1609, 2000, 2256, 2387, 2398) and HIV patients (778, 1692). The most common human health problem associated with *F. verticillioides* is skin lesions (29, 393, 1952, 2277), but it also can infect through wounds (1675, 1812) and has been associated with keratitis (1638). Infections with *F. verticillioides* usually are not associated with hospital settings, but nosocomial outbreaks of diseases attributable to this fungus do occur (774).

Fusarium verticillioides is resistant to most clinical antifungals, *e.g.*, itraconazole, miconazole and flucytosine (1799), with amphotericin B (1748, 1799) and natamycin (1799) reported as the most effective. Fenpropimorph is known to inhibit ergosterol biosynthesis in *F. verticillioides* (458).

Fusarium verticillioides also can cause direct disease in some animals including alligators (647) and freshwater fish (193).

Fumonisins. Fumonisins (691, 1962) are clearly the most important toxin produced by *F. verticillioides*, and some strains may produce these mycotoxins at very high levels (691, 1448, 1546). Fumonisin B_1 is the best known and studied of the fumonisins, but other derivatives are known to occur naturally as well (235, 1307, 1506, 1703, 1946). Members of this family of toxins disrupt sphingolipid metabolism (1425, 1925, 1957, 2288) and cause leukoencephalomalacia in horses (1338), pulmonary edema syndrome in pigs (786, 825, 1639), liver cancer (692, 2270) and liver and kidney toxicity in rats (2269), neurodegeneration in mice (1634), and apoptosis in many types of cells (989). Fumonisins are phytotoxic, but their role in plant diseases caused by *F. verticillioides*, if any, has not been clearly defined (9, 478, 481, 486, 514, 1180), although they have been used successfully to select mutants in *Arabidopsis* with altered pathogen defense reactions (2050). Fumonisins also have some antifungal activity (1043).

Fumonisins have been implicated in human esophageal cancer (1337, 1543, 1803) and in birth defects (1348) in humans. Fumonisins may be recovered from numerous foods, although maize and maize-related products usually are at the highest risk for contamination (262, 2027). Toxin production is higher on degermed maize kernels than it is maize germ tissue (1965). Fumonisin B_1 is sufficiently thermostable that conventional heating during cooking or processing of maize grain does not destroy the compound (528, 2267).

In vitro fumonisin biosynthesis most commonly occurs on cracked maize (1546), but may also be done in liquid media. It requires a well-aerated environment, a pH of 2.5-5.0 (1033) and limited nitrogen (1966). Toxin synthesis is stimulated by the presence of methionine, a fumonisin precursor, in the culture medium (2264). Strains that produce high levels of fumonisins on solid cultures need not produce high levels of fumonisins in liquid cultures and *vice versa* (2264). Production is better at temperatures > 15°C (1856, 2237) and when strains are grown on maize grain rather than on wheat or barley (1355). Reynoso *et al.* (1800) report that female-fertile strains of *F. verticillioides*, on average, produced 20% more fumonisin than did female-sterile strains.

The biochemistry of fumonisin biosynthesis has been examined in some detail, and both the biochemistry and the molecular genetics underlying the pathway generally are well understood (32, 195, 212, 236, 289, 290, 299, 1702, 1737, 1943). The fumonisin biosynthetic gene clusters in *F. proliferatum* and *F. verticillioides* have both been sequenced and have a high degree of synteny, but the flanking regions are completely different (1740, 2276). These data may be interpreted to mean that more than one event resulted in the acquisition of these toxin genes by these fungi and that toxin production is not a homologous character that can be used in phylogenetic studies. Based on microarray studies, > 180 genes may be differentially expressed during fumonisin production (1701).

Other toxins. Feed contaminated with *F. verticillioides* was toxic to rats, causing death, hemorrhage, kidney and liver failure, and diarrhea (4, 2268). *F. verticillioides* also is toxic to numerous other animals including baboons, cattle, chickens, donkeys, ducklings, horses, mice, rabbits, rats, sheep and vervet monkeys (693, 961, 973, 1225, 1233, 1341, 2249). Some, but not all, of these toxic reactions are attributable to fumonisin contamination in feed. *F. verticillioides* was not pathogenic in an experimental test with immunocompetent mice (1392).

Fusarium verticillioides can produce fusaric acid and various derivatives (119, 282, 1341, 1722), and trace levels of beauvericin (1234). Fusarins, particularly fusarin C, are produced by most strains of *F. verticillioides* (308, 601, 694, 942, 944, 1341, 1448, 1904) and the amount of production is sensitive to levels of zinc and manganese (945) and fermentation time (307).

Reports of gibberellin production by *F. verticillioides* probably are attributable to *F. fujikuroi*, especially if the strain in question was recovered from rice. *F. verticillioides* does not produce more than trace levels of moniliformin (1225, 1233, 1349, 1956) and is not known to produce fusaproliferin (1234, 1956), although it probably does produce additional compounds that are toxic at least to ducklings or *in vitro* cell cultures that have not yet been identified (1225, 1452, 1584, 2249). If cultures of *F. verticillioides* are fed isoamylamine, then they can produce nitrosomethylisoamylamine, a carcinogenic nitrosoamine (976).

Fusarium verticillioides is not known to produce trichothecenes (1341, 1452), but it does have genes that encode functional enzymes in the trichothecene biosynthetic pathway (1067). *F. verticillioides* does not produce zearalenone (1341), but can produce 8-bostrycoidin, which may be confused with zearalenone in TLC assays (2005).

Current Species Description: All of the published species descriptions refer to this species in the broad *F. moniliforme sensu lato* sense. There is no current description that takes into account the issues detailed by Seifert *et al.* (1936). The description provided by Gerlach & Nirenberg (697) provides detailed measurements for most of the relevant morphological characters for this species and together with the characters outlined here provides an accurate species description.

References

1. **Abbas, H. K., and C. D. Boyette.** 1996. Control of morning glory species using *Fusarium solani* and its extracts. *International Journal of Pest Management* **42:** 235-239. (*solani*)

2. **Abbas, H. K., and C. J. Mirocha.** 1988. Isolation and purification of a hemorrhagic factor, wortmannin, from *Fusarium oxysporum* N17B. *Applied and Environmental Microbiology* **54:** 1268-1274. (*oxysporum*)

3. **Abbas, H. K., C. J. Mirocha, and R. Gunther.** 1989. Mycotoxins produced by toxic *Fusarium* isolates obtained from agricultural and nonagricultural Arctic areas of Norway. *Mycopathologia* **105:** 143-152. (*acuminatum, avenaceum, oxysporum, sambucinum*)

4. **Abbas, H. K., C. J. Mirocha, T. Kommedahl, P. M. Burnes, R. A. Meronuck, and R. Gunther.** 1988. Toxigenicity of *Fusarium proliferatum* and other *Fusarium* spp. isolated from corn ears in Minnesota, USA. *Phytopathology* **78:** 1258-1260. (*proliferatum, subglutinans, verticillioides*)

5. **Abbas, H. K., and C. J. Mirocha, W. T. Shier, and R. Gunther.** 1992. Bioassay, extraction, and purification procedures for wortmannin, the hemorrhagic factor produced by *Fusarium oxysporum* N17B grown on rice. *Journal of AOAC International* **75:** 474-480. (*oxysporum*)

6. **Abbas, I. H., M. N. Mouhi, J. T. Al-Roubaie, N. N. Hama, and A. H. El-Bahadli.** 1991. *Phomopsis phoenicola* and *Fusarium equiseti*, new pathogens on date palm in Iraq. *Mycological Research* **95:** 509. (*equiseti*)

7. **Abbas, H. K., and C. M. Ocamb.** 1995. First report of production of fumonisin B_1 by *Fusarium polyphialidicum* collected from seeds of *Pinus strobus*. *Plant Disease* **79:** 642. (*polyphialidicum*)

8. **Abbas, H. K., C. M. Ocamb, W. Xie, C. J. Mirocha, and W. T. Shier.** 1995. First report of fumonisin B_1, B_2, and B_3 production by *Fusarium oxysporum* var. *redolens*. *Plant Disease* **79:** 968. (*redolens*)

9. **Abbas, H. K., R. J. Smeda, B. C. Gerwick, and W. T. Shier.** 2000. Fumonisin B_1 from the fungus *Fusarium moniliforme* causes contact toxicity in plants: Evidence from studies with biosynthetically labeled toxin. *Journal of Natural Toxins* **9:** 85-100. (*verticillioides*)

10. **Abdalla, M. Y., A. Al-Rokibah, A. Moretti, and G. Mulè.** 2000. Pathogenicity of toxigenic *Fusarium proliferatum* from date palm in Saudi Arabia. *Plant Disease* **84:** 321-324. (*proliferatum*)

11. **Abergel, C., C. Martinez, J. Fontecilla-Camps, C. Cambillau, P. de Jesus, and M. Lauwereys.** 1990. Crystallization and preliminary X-ray study of a recombinant cutinase from *Fusarium solani pisi*. *Journal of Molecular Biology* **215:** 215-216. (*solani*)

12. **Aboul-Soud, M. A. M., B. W. Yun, L. A. Harrier, and G. J. Loake.** 2004. Transformation of *Fusarium oxysporum* by particle bombardment and characterization of the resulting transformants expressing a GFP transgene. *Mycopathologia* **158:** 475-482. (*oxysporum*)

13. **Abraham, W. R., and H. A. Arfmann.** 1988. Fusalanipyrone, a monoterpenoid from *Fusarium solani*. *Phytochemistry* **27:** 3310-3311. (*solani*)

14. **Abraham, W. R., I. Knoch, and L. Witte.** 1990. Biosynthesis of the terpenoid polyketide fusalanipyrone. *Phytochemistry* **29:** 2877-2878. (*solani*)

15. **Abramson, D., B. McCallum, D. M. Smith, and A. Tekauz.** 2002. Moniliformin in barley inoculated with *Fusarium avenaceum*. *Food Additives and Contaminants* **19:** 765-769. (*avenaceum*)

16. **Achenbach, L. A., J. A. Patrick, and L. E. Gray.** 1997. Genetic homogeneity among isolates of *Fusarium solani* that cause soybean sudden death syndrome. *Theoretical and Applied Genetics* **95:** 474-478. (*solani*)

17. **Achor, D. S., S. Nemec, and R. A. Baker.** 1993. Effects of *Fusarium solani* naphthazarin toxins on the cytology and ultrastructure of rough lemon seedlings. *Mycopathologia* **123:** 117-126. (*solani*)

18. **Adams, G. C., Jr., and E. E. Butler.** 1983. Environmental factors influencing the formation of basidia and basidiospores in *Thanatephorus cucumeris*. *Phytopathology* **73:** 152-155. (6)

19. **Adams, G. C., Jr., W. D. Gubler, and R. G. Grogan.** 1987. Seedling disease of muskmelon and mixed melons in California, USA caused by *Fusarium equiseti*. *Plant Disease* **71:** 370-374. (*equiseti*)

20. **Adams, G., N. Johnson, J. F. Leslie, and L. P. Hart.** 1987. Heterokaryons of *Gibberella zeae* formed following hyphal anastomosis or protoplast fusion. *Experimental Mycology* **11:** 339-353. (5)

21. **Adis, J., O. J. M. Fonseca, and R. Z. Strassen.** 1985. A thrips as vector of the "supersprouting" fungus disease in guarana plants of the Central Amazon. *Fitopatololia Brasiliensis* **10:** 677-679. (*decemcellulare*)

22. **Afanide, B., S. A. Mabadeje, and S. H. Z. Naqvi.** 1976. *Gibberella baccata*, the perfect state of *Fusarium lateritium* in Nigeria. *Mycologia* **68:** 1108-1111. (10)

23. **Agrawal, A. K., and L. K. Sherman.** 2004. Voriconazole-induced musical hallucinations. *Infection* **32:** 293-295. (*solani*)

24. **Ahn, I.-P., H.-S. Chung, and Y.-H. Lee.** 1998. Vegetative compatibility groups and pathogenicity among isolates of *Fusarium oxysporum* f. sp. *cucumerinum*. *Plant Disease* **82:** 244-246. (*oxysporum*)

25. **Aidoo, K. E., A. Anderton, and K. A. Milligan.** 1995. A 2-year survey of the airborne mycoflora in a hospital environment. *International Journal of Environmental Health Research* **5:** 223-228. (3)

26. **Aist, J. R., and C. J. Bayles.** 1991. Ultrastructural basis of mitosis in the fungus *Nectria haematococca* (sexual stage of *Fusarium solani*). I. Asters. *Protoplasma* **161:** 111-122. (*solani*)

27. **Aist, J. R., and C. J. Bayles.** 1991. Ultrastructural basis of mitosis in the fungus *Nectria haematococca* (sexual stage of *Fusarium solani*). II. Spindles. *Protoplasma* **161:** 123-136. (*solani*)

28. **Aist, J. R., and M. W. Berns.** 1981. Mechanics of chromosome separation during mitosis in *Fusarium solani* Fungi Imperfecti: New evidence from ultrastructural and laser mi-

crobeam experiments. *Journal of Cell Biology* **91:** 446-458. (*solani*)

29. **Ajello, L., A. A. Padhye, F. W. Chandler, M. R. McGinnis, L. Morganti, and F. Alberici.** 1985. *Fusarium moniliforme*, a new mycetoma agent restudy of a European case. *European Journal of Epidemiology* **1:** 5-10. (*verticillioides*)

30. **Akinsanmi, O. A., V. Mitter, S. Simpfendorfer, D. Backhouse, and S. Chakraborty.** 2003. Identity and pathogenicity of *Fusarium* spp. isolated from wheat fields in Queensland and northern New South Wales. *Australian Journal of Agricultural Research* **55:** 97-107. (7, *pseudograminearum*)

31. **Albert, B., and C. H. Sellem.** 2002. Dynamics of the mitochondrial genome during *Podospora anserina* aging. *Current Genetics* **40:** 365-373. (4)

32. **Alberts, J. F., W. C. A. Gelderblom, R. Vleggaar, W. F. O. Marasas, and J. P. Rheeder.** 1993. Production of ^{14}C-fumonisin B_1 by *Fusarium moniliforme* MRC 826 in corn cultures. *Applied and Environmental Microbiology* **59:** 2673-2677. (*verticillioides*)

33. **Albisetti, M., R. P. Lauener, T. Guengoer, G. Schaer, F. K. Niggli, D. Nadal.** 2004. Disseminated *Fusarium oxysporum* infection in hemophagocytic lymphohistiocytosis. *Infection* **32:** 364-366. (*oxysporum*)

34. **Albuquerque, P. S. B., and C. N. Bastos.** 1990. Oversprouting and canker of *Cordia alliodora* caused by *Fusarium decemcellulare*. *Fitopatologia Brasileira* **15:** 252-253. (*decemcellulare*)

35. **Alconada, T. M., and M. J. Martinez.** 1994. Purification and characterization of an extracellular endo-1,4-β-xylanase from *Fusarium oxysporum* f. sp. *melonis*. *FEMS Microbiology Letters* **118:** 305-310. (*oxysporum*)

36. **Alconada, T. M., and M. J. Martinez.** 1996. Purification and characterization of a β-glucosidase from the phytopathogenic fungus *Fusarium oxysporum* f. sp. *melonis*. *Letters in Applied Microbiology* **22:** 106-110. (*oxysporum*)

37. **Alejandre-Duran, E., T. Roldan-Arjona, R. R. Ariza, and M. Ruiz-Rubio.** 2003. The photolyase gene from the plant pathogen *Fusarium oxysporum* f. sp. *lycopersici* is induced by visible light and α-tomatine from tomato plant. *Fungal Genetics and Biology* **40:** 159-165. (*oxysporum*)

38. **Alexander, N. J., T. M. Hohn, and S. P. McCormick.** 1998. The *Tri11* gene of *Fusarium sporotrichioides* encodes a cytochrome P-450 monooxygenase required for C-15 hydroxylation in trichothecene biosynthesis. *Applied and Environmental Microbiology* **64:** 221-225. (*sporotrichioides*)

39. **Alexander, N. J., S. P. McCormick, and T. M. Hohn.** 1999. *Tri12*, a trichothecene efflux pump from *Fusarium sporotrichioides*: Gene isolation and expression in yeast. *Molecular and General Genetics* **261:** 977-984. (*sporotrichioides*)

40. **Alexander, N. J., S. P. McCormick, T. M. Larson, and J. E. Jurgenson.** 2004. Expression of *Tri15* in *Fusarium sporotrichioides*. *Current Genetics* **45:** 157-162. (*sporotrichioides*)

41. **Alexandrakis, G., S. Jalali, and P. Gloor.** 1998. Diagnosis of *Fusarium* keratitis in an animal model using the polymerase chain reaction. *British Journal of Ophthalmology* **82:** 306-311. (*solani*)

42. **Ali, H., B. A. Summerell, and L. W. Burgess.** 1991. An evaluation of three media for the isolation of *Fusarium*, *Alternaria* and other fungi from sorghum grain. *Australasian Plant Pathology* **20:** 134-138. (*polyphialidicum*)

44. **Alkan, Y., W. E. Haefeli, J. Burhenne, J. Stein, I. Yaniv, and I. Shalit.** 2004. Voriconazole-induced QT interval prolongation and ventricular tachycardia: A non-concentration-dependent adverse effect. *Clinical Infectious Diseases* **39:** E49-E52, 876. (*solani*)

45. **Aloi, C., and R. P. Baayen.** 1993. Examination of the relationships between vegetative compatibility groups and races in *Fusarium oxysporum* f. sp. *dianthi*. *Plant Pathology* **42:** 839-850. (*oxysporum*)

46. **Altomare, C., A. Logrieco, A. Bottalico, G. Mulè, A. Moretti, and A. Evidente.** 1995. Production of type A trichothecenes and enniatin B by *Fusarium sambucinum* Fückel *sensu lato*. *Mycopathologia* **129:** 177-181. (*sambucinum*, *torulosum*, *venenatum*)

47. **Altomare, C., R. Pengue, M. Favilla, A. Evidente, and A. Visconti.** 2004. Structure-activity relationships of derivatives of fusapyrone, an antifungal metabolite of *Fusarium semitectum*. *Journal of Agricultural and Food Chemistry* **52:** 2997-3001. (*semitectum*)

48. **Altomare, C., G. Perrone, M. C. Zonno, A. Evidente, R. Pengue, F. Fanti, and L. Polonelli.** 2000. Biological characterization of fusapyrone and deoxyfusapyrone, two bioactive secondary metabolites of *Fusarium semitectum*. *Journal of Natural Products* **63:** 1131-1135. (*semitectum*)

49. **Altomare, C., O. Petrini, A. Logrieco, and A. Bottalico.** 1997. Taxonomic relationships among the toxigenic species *Fusarium acuminatum*, *Fusarium sporotrichioides* and *Fusarium tricinctum* by isozyme analysis and RAPD assay. *Canadian Journal of Botany* **75:** 1674-1684. (*armeniacum*)

50. **Alves-Santos, F. M., E. P. Benito, A. P. Eslava, and J. M. Diaz-Minguez.** 1999. Genetic diversity of *Fusarium oxysporum* strains from common bean fields in Spain. *Applied and Environmental Microbiology* **65:** 3335-3340. (*oxysporum*)

51. **Alves-Santos, F. M., B. Ramos, M. A. Garcia-Sanchez, A. P. Eslava, and J. M. Diaz-Minguez.** 2002. A DNA-based procedure for *in planta* detection of *Fusarium oxysporum* f. sp. *phaseoli*. *Phytopathology* **92:** 237-244. (*oxysporum*)

52. **Amalfitano, C., R. Pengue, A. Andolfi, M. Vurro, M. C. Zonno, and A. Evidente.** 2002. HPLC analysis of fusaric acid, 9,10-dehydrofusaric acid and their methyl esters, toxic metabolites from weed pathogenic *Fusarium* species. *Phytochemical Analysis* **13:** 277-282. (*nygamai*, *oxysporum*, *udum*)

53. **Amaral, F. C., R. L. Brandao, J. R. Nicoli, and C. H. D. Ortiz.** 1995. Comparative study of two trehalase activities from *Fusarium oxysporum* var. *lini*. *Canadian Journal of Microbiology* **41:** 1057-1062. (*oxysporum*)

54. **Amir, H., and N. Mahdi.** 1993. Correlation between the ecological aptitudes of various strains of *Fusarium* and their efficacy in the protection of flax against vascular fusariosis. *Canadian Journal of Microbiology* **39:** 234-244. (*oxysporum*)

55. **Ammar, M. S., N. N. Gerber, and L. E. McDaniel.** 1979. New antibiotic pigments related to fusarubin from *Fusarium*

solani (Mart.) Sacc. I. Fermentation, isolation and antimicrobial activities. *Journal of Antibiotics* **32:** 679-684. (*solani*)

56. **Amsellem, Z., S. Barghouthi, B. Cohen, Y. Goldwasser, J. Gressel, L. Hornok, Z. Kerenyi, Y. Kleifeld, O. Klein, J. Kroschel, J. Sauerborn, D. Mueller-Stoever, H. Thomas, M. Vurro, and M. C. Zonno.** 2001. Recent advances in the biocontrol of *Orobanche* (broomrape) species. *BioControl* **46:** 211-228. (*oxysporum*)

57. **Amsellem, Z., Y. Kleifeld, Z. Kerenyi, L. Hornok, Y. Goldwasser, and J. Gressel.** 2001. Isolation, identification, and activity of mycoherbicidal pathogens from juvenile broomrape plants. *Biological Control* **21:** 274-284. (*oxysporum*)

58. **Anagnostakis, S. L.** 1977. Vegetative incompatibility in *Endothia parasitica*. *Experimental Mycology* **1:** 306-316. (5)

59. **Anaissie, E. J., R. Hachem, C. Legrand, P. Legenne, P. Nelson, and G. P. Bodey.** 1992. Lack of activity of amphotericin B in systemic murine fusarial infection. *Journal of Infectious Diseases* **165:** 1155-1157. (*solani*)

60. **Anaissie, E. J., R. T. Kuchar, J. H. Rex, A. Francesconi, M. Kasai, F. M. C. Muller, M. Lozano-Chiu, R. C. Summerbell, M. C. Dignani, S. J. Chanock, and T. J. Walsh.** 2001. Fusariosis associated with pathogenic *Fusarium* species colonization of a hospital water system: A new paradigm for the epidemiology of opportunistic mold infections. *Clinical Infectious Diseases* **33:** 1871-1878. (*oxysporum, solani*)

61. **Anaya, N., and M. I. G. Roncero.** 1995. *Skippy*, a retrotransposon from the fungal plant pathogen *Fusarium oxysporum*. *Molecular and General Genetics* **249:** 637-647. (*oxysporum*)

62. **Anaya, N., and M. I. G. Roncero.** 1996. Stress-induced rearrangement of *Fusarium* retrotransposon sequences. *Molecular and General Genetics* **253:** 89-94. (*oxysporum*)

63. **Anderegg, J., and J. W. Guthrie.** 1981. Seedborne *Fusarium moniliforme* and seedling infection in hybrid sweet corn. *Phytopathology* **71:** 1196-1198. (3, *verticillioides*)

64. **Andersen, B., U. Thrane, A. Svendsen, and I. A. Rasmussen.** 1996. Associated field mycobiota on malt barley. *Canadian Journal of Botany* **74:** 854-858. (*tricinctum*)

65. **Anderson, A. A.** 1958. A new sampler for the collection, sizing and enumeration of viable airborne particles. *Journal of Bacteriology* **76:** 471-484. (3)

66. **Anderson, B., and D. G. White.** 1987. Fungi associated with corn stalks in Illinois, USA in 1982 and 1983. *Plant Disease* **71:** 135-137. (*verticillioides*)

67. **Anderson, J. P., E. Badruzsaufari, P. M. Schenk, J. M. Manners, O. J. Desmond, C. Ehlert, D. J. Maclean, P. R. Ebert, and K. Kazan.** 2004. Antagonistic interaction between abscisic acid and jasmonate-ethylene signaling pathways modulates defense gene expression and disease resistance in *Arabidopsis*. *Plant Cell* **16:** 3460-3479. (*oxysporum*)

68. **Andolfi, A., A. Boari, A. Evidente, and M. Vurro.** 2005. Metabolites inhibiting germination of *Orobanche ramosa* seeds produced by *Myrothecium verrucaria* and *Fusarium compactum*. *Journal of Agricultural and Food Chemistry* **53:** 1598-1603. (*compactum*)

69. **Andrews, G., M. Kurien, V. Anandi, B. Ramakrishna, and R. Raman.** 1996. Nasosinusal fungal granuloma – clinical profile. *Singapore Medical Journal* **37:** 470-474. (*solani*)

70. **Aoki, T., and H. I. Nirenberg.** 1999. *Fusarium globosum* from subtropical Japan and the effect of different light conditions on its conidiogenesis. *Mycoscience* **40:** 1-9. (*globosum*)

71. **Aoki, T., and K. O'Donnell.** 1999. Morphological and molecular characterization of *Fusarium pseudograminearum* sp. nov., formerly recognized as the Group 1 population of *F. graminearum*. *Mycologia* **91:** 597-609. (8, *graminearum, pseudograminearum*)

72. **Aoki, T., and K. O'Donnell.** 1999. Morphological characterization of *Gibberella coronicola* sp. no., obtained through mating experiments of *Fusarium pseudograminearum*. *Mycoscience* **40:** 443-453. (8, *pseudograminearum*)

73. **Aoki, T., K. O'Donnell, Y. Homma, and A. R. Lattanzi.** 2003. Sudden death syndrome of soybean is caused by two morphologically and phylogenetically distinct species within the *Fusarium solani* species complex – *F. virguliforme* in North America and *F. tucumaniae* in South America. *Mycologia* **95:** 660-684. (*solani*)

74. **Aoki, T., K. O'Donnell, and K. Ichikawa.** 2001. *Fusarium fractiflexum* sp. nov. and two other species within the *Gibberella fujikuroi* species complex recently discovered in Japan that form aerial conidia in false heads. *Mycoscience* **42:** 461-478. (*concentricum*)

75. **Appel, D. J., and T. R. Gordon.** 1994. Local and regional variation in populations of *Fusarium oxysporum* from agricultural field soils. *Phytopathology* **84:** 786-791. (5, 11, *oxysporum*)

76. **Appel, D. J., and T. R. Gordon.** 1995. Intraspecific variation within populations of *Fusarium oxysporum* based on RFLP analysis of the intergenic spacer region of the rDNA. *Experimental Mycology* **19:** 120-128. (*oxysporum*)

77. **Applegate, P. J., R. E. Nelson, and R. L. Metzenberg.** 1978. Mutant enrichment by filtration concentration: A variation for the selection of temperature-conditional heterokaryons. *Neurospora Newsletter* **25:** 17. (4)

78. **Arganoza, M. T., J. Ohrnberger, J. Min, and R. A. Akins.** 1994. Suppressor mutants of *Neurospora crassa* that tolerate allelic differences at single or at multiple heterokaryon incompatibility loci. *Genetics* **137:** 731-742. (5)

79. **Arie, T., S. Gouthu, S. Shimazaki, T. Kamakura, M. Kimura, M. Inoue, K. Takio, A. Ozaki, K. Yoneyama, and I. Yamaguchi.** 1998. Immunological detection of endopolygalacturonase secretion by *Fusarium oxysporum* in plant tissue and sequencing of its encoding gene. *Annals of the Phytopathological Society of Japan* **64:** 7-15. (*oxysporum*)

80. **Arie, T., I. Kaneko, T. Yoshida, M. Noguchi, Y. Nomura, and I. Yamaguchi.** 2000. Mating-type genes from asexual phytopathogenic Ascomycetes *Fusarium oxysporum* and *Alternaria alternata*. *Molecular Plant-Microbe Interactions* **13:** 1330-1339. (6, *oxysporum*)

81. **Arney, K. L., R. Tiernan, and M. A. Judson.** 1997. Primary pulmonary involvement of *Fusarium solani* in a lung transplant recipient. *Chest* **112:** 1128-1130. (*solani*)

82. **Arora, D. K., and R. S. Dwivedi.** 1980. Mycoparasitism of *Fusarium* spp. on *Rhizoctonia solani* Kuhn. *Plant and Soil* **55:** 43-53. (*semitectum*)

83. **Arrese, J. E., P. D. Doncker, F. C. Odds, and G. E. Pierard.** 1998. Reduction in the growth of non-dermatophyte molds by itraconazole: Evaluation by corneofungimetry assay. *Mycoses* **41:** 461-465. (*oxysporum*)

84. **Arroyo-Garcia, R., J. L. Cenis, J. Tello, J. M. Martinez-Zapater, and D. Cifuentes.** 2003. Genetic relationships among seven specialized forms of *Fusarium oxysporum* determined by DNA sequencing of the ITS region and AFLPs. *Spanish Journal of Agricultural Research* **1:** 55-63. (*oxysporum*)

85. **Arruda, G. M. T., R. N. G. Miller, M. A. S. V. Ferreira, and A. C. Cafe-Filho.** 2005. Morphological and molecular characterization of the sudden-death syndrome pathogen of soybean in Brazil. *Plant Pathology* **54:** 53-65. (*solani*)

86. **Aruna, C., and S. Audilakshmi.** 2004. Genetic architecture of grain hardness – A durable resistance mechanism for grain moulds in sorghum (*Sorghum bicolor* (L.) Moench). *Indian Journal of Genetics and Plant Breeding* **64:** 35-38. (*thapsinum*)

87. **Asano, Y., S. Ando, Y. Tani, H. Yamada, and T. Ueno.** 1981. Fungal degradation of triacrylonitrile. *Agricultural and Biological Chemistry* **45:** 57-62. (*solani*)

88. **Ashley, J. N., B. C. Hobbs, and H. Raistrick.** 1937. Studies in the biochemistry of micro-organisms. LIII. The crystalline coloring matters of *Fusarium culmorum* (W. G. Smith) Sacc. and related forms. *Biochemistry Journal* **31:** 385-397. (*graminearum*)

89. **Asran, M. R., and H. Buchenauer.** 2002. Virulence of *Fusarium moniliforme* isolates on maize plants in relation to fumonisin and ergosterol levels. *Journal of Phytopathology* **109:** 491-505. (7)

90. **Assabgui, R. A., L. M. Reid, R. I. Hamilton, and J. T. Arnason.** 1993. Correlation of kernel ferulic acid content of maize with resistance to *Fusarium graminearum*. *Phytopathology* **83:** 949-953. (*graminearum*)

91. **Attitalla, I. H., J. Fatehi, J. Levenfors, and S. Brishammar.** 2004. A rapid molecular method for differentiating two special forms (*lycopersici* and *radicis-lycopersici*) of *Fusarium oxysporum*. *Mycological Research* **108:** 787-794. (*oxysporum*)

92. **Audhya, T. K., and D. W. Russell.** 1973. Production of enniatin A. *Canadian Journal of Microbiology* **19:** 1051-1054. (*sambucinum*)

93. **Audhya, T. K., and D. W. Russell.** 1974. Production of enniatins by *Fusarium sambucinum*: Selection of high yield conditions from liquid surface cultures. *Journal of General Microbiology* **82:** 181-190. (*sambucinum*)

94. **Audhya, T. K., and D. W. Russell.** 1975. Enniatin production by *Fusarium sambucinum*: primary, secondary, and unitary metabolism. *Journal of General Microbiology* **86:** 327-332. (*sambucinum*)

95. **Austen, B., H. McCarthy, B. Wilkins, A. Smith, and A. Duncombe.** 2001. Fatal disseminated fusarium infection in acute lymphoblastic leukaemia in complete remission. *Journal of Clinical Pathology* **54:** 488-490. (*dimerum*)

96. **Avalos, J., J. Casadesus, and E. Cerda-Olmedo.** 1985. *Gibberella fujikuroi* mutants obtained with UV radiation and N-methyl-N'-nitro-N-nitrosoguanidine. *Applied and Environmental Microbiology* **49:** 187-191. (4)

97. **Avalos, J., and E. Cerda-Olmedo.** 1987. Carotenoid mutants of *Gibberella fujikuroi*. *Current Genetics* **11:** 505-512. (4)

98. **Avise, J. C.** 1994. *Molecular Markers, Natural History and Evolution*. Chapman and Hall, New York, New York. (9)

99. **Avise, J. C., and R. M. Ball.** 1990. Principles of genealogical concordance in species concepts and biological taxonomy. *Oxford Surveys of Evolutionary Biology* **7:** 45-67. (9)

100. **Awad, A. H. A.** 2002. Environmental study in subway metro stations in Cairo, Egypt. *Journal of Occupational Health* **44:** 112-118. (3)

101. **Baayen, R. P.** 2000. Diagnosis and detection of host-specific forms of *Fusarium oxysporum*. *Bulletin OEPP* **30:** 489-491. (8)

102. **Baayen, R. P., M. G. Forch, C. Waalwijk, P. J. M. Bonants, H. J. M. Loffler, and E. J. A. Roebroeck.** 1998. Pathogenic, genetic and molecular characterization of *Fusarium oxysporum* f. sp. *lilii*. *European Journal of Plant Pathology* **104:** 887-894. (*oxysporum*)

103. **Baayen, R. P., and W. Gams.** 1988. The *Elegans* Fusaria causing wilt disease of carnation. I. Taxonomy. *Netherlands Journal of Plant Pathology* **94:** 273-288. (*redolens*)

104. **Baayen, R. P., and J. Kleijn.** 1989. The *Elegans* Fusaria causing wilt disease of carnation. II. Distinction of vegetative compatibility groups. *Netherlands Journal of Plant Pathology* **95:** 185-194. (*oxysporum*, *redolens*)

105. **Baayen, R. P, K. O'Donnell, P. J. Bonants, E. Cigelnik, L. P. N. M. Kroon, E. J. A. Roebroeck, and C. Waalwijk.** 2000. Gene genealogies and AFLP analyses in the *Fusarium oxysporum* complex identify monophyletic and non-monophyletic *formae speciales* causing wilt and rot disease. *Phytopathology* **90:** 891-900. (5, 8, 9, 11, *oxysporum*, *redolens*)

106. **Baayen, R. P., K. O'Donnell, S. Breeuwsma, D. M. Geiser, and C. Waalwijk.** 2001. Molecular relationships of fungi within the *Fusarium redolens-F. hostae* clade. *Phytopathology* **91:** 1037-1044. (*hostae*, *miscanthi*, *nisikadoi*, *redolens*)

107. **Baayen, R. P., P. H. J. F. van den Boogert, P. J. M. Bonants, J. T. K. Poll, W. J. Blok, and C. Waalwijk.** 2000. *Fusarium redolens* f. sp. *asparagi*, causal agent of asparagus root rot, crown rot and spear rot. *European Journal of Plant Pathology* **106:** 907-912. (*redolens*)

108. **Baayen, R. P., F. van Dreven, M. C. Krijger, and C. Waalwijk.** 1997. Genetic diversity in *Fusarium oxysporum* f. sp. *dianthi* and *Fusarium redolens* f. sp. *dianthi*. *European Journal of Plant Pathology* **103:** 395-408. (*oxysporum*, *redolens*)

109. **Babai, A. A., M. J. Daboussi-Bareyre, and D. Parisot.** 1982. Isolation and genetic analysis of self-sterility and perithecial-pigmentation mutants in a homothallic isolate

of *Nectria haematococca*. *Canadian Journal of Botany* **60:** 79-84. (10)

110. **Bacic, M. K., and D. C. Yoch.** 1998. *In vivo* characterization of dimethylsulfoniopropionate lyase in the fungus *Fusarium lateritium*. *Applied and Environmental Microbiology* **64:** 106-111. (*lateritium*)

111. **Backhouse, D., A. A. Abubakar, L. W. Burgess, J. I. Dennis, G. J. Hollaway, G. B. Wildermuth, H. Wallwork, and F. J. Henry.** 2004. Survey of *Fusarium* species associated with crown rot of wheat and barley in eastern Australia. *Australasian Plant Pathology* **33:** 255-261. (*pseudograminearum*)

112. **Backhouse, D., and L. W. Burgess.** 1995. Mycogeo-graphy of *Fusarium*: Climatic analysis of the distribution within Australia of *Fusarium* species in section *Gibbosum*. *Mycological Research* **99:** 1218-1224. (*acuminatum, compactum, lactis, pseudograminearum, scirpi*)

113. **Backhouse, D., and L. W. Burgess.** 2002. Climactic analysis of the distribution of *Fusarium graminearum*, *Fusarium pseudograminearum*, and *Fusarium culmorum* on cereals in Australia. *Australasian Plant Pathology* **31:** 321-327. (3, *pseudograminearum*)

114. **Backhouse, D, L. W. Burgess, and B. A. Summerell.** 2001. Biogeography of *Fusarium*, p. 122-137. *In* B. A. Summerell, J. F. Leslie, D. Backhouse, W. L. Bryden, and L. W. Burgess (eds.), Fusarium*: Paul E. Nelson Memorial Symposium.* APS Press, St. Paul, Minnesota. (11)

115. **Bacon, C. W., R. M. Bennett, D. N. Hinton, and K. A. Voss.** 1992. Scanning electron microscopy of *Fusarium moniliforme* within asymptomatic corn kernels associated with equine leukoencephalomalacia. *Plant Disease* **76:** 144-148. (*verticillioides*)

116. **Bacon, C. W., and D. M. Hinton.** 1996. Symptomless endophytic colonization of maize by *Fusarium moniliforme*. *Canadian Journal of Botany* **74:** 1195-1202. (*verticillioides*)

117. **Bacon, C. W., D. M. Hinton, and M. D. Richardson.** 1994. A corn seedling assay for resistance to *Fusarium moniliforme*. *Plant Disease* **78:** 302-305. (*verticillioides*)

118. **Bacon, C. W., and P. E. Nelson.** 1994. Fumonisin production in corn by toxigenic strains of *Fusarium moniliforme* and *Fusarium proliferatum*. *Journal of Food Protection* **57:** 514-521. (*proliferatum*)

119. **Bacon, C. W., J. K. Porter, W. P. Norred, and J. F. Leslie.** 1996. Production of fusaric acid by *Fusarium* species. *Applied and Environmental Microbiology* **62:** 4039-4043. (*crookwellense, fujikuroi, heterosporum, napiforme, oxysporum, proliferatum, sambucinum, solani, subglutinans, verticillioides*)

120. **Bacon, C. W., I. E. Yates, D. M. Hinton, and F. Meredith.** 2001. Biological control of *Fusarium moniliforme* in maize. *Environmental Health Perspectives* **109 (Suppl. 2):** 325-332. (*verticillioides*)

121. **Bader, M., A. K. Jafri, T. Krueger, and V. Kumar.** 2003. *Fusarium* osteomyelitis of the foot in a patient with *diabetes mellitus*. *Scandinavian Journal of Infectious Diseases* **35:** 895-896. (*solani*)

122. **Bagga, S., and D. Straney.** 2000. Modulation of cAMP and phosphodiesterase activity by flavonoids which induce spore germination of *Nectria haematococca* MP VI (*Fusarium solani*). *Physiological and Molecular Plant Pathology* **56:** 51-61. (*solani*)

123. **Bai, A. J., E. U. B. Reddi, and C. S. Reddi.** 1981. Vertical profiles of spore concentrations near the ground. *Proceedings of the Indian National Academy of Sciences, Part B Biological Sciences* **47:** 65-77. (3)

124. **Bai, G. H., A. E. Desjardins, and R. D. Plattner.** 2002. Deoxynivalenol-nonproducing *Fusarium graminearum* causes initial infection, but does not cause disease spread in wheat spikes. *Mycopathologia* **153:** 91-98. (*graminearum*)

125. **Bailey, B. A., K. P. Hebbar, M. Strem, L. C. Darlington, and R. D. Lumsden.** 1997. An alginate pril formulation of *Fusarium oxysporum* Schlechtend:Fr. f. sp. *erythroxyli* for biocontrol of *Erythroxylum coca* var. *coca*. *Biocontrol Science and Technology* **7:** 423-435. (*oxysporum*)

126. **Bailey, B. A., K. P. Hebbar, M. Strem, R. D. Lumsden, L. C. Darlington, W. J. Connick, Jr., and D. J. Daigle.** 1998. Formulations of *Fusarium oxysporum* f. sp. *erythroxyli* for biocontrol of *Erythroxylum coca* var. *coca*. *Weed Science* **46:** 682-689. (*oxysporum*)

127. **Bajar, A., G. K. Podila, and P. E. Kolattukudy.** 1991. Identification of a fungal cutinase promoter that is inducible by a plant signal via a phosphorylated trans-acting factor. *Proceedings of the National Academy of Sciences USA* **88:** 8208-8212. (*solani*)

128. **Bajpai, B., and S. Patil.** 1997. Induction of tannin acyl hydrolase (e.c. 3.1.1.20) activity in some members of fungi imperfecti. *Enzyme and Microbial Technology* **20:** 612-614. (*solani*)

129. **Baker, C. J., and D. F. Bateman.** 1978. Cutin degradation by plant pathogenic fungi. *Phytopathology* **68:** 1577-1584. (*solani*)

130. **Baker, C. J., S. L. McCormick, and D. F. Bateman.** 1982. Effects of purified cutin esterase upon the permeability and mechanical strength of cutin membranes. *Phytopathology* **72:** 420-423. (*solani*)

131. **Baker, R., P. Hanchey, and S. D. Dottarar.** 1978. Protection of carnation against Fusarium stem rot by fungi. *Phytopathology* **68:** 1495-1501. (*avenaceum*)

132. **Baker, R. A., J. H. Tatum, and S. Nemec, Jr.** 1990. Antimicrobial activity of naphthoquinones from Fusaria. *Mycopathologia* **111:** 9-15. (*oxysporum, solani*)

133. **Bakshi, S., A. Sztejnberg, and O. Yarden.** 2001. Isolation and characterization of a cold-tolerant strain of *Fusarium proliferatum*, a biocontrol agent of grape downy mildew. *Phytopathology* **91:** 1062-1068. (*proliferatum*)

134. **Balasundaran, M., and R. Gnanaharan.** 1990. Laboratory evaluation of preservative treated rubberwood against fungi. *Journal of Tropical Forest Science* **2:** 303-306. (*decemcellulare*)

135. **Balmas, V., L. Corazza, and A. Magnotta.** 1999. Occurrence of *Fusarium* spp. in uncultivated soils in Italy. *Micologia Italiana* **28:** 5-8. (*polyphialidicum*)

136. **Balmas, V., P. Corda, A. Marcello, and A. Bottalico.** 2000. *Fusarium nygamai* associated with Fusarium foot rot of rice in Sardinia. *Plant Disease* **84:** 807. (*nygamai*)

137. **Bandyopadhyay, R., and L. K. Mughogho.** 1988. Evaluation of field screening techniques for resistance to sorghum grain molds. *Plant Disease* **72:** 500-503. (*thapsinum*)

138. **Bandyopadyhay, R., L. K. Mughogho, M. V. Satyanarayana, and M. E. Kalisz.** 1991. Occurrence of airborne spores of fungi causing grain mould over a sorghum crop. *Mycological Research* **95:** 1315-1320. (3)

139. **Bankole, S. A., and O. O. Mabekoje.** 2004. Occurrence of aflatoxins and fumonisins in preharvest maize from southwestern Nigeria. *Food Additives and Contaminants* **21:** 251-255. (*verticillioides*)

140. **Bao, J. R., D. R. Fravel, N. R. O'Neill, G. Lazarovits, and P. van Berkum.** 2002. Genetic analysis of pathogenic and nonpathogenic *Fusarium oxysporum* from tomato plants. *Canadian Journal of Botany* **80:** 271-279. (*oxysporum*)

141. **Bao, J. R., and G. Lazarovits.** 2001. Differential colonization of tomato roots by nonpathogenic and pathogenic *Fusarium oxysporum* strains may influence Fusarium wilt control. *Phytopathology* **91:** 449-456. (*oxysporum*)

142. **Bao, J. R., and G. Lazarovits.** 2002. Evaluation of three procedures for recovery of GUS enzyme and colony forming units of a nonpathogenic strain of *Fusarium oxysporum*, 70T01, from inoculated tomato roots. *Canadian Journal of Plant Pathology* **24:** 340-348. (*oxysporum*)

143. **Bao, J. R., J. Velema, K. F. Dobinson, and G. Lazarovits.** 2000. Using GUS expression in a nonpathogenic *Fusarium oxysporum* strain to measure fungal biomass. *Canadian Journal of Plant Pathology* **22:** 70-78. (*oxysporum*)

144. **Bapat, S., and A. K. Shah.** 2000. Biological control of fusarial wilt of pigeon pea by *Bacillus brevis*. *Canadian Journal of Microbiology* **46:** 125-132. (*udum*)

145. **Baran, R. A. T., and B. M. Piraaccini.** 1997. Uncommon clinical patterns of *Fusarium* nail infection: Report of three cases. *British Journal of Dermatology* **136:** 424-427. (*oxysporum*)

146. **Barata, R. A., M. H. G. Andrade, R. D. Rodrigues, and I. M. Castro.** 2002. Purification and characterization of an extracellular trypsin-like protease of *Fusarium oxysporum* var. *lini*. *Journal of Bioscience and Bioengineering* **94:** 304-308. (*oxysporum*)

147. **Barbosa, I. P., and C. Kemmelmeier.** 1993. Chemical composition of the hyphal wall from *Fusarium graminearum*. *Experimental Mycology* **17:** 274-283. (*graminearum*)

148. **Barbosa-Tessmann, I. P., D. A. da Silva, R. M. Peralta, and C. Kemmelmeier.** 2001. A new species of *Fusarium* producer of galactose oxidase. *Journal of Basic Microbiology* **41:** 143-148. (*subglutinans*)

149. **Barclay, M., J. C. Day, I. P. Thompson, C. J. Knowles, and M. J. Bailey.** 2002. Substrate-regulated cyanide hydratase (*chy*) gene expression in *Fusarium solani*: the potential of a transcription-based assay for monitoring the biotransformation of cyanide complexes. *Environmental Microbiology* **4:** 183-189. (*solani*)

150. **Barclay, M., A. Hart, C. J. Knowles, J. C. L. Meeussen, and V. A. Tett.** 1998. Biodegradation of metal cyanides by mixed and pure cultures of fungi. *Enzyme and Microbial Technology* **22:** 223-231. (*oxysporum, solani*)

151. **Barclay, M., V. A. Tett, and C. J. Knowles.** 1998. Metabolism and enzymology of cyanide/metallocyanide biodegradation by *Fusarium solani* under neutral and acidic conditions. *Enzyme and Microbial Technology* **23:** 321-330. (*solani*)

152. **Barua, A. S., J. Saha, S. Chaudhuri, A. Chowdhury, and N. Adityachaudhury.** 1990. Degradation of pendimethalin by soil fungi. *Pesticide Science* **29:** 419-426. (*solani*)

153. **Barve, M. P., M. P. Haware, M. N. Sainani, P. K. Ranjekar, and V. S. Gupta.** 2001. Potential of microsatellites to distinguish four races of *Fusarium oxysporum* f. sp. *ciceri* prevalent in India. *Theoretical and Applied Genetics* **102:** 138-147. (*oxysporum*)

154. **Bastos, C. N., and A. O. Santos.** 2001. Oversprouting of inflorescences of the limao de Caiena caused by *Fusarium decemcellulare*. *Fitopatologia Brasileira* **26:** 222. (*decemcellulare*)

155. **Bateman, G. L., H. Kwasna, and E. Ward.** 1996. Relationships among *Fusarium* spp. estimated by comparing restriction fragment length polymorphisms in polymerase chain reaction-amplified nuclear rDNA. *Canadian Journal of Microbiology* **42:** 1232-1240. (7)

156. **Bayman, P., and P. J. Cotty.** 1991. Improved media for selecting nitrate nonutilizing mutants in *Aspergillus flavus*. *Mycologia* **83:** 311-316. (2, 5)

157. **Beatty, N. P., M. L. Smith, and N. L. Glass.** 1994. Molecular characterization of mating-type loci in selected homothallic species of *Neurospora*, *Gelasinospora*, and *Anixiella*. *Mycological Research* **98:** 1309-1316. (6)

158. **Becker, P., and K. M. Weltring.** 1998. Purification and characterization of α-chaconinase of *Gibberella pulicaris*. *FEMS Microbiology Letters* **167:** 197-202. (*sambucinum*)

159. **Beckman, C. H.** 1987. *The Nature of Wilt Diseases in Plants*. APS Press, St. Paul, Minnesota. (*oxysporum*)

160. **Belabid, L., and Z. Fortas.** 2002. Virulence and vegetative compatibility of Algerian isolates of *Fusarium oxysporum* f. sp. *lentis*. *Phytopathologia Mediterranea* **41:** 179-187. (*oxysporum*)

161. **Belien, T., S. van Campenhout, M. van Acker, and G. Volckaert.** 2005. Cloning and characterization of two endoxylanases from the cereal phytopathogen *Fusarium graminearum* and their inhibition profile against endoxylanase inhibitors from wheat. *Biochemical and Biophysical Research Communications* **327:** 407-414. (*graminearum*)

162. **Bell, A. A., M. H. Wheeler, J. Liu, R. D. Stipanovic, L. S. Puckhaber, and H. Orta.** 2003. United States Department of Agriculture-Agricultural Research Service studies on polyketide toxins of *Fusarium oxysporum* f. sp. *vasinfectum*: Potential targets for disease control. *Pest Management Science* **59:** 736-747. (*oxysporum*)

163. **Bell, T. A., D. V. Lightner, and R. R. Williams.** 1990. A promising new chemotherapeutant for use in treatment of *Fusarium solani* infections in penaeid shrimp. *Journal of the World Aquaculture Society* **21:** 131-136. (*solani*)

164. **Benhamou, N., C. Garand, and A. Goulet.** 2002. Ability of nonpathogenic *Fusarium oxysporum* strain Fo47 to induce resistance against *Pythium ultimum* infection in cucumber. *Applied and Environmental Microbiology* **68:** 4044-4060. (*oxysporum*)

165. **Benigno, D. R. A., and F. C. Querbal.** 1977. *Host Index of Plant Diseases in the Philippines*. University of the Philippines, Los Baños, the Philippines. (11)

166. **Bentley, S., and B. J. Bassam.** 1996. A robust DNA amplification fingerprinting system applied to analysis of genetic variation within *Fusarium oxysporum* f. sp. *cubense*. *Journal of Phytopathology* **144:** 207-213. (*oxysporum*)

167. **Bentley, S., K. G. Pegg, N. Y. Moore, R. D. Davis, and I. W. Buddenhagen.** 1998. Genetic variation among vegetative compatibility groups of *Fusarium oxysporum* f. sp. *cubense* analyzed by DNA fingerprinting. *Phytopathology* **88:** 1283-1293. (5)

168. **Benyon, F. H. L., and L. W. Burgess.** 1997. Characterization of *Fusarium avenaceum* subspecies *avenaceum, nurragi*, and *aywerte*. *Cereal Research Communications* **25:** 579-580. (*armeniacum*)

169. **Benyon, F. H. L., L. W. Burgess, and P. J. Sharp.** 2000. Molecular genetic investigations and reclassification of *Fusarium* species in sections *Fusarium* and *Roseum*. *Mycological Research* **104:** 1164-1174. (8, 9, *armeniacum, avenaceum, aywerte, nurragi, pseudograminearum*)

170. **Benyon, F., B. A. Summerell, and L. W. Burgess.** 1996. Association of *Fusarium* species with root rot of *Cymbidium* orchids. *Australasian Plant Pathology* **25:** 226-228. (*proliferatum, solani, subglutinans*)

171. **Beremand, M. N.** 1987. Isolation and characterization of mutants blocked in T-2 toxin biosynthesis. *Applied and Environmental Microbiology* **53:** 1855-1859. (*sporotrichioides*)

172. **Beremand, M. N.** 1989. Genetic and mutational research tools for investigating the genetics and molecular biology of trichothecene production in *Gibberella pulicaris* (*Fusarium sambucinum*). *Mycopathologia* **107:** 67-74. (*sambucinum*)

173. **Beremand, M. N., and A. E. Desjardins.** 1988. Trichothecene biosynthesis in *Gibberella pulicaris*: Inheritance of C-8 hydroxylation. *Journal of Industrial Microbiology* **3:** 167-174. (*sambucinum*)

174. **Beremand, M. N., A. E. Desjardins, T. M. Hohn, and F. L. van Middlesworth.** 1991. Survey of *Fusarium sambucinum* (*Gibberella pulicaris*) for mating type, trichothecene production, and other selected traits. *Phytopathology* **81:** 1452-1458. (*sambucinum*)

175. **Beremand, M. N., F. van Middlesworth, S. Taylor, R. D. Plattner, and D. Weisleder.** 1988. Leucine auxotrophy specifically alters the pattern of trichothecene production by a T-2 toxin-producing strain of *Fusarium sporotrichioides*. *Applied and Environmental Microbiology* **54:** 2759-2766. (*sporotrichioides*)

176. **Berka, R. M., M. W. Rey, K. M. Brown, T. Byun, and A. V. Klotz.** 1998. Molecular characterization and expression of a phytase gene from the thermophilic fungus *Thermomyces lanuginosus*. *Applied and Environmental Microbiology* **64:** 4423-4427. (*venenatum*)

177. **Berkeley, M. J.** 1875. Notices of the North American Fungi. *Grevillea* **3:** 97-112. (*semitectum*)

178. **Berrocal-Lobo, M., and A. Molina.** 2004. Ethylene response factor 1 mediates *Arabidopsis* resistance to the soil-borne fungus *Fusarium oxysporum*. *Molecular Plant-Microbe Interactions* **17:** 763-770. (*oxysporum*)

179. **Beyer, M., S. Roeding, A. Ludewig, and J. A. Verreet.** 2004. Germination and survival of *Fusarium graminearum* macroconidia as affected by environmental factors. *Journal of Phytopathology* **152:** 92-97. (*graminearum*)

180. **Beyer, M., J. A. Verreet, and W. S. M. Ragab.** 2005. Effect of relative humidity on germination of ascospores and macroconidia of *Gibberella zeae* and deoxynivalenol production. *International Journal of Food Microbiology* **98:** 233-240. (*graminearum*)

181. **Bigley, V. H., R. F. Duarte, R. D. Gosling, C. C. Kibbler, S. Seaton, and M. Potter.** 2004. *Fusarium dimerum* infection in a stem cell transplant recipient treated successfully with voriconazole. *Bone Marrow Transplantation* **34:** 815-817. (*dimerum*)

182. **Bilai, V. I.** 1952. Taxonomy of the *Sporotrichiella* section of the genus *Fusarium*. *Dopov. Akad. Nauk. Ukr. SSR* **5:** 415-419. (8)

183. **Bilai, V. I.** 1955. *The Fusaria (Biology and Systematics)*. Publ. Acad. Sci. Ukr. SSR, Kiev, USSR. (8)

184. **Bilai, V. I.** 1970. Experimental morphogenesis in the fungi of the genus *Fusarium* and their taxonomy. *Ann. Acad. Sciences Fenn. A, IV Biologica* **168:** 7-18. (8)

185. **Bilai, V. I.** 1978. Principles of systematics and structure of phytopathogenic species of the genus *Fusarium*. *Mikrobiologicheskii Zhurnal* **40:** 148-156. (8)

186. **Billich, A., U. Keller, H. Kleinhauf, and R. Zocher.** 1988. Production of protoplasts from *Fusarium scirpi* by lytic enzymes from *Streptomyces tsusimaensis*. *Applied Microbiology and Biotechnology* **28:** 442-444. (*scirpi*)

187. **Billich, A., and R. Zocher.** 1988. Constitutive expression of enniatin synthetase during fermentative growth of *Fusarium scirpi*. *Applied and Environmental Microbiology* **54:** 2504-2509. (*scirpi*)

188. **Bisakowski, B., S. Kermasha, and E. Spinnler.** 1998. Characterization of purified lipoxygenase extracts from *Fusarium proliferatum*. *Journal of Agricultural and Food Chemistry* **46:** 2382-2388. (*proliferatum*)

189. **Bisen, P. S., K. Ghosh, and G. P. Agarwal.** 1982. Induction and inhibition of cellulase complex in *Fusarium solani*. *Biochemie und Physiologie der Pflanzen* **177:** 593-599. (*solani*)

190. **Bishop, D. L.** 2002. Gene expression of a vacuolar peroxidase with stress-induced pathogenesis in wheat sheaths. *Physiological and Molecular Plant Pathology* **61:** 65-71. (*proliferatum*)

191. **Bishop, D. L., N. J. Chatterton, P. A. Harrison, and R. D. Hatfield.** 2002. Changes in carbohydrate coordinated partitioning and cell wall remodeling with stress-induced pathogenesis in wheat sheaths. *Physiological and Molecular Plant Pathology* **61:** 53-63. (*proliferatum*)

192. **Bishop, G. C., and A. H. Ilsley.** 1978. Production of enniatin as a criterion for confirming the identity of *Fusarium lateritium* isolates. *Australian Journal of Biological Sciences* **31:** 93-96. (*lateritium*)

193. **Bisht, D., G. S. Bisht, and R. D. Khulbe.** 2000. *Fusarium*: A new threat to fish population in reservoirs of Kumaun, India. *Current Science* **78:** 1241-1245. (*udum, verticillioides*)

194. **Bistis, G.** 1981. Chemotropic interactions between trichogynes and conidia of opposite mating type in *Neurospora crassa*. *Mycologia* **73:** 959-975. (6, 10)

195. **Blackwell, B. A., J. D. Miller, and M. E. Savard.** 1994. Production of carbon 14-labeled fumonisin in liquid culture. *Journal of AOAC International* **77:** 506-511. (*verticillioides*)

196. **Blakeslee, A. F.** 1904. Sexual reproduction in the Mucorineae. *Proceedings of the American Academy of Arts and Sciences* **40:** 205-319. (6)

197. **Blaney, B. J., and R. L. Dodman.** 2002. Production of zearalenone, deoxynivalenol, nivalenol, and acetylated derivatives by Australian isolates of *Fusarium graminearum* and *F. pseudograminearum* in relation to source and culturing conditions. *Australian Journal of Agricultural Research* **53:** 1317-1326. (*pseudograminearum*)

198. **Blaney, B. J., P. E. Green, and M. D. Connole.** 1985. Fungal metabolites with insecticidal activity: relative toxicity of extracts of fungal cultures to sheep blowfly, *Lucilia cuprina* (Wied.). *General and Applied Entomology* **17:** 42-46. (*semitectum*)

199. **Blinkovsky, A. M., T. Byun, K. M. Brown, and E. J. Golightly.** 1999. Purification, characterization, and heterologous expression in *Fusarium venenatum* of a novel serine carboxypeptidase from *Aspergillus oryzae*. *Applied and Environmental Microbiology* **65:** 3298-3303. (*venenatum*)

200. **Blinkovsky, A. M., T. Byun, K. M. Brown, E. J. Golightly, and A. V. Klotz.** 2000. A non-specific aminopeptidase from *Aspergillus*. *Biochimica et Biophysica Acta* **1480:** 171-181. (*venenatum*)

201. **Blish, B. J., M. L. Carson, M. A. Cubeta, and W. M. Hagler, G. A. Payne.** 2004. Infection and fumonisin production by *Fusarium verticillioides* in developing maize kernels. *Phytopathology* **94:** 88-93. (*verticillioides*)

202. **Blodgett, J. T., W. J. Swart, and S. Louw.** 1998. First report of *Fusarium sambucinum*, *F. oxysporum*, and *F. subglutinans* associated with stem decay of *Amaranthus hybridus* in South Africa. *Plant Disease* **82:** 1062. (*sambucinum*)

203. **Blodgett, J. T., W. J. Swart, and S. Louw**. 2004. Identification of fungi and fungal pathogens associated with *Hypolixus haerens* and decayed and cankered stems of *Amaranthus hybridus*. *Plant Disease* **88:** 333-337. (*sambucinum*)

204. **Blok, W. J., and G. J. Bollen.** 1997. Host specificity and vegetative compatibility of Dutch isolates of *Fusarium oxysporum* f. sp. *asparagi*. *Canadian Journal of Botany* **75:** 383-393. (*oxysporum*)

205. **Blok, W. J., M. J. Zwankhuizen, and G. J. Bollen.** 1997. Biological control of *Fusarium oxysporum* f. sp. *asparagi* by applying nonpathogenic isolates of *F. oxysporum*. *Biocontrol Science and Technology* **7:** 527-541. (*oxysporum*)

206. **Bloomberg, W. J.** 1979. Model simulations of infection of Douglas fir (*Pseudotsuga menziesii*) seedlings by *Fusarium oxysporum*. *Phytopathology* **69:** 1072-1077. (11)

207. **Bloomberg, W. J.** 1988. Modeling applied to the analysis of crop pest management interactions. *Memoirs of the Entomological Society of Canada* **143:** 29-38. (11)

208. **Bluhm, B. H., M. A. Cousin, and C. P. Woloshuk.** 2004. Multiplex real-time PCR detection of fumonisin-producing and trichothecene-producing groups of *Fusarium* species. *Journal of Food Protection* **67:** 536-543. (*graminearum*, *verticillioides*)

209. **Boari, A., and M. Vurro.** 2004. Evaluation of *Fusarium* spp. and other fungi as biological control agents of broomrape (*Orobanche ramosa*). *Biological Control* **30:** 212-219. (*oxysporum*)

210. **Bodey, G. P., M. Boktour, S. Mays, M. Duvic, D. Kontoyiannis, R. Hachem, and I. Raad.** 2002. Skin lesions associated with *Fusarium* infection. *Journal of the American Academy of Dermatology* **47:** 659-666. (*dimerum*, *oxysporum*, *proliferatum*, *semitectum*, *solani*, *verticillioides*)

211. **Boehm, E. W. A., R. C. Ploetz, and H. C. Kistler.** 1994. Statistical analysis of electrophoretic karyotype variation among vegetative compatibility groups of *Fusarium oxysporum* f. sp. *cubense*. *Molecular Plant-Microbe Interactions* **7:** 196-207. (5, 7, *oxysporum*)

212. **Bojja, R. S., R. L. Cerny, R. H. Proctor, and L. Du.** 2004. Determining the biosynthetic sequence in the early steps the fumonisin pathway by use of three gene-disruption mutants of *Fusarium verticillioides*. *Journal of Agricultural and Food Chemistry* **52:** 2855-2860. (*verticillioides*)

213. **Boland, G. J., and G. D. Inglis.** 1989. Antagonism of white mold, *Sclerotinia sclerotiorum*, of bean by fungi from bean and rapeseed flowers. *Canadian Journal of Botany* **67:** 1775-1781. (*heterosporum*)

214. **Bolkan, H. A., J. C. Dianese, and F. P. Cupertino.** 1979. Survival and colonization potential of *Fusarium moniliforme* var. *subglutinans* in soil. *Phytopathology* **69:** 1298-1300. (*guttiforme*)

215. **Bolkan, H. A., J. C. Dianese, and F. P. Cupertino.** 1979. Pineapple flowers as principal infection sites for *Fusarium moniliforme* var. *subglutinans*. *Plant Disease Reporter* **63:** 655-657. (*guttiforme*)

216. **Bolkan, H. A., J. C. Dianese, C. B. da Silva, J. C. A. D. Araujo.** 1982. Influence of carbon source, light, water potential and temperature on growth and sporulation of *Fusarium moniliforme* var. *subglutinans*. *Revista de Microbiologia* **13:** 264-271. (*guttiforme*)

217. **Bonnin, E., A. le Goff, R. Korner, G. J. W. M. van Alebeek, T. M. I. E. Christensen, A. G. J. Voragen, P. Roepstorff, C. Caprari, and J. F. Thibault.** 2001. Study of the mode of action of endopolygalacturonase from *Fusarium moniliforme*. *Biochimica et Biophysica Acta* **1526:** 301-309. (*verticillioides*)

218. **Bonnin, E., A. le Goff, R. Korner, J. Vigouroux, P. Roepstorff, and J. F. Thibault.** 2002. Hydrolysis of pectins with different degrees and patterns of methylation by the endopolygalacturonase of *Fusarium moniliforme*. *Biochimica et Biophysica Acta* **1596:** 83-94. (*verticillioides*)

219. **Bonnin, E., A. le Goff, G. J. W. M. van Alebeek, A. G. J. Voragen, and J. F. Thibault.** 2003. Mode of action of *Fusarium moniliforme* endopolygalacturonase towards acetylated pectin. *Carbohydrate Polymers* **52:** 381-388. (*verticillioides*)

220. **Boominathan, K., and A. Mahadevan.** 1989. Dissimilation of aromatic substances by fungi. *Zentralblatt für Mikrobiologie* **144:** 37-45. (*oxysporum*)

221. **Boonpasart, S., N. Kasetsuwan, V. Puangsricharern, L. Pariyakanok, and T. Jittpoonkusol.** 2002. Infectious kera-titis

at King Chulalongkorn Memorial Hospital: A-12-year retrospective study of 391 cases. *Journal of the Medical Association of Thailand* **85(Suppl. 1):** S217-S230. (1)

222. **Booth, C.** 1971. *The Genus* Fusarium. Commonwealth Mycological Institute, Kew, Surrey, United Kingdom. (4, 8, 10, *acuminatum, graminearum, heterosporum, lateritium, merismoides, oxysporum, poae, redolens, sacchari, sambucinum, scirpi, semitectum, solani, verticillioides*)

223. **Booth, C., and B. C. Sutton.** 1984. *Fusarium pallidoroseum*, the correct name for *F. semitectum* Auct. *Transactions of the British Mycological Society* **83:** 702-704. (*semitectum*)

224. **Borras, O., R. Santos, A. P. Matos, R. S. Cabral, and M. Arzola.** 2001. A first attempt to use a *Fusarium subglutinans* culture filtrate for the selection of pineapple cultivars resistant to fusariose disease. *Plant Breeding* **120:** 435-438. (*guttiforme*)

225. **Boshoff, W. H. P., Z. A. Pretorius, and W. J. Swart.** 1999. A comparison of head infection and blight development caused by *Fusarium graminearum* and *Fusarium crookwellense* in wheat. *South African Journal of Plant and Soil* **16:** 79-84. (*crookwellense*)

226. **Bossi, P., E. Mortier, C. Michon, H. Gaudin, A. M. Simonpoli, J. Pouchot, and P. Vinceneux.** 1995. *Fusarium solani* sinusitis in a patient with AIDS. *Journal de Mycologie Medicale* **5:** 56-57. (*solani*)

227. **Bourett, T. M., J. A. Sweigard, K. J. Czymmek, A. Carroll, and R. J. Howard.** 2002. Reef coral fluorescent proteins for visualizing fungal pathogens. *Fungal Genetics and Biology* **37:** 211-220. (*verticillioides*)

228. **Bourne, B. A.** 1953. Studies on the dissemination of sugar cane diseases. *Sugar Journal* **16:** 19, 22 (*sacchari*)

229. **Bowden, R. L., and J. F. Leslie.** 1992. Nitrate-nonutilizing mutants of *Gibberella zeae* (*Fusarium graminearum*) and their use in determining vegetative compatibility. *Experimental Mycology* **16:** 307-315. (5, 6, 9, *graminearum*)

230. **Bowden, R. L., and J. F. Leslie.** 1999. Sexual recombination in *Gibberella zeae* (*Fusarium graminearum*). *Phytopathology* **89:** 182-188. (2, 6, 9, 11, *graminearum*)

231. **Boyette, C. D., G. E. Templeton, and L. R. Oliver.** 1984. Texas Gourd (*Cucurbita texana*) control with *Fusarium solani* f. sp. *cucurbitae*. *Weed Science* **32:** 649-655. (*solani*)

232. **Bragulat, M. R., E. Martinez, G. Castellá, and F. J. Cabañes.** 2004. Selective efficacy of culture media recommended for isolation and enumeration of *Fusarium* spp. *Journal of Food Protection* **67:** 207-211. (2)

233. **Bramel-Cox, P. J., I. S. Stein, D. M. Rodgers, and L. E. Claflin.** 1988. Inheritance of resistance to *Macrophomina phaseolina* Tassi Goid. and *Fusarium moniliforme* Sheldon in sorghum. *Crop Science* **28:** 37-40. (*thapsinum*)

234. **Brandao, R. L., J. R. Nicol, and A. F. D. S. Figueiredo.** 1987. Purification and characterization of a β-galactosidase from *Fusarium oxysporum* var. *lini*. *Journal of Dairy Science* **70:** 1331-1337. (*oxysporum*)

235. **Branham, B. E., and R. D. Plattner.** 1993. Isolation and characterization of a new fumonisin from liquid cultures of *Fusarium moniliforme*. *Journal of Natural Products* **56:** 1630-1633. (*verticillioides*)

236. **Branham, B. E., and R. D. Plattner.** 1993. Alanine is a precursor in the biosynthesis of fumonisin B_1 by *Fusarium moniliforme*. *Mycopathologia* **124:** 99-104. (*verticillioides*)

237. **Brewbaker, J. L.** 1998. Disease-resistant tropical supersweet corn populations. *Hortscience* **33:** 1262-1264. (*verticillioides*)

238. **Bridge, P. D., L. A. Hopkinson, and M. A. Rutherford.** 1995. Rapid mitochondrial probes for analysis of polymorphisms in *Fusarium oxysporum* special forms. *Letters in Applied Microbiology* **21:** 198-201. (*oxysporum*)

239. **Bridge, P. D., M. A. Ismail, and M. A. Rutherford.** 1993. An assessment of aesculin hydrolysis, vegetative compatibility and DNA polymorphism as criteria for characterizing pathogenic races within *Fusarium oxysporum* f. sp. *vasinfectum*. *Plant Pathology* **42:** 264-269. (*oxysporum*)

240. **Brill, G. M., W. M. Kati, D. Montgomery, J. P. Karwowski, P. E. Humphrey, M. Jackson, J. J. Clement, S. Kadam, R. H. Chen, and J. B. McAlpine.** 1996. Novel triterpene sulfates from *Fusarium compactum* using a rhinovirus 3C protease inhibitor screen. *Journal of Antibiotics* **49:** 541-546. (*compactum*)

241. **Britz, H., T. A. Coutinho, B. D. Wingfield, W. F. O. Marasas, and M. J. Wingfield.** 2005. Diversity and differentiation in two populations of *Gibberella circinata* in South Africa. *Plant Pathology* **54:** 46-52. (*circinatum*)

242. **Britz, H., T. A. Coutinho, M. J. Wingfield, and W. F. O. Marasas.** 2002. Validation of the description of *Gibberella circinata* and morphological differentiation of the anamorph *Fusarium circinatum*. *Sydowia* **54:** 9-22. (8, 9, *circinatum, subglutinans*)

243. **Britz, H., T. A. Coutinho, M. J. Wingfield, W. F. O. Marasas, T. R. Gordon, and J. F. Leslie.** 1999. *Fusarium subglutinans* f. sp. *pini* represents a distinct mating population in the *Gibberella fujikuroi* species complex. *Applied and Environmental Microbiology* **65:** 1198-1201. (6, 8, 9, *circinatum, sacchari*)

244. **Britz, H., E. T. Steenkamp, T. A. Coutinho, B. D. Wingfield, W. F. O. Marasas, and M. J. Wingfield.** 2002. Two new species of *Fusarium* section *Liseola* associated with mango malformation. *Mycologia* **94:** 722-730. (8, *mangiferae, pseudocircinatum, sterilihyphosum, subglutinans*)

245. **Britz, H., B. D. Wingfield, T. A. Coutinho, and M. J. Wingfield.** 2002. Sequence characterized amplified polymorphic markers for the pitch canker pathogen, *Fusarium circinatum*. *Molecular Ecology Notes* **2:** 577-580. (*circinatum*)

246. **Britz, H., M. J. Wingfield, T. A. Coutinho, W. F. O. Marasas, and J. F. Leslie.** 1998. Female fertility and mating type distribution in a South African population of *Fusarium subglutinans* f. sp. *pini*. *Applied and Environmental Microbiology* **64:** 2094-2095. (6, 9, *circinatum*)

247. **Broadhurst, P. G.** 1990. *Fusarium graminearum* causing stub dieback of carnations in New Zealand. *New Zealand Journal of Crop and Horticultural Science* **18:** 137-140. (*avenaceum, culmorum, graminearum*)

248. **Broadhurst, P. G., and P. R. Johnston.** 1994. *Gibber-ella tumida sp. nov.* – teleomorph of *Fusarium tumidum* from gorse in New Zealand. *Mycological Research* **98:** 729-732. (10)

249. **Brody, S.** 1992. Circadian rhythms in *Neurospora crassa*: The role of mitochondria. *Chronobiology Interna-tional* **9:** 222-230. (4)

250. **Brooker, N. L., J. F. Leslie, and M. B. Dickman.** 1991. Nitrate nonutilizing mutants of *Colletotrichum* and their use in studies of vegetative compatibility and genetic relatedness. *Phytopathology* **81:** 672-677. (2)

251. **Broomhead, J. N., D. R. Ledoux, A. J. Bermudez, and G. E. Rottinghaus.** 2002. Chronic effects of moniliformin in broilers and turkeys fed dietary treatments to market age. *Avian Diseases* **46:** 901-908. (*fujikuroi*)

252. **Brown, D. T., P. D. Turner, and C. O'Reilly.** 1995. Expression of the cyanide hydratase enzyme from *Fusarium lateritium* in *Escherichia coli* and identification of an essential cysteine residue. *FEMS Microbiology Letters* **134:** 143-146. (*lateritium*)

253. **Brown, D. W., R. B. Dyer, S. P. McCormick, D. F. Kendra, and R. D. Plattner.** 2004. Functional demarcation of the *Fusarium* core trichothecene gene cluster. *Fungal Genetics and Biology* **41:** 454-462. (*graminearum, sporotrichioides*)

254. **Brown, D. W., S. P. McCormick, N. J. Alexander, R. H. Proctor, and A. E. Desjardins.** 2002. Inactivation of a cytochrome P-450 is a determinant of trichothecene diversity in *Fusarium* species. *Fungal Genetics and Biology* **36:** 224-233. (*sporotrichioides*)

255. **Brown, D. W., R. H. Proctor, R. B. Dyer, and R. D. Plattner.** 2003. Characterization of a *Fusarium* two-gene cluster involved in trichothecene C-8 modification. *Journal of Agricultural and Food Chemistry* **51:** 7936-7944. (*sporotrichioides*)

256. **Brown, R. L., T. E. Cleveland, C. P. Woloshuk, G. A. Payne, and D. Bhatnagar.** 2001. Growth inhibition of a *Fusarium verticillioides* GUS strain in corn kernels of aflatoxin-resistant genotypes. *Applied Microbiology and Biotechnology* **57:** 708-711. (*verticillioides*)

257. **Bryden, W. L., A. Logrieco, H. K. Abbas, J. K. Porter, R. F. Vesonder, J. L. Richard, and R. J. Cole.** 2001. Other significant *Fusarium* mycotoxins, p. 360-392. *In* B. A. Summerell, J. F. Leslie, D. Backhouse, W. L. Bryden, and L. W. Burgess (eds.), Fusarium*: Paul E. Nelson Memorial Symposium.* APS Press, St. Paul, Minnesota. (*thapsinum*)

258. **Brückner, B., S. E. Unkles, K. Weltring, and J. R. Kinghorn.** 1992. Transformation of *Gibberella fujikuroi*: Effect of the *Aspergillus nidulans AMA1* sequence on frequency and integration. *Current Genetics* **22:** 313-316. (*fujikuroi*)

259. **Buerstmayr, H., L. Legzdina, B. Steiner, and M. Lemmens.** 2004. Variation for resistance to Fusarium head blight in spring barley. *Euphytica* **137:** 279-290. (*graminearum*)

260. **Bu'Lock, J. D., and S. Chulze.** 1990. Effect of a regulatory mutation on trichothecene production by *Fusarium graminearum* strain NRRL 3198. *Mycological Research* **94:** 851-853. (*graminearum*)

261. **Bu'Lock, J. D., C. E. Wright, and J. E. Mooney.** 1986. Use of a protoplast fusion test to establish the status of mycotoxin genes in an edible *Fusarium*. *Biotechnology Letters* **8:** 621-624. (*venenatum*)

262. **Bullerman, L. B., and W. Y. J. Tsai.** 1994. Incidence and levels of *Fusarium moniliforme*, *Fusarium proliferatum* and fumonisins in corn and corn-based foods and feeds. *Journal of Food Protection* **57:** 541-546. (*verticillioides*)

263. **Burge, H. P., J. R. Boise, J. A. Rutherford, and W. R. Solomon.** 1977. Comparative recoveries of airborne fungus spores by viable and nonviable modes of volumetric collection. *Mycopathologia* **61:** 27-34. (3)

264. **Burge, H. P., J. R. Boise, W. R. Solomon, and E. Bandera.** 1978. Fungi in libraries – An aerometric study. *Mycopathologia* **64:** 67-72. (3)

265. **Burgess, L. W.** 1981. General ecology, p. 225-235. *In* P. E. Nelson, T. A. Toussoun, and R. J. Cook (eds.), Fusarium*: Diseases, Biology and Taxonomy*. Pennsylvania State University Press, University Park, Pennsylvania. (9)

266. **Burgess, L. W., D. Backhouse, B. A. Summerell, A. B. Pattison, T. A. Klein, R. J. Esdaile, and G. Ticehurst.** 1993. Long-term effects of stubble management on the incidence of infection of wheat by *Fusarium graminearum* Schw. Group 1. *Australian Journal of Experimental Agriculture* **33:** 451-456. (*pseudograminearum*)

267. **Burgess, L. W., D. Backhouse, B. A. Summerell, and L. J. Swan.** 2001. Crown rot of wheat, p. 271-294. *In* B. A. Summerell, J. F. Leslie, D. Backhouse, W. L. Bryden, and L. W. Burgess (eds.), Fusarium*: Paul E. Nelson Memorial Symposium.* APS Press, St. Paul, Minnesota. (*pseudograminearum*)

268. **Burgess, L. W., D. Backhouse, L. J. Swan, and R. J. Esdaile.** 1996. Control of Fusarium crown rot of wheat by late stubble burning and rotation with sorghum. *Australasian Plant Pathology* **25:** 229-233. (2)

269. **Burgess, L. W., R. L. Dodman, W. Pont, and P. Mayers.** 1981. Fusarium diseases of wheat, maize and grain sorghum in eastern Australia, p. 64-76. *In* P. E. Nelson, T. A. Toussoun, and R. J. Cook (eds.), Fusarium*: Diseases, Biology and Taxonomy*. Pennsylvania State University Press, University Park, Pennsylvania. (*pseudograminearum*)

270. **Burgess, L. W., G. A. Forbes, C. Windels, P. E. Nelson, W. F. O. Marasas, and K. P. Gott.** 1993. Characterization and distribution of *Fusarium acuminatum* subsp. *armeniacum*, subsp. nov. *Mycologia* **85:** 119-124. (8, *acuminatum, armeniacum*)

271. **Burgess, L. W., T. A. Klein, W. L. Bryden, and N. F. Tobin.** 1987. Head blight of wheat caused by *Fusarium graminearum* Group 1 in New South Wales in 1983. *Australasian Plant Pathology* **16:** 72-78. (*pseudograminearum*)

272. **Burgess, L. W., P. E. Nelson, and B. A. Summerell.** 1989. Variability and stability of morphological characters in *Fusarium oxysporum*. *Mycologia* **81:** 818-822. (2)

273. **Burgess, L. W., P. E. Nelson, and T. A. Toussoun.** 1982. Characterization, geographic distribution, and ecology of *Fusarium crookwellense sp. nov. Transactions of the British Mycological Society* **79:** 497-505. (*crookwellense*)

274. **Burgess, L. W., P. E. Nelson, and T. A. Toussoun.** 1989. Stability of morphological characters of *Fusarium nygamai*. *Mycologia* **81:** 480-482. (*nygamai*)

275. **Burgess, L. W., P. E. Nelson, T. A. Toussoun, and G. A. Forbes.** 1988. Distribution of *Fusarium* species in sections *Roseum*, *Arthrosporiella*, *Gibbosum*, and *Discolor* recovered from grassland, pasture, and pine nursery soils of eastern Australia. *Mycologia* **80:** 815-824. (*armeniacum, longipes, semitectum*)

276. **Burgess, L. W., P. E. Nelson, T. A. Toussoun, and W. F. O. Marasas.** 1985. *Fusarium scirpi*: Emended description and notes on geographic distribution. *Mycologia* **77:** 212-218. (*equiseti, scirpi*)

277. **Burgess, L. W., and B. A. Summerell.** 1992. Mycogeography of *Fusarium*: Survey of *Fusarium* species from subtropical and semi-arid grassland soils from Queensland, Australia. *Mycological Research* **96:** 780-784. (*chlamydosporum, compactum, longipes, napiforme, nygamai, scirpi, solani*)

278. **Burgess, L. W., and B. A. Summerell.** 2000. Taxonomy of *Fusarium*: *Fusarium acuminatum* stat. & comb. nov. *Mycotaxon* **75:** 347-348. (8, *acuminatum*)

279. **Burgess, L. W., B. A. Summerell, S. Bullock, K. P. Gott, and D. Backhouse.** 1994. *Laboratory Manual for Fusarium Research*, 3rd ed. University of Sydney/Royal Botanic Gardens, Sydney, Australia. (P, 4, 8, 9)

280. **Burgess, L. W., B. A. Summerell, and P. E. Nelson.** 1991. An evaluation of several media for use in identification of some *Fusarium* spp. *Australasian Plant Pathology* **20:** 86-88. (2)

281. **Burgess, L. W., and D. Trimboli.** 1986. Characterization and distribution of *Fusarium nygamai*, sp. nov. *Mycologia* **78:** 223-229. (8, *nygamai, verticillioides*)

282. **Burmeister, H. R., M. D. Grove, R. E. Peterson, D. Weisleder, and R. D. Plattner.** 1985. Isolation and characterization of two new fusaric acid analogs from *Fusarium moniliforme* NRRL 13163. *Applied and Environmental Microbiology* **50:** 311-314. (*verticillioides*)

283. **Burmeister, H. R., and R. D. Plattner.** 1987. Enniatin production by *Fusarium tricinctum* and its effect on germinating wheat seeds. *Phytopathology* **77:** 1483-1487. (*tricinctum*)

284. **Burmeister, H. R., and R. F. Vesonder.** 1990. Production of a novel steroid sulfate metabolite 4,4,24-trimethyl-cholesta-8,14,24-28-trien-2-α-3-β-11-α-12-β-tetrol-12-acetate-3-sulfate by *Fusarium* species and its biological activity. *Applied and Environmental Microbiology* **56:** 3209-3212. (*acuminatum, chlamydosporum, culmorum, equiseti, graminearum, sambucinum, sporotrichioides*)

285. **Burmester, J., A. Haese, and R. Zocher.** 1995. Highly conserved methyltransferases as an integral part of peptide synthetases. *Biochemistry and Molecular Biology International* **37:** 201-207. (*sambucinum, scirpi*)

286. **Burnett, J. H.** 1975. *Mycogenetics*. John Wiley, London. (4)

287. **Bushelman, S. J., J. P. Callen, D. N. Roth, L. M. Cohen.** 1995. Disseminated *Fusarium solani* infection. *Journal of the American Academy of Dermatology* **32:** 346-351. (*solani*)

288. **Bushnell, G. W., G. A. Poulton, and Y. L. Li.** 1984. Pyrones 10. Lateropyrone, a new antibiotic from the fungus *Fusarium lateritium*. *Canadian Journal of Chemistry* **62:** 2101-2106. (*lateritium*)

289. **Butchko, R. E., R. D. Plattner, and R. H. Proctor.** 2003. *FUM9* is required for C-5 hydroxylation of fumonisins and complements the meitotically defined *Fum3* locus in *Gibberella moniliformis*. *Applied and Environmental Microbiology* **69:** 6935-6937. (*verticillioides*)

290. **Butchko, R. A. E., R. D. Plattner, and R. H. Proctor.** 2003. *FUM13* encodes a short chain dehydrogenase/reductase required for C-3 carbonyl reduction during fumonisin biosynthesis in *Gibberella moniliformis*. *Journal of Agricultural and Food Chemistry* **51:** 3000-3006. (*verticillioides*)

291. **Butler, E. J., and A. H. Khan.** 1913. Some new sugarcane diseases. *Memoirs of the Department of Agriculture in India, Botanical series* **6:** 185-190. (*sacchari*)

292. **Cabañes, F. J., J. M. Alonso, G. Castellá, F. Alegre, M. Domingo, and S. Pont.** 1997. Cutaneous hyalohyphomycosis caused by *Fusarium solani* in a loggerhead sea turtle (*Caretta caretta* L.). *Journal of Clinical Microbiology* **35:** 3343-3345. (*solani*)

293. **Caesar, A. J.** 1996. Identity, pathogenicity, and comparative virulence of *Fusarium* spp. related to stand declines of leafy spurge (*Euphorbia esula*) in the Northern Plains. *Plant Disease* **80:** 1395-1398. (*oxysporum, proliferatum, solani*)

294. **Caesar, A. J., G. Campobasso, and G. Terragitti.** 1998. Identification, pathogenicity and comparative virulence of *Fusarium* spp. associated with diseased *Euphorbia* spp. in Europe. *Biocontrol Science and Technology* **8:** 313-319. (*proliferatum, solani*)

295. **Cai, G., L. R. Gale, R. W. Schneider, H. C. Kistler, R. M. Davis, K. S. Elias, E. M. Miyao.** 2003. Origin of race 3 of *Fusarium oxysporum* f. sp. *lycopersici* at a single site in California. *Phytopathology* **93:** 1014-1022. (11)

296. **Calado, C. R. C., G. E. Hamilton, J. M. S. Cabral, L. P. Fonseca, and A. Lyddiatt.** 2001. Direct product sequestration of a recombinant cutinase from batch fermentations of *Saccharomyces cerevisiae*. *Bioseparation* **10:** 87-97. (*solani*)

297. **Calado, C. R. C., S. M. S. Monteiro, J. M. S. Cabral, and L. P. Fonseca.** 2002. Effect of pre-fermentation on the production of cutinase by a recombinant *Saccharomyces cerevisiae*. *Journal of Bioscience and Bioengineering* **93:** 354-359. (*solani*)

298. **Calado, C. R. C., M. A. Taipa, J. M. S. Cabral, and L. P. Fonseca.** 2002. Optimization of culture conditions and characterization of cutinase produced by recombinant *Saccharomyces cerevisiae*. *Enzyme and Microbial Technology* **31:** 161-170. (*solani*)

299. **Caldas, E. D., K. Sadilkova, B. L. Ward, A. D. Jones, C. K. Winter, and D. G. Gilchrist.** 1998. Biosynthetic studies of fumonisin B_1 and AAL toxins. *Journal of Agricultural and Food Chemistry* **46:** 4734-4743. (*verticillioides*)

300. **Camargo de Morais, M. M., M. M. D. Maia, F. F. S. Borba, K. G. Melo, C. M. S. O. Santos, E. R. A. Reis, M. A. Morais, Jr., and J. L. Lima-Filho.** 2003. Oil/mineral-salts medium designed for easy recovery of extracellular lipase from *Fusarium oxysporum* AM3. *World Journal of Microbiology and Biotechnology* **19:** 17-20. (*oxysporum*)

301. **Camin, A. M., C. Michelet, T. Langanay, C. de Place, S. Chevrier, E. Gueho, and C. Guiguen.** 1999. Endocarditis due to *Fusarium dimerum* four years after coronary artery bypass grafting. *Clinical and Infectious Diseases* **28:** 150. (*dimerum*)

302. **Campbell, J. L., and B. C. Turner.** 1987. Recombination block in the spore killer region of *Neurospora*. *Genome* **29:** 129-135. (10)

303. **Candau, R., J. Avalos, and E. Cerda-Olmedo.** 1992. Regulation of gibberellin biosynthesis in *Gibberella fujikuroi*. *Plant Physiology* **100:** 1184-1188. (*fujikuroi*)

304. **Cane, D. E., J.-H. Shim, Q. Xue, B. C. Fitzsimmons, and T. M. Hohn.** 1995. Trichodiene synthase: Identification of active site residues by site-directed mutagenesis. *Biochemistry* **34:** 2480-2488. (*sambucinum*)

305. **Cane, D. E., Z. Wu, J. S. Oliver, and T. M. Hohn.** 1993. Overproduction of soluble trichodiene synthase from *Fusarium sporotrichioides* in *Escherichia coli*. *Archives of Biochemistry and Biophysics* **300:** 416-422. (*sporotrichioides*)

306. **Cane, D. E., Q. Xue, and B. C. Fitzsimmons.** 1996. Trichodiene synthase: Probing the role of the highly conserved aspartate-rich region by site-directed mutagenesis. *Biochemistry* **35:** 12369-12376. (*sporotrichioides*)

307. **Cantalejo, M. J., J. M. Carrasco, and E. Hernandez.** 1997. Relationship between the stages of development of *Fusarium moniliforme* ATCC 38932 and production of fusarin C. *Journal of Food Protection* **60:** 433-435. (*verticillioides*)

308. **Cantalejo, M. J., J. M. Carrasco, and E. Hernandez.** 1997. Fusarin C production by *Fusarium* spp. from Spain. *Journal of Food Protection* **60:** 837-842. (*oxysporum, poae, verticillioides*)

309. **Capasso, R., A. Evidente, A. Cutignano, M. Vurro, M. C. Zonno, and A. Bottalico.** 1996. Fusaric and 9,10-dehydrofusaric acids and their methyl esters from *Fusarium nygamai*. *Phytochemistry* **41:** 1035-1039. (*nygamai*)

310. **Caprari, C., C. Bergmann, Q. Migheli, G. Salvi, P. Albersheim, A. Darvill, F. Cervone, and G. de Lorenzo.** 1993. *Fusarium moniliforme* secretes four endopolygalacturonases derived from a single gene product. *Physiological and Molecular Plant Pathology* **43:** 453-462. (*verticillioides*)

311. **Caprari, C., B. Mattei, M. L. Basile, G. Salvi, V. Crescenzi, G. de Lorenzo, and F. Cervone.** 1996. Mutagenesis of endopolygalacturonase from *Fusarium moniliforme*: Histidine residue 234 is critical for enzymatic and macerating activities and not for binding to polygalacturonase-inhibiting protein (PGIP). *Molecular Plant-Microbe Interactions* **9:** 617-624. (*verticillioides*)

312. **Caprari, C., A. Richter, C. Bergmann, S. lo Cicero, G. Salvi, F. Cervone, and G. de Lorenzo.** 1993. Cloning and characterization of a gene encoding the endopolygalacturonase of *Fusarium moniliforme*. *Mycological Research* **97:** 497-505. (*verticillioides*)

313. **Carbone, I., and L. M. Kohn.** 1999. A method for designing primer sets for speciation studies in filamentous ascomycetes. *Mycologia* **91:** 553-556. (7, *solani*)

314. **Caracuel, Z., C. Casanova, M. I. G. Roncero, A. di Pietro, and J. Ramos.** 2003. pH response transcription factor PacC controls salt stress tolerance and expression of the P-type Na+-ATPase Ena1 in *Fusarium oxysporum*. *Eukaryotic Cell* **2:** 1246-1252. (*oxysporum*)

315. **Cardenas, F., M. S. de Castro, J. M. Sanchez-Montero, J. V. Sinisterra, M. Valmaseda, S. W. Elson, and E. Alvarez.** 2001. Novel microbial lipases: Catalytic activity in reactions in organic media. *Enzyme and Microbial Technology* **28:** 145-154. (*oxysporum, poae*)

316. **Cardinale, F., and A. Matta.** 2001. Basic xylanases from the fungal tomato pathogen *Fusarium oxysporum* f. sp. *lycopersici*. *Journal of Plant Pathology* **83:** 27-36. (*oxysporum*)

317. **Carlile, M. J.** 1996. The discovery of fungal sex horones. II. Antheridiol. *The Mycologist* **10:** 113-117. (6)

318. **Carrion, G., and A. Bonet.** 2004. Mycobiota associated with the coffee berry borer (Coleoptera: Scolytidae) and its galleries in fruit. *Annals of the Entomological Society of America* **97:** 492-499. (*heterosporum*)

319. **Carter, J. P., H. N. Rezanoor, D. Holden, A. E. Desjardins, R. D. Plattner, and P. Nicholson.** 2002. Variation in pathogenicity associated with the genetic diversity of *Fusarium graminearum*. *European Journal of Plant Pathology* **108:** 573-583. (*graminearum*)

320. **Carter, M. V.** 1983. Biological control of *Eutypa armeniacae*. 5. Guidelines for establishing routine wound protection in commercial apricot orchards. *Australian Journal of Experimental Agriculture and Animal Husbandry* **23:** 429-436. (*lateritium*)

321. **Carter, W. W.** 1979. Corky dry rot of cantaloupe caused by *Fusarium roseum* 'Semitectum'. *Plant Disease Reporter* **63:** 1080-1084. (*semitectum*)

322. **Carter, W. W.** 1981. Reevaluation of heated water dip as a post-harvest treatment for controlling surface and decay fungi of muskmelon fruits (*Cucumis melo*). *Hortscience* **16:** 334-335. (3)

323. **Carvalho, C. M. L., M. R. Aires-Barros, and J. M. S. Cabral.** 1999. Cutinase: From molecular level to bioprocess development. *Biotechnology and Bioengineering* **66:** 17-34. (*solani*)

324. **Casselton, L. A., and N. S. Olesnicky.** 1998. Molecular genetics of mating recognition in basidiomycete fungi. *Microbiology and Molecular Biology Reviews* **62:** 55-70. (6)

325. **Castagnola, E., A. Garaventa, M. Conte, A. Barretta, E. Faggi, and C. Viscoli.** 1993. Survival after fungaemia due to *Fusarium moniliforme* in a child with neuroblastoma. *European Journal of Clinical Microbiology and Infectious Diseases* **12:** 308-309. (*verticillioides*)

326. **Castellá, G., M. R. Bragulat, M. V. Rubiales, and F. J. Cabañes.** 1997. Malachite green agar, a new selective medium for *Fusarium*. *Mycopathologia* **137:** 173-178. (2)

327. **Castellá, G., G. P. Munkvold, P. Imerman, and W. G. Hyde.** 1999. Effects of temperature, incubation period and substrate on production of fusaproliferin by *Fusarium subglutinans* ITEM 2404. *Natural Toxins* **7:** 129-132. (*subglutinans*)

328. **Castor, L. L., and R. A. Frederiksen.** 1980. Fusarium head blight occurrence and effects on sorghum (*Sorghum bicolor*) yield and grain characteristics in Texas, USA. *Plant Disease* **64:** 1017-1019. (*thapsinum*)

329. **Castro, I. M., and M. C. Loureiro-Dias.** 1991. Glycerol utilization in *Fusarium oxysporum* var. *lini* regulation of transport and metabolism. *Journal of General Microbiology* **137:** 1497-1502. (*oxysporum*)

330. **Catcheside, D. E. A., and D. G. Catcheside.** 1979. Survival of *Neurospora* conidia on silica gel. *Neurospora Newsletter* **26:** 24-25. (4)

331. **Cerda-Olmedo, E., R. Fernandez-Martin, and J. Avalos.** 1994. Genetics and gibberellin production in *Gibberella fujikuroi*. *Antonie van Leeuwenhoek* **65:** 217-225. (*fujikuroi*)

332. **Chacon-Martinez, C. A., J. M. Anzola, A. Rojas, F. Hernandez, H. Junca, W. Ocampo, and P. del Portillo.** 2004. Identification and characterization of the α-L-arabinofuranosidase B of *Fusarium oxysporum* f. sp. *dianthi*. *Physiological and Molecular Plant Pathology* **64:** 201-208. (*oxysporum*)

333. **Chaisrisook, C., and J. F. Leslie.** 1990. A nuclear gene controlling perithecial pigmentation in *Gibberella fujikuroi* (*Fusarium moniliforme*). *Journal of Heredity* **81:** 189-192. (6, 10, *verticillioides*)

334. **Chakrabarti, A., P. K. Mukherjee, P. D. Sherkhane, A. S. Bhagwat, and N. B. K. Murthy.** 2001. A simple and rapid molecular method for distinguishing between races of *Fusarium oxysporum* f. sp. *ciceris* from India. *Current Science* **80:** 571-575. (*oxysporum*)

335. **Chakraborty, A., and P. K. S. Gupta.** 1995. Factors affecting cross protection of Fusarium wilt of pigeon pea by soilborne nonpathogenic fungi. *Phytoparasitica* **23:** 323-334. (*oxysporum, sacchari*)

336. **Chakraborty, A., and P. K. S. Gupta.** 1998. Management of Fusarium wilt of pigeonpea by pre-inoculation of plants with non-pathogenic *Fusaria*. *Annals of Plant Protection Sciences* **6:** 121-126. (*sacchari*)

337. **Chakraborty, S. K., and A. Bhattacharyyaa.** 1991. Degradation of butachlor by two soil fungi. *Chemosphere* **23:** 99-106. (*solani*)

338. **Chakraborty, S. K., A. Chowdhury, A. Bhattacharyya, S. Ghosh, S. Pan, R. Waters, and N. Adityachaudhury.** 1995. Microbial degradation of oxadiazon by soil fungus *Fusarium solani*. *Journal of Agricultural and Food Chemistry* **43:** 2964-2969. (*solani*)

339. **Chalvet, F., C. Grimaldi, F. Kaper, T. Langin, and M.-J. Daboussi.** 2003. *Hop*, an active Mutator-like element in the genome of the fungus *Fusarium oxysporum*. *Molecular Biology and Evolution* **20:** 1362-1375. (*oxysporum*)

340. **Champeil, A., J. F. Fourbet, and T. Dore.** 2004. Effects of grain sampling procedures on *Fusarium* mycotoxin assays in wheat grains. *Journal of Agricultural and Food Chemistry* **52:** 6049-6054. (*graminearum*)

341. **Chandler, E. A., D. R. Simpson, M. A. Thomsett, and P. Nicholson.** 2003. Development of PCR assays to *Tri7* and *Tri13* trichothecene biosynthetic genes, and characterization of chemotypes of *Fusarium graminearum, Fusarium culmorum* and *Fusarium cerealis*. *Physiological and Molecular Plant Pathology* **62:** 355-367. (*crookwellense, graminearum*)

342. **Chary, S. J., and S. M. Reddy.** 1985. Starch degrading enzymes of two species of *Fusarium*. *Folia Microbiologica* **30:** 452-457. (*oxysporum, scirpi*)

343. **Chattopadhyay, N. C., and B. Nandi.** 1978. Changes in nitrogen content in malformed inflorescence of mango caused by *Fusarium moniliforme* var. *subglutinans*. *Giornale Botanico Italiano* **112:** 343-346. (*mangiferae*)

344. **Chattopadhyay, N. C., and B. Nandi.** 1981. Nutrition in *Fusarium moniliforme* var. *subglutinans* causing mango malformation. *Mycologia* **73:** 407-414. (*mangiferae*)

345. **Chelkowski, J., G. L. Bateman, and C. J. Mirocha.** 1999. Identification of toxigenic *Fusarium* species using PCR assays. *Journal of Phytopathology* **147:** 307-311. (7)

346. **Chelkowski, J., M. Manka, H. Kwasna, A. Visconti, and P. Golinski.** 1989. *Fusarium sporotrichioides* (Sherb.), *Fusarium tricinctum* (Corda) Sacc. and *Fusarium poae* Wollenw.: Cultural characteristics, toxigenicity and pathogenicity towards cereals. *Journal of Phytopathology* **124:** 155-161. (*tricinctum*)

347. **Chelkowski, J., M. Zawadzki, P. Zajkowski, A. Logrieco, and A. Bottalico.** 1990. Moniliformin production by *Fusarium* species. *Mycotoxin Research* **6:** 41-45. (*acuminatum, anthophilum, avenaceum, dlamini, oxysporum, proliferatum, solani, subglutinans, tricinctum*)

348. **Chen, F.-J.** 1994. Genetic analysis of pyriform microconidium production in mating population B of *Gibberella fujikuroi*. *Canadian Journal of Botany* **72:** 1209-1210. (*sacchari*)

349. **Chen, W.-Q., and W. J. Swart.** 2001. Genetic variation among *Fusarium oxysporum* isolates associated with root rot of *Amaranthus hybridus* in South Africa. *Plant Disease* **85:** 1076-1080. (*oxysporum*)

350. **Chimwamurombe, P. M., B. D. Wingfield, A. M. Botha, and M. J. Wingfield.** 2001. Cloning and sequence analysis of the endopolygalacturonase gene from the pitch canker fungus, *Fusarium circinatum*. *Current Microbiology* **42:** 350-352. (*circinatum*)

351. **Chinain, M., and A. Vey.** 1988. Experimental study of *Fusarium solani* infections in *Astacus leptodactylus* and *Pacifastacus leniusculus* Crustacea Decapoda. *Diseases of Aquatic Organisms* **5:** 215-224. (*solani*)

352. **Chitarra, G. S., P. Breeuwer, F. M. Rombouts, T. Abee, and J. Dijksterhuis.** 2005. Differentiation inside multicelled macroconidia of *Fusarium culmorum* during early germination. *Fungal Genetics and Biology* **42:** 694-703. (*culmorum*)

353. **Cho, J. H., J. C. Rupe, M. S. Cummings, and E. E. Gbur, Jr.** 2001. Isolation and identification of *Fusarium solani* f. sp. *glycines* from soil on Modified Nash and Snyder's medium. *Plant Disease* **85:** 256-260. (*solani*)

354. **Choi, G. H., E. T. Marek, C. L. Schardl, M. G. Richey, S. Chang, and D. A. Smith.** 1990. *sti35*, A stress-responsive gene in *Fusarium* spp. *Journal of Bacteriology* **172:** 4522-4528. (*oxysporum, solani*)

355. **Choi, G. H., M. G. Richey, C. S. M. Turbek, and D. A. Smith.** 1987. Comparative characteristics of a kievitone hydratase-producing isolate of *Fusarium solani* f. sp. *phaseoli* and of two strains deficient in this enzyme. *Physiological and Molecular Plant Pathology* **30:** 215-224. (*solani*)

356. **Chotena, M., D. J. Makus, and W. R. Simpson.** 1980. Effect of water stress on production and quality of sweet corn (*Zea mays*) seed. *Journal of the American Society for Horticultural Science* **105:** 289-293. (*verticillioides*)

357. **Chow, L.-P., M. Kamo, Y. Ueno, and A. Tsugita.** 1996. Amino acid sequence of peptidyl-prolyl isomerase a of *Fusarium sporotrichioides*. *Natural Toxins* **4:** 149-155. (*sporotrichioides*)

358. **Christakopoulos, P., P. W. Goodenough, D. Kekos, B. J. Macris, M. Claeyssens, and M. K. Bhat.** 1994. Purification and characterization of an extracellular β-glucosidase with transglycosylation and exo-glucosidase activities from *Fusarium oxysporum*. *European Journal of Biochemistry* **224:** 379-385. (*oxysporum*)

359. **Christakopoulos, P., P. Katapodis, D. G. Hatzinikolaou, D. Kekos, and B. J. Macris.** 2000. Purification and characterization of an extracellular α-L-arabinofuranosidase from *Fusarium oxysporum*. *Applied Biochemistry and Biotechnology* **87:** 127-133. (*oxysporum*)

360. **Christakopoulos, P., D. Kekos, F. N. Kolisis, and B. J. Macris.** 1995. Controlling simultaneous production of endoglucanase and β-glucosidase by *Fusarium oxysporum* in submerged culture. *Biotechnology Letters* **17:** 883-888. (*oxysporum*)

361. **Christakopoulos, P., D. Kekos, B. J. Macris, M. Claeyssens, and M. K. Bhat.** 1996. Purification and characterization of a major xylanase with cellulase and transferase activities from *Fusarium oxysporum*. *Carbohydrate Research* **289:** 91-104. (*oxysporum*)

362. **Christakopoulos, P., D. P. Koullas, D. Kekos, E. G. Koukios, and B. J. Macris.** 1991. Direct ethanol conversion of pretreated straw by *Fusarium oxysporum*. *Bioresource Technology* 35: 297-300. (*oxysporum*)

363. **Christakopoulos, P., D. P. Koullas, D. Kekos, E. G. Koukios, and B. J. Macris.** 1991. Direct conversion of straw to ethanol by *Fusarium oxysporum*: Effect of cellulose crystallinity. *Enzyme and Microbial Technology* **13:** 272-274. (*oxysporum*)

364. **Christakopoulos, P., B. J. Macris, and D. Kekos.** 1990. On the mechanism of direct conversion of cellulose to ethanol by *Fusarium oxysporum*: Effect of cellulase and β-glucosidase. *Applied Microbiology and Biotechnology* **33:** 18-20. (*oxysporum*)

365. **Christakopoulos, P., D. Mamma, D. Kekos, and B. J. Macris.** 1999. Enhanced acetyl esterase production by *Fusarium oxysporum*. *World Journal of Microbiology and Biotechnology* **15:** 443-446. (*oxysporum*)

366. **Christakopoulos, P., W. Nerinckx, D. Kekos, B. Macris, and M. Claeyssens.** 1996. Purification and characterization of two low molecular mass alkaline xylanases from *Fusarium oxysporum* F3. *Journal of Biotechnology* **51:** 181-189. (*oxysporum*)

367. **Christakopoulos, P., W. Nerinckx, D. Kekos, B. Macris, and M. Claeyssens.** 1997. The alkaline xylanase III from *Fusarium oxysporum* F3 belongs to family F/10. *Carbohydrate Research* **302:** 191-195. (*oxysporum*)

368. **Chrzanowska, J., M. Kolaczkowska, and A. Polanowski.** 1990. Proteolysis of casein by a proteinase from *Fusarium moniliforme* in solution and in Emmental cheese. *Milchwissenschaft* **45:** 164-167. (*verticillioides*)

369. **Chu, Q., W. Wu, M. E. Cook, and E. B. Smalley.** 1995. Induction of tibial dyschondroplasia and suppression of cell-mediated immunity in chickens by *Fusarium oxysporum* grown on sterile corn. *Avian Diseases* **39:** 100-107. (*oxysporum*)

370. **Chu, Y.-M., J.-J. Jeon, S.-J. Yea, Y.-H. Kim, S.-H. Yun, Y. W. Lee, and K.-H. Kim.** 2002. Double-stranded RNA mycovirus from *Fusarium graminearum*. *Applied and Environmental Microbiology* **68:** 2529-2534. (*graminearum*)

371. **Chu, Y.-M., W.-S. Lim, S.-J. Yea, J.-D. Cho, Y.-W. Lee, and K.-H. Kim.** 2004. Complexity of dsRNA mycovirus isolated from *Fusarium graminearum*. *Virus Genes* **28:** 135-143. (*graminearum*)

372. **Chulze, S. N., A. Torres, M. L. Ramirez, and J. F. Leslie.** 2000. Genetic variation in *Fusarium* section *Liseola* from no-till maize in Argentina. *Applied and Environmental Microbiology* **66:** 5312-5315. (5, 6, 9, 11, *verticillioides*)

373. **Ciarmela, M. L., M. C. Minvielle, G. Lori, and J. A. Basualdo.** 2002. Biological interaction between soil fungi and *Toxocara canis* eggs. *Veterinary Parasitology* **103:** 251-257. (*semitectum*)

374. **Cieslinski, G., and E. Lis.** 1989. Studies on the infection of strawberry roots *Fragaria grandiflora* Duch. by fungi of the genus *Fusarium* in *in vitro* conditions. *Fruit Science Reports* **16:** 7-16. (*culmorum*)

375. **Ciotola, M., A. K. Watson, and S. G. Hallett.** 1995. Discovery of the isolate of *Fusarium oxysporum* with potential to control *Striga hermonthica* in Africa. *Weed Research* **35:** 303-309. (*oxysporum*)

376. **Clancy, C. J., and M. H. Nguyen.** 1998. *In vitro* efficacy and fungicidal activity of voriconazole against *Aspergillus* and *Fusarium* species. *European Journal of Clinical Microbiology and Infectious Diseases* **17:** 573-575. (*oxysporum*)

377. **Clark, C. A., M. W. Hoy, and P. E. Nelson.** 1995. Variation among isolates of *Fusarium lateritium* from sweet potato for pathogenicity and vegetative compatibility. *Phytopathology* **85:** 624-629. (*denticulatum*)

378. **Clark, C. A., J. W. Hyun, and M. W. Hoy.** 1998. Relationships among wilt-inducing isolates of *Fusarium oxysporum* from sweet potato and tobacco. *Plant Disease* **82:** 530-536. (11)

379. **Clarke, D. C., and M. Christensen.** 1981. The soil micro fungal community of a South Dakota, USA grassland. *Canadian Journal of Botany* **59:** 1950-1960. (*solani*)

380. **Clarkson, J. D. S.** 1978. Pathogenicity of *Fusarium* spp. associated with foot rots of peas and beans. *Plant Pathology* **27:** 110-117. (*redolens*)

381. **Clausen, K.** 2001. Enzymatic oil-degumming by a novel microbial phospholipase. *European Journal of Lipid Science and Technology* **103:** 333-340. (*oxysporum*)

382. **Clear, R. M., and S. K. Patrick.** 1992. A simple medium to aid the identification of *Fusarium moniliforme*, *Fusarium proliferatum*, and *Fusarium subglutinans*. *Journal of Food Protection* 55: 120-122. (*proliferatum*)

383. **Clements, M. J., K. W. Campbell, C. M. Maragos, C. Pilcher, J. M. Headrick, J. K. Pataky, and D. G. White.** 2003. Influence of Cry1Ab protein and hybrid genotype on fumonisin contamination and fusarium ear rot of corn. *Crop Science* **43:** 1283-1293. (*verticillioides*)

384. **Clements, M. J., C. E. Kleinschmidt, C. M. Maragos, J. K. Pataky, and D. G. White.** 2003. Evaluation of inoculation techniques for Fusarium ear rot and fumonisin contamination of corn. *Plant Disease* **87:** 147-153. (*proliferatum, verticillioides*)

385. **Cleveland, T. E., and D. A. Smith.** 1983. Partial purification and further characterization of kievitone hydratase from cell-free culture filtrates of *Fusarium solani* f. sp. *phaseoli*. *Physiological Plant Pathology* **22:** 129-142. (*solani*)

386. **Cluness, M. J., P. D. Turner, E. Clements, D. T. Brown, and C. O'Reilly.** 1993. Purification and properties of cyanide hydratase from *Fusarium lateritium* and analysis of the corresponding *chy1* gene. *Journal of General Microbiology* **139:** 1807-1815. (*lateritium*)

387. **Coenen, A., F. Debets, and R. Hoekstra.** 1994. Additive action of partial heterokaryon incompatibility (partial-*het*) genes in *Aspergillus nidulans*. *Current Genetics* **26:** 233-237. (5)

388. **Cohen, B. A., Z. Amsellem, S. Lev-Yadun, and J. Gressel.** 2002. Infection of tubercles of the parasitic weed *Orobanche aegyptiaca* by mycoherbicidal *Fusarium* species. *Annals of Botany* **90:** 567-578. (*oxysporum*)

389. **Cohen, B. A., Z. Amsellem, R. Maor, A. Sharon, and J. Gressel.** 2002. Transgenically enhanced expression of indole-3-acetic acid confers hypervirulence to plant pathogens. *Phytopathology* **92:** 590-596. (*oxysporum*)

390. **Cole, R. J., J. W. Dorner, R. H. Cox, B. M. Cunfer, H. G. Cutler, and B. P. Stuart.** 1981. The isolation and identification of several trichothecene mycotoxins from *Fusarium heterosporum*. *Journal of Natural Products* **44:** 324-330. (*heterosporum*)

391. **Cole, R. J., J. W. Dorner, J. Gilbert, D. N. Mortimer, C. Crews, J. C. Mitchell, R. M. Windingstad, P. E. Nelson, and H. G. Cutler.** 1988. Isolation and identification of trichothecenes from *Fusarium compactum* suspected in the aetiology of a major intoxication of sandhill cranes. *Journal of Agricultural & Food Chemistry* **36:** 1163-1167. (*compactum*)

392. **Coleman, J. O. D., S. J. Hiscock, and F. M. Dewey.** 1993. Monoclonal antibodies to purified cutinase form *Fusarium solani* f. sp. *pisi*. *Physiological and Molecular Plant Pathology* **43:** 391-401. (*solani*)

393. **Collins, M. S., and M. G. Rinaldi.** 1977. Cutaneous infection in man caused by *Fusarium moniliforme*. *Sabouraudia* **15:** 151-160. (*verticillioides*)

394. **Colorni, A.** 1989. Fusariosis in the shrimp *Penaeus semisulcatus* cultured in Israel. *Mycopathologia* **108:** 145-147. (13)

395. **Compel, P., I. Papp, M. Bibo, C. Fekete, and L. Hornok.** 1999. Genetic interrelationships and genome organization of double-stranded RNA elements of *Fusarium poae*. *Virus Genes* **18:** 49-56. (*poae*)

396. **Connick, W. J., Jr., D. J. Daigle, A. B. Pepperman, K. P. Hebbar, R. D. Lumsden, T. W. Anderson, and D. C. Sands.** 1998. Preparation of stable, granular formulations containing *Fusarium oxysporum* pathogenic to narcotic plants. *Biological Control* **13:** 79-84. (*oxysporum*)

397. **Cook, R. J.** 1980. Fusarium foot rot of wheat and its control in the Pacific Northwest. *Plant Disease* **64:** 1061-1066. (*pseudograminearum*)

398. **Cooper, K. W.** 1940. Relations of *Pediculopsis graminum* and *Fusarium poae* to central bud rot of carnations. *Phytopathology* **30:** 853-859. (*poae*)

399. **Coppin, E., R. Debuchy, S. Arnaise, and M. Picard.** 1997. Mating types and sexual development in filamentous ascomycetes. *Microbiology and Molecular Biology Reviews* **61:** 411-428. (6)

400. **Correll, J. C., T. R. Gordon, and A. H. McCain.** 1992. Genetic diversity in California and Florida populations of the pitch canker fungus *Fusarium subglutinans* f. sp. *pini*. *Phytopathology* **82:** 415-420. (*circinatum*)

401. **Correll, J. C., T. R. Gordon, A. H. McCain, J. W. Fox, C. S. Koehler, D. L. Wood, and M. E. Schultz.** 1991. Pitch canker disease in California, USA: Pathogenicity, distribution and canker development on Monterey Pine, *Pinus radiata*. *Plant Disease* **75:** 676-682. (*circinatum*)

402. **Correll, J. C., C. J. R. Klittich, and J. F. Leslie.** 1987. Nitrate non-utilizing mutants of *Fusarium oxysporum* and their use in vegetative compatibility tests. *Phytopathology* **77:** 1640-1646. (2, 5)

403. **Correll, J. C., C. J. R. Klittich, and J. F. Leslie.** 1989. Heterokaryon self-incompatibility in *Gibberella fujikuroi* (*Fusarium moniliforme*). *Mycological Research* **93:** 21-27. (5, *verticillioides*)

404. **Correll, J. C., J. E. Puhalla, and R. W. Schneider.** 1986. Vegetative compatibility groups among non-pathogenic root-colonizing strains of *Fusarium oxysporum*. *Canadian Journal of Botany* **64:** 2358-2361. (5, *oxysporum*)

405. **Correll, J. C., J. E. Puhalla, and R. W. Schneider.** 1986. Identification of *Fusarium oxysporum* f. sp. *apii* on the basis of colony size, virulence and vegetative compatibility. *Phytopathology* **76:** 396-400. (5)

406. **Cosentino, S., P. L. Pisano, M. E. Fadda, and F. Palmas.** 1990. Pollen and mold allergy aerobiologic survey in the atmosphere of Cagliari, Italy 1986-1988. *Annals of Allergy* **65:** 393-400. (3)

407. **Costa, M. J. L., M. T. Cunha, J. M. S. Cabral, and M. R. Aires-Barros.** 2000. Scale-up of recombinant cutinase recovery by whole broth extraction with PEG-phosphate aqueous two-phase. *Bioseparation* **9:** 231-238. (*solani*)

408. **Costello, B. P. J. D. L., P. Evans, R. J. Ewen, H. E. Gunson, P. R. H. Jones, N. M. Ratcliffe, and P. T. N. Spencer-Phillips.** 2001. Gas chromatography-mass spectrometry analyses of volatile organic compounds from potato tubers inoculated with *Phytophthora infestans* or *Fusarium coeruleum*. *Plant Pathology* **50:** 489-496. (*torulosum*)

409. **Cotillon, A. C., and R. Morfin.** 1999. Transformation of 3-OH-steroids by *Fusarium moniliforme* 7-α-hydroxylase. *Journal of Steroid Biochemistry and Molecular Biology* **68:** 229-237. (*verticillioides*)

410. **Cotten, T. K., and G. P. Munkvold.** 1998. Survival of *Fusarium moniliforme*, *F. proliferatum*, and *F. subglutinans* in maize stalk residue. *Phytopathology* **88:** 550-555. (*proliferatum, subglutinans, verticillioides*)

411. **Couteaudier, Y., M.-J. Daboussi, A. Eparvier, T. Langin, and J. Orcival.** 1993. The GUS gene fusion system (*Es-*

cherichia coli β-D-glucuronidase gene), a useful tool in studies of root colonization by *Fusarium oxysporum*. *Applied and Environmental Microbiology* **59:** 1767-1773. (*oxysporum*)

412. **Covarelli, L., A. S. Turner, and P. Nicholson.** 2004. Repression of deoxynivalenol accumulation and expression of *tri* genes in *Fusarium culmorum* by fungicides *in vitro*. *Plant Pathology* **53:** 22-28. (*culmorum*)

413. **Cove, D. J.** 1976. Chlorate toxicity in *Aspergillus nidulans*. Studies of mutants altered in nitrate assimilation. *Molecular and General Genetics* **146:** 147-159. (5)

414. **Covert, S. F., A. Briley, M. M. Wallace, and V. T. McKinney.** 1999. Partial *MAT-2* gene structure and the influence of temperature on mating success in *Gibberella circinata*. *Fungal Genetics and Biology* **28:** 43-54. (6, 9, *circinatum*)

415. **Crawford, M. S., and P. E. Kolattukudy.** 1987. Pectate lyase from *Fusarium solani* f. sp. *pisi*: Purification, characterization, and *in vitro* translation of the messenger RNA and involvement in pathogenicity. *Archives of Biochemistry and Biophysics* **258:** 196-205. (*solani*)

416. **Creveld, L. D., A. Amadei, R. C. van Schaik, H. A. M. Pepermans, J. de Vlieg, and H. J. C. Berendsen.** 1998. Identification of functional and unfolding motions of cutinase as obtained from molecular dynamics computer simulations. *Proteins* **33:** 253-264. (*solani*)

417. **Creveld, L. D., W. Meijberg, H. J. C. Berendsen, and H. A. M. Pepermans.** 2001. DSC studies of *Fusarium solani pisi* cutinase: Consequences for stability in the presence of surfactants. *Biophysical Chemistry* **92:** 65-75. (*solani*)

418. **Crisp, L. M., and C. E. Bland.** 1990. Potential use of ozone to disinfect sea water of fungi causing diseases of cultured marine crustacea. *Journal of Invertebrate Pathology* **55:** 380-386. (*solani*)

419. **Crowhurst, R. N., S. J. Binnie, J. K. Bowen, B. T. Hawthorne, K. M. Plummer, J. Rees-George, E. H. A. Rikkerink, and M. D. Templeton.** 1997. Effect of disruption of a cutinase gene (*cutA*) on virulence and tissue specificity of *Fusarium solani* f. sp. *cucurbitae* race 2 toward *Cucurbita maxima* and *C. moschata*. *Molecular Plant-Microbe Interactions* **10:** 355-368. (*solani*)

420. **Crowhurst, R. N., G. J. Rees, E. H. A. Rikkerink, and M. D. Templeton.** 1992. High efficiency transformation of *Fusarium solani cucurbitae* race 2 mating population V. *Current Genetics* **21:** 463-469. (*solani*)

421. **Cuero, R., and T. Ouellet.** 2005. Metal ions modulate gene expression and accumulation of the mycotoxins aflatoxin and zearalenone. *Journal of Applied Microbiology* **98:** 598-605. (*graminearum*)

422. **Cuero, R. G.** 1980. Ecological distribution of *Fusarium solani* and its opportunistic action related to mycotic keratitis in Cali, Colombia. *Journal of Clinical Microbiology* **12:** 455-461. (*solani*)

423. **Cullen, D., R. W. Caldwell, and E. B. Smalley.** 1982. Cultural characteristics, pathogenicity and zearalenone production by strains of *Gibberella zeae* isolated from corn. *Phytopathology* **72:** 1415-1418. (*graminearum*)

424. **Cullen, D., and E. B. Smalley.** 1983. Heterokaryosis in *Fusarium tricinctum* and *Fusarium sporotrichioides*. *Journal of General Microbiology* **129:** 3035-3042. (*sporotrichioides*)

425. **Cumagun, C. J. R., R. L. Bowden, J. E. Jurgenson, J. F. Leslie, and T. Miedaner.** 2004. Genetic mapping of pathogenicity and aggressiveness of *Gibberella zeae* (*Fusarium graminearum*) towards wheat. *Phytopathology* **94:** 520-526. (7, *graminearum*)

426. **Cumagun, C. J. R., and T. Miedaner.** 2004. Segregation for aggressiveness and deoxynivalenol production of a population of *Gibberella zeae* causing head blight of wheat. *European Journal of Plant Pathology* **110:** 789-799. (*graminearum*)

427. **Cunha, M. T., M. J. L. Costa, C. R. C. Calado, L. P. Fonseca, M. R. Aires-Barros, and J. M. S. Cabral.** 2003. Integration of production and aqueous two-phase systems extraction of extracellular *Fusarium solani pisi* cutinase fusion proteins. *Journal of Biotechnology* **100:** 55-64. (*solani*)

428. **Curir, P., B. Danieli, M. Dolci, C. Pasini, L. Guglieri, and M. Sacco.** 2000. Reductive detoxification of the acetophenone skeleton of the carnation phytoanticipin by *Fusarium oxysporum* f. sp. *dianthi*. *Plant Pathology* **49:** 742-747. (*oxysporum*)

429. **Curir, P., L. Guglieri, M. Dolci, A. Capponi, and G. Aurino.** 2000. Fusaric acid production by *Fusarium oxysporum* f. sp. *lilii* and its role in the lily basal rot disease. *European Journal of Plant Pathology* **106:** 849-856. (*oxysporum*)

430. **Curragh, H. J., H. Mooibroek, J. G. H. Wessels, R. Marchant, and E. Mullan.** 1993. Protoplast formation and DNA-mediated transformation of *Fusarium culmorum* to hygromycin B resistance. *Mycological Research* **97:** 313-317. (*culmorum*)

431. **Daboussi, M.-J., J. M. Daviere, S. Graziani, and T. Langin.** 2002. Evolution of the *Fot1* transposons in the genus *Fusarium*: Discontinuous distribution and epigenetic inactivation. *Molecular Biology and Evolution* **19:** 510-520. (*oxysporum*)

432. **Daboussi, M.-J., and C. Gerlinger.** 1992. Parasexual cycle and genetic analysis following protoplast fusion in *Nectria haematococca*. *Current Genetics* **21:** 385-392. (*solani*)

433. **Daboussi, M.-J., and T. Langin.** 1994. Transposable elements in the fungal plant pathogen *Fusarium oxysporum*. *Genetica* **93:** 49-59. (*oxysporum*)

434. **Daboussi, M. J., T. Langin, and Y. Brygoo.** 1992. *Fot1*, a new family of fungal transposable elements. *Molecular and General Genetics* **232:** 12-16. (*oxysporum*)

435. **Daboussi, M. J., T. Langin, F. Deschamps, Y. Brygoo, C. Scazzocchio, and G. Burger.** 1991. Heterologous expression of the *Aspergillus nidulans* regulatory gene *nirA* in *Fusarium oxysporum*. *Gene* **109:** 155-160. (*oxysporum*)

436. **Daiber, A., H. Shoun, and V. Ullrich.** 2005. Nitric oxide reductase (P450nor) from *Fusarium oxysporum*. *Journal of Inorganic Biochemistry* **99:** 185-193. (*oxysporum*)

437. **Dalgaard, J. Z., and A. J. S. Klar.** 2001. Does *Schizosaccharomyces pombe* exploit the intrinsic asymmetry of DNA synthesis to imprint daughter cells for mating type switching? *Trends in Genetics* **17:** 153-157. (6)

438. **Damicone, J. P., D. R. Cooley, and W. J. Manning.** 1981. Benomyl in acetone eradicates *Fusarium moniliforme* and *F. oxysporum* from asparagus seed. *Plant Disease* **65:** 892-893. (3)

439. **Daniels, B. A.** 1983. Elimination of *Fusarium moniliforme* from corn seed. *Plant Disease* **67:**609-611. (3)

440. **Dantzer, W. R., A. L. I. Pomettoi, and P. A. Murphy.** 1996. Fumonisin B_1 production by *Fusarium proliferatum* strain M5991 in a modified Myro liquid medium. *Natural Toxins* **4:** 168-173. (*proliferatum*)

441. **Dantzig, A. H., S. H. Zuckerman, and M. M. Andonov-Ronald.** 1986. Isolation of a *Fusarium solani* mutant reduced in cutinase activity and virulence. *Journal of Bacteriology* **168:** 911-916. (*solani*)

442. **Daroda, L., K. Hahn, D. Pashkoulov, and E. Benvenuto.** 2001. Molecular characterization and *in planta* detection of *Fusarium moniliforme* endopolygalacturonase isoforms. *Physiological and Molecular Plant Pathology* **59:** 317-325. (*verticillioides*)

443. **Darvas, J. M., and J. M. Kotze.** 1987. Fungi associated with pre- and postharvest diseases of avocado fruit at Westfalia Estate, South Africa. *Phytophylactica* **19:** 83-86. (*decemcellulare*)

444. **Darvas, J. M., J. M. Kotze, and F. C. Wehner.** 1987. Field occurrence and control of fungi causing postharvest decay of avocados. *Phytophylactica* **19:** 453-456. (*decemcellulare*)

445. **Darvas, J. M., J. M. Kotze, and F. C. Wehner.** 1987. Pathogenicity of fungi causing pre- and postharvest diseases of avocado fruit. *Phytophylactica* **19:** 489-493. (*sambucinum, solani*)

446. **Das, J.** 1991. Influence of potassium in the agar medium on the growth pattern of the filamentous fungus *Fusarium solani*. *Applied and Environmental Microbiology* **57**: 3033-3036. (*solani*)

447. **Das, J., and H. G. Busse.** 1990. Light-driven diurnal zonation in the filamentous fungus *Fusarium solani*. *International Journal of Developmental Biology* **34:** 319-322. (*solani*)

448. **Daubner, S. C., G. Gadda, M. P. Valley, and P. F. Fitzpatrick.** 2002. Cloning of nitroalkane oxidase from *Fusarium oxysporum* identifies a new member of the acyl-CoA dehydrogenase superfamily. *Proceedings of the National Academy of Sciences USA* **99:** 2702-2707. (*oxysporum*)

449. **Daviere, J. M., T. Langin, and M.-J. Daboussi.** 2001. Potential role of transposable elements in the rapid reorganization of the *Fusarium oxysporum* genome. *Fungal Genetics and Biology* **34:** 177-192. (*oxysporum*)

450. **Davis, G. R. F., N. D. Westcott, J. D. Smith, G. A. Neish, and H. B. Schiefer.** 1982. Toxigenic isolates of *Fusarium sporotrichioides* obtained from hay in Saskatchewan, Canada. *Canadian Journal of Microbiology* **28:** 259-261. (*sporotrichioides*)

451. **Davis, P. J., D. Wiese, and J. R. Rosazza.** 1977. Microbial transformations of glaucine. *Journal of the Chemical Society Perkin Transactions* **1977:** 1-6. (*solani*)

452. **Davis, R. D., N. Y. Moore, and J. K. Kochman.** 1996. Characterization of a population of *Fusarium oxysporum* f. sp. *vasinfectum* causing wilt of cotton in Australia. *Australian Journal of Agricultural Research* **47:** 1143-1156. (*oxysporum*)

453. **de Groot, M. J. A., P. Bundock, P. J. J. Hooykaas, and A. G. M. Beijersbergen.** 1998. *Agrobacterium tumefaciens*-mediated transformation of filamentous fungi. *Nature Biotechnology* **16:** 839-842. (*venenatum*)

454. **de Haan, L. A. M., A. Numansen, E. J. A. Roebroeck, and J. van-Doorn.** 2000. PCR detection of *Fusarium oxysporum* f. sp. *gladioli* race 1, causal agent of Gladiolus yellows disease, from infected corms. *Plant Pathology* **49:** 89-100. (*oxysporum*)

455. **de las Heras, A., B. Patino, M. L. Posada, M. J. Martinez, C. Vazquez, and M. T. Gonzalez-Jaen.** 2003. Characterization and *in vitro* expression patterns of an exopolygalacturonase encoding gene from *Fusarium oxysporum* f. sp. *radicis-lycopersici*. *Journal of Applied Microbiology* **94:** 856-864. (*oxysporum*)

456. **de Lorenzo, G., G. Salvi, L. Degra, R d'Ovidio, and F. Cervone.** 1987. Induction of extracellular polygalacturonase and its messenger RNA in the phytopathogenic fungus *Fusarium moniliforme*. *Journal of General Microbiology* **133:** 3365-3374. (*verticillioides*)

457. **de Wolf, E. D., L. V. Madden, and P. E. Lipps.** 2003. Risk assessment models for wheat Fusarium head blight epidemics based on within-season weather data. *Phytopathology* **93:** 428-435. (11)

458. **Debieru, D., C. Gall, M. Credt, J. Bach, C. Malosse, and P. Leroux.** 1992. Ergosterol biosynthesis and its inhibition by fenpropimorph in *Fusarium* species. *Phytochemistry* **31:** 1223-1233. (*decemcellulare, lateritium, oxysporum, solani, verticillioides*)

459. **Defago, G., and H. Kern.** 1983. Induction of *Fusarium solani* mutants insensitive to tomatine: Their pathogenicity and aggressiveness to tomato fruits and pea plants. *Physiological Plant Pathology* **22:** 29-38. (*solani*)

460. **Defago, G., H. Kern, and L. Sedlar.** 1983. Genetic analysis of tomatine insensitivity, sterol content and pathogenicity for green tomato fruits in mutants of *Fusarium solani*. *Physiological Plant Pathology* **22:** 39-44. (*solani*)

461. **Delgado-Jarana, J., A. L. Martinez-Rocha, R. Roldan-Rodriguez, M. I. G. Roncero, and A. di Pietro.** 2005. *Fusarium oxysporum* G protein β subunit Fgb1 regulates hyphal growth, development, and virulence through multiple signaling pathways. *Fungal Genetics and Biology* **42:** 61-72. (*oxysporum*)

462. **Delserone, L. M., K. McCluskey, D. E. Matthews, and H. D. VanEtten.** 1999. Pisatin demethylation by fungal pathogens and nonpathogens of pea: Association with pisatin tolerance and virulence. *Physiological and Molecular Plant Pathology* **55:** 317-326. (*solani*)

463. **Denison, S. H., and G. S. May.** 1994. Mitotic catastrophe is the mechanism of lethality for mutations that confer mutagen sensitivity in *Aspergillus nidulans*. *Mutation Research* **304:** 193-202. (4)

464. **Denny, T. P., and H. D. VanEtten.** 1982. Metabolism of the phytoalexins medicarpin and maackiain by *Fusarium solani*. *Phytochemistry* **21:** 1023-1028. (*solani*)

465. **Deschamps, F., T. Langin, P. Maurer, C. Gerlinger, B. Felenbok, and M.-J. Daboussi.** 1999. Specific expression of the *Fusarium* transposon *Fot1* and effects on target gene transcription. *Molecular Microbiology* **31:** 1373-1383. (*oxysporum*)

466. **Desjardins, A. E.** 1995. Population structure of *Gibberella pulicaris* (anamorph *Fusarium sambucinum*) from po-

tato tuber dry rot in North America and Europe. *American Potato Journal* **72:** 145-156. (*sambucinum*)

467. **Desjardins, A. E.** 2003. *Gibberella* from *a*(*venaceae*) to *z*(*eae*). *Annual Review of Phytopathology* **41:** 171-198. (10)

468. **Desjardins, A. E.** 2005. Fusarium *mycotoxins: Chemistry, genetics and biology*. APS Press, St. Paul, Minnesota. (P, 8, 9)

469. **Desjardins, A. E., and M. Beremand.** 1987. A genetic system for trichothecene toxin production in *Gibberella pulicaris* (*Fusarium sambucinum*). *Phytopathology* **77:** 678-683. (2, 6, *sambucinum*)

470. **Desjardins, A. E., E. A. Christ-Harned, S. P. McCormick, and G. A. Secor.** 1993. Population structure and genetic analysis of field resistance to thiabendazole in *Gibberella pulicaris* from potato tubers. *Phytopathology* **83:** 164-170. (*sambucinum*)

471. **Desjardins, A. E., and H. W. Gardner.** 1989. Genetic analysis in *Gibberella pulicaris*: Rishitin tolerance, rishitin metabolism, and virulence on potato tubers. *Molecular Plant-Microbe Interactions* **2:** 26-34. (*sambucinum*)

472. **Desjardins, A. E., and H. W. Gardner.** 1991. Virulence of *Gibberella pulicaris* on potato tubers and its relationship to a gene for rishitin metabolism. *Phytopathology* **81:** 429-435. (*sambucinum*)

473. **Desjardins, A. E., H. W. Gardner, and R. D. Plattner.** 1989. Detoxification of the potato phytoalexin lubimin by *Gibberella pulicaris*. *Phytochemistry* **28:** 431-438. (*sambucinum*)

474. **Desjardins, A. E., H. W. Gardner, and K. M. Welt-ring.** 1992. Detoxification of sesquiterpene phytoalexins by *Gibberella pulicaris* (*Fusarium sambucinum*) and its importance for virulence on potato tubers. *Journal of Industrial Microbiology* **9:** 201-211. (*sambucinum*)

475. **Desjardins, A. E., T. M. Hohn, and S. P. McCormick.** 1992. Effect of gene disruption of trichodiene synthase on the virulence of *Gibberella pulicaris*. *Molecular Plant-Microbe Interactions* **5:** 214-222. (*sambucinum*)

476. **Desjardins, A. E., A. M. Jarosz, R. D. Plattner, N. J. Alexander, D. W. Brown, and J. E. Jurgenson.** 2004. Patterns of trichothecene production, genetic variability, and virulence to wheat of *Fusarium graminearum* from smallholder farms in Nepal. *Journal of Agricultural and Food Chemistry* **52:** 6341-6346. (*graminearum*)

477. **Desjardins, A. E., H. K. Manandhar, R. D. Plattner, G. G. Manandhar, S. M. Poling, and C. M. Maragos.** 2000. *Fusarium* species from Nepalese rice and production of mycotoxins and gibberellic acid by selected species. *Applied and Environmental Microbiology* **66:** 1020-1025. (*proliferatum, torulosum*)

478. **Desjardins, A. E., G. P. Munkvold, R. D. Plattner, and R. H. Proctor.** 2002. *FUM1*: A gene required for fumonisin biosynthesis but not for maize ear rot and ear infection by *Gibberella moniliformis* in field tests. *Molecular Plant Microbe Interactions* **15:** 1157-1164. (*verticillioides*)

479. **Desjardins, A. E., and P. E. Nelson.** 1995. Sexual fertility of forty *Fusarium* strains from the European *Fusarium sambucinum* project. *Mycopathologia* **129:** 149-151. (*sambucinum*)

480. **Desjardins, A. E., and R. D. Plattner.** 1989. Trichothecene toxin production by strains of *Gibberella pulicaris* (*Fusarium sambucinum*) in liquid culture and in potato tubers. *Journal of Agricultural and Food Chemistry* **37:** 388-392. (*sambucinum*)

481. **Desjardins, A. E., and R. D. Plattner.** 2000. Fumonisin B_1-nonproducing strains of *Fusarium verticillioides* cause maize (*Zea mays*) ear infection and ear rot. *Journal of Agricultural and Food Chemistry* **48:** 5773-5780. (*verticillioides*)

482. **Desjardins, A. E., and R. D. Plattner.** 2003. Diverse traits for pathogen fitness in *Gibberella zeae*. *Canadian Journal of Plant Pathology* **25:** 21-27. (*graminearum*)

483. **Desjardins, A. E., R. D. Plattner, and M. N. Beremand.** 1987. Ancymidol blocks trichothecene biosymthesis and leads to accumulation of trichodiene in *Fusarium sporotrichioides* and *Gibberella pulicaris*. *Applied and Environmental Microbiology* **53:** 1860-1865. (*sambucinum, sporotrichioides*)

484. **Desjardins, A. E., R. D. Plattner, and T. R. Gordon.** 2000. *Gibberella fujikuroi* mating population A and *Fusarium subglutinans* from teosinte species and maize from Mexico and Central America. *Mycological Research* **104:** 865-872. (*subglutinans, verticillioides*)

485. **Desjardins, A. E., R. D. Plattner, M. Lu, and L. E. Claflin.** 1998. Distribution of fumonisins in maize ears infected with strains of *Fusarium moniliforme* that differ in fumonisin production. *Plant Disease* **82:** 953-958. (*verticillioides*)

486. **Desjardins, A. E., R. D. Plattner, T. C. Nelsen, and J. F. Leslie.** 1995. Genetic analysis of fumonisin production and virulence of *Gibberella fujikuroi* mating population A (*Fusarium moniliforme*) on maize (*Zea mays*) seedlings. *Applied and Environmental Microbiology* **61:** 79-86. (*verticillioides*)

487. **Desjardins, A. E., G. F. Spencer, and R. D. Plattner.** 1989. Tolerance and metabolism of furanocoumarins by the phytopathogenic fungus *Gibberella pulicaris* (*Fusarium sambucinum*). *Phytochemistry* **28:** 2963-2969. (*sambucinum*)

488. **Dettman, J. R., D. J. Jacobson, and J. W. Taylor.** 2003. A multi-locus genealogical approach to phylogenetic species recognition in the model eukaryote *Neurospora*. *Evolution* **57:** 2703-2720. (9)

489. **Dettman, J. R., D. J. Jacobson, E. Turner, A. Pringle, and J. W. Taylor.** 2003. Reproductive isolation and phylogenetic divergence in *Neurospora*: Comparing methods of species recognition in a model eukaryote. *Evolution* **57:** 2721-2741. (9)

490. **Dharmaputra, O. S., and I. Retnowati.** 1996. Fungi isolated from groundnuts in some locations of West Java. *Biotropia* **9:** 15-25. (*longipes*)

491. **Dhingra, O. D., and J. J. Muchovej.** 1979. Pod rot, seed rot, and root rot of snap bean and dry bean caused by *Fusarium semitectum*. *Plant Disease Reporter* **63:** 84-87. (*semitectum*)

492. **di Menna, M. E., D. R. Lauren, and W. A. Smith.** 1991. An effect of incubation temperature on zearalenone production by strains of *Fusarium crookwellense*. *Mycopathologia* **116:** 81-86. (*crookwellense*)

493. **di Pietro, A., N. Anaya, and M. I. G. Roncero.** 1994. Occurrence of a retrotransposon-like sequence among different *formae speciales* and races of *Fusarium oxysporum*. *Mycological Research* **98:** 993-996. (*oxysporum*)

494. **di Pietro, A., F. I. Garcia-Maceira, M. D. Huertas-Gonzalez, M. C. Ruiz-Roldan, Z. Caracuel, A. S. Barbieri, and M.**

I. G. Roncero. 1998. Endopolygalacturonase PG1 in different *formae speciales* of *Fusarium oxysporum*. *Applied and Environmental Microbiology* **64:** 1967-1971. (*oxysporum*)

495. **di Pietro, A., F. I. Garcia-Maceira, E. Meglecz, and M. I. G. Roncero.** 2001. A MAP kinase of the vascular wilt fungus *Fusarium oxysporum* is essential for root penetration and pathogenesis. *Molecular Microbiology* **39:** 1140-1152. (*oxysporum*)

496. **di Pietro, A., M. D. Huertas-Gonzalez, J. F. Gutierrez-Corona, G. Martinez-Cadena, E. Meglecz, and M. I. G. Roncero.** 2001. Molecular characterization of a subtilase from the vascular wilt fungus *Fusarium oxysporum*. *Molecular Plant-Microbe Interactions* **14:** 653-662. (*oxysporum*)

497. **di Pietro, A., and M. I. G. Roncero.** 1996. Purification and characterization of an exo-polygalacturonase from the tomato vascular wilt pathogen *Fusarium oxysporum* f. sp. *lycopersici*. *FEMS Microbiology Letters* **145:** 295-299. (*oxysporum*)

498. **di Pietro, A., and M. I. G. Roncero.** 1996. Endopolygalacturonase from *Fusarium oxysporum* f. sp. *lycopersici*: Purification, characterization, and production during infection of tomato plants. *Phytopathology* **86:** 1324-1330. (*oxysporum*)

499. **di Pietro, A., and M. I. G. Roncero.** 1996. Purification and characterization of a pectate lyase from *Fusarium oxysporum* f. sp. *lycopersici* produced on tomato vascular tissue. *Physiological and Molecular Plant Pathology* **49:** 177-185. (*oxysporum*)

500. **di Pietro, A., and M. I. G. Roncero.** 1998. Cloning, expression, and role in pathogenicity of *pg1* encoding the major extracellular endopolygalacturonase of the vascular wilt pathogen *Fusarium oxysporum*. *Molecular Plant-Microbe Interactions* **11:** 91-98. (*oxysporum*)

501. **di Primo, P., C. Cappelli, and T. Katan.** 2002. Vegetative compatibility grouping of *Fusarium oxysporum* f. sp. *gladioli* from saffron. *European Journal of Plant Pathology* **108:** 869-875. (*oxysporum*)

502. **di Primo, P., G. Cartia, and T. Katan.** 2001. Vegetative compatibility and heterokaryon stability in *Fusarium oxysporum* f. sp. *radicis-lycopersici* from Italy. *Plant Pathology* **50:** 371-382. (5)

503. **Dianese, J. C., H. A. Bolkan, C. B. da Silva, and F. A. A. Couto.** 1981. Pathogenicity of epiphytic *Fusarium moniliforme* var. *subglutinans* to pineapple (*Ananas comosus*). *Phytopathology* **71:** 1145-1149. (*guttiforme*)

504. **Dick, M.** 1998. Pine pitch canker – The threat to New Zealand. *New Zealand Forestry* **42:** 30-34. (*circinatum*)

505. **Dickman, M. B., and J. F. Leslie.** 1992. Complementation of a *nnu* mutant of *Gibberella zeae* (*Fusarium graminearum*) by the regulatory *nit-2* gene of *Neurospora crassa*. *Molecular and General Genetics* **235:** 458-462. (5, *graminearum*)

506. **Dietert, M. F., H. D. VanEtten, and P. S. Matthews.** 1983. *Nectria haematococca* mating population VI: Cultural parameters affecting growth, conidiation and perithecial formation. *Canadian Journal of Botany* **61:** 1178-1184. (6)

507. **Diniz-Maia, M. D. M., M. M. Camargo de Morais, M. A. de Morais, Jr., E. H. Magalhaes-Melo, and J. L. de Lima Filho.** 1999. Production of extracellular lipase by the phytopathogenic fungus *Fusarium solani* FS1. *Revista de Microbiologia* **30:** 304-309. (*solani*)

508. **Diolez, A., T. Langin, C. Gerlinger, Y. Brygoo, and M.-J. Daboussi.** 1993. The *Nia* gene of *Fusarium oxysporum*: Isolation, sequence and development of a homologous transformation system. *Gene* **131:** 61-67. (*oxysporum*)

509. **Dixit, A., W. Lewis, J. Baty, W. Crozier, and J. Wed-ner.** 2000. Deuteromycete aerobiology and skin reactivity patterns: A two year concurrent study in Corpus Christi, Texas, USA. *Grana* **39:** 209-218. (3)

510. **Dixit, V. S., A. R. Kumar, A. Pant, and M. I. Khan.** 2004. Low molecular mass pectate lyase from *Fusarium moniliforme*: Similar modes of chemical and thermal denaturation. *Biochemical and Biophysical Research Communications* **315:** 477-484. (*verticillioides*)

511. **D'Mello, J. P. F., A. M. C. Macdonald, D. Postel, W. T. P. Dijksma, A. Dujardin, and C. M. Placinta.** 1998. Pesticide use and mycotoxin production in *Fusarium* and *Aspergillus* phytopathogens. *European Journal of Plant Pathology* **104:** 741-751. (*culmorum*)

512. **D'Mello, J. P. F., A. M. C. Macdonald, and R. Rinna.** 2001. Effects of azoxystrobin on mycotoxin production in a carbendazim-resistant strain of *Fusarium sporotrichioides*. *Phytoparasitica* **29:** 431-440. (*sporotrichioides*)

513. **Dobzhansky, T.** 1950. Mendelian populations and their evolution. *American Naturalist* **84:** 401-418. (9)

514. **Doehlert, D. C., C. A. Knutson, and R. F. Vesonder.** 1994. Phytotoxic effects of fumonisin B_1 on maize seedling growth. *Mycopathologia* **127:** 117-121. (*verticillioides*)

515. **Dohare, S., M. M. Mishra, and B. Kumar.** 2003. Effect of wilt on juice quality of sugarcane. *Annals of Biology* **19:** 183-186. (*sacchari*)

516. **Donaldson, G. C., L. A. Ball, P. E. Axelrood, and N. L. Glass.** 1995. Primer sets developed to amplify conserved genes from filamentous ascomycetes are useful in differentiating *Fusarium* species associated with conifers. *Applied and Environmental Microbiology* **61:** 1331-1340. (*oxysporum*, *proliferatum*)

517. **Doohan, F. M., D. W. Parry, P. Jenkinson, and P. Nicholson.** 1998. The use of species-specific PCR-based assays to analyze Fusarium ear blight of wheat. *Plant Pathology* **47:** 197-205. (7, *poae*)

518. **Doohan, F. M., D. W. Parry, and P. Nicholson.** 1999. Fusarium ear blight of wheat: The use of quantitative PCR and visual disease assessment in studies of disease control. *Plant Pathology* **48:** 209-217. (*poae*)

519. **Doohan, F. M., P. Smith, D. W. Parry, and P. Nicholson.** 1998. Transformation of *Fusarium culmorum* with the β-D-glucuronidase (GUS) reporter gene: A system for studying host-pathogen relationships and disease control. *Physiological and Molecular Plant Pathology* **53:** 253-268. (*culmorum*)

520. **Doval, S. L, C. K. Kaul, and B. P. Mathur.** 1976. Note on control of mango malformation. *Indian Journal of Agricultural Sciences* **46:** 545-546. (*mangiferae*)

521. **Doyle, J. J.** 1997. Trees within trees: Genes and species, molecules and morphology. *Systematic Biology* **46:** 537-553. (9)

522. **Dubey, R. C., and R. S. Dwivedi.** 1986. Destructive mycoparasitic behavior of *Fusarium solani* Mart. App. & Woll.

against *Mucor spinosus* van Tieghem. *Microbios Letters* **32:** 123-128. (*solani*)

523. **Duhart, B. T., Jr., D. Zhang, J. Deck, J. P. Freeman, and C. E. Cerniglia.** 1999. Biotransformation of protriptyline by filamentous fungi and yeasts. *Xenobiotica* **29:** 733-746. (*oxysporum*)

524. **Duijff, B. J., D. Pouhair, C. Olivain, C. Alabouvette, and P. Lemanceau.** 1998. Implication of systemic induced resistance in the suppression of Fusarium wilt of tomato by *Pseudomonas fluorescens* WCS417r and by nonpathogenic *Fusarium oxysporum* Fo47. *European Journal of Plant Pathology* **104:** 903-910. (*oxysporum*)

525. **Dukes, G. H., Jr., G. M. McGuire, II, D. Jackson, and S. L. McGuire.** 2003. The occurrence of *Fusarium merismoides* var. *chlamydosporale* (Tuberculariaceae) in Rankin County, Mississippi. *Journal of the Mississippi Academy of Sciences* **48:** 115-116. (*merismoides*)

526. **Dumestre, A., T. Chone, J. M. Portal, M. Gerard, and J. Berthelin.** 1997. Cyanide degradation under alkaline conditions by a strain of *Fusarium solani* isolated from contaminated soils. *Applied and Environmental Microbiology* **63:** 2729-2734. (*solani*)

527. **Dunlap, J. C., J. J. Loros, Y. Liu, and S. K. Crosthwaite.** 1999. Eukaryotic circadian systems: Cycles in common. *Genes to Cells* **4:** 1-10. (4)

528. **Dupuy, J., P. Le Bars, H. Boudra, and J. Le Bars.** 1993. Thermostability of fumonisin B_1, a mycotoxin from *Fusarium moniliforme*, in corn. *Applied and Environmental Microbiology* **59:** 2864-2867. (*verticillioides*)

529. **Duran-Paramo, E., H. Molina-Jimenez, M. A. Brito-Arias, and F. Robles-Martinez.** 2004. Gibberellic acid production by free and immobilized cells in different culture systems. *Applied Biochemistry and Biotechnology* **113-116**: 381-388. (*fujikuroi*)

530. **DuTeau, N. M., and J. F. Leslie.** 1991. A simple, rapid procedure for the isolation of DNA for PCR from *Gibberella fujikuroi* (*Fusarium* section *Liseola*). *Fungal Genetics Newsletter* **38:** 72. (7)

531. **Duttamajumder, S. K., B. L. Srivastava, and R. Kapur.** 2004. Wilt disease: a major deterrent in breeding sugarcane genotypes for high sugar accumulation potential in subtropical India. *Indian Journal of Agricultural Sciences* **74:** 80-83. (*sacchari*)

532. **Dwinell, L. D., S. W. Fraedrich, and D. Adams.** 2001. Diseases of pines caused by the pitch canker fungus, p. 225-232. *In* B. A. Summerell, J. F. Leslie, D. Backhouse, W. L. Bryden, and L. W. Burgess (eds.), Fusarium*: Paul E. Nelson Memorial Symposium.* APS Press, St. Paul, Minnesota. (*circinatum*)

533. **Dwivedi, S. K.** 1996. Studies on some aspects of wilt disease of guava (*Psidium guajava* L.). *Journal of Mycopathological Research* **34:** 99-103. (*longipes*)

534. **Dwivedi, S. K., R. C. Mishra, and R. S. Dwivedi.** 1988. Incidence of wilt disease of guava (*Psidium guajava* L.) in Varanasi, India. *International Journal of Tropical Plant Diseases* **6:** 213-216. (*longipes*)

535. **Dyer, P. S., D. S. Ingram, and K. Johnstone.** 1993. Evidence for the involvement of linoleic acid and other endogenous lipid factors in perithecial development of *Nectria haematococca* mating population VI. *Mycological Research* **97:** 485-496. (2, 6)

536. **Eddine, A. N., F. Hannemann, and W. Schaefer.** 2001. Cloning and expression analysis of *NhL1*, a gene encoding an extracellular lipase, from the fungal pea pathogen *Nectria haematococca* MP VI (*Fusarium solani* f. sp. *pisi*) that is expressed in planta. *Molecular Genetics and Genomics* **265:** 215-224. (*solani*)

537. **Edel, V., C. Steinberg, I. Avelange, G. Laguerre, and C. Alabouvette.** 1995. Comparison of three molecular methods for the characterization of *Fusarium oxysporum* strains. *Phytopathology* **85:** 579-585. (*oxysporum*)

538. **Edel, V., C. Steinberg, N. Gautheron, and C. Alabouvette.** 1997. Evaluation of restriction analysis of polymerase chain reaction (PCR)-amplified ribosomal DNA for the identification of *Fusarium* species. *Mycological Research* **101:** 179-187. (7)

539. **Edel, V., C. Steinberg, N. Gautheron, and C. Alabouvette.** 2000. Ribosomal DNA-targeted oligonucleotide probe and PCR assay specific for *Fusarium oxysporum*. *Mycological Research* **104:** 518-526. (*oxysporum*)

540. **Edel, V., C. Steinberg, N. Gautheron, G. Recorbet, and C. Alabouvette.** 2001. Genetic diversity of *Fusarium oxysporum* populations isolated from different soils in France. *FEMS Microbiology Ecology* **36:** 61-71. (*oxysporum*)

541. **Edwards, D. G.** 1993. The nutritional evaluation of mycoprotein. *International Journal of Food Sciences and Nutrition* **44(Suppl. 1):** S37-S43. (*venenatum*)

542. **Edwards, E. T.** 1935. Studies on *Gibberella fujikuroi* var. *subglutinans* the hitherto undescribed ascigerous stage of *Fusarium moniliforme* var. *subglutinans* and its pathogenicity on maize in New South Wales. *Department of Agriculture of New South Wales Scientific Bulletin* **49:** 1–68. (*subglutinans*)

543. **Edwards, S. G., S. R. Pirgozliev, M. C. Hare, and P. Jenkinson.** 2001. Quantification of trichothecene-producing *Fusarium* species in harvested grain by competitive PCR to determine efficacies of fungicides against fusarium head blight of winter wheat. *Applied and Environmental Microbiology* **67:** 1575-1580. (*sporotrichioides*)

544. **Edwards, S. V., and P. Beerli.** 2000. Perspective: Gene divergence, population divergence, and the variance in coalescence time in phylogeographic studies. *Evolution* **54:** 1839-1854. (9)

545. **Egan, B. T., R. C. Magarey, and B. J. Croft.** 1997. Sugar cane, p. 277-302. In: R. J. Hillocks and J. M. Walker (eds.), *Soilborne Diseases of Tropical Crops*. CAB International, Wallingford, Oxon, UK. (*sacchari*)

546. **Eged, S.** 2002. Changes in the content of fusaric acid during *Fusarium oxysporum* ontogenesis. *Biologia* **57:** 725-728. (*oxysporum*)

547. **Egmond, M. R., and J. de Vlieg.** 2000. *Fusarium solani pisi* cutinase. *Biochimie* **82:** 1015-1021. (*solani*)

548. **El Ani, A. S.** 1990. *In vivo* and *in vitro* nuclear conditions in the hyphae of *Fusarium solani* and *Aspergillus flavus*. *Mycologia* **82:** 566-568. (13)

549. **El Forgany, M., and D. J. Makus.** 1979. Effect of water stress on seed yield and quality of the sweet corn inbred "Luther Hill". *Journal of the American Society for Horticultural Science* **104:** 102-104. (*verticillioides*)

550. **El Gholl, N. E., J. J. McRitchie, C. L. Schoulties, and W. H. Ridings.** 1978. The identification and induction of perithecia and pathogenicity of *Gibberella tricincta*, new species. *Canadian Journal of Botany* **56:** 2203-2206. (*tricinctum*)

551. **El Gholl, N. E., C. L. Schoulties, and W. H. Ridings.** 1979. Factors affecting perithecial production in *Gibberella tricincta*. *Canadian Journal of Botany* **57:** 2497-2500. (*tricinctum*)

552. **El Gindy, A. A., and R. R. Saad.** 1990. Fungi of virgin and cultivated soil of Salhiah desert Egypt. *Zentralblatt für Mikrobiologie* **145:** 547-551. (*solani*)

553. **El Masry, H. G.** 1991. Optimization of dextranase synthesis by a locally isolated *Fusarium moniliforme* 12. *Zentralblatt für Mikrobiologie* **146:** 185-192. (*verticillioides*)

554. **El Meleigi, M. A., L. E. Claflin, and R. J. Raney.** 1983. Effect of seedborne *Fusarium moniliforme* and irrigation scheduling on colonization of root and stalk tissue: Stalk rot incidence and grain yields. *Crop Science* **23:** 1025-1028. (*verticillioides*)

555. **Elena, K., and A. C. Pappas.** 2002. Pathogenicity and vegetative compatibility of *Fusarium oxysporum* f. sp. *phaseoli* in Greece. *Journal of Phytopathology* **150:** 495-499. (*oxysporum*)

556. **Elena, K., and E. C. Tjamos.** 1995. Vegetative compatibility groups of *Fusarium oxysporum* f. sp. *dianthi* from plants and the rhizosphere of carnation in Greece. *Plant Pathology* **44:** 148-152. (*oxysporum*)

557. **Elias, K. S., and P. J. Cotty.** 1995. A rose bengal amended medium for selecting nitrate metabolism mutants from fungi. *Canadian Journal of Botany* **73:** 680-682. (2, 5)

558. **Elias, K. S., and R. W. Schneider.** 1991. Vegetative compatibility groups in *Fusarium oxysporum* f. sp. *lycopersici*. *Phytopathology* **81:** 159-162. (*oxysporum*)

559. **Elias, K. S., and R. W. Schneider.** 1992. Genetic diversity within and among races and vegetative compatibility groups of *Fusarium oxysporum* f. sp. *lycopersici* as determined by isozyme analysis. *Phytopathology* **82:** 1421-1427. (*oxysporum*)

560. **Elias, K. S., R. W. Schneider, and M. M. Lear.** 1991. Analysis of vegetative compatibility groups in nonpathogenic populations of *Fusarium oxysporum* isolated from symptomless tomato roots. *Canadian Journal of Botany* **69:** 2089-2094. (5, *oxysporum*)

561. **Elias, K. S., D. Zamir, T. Lichtman-Pleban, and T. Katan.** 1993. Population structure of *Fusarium oxysporum* f. sp. *lycopersici*: Restriction fragment length polymorphisms provide genetic evidence that vegetative compatibility group is an indicator of evolutionary origin. *Molecular Plant-Microbe Interactions* **6:** 565-572. (*oxysporum*)

562. **Eljaschewitsch, J., J. Sandfort, K. Tintelnot, I. Horbach, and B. Ruf.** 1996. Port-a-cath-related *Fusarium oxysporum* infection in an HIV-infected patient: Treatment with liposomal amphotericin B. *Mycoses* **39:** 115-119. (1, *oxysporum*)

563. **Ellanskaya, I. A., P. A. Volz, E. Nevo, S. P. Wasser, and E. V. Sokolova.** 1998. Soil micromycetes of Evolution Canyon, Lower Nahal Oren, Mt. Carmel National Park, Israel. *Microbios* **92:** 19-33. (*solani*)

564. **Ellis, J. B., and B. M. Everhart.** 1895. New species of fungi from various localities. *Proceedings of the Academy of Natural Sciences of Philadelphia*, p. 413-441. (*acuminatum*)

565. **Elmer, W. H.** 1991. Vegetative compatibility groups of *Fusarium proliferatum* from asparagus and comparisons of virulence, growth rates, and colonization of asparagus residues among groups. *Phytopathology* **81:** 852-857. (5, *proliferatum*)

566. **Elmer, W. H.** 1995. A single mating population of *Gibberella fujikuroi* (*Fusarium proliferatum*) predominates in asparagus fields in Connecticut, Massachusetts, and Michigan. *Mycologia* **87:** 68-71. (*proliferatum*)

567. **Elmer, W. H.** 1996. Fusarium fruit rot of pumpkin in Connecticut. *Plant Disease* **80:** 131-135. (*acuminatum*, *equiseti*)

568. **Elmer, W. H.** 2001. Fusarium diseases of asparagus, p. 248-262. *In* B. A. Summerell, J. F. Leslie, D. Backhouse, W. L. Bryden, and L. W. Burgess (eds.), *Fusarium: Paul E. Nelson Memorial Symposium*. APS Press, St. Paul, Minnesota. (11, *proliferatum*)

569. **Elmer, W. H., and C. T. Stephens.** 1989. Classification of *Fusarium oxysporum* f. sp. *asparagi* into vegetatively compatible groups. *Phytopathology* **79:** 88-93. (*oxysporum*)

570. **Elmer, W. H., B. A. Summerell, L. W. Burgess, D. Backhouse, and A. A. Abubaker.** 1997. *Fusarium* species associated with asparagus crowns and soil in Australia and New Zealand. *Australasian Plant Pathology* **26:** 255-261. (*nygamai*, *proliferatum*, *scirpi*)

571. **Elmer, W. H., B. A. Summerell, L. W. Burgess, and E. L. Nigh, Jr.** 1999. Vegetative compatibility groups in *Fusarium proliferatum* from asparagus in Australia. *Mycologia* **91:** 650-654. (*proliferatum*)

572. **Elmer, W. H., R. L. Wick, and P. Haviland.** 1994. Vegetative compatibility among *Fusarium oxysporum* f. sp. *basilicum* isolates recovered from basil seed and infected plants. *Plant Disease* **78:** 789-791. (*oxysporum*)

573. **Elzein, A., and J. Kroschel.** 2004. *Fusarium oxysporum* Foxy2 shows potential to control both *Striga hermonthica* and *S. asiatica*. *Weed Research* **44:** 433-438. (*oxysporum*)

574. **Elzein, A., and J. Kroschel.** 2004. Influence of agricultural by-products in liquid culture on chlamydospore production by the potential mycoherbicide *Fusarium oxysporum* Foxy2. *Biocontrol Science and Technology* **14:** 823-836. (*oxysporum*)

575. **Elzein, A., J. Kroschel, D. Müller-Stoever.** 2004. Optimization of storage conditions for adequate shelf-life of 'Pesta' formulation of *Fusarium oxysporum* 'Foxy 2', a potential mycoherbicide for *Striga*: Effects of temperature, granule size and water activity. *Biocontrol Science and Technology* **14:** 545-559. (*oxysporum*)

576. **Elzein, A., J. Kroschel, D. Müller-Stoever.** 2004. Effects of inoculum type and propagule concentration on shelf life of Pesta formulations containing *Fusarium oxysporum* Foxy 2, a potential mycoherbicide agent for *Striga* spp. *Biological Control* **30:** 203-211. (*oxysporum*)

577. **Enebak, S. A., and G. R. Stanosz.** 2003. Responses of conifer species of the Great Lakes region of North America to inoculation with the pitch canker pathogen *Fusarium circinatum*. *Forest Pathology* **33:** 333-338. (*circinatum*)

578. **Engelbrecht, M. C., W. A. Smit, and P. S. Knox-Davies.** 1983. Damping off of rooibos tea, *Aspalathus linearis*. *Phytophylactica* **15:** 121-124. (*chlamydosporum*)

579. **Engelhardt, J. A., W. W. Carlton, and J. F. Tuite.** 1989. Toxicity of *Fusarium moniliforme* var. *subglutinans* for chicks, ducklings and turkey poults. *Avian Diseases* **33:** 357-360. (*subglutinans*)

580. **Engels, R., and J. Kramer.** 1996. The incidence of Fusaria and occurrence of selected *Fusarium* mycotoxins on *Lolium* spp. in Germany. *Mycotoxin Research* **12:** 31-40. (*anthophilum*)

581. **Enkerli, J., G. Bhatt, and S. F. Covert.** 1997. *Nht1*, a transposable element cloned from a dispensable chromosome in *Nectria haematococca. Molecular Plant-Microbe Interactions* **10:** 742-749. (*solani*)

582. **Epple, P., K. Apel, and H. Bohlmann.** 1997. Overexpression of an endogenous thionin enhances resistance of *Arabidopsis* against *Fusarium oxysporum. Plant Cell* **9:** 509-520. (*oxysporum*)

583. **Epple, P., A. Vignutelli, K. Apel, and H. Bohlmann.** 1998. Differential induction of the *Arabidopsis thaliana Thi2.1* gene by *Fusarium oxysporum* f. sp. *matthiolae. Molecular Plant-Microbe Interactions* **11:** 523-529. (*oxysporum*)

584. **Epstein, L., Y.-H. Kwon, D. E. Almond, L. M. Schached, and M. J. Jones.** 1994. Genetic and biochemical characterization of *Nectria haematococca* strains with adhesive and adhesion-reduced macroconidia. *Applied and Environmental Microbiology* **60:** 524-530. (*solani*)

585. **Erdey, D. P., D. J. Mycock, and P. Berjak.** 1997. The elimination of *Fusarium moniliforme* (Sheldon) infection in maize caryopses by hot water treatments. *Seed Science and Technology* **25:** 485-501. (3)

586. **Esaki, N., M. Watanabe, T. Kurihara, and K. Soda.** 1994. Fungal thermostable α-dialkylamino acid aminotransferase: Occurrence, purification and characterization. *Archives of Microbiology* **161:** 110-115. (*solani*)

587. **Espionel-Ingroff, A.** 1998. *In vitro* activity of the new triazole voriconazole (UK-109,496) against opportunistic filamentous and dimorphic fungi and common and emerging yeast pathogens. *Journal of Clinical Microbiology* **36:** 198-202. (*oxysporum, solani*)

588. **Espionel-Ingroff, A., K. Boyle, and D. J. Sheehan.** 2001. *In vitro* antifungal activities of voriconazole and reference agents as determined by NCCLS methods: Review of the literature. *Mycopathologia* **150:** 101-115. (*oxysporum*)

589. **Esser, K., and R. Kuenen.** 1967. *Genetics of Fungi.* Springer-Verlag, New York. (4)

590. **Etcheverry, M., A. Nesci, G. Barros, A. Torres, and S. Chulze.** 1999. Occurrence of *Aspergillus* section *Flavi* and aflatoxin B_1 in corn genotypes and corn meal in Argentina. *Mycopathologia* **147:** 37-41. (*nygamai*)

591. **Etcheverry, M., A. Torres, M. L. Ramirez, S. Chulze, and N. Magan.** 2002. In vitro control of growth and fumonisin production by *Fusarium verticillioides* and *F. proliferatum* using antioxidants under different water availability and temperature regimes. *Journal of Applied Microbiology* **92:** 624-632. (*proliferatum, verticillioides*)

592. **Evidente, A., C. Amalfitano, R. Pengue, and C. Altomare.** 1999. High performance liquid chromatography for the analysis of fusapyrone and deoxyfusapyrone, two antifungal α-pyrones from *Fusarium semitectum. Natural Toxins* **7:** 133-137. (*semitectum*)

593. **Evidente, A., L. Conti, C. Altomare, A. Bottalico, G. Sindona, A. L. Segre, A. Logrieco.** 1994. Fusapyrone and deoxyfusapyrone, two antifungal α-pyrones from *Fusarium semitectum. Natural Toxins* **2:** 4-13. (*semitectum*)

594. **Fadda, M. E., S. Cosentino, C. Atzori, M. Deplano, P. Cardia, and F. Palmas.** 1990. Mold allergy fungal airborne spores in the homes of allergic and non-allergic subjects. *Folia Allergolica Immunologia Clinica* **37:** 335-342. (3)

595. **Fajola, A. O.** 1983. Seedborne fungi of tobacco (*Nicotiana tabacum*) in Nigeria. *Seed Research* **11:** 77-81. (*chlamydosporum*)

596. **Falcon, M. A., A. Rodriguez, A. Carnicero, V. Regalado, F. Perestelo, O. Milstein, and G. de la Fuente.** 1995. Isolation of microorganisms with lignin transformation potential from soil of Tenerife Island. *Soil Biology and Biochemistry* **27:** 121-126. (*oxysporum*)

597. **Falk, S. P., R. C. Pearson, D. M. Gadoury, R. C. Seem, and A. Sztejnberg.** 1996. *Fusarium proliferatum* as a biocontrol agent against grape downy mildew. *Phytopathology* **86:** 1010-1017. (*proliferatum*)

598. **Falloon, R. E.** 1985. Fungi pathogenic to ryegrass *Lolium* spp. seedlings. *Plant and Soil* **86:** 79-86. (*culmorum*)

599. **☑Fang, C.-T., S.-C. Chang, J.-L. Tang, P.-R. Hsueh, Y.-L. Chang, and C.-C. Hung, and Y.-C. Chen.** 1997. *Fusarium solani* fungemia in a bone marrow transplant recipient. *Journal of the Formosan Medical Association* **96:** 129-133. (*solani*)

600. **Farber, J. M., and G. W. Sanders.** 1986. Production of fusarin C by *Fusarium* spp. *Journal of Agricultural and Food Chemistry* **34:** 963-966. (*avenaceum, culmorum, graminearum, poae, sambucinum, sporotrichioides*)

601. **Farber, J. M., and G. W. Sanders.** 1986. Fusarin C production by North American isolates of *Fusarium moniliforme. Applied and Environmental Microbiology* **51:** 381-384. (*verticillioides*)

602. **Farber, J. M., G. W. Sanders, G. A. Lawrence, and P. M. Scott.** 1988. Production of moniliformin by Canadian isolates of *Fusarium. Mycopathologia* **101:** 187-190. (*graminearum, subglutinans*)

603. **Farina, C., F. Vailati, A. Manisco, and A. Goglio.** 1999. Fungaemia survey: A 10-year experience in Bergamo, Italy. *Mycoses* **42:** 543-548. (*verticillioides*)

604. **Farkya, S., A. K. Pandey, and R. C. Rajak.** 2001. Effect of edaphic and climatic factors on the mycoherbicidal potential of *Fusarium* spp. against *Parthenium hysterophorus* L. *Journal of Biological Control* **15:** 177-182. (*oxysporum*)

605. **Farnworth, N. E., G. D. Robson, A. P. J. Trinci, and M. G. Wiebe.** 2003. Trypsin-like protease (TLP) production in *Fusarium oxysporum* and *Fusarium venenatum* and use of the TLP promoter for recombinant protein (glucoamylase) production. *Enzyme and Microbial Technology* **33:** 85-91. (*venenatum*)

606. **Fawzi, E. M.** 2003. Production and purification of β-glucosidase and protease by *Fusarium proliferatum* NRRL 26517 grown on *Ficus nitida* wastes. *Annals of Microbiology* **53:** 463-476. (*proliferatum*)

607. **Federici, L., B. Mattei, C. Caprari, C. Savino, F. Cervone, and D. Tsernoglou.** 1999. Crystallization and preliminary X-ray diffraction study of the endo-polygalacturonase from

Fusarium moniliforme. *Acta Crystallographica Section D: Biological Crystallography* **55:** 1359-1361. (*verticillioides*)

608. **Fekete, C., G. Giczey, I. Papp, L. Szabo, and L. Hornok.** 1995. High-frequency occurrence of virus-like particles with double-stranded RNA genome in *Fusarium poae*. *FEMS Microbiology Letters* **131:** 295-299. (*poae*)

609. **Fekete, C. and L. Hornok.** 1997. A repetitive DNA sequence associated with karyotype variability in *Fusarium poae*. *Acta Phytopathologica et Entomologica Hungarica* **32:** 29-38. (*poae*)

610. **Fekete, C., A. Logrieco, G. Giczey, and L. Hornok.** 1997. Screening of fungi for the presence of the trichodiene synthase encoding sequence by hybridization to the *Tri5* gene cloned from *Fusarium poae*. *Mycopathologia* **138:** 91-97. (*sporotrichioides*, *tricinctum*)

611. **Fekete, C., R. Nagy, A. J. M. Debets, and L. Hornok.** 1993. Electrophoretic karyotypes and gene mapping in eight species of the *Fusarium* sections *Arthrosporiella* and *Sporotrichiella*. *Current Genetics* **24:** 500-504. (*camptoceras*, *poae*)

612. **Fernandez, D., K. Assigbetse, M. P. Dubois, and J. P. Geiger.** 1994. Molecular characterization of races and vegetative compatibility groups in *Fusarium oxysporum* f. sp. *vasinfectum*. *Applied and Environmental Microbiology* **60:** 4039-4046. (*oxysporum*)

613. **Fernandez, D., M. Ouinten, A. Tantaoui, and J. P. Geiger.** 1997. Molecular records of micro-evolution within the Algerian population of *Fusarium oxysporum* f. sp. *albedinis* during its spread to new oases. *European Journal of Plant Pathology* **103:** 485-490. (*oxysporum*)

614. **Fernandez, D., M. Ouinten, A. Tantaoui, J. P. Geiger, M.-J. Daboussi, and T. Langin.** 1998. *Fot1* insertions in the *Fusarium oxysporum* f. sp. *albedinis* genome provide diagnostic PCR targets for detection of the date palm pathogen. *Applied and Environmental Microbiology* **64:** 633-636. (*oxysporum*)

615. **Fernandez, J. A., D. S. Wofford, and J. L. Horton.** 1985. Interactive effects of freezing and common rot fungi on winter wheat. *Phytopathology* **75:** 845-847. (*acuminatum*)

616. **Fernandez, N., B. Patino, and C. Vazquez.** 1993. Pectin degrading enzymes secreted by six isolates of *Fusarium oxysporum*. *Mycological Research* **97:** 461-466. (*oxysporum*)

617. **Fernandez-Martin, R., E. Cerda-Olmedo, and J. Avalos.** 2000. Homologous recombination and allele replacement in transformants of *Fusarium fujikuroi*. *Molecular and General Genetics* **263:** 838-845. (*fujikuroi*)

618. **Fernandez-Martin, R., F. Reyes, C. E. Domenech, E. Cabrera, P. M. Bramley, A. F. Barrero, J. Avalos, and E. Cerda-Olmedo.** 1995. Gibberellin biosynthesis in *gib* mutants of *Gibberella fujikuroi*. *Journal of Biological Chemistry* **270:** 14970-14974. (*fujikuroi*)

619. **Fernando, G., W. Zimmermann, and P. E. Kolattukudy.** 1984. Suberin grown *Fusarium solani* f. sp. *pisi* generates a cutinase-like esterase which depolymerizes the aliphatic components of suberin. *Physiological Plant Pathology* **24:** 143-156. (*solani*)

620. **Fernando, W. G. D., T. C. Paulitz, W. L. Seaman, P. Dutilleul, and J. D. Miller.** 1997. Head blight gradients caused by *Gibberella zeae* from area sources of inoculum in wheat field plots. *Phytopathology* **87:** 414-421. (3)

621. **Ferreira, B. S., C. R. C. Calado, F. van Keulen, L. P. Fonseca, J. M. S. Cabral, and M. M. R. da Fonseca.** 2003. Towards a cost effective strategy for cutinase production by a recombinant *Saccharomyces cerevisiae*: Strain physiological aspects. *Applied Microbiology and Biotechnology* **61:** 69-76. (*solani*)

622. **Ferreira, B. S., C. R. C. Calado, F. van Keulen, L. P. Fonseca, J. M. S. Cabral, and M. M. R. da Fonseca.** 2004. Recombinant *Saccharomyces cerevisiae* strain triggers acetate production to fuel biosynthetic pathways. *Journal of Biotechnology* **109:** 159-167. (*solani*)

623. **Fiely, M. B., J. C. Correll, and T. E. Morelock.** 1995. Vegetative compatibility, pathogenicity, and virulence diversity of *Fusarium oxysporum* recovered from spinach. *Plant Disease* **79:** 990-993. (*oxysporum*)

624. **Filion, M., M. St. Arnaud, and S. H. Jabaji-Hare.** 2003. Direct quantification of fungal DNA from soil substrate using real-time PCR. *Journal of Microbiological Methods* **53:** 67-76. (*solani*)

625. **Fisher, N. L., L. W. Burgess, T. A. Toussoun, and P. E. Nelson.** 1982. Carnation leaves as a substrate and for preserving cultures of *Fusarium* species. *Phytopathology* **72:** 151-153. (2, 4)

626. **Fisher, N. L., W. F. O. Marasas, and T. A. Toussoun.** 1983. Taxonomic importance of microconidial chains in *Fusarium* section *Liseola* and effects of water potential on their formation. *Mycologia* **75:** 693-698. (2, *proliferatum*)

627. **Fitzpatrick, P. F., A. M. Orville, A. Nagpal, and M. P. Valley.** 2005. Nitroalkane oxidase, a carbanion-forming flavoprotein homologous to acyl-CoA dehydrogenase. *Archives of Biochemistry and Biophysics* **433:** 157-165. (*oxysporum*)

628. **Flaherty, J. E., A. M. Pirttila, B. H. Bluhm, and C. P. Woloshuk.** 2003. *PAC1*, a pH-regulatory gene from *Fusarium verticillioides*. *Applied and Environmental Microbiology* **69:** 5222-5227. (*verticillioides*)

629. **Fleissner, A., C. Sopalla, and K. M. Weltring.** 2002. An ATP-binding cassette multidrug-resistance transporter is necessary for tolerance of *Gibberella pulicaris* to phytoalexins and virulence on potato tubers. *Molecular Plant-Microbe Interactions* **15:** 102-108. (*sambucinum*)

630. **Fletcher, J. T.** 1994. Fusarium stem and fruit rot of sweet peppers in the glasshouse. *Plant Pathology* **43:** 225-227. (*solani*)

631. **Fletcher, J. T., and K. L. Lord.** 1985. A stem rot of tomato caused by *Fusarium merismoides*. *Plant Pathology* **34:** 443-445. (*merismoides*)

632. **Flipsen, J. A. C., A. C. M. Appel, H. T. M. W. M van der Hijden, and C. T. Verrips.** 1998. Mechanims of removal of immobilized triacylglycerol by lipolytic enzymes in a sequential laundry wash process. *Enzyme and Microbial Technology* **23:** 274-280. (*solani*)

633. **Flipsen, J. A. C., M. A. van Schaick, R. Dijkman, H. T. W. M. van der Hijden, H. M. Verheij, and M. R. Egmond.** 1999. Interfacial binding of cutinase rather than its catalytic activity determines the steady state interfacial tension during oil drop lipid hydrolysis. *Chemistry and Physics of Lipids* **97:** 181-191. (*solani*)

634. **Florakis, G. J., G. Moazami, H. Schubert, C. J. Koester, and J. D. Auran.** 1997. Scanning slit confocal microscopy of fungal keratitis. *Archives of Ophthalmology* **115:** 1461-1463. (*solani*)

635. **Fong, Y. K., S. Anuar, H. P. Lim, F. Y. Tham, and F. R. Sanderson.** 2000. A modified filter paper technique for long-term preservation of some fungal cultures. *Mycologist* **14:** 127-130. (4)

636. **Forbes, G. A.** 1984. Studies on the *Fusarium* species in the sections *Arthrosporiella*, *Discolor*, *Gibbosum* and *Roseum*. M. Sc. Agr. Thesis, University of Sydney, Sydney, Australia. 112 pp. (*armeniacum*)

637. **Foroozan, R., R. C. Eagle, Jr., and E. J. Cohen.** 2000. Fungal keratitis in a soft contact lens wearer. *CLAO Journal* **26:** 166-168. (*solani*)

638. **Fotso, J., J. F. Leslie, and J. S. Smith.** 2002. Production of beauvericin, moniliformin, fusaproliferin, and fumonisins B_1, B_2 and B_3 by *ex*-type strains of fifteen *Fusarium* species. *Applied and Environmental Microbiology* **68:** 5195-5197. (*acutatum, begoniae, brevicatenulatum, bulbicola, circinatum, concentricum, denticulatum, guttiforme, lactis, nisikadoi, phyllophilum, pseudoanthophilum, pseudocircinatum, pseudonygamai, ramigenum*)

639. **Francis, R. G., and L. W. Burgess.** 1977. Characteristics of two populations of *Fusarium roseum* 'Graminearum' in eastern Australia. *Transactions of the British Mycological Society* **68:** 421-427. (*pseudograminearum*)

640. **Fravel, D., C. Olivain, and C. Alabouvette.** 2003. *Fusarium oxysporum* and its biocontrol. *New Phytologist* **157:** 493-502. (*oxysporum*)

641. **Frederiksen, R. A., and G. N. Odvody, eds.** 2000. *Compendium of Sorghum Diseases*, 2nd ed. APS Press, St. Paul, Minnesota. (*thapsinum*)

642. **Freeman, S., C. Ginzburg, and J. Katan.** 1989. Heat shock protein synthesis in propagules of *Fusarium oxysporum* f. sp. *niveum*. *Phytopathology* **79:** 1054-1058. (*oxysporum*)

643. **Freeman, S., D. Klein-Gueta, and N. Korolev.** 2004. Epidemiology and survival of *Fusarium mangiferae*, the causal agent of mango malformation disease. *Acta Horticulturae* **645:** 487-491. (*mangiferae*)

644. **Freeman, S., M. Maimon, and Y. Pinkas.** 1999. Use of GUS transformants of *Fusarium subglutinans* for determining etiology of mango malformation disease. *Phytopathology* **89:** 456-461. (*mangiferae*)

645. **Freeman, S., A. Zveibil, H. Vintal, and M. Maymon.** 2002. Isolation of nonpathogenic mutants of *Fusarium oxysporum* f. sp. *melonis* for biological control of Fusarium wilt in cucurbits. *Phytopathology* **92:** 164-168. (*oxysporum*)

646. **Freidank, H.** 1995. Hyalohyphomycoses due to *Fusarium* spp.: Two case reports and review to the literature. *Mycoses* **38:** 69-74. (*solani, verticillioides*)

647. **Frelier, P. F., L. Sigler, and P. E. Nelson.** 1985. Mycotic pneumonia caused by *Fusarium moniliforme* in an alligator (*Alligator mississippiensis*). *Sabouraudia* **23:** 399-402. (*verticillioides*)

648. **Frisullo, S., A. Logrieco, A. Moretti, G. Grammatikaki, and A. Bottalico.** 1994. Banana corm and root rot by *Fusarium compactum* in Crete. *Phytopathologia Mediterranea* **33:** 78-82. (*compactum*)

649. **Fu, Y. H., and G. A. Marzluf.** 1990. *nit-2*, the major positive-acting nitrogen regulatory gene of *Neurospora crassa* encodes a sequence-specific DNA binding protein. *Proceedings of the National Academy of Sciences (USA)* **87:** 5331-5335. (5)

650. **Fuchs, J. G., Y. Moenne-Loccoz, and G. Defago.** 1997. Nonpathogenic *Fusarium oxysporum* strain Fo47 induces resistance to Fusarium wilt in tomato. *Plant Disease* **81:** 492-496. (*oxysporum*)

651. **Fuchs, J. G., Y. Moenne-Loccoz, and G. Defago.** 1999. Ability of nonpathogenic *Fusarium oxysporum* Fo47 to protect tomato against Fusarium wilt. *Biological Control* **14:** 105-110. (*oxysporum*)

652. **Fuchs, U., K. J. Czymmek, and J. A. Sweigard.** 2004. Five hydrophobin genes in *Fusarium verticillioides* include two required for microconidial chain formation. *Fungal Genetics and Biology* **41:** 852-864. (*verticillioides*)

653. **Fugro, P. A.** 1999. A new disease of okra (*Abelmoschus esculentus* L.) in India. *Journal of Mycology and Plant Pathology* **29:** 264. (*chlamydosporum*)

654. **Fukaya, N., L.-P. Chow, Y. Sugiura, A. Tsugita, Y. Ueno, and K. Tabuchi.** 1995. Partial amino acid sequence of peptidyl-prolyl isomerases of *Fusarium sporotrichioides*. *Journal of Biomedical Science* **2:** 353-356. (*sporotrichioides*)

655. **Funnell, D. L., P. S. Matthews, H. D. VanEtten.** 2001. Breeding for highly fertile isolates of *Nectria haematococca* MP VI that are highly virulent on pea and *in planta* selection for virulent recombinants. *Phytopathology* **91:** 92-101. (*solani*)

656. **Funnell, D. L., and H. D. VanEtten.** 2002. Pisatin demethylase genes are on dispensable chromosomes while genes for pathogenicity on carrot and ripe tomato are on other chromosomes in *Nectria haematococca*. *Molecular Plant-Microbe Interactions* **15:** 840-846. (*solani*)

657. **Furlong, E. B., L. M. V. Soares, C. C. Lasca, and E. Y. Kohara.** 1995. Mycotoxins and fungi in wheat stored in elevators in the state of Rio Grande do Sul, Brazil. *Food Additives and Contaminants* **12:** 683-688. (*dimerum*)

658. **Gabler M., and L. Fischer.** 1999. Production of a new D-amino acid oxidase from the fungus *Fusarium oxysporum*. *Applied and Environmental Microbiology* **65:** 3750-3753. (*oxysporum*)

659. **Gabriele, P., and R. K. Hutchins.** 1996. Fusarium endophthalmitis in an intravenous drug abuser. *American Journal of Ophthalmology* **122:** 119-121. (*dimerum*)

660. **Gadda, G., A. Banerjee, and P. F. Fitzpatrick.** 2000. Identification of an essential tyrosine residue in nitroalkane oxidase by modification with tetranitromethane. *Biochemistry* **39:** 1162-1168. (*oxysporum*)

661. **Gadda, G., D. Y. Choe, and P. F. Fitzpatrick.** 2000. Use of pH and kinetic isotope effects to dissect the effects of substrate size on binding and catalysis by nitroalkane oxidase. *Archives of Biochemistry and Biophysics* **382:** 138-144. (*oxysporum*)

662. **Gadda, G., R. D. Edmondson, D. H. Russell, and P. F. Fitzpatrick.** 1997. Identification of the naturally occurring flavin of nitroalkane oxidase from *Fusarium oxysporum* as a 5-nitrobutyl-FAD and conversion of the enzyme to the active FAD-containing form. *Journal of Biological Chemistry* **272:** 5563-5570. (*oxysporum*)

663. **Gadda, G., and P. F. Fitzpatrick.** 1998. Biochemical and physical characterization of the active FAD-containing form of nitroalkane oxidase from *Fusarium oxysporum. Biochemistry* **37:** 6154-6164. (*oxysporum*)

664. **Gadda, G., and P. F. Fitzpatrick.** 1999. Substrate specificity of a nitroalkane-oxidizing enzyme. *Archives of Biochemistry and Biophysics* **363:** 309-313. (*oxysporum*)

665. **Gagkaeva, T. Y., and T. Yli-Mattila.** 2004. Genetic diversity of *Fusarium graminearum* in Europe and Asia. *European Journal of Plant Pathology* **110:** 551-562. (*graminearum*)

666. **Gajdusek, D. C.** 1953. Acute infectious hemorrhagic fevers and mycotoxicosis in the Union of Soviet Socialist Republics, p. 82-106. In *Medical Science Publication* No. 2, Walter Reed Army Medical Center, Washington, D.C. (1)

667. **Gale, L. R., J. D. Bryant, H. Giese, T. Katan, K. O'Donnell, H. Suga, T. R. Usgaard, T. J. Ward, and H. C. Kistler.** 2005. A genetic map of *Gibberella zeae* using sequence-tagged sites and AFLPs. *Fungal Genetics Newsletter* **52(Suppl.):** 82. (7, *graminearum*)

668. **Gale, L. R., L. F. Chen, C. A. Hernick, K. Takamura, and H. C. Kistler.** 2002. Population analysis of *Fusarium graminearum* from wheat fields in eastern China. *Phytopathology* **92:** 1315-1322. (9, 11, *graminearum*)

669. **Gale, L. R., T. Katan, and H. C. Kistler.** 2003. The probable center of origin of *Fusarium oxysporum* f. sp. *lycopersici* VCG 0033. *Plant Disease* **87:** 1433-1438. (11)

670. **Gallegos, A., D. J. Jacobson, N. B. Raju, M. P. Skupinski, and D. O. Natvig.** 2000. Suppressed recombination and a pairing anomaly on the mating-type chromosome of *Neurospora tetrasperma. Genetics* **154:** 623-633. (6)

671. **Gams, W.** 1971. Cephalosporium-*artige Schimmelpilze (Hyphomycetes)*. Gustav Fischer Verlag, Stuttgart, Germany. (*sacchari*)

672. **Gams, W., M. Klamer, and K. O'Donnell.** 1999. *Fusarium miscanthi sp. nov.* from Miscanthus litter. *Mycologia* **91:** 263-268. (8, *miscanthi*, *nisikadoi*)

673. **Gams, W., H. I. Nirenberg, K. A. Seifert, D. Brayford, and U. Thrane.** 1997. Proposal to conserve the name *Fusarium sambucinum* (Hyphomycetes). *Taxon* **46:** 111-113. (*sambucinum*)

674. **Ganetzky, B.** 2000. Tracking down a cheating gene. *American Scientist* **88:** 128-135. (10)

675. **Garbayo, I., C. Vilchez, J. E. Nava-Saucedo, and J. N. Barbotin.** 2003. Nitrogen, carbon and light-mediated regulation studies of carotenoid biosynthesis in immobilized mycelia of *Gibberella fujikuroi. Enzyme and Microbial Technology* **33:** 629-634. (*fujikuroi*)

676. **Garcia, M. L., J. M. Herreras, E. Dios, P. Argueso, and A. Almaraz.** 2002. Evaluation of lectin staining in the diagnosis of fungal keratitis in an experimental rabbit model. *Molecular Vision* **8:** 10-16. (*solani*)

677. **Garcia-Maceira, F. I., A. di Pietro, M. D. Huertas-Gonzalez, M. C. Ruiz-Roldan, and M. I. G. Roncero.** 2001. Molecular characterization of an endopolygalacturonase from *Fusarium oxysporum* expressed during early stages of infection. *Applied and Environmental Microbiology* **67:** 2191-2196. (*oxysporum*)

678. **Garcia-Pedrajas, M. D., B. W. Bainbridge, J. B. Heale, E. Perez-Artes, and R. M. Jimenez-Diaz.** 1999. A simple PCR-based method for the detection of the chickpea-wilt pathogen *Fusarium oxysporum* f. sp. *ciceris* in artificial and natural soils. *European Journal of Plant Pathology* **105:** 251-259. (*oxysporum*)

679. **Garcia-Pedrajas, M. D., and M. I. G. Roncero.** 1996. A homologous and self-replicating system for efficient transformation of *Fusarium oxysporum. Current Genetics* **29:** 191-198. (*oxysporum*)

680. **Gardner, H. W., A. E. Desjardins, S. P. McCormick, and D. Weisleder.** 1994. Detoxification of the potato phytoalexin rishitin by *Gibberella pulicaris. Phytochemistry* **37:** 1001-1005. (*sambucinum*)

681. **Gardner, H. W., A. E. Desjardins, and R. D. Plattner.** 1988. Biotransformation of the potato phytoalexin lubimin by *Gibberella pulicaris*: Identification of major products. *Biochimica et Biophysica Acta* **966:** 347-356. (*sambucinum*)

682. **Gari-Toussaint, M., J. M. Leguay, C. Zur, J. F. Michiels, L. Ferrara, F. Negre, and Y. le Fichoux.** 1997. *Fusarium solani* keratitis in a diabetic patient. *Journal de Mycologie Medicale* **7:** 227-231. (*solani*)

683. **Garibaldi, A., G. Gilardi, and M. L. Gullino.** 2004. Seed transmission of *Fusarium oxysporum* f. sp. *lactucae. Phytoparasitica* **32:** 61-65. (*oxysporum*)

684. **Garnjobst, L.** 1953. Genetic control of heterokaryosis in *Neurospora crassa. American Journal of Botany* **42:** 444-448. (5)

685. **Gatch, E. W., and G. P. Munkvold.** 2002. Fungal species composition in maize stalks in relation to European corn borer injury and transgenic insect protection. *Plant Disease* **86:** 1156-1162 (*graminearum*, *proliferatum*, *subglutinans*, *verticillioides*)

686. **Gaur, V. K., and L. C. Sharma.** 1989. Variability in single spore isolates of *Fusarium udum* Butler. *Mycopathologia* **107:** 9-16. (*udum*)

687. **Gaylarde, C., and J. Kelley.** 1995. Genetic variation in *Fusarium merismoides* preserved by two different methods. *World Journal of Microbiology and Biotechnology* **11:** 319-321. (4, *merismoides*)

688. **Geiser, D. M., M. Jimenez-Gasco, S. Kang, I. Makalowska, N. Veeraraghavan, T. J. Ward, N. Zhang, G. A. Kuldau, and K. O'Donnell.** 2004. FUSARIUM-ID v. 1.0: A DNA sequence database for identifying *Fusarium. European Journal of Plant Pathology* **110:** 473-479. (7)

689. **Geiser, D. M., J. H. Juba, B. Wang, and S. N. Jeffers.** 2001. *Fusarium hostae* sp. nov., a relative of *F. redolens* with a *Gibberella* teleomorph. *Mycologia* **93:** 670-678. (3, 8, *hostae*)

690. **Geiser, D. M., M. L. Lewis-Ivey, G. Hakiza, J. H. Juba, and S. A. Miller.** 2005. *Gibberella xylarioides* (anamorph *Fusarium xylarioides*), a causative agent of coffee wilt disease in Africa, is a previously unrecognized member of the *G. fujikuroi* species complex. *Mycologia* **97:** 191-201. (4, 11)

691. **Gelderblom, W. C. A., K. Jaskiewicz, W. F. O. Marasas, P. G. Thiel, R. M. Horak, M. Vleggaar, and N. P. J. Kriek.** 1988. Fumonisins – Novel mycotoxins with cancer-promoting activity produced by *Fusarium moniliforme. Applied and Environmental Microbiology* **54:** 1806-1811. (*verticillioides*)

692. **Gelderblom, W. C. A., N. P. J. Kriek, W. F. O. Marasas, and P. G. Thiel.** 1991. Toxicity and carcinogenicity of the *Fusarium moniliforme* metabolite fumonisin B_1 in rats. *Carcinogenesis* **12:** 1247-1252. (*verticillioides*)

693. **Gelderblom, W. C. A., J. V. Seier, P. W. Snijman, D. J. van Schalkwyk, G. S. Shephard, and W. F. O. Marasas.** 2001. Toxicity of culture material of *Fusarium verticillioides* strain MRC 826 to nonhuman primates. *Environmental Health Perspectives* **109(Suppl. 2):** 267-276. (*verticillioides*)

694. **Gelderblom, W. C. A., P. G. Thiel, W. F. O. Marasas, and K. J. van der Merwe.** 1984. Natural occurrence of fusarin C, a mutagen produced by *Fusarium moniliforme* in corn. *Journal of Agricultural and Food Chemistry* **32:** 1064-1067. (*graminearum*, *verticillioides*)

695. **Gerber, N. N., and M. S. Ammar.** 1979. New antibiotic pigments related to fusarubin from *F. solani* (Mart.) Sacc. II. Structure elucidations. *Journal of Antibiotics* **32:** 685-688. (*solani*)

696. **Gerlach, K. S., S. Bentley, N. Y. Moore, K. G. Pegg, and E. A. B. Aitken.** 2001. Characterization of Australian isolates of *Fusarium oxysporum* f. sp. *cubense* by DNA fingerprinting analysis. *Australian Journal of Agricultural Research* **51:** 945-953. (*oxysporum*)

697. **Gerlach, W., and H. Nirenberg.** 1982. The genus *Fusarium* – A pictorial atlas. *Mitteilungen aus der Biologischen Bundesanstalt Für Land- und Forstwirtschaft (Berlin – Dahlem)* **209:** 1-405. (8, 9, *anthophilum*, *camptoceras*, *equiseti*, *heterosporum*, *lactis*, *oxysporum*, *poae*, *proliferatum*, *redolens*, *sporotrichioides*, *sacchari*, *sambucinum*, *scirpi*, *semitectum*, *solani*, *subglutinans*, *succisae*, *tricinctum*, *udum*, *verticillioides*)

698. **Ghareib, M., H. S. Hamdy, and A. A. Khalil.** 2001. Production of intracellular milk-clotting enzyme in submerged cultures of *Fusarium subglutinans*. *Acta Microbiologica Polonica* **50:** 139-147. (*subglutinans*)

699. **Gianni, C., A. Cerri, and C. Crosti**. 1997. Unusual clinical features of fingernail infection by *Fusarium oxysporum*. *Mycoses* **40:** 455-459. (*oxysporum*)

700. **Gibb, E., and J. H. Walsh.** 1980. Effect of nutritional factors and carbon dioxide on growth of *Fusarium moniliforme* and other fungi in reduced oxygen concentrations. *Transactions of the British Mycological Society* **74:** 111-118. (*solani*, *verticillioides*)

701. **Gilbert, J., R. L. Conner, M. R. Fernandez, D. McLaren, and S. M. Woods.** 2003. Role of spring wheat seed infested with *Fusarium graminearum* in spread and development of Fusarium head blight and effects on agronomic performance. *Canadian Journal of Plant Pathology* **25:** 73-81. (*culmorum*, *graminearum*)

702. **Gilbert, J., and W. G. D. Fernando.** 2004. Epidemiology and biological control of *Gibberella zeae*/*Fusarium graminearum*. *Canadian Journal of Plant Pathology* **26:** 464-472. (*graminearum*)

703. **Gilbertson, R. L, W. M. Brown, Jr., E. G. Ruppel, and J. L. Capinera.** 1986. Association of corn stalk rot *Fusarium* spp. and western corn rootworm beetles in Colorado, USA. *Phytopathology* **76:** 1309-1314. (*verticillioides*)

704. **Gillespie, D. R., and J. G. Menzies.** 1993. Fungus gnats vector *Fusarium oxysporum* f. sp. *radicis-lycopersici*. *Annals of Applied Biology* **123:** 539-544. (*oxysporum*)

705. **Giordano, W., J. Avalos, E. Cerda-Olmedo, and C. E. Domenech.** 1999. Nitrogen availability and production of bikaverin and gibberellins in *Gibberella fujikuroi*. *FEMS Microbiology Letters* **173:** 389-393. (*fujikuroi*)

706. **Giraud, T., E. Fournier, D. Vautrin, M. Solignac, E. Vercken, B. Bakan, and Y. Brygoo.** 2002. Isolation of eight polymorphic microsatellite loci, using an enrichment protocol, in the phytopathogenic fungus *Fusarium culmorum*. *Molecular Ecology Notes* **2:** 121-123. (*culmorum*)

707. **Gkargkas, K., D. Mamma, G. Nedev, E. Topakas, P. Christakopoulos, D. Kekos, and B. J. Macris.** 2004. Studies on a N-acetyl-β-D-glucosaminidase produced by *Fusarium oxysporum* F3 grown in solid-state fermentation. *Process Biochemistry* **39:** 1599-1605. (*oxysporum*)

708. **Glass, N. L., J. Grotelueschen, and R. L. Metzenberg.** 1990. *Neurospora crassa A* mating type region. *Proceedings of the National Academy of Sciences (USA)* **87:** 4912-4916. (6)

709. **Glass, N. L., D. J. Jacobson, and P. K. T. Shiu.** 2000. The genetics of hyphal fusion and vegetative incompatibility in filamentous ascomycete fungi. *Annual Review of Genetics* **34:** 165-186. (5)

710. **Glass, N. L., and M. L. Smith.** 1994. Structure and function of a mating-type gene from the homothallic species *Neurospora africana*. *Molecular and General Genetics* **244:** 401-409. (6)

711. **Glenn, A. E., S. E. Gold, and C. W. Bacon.** 2002. *Fdb1* and *Fdb2*, *Fusarium verticillioides* loci necessary for detoxification of preformed antimicrobials from corn. *Molecular Plant-Microbe Interactions* **15:** 91-101. (*verticillioides*)

712. **Glenn, A. E., D. M. Hinton, I. E. Yates, and C. W. Bacon.** 2001. Detoxification of corn antimicrobial compounds as the basis for isolating *Fusarium verticillioides* and some other *Fusarium* species from corn. *Applied and Environmental Microbiology* **67:** 2973-2981. (2, *graminearum*, *subglutinans*, *verticillioides*)

713. **Glenn, A. E., F. I. Meredith, W. H. Morrison III, and C. W. Bacon.** 2003. Identification of intermediate and branch metabolites resulting from biotransformation of 2-benzoxazolinone by *Fusarium verticillioides*. *Applied and Environmental Microbiology* **69:** 3165-3169. (*verticillioides*)

714. **Glenn, A. E., E. A. Richardson, and C. W. Bacon.** 2004. Genetic and morphological characterization of a *Fusarium verticillioides* conidiation mutant. *Mycologia* **96:** 968-980. (*verticillioides*)

715. **Glinski, M., C. Urbanke, T. Hornbogen, and R. Zocher.** 2002. Enniatin synthetase is a monomer with extended structure: Evidence for an intramolecular reaction mechanism. *Archives of Microbiology* **178:** 267-273. (*scirpi*)

716. **Godoy, P., F. Nunes, V. Silva, J. Tomimori-Yamashita, L. Zaror, and O. Fischman.** 2004. Onychomycosis caused by *Fusarium solani* and *Fusarium oxysporum* in Sao Paulo, Brazil. *Mycopathologia* **157:** 287-290. (*oxysporum*, *solani*)

717. **Goldlust, A., and Z. Bohak.** 1989. Induction, purification and characterization of the nitrilase of *Fusarium oxysporum* f. sp. *melonis*. *Biotechnology and Applied Biochemistry* **11:** 581-601. (*oxysporum*)

718. **Goldschmied, R. A., J. Friedman, and C. S. Block.** 1993. *Fusarium* spp. isolated from non-ocular sites: A 10 year experience at an Israeli general hospital. *Journal de Mycologie Medicale* **3:** 99-102. (1, *equiseti*, *oxysporum*)

719. **Golebniak, B.** 2001. The response of meadow fescue, perennial ryegrass and Italian ryegrass to infection by *Fusa-*

rium avenaceum, F. culmorum and *F. graminum. Journal of Plant Protection Research* **41:** 395-401. (*culmorum*)

720. **Golinski, P., Z. Kaczmarek, I. Kiecana, H. Wisniewska, P. Kaptur, M. Kostecki, and J. Chelkowski.** 2002. Fusarium head blight of common Polish winter wheat cultivars – comparison of effects of *Fusarium avenaceum* and *Fusarium culmorum* on yield components. *Journal of Phytopathology* **150:** 135-141. (*avenaceum*)

721. **Golinski, P., J. Perkowski, M. Kostecki, J. Grabarkiewicz-Szczesna, and J. Chelkowski.** 1996. *Fusarium* species and *Fusarium* toxins in wheat in Poland: A comparison with neighbour countries. *Sydowia* **48:** 12-22. (*tricinctum*)

722. **Gomez-Gomez, E., N. Anaya, M. I. G. Roncero, and C. Hera.** 1999. *Folyt1*, a new member of the *hAT* family, is active in the genome of the plant pathogen *Fusarium oxysporum. Fungal Genetics and Biology* **27:** 67-76. (*oxysporum*)

723. **Gomez-Gomez, E., M. I. G. Roncero, A. di Pietro, and C. Hera.** 2001. Molecular characterization of a novel endo-β-1,4-xylanase gene from the vascular wilt fungus *Fusarium oxysporum. Current Genetics* **40:** 268-275. (*oxysporum*)

724. **Gomez-Gomez, E., M. C. Ruiz-Roldan, A. di Pietro, M. I. G. Roncero, and C. Hera.** 2002. Role in pathogenesis of two endo-β-1,4-xylanase genes from the vascular wilt fungus *Fusarium oxysporum. Fungal Genetics and Biology* **35:** 213-222. (*oxysporum*)

726. **Goncalves, A. M. D., M. R. Aires-Barros, and J. M. S. Cabral.** 2003. Interaction of an anionic surfactant with a recombinant cutinase from *Fusarium solani pisi*: A spectroscopic study. *Enzyme and Microbial Technology* **32:** 868-879. (*solani*)

727. **Goncalves, A. P. V., J. M. S. Cabral, and M. R. Aires-Barros.** 1996. Immobilization of a recombinant cutinase by entrapment and by covalent binding: Kinetic and stability studies. *Applied Biochemistry and Biotechnology* **60:** 217-228. (*solani*)

728. **Goncalves, A. M., A. P. Serro, M. R. Aires-Barros, and J. M. S. Cabral.** 2000. Effects of ionic surfactants used in reversed micelles on cutinase activity and stability. *Biochimica et Biophysica Acta* **1480:** 92-106. (*solani*)

729. **Gonzalez, H. H. L., E. J. Martinez, and S. L. Resnik.** 1997. Fungi associated with sorghum grain from Argentina. *Mycopathologia* **139:** 35-41. (*thapsinum*)

730. **Gonzalez, M. S., F. Fuenmayor, F. Godoy, and R. Navas.** 2003. First report of *Fusarium denticulatum* from sweet potato in Venezuela. *Plant Disease* **87:** 202. (*denticulatum*)

731. **Gonzalez-Candelas, L., and P. E. Kolattukudy.** 1992. Isolation and analysis of a novel inducible pectate lyase gene from the phytopathogenic fungus *Fusarium solani* f. sp. *pisi* (*Nectria haematococca* mating population VI). *Journal of Bacteriology* **174:** 6343-6349. (*solani*)

732. **Gonzalez-Jaen, M. T., S. Mirete, B. Patino, E. Lopez-Errasquin, and C. Vazquez.** 2004. Genetic markers for the analysis of variability and for production of specific diagnostic sequences in fumonisin-producing strains of *Fusarium verticillioides. European Journal of Plant Pathology* **110:** 525-532. (*verticillioides*)

733. **Gooch, V. D., R. A. Wehseler, and C. G. Gross.** 1994. Temperature effects on the resetting of the phase of the *Neurospora* circadian rhythm. *Journal of Biological Rhythms* **9:** 83-94. (4)

734. **Goodman, D. M., and L. L. Burpee.** 1991. Biological control of Dollar Spot disease of creeping bentgrass. *Phytopathology* **81:** 1438-1446. (*heterosporum*)

735. **Gopinath, A., H. S. Prakash, and H. S. Shetty.** 1985. Grain mold of sorghum: Role of associated fungi on the health and viability of seeds. *International Journal of Tropical Plant Diseases* **3:** 177-182. (*semitectum*)

736. **Gopinath, A., H. S. Prakash, and H. S. Shetty.** 1988. Effect of sorghum and pearl millet pollen on the conidial germination of *Fusarium moniliforme* and *Fusarium semitectum. Geobios* **15:** 14-18. (*semitectum*)

737. **Gopinathan, U., T. Ramakrishna, M. Willcox, C. M. Rao, D. Balasubramanian, A. Kulkarni, G. K. Vemuganti, and G. N. Rao.** 2001. Enzymatic, clinical and histologic evaluation of corneal tissues in experimental fungal keratitis in rabbits. *Experimental Eye Research* **72:** 433-442. (*solani*)

738. **Gordon, C., S. Thomas, A. Griffen, G. D. Robson, A. P. J. Trinci, and M. G. Wiebe.** 2001. Combined use of growth rate correlated and growth rate independent promoters for recombinant glucoamylase production in *Fusarium venenatum. FEMS Microbiology Letters* **194:** 229-234. (*venenatum*)

739. **Gordon, T. R., and D. Okamoto.** 1991. Vegetative compatibility groupings in a local population of *Fusarium oxysporum. Canadian Journal of Botany* **69:** 168-172. (*oxysporum*)

740. **Gordon, T. R., and D. Okamoto.** 1992. Variation in mitochondrial DNA among vegetatively compatible isolates of *Fusarium oxysporum. Experimental Mycology* **16:** 245-250. (5, *oxysporum*)

741. **Gordon, T. R., and D. Okamoto.** 1992. Variation within and between populations of *Fusarium oxysporum* based on vegetative compatibility and mitochondrial DNA. *Canadian Journal of Botany* **70:** 1211-1217. (*oxysporum*)

742. **Gordon, T. R., and D. Okamoto.** 1992. Population structure and the relationship between pathogenic and nonpathogenic strains of *Fusarium oxysporum. Phytopathology* **82:** 73-77. (*oxysporum*)

743. **Gordon, T. R., D. Okamoto, A. J. Storer, and D. L. Wood.** 1998. Susceptibility of five landscape pines to pitch canker disease, caused by *Fusarium subglutinans* f. sp. *pini. HortScience* **33:** 868-871. (*circinatum*)

744. **Gordon, T. R., A. J. Storer, and D. Okamoto.** 1996. Population structure of the pitch canker pathogen, *Fusarium subglutinans* f. sp. *pini*, in California. *Mycological Research* **100:** 850-854. (*circinatum*)

745. **Gordon, W. L.** 1944. The occurrence of *Fusarium* species in Canada. I. Species of *Fusarium* isolated from farm samples of cereal seed in Manitoba. *Canadian Journal of Research, C* **22:** 282-286. (8)

746. **Gordon, W. L.** 1952. The occurrence of *Fusarium* species in Canada. II. Prevalence and taxonomy of *Fusarium* species in cereal seed. *Canadian Journal of Botany* **30:** 209-251. (8, *acuminatum, scirpi*)

747. **Gordon, W. L.** 1954. The occurrence of *Fusarium* species in Canada. III. Taxonomy of *Fusarium* species in the seed of vegetable, forage, and miscellaneous crops. *Canadian Journal of Botany* **32:** 576-590. (8)

748. **Gordon, W. L.** 1954. The occurrence of *Fusarium* species in Canada. IV. Taxonomy and prevalence of *Fusarium* species

in the soil of cereal plots. *Canadian Journal of Botany* **32:** 622-629. (8)

749. **Gordon, W. L.** 1956. The occurrence of *Fusarium* species in Canada. V. Taxonomy and geographic distribution of *Fusarium* species in soil. *Canadian Journal of Botany* **34:** 833-846. (8)

750. **Gordon, W. L.** 1956. The taxonomy and habitats of the *Fusarium* species in Trinidad, B.W.I. *Canadian Journal of Botany* **34:** 847-864. (8)

751. **Gordon, W. L.** 1959. The occurrence of *Fusarium* species in Canada. VI. Taxonomy and geographic distribution of *Fusarium* species on plants, insects, and fungi. *Canadian Journal of Botany* **37:** 257-290. (8)

752. **Gordon, W. L.** 1960. The taxonomy and habitats of *Fusarium* species from tropical and temperate regions. *Canadian Journal of Botany* **38:** 643-658. (8, 11)

753. **Gordon, W. L.** 1961. Sex and mating types in relation to the production of perithecia by certain species of *Fusarium*. *Proceedings of the Canadian Phytopathological Society* **28:** 11. (*heterosporum*)

754. **Gordon-Lennox, G., and D. Gindrat.** 1987. Relationship of root necrosis potential to asparagus (*Asparagus officinalis*) decline in Switzerland. *Biology and Fertility of Soils* **3:** 195-198. (*redolens*)

755. **Goswami, R. S., and H. C. Kistler.** 2004. Heading for disaster: *Fusarium graminearum* on cereal crops. *Molecular Plant Pathology* **5:** 515-525. (*graminearum*)

756. **Gott, K. P., L. W. Burgess, V. Balmas, and J. Duff.** 1994. Mycogeography of *Fusarium*: *Fusarium* species in soils from Palm Valley, central Australia. *Australasian Plant Pathology* **23:** 112-117. (*polyphialidicum, scirpi*)

757. **Gottwald, S., C. U. Germeier, and W. Ruhmann.** 2001. Computerized image analysis in *Fusarium* taxonomy. *Mycological Research* **105:** 206-214. (11)

758. **Gourley, L. M., C. H. Andrews, L. L. Singleton, and L. Araujo.** 1977. Effects of *Fusarium moniliforme* on seedling development of sorghum cultivars. *Plant Disease Reporter* **61:** 616-618. (*verticillioides*)

759. **Govind, N. S., and E. Cerda-Olmedo.** 1986. Sexual activation of carotenogenesis in *Phycomyces blakesleeanus*. *Journal of General Microbiology* **132:** 2775-2780. (6)

760. **Grabarkiewicz-Szczesna, J., E. Foremska, M. Kostecki, P. Golinski, and J. Chelkowski.** 1999. Trichothecene accumulation in kernels of corn inoculated with *Fusarium poae* (Peck) Wollenw. *Nahrung* **43:** 330-332. (*poae*)

761. **Gracia-Garza, J. A., and D. R. Fravel.** 1998. Effect of relative humidity on sporulation of *Fusarium oxysporum* in various formulations and effect of water on spore movement through soil. *Phytopathology* **88:** 544-549. (*oxysporum*)

762. **Gracia-Garza, J. A., D. R. Fravel, B. A. Bailey, and P. K. Hebbar.** 1998. Dispersal of formulations of *Fusarium oxysporum* f. sp. *erythroxyli* and *F. oxysporum* f. sp. *melonis* by ants. Phytopathology **88:** 185-189. (*oxysporum*)

763. **Gracia-Garza, J. A., D. R. Fravel, A. J. Nelson, K. S. Elias, B. A. Bailey, E. Arevalo-Gardini, and L. C. Darlington.** 1999. Potential for dispersal of *Fusarium oxysporum* f. sp. *erythroxyli* by infested seed. *Plant Disease* **83:** 451-455. (*oxysporum*)

764. **Granchinho, S. C. R., C. M. Franz, E. Polishchuk, W. R. Cullen, and K. J. Reimer.** 2002. Transformation of arsenic (V) by the fungus *Fusarium oxysporum melonis* isolated from the alga *Fucus gardneri*. *Applied Organometallic Chemistry* **16:** 721-726. (*oxysporum*)

765. **Granshaw, T., M. Tsukamoto, and S. Brody.** 2003. Circadian rhythms in *Neurospora crassa*: Farnesol and geraniol allow expression of rhythmicity in the otherwise arrhythmic strains *frq-10*, *wc-1* and *wc-2*. *Journal of Biological Rhythms* **18:** 287-296. (4)

766. **Gray, J. S., G. C. J. Martin, and W. Rigby.** 1967. Aurofusarin. *Journal of the Chemical Society* **1967(C):** 2580-2587. (*graminearum*)

767. **Graybill, J. R., L. K. Najvar, G. M. Gonzalez, S. Hernandez, and R. Bocanegra.** 2003. Improving the mouse model for studying the efficacy of voriconazole. *Journal of Antimicrobial Chemotherapy* **51:** 1373-1376. (*solani*)

768. **Greene, A. V., N. Keller, H. Haas, and D. Bell-Pedersen.** 2003. A circadian oscillator in *Aspergillus* spp. regulates daily development and gene expression. *Eukaryotic Cell* **2:** 231-237. (4)

769. **Grell, R. F.** 1962. A new model for secondary nondisjunction: the role of distributive pairing. *Genetics* **47:** 1737-1754. (7)

770. **Greuter, W., J. McNeill, F. R. Barrie, H. M. Burdet, V. Demoulin, T. S. Filgueiras, D. H. Nicolson, P. C. Silva, J. E. Skog, P. Trehane, N. J. Turland, and D. L. Hawksworth, eds.** 2000. *International Code of Botanical Nomenclature (Saint Louis Code) adopted by the Sixteenth International Botanical Congress St. Louis, Missouri, July – August 1999*. Koeltz Scientific Books, Königstein, Germany. (4, 8, 9, *circinatum*)

771. **Griffee, P. J.** 1976. Pathogenicity of some fungi isolated from diseased crowns of banana hands. *Phytopathologische Zeitschrift* **85:** 206-216. (*semitectum*)

772. **Griffee, P. J., and O. J. Burden.** 1976. Fungi associated with crown rot of boxed bananas in the Windward Islands. *Phytopathologische Zeitschrift* **85:** 149-158. (*semitectum*)

773. **Griffiths, A. J. F.** 1992. Fungal senescence. *Annual Review of Genetics* **26:** 351-372. (4)

774. **Grigis, A., C. Farina, F. Symoens, N. Nolard, and A. Goglio.** 2000. Nosocomial pseudo-outbreak of *Fusarium verticillioides* associated with sterile plastic containers. *Infection Control and Hospital Epidemiology* **21:** 50-52. (*verticillioides*)

775. **Griswold, K. E., N. A. Mahmood, B. L. Iverson, and G. Georgiou.** 2003. Effects of codon usage versus putative 5'-mRNA structure on the expression of *Fusarium solani* cutinase in the *Escherichia coli* cytoplasm. *Protein Expression and Purification* **27:** 134-142. (*solani*)

776. **Grunwald, N. J., V. A. Coffman, and J. M. Kraft.** 2003. Sources of partial resistance to Fusarium root rot in the *Pisum* core collection. *Plant Disease* **87:** 1197-1200. (*solani*)

777. **Guarro, J., and J. Gene.** 1995. Opportunistic Fusarial infections in humans. *European Journal of Clinical Microbiology and Infectious Diseases* **14:** 741-754. (*oxysporum*)

778. **Guarro, J., M. Nucci, T. Akiti, and J. Gene.** 2000. Mixed infection caused by two species of *Fusarium* in a human immunodeficiency virus-positive patient. *Journal of Clinical Microbiology* **38:** 3460-3462. (1, *solani*, *verticillioides*)

779. **Guarro, J., M. Nucci, T. Akiti, J. Gene, M. D. G. C. Barreiro, and R. T. Goncalves.** 2000. Fungemia due to *Fusarium sacchari* in an immunosuppressed patient. *Journal of Clinical Microbiology* **38:** 419-421. (*sacchari*)

780. **Guarro, J., I. Pujol, and E. Mayayo.** 1999. *In vitro* and *in vivo* experimental activities of antifungal agents against *Fusarium solani. Antimicrobial Agents and Chemotherapy* **43:** 1256-1257. (*solani*)

781. **Guenther, J. C., and F. Trail.** 2005. The development and differentiation of *Gibberella zeae* (anamorph: *Fusarium graminearum*) during colonization of wheat. *Mycologia* **97:** 229-237. (10)

782. **Guevara, M. A., M. T. Gonzalez-Jaen, and P. Estevez.** 1997. Multiple forms of pectic lyases and polygalacturonase from *Fusarium oxysporum* f. sp. *radicis-lycopersici*: Regulation of their synthesis by galacturonic acid. *Canadian Journal of Microbiology* **43:** 245-253. (*oxysporum*)

783. **Guinvarc'h, A., L. Guilbert, A. Marmorat-Khuong, V. Lavarde, P. Chevalier, C. Amrein, R. Guillemain, and A. Berrebi.** 1998. Disseminated *Fusarium solani* infection with endocarditis in a lung transplant recipient. *Mycoses* **41:** 59-61. (*solani*)

784. **Gulya, T. J., Jr., C. A. Martinson, and P. J. Loesch, Jr.** 1980. Evaluation of inoculation techniques and rating dates for *Fusarium moniliforme* ear rot of *opaque-2* maize. *Phytopathology* **70:** 1116-1118. (*verticillioides*)

785. **Gulya, T. J., Jr., C. A. Martinson, and L. H. Tiffany.** 1979. Ear-rotting fungi associated with *opaque-2* maize. *Plant Disease Reporter* **63:** 370-373. (*verticillioides*)

786. **Gumprecht, L. A., V. R. Beasley, R. M. Weigel, H. M. Parker, M. E. Tumbleson, C. W. Bacon, F. I. Meredith, and W. M. Haschek.** 1998. Development of fumonisin-induced hepatotoxicity and pulmonary edema in orally dosed swine: Morphological and biochemical alterations. *Toxicologic Pathology* **26:** 777-788. (*verticillioides*)

787. **Gunn, L. V., and B. A. Summerell.** 2002. Differentiation of *Fusarium oxysporum* isolates from *Phoenix canariensis* (Canary Island Date Palm) by vegetative compatibility grouping and molecular analysis. *Australasian Plant Pathology* **31:** 351-358. (5)

788. **Guo, W., L. Gonzalez-Candelas, and P. E. Kolattukudy.** 1995. Cloning of a new pectate lyase gene *pelC* from *Fusarium solani* f. sp. *pisi* (*Nectria haematococca*, mating type VI) and characterization of the gene product expressed in *Pichia pastoris*. *Archives of Biochemistry and Biophysics* **323:** 352-360. (*solani*)

789. **Guo, W., L. Gonzalez-Candelas, and P. E. Kolattukudy.** 1995. Cloning of a novel constitutively expressed pectate lyase gene *pelB* from *Fusarium solani* f. sp. *pisi* (*Nectria haematococca*, mating type VI) and characterization of the gene product expressed in *Pichia pastoris*. *Journal of Bacteriology* **177:** 7070-7077. (*solani*)

790. **Guo, W., L. Gonzalez-Candelas, and P. E. Kolattukudy.** 1996. Identification of a novel *pelD* gene expressed uniquely *in planta* by *Fusarium solani* f. sp. *pisi* (*Nectria haematococca*, mating type VI) and characterization of its protein product as an endo-pectate lyase. *Archives of Biochemistry and Biophysics* **332:** 305-312. (*solani*)

791. **Gupta, A. K., and I. S. Bhatia.** 1982. Gluco-fructosan biosynthesis in *Fusarium oxysporum*: Regulation and substrate specificity of fructosyl transferase E.C. 2.4.1.9 and invertase E.C. 3.2.1.26. *Phytochemistry* **21:** 1249-1254. (*oxysporum*)

792. **Gupta, A. K., B. Nagpal, N. Kaur, and R. Singh.** 1988. Mycelial and extracellular inulinases from *Fusarium oxysporum* grown on aqueous extract of *Cichorium intybus* roots. *Journal of Chemical Technology and Biotechnology* **42:** 69-76. (*oxysporum*)

793. **Gupta, A. K., and V. N. Pathak.** 1988. Some new fungi associated with papaya fruits. *Indian Phytopathology* **41:** 479-480. (*scirpi*)

794. **Gupta, R. C., R. S. Upadhyay, and B. Rai.** 1979. Hyphal parasitism and chlamydospore formation by *Fusarium oxysporum* on *Rhizoctonia solani*. *Mycopathologia* **67:** 147-152. (*oxysporum*)

795. **Gupta, S.** 1991. Newer evidence to demonstrate mycovirus of *Fusarium moniliforme* var. *subglutinans* as causal agent of mango shoot malformation. *Journal of Entomological Research* **15:** 222-228. (*mangiferae*)

796. **Gupta, S., S. B. Krasnoff, N. L. Underwood, J. A. A. Renwick, and D. W. Roberts.** 1991. Isolation of beauvericin as an insect toxin from *Fusarium semitectum* and *Fusarium moniliforme* var. *subglutinans*. *Mycopathologia* **115:** 185-189. (*concentricum, semitectum, subglutinans*)

797. **Guzman-Cottrill, J. A., X. Zheng, and E. G. Chadwick.** 2004. *Fusarium solani* endocarditis successfully treated with liposomal amphotericin B and voriconazole. *Pediatric Infectious Disease Journal* **23:** 1059-1061. (*solani*)

798. **Hacker, C., M. Glinski, T. Hornbogen, A. Doller, and R. Zocher.** 2000. Mutational analysis of the N-methyltransferase domain of the multifunctional enzyme enniatin synthetase. *Journal of Biological Chemistry* **275:** 30826-30832. (*scirpi*)

799. **Hadar, E., J. Katan, and T. Katan.** 1989. The use of nitrate-nonutilizing mutants and a selective medium for studies of pathogenic strains of *Fusarium oxysporum*. *Plant Disease* **73:** 800-803. (*oxysporum*)

800. **Hadwiger, L. A., M.-M. Chang, and M. A. Parsons.** 1995. *Fusarium solani* DNase is a signal for increasing expression of nonhost disease resistance response genes, hypersensitivity, and pisatin production. *Molecular Plant-Microbe Interactions* **8:** 871-879. (*solani*)

801. **Hadwiger, L. A., and W. Wagoner.** 1983. Electrophoretic patterns of pea and *Fusarium solani* proteins synthesized *in vitro* or *in vivo* which characterize the compatible and incompatible interactions. *Physiological Plant Pathology* **23:** 153-162. (*solani*)

802. **Haefner, K.** 1967. A simple apparatus for producing agar layers of uniform thickness for microbiological micromanipulator work. *Zeitschrift für Allgemeine Mikrobiologie* **7:** 229-231. (4)

803. **Haese, A., R. Pieper, T. von Ostrowski, and R. Zocher.** 1994. Bacterial expression of catalytically active fragments of the multifunctional enzyme enniatin synthetase. *Journal of Molecular Biology* **243:** 116-122. (*scirpi*)

804. **Haese, A., M. Schubert, M. Herrmann, and R. Zocher.** 1993. Molecular characterization of the enniatin synthetase gene encoding a multifunctional enzyme catalyzing N-

methyldepsipeptide formation in *Fusarium scirpi. Molecular Microbiology* **7:** 905-914. (*scirpi*)

805. **Hagen, B., and C. Hagen.** 1995. Computer-aided morphometrical analysis of macroconidia in *Fusarium sambucinum* Fückel *sensu lato* strains. *Mycopathologia* **129:** 143-147. (*sambucinum*)

806. **Hajek, A. E., J. S. Elkinton, and R. A. Humber.** 1997. Entomopathogenic hyphomycetes associated with gypsy moth larvae. *Mycologia* **89:** 825-829. (*polyphialidicum*)

807. **Hall, R. & J. C. Sutton.** 1998. Relation of weather, crop and soil variables to the prevalence, incidence and severity of basal infections of winter wheat in Ontario. *Canadian Journal of Plant Pathology* **20:** 69-80. (*avenaceum*)

808. **Hamann, A., and H. D. Osiewacz.** 2004. Genetic anal-ysis of spore killing in the filamentous ascomycete *Podospora anserina. Fungal Genetics and Biology* **41:** 1088-1098. (10)

809. **Han, Y.-K., T. Lee, K.-H. Han, S.-H. Yun, and Y.-W. Lee.** 2004. Functional analysis of the homoserine O-acetyltransferase gene and its identification as a selectable marker in *Gibberella zeae. Current Genetics* **46:** 205-212. (*graminearum*)

810. **Hancock, J. G.** 1983. Seedling diseases of alfalfa in California. *Plant Disease* **67:** 1203-1208. (*acuminatum*)

811. **Hansen, H. N., and R. E. Smith.** 1932. The mechanisms of variation in imperfect fungi: *Botrytis cinerea. Phytopathology* **37:** 369-371. (4)

812. **Hara, H., and J. W. Beardsley, Jr.** 1979. The biology of the black twig borer *Xylosandrus compactus* in Hawaii, USA. *Proceedings of the Hawaiian Entomological Society* **23:** 55-70. (*solani*)

813. **Harbaugh, B. K., and R. J. McGovern.** 2000. Susceptibility of 46 *Lisianthus* cultivars to Fusarium crown and stem rot. *HortTechnology* **10:** 816-189. (*avenaceum*)

814. **Hareendrath, V., K. P. V. Nair, and S. Paulose.** 1987. *Fusarium pallidioroseum* (Cooke) Sacc. as a fungal pathogen of *Aphis craccivora* Koch. *Entomon* **12:** 392-394. (*semitectum*)

815. **Hargreaves, A. J., and W. R. Jarvis.** 1972. Raspberry bud death and lateral wilt. *Scottish Agriculture* **51:** 384-387. (*avenaceum*)

816. **Hargreaves, M., S. Parappukkaran, L. Morawska, J. Hitchins, C. He, and D. Gilbert.** 2003. A pilot investigation into associations between indoor airborne fungal and non-biological particle concentrations in residential houses in Brisbane, Australia. *Science of the Total Environment* **312:** 89-101. (3)

817. **Harman, G. E., L. R. Mattick, G. Nash, and B. L. Nedrow.** 1980. Stimulation of fungal spore germination and inhibition of sporulation in fungal vegetative thalli by fatty acids and their volatile peroxidation products. *Canadian Journal of Botany* **58:** 1541-1547. (*solani*)

818. **Harper, D. B.** 1977. Fungal degradation of aromatic nitriles: Enzymology of carbon-nitrogen bond cleavage by *Fusarium solani. Biochemical Journal* **167:** 685-692. (*solani*)

819. **Harrington, T. C., and D. M. Rizzo.** 1999. Defining species in the fungi, p. 43-71. *In* Worrall, J. J., ed. *Structure and Dynamics of Fungal Populations,* Kluwer Academic Publishers: Boston Massachusetts. (9)

820. **Hart, L. P., E. Gendloff, E. C. Rossman.** 1984. Effect of corn genotypes on ear rot infection by *Gibberella zeae. Plant Disease* **68:** 296-298. (*graminearum*)

821. **Hartill, W. F. T., and P. G. Broadhurst.** 1989. *Fusarium avenaceum* as a pathogen of stonefruit in New Zealand. *New Zealand Journal of Crop and Horticultural Science* **17:** 293-295. (*avenaceum*)

822. **Hartung, C., and M. Reyes-Lugo.** 1996. *Fusarium solani* invader of the eggs of the insect *Panstrongylus geniculatus* in a vivarium. *Mycopathologia* **135:** 183-185. (*solani*)

823. **Harveson, R. M., and C. M. Rush.** 1997. Genetic variation among *Fusarium oxysporum* isolates from sugar beet as determined by vegetative compatibility. *Plant Disease* **81:** 85-88. (5, *oxysporum*)

824. **Harvey, R. B., T. S. Edrington, L. F. Kubena, G. E. Rottinghaus, J. R. Turk, K. J. Genovese, and D. J. Nisbet.** 2001. Toxicity of moniliformin from *Fusarium fujikuroi* culture material to growing barrows. *Journal of Food Protection* **64:** 1780-1784. (*fujikuroi*)

825. **Haschek, W. M., L. A. Gumprecht, G. Smith, M. E. Tumbleson, and P. D. Constable.** 2001. Fumonisin toxicosis in swine: An overview of porcine pulmonary edema and current perspectives. *Environmental Health Perspectives* **109 (Suppl. 2):** 251-257. (*verticillioides*)

826. **Haseltine, W. A.** 1983. Ultraviolet light and mutagenesis revisited. *Cell* **33:** 13-17. (4)

827. **Hatsch, D., V. Phalip, and J. M. Jeltsch.** 2004. Use of genes encoding cellobiohydrolase C and topoisomerase II as targets for phylogenetic analysis and identification of *Fusarium. Research in Microbiology* **155:** 290-296. (7, *solani*)

828. **Hawley, R. S., K. S. McKim, and T. Arbel.** 1993. Meiotic segregation in *Drosophila melanogaster* females: Molecules, mechanisms, and myths. *Annual Review of Genetics* **27:** 281-317. (7)

829. **Hawthorne, B. T., R. D. Ball, and J. Rees-George.** 1994. Genetic analysis of variation of pathogenicity in *Nectria haematococca* (*Fusarium solani*) on *Cucurbita* spp. *Mycological Research* **98:** 1183-1191. (*solani*)

830. **Hawthorne, B. T., and J. Rees-George.** 1996. Use of nitrate non-utilizing mutants to study vegetative incompatibility in *Fusarium solani* (*Nectria haematococca*), especially members of mating populations I, V and VI. *Mycological Research* **100:** 1075-1081. (*solani*)

831. **Hawthorne, B. T., J. Rees-George, and P. G. Broadhurst.** 1992. Mating behaviour and pathogenicity in New Zealand isolates of *Nectria haematococca* (*Fusarium solani*). *New Zealand Journal of Crop and Horticultural Science* **20:** 51-57. (*solani*)

832. **Hawthorne, B. T., J. Rees-George, and R. N. Crowhurst.** 2001. Induction of cutinolytic esterase activity during saprophytic growth of cucurbit pathogens, *Fusarium solani* f. sp. *cucurbitae* races one and two (*Nectria haematococca* MPI and MPV, respectively). *FEMS Microbiology Letters* **194:** 135-141. (*solani*)

833. **He, J., Y. Ruan, and D. Straney.** 1996. Analysis of determinants of binding and transcriptional activation of the pisatin-responsive DNA binding factor of *Nectria haematococca. Molecular Plant-Microbe Interactions* **9:** 171-179. (*solani*)

834. **He, Z., J. M. S. Wong, H. S. Maniar, S. J. Brill, and C. J. Ingles.** 1996*b*. Assessing the requirements for nucleotide excision repair proteins of *Saccharomyces cerevisiae* in an *in vitro* system. *Journal of Biological Chemistry* **271:** 28243-28249. (4)

835. **Headrick, J. M., D. A. Glawe, and J. K. Pataky.** 1988. Ascospore polymorphism in *Gibberella zeae*. *Mycologia* **80:** 679-684. (10)

836. **Headrick, J. M., and J. K. Pataky.** 1989. Resistance to kernel infection by *Fusarium moniliforme* in inbred lines of sweet corn and the effect of infection on emergence. *Plant Disease* **73:** 887-892. (*verticillioides*)

837. **Heaton, L. A., and J. F. Leslie.** 2004. Double-stranded RNAs associated with *Fusarium proliferatum* mitochondria. *Mycological Progress* **3:** 193-198. (*proliferatum*)

838. **Hebbar, K. P., B. A. Bailey, S. M. Poch, J. A. Lewis, and R. D. Lumsden.** 1999. An improved granular formulation for a mycoherbicidal strain of *Fusarium oxysporum*. *Weed Science* **47:** 473-478. (*oxysporum*)

839. **Hebbar, K. P., J. A. Lewis, S. M. Poch, and R. D. Lumsden.** 1996. Agricultural by-products as substrates for growth, conidiation and chlamydospore formation by a potential mycoherbicide, *Fusarium oxysporum* strain EN4. *Biocontrol Science and Technology* **6:** 263-275. (*oxysporum*)

840. **Hebbar, K. P., R. D. Lumsden, J. A. Lewis, S. M. Poch, and B. A. Bailey.** 1998. Formulation of mycoherbicidal strains of *Fusarium oxysporum*. *Weed Science* **46:** 501-507. (*oxysporum*)

841. **Hebbar, K. P., R. D. Lumsden, S. M. Poch, and J. A. Lewis.** 1997. Liquid fermentation to produce biomass of mycoherbicidal strains of *Fusarium oxysporum*. *Applied Microbiology and Biotechnology* **48:** 714-719. (*oxysporum*)

842. **Hedden, P., A. L. Phillips, M. C. Rojas, E. Carrera, and B. Tudzynski.** 2001. Gibberellin biosynthesis in plants and fungi: A case of convergent evolution? *Journal of Plant Growth Regulation* **20:** 319-331. (*fujikuroi*)

843. **Heiser, I., W. Osswald, R. Baker, S. Nemec, and E. F. Elstner.** 1998. Activation of *Fusarium* naphthazarin toxins and other *p*-quinones by reduced thioctic acid. *Journal of Plant Physiology* **153:** 276-280. (*solani*)

844. **Heiser, I., W. Osswald, and E. F. Elstner.** 1998. The formation of reactive oxygen species by fungal and bacterial phytotoxins. *Plant Physiology and Biochemistry* **36:** 703-713. (*solani*)

845. **Helbig, J. B., and R. B. Carroll.** 1984. Dicotyledonous weeds as a source of *Fusarium oxysporum* pathogenic on soybean. *Plant Disease* **68:** 694-696. (*oxysporum*)

846. **Helm, T. N., D. L. Longworth, G. S. Hall, B. J. Bolwell, B. Fernandez, and K. J. Tomecki.** 1990. Case report and review of resolved fusariosis. *Journal of the American Academy of Dermatology* **23:** 393-398. (*proliferatum*)

847. **Hemo, I., J. Pe'Er, and I. Polacheck.** 1989. *Fusarium oxysporum* keratitis. *Ophthalmologica* **198:** 3-7. (*oxysporum*)

848. **Hennequin, C., E. Abachin, F. Symoens, V. Lavarde, G. Reboux, N. Nolard, and P. Berche.** 1999. Identification of *Fusarium* species involved in human infections by 28S rRNA gene sequencing. *Journal of Clinical Microbiology* **37:** 3586-3589. (*oxysporum*)

849. **Hennequin, C., V. Lavarde, and J. L. Poirot, M. Rabodonirina, A. Datry, S. Aractingi, J. Dupouy-Camet, D. Caillot, F. Grange, L. Kures, O. Morin, B. Lebeau, S. Bretagne, C. Guigen, D. Basset, and R. Grillot.** 1997. Invasive *Fusarium* infections: A retrospective survey of 31 cases. *Journal of Medical and Veterinary Mycology* **35:** 107-114. (*oxysporum, solani, verticillioides*)

850. **Henriksson, G., D. E. Akin, R. T. Hanlin, C. Rodriguez, D. D. Archibald, L. L. Rigsby, and K. E. L. Eriksson.** 1997. Identification and retting efficiencies of fungi isolated from dew-retted flax in the United States and Europe. *Applied and Environmental Microbiology* **63:** 3950-3956. (*lateritium*)

851. **Herbrecht, R., R. Kessler, C. Kravanja, M. H. Meyer, J. Waller, and V. Letscher-Bru.** 2004. Successful treatment of *Fusarium proliferatum* pneumonia with posaconazole in a lung transplant recipient. *Journal of Heart and Lung Transplantation* **23:** 1451-1454. (*proliferatum*)

852. **Herrman, M., R. Zocher, and A. Haese.** 1996. Effect of disruption of the enniatin synthetatse gene on the virulence of *Fusarium avenaceum*. *Molecular Plant-Microbe Interactions* **9:** 226-232. (*avenaceum*)

853. **Herskowitz, I.** 1988. Life cycle of the budding yeast *Saccharomyces cerevisiae*. *Microbiological Reviews* **52:** 536-553. (6)

854. **Hestbjerg, H., K. F. Nielsen, U. Thrane, and S. Elmholt.** 2002. Production of trichothecenes and other secondary metabolites by *Fusarium culmorum* and *Fusarium equiseti* on common laboratory media and a soil organic matter agar: An ecological interpretation. *Journal of Agricultural and Food Chemistry* **50:** 7593-7599. (*equiseti*)

855. **Heyndrickx, A., N. Sookvanichsilp, and M. van den Heede.** 1989. Detection of trichothecene mycotoxins yellow rain in blood, urine and feces of Iranian soldiers treated as victims of a gas attack. *Rivista di Tossicologia Sperimentale e Clinica* **19:** 7-12. (1)

856. **Hidalgo, O. B., A. Pires de Matos, R. S. Cabral, R. T. Tussel, M. Arzola, R. Santos, and M. C. Perez.** 1998. Phytotoxic effect of culture filtrate from *Fusarium subglutinans* the causal agent of fusariose of pineapple (*Ananas comosus* (L.) Merr.). *Euphytica* **104:** 73-77. (*guttiforme*)

857. **Hidalgo, O. B., R. Santos, R. T. Tussel, A. Pires-de-Matos, R. S. Cabral, M. Arzola, and M. C. Perez.** 1999. Phytotoxity of *Fusarium subglutinans* culture filtrates on *in vitro* plantlets and calli of resistant and susceptible pineapple (*Ananas comosus*). *Plant Pathology* **48:** 756-758. (*guttiforme*)

858. **Hidy, P. H., R. S. Baldwin, R. L. Greasham, C. L. Keith, and J. R. McMullen.** 1977. Zearalenone and some derivatives: Production and biological activity. *Advances in Applied Microbiology* **22:** 59-82. (1, 4, *graminearum*)

859. **Hiemenz, J. W., B. Kennedy, and K. J. Kwon-Chung.** 1990. Invasive Fusariosis associated with an injury by a stingray barb. *Journal of Medical and Veterinary Mycology* **28:** 209-214. (*solani*)

860. **Hildebrand, D. C., and A. H. McCain.** 1978. The use of various substrates for large-scale production of *Fusarium oxysporum* f. sp. *cannabis* inoculum. *Phytopathology* **68:** 1099-1101. (*oxysporum*)

861. **Hillocks, R. J., E. Minja, A. Mwaga, M. Silim-Nahdy, and P. Subrahmanyam.** 2000. Diseases and pests of pigeon-

pea in eastern Africa: A review. *International Journal of Pest Management* **46:** 7-18. (*udum*)

862. **Hinojo, M. J., A. Llorens, R. Mateo, B. Patino, M. T. Gonzalez-Jaen, and M. Jimenez.** 2004. Utility of the polymerase chain reaction-restriction fragment length polymorphisms of the intergenic spacer region of the rDNA for characterizing *Gibberella fujikuroi* isolates. *Systematic and Applied Microbiology* **27:** 681-688. (*verticillioides*)

863. **Hirata, T., E. Kimishima, T. Aoki, H. I. Nirenberg, and K. O'Donnell.** 2001. Morphological and molecular characterization of *Fusarium verticillioides* from rotten banana imported into Japan. *Mycoscience* **42:** 155-166. (*verticillioides*)

864. **Hirose, H., H. Terasaki, S. Awaya, and T. Yasuma.** 1997. Treatment of fungal corneal ulcers with amphotericin B ointment. *American Journal of Ophthalmology* **124:** 836-838. (*solani*)

865. **Hirota, A., Y. Ando, S. Monma, and H. Hirota.** 1994. FCRR-toxin, a novel phytotoxin from *Fusarium oxysporum* f. sp. *radicis-lycopersici*. *Bioscience Biotechnology and Biochemistry* **58:** 1931-1932. (*oxysporum*)

866. **Hirsch, D. J., S. R. Hirsch, and J. H. Kalbfleisch.** 1978. Effect of central air conditioning and meteorological factors on indoor spore counts. *Journal of Allergy and Clinical Immunology* **62:** 22-26. (3)

867. **Hirschi, K., and H. VanEtten.** 1996. Expression of the pisatin detoxifying genes (*PDA*) of *Nectria haematococca in vitro* and *in planta*. *Molecular Plant-Microbe Interactions* **9:** 483-491. (*solani*)

868. **Hirst, J. M.** 1952. An automatic volumetric spore trap. *Annals of Applied Biology* **39:** 257-265. (3)

869. **Ho, Y. W., and G. Varghese.** 1986. Pathogenic potential of soil Fusaria from Malaysian oil palm habitats. *Journal of Phytopathology* **115:** 325-331. (*heterosporum*)

870. **Hoch, H. C., and G. S. Abawi.** 1979. Mycoparasitism of oospores of *Pythium ultimum* by *Fusarium merismoides*. *Mycologia* **71:** 621-625. (*merismoides*)

871. **Hochlowski, J. E., D. N. Whittern, A. Buko, L. Alder, and J. B. McAlpine.** 1995. Fusacandins A and B; novel antifungal antibiotics of the papulacandin class from *Fusarium sambucinum*. II. Isolation and structural elucidation. *Journal of Antibiotics* **48:** 614-618. (*sambucinum*)

872. **Hoelker, U., R. M. Fakoussa, and M. Hoefer.** 1995. Growth substrates control the ability of *Fusarium oxysporum* to solubilize low rank coal. *Applied Microbiology and Biotechnology* **44:** 351-355. (*oxysporum*)

873. **Hoelker, U., S. Ludwig, T. Scheel, and M. Hoefer.** 1999. Mechanisms of coal solubilization by the deuteromycetes *Trichoderma atroviride* and *Fusarium oxysporum*. *Applied Microbiology and Biotechnology* **52:** 57-59. (*oxysporum*)

874. **Hoff, M., R. M. Trueb, B. K. Ballmer-Weber, S. Vieths, and B. Wuethrich.** 2003. Immediate-type hypersensitivity reaction to ingestion of mycoprotein (Quorn) in a patient allergic to molds caused by acidic ribosomal protein P2. *Journal of Allergy and Clinical Immunology* **111:** 1106-1110. (*venenatum*)

875. **Hoffmann, B., S. E. Eckert, S. Krappmann, and G. H. Braus.** 2001. Sexual diploids of *Aspergillus nidulans* do not form by random fusion of nuclei in the heterokaryon. *Genetics* **157:** 141-147. (6)

876. **Hohn, T. M., and M. N. Beremand.** 1989. Regulation of trichodiene synthase in *Fusarium sporotrichioides* and *Gibberella pulicaris* (*Fusarium sambucinum*). *Applied and Environmental Microbiology* **55:** 1500-1503. (*sambucinum, sporotrichioides*)

877. **Hohn, T. M., and M. N. Beremand.** 1989. Isolation and nucleotide sequence of a sesquiterpene cyclase gene from the trichothecene-producing fungus *Fusarium sporotrichioides*. *Gene* **79:** 131-138. (*sporotrichioides*)

878. **Hohn, T. M., and A. E. Desjardins.** 1992. Isolation and gene disruption of the *tox5* gene encoding trichodiene synthase in *Gibberella pulicaris*. *Molecular Plant-Microbe Interactions* **5:** 249-256. (*sambucinum*)

879. **Hohn, T. M., A. E. Desjardins, and S. P. McCormick.** 1993. Analysis of *Tox5* gene expression in *Gibberella pulicaris* strains with different trichothecene production phenotypes. *Applied and Environmental Microbiology* **59:** 2359-2363. (*sambucinum*)

880. **Hohn, T. M., A. E. Desjardins, and S. P. McCormick.** 1995. The *Tri4* gene of *Fusarium sporotrichioides* encodes a cytochrome P450 monooxygenase involved in trichothecene biosynthesis. *Molecular and General Genetics* **248:** 95-102. (*sporotrichioides*)

881. **Hohn, T. M., R. Krishna, and R. H. Proctor.** 1999. Characterization of a transcriptional activator controlling trichothecene toxin biosynthesis. *Fungal Genetics and Biology* **26:** 224-235. (*sporotrichioides*)

882. **Hohn, T. M., S. P. McCormick, and A. E. Desjardins.** 1993. Evidence for a gene cluster involving trichothecene-pathway biosynthetic genes in *Fusarium sporotrichioides*. *Current Genetics* **24:** 291-295. (*sporotrichioides*)

883. **Hohn, T. M., and R. D. Plattner.** 1989. Expression of the trichodiene synthase gene of *Fusarium sporotrichioides* in *Escherichia coli* results in sesquiterpene production. *Archives of Biochemistry and Biophysics* **275:** 92-97. (*sporotrichioides*)

884. **Hohn, T. M., and F. van Middlesworth.** 1986. Purification and characterization of the sesquiterpene cyclase trichodiene synthetase from *Fusarium sporotrichioides*. *Archives of Biochemistry and Biophysics* **251:** 756-761. (*sporotrichioides*)

885. **Holenstein, J., and A. G. Defago.** 1983. Inheritance of naphthazarin production and pathogenicity to pea in *Nectria haematococca*. *Journal of Experimental Botany* **34:** 927-935. (*solani*)

886. **Holley, R. N., P. B. Hamilton, and M. M. Goodman.** 1989. Evaluation of tropical maize germplasm for resistance to kernel colonization by *Fusarium moniliforme*. *Plant Disease* **73:** 578-580. (*verticillioides*)

887. **Holmes, S. J. I.** 1979. Effects of *Fusarium nivale* and *Fusarium culmorum* on the establishment of four species of pasture grass. *Annals of Applied Biology* **91:** 243-250. (*culmorum*)

888. **Holubova-Jechova, V., W. Gams, and H. I. Nirenberg.** 1994. Revisiones Generum Obscurorum *Hyphomycetum*: A revision of the *Selenosporium* species described by A. C. J. Corda. *Sydowia* **46:** 247-256. (*tricinctum*)

889. **Honda, K., M. Kataoka, and S. Shimizu.** 2002. Functional analyses and application of microbial lactonohydrolases. *Biotechnology and Bioprocess Engineering* **7:** 130-137. (*oxysporum*)

890. **Honda, K., H. Tsuboi, T. Minetoki, H. Nose, K. Sakamoto, M. Kataoka, and S. Shimizu.** 2005. Expression of the *Fusa-*

rium oxysporum lactonase gene in *Aspergillus oryzae*: Molecular properties of the recombinant enzyme and its application. *Applied Microbiology and Biotechnology* **66:** 520-526. (*oxysporum*)

891. **Hooker, D. C., A. W. Schaafsma, and L. Tamburic-Ilincic.** 2002. Using weather variables pre- and post-heading to predict deoxynivalenol content in winter wheat. *Plant Disease* **86:** 611-619. (11)

892. **Hoover, K., D. L. Wood, A. J. Storer, J. W. Fox, and W. E. Bros.** 1996. Transmission of the pitch canker fungus, *Fusarium subglutinans* f. sp. *pini*, to Monterey pine, *Pinus radiata*, by cone-and twig-infesting beetles. *Canadian Entomologist* **128:** 981-994. (*circinatum*)

893. **Hope, R., D. Aldred, and N. Magan.** 2005. Comparison of environmental profiles for growth and deoxynivalenol production by *Fusarium culmorum* and *F. graminearum* on wheat grain. *Letters in Applied Microbiology* **40:** 295-300. (*culmorum, graminearum*)

894. **Horgan, G., and M. Dick.** 2004. Pitch canker – Terminator of economic viability for New Zealand's commercial forest estate? *New Zealand Journal of Forestry* **49:** 28-30. (*circinatum*)

895. **Hornbogen, T., R. Pieper, K. Hoffmann, H. Kleinkauf, and R. Zocher.** 1992. Two new cyclophilins from *Fusarium sambucinum* and *Aspergillus niger*: Resistance of cyclophilin-cyclosporin A complexes against proteolysis. *Biochemical and Biophysical Research Communications* **187:** 791-796. (*sambucinum*)

896. **Hornok, L., C. Fekete, and G. Giczey.** 1996. Molecular characterization of *Fusarium poae*. *Sydowia* **48:** 23-31. (*poae*)

897. **Hose, J. E., D. V. Lightner, R. M. Redman, and D. A. Danald.** 1984. Observations on the pathogenesis of the imperfect fungus *Fusarium solani* in the California Brown Shrimp *Penaeus californiensis*. *Journal of Invertebrate Pathology* **44:** 292-303. (*solani*)

898. **Hose, J. E., and G. G. Martin.** 1989. Defense functions of granulocytes in the ridgeback prawn *Sicyonia ingentis*. *Journal of Invertebrate Pathology* **53:** 335-346. (*solani*)

899. **Hoshino, T., A. Mizutani, T. Sasaki, H. Hidaka, and T. Yamane.** 1994. Purification and partial amino acid sequence of calmodulin from *Fusarium oxysporum*. *Journal of General and Applied Microbiology* **40:** 43-51. (*oxysporum*)

900. **Hoshino, T., T. Sasaki, Y. Watanabe, T. Nagasawa, and T. Yamane.** 1992. Purification and some characteristics of extracellular lipase from *Fusarium oxysporum* f. sp. *lini*. *Bioscience Biotechnology and Biochemistry* **56:** 660-664. (*oxysporum*)

901. **Hottiger, T, and T. Boller.** 1991. Ethylene biosynthesis in *Fusarium oxysporum* f. sp. *tulipae* proceeds from glutamate-2-oxoglutarate and requires oxygen and ferrous ions *in vivo*. *Archives of Microbiology* **157:** 18-22. (*oxysporum*)

902. **Hou, Z, C. Xue, Y. Peng, T. Katan, H. C. Kistler, and J.-R. Xu.** 2002. A mitogen-activated protein kinase gene (*MGV1*) in *Fusarium graminearum* is required for female fertility, heterokaryon formation and plant infection. *Molecular Plant-Microbe Interactions* **15:** 1119-1127. (10, *graminearum*)

903. **Howlett, B. J., J. F. Leslie, and D. D. Perkins.** 1993. Putative multiple alleles at the vegetative (heterokaryon) incompatibility loci *het-c* and *het-8* in *Neurospora crassa*. *Fungal Genetics Newsletter* **40:** 40-42. (5)

904. **Howson, W. T., R. C. McGinnis, and W. L. Gordon.** 1963. Cytological studies on the perfect stages of some species of *Fusarium*. *Canadian Journal of Genetics and Cytology* **5:** 60-64. (7)

905. **Hsieh, W. H., S. N. Smith, and W. C. Snyder.** 1977. Mating groups in *Fusarium moniliforme*. *Phytopathology* **67:** 1041-1043. (2, 6, 8)

906. **Hsieh, W. H., W. C. Snyder, and S. N. Smith.** 1979. Influence of carbon sources, amino acids and water potential on growth and sporulation of *Fusarium moniliforme*. *Phytopathology* **69:** 602-604. (*verticillioides*)

907. **Hua, L., Z.-H. Sun, Y. Leng, and Z.-Q. Hu.** 2005. Continuous biocatalytic resolution of DL-pantolactone by cross-linked cells in a membrane bioreactor. *Process Biochemistry* **40:** 1137-1142. (*oxysporum*)

908. **Hua, L., Z.-H. Sun, P. Zheng, and Y. Xu.** 2004. Biocatalytic resolution of DL-pantolactone by glutaraldehyde cross-linked cells of *Fusarium moniliforme* CGMCC 0536. *Enzyme and Microbial Technology* **35:** 161-166. (*verticillioides*)

909. **Hua-Van, A., J. M. Daviere, F. Kaper, T. Langin, and M.-J. Daboussi.** 2000. Genome organization in *Fusarium oxysporum*: Clusters of class II transposons. *Current Genetics* **37:** 339-347. (*oxysporum*)

910. **Hua-Van, A., F. Hericourt, P. Capy, M.-J. Daboussi, and T. Langin.** 1998. Three highly divergent subfamilies of the *impala* transposable element coexist in the genome of the fungus *Fusarium oxysporum*. *Molecular and General Genetics* **259:** 354-362. (*oxysporum*)

911. **Hua-Van, A., T. Langin, and M.-J. Daboussi.** 2001. Evolutionary history of the *impala* transposon in *Fusarium oxysporum*. *Molecular Biology and Evolution* **18:** 1959-1969. (*oxysporum*)

912. **Hua-Van, A., T. Langin, and M. J. Daboussi.** 2002. Aberrant transposition of a *Tc1-mariner* element, *impala*, in the fungus *Fusarium oxysporum*. *Molecular Genetics and Genomics* **267:** 79-87. (*oxysporum*)

913. **Hua-Van, A., J. A. Pamphile, T. Langin, and M.-J. Daboussi.** 2001. Transposition of autonomous and engineered *impala* transposons in *Fusarium oxysporum* and a related species. *Molecular and General Genetics* **264:** 724-731. (*oxysporum, verticillioides*)

914. **Hue, F. X., M. Huerre, M. A. Rouffault, and C. de Bievre.** 1999. Specific detection of *Fusarium* species in blood and tissues by a PCR technique. *Journal of Clinical Microbiology* **37:** 2434-2438. (*oxysporum*)

915. **Huertas-Gonzalez, M. D., M. C. Ruiz-Roldan, F. I. Garcia-Maceira, M. I. G. Roncero, and A. di Pietro.** 1999. Cloning and characterization of *pl1* encoding an *in planta*-secreted pectate lyase of *Fusarium oxysporum*. *Current Genetics* **35:** 36-40. (*oxysporum*)

916. **Hughes, J. B., J. J. Hellmann, T. H. Rickets, and B. J. M. Bohannan.** 2001. Counting the uncountable: Statistical approaches to estimating microbial diversity. *Applied and Environmental Microbiology* **67:** 4399-4406. (11)

917. **Hughes, R., J. Dart, and S. Kilvington.** 2003. Activity of the amidoamine myristamidopropyl dimethylamine against

keratitis pathogens. *Journal of Antimicrobial Chemotherapy* **51:** 1415-1418. (*solani*)

918. **Hunter, T., and V. W. L. Jordan.** 1974. Fruit blotch of strawberry caused by *Fusarium sambucinum* Fückel. *Plant Pathology* **23:** 118. (*sambucinum*)

919. **Husson, F., A. Couturier, S. Kermasha, and J. M. Belin.** 1998. Induction and localization of a lipoxygenase from *Fusarium proliferatum*. *Journal of Molecular Catalysis B Enzymatic* **5:** 159-163. (*proliferatum*)

920. **Husson, F., Y. Pagot, S. Kermasha, and J. M. Belin.** 1998. *Fusarium proliferatum*: Induction and intracellular location of a lipoxygenase. *Enzyme and Microbial Technology* **23:** 42-48. (*proliferatum*)

921. **Hwang, S. F., B. D. Gossen, G. D. Turnbull, K. F. Chang, R. J. Howard, and A. G. Thomas.** 2000. Effect of temperature, seeding date, fungicide seed treatment and inoculation with *Fusarium avenaceum* on seedling survival, root rot severity, and yield of lentil. *Canadian Journal of Plant Science* **80:** 899-907. (*avenaceum*)

922. **Hyun, J.-W., S.-C. Lee, D.-H. Kim, S.-W. Ko, and K.-S. Kim.** 2000. Fusarium fruit rot of citrus in Jeju Island. *Mycobiology* **28:** 158-162. (*proliferatum*)

923. **Iannelli, D., R. Capparelli, G. Cristinzio, F. Marziano, F. Scala, and C. Noviello.** 1982. Serological differentiation among *formae speciales* and physiological races of *Fusarium oxysporum*. *Mycologia* **74:** 313-319. (*oxysporum*)

924. **Ichikawa, K., and T. Aoki.** 2000. New leaf spot disease of *Cymbidium* species caused by *Fusarium subglutinans* and *Fusarium proliferatum*. *Journal of General Plant Pathology* **66:** 213-218. (*proliferatum*, *sacchari*, *subglutinans*)

925. **Ichikawa, K., S. Kawasaki, C. Tanaka, and M. Tsuda.** 2003. Induced resistance against Fusarium diseases of *Cymbidium* species by weakly virulent strain HPF-1 (*Fusarium* sp.). *Journal of General Plant Pathology* **69:** 400-405. (*oxysporum*)

926. **Ikediobi, C. O., E. C. Ogundu, and A. I. Ukoha.** 1985. Production of linamarase by *Aspergillus sydowi* and *Fusarium equiseti*. *Process Biochemistry* **20:** 99-102. (*equiseti*)

927. **Imwidthaya, S., C. Chuntrasakul, and N. Chantarakul.** 1984. Opportunistic fungal infection of the burn wound. *Journal of the Medical Association of Thailand* **67:** 242-248. (*semitectum*)

928. **Inch, S., and J. Gilbert.** 2003. The incidence of *Fusarium* species recovered from inflorescences of wild grasses in southern Manitoba. *Canadian Journal of Plant Pathology* **25:** 379-383. (*poae*, *sporotrichioides*)

929. **Inglis, D. A., and R. J. Cook.** 1986. Persistence of chlamydospores of *Fusarium culmorum* in wheat soils of eastern Washington USA. *Phytopathology* **76:** 1205-1208. (*culmorum*)

930. **Inglis, G. D., and G. J. Boland.** 1992. Evaluation of filamentous fungi isolated from petals of bean and rapeseed for suppression of white mold. *Canadian Journal of Microbiology* **38:** 124-129. (*heterosporum*)

931. **Inoue, I., F. Namiki, and T. Tsuge.** 2002. Plant colonization by the vascular wilt fungus *Fusarium oxysporum* requires *FOW1*, a gene encoding a mitochondrial protein. *Plant Cell* **14:** 1869-1883. (*oxysporum*)

932. **Inoue, I., T. Ohara, F. Namiki, and T. Tsuge.** 2001. Isolation of pathogenicity mutants of *Fusarium oxysporum* f. sp. *melonis* by insertional mutagenesis. *Journal of General Plant Pathology* **67:** 191-199. (13)

933. **Irena, K., and K. Elzbieta.** 1999. Pathogenicity of *Fusarium* spp. on oats seedlings (*Avena sativa* L.). *Plant Breeding and Seed Science* **43:** 90-99. (*crookwellense*)

934. **Ishibashi, Y., S. Hommura, and Y. Matsumoto.** 1987. Direct examination vs. culture of biopsy specimens for the diagnosis of keratomycosis. *American Journal of Ophthalmology* **103:** 636-640. (*solani*)

935. **Isogai, T., H. Ono, Y. Ishitani, H. Kojo, Y. Ueda, and M. Kohsaka.** 1990. Structure and expression of complementary DNA for D-amino acid oxidase active against cephalosporin C from *Fusarium solani*. *Journal of Biochemistry* **108:** 1063-1069. (*solani*)

936. **Itakura, H., and R. Kinosita.** 1975. Toxic fungi isolated from Uganda foodstuffs. A histopathological study of acute toxicity of fungal cultural filtrates. *Tropical Medicine* **17:** 73-90. (*decemcellulare*)

937. **Ito, S., T. Eto, S. Tanaka, N. Yamauchi, H. Takahara, and T. Ikeda.** 2004. Tomatidine and lycotetraose, hydrolysis products of α-tomatine by *Fusarium oxysporum* tomatinase, suppress induced defense responses in tomato cells. *FEBS Letters* **571:** 31-34. (*oxysporum*)

938. **Ito, S., T. Kawaguchi, A. Nagata, H. Tamura, H. Matsushita, H. Takahara, S. Tanaka, and T. Ikeda.** 2004. Distribution of the *FoToml* gene encoding tomatinase in *formae speciales* of *Fusarium oxysporum* and identification of a novel tomatinase from *F. oxysporum* f. sp. *radicis-lycopersici*, the causal agent of Fusarium crown and root rot of tomato. *Journal of General Plant Pathology* **70:** 195-201. (*oxysporum*)

939. **Ito, S., H. Takahara, T. Kawaguchi, S. Tanaka, and M. Kameya-Iwaki.** 2002. Post-transcriptional silencing of the tomatinase gene in *Fusarium oxysporum* f. sp *lycopersici*. *Journal of Phytopathology* **150:** 474-480. (*oxysporum*)

940. **Iwahara, S., T. Nishihara, T. Jomori, M. Kuwahara, and T. Higuchi.** 1980. Enzymic oxidation of α, β unsaturated alcohols in the side chains of lignin related compounds. *Journal of Fermentation Technology* **58:** 183-187. (*solani*)

941. **Jackowiak, H., D. Packa, M. Wiwart, and J. Perkowski.** 2005. Scanning electron microscopy of *Fusarium*-damaged kernels of spring wheat. *International Journal of Food Microbiology* **98:** 113-123. (*culmorum*)

942. **Jackson, M. A., and S. N. Freer.** 1991. The influence of carbon and nitrogen nutrition on fusarin C biosynthesis by *Fusarium moniliforme*. *FEMS Microbiology Letters* **82:** 323-328. (*verticillioides*)

943. **Jackson, M. A., D. J. Frost, J. P. Karwowski, P. E. Humphrey, S. K. Dahod, W. S. Choi, K. Brandt, L. H. Malmberg, R. R. Rasmussen, M. H. Scherr, R. K. Flamm, S. Kadam, and J. B. McAlpine.** 1995. Fusacandins A and B; novel antifungal antibiotics of the papulacandin class from *Fusarium sambucinum*. I. Identity of the producing organism, fermentation and biological activity. *Journal of Antibiotics* **48:** 608-613. (*sambucinum*)

944. **Jackson, M. A., and A. C. Lanser.** 1993. Glucose and zinc concentration influence fusarin C synthesis, ethanol synthesis and lipid composition in *Fusarium moniliforme* submerged cultures. *FEMS Microbiology Letters* **108:** 69-74. (*verticillioides*)

945. **Jackson, M. A., P. J. Slininger, and R. J. Bothast.** 1989. Effects of zinc, iron, cobalt, and manganese on *Fusarium moniliforme* NRRL 13616 growth and fusarin C biosynthesis in submerged cultures. *Applied and Environmental Microbiology* **55:** 649-655. (*verticillioides*)

946. **Jacobson, D. J.** 1992. Control of mating-type heterokaryon incompatibility by the *tol* gene in *Neurospora crassa* and *N. tetrasperma*. *Genome* **35:** 347-353. (5)

947. **Jacobson, D. J., and T. R. Gordon.** 1988. Vegetative compatibility and self-incompatibility within *Fusarium oxysporum* f. sp. *melonis*. *Phytopathology* **78:** 668-672. (5, *oxysporum*)

948. **Jacobson, D. J., and T. R. Gordon.** 1990. Further investigations of vegetative compatibility within *Fusarium oxysporum* f. sp. *melonis*. *Canadian Journal of Botany* **68:** 1245-1248. (*oxysporum*)

949. **Jacobson, D. J., and T. R. Gordon.** 1990. Variability of mitochondrial DNA as an indicator of relationships between populations of *Fusarium oxysporum* f. sp. *melonis*. *Mycological Research* **94:** 734-744. (*oxysporum*)

950. **Jacobson, D. J., and T. R. Gordon.** 1991. *Fusarium oxysporum* f. sp. *melonis*: A case study of diversity within a *forma specialis*. *Phytopathology* **81:** 1064-1067. (5)

951. **Jacobson, M. F.** 2003. Adverse reactions linked to Quorn-brand foods. *Allergy* **58:** 455-456. (*venenatum*)

952. **Jaeger, E. E. M., N. M. Carroll, S. Choudhury, A. A. S. Dunlop, H. M. A. Towler, M. M. Matheson, P. Adamson, N. Okhravi, and S. Lightman.** 2000. Rapid detection and identification of *Candida*, *Aspergillus*, and *Fusarium* species in ocular samples using nested PCR. *Journal of Clinical Microbiology* **38:** 2902-2908. (7, *solani*)

953. **Jain, K. C., and M. V. Reddy.** 1995. Inheritance of resistance to Fusarium wilt in pigeonpea [*Cajanus cajan* (L.) Millsp.]. *Indian Journal of Genetics and Plant Breeding* **55:** 434-437. (*udum*)

954. **Jain, S., K. Akiyama, T. Kan, T. Ohguchi, and R. Takata.** 2003. The G protein β subunit FGB1 regulates development and pathogenicity in *Fusarium oxysporum*. *Current Genetics* **43:** 79-86. (*oxysporum*)

955. **Jain, S., K. Akiyama, K. Mae, T. Ohguchi, and R. Takata.** 2002. Targeted disruption of a G protein α subunit gene results in reduced pathogenicity in *Fusarium oxysporum*. *Current Genetics* **41:** 407-413. (*oxysporum*)

956. **Jain, S., K. Akiyama, R. Takata, and T. Ohguchi.** 2005. Signaling via the G protein α subunit FGA2 is necessary for pathogenesis in *Fusarium oxysporum*. *FEMS Microbiology Letters* **243:** 165-172. (*oxysporum*)

957. **Jamil, K., J. Narasaiah, and G. Thyagarajan.** 1983. Studies on the evaluation of naturally occurring fungal pathogens of water hyacinth (*Eichhornia crassipes*) for biological control of the weed. *Indian Journal of Botany* **6:** 185-189. (*solani*)

958. **Jardine, D. J., and J. F. Leslie.** 1992. Aggressiveness of *Gibberella fujikuroi* (*Fusarium moniliforme*) isolates to grain sorghum under greenhouse conditions. *Plant Disease* **76:** 897-900. (3, *thapsinum*)

959. **Jardine, D. J., and J. F. Leslie**. 1999. Aggressiveness to mature maize plants of *Fusarium* strains differing in the ability to produce fumonisin. *Plant Disease* **83:** 690-693. (3, *thapsinum*)

960. **Jarvis, W. R., and R. A. Shoemaker.** 1978. Taxonomic status of *Fusarium oxysporum* causing foot and root rot of tomato. *Phytopathology* **68:** 1679-1680. (*oxysporum*)

961. **Jaskiewicz, K., W. F. O. Marasas, and J. J. F. Taljaard.** 1987. Hepatitis in vervet monkeys caused by *Fusarium moniliforme*. *Journal of Comparative Pathology* **97:** 281-292. (*verticillioides*)

962. **Javaid, I., and M. Ashraf.** 1977. Wheat diseases in Zambia 1973-1976. *Plant Disease Reporter* **61:** 953-954. (*longipes*)

963. **Jayalakshmi, S. K., K. Sreeramulu, and V. I. Benigi.** 2003. Efficacy of *Trichoderma* spp. against pigeonpea wilt caused by *Fusarium udum* Butler. *Journal of Biological Control* **17:** 75-78. (*udum*)

964. **Jelen, H. H., C. J. Mirocha, E. Wasowicz, and E. Kaminski.** 1995. Production of volatile sesquiterpenes by *Fusarium sambucinum* strains with different abilities to synthesize trichothecenes. *Applied and Environmental Microbiology* **61:** 3815-3820. (*sambucinum*)

965. **Jelsch, C., S. Longhi, and C. Cambillau.** 1998. Packing forces in nine crystal forms of cutinase. *Proteins* **31:** 320-333. (*solani*)

966. **Jenczmionka, N. J., and W. Schaefer.** 2005. The *Gpmk1* MAP kinase of *Fusarium graminearum* regulates the induction of specific secreted enzymes. *Current Genetics* **47:** 29-36. (*graminearum*)

967. **Jeney, A., E. Beki, G. Mulè, and L. Hornok.** 2004. Identification of growth stage specific transcript profiles in *Fusarium proliferatum* (*Gibberella fujikuroi*, mating population D) by cDNA-AFLP analysis. *European Journal of Plant Pathology* **110:** 619-625. (*proliferatum*)

968. **Jenkinson, P., and D. W. Parry.** 1994. Splash dispersal of conidia of *Fusarium culmorum* and *Fusarium avenaceum*. *Mycological Research* **98:** 506-510. (*avenaceum*, *culmorum*)

969. **Jennings, J. C., P. C. Apel-Birkhold, B. A. Bailey, and J. D. Anderson.** 2000. Induction of ethylene biosynthesis and necrosis in weed leaves by a *Fusarium oxysporum* protein. *Weed Science* **48:** 7-14. (*oxysporum*)

970. **Jennings, J. C., P. C. Apel-Birkhold, N. M. Mock, C. J. Baker, J. D. Anderson, and B. A. Bailey.** 2001. Induction of defense responses in tobacco by the protein Nep1 from *Fusarium oxysporum*. *Plant Science* **161:** 891-899. (*oxysporum*)

971. **Jennings, P., M. E. Coates, K. Walsh, J. A. Turner, and P. Nicholson.** 2004. Determination of deoxynivalenol- and nivalenol-producing chemotypes of *Fusarium graminearum* isolated from wheat crops in England and Wales. *Plant Pathology* **53:** 643-652. (*graminearum*)

972. **Jensen, C. G., J. R. Aist, C. J. Bayles, S. M. Bollard, and L. C. W. Jensen.** 1991. Ultrastructural basis of mitosis in the fungus *Nectria haematococca* (sexual stage of *Fusarium solani*). III. Intermicrotubule bridges. *Protoplasma* **161:** 137-149. (*solani*)

973. **Jeschke, N., P. E. Nelson, and W. F. O. Marasas.** 1987. Toxicity to ducklings of *Fusarium moniliforme* isolated from corn intended for use in poultry feed. *Poultry Science* **66:** 1619-1623. (*verticillioides*)

974. **Jeschke, N., P. E. Nelson, and W. F. O. Marasas.** 1990. *Fusarium* spp. isolated from soil samples collected at different altitudes in the Transkei, southern Africa. *Mycologia* **82:** 727-733. (*dlamini*, *napiforme*, *solani*)

975. **Jestoi, M., M. Rokka, T. Yli-Mattila, P. Parikka, A. Rizzo, and K. Peltonen.** 2004. Presence and concentrations of the *Fusarium*-related mycotoxins beauvericin, enniatins and moniliformin in Finnish grain samples. *Food Additives and Contaminants* **21:** 794-802. (*avenaceum*)

976. **Ji, C., M.-H. Li, J.-J. Li, and S. J. Lu.** 1986. Synthesis of nitrosomethylisoamylamine from isoamylamine and sodium nitrite by fungi. *Carcinogenesis* **7:** 301-304. (*verticillioides*)

977. **Jimenez, M., T. Huerta, and R. Mateo.** 1997. Mycotoxin production by *Fusarium* species isolated from bananas. *Applied and Environmental Microbiology* **63:** 364-369. (*camptoceras*)

978. **Jimenez, M., A. Logrieco, and A. Bottalico.** 1993. Occurrence and pathogenicity of *Fusarium* species in banana fruits. *Journal of Phytopathology* **137:** 214-220. (*camptoceras, proliferatum, subglutinans*)

979. **Jimenez-Gasco, M., and R. M. Jimenez-Diaz.** 2003. Development of a specific polymerase chain reaction-based assay for the identification of *Fusarium oxysporum* f. sp. *ciceris* and its pathogenic races 0, 1A, 5, and 6. *Phytopathology* **93:** 200-209. (*oxysporum*)

980. **Jimenez-Gasco, M. M., M. G. Milgroom, and R. M. Jimenez-Diaz.** 2002. Gene genealogies support *Fusarium oxysporum* f. sp. *ciceris* as a monophyletic group. *Plant Pathology* **51:** 72-77. (*oxysporum*)

981. **Jimenez-Gasco, M. M., M. G. Milgroom, and R. M. Jimenez-Diaz.** 2004. Stepwise evolution of races in *Fusarium oxysporum* f. sp. *ciceris* inferred from fingerprinting with repetitive DNA sequences. *Phytopathology* **94:** 228-235. (*oxysporum*)

982. **Joffe, A. Z.** 1983. Environmental conditions conducive to *Fusarium* toxin formation causing serious outbreaks in animals and man. *Veterinary Research Communications* **7:** 187-193. (1)

983. **Joffe, A. Z.** 1986. Fusarium *Species: Their Biology and Toxicology*. John Wiley and Sons, New York City. (8)

984. **Joffe, A. Z., and J. Palti.** 1977. Species of *Fusarium* found in uncultivated desert-type soils in Israel. *Phytoparasitica* **5:** 119-122. (*avenaceum, equiseti, oxysporum, semitectum, solani, verticillioides*)

985. **Joffe, A. Z., J. Palti, and R. Arbel-Sherman.** 1973. *Fusarium moniliforme* Sheld. In Israel [*Gibberella fujikuroi* (Saw.) Wollenw.]. *Mycopathologia Mycologia Applicata* **50:** 85-107. (*bulbicola, subglutinans*)

986. **Johnson, E. M., A. Szekely, and D. W. Warnock.** 1998. *In vitro* activity of voriconazole, itraconazole and amphotericin B against filamentous fungi. *Journal of Antimicrobial Chemotherapy* **42:** 741-745. (*solani*)

987. **Johnson, G. I.** 1976. Brown etch or zonate ring spot of Butternut gramma. *Australasian Plant Pathology Society Newsletter* **5:** 48. (*equiseti*)

988. **Johnson, S. W., and R. C. Coolbaugh.** 1990. Light stimulated gibberellin biosynthesis in *Gibberella fujikuroi*. *Plant Physiology* **94:** 1696-1701. (*fujikuroi*)

989. **Jones, C., J. R. Ciacci-Zanella, Y. Zhang, G. Henderson, and M. Dickman.** 2001. Analysis of fumonisin B_1-induced apoptosis. *Environmental Health Perspectives* **109 (Suppl. 2):** 315-320. (*verticillioides*)

990. **Jones, M. J., and L. Epstein.** 1989. Adhesion of *Nectria haematococca* macroconidia. *Physiological and Molecular Plant Pathology* **35:** 453-461. (*solani*)

991. **Jones, M. J., and L. Epstein.** 1990. Adhesion of macroconidia to the plant surface and virulence of *Nectria haematococca*. *Applied and Environmental Microbiology* **56:** 3772-3778. (13)

992. **Joshi, A., S. K. Garg, and J. Verma.** 1990. Production of ethanol from sugars in wood hydrolysate by *Fusarium oxysporum*. *World Journal of Microbiology and Biotechnology* **6:** 10-14. (*oxysporum*)

993. **Julien, J., H. S. Poirier, and Y. Brygoo.** 1992. *Foret1*, a reverse transcriptase-like sequence in the filamentous fungus *Fusarium oxysporum*. *Nucleic Acids Research* **20:** 3933-3937. (*oxysporum*)

994. **Jurgenson, J. E., R. L. Bowden, K. A. Zeller, J. F. Leslie, N. J. Alexander, and R. D. Plattner.** 2002. A genetic map of *Gibberella zeae* (*Fusarium graminearum*). *Genetics* **160:** 1452-1460. (6, 7, 9, 10, *graminearum*)

995. **Jurgenson, J. E., K. A. Zeller, and J. F. Leslie.** 2002. An expanded genetic map of *Gibberella moniliformis* (*Fusarium verticillioides*). *Applied and Environmental Microbiology* **68:** 1972-1979. (6, 7, *verticillioides*)

996. **Kabeere, F., M. J. Hill, and J. G. Hampton.** 1997. The transmission of *Fusarium subglutinans* from maize seeds to seedlings. *Australasian Plant Pathology* **26:** 126-130. (*subglutinans*)

997. **Kacprzak, M., F. O. Asiegbu, G. Daniel, J. Stenlid, M. Manka, and M. Johansson.** 2001. Resistance reaction of conifer species (European larch, Norway spruce, Scots pine) to infection by selected necrotrophic damping-off pathogens. *European Journal of Plant Pathology* **107:** 191-207. (*culmorum*)

998. **Kaiser, S. A. K. M., and P. K. S. Gupta.** 1976. Serological and electrophoretic studies of three special forms of *Fusarium oxysporum*. *Transactions of the British Mycological Society* **67:** 33-37. (*oxysporum*)

999. **Kamaya, Y., F. Nakatsubo, and T. Higuchi.** 1983. Degradation of trimeric lignin model compounds aryl glycerol β-syringaresinol ethers by *Fusarium solani* M-13-1. *Agricultural and Biological Chemistry* **47:** 299-308. (*solani*)

1000. **Kamaya, Y., F. Nakatsubo, T. Higuchi, and S. Iwahara.** 1981. Degradation of racemic syringaresinol, a β-β' linked lignin model compound by *Fusarium solani* M-13-1. *Archives of Microbiology* **129:** 305-309. (*solani*)

1001. **Kamper, J. T., U. Kamper, L. M. Rogers, and P. E. Kolattukudy.** 1994. Identification of regulatory elements in the cutinase promoter from *Fusarium solani* f. sp. *pisi* (*Nectria haematococca*). *Journal of Biological Chemistry* **269:** 9195-9204. (*solani*)

1002. **Kanaan, Y. M., and A. H. Bahkali.** 1993. Frequency and cellulolytic activity of seed-borne *Fusarium* species isolated from Saudi Arabian cereal cultivars. *Zeitschrift für Pflanzenkrankheiten und Pflanzenschutz* **100:** 291-298. (*chlamydosporum*)

1003. **Kanaya, S., and H. Yoshida.** 1979. Phosphodiesterase phosphomonoesterases from *Fusarium moniliforme*: Separation and properties for four isozymes. *Journal of Biochemistry* **85:** 791-798. (*verticillioides*)

1004. **Kanaya, S., and H. Yoshida.** 1989. PDM phosphatase of *Fusarium moniliforme* possesses two phosphomonoesterase active sites with different affinities for *p*-nitrophenyl phosphate. *Biochemistry International* **18:** 1121-1128. (*verticillioides*)

1005. **Kang, Z., I. Zingen-Sell, and H. Buchenauer.** 2005. Infection of wheat spikes by *Fusarium avenaceum* and alterations of cell wall components in the infected tissue. *European Journal of Plant Pathology* **111:** 19-28. (*avenaceum*)

1006. **Kannaiyan, J., Y. L. Nene, M. V. Reddy, J. G. Ryan, and T. N. Raju.** 1984. Prevalence of pigeonpea (*Cajanus cajan*) diseases and associated crop losses in Asia, Africa and the Americas. *Tropical Pest Management* **30:** 62-71. (*udum*)

1007. **Kano, R., T. Okayama, M. Hamamoto, T. Nagata, K. Ohno, H. Tsujimoto, H. Nakayama, K. Doi, K. Fujiwara, and A. Hasegawa.** 2002. Isolation of *Fusarium solani* from a dog: Identification by molecular analysis. *Medical Mycology* **40:** 435-437. (*solani*)

1008. **Katan, T.** 1999. Current status of vegetative compatibility groups in *Fusarium oxysporum*. *Phytoparasitica*. **27:** 51-64. (5, *oxysporum*)

1009. **Katan, T., R. Berliner, and J. Katan.** 1994. Vegetative compatibility in populations of *Fusarium oxysporum* from wild carnation. *Mycological Research* **98:** 1415-1418. (*oxysporum*)

1010. **Katan, T., and P. di Primo.** 1999. Current status of vegetative compatibility groups in *Fusarium oxysporum*: Supplement (1999). *Phytoparasitica* **27:** 273-277. (*oxysporum*)

1011. **Katan, T., A. Gamliel, and J. Katan.** 1996. Vegetative compatibility of *Fusarium oxysporum* from sweet basil in Israel. *Plant Pathology* **45:** 656-661. (*oxysporum*)

1012. **Katan, T., E. Hadar, and J. Katan.** 1989. Vegetative compatibility of *Fusarium oxysporum* f. sp. *dianthi* from carnation in Israel. *Plant Pathology* **38:** 376-381. (*oxysporum*)

1013. **Katan, T., and J. Katan.** 1988. Vegetative compatibility grouping of *Fusarium oxysporum* f. sp. *vasinfectum* from tissue and the rhizosphere of cotton plants. *Phytopathology* **78:** 852-855. (*oxysporum*)

1014. **Katan, T., and J. Katan.** 1999. Vegetative compatibility grouping in *Fusarium oxysporum* f. sp. *radicis-lycopersici* from the UK, the Netherlands, Belgium and France. *Plant Pathology* **48:** 541-549. (5, *oxysporum*)

1015. **Katan, T., D. Zamir, M. Sarfatti, and J. Katan.** 1991. Vegetative compatibility groups and subgroups in *Fusarium oxysporum* f. sp. *radicis-lycopersici*. *Phytopathology* **81:** 255-262. (5, *oxysporum*)

1016. **Kataoka, M., K. Shimizu, K. Sakamoto, H. Yamada, and S. Shimizu.** 1995. Optical resolution of racemic pantolactone with a novel fungal enzyme, lactonohydrolase. *Applied Microbiology and Biotechnology* **43:** 974-977. (*oxysporum*)

1017. **Kataoka, M., K. Shimizu, K. Sakamoto, H. Yamada, and S. Shimizu.** 1995*b*. Lactonohydrolase-catalyzed optical resolution of pantoyl lactone: Selection of a potent enzyme producer and optimization of culture and reaction conditions for practical resolution. *Applied Microbiology and Biotechnology* **44:** 333-338. (*oxysporum*)

1018. **Katayama, T., F. Nakatsubo, and T. Higuchi.** 1980. Initial reactions in the fungal degradation of guaiacyl glycerol β-coniferyl ether, a lignin substrate model. *Archives of Microbiology* **126:** 127-132. (*solani*)

1019. **Katayama, T., F. Nakatsubo, and T. Higuchi.** 1981. Degradation of aryl glycerol β-aryl ethers, lignin substructure models by *Fusarium solani*. *Archives of Microbiology* **130:** 198-203. (*solani*)

1020. **Kathariou, S.** 1981. Gene pool organization in *Fusarium moniliforme*. Ph.D. dissertation, University of California-Berkeley. (8)

1021. **Kathariou, S., and P. T. Spieth.** 1982. Spore killer polymorphism in *Fusarium moniliforme*. *Genetics* **102:** 19-24. (10, *verticillioides*)

1022. **Katona, S. J., and E. R. Kaminski.** 2002. Sensitivity to Quorn mycoprotein (*Fusarium venenatum*) in a mold-allergic patient. *Journal of Clinical Pathology* **55:** 876-877. (*venenatum*)

1023. **Katta, S. K., A. E. Cagampang, L. S. Jackson, and L. B. Bullerman.** 1997. Distribution of *Fusarium* molds and fumonisins in dry-milled corn fractions. *Cereal Chemistry* **74:** 858-863. (*verticillioides*)

1024. **Kaufman, L., P. G. Standard, M. Jalbert, and D. E. Kraft.** 1997. Immunohistologic identification of *Aspergillus* spp. and other hyaline fungi by using polyclonal fluorescent antibodies. *Journal of Clinical Microbiology* **35:** 2206-2209. (*solani*)

1025. **Kavelman, R., and B. Kendrick.** 1978. Degradation of a plastic polyepsilon caprolactone by hyphomycetes. *Mycologia* **70:** 87-103. (*solani*)

1026. **Kawabe, M., K. Mizutani, T. Yoshida, T. Teraoka, K. Yoneyama, I. Yamaguchi, and T. Arie.** 2004. Cloning of the pathogenicity-related gene *FPD1* in *Fusarium oxysporum* f. sp. *lycopersici*. *Journal of General Plant Pathology* **70:** 16-20. (*oxysporum*)

1027. **Kawchuk, L. M., L. J. Hutchison, C. A. Verhaeghe, D. R. Lynch, P. S. Bains, and J. D. Holley.** 2002. Isolation of the β-tubulin gene and characterization of thiabendazole resistance in *Gibberella pulicaris*. *Canadian Journal of Plant Pathology* **24:** 233-238. (*sambucinum*)

1028. **Kayali, H. A., and L. Tarhan.** 2005. Role of pyruvate and ascorbate production in regulation of antioxidant enzymes and membrane LPO levels in *Fusarium acuminatum*. *Applied Biochemistry and Biotechnology* **120:** 15-27. (*acuminatum*)

1029. **Kedera, C. J., J. F. Leslie, and L. E. Claflin.** 1994. Genetic diversity of *Fusarium* section *Liseola* (*Gibberella fujikuroi*) in individual maize plants. *Phytopathology* **84:** 603-607. (3, *proliferatum*, *subglutinans*, *verticillioides*)

1030. **Kedera, C. J., R. D. Plattner, and A. E. Desjardins.** 1999. Incidence of *Fusarium* spp. and levels of fumonisin B_1 in maize in western Kenya. *Applied and Environmental Microbiology* **65:** 41-44. (*verticillioides*)

1031. **Keith, C. L.** 1972. Process for producing zearalenone. U. S. Patent No. 2,272,578. (4)

1032. **Keller, S. E., and T. M. Sullivan.** 1998. Identification of a variant of *Fusarium proliferatum* that hydrolyzes fumonisin B_1. *Journal of Agricultural and Food Chemistry* **46:** 2823-2826. (*proliferatum*)

1033. **Keller, S. E., T. M. Sullivan, and S. Chirtel.** 1997. Factors affecting the growth of *Fusarium proliferatum* and the production of fumonisin B_1: Oxygen and pH. *Journal of Industrial Microbiology and Biotechnology* **19:** 305-309. (*verticillioides*)

1034. **Kellock, A. W., L. L. Stubbs, and D. G. Parbery.** 1978. Seed-borne *Fusarium* species on subterranean clover and other pasture legumes. *Australian Journal of Agricultural Research* **29:** 975-982. (*avenaceum*)

1035. **Kelly, A. G., B. W. Bainbridge, J. B. Heale, E. Perez-Artes, and R. M. Jimenez-Diaz.** 1998. *In planta* polymerase chain reaction detection of the wilt-inducing pathotype of *Fusarium oxysporum* f. sp. *ciceris* in chickpea (*Cicer arietinum* L.). *Physiological and Molecular Plant Pathology* **52:** 397-409. (*oxysporum*)

1036. **Kerényi, Z., and L. Hornok.** 2002. Structure and function of mating type genes in *Fusarium* species. *Acta Microbiologica et Immunologica Hungarica* **49:** 313-314. (*avenaceum, camptoceras, crookwellense, culmorum, semitectum, sporotrichioides*)

1037. **Kerényi, Z., A. Moretti, C. Waalwijk, B. Oláh, and L. Hornok.** 2004. Mating type sequences in asexually reproducing *Fusarium* species. *Applied and Environmental Microbiology* **70:** 4419-4423. (6, 7, 9, 11, *avenaceum, culmorum, poae, semitectum*)

1038. **Kerényi, Z., E. Taborhegyi, A. Pomazi, and L. Hornok.** 1997. Variability amongst strains of *Fusarium poae* assessed by vegetative compatibility and RAPD polymorphism. *Plant Pathology* **46:** 882-889. (5, *poae*)

1039. **Kerényi, Z., K. Zeller, L. Hornok, and J. F. Leslie.** 1999. Standardization of mating-type terminology in the *Gibberella fujikuroi* species complex. *Applied and Environmental Microbiology* **65:** 4071-4076. (6, 7, 9, 11)

1040. **Kerr, K. M., and P. J. Davis.** 1983. Microbial transformations in organic synthesis. 3. Synthesis, characterization and fungal metabolism of cis-7-methylglaucine and trans-7-methylglaucine. *Journal of Organic Chemistry* **48:** 928-932. (*solani*)

1041. **Kerr, K. M., and P. J. Davis.** 1986. Microbial transformations in organic synthesis. 4. Stereoselective fungal metabolism of 7-methylglaucine. *Journal of Organic Chemistry* **51:** 1741-1744. (*solani*)

1042. **Kessler, K. J., Jr.** 1974. An apparent symbiosis between *Fusarium* fungi and ambrosia beetles causes canker on Black Walnut stems. *Plant Disease Reporter* **58:** 1044-1047. (*lateritium*)

1043. **Keyser, Z., H. F. Vismer, J. A. Klaasen, P. W. Snijman, and W. F. O. Marasas.** 1999. The antifungal effect of fumonisin B_1 on *Fusarium* and other fungal species. *South African Journal of Science* **95:** 455-458. (*verticillioides*)

1044. **Khan, R., and D. C. Straney.** 1999. Regulatory signals influencing expression of the *PDA1* gene of *Nectria haematococca* MP VI in culture and during pathogenesis of pea. *Molecular Plant-Microbe Interactions* **12:** 733-742. (*solani*)

1045. **Khan, R., R. Tan, A. Galvez-Mariscal, and D. Straney.** 2003. A binuclear zinc transcription factor binds the host isoflavonoid-responsive element in a fungal cytochrome p450 gene responsible for detoxification. *Molecular Microbiology* **49:** 117-130. (*solani*)

1046. **Khang, C. K., S.-Y. Park, Y.-H. Lee, and S. Kang.** 2005. A dual selection based, targeted gene replacement tool for *Magnaporthe grisea* and *Fusarium oxysporum*. *Fungal Genetics and Biology* **42:** 483-492. (*oxysporum*)

1047. **Khonga, E. B., and J. C. Sutton.** 1988. Inoculum production and survival of *Gibberella zeae* in maize and wheat residues. *Canadian Journal of Plant Pathology* **10:** 232-239. (*graminearum*)

1048. **Kido, T., K. Hashizume, and K. Soda.** 1978. Purification and properties of nitroalkane oxidase from *Fusarium oxysporum*. *Journal of Bacteriology* **133:** 53-58. (*oxysporum*)

1049. **Kienh, T. E., P. E. Nelson, E. M. Bernard, F. F. Edwards, B. Koziner, and D. Armstrong.** 1985. Catheter-associated fungemia caused by *Fusarium chlamydosporum* in a patient with lymphocytic lymphoma. *Journal of Clinical Microbiology* **21:** 501-504. (*chlamydosporum*)

1050. **Kilic, O, and G. J. Griffin.** 1998. Effect of dsRNA-containing and dsRNA-free hypovirulent isolates of *Fusarium oxysporum* on severity of Fusarium seedling disease of soybean in naturally infested soil. *Plant and Soil* **201:** 125-135. (*oxysporum*)

1051. **Killham, K., and M. Wainwright.** 1981. Microbial release of sulfur ions from atmospheric pollution deposits. *Journal of Applied Ecology* **18:** 889-896. (*solani*)

1052. **Kim, D.-H.** 1997. Induced change in DNA methylation of *Fusarium oxysporum* f. sp. *niveum* due to successive transfer. *Journal of Biochemistry and Molecular Biology* **30:** 216-221. (*oxysporum*)

1053. **Kim, D.-H., R. D. Martyn, and C. W. Magill.** 1992. Restriction fragment length polymorphism groups and physical map of mitochondrial DNA from *Fusarium oxysporum* f. sp. *niveum*. *Phytopathology* **82:** 346-353. (*oxysporum*)

1054. **Kim, D.-H., R. D. Martyn, and C. W. Magill.** 1993. Chromosomal polymorphism in *Fusarium oxysporum* f. sp. *niveum*. *Phytopathology* **83:** 1209-1216. (*oxysporum*)

1055. **Kim, H. J., and B. R. Min.** 2004. Nucleotide divergence analysis of IGS region in *Fusarium oxysporum* and its *formae speciales* based on the sequence. *Mycobiology* **32:** 119-122. (*oxysporum*)

1056. **Kim, H.-S., T. Lee, M. Dawlatana, S.-H. Yun, and Y.-W. Lee.** 2003. Polymorphism of trichothecene biosynthesis genes in deoxynivalenol- and nivalenol-producing *Fusarium graminearum* isolates. *Mycological Research* **107:** 190-197. (*graminearum*)

1057. **Kim, J.-C., and Y.-W. Lee.** 1994. Sambutoxin, a new mycotoxin produced by toxic *Fusarium* isolates obtained from rotted potato tubers. *Applied and Environmental Microbiology* **60:** 4380-4386. (*oxysporum, sambucinum*)

1058. **Kim, J.-C., Y.-W. Lee, H. Tamura, and T. Yoshizawa.** 1995. Sambutoxin: A new mycotoxin isolated from *Fusarium sambucinum*. *Tetrahedron Letters* **36:** 1047-1050. (*sambucinum*)

1059. **Kim, J.-C., Y.-W. Lee, and S. H. Yu.** 1995. Sambutoxin-producing isolates of *Fusarium* species and occurrence of sambutoxin in rotten potato tubers. *Applied and Environmental Microbiology* **61:** 3750-3751. (*oxysporum, sambucinum, semitectum*)

1060. **Kim, J.-C., J.-B. Park, G.-W. Kim, W. B. Kim, and Y.-W. Lee.** 1998. *In vitro* cytotoxicity of sambutoxin. *Agricultural Chemistry and Biotechnology* **41:** 273-274. (*sambucinum*)

1061. **Kim, J.-E., K.-H. Han, J. Jin, H. Kim, J.-C. Kim, S.-H. Yun, and Y.-W. Lee.** 2005. Putative polyketide synthase and laccase genes for biosynthesis of aurofusarin in *Gib-*

berella zeae. *Applied and Environmental Microbiology* **71:** 1701-1708. (*graminearum*)

1062. **Kim, J. E., S. L. Perkins, and G. J. Harris.** 2003. Voriconazole treatment of fungal scleritis and epibulbar abscess resulting from scleral buckle infection. *Archives of Ophthalmology* **121:** 735-737. (*solani*)

1063. **Kim, K.-H., Y.-W. Lee, C. J. Mirocha, and R. J. Pawlosky.** 1990. Isoverrucarol production by *Fusarium oxysporum* CJS-12 isolated from corn. *Applied and Environmental Microbiology* **56:** 260-263. (*oxysporum*)

1064. **Kim, Y.-H., J. Lee, J.-Y. Ahn, M.-B. Gu, and S.-H. Moon.** 2002. Enhanced degradation of an endocrine-disrupting chemical, butyl benzyl phthalate, by *Fusarium oxysporum* f. sp. *pisi* cutinase. *Applied and Environmental Microbiology* **68:** 4684-4688. (*oxysporum*)

1065. **Kim, Y. H., J. Lee, and S.-H. Moon.** 2003. Uniqueness of microbial cutinases in hydrolysis of *p*-nitrophenyl esters. *Journal of Microbiology and Biotechnology* **13:** 57-63. (*solani*)

1066. **Kimura, M., G. Matsumoto, Y. Shingu, K. Yoneymama, and I. Yamaguchi.** 1998. The mystery of the trichothecene 3-O-acetyltransferase gene: Analysis of the region around *Tri101* and characterization of its homologue form *Fusarium sporotrichioides*. *FEBS Letters* **435:** 163-168. (*sporotrichioides*)

1067. **Kimura, M., T. Tokai, G. Matsumoto, M. Fujimura, H. Hamamoto, K. Yoneyama, T. Shibata, and I. Yamaguchi.** 2003. Trichothecene nonproducer *Gibberella* species have both functional and nonfunctional 3-O-acetyltransferase genes. *Genetics* **163:** 677-684. (*oxysporum*, *verticillioides*)

1068. **Kimura, M, and I. Yamaguchi.** 1999. The mystery of the trichothecene 3-O-acetyltransferase gene: *Tri101* evolved independently of other trichothecene biosynthetic genes in the gene cluster. *Pesticide Science* **55:** 372-374. (*sporotrichioides*)

1069. **King, S. B., and G. E. Scott.** 1981. Genotypic differences in maize (*Zea mays*) to kernel infection by *Fusarium moniliforme*. *Phytopathology* **71:** 1245-1247. (*verticillioides*)

1070. **Kinzel, J. J., and J. F. Leslie.** 1985. NADPH-dependent glutamate dehydrogenase activity in *Gibberella zeae*. *Experimental Mycology* **9:** 170-173. (5)

1071. **Kiprop, E. K., J. P. Baudoin, A. W. Mwang'ombe, P. M. Kimani, G. Mergeai, and A. Maquet.** 2002. Characterization of Kenyan isolates of *Fusarium udum* from pigeonpea [*Cajanus cajan* (L.) Millsp.] by cultural characteristics, aggressiveness and AFLP analysis. *Journal of Phytopathology* **150:** 517-525. (*udum*)

1072. **Kiprop, E. K., A. W. Mwang'ombe, J. P. Baudoin, P. M. Kimani, and G. Mergeai.** 2002. Cultural characteristics, pathogenicity and vegetative compatibility of *Fusarium udum* isolates from pigeonpea [*Cajanus cajan* (L.) Millsp.] in Kenya. *European Journal of Plant Pathology* **108:** 147-154. (*udum*)

1073. **Kiryu, H., S. Yoshida, Y. Suenaga, and M. Asahi.** 1991. Invasion and survival of *Fusarium solani* in the dexamethasone-treated cornea of rabbits. *Journal of Medical and Veterinary Mycology* **29:** 395-406. (*solani*)

1074. **Kishi, K., T. Furukawa, and T. Aoki.** 1999. Purple spot of aloe (*Aloe arborescens* Mill.) caused by *Fusarium phyllophilum* Nirenberg *et* O'Donnell (new disease). *Annals of the Phytopathological Society of Japan* **65:** 576-587. (*phyllophilum*)

1075. **Kistler, H. C.** 2001. Evolution of host specificity in *Fusarium oxysporum*. p. 70-96. *In* B. A. Summerell, J. F. Leslie, D. Backhouse, W. L. Bryden, and L. W. Burgess (eds.), *Fusarium: Paul E. Nelson Memorial Symposium*. APS Press, St. Paul, Minnesota. (7, 8, 11)

1076. **Kistler, H. C., C. Alabouvette, R. P. Baayen, S. Bentley, D. Brayford, A. Coddington, J. C. Correll, M.-J. Daboussi, K. Elias, D. Fernandez, T. R. Gordon, T. Katan, H. G. Kim, J. F. Leslie, R. D. Martyn, Q. Migheli, N. Y. Moore, K. O'Donnell, R. C. Ploetz, M. A. Rutherford, B. Summerell, C. Waalwijk, and S. Woo.** 1998. Systematic numbering of vegetative compatibility groups in the plant pathogenic fungus *Fusarium oxysporum*. *Phytopathology* **88:** 30-32. (5, *oxysporum*)

1077. **Kistler, H. C., and U. Benny.** 1989. The mitochondrial genome of *Fusarium oxysporum*. *Plasmid* **22:** 86-89. (*oxysporum*)

1078. **Kistler, H. C., U. Benny, E. W. A. Boehm, and T. Katan.** 1995. Genetic duplication in *Fusarium oxysporum*. *Current Genetics* **28:** 173-176. (*oxysporum*)

1079. **Kistler, H. C., U. Benny, and W. A. Powell.** 1997. Linear mitochondrial plasmids of *Fusarium oxysporum* contain genes with sequence similarity to genes encoding a reverse transcriptase from *Neurospora* spp. *Applied and Environmental Microbiology* **63:** 3311-3313. (*oxysporum*)

1080. **Kistler, H. C., and S. A. Leong.** 1986. Linear plasmid-like DNA in the plant pathogenic fungus *Fusarium oxysporum* f. sp. *conglutinans*. *Journal of Bacteriology* **167:** 587-593. (*oxysporum*)

1081. **Kistler, H. C., L. W. Meinhardt, and U. Benny.** 1996. Mutants of *Nectria haematococca* created by a site-directed chromosome breakage are greatly reduced in virulence toward pea. *Molecular Plant-Microbe Interactions* **9:** 804-809. (*solani*)

1082. **Kistler, H. C., E. A. Momol, and U. Benny.** 199[illegible]. Repetitive genomic sequences for determining relatedness among strains of *Fusarium oxysporum*. *Phytopathology* **81:** 331-336. (*oxysporum*)

1083. **Kistler, H. C., and H. D. VanEtten.** 1981. Phaseolin metabolism and tolerance in *Fusarium solani* f. sp. *phaseoli*. *Physiological Plant Pathology* **19:** 257-272. (*solani*)

1084. **Kitamura, Y., Y. Abe, and T. Yasui.** 1991. Metabolism of levoglucosan, 1,6-anhydro-β-D-glucopyranose, in microorganisms. *Agricultural and Biological Chemistry* **55:** 515-522. (*solani*)

1085. **Klaasen, J. A., and P. E. Nelson.** 1996. Identification of a mating population, *Gibberella nygamai sp. nov.*, within the *Fusarium nygamai* anamorph. *Mycologia* **88:** 965-969. (2, 6, 8, 9, *nygamai*)

1086. **Klaasen, J. A., and P. E. Nelson.** 1998. Fumonisin production by field strains of *Fusarium nygamai* (*Gibberella nygamai*) and ascospore progeny of laboratory crosses. *World Journal of Microbiology and Biotechnology* **14:** 873-877. (*nygamai*)

1087. **Klein, K. K., and J. C. Correll.** 2001. Vegetative compatibility group diversity in *Fusarium*, p. 83-96. *In* B. A. Summerell, J. F. Leslie, D. Backhouse, W. L. Bryden, and L. W. Burgess (eds.), *Fusarium: Paul E. Nelson Memorial Symposium*. APS Press, St. Paul, Minnesota. (11, *oxysporum*)

1088. **Klittich, C. J. R., and C. R. Bronson.** 1986. Reduced fitness associated with *TOX1* of *Cochliobolus heterostrophus*. *Phytopathology* **76:** 1294-1298. (6)

1089. **Klittich, C. J. R., J. C. Correll, and J. F. Leslie.** 1988. Inheritance of sectoring frequency in *Fusarium moniliforme* (*Gibberella fujikuroi*). *Experimental Mycology* **12:** 289-294. (5)

1090. **Klittich, C. J. R., and J. F. Leslie.** 1988. Nitrate reduction mutants of *Fusarium moniliforme* (*Gibberella fujikuroi*). *Genetics* **118:** 417-423. (2, 5, 6, *verticillioides*)

1091. **Klittich, C. J. R., and J. F. Leslie.** 1989. Chlorate-resistant, nitrate-utilizing (*crn*) mutants of *Fusarium moniliforme* (*Gibberella fujikuroi*). *Journal of General Microbiology* **135:** 721-727. (5)

1092. **Klittich, C. J. R., and J. F. Leslie.** 1992. Identification of a second mating population within the *Fusarium moniliforme* anamorph of *Gibberella fujikuroi*. *Mycologia* **84:** 541-547. (6, 8, 9, *thapsinum*)

1093. **Klittich, C. J. R., J. F. Leslie, P. E. Nelson, and W. F. O. Marasas.** 1997. *Fusarium thapsinum* (*Gibberella thapsina*): a new species in section *Liseola* from sorghum. *Mycologia* **89:** 643-652. (8, 9, *thapsinum*)

1094. **Klotz, L. V., P. E. Nelson, and T. A. Toussoun.** 1988. A medium for enhancement of chlamydospore formation in *Fusarium* species. *Mycologia* **80:** 108-109. (2)

1095. **Knight, C.** 1982. Pathogenicity of some fungi associated with crown rot of bananas. *Phytopathologische Zeitschrift* **104:** 13-18. (*subglutinans*)

1096. **Knight, C., D. F. Cutts, and J. Colhoun.** 1977. The role of *Fusarium semitectum* in causing crown rot of bananas. *Phytopathologische Zeitschrift* **89:** 170-176. (*semitectum*)

1097. **Knoll, S., S. Mulfinger, L. Niessen, and R. F. Vogel.** 2002. Rapid preparation of *Fusarium* DNA from cereals for diagnostic PCR using sonification and an extraction kit. *Plant Pathology* **51:** 728-734. (7)

1098. **Knutsen, A. K., M. Torp, and A. Holst-Jensen.** 2004. Phylogenetic analyses of the *Fusarium poae*, *Fusarium sporotrichioides* and *Fusarium langsethiae* species complex based on partial sequences of the translation elongation factor-1α gene. *International Journal of Food Microbiology* **95:** 287-295. (*poae*)

1099. **Kobayashi, M., Y. Matsuo, A. Takimoto, S. Suzuki, F. Maruo, and H. Shoun.** 1996. Denitrification, a novel type of respiratory metabolism in fungal mitochondrion. *Journal of Biological Chemistry* **271:** 16263-16267. (*oxysporum*)

1100. ☑**Kobayashi, M., M. Shinohara, C. Sakoh, M. Kataoka, and S. Shimizu.** 1998. Lactone ring cleaving enzyme: Genetic analysis, novel RNA editing, and evolutionary implications. *Proceedings of the National Academy of Sciences USA* **95:** 12787-12792. (*oxysporum*)

1101. **Kobayashi, M., and H. Shoun.** 1995. The copper-containing dissimilatory nitrite reductase involved in the denitrifying system of the fungus *Fusarium oxysporum*. *Journal of Biological Chemistry* **270:** 4146-4151. (*oxysporum*)

1102. **Koch, N., and W. Huth.** 1997. Interaction of barley yellow dwarf virus and *Fusarium culmorum* (W. G. Sm.) Sacc. in winter wheat. *Journal of Phytopathology* **145:** 425-428. (*culmorum*)

1103. **Kodo, N., C. Nakamura, H. Kato, T. Yoshizawa, N. Mori, and C. Kaneda.** 1995. Restriction fragment length polymorphisms of mitochondrial DNAs from seven *Fusarium* species causing fusarium head blight. *Japanese Journal of Genetics* **70:** 435-451. (*culmorum*, *sporotrichioides*)

1104. **Koenig, R. L., R. C. Ploetz, and H. C. Kistler.** 1997. *Fusarium oxysporum* f. sp. *cubense* consists of a small number of divergent and globally distributed clonal lineages. *Phytopathology* **87:** 915-923. (5, 11, *oxysporum*)

1105. **Koenig, T., A. Kapus, and B. Sarkadi.** 1993. Effects of equisetin on rat liver mitochondria: Evidence for inhibition of substrate anion carriers of the inner membrane. *Journal of Bioenergetics and Biomembranes* **25:** 537-545. (*equiseti*)

1106. **Koike, S. T., T. R. Gordon, and B. J. Aegerter.** 2003. Root and basal rot of leek caused by *Fusarium culmorum* in California. *Plant Disease* **87:** 601. (*culmorum*)

1107. **Kolaczkowska, M., J. Chrzanowska, A. Jacyk, K. Szoltysek, and A. Polanowski.** 1988. Factors affecting rennin-like proteinase production by *Fusarium moniliforme*. *Milchwissenschaft* **43:** 83-86. (*verticillioides*)

1108. **Kolaczkowska, M. K., M. Wieczorek, and A. Polanowski.** 1983. An aspartic proteinase (e.c. 3.4.23.6) from *Fusarium moniliforme*: Purification and general properties. *European Journal of Biochemistry* **132:** 557-562. (*verticillioides*)

1109. **Kolattukudy, P. E., L. M. Rogers, D. Li, C.-S. Hwang, and M. A. Flaishman.** 1995. Surface signaling in pathogenesis. *Proceedings of the National Academy of Sciences USA* **92:** 4080-4087. (*solani*)

1110. **Koller, W., C. R. Allan, and P. E. Kolattukudy.** 1982. Inhibition of cutinase and prevention of fungal penetration into plants by Benomyl: A possible protective mode of action. *Pesticide Biochemistry and Physiology* **18:** 15-25. (*solani*)

1111. **Koller, W., C. R. Allan, and P. E. Kolattukudy.** 1982. Role of cutinase and cell wall degrading enzymes in infection of *Pisum sativum* by *Fusarium solani* f. sp. *pisi*. *Physiological Plant Pathology* **20:** 47-60. (*solani*)

1112. **Koller, W., C. P. Allan, and P. E. Kolattukudy.** 1982. Protection of *Pisum sativum* from *Fusarium solani* f. sp. *pisi* by inhibition of cutinase with organo-phosphorus pesticides. *Phytopathology* **72:** 1425-1430. (*solani*)

1113. **Koller, W., and P. E. Kolattukudy.** 1982. Mechanism of action of cutinase: Chemical modification of the catalytic triad characteristic for serine hydrolases. *Biochemistry* **21:** 3083-3090. (*solani*)

1114. **Kolmark, H. G.** 1984. Mutants with continuous microcycle conidiation in the filamentous fungus *Fusarium solani* f. sp. *pisi*. *Molecular and General Genetics* **198:** 12-18. (*solani*)

1115. **Komada, H.** 1975. Development of a selective medium for quantitative isolation of *Fusarium oxysporum* from natural soil. *Review of Plant Protection Research* **8:** 114-125. (2)

1116. **Kommedahl, T., H. K. Abbas, P. M. Burnes, and C. J. Mirocha.** 1988. Prevalence and toxigenicity of *Fusarium* spp. from soils of Norway near the Arctic Circle. *Mycologia* **80:** 790-794. (*equiseti*, *merismoides*, *oxysporum*, *sambucinum*, *semitectum*, *solani*)

1117. **Kommedahl, T., C. E. Windels, and D. S. Lang.** 1975. Comparison of *Fusarium* populations in grasslands of Minnesota, USA and Iceland. *Mycologia* **67:** 38-44. (*sambucinum*)

1118. **Kommedahl, T., C. E. Windels, and R. E. Stucker.** 1979. Occurrence of *Fusarium* species in roots and stalks of symptomless corn plants during the growing season. *Phytopathology* **69:** 961-966. (*armeniacum, oxysporum*)

1119. **Kommedahl, T., C. E. Windels, and H. B. Wiley.** 1978. *Fusarium* infected stalks and other diseases of corn in Minnesota USA in 1977. *Plant Disease Reporter* **62:** 692-694. (*oxysporum, verticillioides*)

1120. **Kondo, E., and T. Mitsugi.** 1966. Microbiological synthesis of 16-keto steroids from steroidal sapogenins. *Journal of the American Chemical Society* **88:** 4737-4738. (*solani*)

1121. **Kondo, N., F. Kodama, and A. Ogoshi.** 1997. Vegetative compatibility groups of *Fusarium oxysporum* f. sp. *adzukicola* and nonpathogenic *Fusarium oxysporum* on Adzuki bean isolated from adzuki bean fields in Hokkaido. *Annals of the Phytopathological Society of Japan* **63:** 8-12. (*oxysporum*)

1122. **Kono, Y., H. Yamamoto, M. Takeuchi, and H. Komada.** 1995. Alterations in superoxide dismutase and catalase in *Fusarium oxysporum* during starvation-induced differentiation. *Biochimica et Biophysica Acta* **1268:** 35-40. (*oxysporum*)

1123. **Konstantinova, P., and T. Yli-Mattila.** 2004. IGS-RFLP analysis and development of molecular markers for identification of *Fusarium poae, Fusarium langsethiae, Fusarium sporotrichioides* and *Fusarium kyushuense*. *International Journal of Food Microbiology* **95:** 321-331. (*poae*)

1124. **Kordali, S., and E. Demirci.** 1998. *Fusarium* species from various vegetables in Erzincan, Turkiye. *Journal of Turkish Phytopathology* **27:** 131-136. (*scirpi*)

1125. **Kornerup, A., and J. H. Wanscher.** 1978. *Methuen Handbook of Color*, 3rd ed. Methuen and Co., London. (12)

1126. **Korolev, N., and T. Katan.** 1997. Improved medium for selecting nitrate non-utilizing (*nit*) mutants of *Verticillium dahliae*. *Phytopathology* **87:** 1067-1070. (2, 5)

1127. **Kosiak, B., M. Torp, E., Skjerve, and B. Andersen.** 2004. *Alternaria* and *Fusarium* in Norwegian grains of reduced quality – a matched pair sample study. *International Journal of Food Microbiology* **93:** 51-62. (*poae*)

1128. **Kosiak, B., M. Torp, E. Skjerve, and U. Thrane.** 2003. The prevalence and distribution of *Fusarium* species in Norwegian cereals: A survey. *Acta Agriculturae Scandinavica Section B: Soil and Plant Science* **53:** 168-176. (*poae, tricinctum*)

1129. **Kostecki, M., H. Wisniewska, G. Perrone, A. Ritieni, P. Golinski, J. Chelkowski, and A. Logrieco.** 1999. The effects of cereal substrate and temperature on production of beauvericin, moniliformin and fusaproliferin by *Fusarium subglutinans* ITEM-1434. *Food Additives and Contaminants* **16:** 361-365. (*subglutinans*)

1130. **Koti-Babu, A. C. M., and K. S. Rao.** 1999. Histopathology of floral organs of *Mangifera indica* L. (Anacardiaceae) as affected by *Fusarium moniliforme* J. Sheld. *Phyton* **39:** 239-249. (*mangiferae*)

1131. **Kraft, J. M., and G. C. Papavizas.** 1983. Use of host resistance, *Trichoderma* and fungicides to control soilborne diseases and increase seed yields of peas (*Pisum sativum*). *Plant Disease* **67:** 1234-1237. (*solani*)

1132. **Kraft, J. M., and D. W. Wilkins.** 1989. The effects of pathogen numbers and tillage on root disease severity, root length, and seed yields in green peas. *Plant Disease* **73:** 884-887. (*solani*)

1133. **Krasnopol'skaya, L. M., A. P. Makeeva, L. M. Sokolova, and V. L. Sadovskaya.** 1997. Production of gibberellins A_4 and A_7 by *Fusarium moniliforme* strains. *Microbiology* **66:** 419-422. (*verticillioides*)

1134. **Krause, M., A. Lindemann, M. Glinski, T. Hornbogen, G. Bonse, P. Jeschke, G. Thielking, W. Gau, H. Kleinkauf, and R. Zocher.** 2001. Directed biosynthesis of new enniatins. *Journal of Antibiotics* **54:** 797-804. (*sambucinum, semitectum*)

1135. **Krcmery, V., Jr., P. Fuchsberger, J. Trupl, M. Blahova, A. Danisovicova, J. Svec, and L. Drgona.** 1993. Fungal pathogens in etiology of septic shock in neutropenic patients with cancer. *Zentralblatt für Bakteriologie* **278:** 562-565. (*dimerum, solani*)

1136. **Krcmery, V., Jr., Z. Jesenska, S. Spanik, J. Gyarfas, J. Nogova, R. Botek, J. Mardiak, J. Sufliarsky, J. Sisolakova, M. Vanickova, A. Kunova, M. Studena, and J. Trupl.** 1997. Fungaemia due to *Fusarium* spp. in cancer patients. *Journal of Hospital Infection* **36:** 223-228. (1, *oxysporum, solani*)

1137. **Kriek, N. P. J., W. F. O. Marasas, P. S. Steyn, S. J. van Rensberg, and M. Steyn.** 1977. Toxicity of a moniliformin-producing strain of *Fusarium moniliforme* var. *subglutinans*. *Food, Cosmetics and Toxicology* **15:** 579-587. (*subglutinans*)

1138. **Krishnaveni, S., J. M. Jeoung, S. Muthukrishnan, and G. H. Liang.** 2001. Transgenic sorghum plants constitutively expressing a rice chitinase gene show improved resistance to stalk rot. *Journal of Genetics and Breeding* **55:** 151-158. (*thapsinum*)

1139. **Kristensen, R., M. Torp, B. Kosiak, and A. Holst-Jensen.** 2005. Phylogeny and toxigenic potential is correlated in *Fusarium* species as revealed by partial translocation elongation factor 1α gene sequences. *Mycological Research* **109:** 173-186. (7)

1140. **Kritzinger, Q., T. A. S. Aveling, W. F. O. Marasas, J. P. Rheeder, L. van der Westhuizen, and G. S. Shephard.** 2003. Mycoflora and fumonisin mycotoxins associated with cowpea [*Vigna unguiculata* (L.) Walp] seeds. *Journal of Agricultural and Food Chemistry* **51:** 2188-2192. (*sambucinum, subglutinans*)

1141. **Krivobok, S., E. Miriouchkine, F. Seigle-Murandi, and J. L. Benoit-Guyod.** 1998. Biodegradation of anthracene by soil fungi. *Chemosphere* **37:** 523-530. (*subglutinans*)

1142. **Kroes, G. M. L. W., R. P. Baayen, and W. Lange.** 1998. Histology of root rot of flax seedlings (*Linum usitatissimum*) infected by *Fusarium oxysporum* f. sp. *lini*. *European Journal of Plant Pathology* **104:** 725-736. (*oxysporum*)

1143. **Kroon, B. A. M., and D. M. Elgersma.** 1991. Induction of mutants of *Fusarium oxysporum* f. sp. *lycopersici* with altered virulence. *Netherlands Journal of Plant Pathology* **97:** 409-416. (5)

1144. **Kroschel, J., A. Hundt, A. A. Abbasher, and J. Sauerborn.** 1996. Pathogenicity of fungi collected in northern Ghana to *Striga hermonthica*. *Weed Research* **36:** 515-520. (*solani*)

1145. **Krulder, J. W. M., R. W. Brimicombe, P. W. Wijermans, and W. Gams.** 1996. Systemic *Fusarium nygamai* infection in a patient with lymphoblastic non-Hodgkin's lymphoma. *Mycoses* **39:** 121-123. (*nygamai*)

1146. **Kuhad, R. C., M. Kumar, and S. Singh.** 1994. A hypercellulolytic mutant of *Fusarium oxysporum*. *Letters in Applied Microbiology* **19:** 397-400. (*oxysporum*)

1147. **Kuhlman, E. G.** 1982. Varieties of *Gibberella fujikuroi* with anamorphs in *Fusarium* section *Liseola*. *Mycologia* **74:** 759-768. (8)

1148. **Kuhlman, E. G., L. D. Dwinell, P. E. Nelson, and C. Booth.** 1978. Characterization of the *Fusarium* causing pitch canker of southern pines. *Mycologia* **70:** 1131-1143. (*circinatum*)

1149. **Kuhn, P. J., and D. A. Smith.** 1979. Isolation from *Fusarium solani* f. sp. *phaseoli* of an enzymic system responsible for kievitone and phaseolidin detoxification. *Physiological Plant Pathology* **14:** 179-190. (*solani*)

1150. **Kuldau, G. A., and I. E. Yates.** 2000. Evidence for *Fusarium* endophytes in cultivated and wild plants, p. 85-117. *In* C. W. Bacon and J. F. White, Jr. (eds.), *Microbial Endophytes*. Marcel Dekker, New York City, New York. (9)

1151. **Kulik, T., G. Fordonski, A. Pszczolkowska, K. Plodzien, and M. Lapinski.** 2004. Development of PCR assay based on *ITS2* rDNA polymorphism for the detection and differentiation of *Fusarium sporotrichioides*. *FEMS Microbiology Letters* **239:** 181-186. (*sporotrichioides*)

1152. **Kulprecha, S., T. Ueda, T. Nihira, T. Yoshida, and H. Taguchi.** 1985. Optimum conditions for ursodeoxycholic acid production from lithocholic acid by *Fusarium equiseti* M-41. *Applied and Environmental Microbiology* **49:** 338-344. (*equiseti*)

1153. **Kumar, J., U. S. Singh, and S. P. S. Beniwal.** 1993. Mango malformation: One hundred years of research. *Annual Review of Phytopathology* **31:** 217-232. (*mangiferae*)

1154. **Kumar, P. K. R., and B. K. Lonsane.** 1987. Gibberellic acid by solid state fermentation: Consistent and improved yields. *Biotechnology and Bioengineering* **30:** 267-271. (*fujikuroi*)

1155. **Kumar, P. K. R., A. Singh, and K. Schuegerl.** 1991. Formation of acetic acid from cellulosic substrates by *Fusarium oxysporum*. *Applied Microbiology and Biotechnology* **34:** 570-572. (*oxysporum*)

1156. **Kunert, J., P. Chmelik, and V. Bic.** 1993. *Fusarium solani*: Invader of the ophidian eggs of *Elaphe guttata* in captivity. *Mycopathologia* **122:** 65-68. (*solani*)

1157. **Kuninaga, S., and R. Yokosawa.** 1989. Genetic relatedness within and between *formae speciales* of *Fusarium oxysporum* as measured by DNA-DNA reassociation kinetics. *Annals of the Phytopathological Society of Japan* **55:** 216-223. (*oxysporum*)

1158. **Kunz, B. A., A. F. L. Straffon, and E. J. Vonarx.** 2000. DNA damage-induced mutation: Tolerance via translesion synthesis. *Mutation Research* **451:** 169-185. (4)

1159. **Kurek, E., and M. Majewska.** 2004. *In vitro* remobilization of Cd immobilized by fungal biomass. *Geoderma* **122:** 235-246. (*culmorum*)

1160. **Kurmut, A. M., H. I. Nirenberg, H. Bochow, and C. Buttner.** 2002. *Fusarium nygamai*, causal agent of root rot of *Vicia faba* L. in the Sudan. *Mededelingen Faculteit Landbouwkundige en Toegepaste Biologische Wetenschappen Universiteit Gent* **67:** 269-274. (*nygamai*)

1161. **Kuske, C. R., K. L. Banton, D. L. Adorada, P. C. Stark, K. K. Hill, and P. J. Jackson.** 1998. Small-scale DNA sample preparation method for field PCR detection of microbial cells and spores in soil. *Applied and Environmental Microbiology* **64:** 2463-2472. (7, *verticillioides*)

1162. **Kuwahara, M., H. Yanase, Y. Ishida, and Y. Kikuchi.** 1980. Metabolism of aliphatic nitriles in *Fusarium solani*. *Journal of Fermentation Technology* **58:** 573-578. (*solani*)

1163. **Kuwazaki, S., N. Takaya, A. Nakamura, and H. Shoun.** 2003. Formate-forming fungal catabolic pathway to supply electrons to nitrate respiration. *Bioscience Biotechnology and Biochemistry* **67:** 937-939. (*oxysporum*)

1164. **Kwasna, H., Chelkowski, J., and P. Zajkowski.** 1991. Grzyby (Mycota), tom XXII. Sierpik (*Fusarium*). Polska Akademia Nauk, Flora Polska, Warsaw, Poland. (8)

1165. **Kwon, S.-I., and A. J. Anderson.** 2001. Laccase isozymes: Production by an opportunistic pathogen, a *Fusarium proliferatum* isolate from wheat. *Physiological and Molecular Plant Pathology* **59:** 235-242. (*proliferatum*)

1166. **Kwon, S.-I., and A. J. Anderson.** 2001. Differential production of superoxide dismutase and catalase isozymes during infection of wheat by a *Fusarium proliferatum*-like fungal isolate. *Physiological and Molecular Plant Pathology* **58:** 73-81. (*proliferatum*)

1167. **Kwon, S.-I., and A. J. Anderson.** 2002. Genes for multicopper proteins and laccase activity: Common features in plant-associated *Fusarium* isolates. *Canadian Journal of Botany* **80:** 563-570. (*proliferatum*)

1168. **Kwon, S.-I., C. D. von Dohlen, and A. J. Anderson.** 2001. Gene sequence analysis of an opportunistic wheat pathogen, an isolate of *Fusarium proliferatum*. *Canadian Journal of Botany* **79:** 1115-1121. (*proliferatum*)

1169. **Laday, M., F. Bagi, A. Mesterhazy, and A. Szecsi.** 2000. Isozyme evidence for two groups of *Fusarium graminearum*. *Mycological Research* **104:** 788-793. (*pseudograminearum*)

1170. **Laday, M., G. Mulè, A. Moretti, Z. Hamari, A. Juhasz, A. Szecsi, and A. Logrieco.** 2004. Mitochondrial DNA variability in *Fusarium proliferatum* (*Gibberella intermedia*). *European Journal of Plant Pathology* **110:** 563-571. (*proliferatum*)

1171. **Laday, M., and A. Szecsi.** 2001. Distinct electrophoretic isozyme profiles of *Fusarium graminearum* and closely related species. *Systematic and Applied Microbiology* **24:** 67-75. (*pseudograminearum*)

1172. **Lagopodi, A. L., A. F. J. Ram, G. E. M. Lamers, P. J. Punt, C. A. M. J. J. van den Hondel, B. J. J. Lugtenberg, and G. V. Bloemberg.** 2002. Novel aspects of tomato root colonization and infection by *Fusarium oxysporum* f. sp. *radicis-lycopersici* revealed by confocal laser scanning microscopic analysis using the green fluorescent protein as a marker. *Molecular Plant-Microbe Interactions* **15:** 172-179. (*oxysporum*)

1173. **Lairini, K., A. Perez-Espinosa, M. Pineda, and M. Ruiz-Rubio.** 1996. Purification and characterization of tomatinase from *Fusarium oxysporum* f. sp. *lycopersici*. *Applied and Environmental Microbiology* **62:** 1604-1609. (*oxysporum*)

1174. **Lairini, K., A. Perez-Espinosa, and M. Ruiz-Rubio.** 1997. Tomatinase induction in *formae speciales* of *Fusarium oxysporum* non-pathogenic of tomato plants. *Physiological and Molecular Plant Pathology* **50:** 37-52. (*oxysporum*)

1175. **Lairini, K., and M. Ruiz-Rubio.** 1997. Detection of tomatinase from *Fusarium oxysporum* f. sp. *lycopersici* in infected tomato plants. *Phytochemistry* **45:** 1371-1376. (*oxysporum*)

1176. **Lairini, K., and M. Ruiz-Rubio.** 1998. Detoxification of α-tomatine by *Fusarium solani*. *Mycological Research* **102:** 1375-1380. (*oxysporum, solani*)

1177. **Lamare, S., R. Lortie, and M. D. Legoy.** 1997. Kinetic studies of *Fusarium solani pisi* cutinase used in a gas/solid system: Transesterification and hydrolysis reactions. *Biotechnology and Bioengineering* **56:** 1-8. (*solani*)

1178. **Lamondia, J. A., and W. H. Elmer.** 1989. Pathogenicity and vegetative compatibility among isolates of *Fusarium oxysporum* and *Fusarium moniliforme* colonizing asparagus tissues. *Canadian Journal of Botany* **67:** 2420-2424. (*oxysporum*)

1179. **Lamprecht, S. C., P. S. Knox-Davies, and W. F. O. Marasas.** 1988. Fungi associated with root rot of annual *Medicago* spp. in South Africa. *Phytophylactica* **20:** 281-286. (*acuminatum, avenaceum*)

1180. **Lamprecht, S. C., W. F. O. Marasas, J. F. Alberts, M. E. Cawood, W. C. A. Gelderblom, G. S. Shephard, P. G. Thiel, and F. J. Calitz.** 1994. Phytotoxicity of fumonisins and TA-toxin to corn and tomato. *Phytopathology* **84:** 383-391. (*verticillioides*)

1181. **Lamprecht, S. C., W. F. O. Marasas, P. G. Thiel, D. J. Schneider, and P. S. Knox-Davies.** 1986. Incidence and toxigenicity of seedborne *Fusarium* species from annual *Medicago* species in South Africa. *Phytopathology* **76:** 1040-1042. (*acuminatum, avenaceum, sambucinum*)

1182. **Lamprecht, S. C., W. F. O. Marasas, P. S. van Wyk, and P. S. Davies.** 1988. *Fusarium tricinctum* (Fungi, Hyphomycetes) in South Africa: Morphology and pathogenicity. *Bothalia* **18:** 189-194. (*tricinctum*)

1183. **Langevin, F., F. Eudes, and A. Comeau.** 2004. Effect of trichothecenes produced by *Fusarium graminearum* during Fusarium head blight development in six cereal species. *European Journal of Plant Pathology* **110:** 735-746. (*graminearum*)

1184. **Langin, T., P. Capy, and M.-J. Daboussi.** 1995. The transposable element *impala*, a fungal member of the *Tc1-mariner* superfamily. *Molecular and General Genetics* **246:** 19-28. (*oxysporum*)

1185. **Langin, T., M.-J. Daboussi, C. Gerliner, and Y. Brygoo.** 1990. Influence of biological parameters and gene transfer technique on transformation of *Fusarium oxysporum*. *Current Genetics* **17:** 313-320. (*oxysporum*)

1186. **Lappalainen, S., M. Nikulin, S. Berg, P. Parikka, E. L. Hintikka, and A. L. Pasanen.** 1996. *Fusarium* toxins and fungi associated with handling of grain on eight Finnish farms. *Atmospheric Environment* **30:** 3059-3065. (3)

1187. **Larkin, R. P., and D. R. Fravel.** 1998. Efficacy of various fungal and bacterial biocontrol organisms for control of Fusarium wilt of tomato. *Plant Disease* **82:** 1022-1028. (*oxysporum, solani*)

1188. **Larkin, R. P., and D. R. Fravel.** 1999. Mechanisms of action and dose-response relationships governing biological control of Fusarium wilt of tomato by nonpathogenic *Fusarium* spp. *Phytopathology* **89:** 1152-1161. (*oxysporum*)

1189. **Larkin, R. P., and D. R. Fravel.** 2002. Effects of varying environmental conditions on biological control of Fusarium wilt of tomato by nonpathogenic *Fusarium* spp. *Phytopathology* **92:** 1160-1166. (*oxysporum, solani*)

1190. **Larkin, R. P., D. L. Hopkins, and F. N. Martin.** 1990. Vegetative compatibility within *Fusarium oxysporum* f. sp. *niveum* and its relationship to virulence, aggressiveness and race. *Canadian Journal of Microbiology* **36:** 352-358. (*oxysporum*)

1191. **Larkin, R. P., D. L. Hopkins, and F. N. Martin.** 1996. Suppression of Fusarium wilt of watermelon by nonpathogenic *Fusarium oxysporum* and other microorganisms recovered from a disease-suppressive soil. *Phytopathology* **86:** 812-819. (*oxysporum*)

1192. **Larson, C., L. J. Francl, and T. Friesen.** 2001. Evaluation of the Burkard cyclonic spore sampler for collection efficiency of ascospores. *Plant Disease* **85:** 1249-1252. (3)

1193. **Larsson, M., and J. Olofsson.** 1994. Prevalence and pathogenicity of spinach root pathogens of the genera *Aphanomyces, Phytophthora, Fusarium, Cylindrocarpon*, and *Rhizoctonia* in Sweden. *Plant Pathology* **43:** 251-260. (*redolens*)

1194. **Lawrence, E. B., P. E. Nelson, and J. E. Ayers.** 1981. Histopathology of sweet corn (*Zea mays*) seed and plants infected with *Fusarium moniliforme* and *Fusarium oxysporum*. *Phytopathology* **71:** 379-386. (*verticillioides*)

1195. **Lawrence, E. B., P. E. Nelson, and T. A. Toussoun.** 1985. Inheritance of compatibility and sex in *Gibberella baccata*. *Phytopathology* **75:** 322-324. (*lateritium*)

1196. **Lawrence, E. B., P. E. Nelson, and T. A. Toussoun.** 1985. Genetics of certain morphological characteristics of *Gibberella baccata*. *Phytopathology* **75:** 741-747. (*lateritium*)

1197. **Lawrence, E. B., P. E. Nelson, and T. A. Toussoun.** 1986. Spontaneous mutability and colony morphology of *Gibberella baccata*. *Canadian Journal of Genetics and Cytology* **28:** 932-941. (*lateritium*)

1198. **Leach, C. M.** 1962. Sporulation of diverse species of fungi under near-ultraviolet radiation. *Canadian Journal of Botany* **40:** 151-161. (4)

1199. **Leach, J., B. R. Lang, and O. C. Yoder.** 1982. Methods for selection of mutants and *in vitro* culture of *Cochliobolus heterostrophus*. *Journal of General Microbiology* **128:** 1719-1729. (4)

1200. **Leach, J. G.** 1940. *Insect Transmission of Plant Diseases*. McGraw-Hill, London. (*poae*)

1201. **Lee, B.-D., W.-G. Kim, W.-D. Cho, and J.-M. Sung.** 2002. Occurrence of dry rot on *Cymbidium* orchids caused by *Fusarium* spp. in Korea. *Plant Pathology Journal* **18:** 156-160. (*proliferatum*)

1202. **Lee, C., H. Goerisch, H. Kleinkauf, and R. Zocher.** 1992. A highly specific D-hydroxyisovalerate dehydrogenase from the enniatin producer *Fusarium sambucinum*. *Journal of Biological Chemistry* **267:** 11741-11744. (*sambucinum*)

1203. **Lee, C., H. Goerisch, and R. Zocher.** 2000. The kinetic investigation of D-hydroxyisovalerate dehydrogenase from *Fusarium sambucinum*. *Journal of Biochemistry and Molecular Biology* **33:** 228-233. (*sambucinum*)

1204. **Lee, C., and R. Zocher.** 1996. The biochemical characterization of D-hydroxyisovalerate dehydrogenase, a key enzyme in the biosynthesis of enniatins. *Journal of Biochemistry and Molecular Biology* **29:** 493-499. (*sambucinum*)

1205. **Lee, J., J. E. Jurgenson, J. F. Leslie, and R. L. Bowden.** 2005. The alignment between physical and genetic maps of *Gibberella zeae*. *Fungal Genetics Newsletter* **52 (Suppl.):** 210. (7, *graminearum*)

1206. **Lee, J., T. Lee, Y.-W. Lee, S. H. Yun, and B. G. Turgeon.** 2003. Shifting fungal reproductive mode by manipulation of *mating type* genes: Obligatory heterothallism of *Gibberella zeae*. *Molecular Microbiology* **50:** 145-152. (6, *graminearum*)

1207. **Lee, S. B., M. G. Milgroom, and J. W. Taylor.** 1988. A rapid, high yield mini-prep method for isolation of total genomic DNA from fungi. *Fungal Genetics Newsletter* **35:** 23-24. (7)

1208. **Lee, S. K., S. S. Kim, D. H. Nahm, H. S. Park, Y. J. Oh, K. J. Park, S. O. Kim, and S. J. Kim.** 2000. Hypersensitivity pneumonitis caused by *Fusarium napiforme* in a home environment. *Allergy* **55:** 1190-1193. (*napiforme*)

1209. **Lee, T., Y.-K. Han, K.-H. Kim, S.-H. Yun, and Y.-W. Lee.** 2002. *Tri13* and *Tri7* determine deoxynivalenol- and nivalenol-producing chemotypes of *Gibberella zeae*. *Applied and Environmental Microbiology* **68:** 2148-2154. (*graminearum*)

1210. **Lee, T., D. W. Oh, H.-S. Kim, J. Lee, Y.-H. Kim, S.-H. Yun, and Y.-W. Lee.** 2001. Identification of deoxynivalenol- and nivalenol-producing chemotypes of *Gibberella zeae* by using PCR. *Applied and Environmental Microbiology* **67:** 2966-2972. (*graminearum*)

1211. **Lee, Y. M., Y. K. Choi, and B.-R. Min.** 2000. Molecular characterization of *Fusarium solani* and its *formae speciales* based on sequences analysis of the internal transcribed spacer (ITS) region of ribosomal DNA. *Mycobiology* **28:** 82-88. (*solani*)

1212. **Legrand, C., E. Anaissie, R. Hashem, P. Nelson, G. P. Bodey, and J. Ro.** 1991. Experimental Fusarial hyalohyomycosis in a murine model. *Journal of Infectious Diseases* **164:** 944-948. (*solani*)

1213. **Lekanda, J. S., and J. R. Perez-Correa.** 2004. Energy and water balances using kinetic modeling in a pilot-scale SSF bioreactor. *Process Biochemistry* **39:** 1793-1802. (*fujikuroi*)

1214. **Lemanceau, P., P. A. H. M. Bakker, W. J. de Kogel, C. Alabouvette, and B. Schippers.** 1993. Antagonistic effect of nonpathogenic *Fusarium oxysporum* Fo47 and pseudobactin 358 upon pathogenic *Fusarium oxysporum* f. sp. *dianthi*. *Applied and Environmental Microbiology* **59:** 74-82. (*oxysporum*)

1215. **Leslie, J. F.** 1981. Inbreeding for isogeneity by backcrossing to a fixed parent in haploid and diploid eukaryotes. *Genetical Research* **37:** 239-252. (6)

1216. **Leslie, J. F.** 1983. Some genetic techniques for *Gibberella zeae*. *Phytopathology* **73:** 1005-1008. (2, 4, *graminearum*)

1217. **Leslie, J. F.** 1986. Utilization of nitrogen sources by *Gibberella zeae*. *Mycologia* **78:** 568-576. (*graminearum*)

1218. **Leslie, J. F.** 1987. A nitrate nonutilizing mutant of *Gibberella zeae*. *Journal of General Microbiology* **133:** 1279-1287. (5, *graminearum*)

1219. **Leslie, J. F.** 1991. Mating populations in *Gibberella fujikuroi* (*Fusarium* section *Liseola*). *Phytopathology* **81:** 1058-1060. (6, 8, 9, 11)

1220. **Leslie, J. F.** 1993. Fungal vegetative compatibility. *Annual Review of Phytopathology* **31:** 127-151. (5, 11, *oxysporum*)

1221. **Leslie, J. F.** 1995. *Gibberella fujikuroi*: Available populations and variable traits. *Canadian Journal of Botany* **73 (Suppl. 1):** S282-S291. (6, 9, 11)

1222. **Leslie, J. F.** 1999. Genetic status of the *Gibberella fujikuroi* species complex. *Plant Pathology Journal* **15:** 259-269. (6, 9, 11)

1223. **Leslie, J. F., and M. B. Dickman.** 1991. Fate of DNA encoding hygromycin resistance after meiosis in transformed strains of *Gibberella fujikuroi* (*Fusarium moniliforme*). *Applied and Environmental Microbiology* **57:** 1423-1429. (*verticillioides*)

1224. **Leslie, J. F., and K. K. Klein.** 1996. Female fertility and mating-type effects on effective population size and evolution in filamentous fungi. *Genetics* **144:** 557-567. (6, 9, 10, 11, *proliferatum*, *thapsinum*, *verticillioides*)

1225. **Leslie, J. F., W. F. O. Marasas, G. S. Shephard, E. W. Sydenham, S. Stockenström, and P. G. Thiel.** 1996. Duckling toxicity and the production of fumonisin and moniliformin by isolates in the A and F mating populations of *Gibberella fujikuroi*. *Applied and Environmental Microbiology* **62:** 1182-1187. (*thapsinum*, *verticillioides*)

1226. **Leslie, J. F., C. A. S. Pearson, P. E. Nelson, and T. A. Toussoun.** 1990. *Fusarium* species from corn, sorghum, and soybean fields in the central and eastern United States. *Phytopathology* **80:** 343-350. (3, *oxysporum*, *proliferatum*, *semitectum*, *solani*)

1227. **Leslie, J. F., and R. D. Plattner.** 1991. Fertility and fumonisin B_1 production by strains of *Fusarium moniliforme* (*Gibberella fujikuroi*). *Proceedings of the 17th Sorghum Improvement Conference of North America (Lubbock, Texas)*, pp. 80-84. (*thapsinum*, *verticillioides*)

1228. **Leslie, J. F., and N. B. Raju.** 1985. Recessive mutations from natural populations of *Neurospora crassa* that are expressed in the sexual diplophase. *Genetics* **111:** 759-777. (6)

1229. **Leslie, J. F., B. A. Summerell, S. Bullock, and F. J. Doe.** 2005. *Gibberella sacchari*: The teleomorph of *Fusarium sacchari*. *Mycologia* **97:** 718-724. (6, 8, 9, 10, *sacchari*)

1230. **Leslie, J. F., and C. T. Yamashiro.** 1997. Effects of the *tol* mutation on allelic interactions at *het* loci in *Neurospora crassa*. *Genome* **40:** 834-840. (5)

1231. **Leslie, J. F., and K. A. Zeller.** 1996. Heterokaryon compatibility in fungi – More than just another way to die. *Journal of Genetics* **75:** 415-424. (5)

1232. **Leslie, J. F., and K. A. Zeller.** 1997. Mutants that blur the line between biological species and vegetative compatibility groups. *Cereal Research Communications* **25:** 539-542. (5)

1233. **Leslie, J. F., K. A. Zeller, S. C. Lamprecht, J. P. Rheeder, and W. F. O. Marasas.** 2005. Toxicity, pathogenicity and genetic differentiation of five species of *Fusarium* from sorghum and millet. *Phytopathology* **95:** 275-283. (3, 9, *andiyazi*, *nygamai*, *pseudonygamai*, *thapsinum*, *verticillioides*)

1234. **Leslie, J. F., K. A. Zeller, A. Logrieco, G. Mulè, A. Moretti, and A. Ritieni.** 2004. Species diversity and toxin production by strains in the *Gibberella fujikuroi* species complex isolated from native prairie grasses in Kansas. *Applied and Environmental Microbiology* **70:** 2254-2262. (9, *fujikuroi*, *konzum*, *proliferatum*, *subglutinans*, *thapsinum*, *verticillioides*)

1235. **Leslie, J. F., K. A. Zeller, and B. A. Summerell.** 2001. Icebergs and species in populations of *Fusarium*. *Physiological and Molecular Plant Pathology* **59:** 107-117. (P, 9)

1236. **Leslie, J. F., K. A. Zeller, M. Wohler, and B. A. Summerell.** 2004. Interfertility of two mating populations in the *Gibberella fujikuroi* species complex. *European Journal of Plant Pathology* **110:** 610-618. (6, 7, 8, 9, 11, *anthophilum*, *fujikuroi*, *proliferatum*, *succisae*)

1237. **Letscher-Bru, V., F. Campos, J. Waller, R. Randriamahazaka, E. Candolfi, and R. Herbrecht.** 2002. Successful outcome of treatment of a disseminated infection due to *Fusarium dimerum* in a leukemia patient. *Journal of Clinical Microbiology* **40:** 1100-1102. (*dimerum*)

1238. **Leu, H. S., A. Y. S. Lee, and T. T. Kuo.** 1995. Recurrence of *Fusarium solani* abscess formation in an otherwise healthy patient. *Infection* **23:** 303-305. (*solani*)

1239. **Lew, H., J. Chelkowski, P. Pronczuk, and W. Edinger.** 1996. Occurrence of the mycotoxin moniliformin in maize (*Zea mays* L.) ears infected by *Fusarium subglutinans* (Wollenweber and Reinking) Nelson *et al.* *Food Additives and Contaminants* **13:** 321-324. (*subglutinans*)

1240. **Lewis, C. E.** 1913. Comparative studies of certain disease-producing species of *Fusarium*. *Bulletin of the Maine Agricultural Experiment Station* **219:** 203-258. (*poae*)

1241. **Lewis, R. E., N. P. Wiederhold, and M. E. Klepser.** 2005. *In vitro* pharmacodynamics of amphotericin B, itraconazole, and voriconazole against *Aspergillus*, *Fusarium*, and *Scedosporium* spp. *Antimicrobial Agents and Chemotherapy* **49:** 945-951. (*oxysporum*, *solani*)

1242. **Lewis, Z. A., A. Correa, C. Schwerdtfeger, K. L. Link, X. Xie, R. H. Gomer, T. Thomas, D. J. Ebbole, and D. Bell-Pedersen.** 2002. Overexpression of white collar-1 (*wc-1*) activates circadian clock-associated genes, but is not sufficient to induce most light-regulated gene expression in *Neurospora crassa*. *Molecular Microbiology* **45:** 917-931. (4)

1243. **Lezinou, V., P. Christakopoulos, D. Kekos, and B. J. Macris.** 1994. Simultaneous saccharification and fermentation of sweet sorghum carbohydrates to ethanol in a fed-batch process. *Biotechnology Letters* **16:** 983-988. (*oxysporum*)

1244. **Lezinou, V., P. Christakopoulos, L. W. Li, D. Kekos, and B. J. Macris.** 1995. Study of a single and mixed culture for the direct bioconversion of sorghum carbohydrates to ethanol. *Applied Microbiology and Biotechnology* **43:** 412-415. (*oxysporum*)

1245. **Li, D., K.-R. Chung, D. A. Smith, and C. L. Schardl.** 1995. The *Fusarium solani* gene encoding kievitone hydratase, a secreted enzyme that catalyzes detoxification of a bean phytoalexins. *Molecular Plant-Microbe Interactions* **8:** 388-397. (*solani*)

1246. **Li, D., and P. E. Kolattukudy.** 1995. Cloning and expression of cDNA encoding a protein that binds a palindromic promoter element essential for induction of fungal cutinase by plant cutin. *Journal of Biological Chemistry* **270:** 11753-11756. (*solani*)

1247. **Li, D., and P. E. Kolattukudy.** 1997. Cloning of cutinase transcription factor 1, a transactivating protein containing Cys_6Zn_2 binuclear cluster DNA-binding motif. *Journal of Biological Chemistry* **272:** 12462-12467. (*solani*)

1248. **Li, D., L. Rogers, and P. E. Kolattukudy.** 1997. Cloning and expression of cDNA encoding a mitogen-activated protein kinase from a phytopathogenic filamentous fungus. *Gene* **195:** 161-166. (*solani*)

1249. **Li, D, T. Sirakova, L. Rogers, W. F. Ettinger, and P. E. Kolattukudy.** 2002. Regulation of constitutively expressed and induced cutinase genes by different zinc finger transcription factors in *Fusarium solani* f. sp. *pisi* (*Nectria haematococca*). *Journal of Biological Chemistry* **277:** 7905-7912. (*solani*)

1250. **Li, H.-P., A.-B. Wu, C.-S. Zhao, O. Scholten, H. Loffler, and Y.-C. Liao.** 2005. Development of a generic PCR detection of deoxynivalenol- and nivalenol-chemotypes of *Fusarium graminearum*. *FEMS Microbiology Letters* **243:** 505-511. (*graminearum*)

1251. **Li, R.-Y., Z. Wan, A.-P. Wang, Y. N. Shen, C.-M. Lu, M. Li, L.-Y. Xi, W.-D. Liu, and F.-Q. Zeng.** 2004. *In vitro* susceptibility testing of amorolfine in pathogenic fungi isolated from dermatomycosis patients in China *Mycoses* **47:** 402-406. (*solani*)

1252. **Li, S., Y.-K. Tam, and G. L. Hartman.** 2000. Molecular differentiation of *Fusarium solani* f. sp. *glycines* from other *F. solani* based on mitochondrial small subunit rDNA sequences. *Phytopathology* **90:** 491-497. (*solani*)

1253. **Liddell, C. M.** 1985. The comparative pathogenicity of *Fusarium graminearum* Group 1, *Fusarium culmorum*, and *Fusarium crookwellense* as crown, foot and root rot of pathogens of wheat. *Australasian Plant Pathology* **14:** 29-31. (2, *crookwellense*)

1254. **Liddell, C. M., and L. W. Burgess.** 1985. Survival of *Fusarium moniliforme* at controlled temperature and relative humidity. *Transactions of the British Mycological Society* **84:** 121-130. (*verticillioides*)

1255. **Liddell, C. M., L. W. Burgess, and P. W. J. Taylor.** 1986. Reproduction of crown rot of wheat caused by *Fusarium graminearum* Group 1 in the greenhouse. *Plant Disease* **70:** 632-635. (2)

1256. **Lieberman, T. W., A. P. Ferry, and E. J. Bottone.** 1979. *Fusarium solani* endopthalmitis without primary corneal involvement. *American Journal of Ophthalmology* **88:** 764-767. (*solani*)

1257. **Lievens, B., M. Brouwer, A. C. R. C. Vanachter, C. A. Lévesque, B. P. A. Cammune, and B. P. H. J. Thomma.** 2003. Design and development of a DNA array for rapid detection and identification of multiple tomato vascular wilt pathogens. *FEMS Microbiology Letters* **223:** 113-122. (7, *oxysporum*)

1258. **Lim, G.** 1972. *Fusarium* populations in intensively cultivated soils. *Tropical Agriculture* **49:** 77-80. (3)

1259. **Lim, G., T. K. Tan, and A. Toh.** 1989. The fungal problem in buildings in the humid tropics. *International Biodeterioration* **25:** 27-38. (*decemcellulare*)

1260. **Lin, T. S., and P. E. Kolattukudy.** 1980. Isolation and characterization of a cuticular polyester cutin hydrolyzing enzyme from phytopathogenic fungi. *Physiological Plant Pathology* **17:** 1-16. (*sambucinum, solani*)

1261. **Lin, Y.-S., and R. J. Cook.** 1977. Root rot of lentils caused by *Fusarium roseum* 'Avenaceum'. *Plant Disease Reporter* **61:** 72-75. (*avenaceum*)

1262. **Line, M. A.** 1977. Microorganisms tolerant to creosote in creosote-treated soils. *International Biodeterioration Bulletin* **13:** 102-107. (*solani*)

1263. **Link, H. F.** 1809. Observationes in ordines plantarum naturals, Dissetatio I. *Mag. Ges. Naturf. Freunde, Berlin* **3:** 3-42. (1, 8, 9)

1264. **Linnabary, R. D., and M. P. Tarrier.** 1988. Acute bovine pulmonary emphysema caused by the fungus *Fusarium semitectum*. *Veterinary and Human Toxicology* **30:** 255-256. (*semitectum*)

1265. **Linnemannstons, P., M. M. Prado, R. Fernandez-Martin, B. Tudzynski, and J. Avalos.** 2002. A carotenoid biosynthesis gene cluster in *Fusarium fujikuroi*: the genes *carB* and *carRA*. *Molecular Genetics and Genomics* **267:** 593-602. (*fujikuroi*)

1266. **Linnemannstons, P., T. Voss, P. Hedden, P. Gaskin, and B. Tudzynski.** 1999. Deletions in the gibberellin biosynthesis gene cluster of *Gibberella fujikuroi* by restriction enzyme-mediated integration and conventional transformation-mediated mutagenesis. *Applied and Environmental Microbiology* **65:** 2558-2564. (*fujikuroi*)

1267. **Lionakis, M. S., R. E. Lewis, G. Samonis, and D. P. Kontoyiannis.** 2003. Pentamidine is active *in vitro* against *Fusarium* species. *Antimicrobial Agents and Chemotherapy* **47:** 3252-3259. (*oxysporum, solani*)

1268. **Little, C. R., and C. W. Magill.** 2003. Elicitation of defense response genes in sorghum floral tissues infected by *Fusarium thapsinum* and *Curvularia lunata* at anthesis. *Physiological and Molecular Plant Pathology* **63:** 271-279. (*thapsinum*)

1269. **Little, C. R., and C. W. Magill.** 2004. Colonization of sorghum peduncles by *Fusarium thapsinum* and *Curvularia lunata*: Subsequent pigment accumulation. *International Sorghum and Millets Newsletter* **45:** 28-30. (*thapsinum*)

1270. **Liu, W., and L. Sundheim.** 1996. Nitrate nonutilizing mutants and vegetative compatibility groups in *Fusarium poae*. *Fungal Genetics and Biology* **20:** 12-17. (*poae*)

1271. **Liu, W., L. Sundheim, and W. Langseth.** 1997. Trichothecene production and the relationship to vegetative compatibility groups in *Fusarium poae*. *Mycopathologia* **140:** 105-114. (*poae*)

1272. **Liu, Y., C. Heintzen, J. Loros, and J. C. Dunlap.** 1999. Regulation of clock genes. *Cellular and Molecular Life Sciences* **55:** 1195-1205. (4)

1273. **Liu, Z., and Z. Sun.** 2004. Cloning and expression of D-lactonohydrolase cDNA from *Fusarium moniliforme* in *Saccharomyces cerevisiae*. *Biotechnology Letters* **26:** 1861-1865. (*verticillioides*)

1274. **Llorens, A., R. Mateo, M. J. Hinojo, A. Logrieco, and M. Jimenez.** 2004. Influence of the interactions among ecological variables in the characterization of zearalenone producing isolates of *Fusarium* spp. *Systematic and Applied Microbiology* **27:** 253-260. (*culmorum, graminearum*)

1275. **Llorens, A., R. Mateo, M. J. Hinojo, F. M. Valle-Algarra, and M. T. I. Jimenez.** 2004. Influence of environmental factors on the biosynthesis of type B trichothecenes by isolates of *Fusarium* spp. from Spanish crops. *International Journal of Food Microbiology* **94:** 43-54. (*culmorum, graminearum*)

1276. **Lodwig, E. M., P. D. Bridge, M. A. Rutherford, J. Kung'u, and P. Jeffries.** 1999. Molecular differences distinguish clonal lineages within East African populations of *Fusarium oxysporum* f. sp. *cubense*. *Journal of Applied Microbiology* **86:** 71-77. (*oxysporum*)

1277. **Loesch, P. J., Jr., D. C. Foley, and D. F. Cox.** 1976. Comparative resistance of *opaques-2* and normal inbred lines of maize to ear-rotting pathogens. *Crop Science* **16:** 841-842. (*verticillioides*)

1278. **Loffler, H. J. M., and B. Schippers.** 1984. Ammonia-induced mycostasis is not mediated by enhanced release of carbon compounds. *Canadian Journal of Microbiology* **30:** 1038-1041. (*solani*)

1279. **Logrieco, A., C. Altomare, A. Moretti, and A. Bottalico.** 1992. Cultural and toxigenic variability in *Fusarium acuminatum*. *Mycological Research* **96:** 518-523. (*acuminatum, armeniacum*)

1280. **Logrieco, A., A. Bottalico, G. Mulè, A. Moretti, and G. Perrone.** 2003. Epidemiology of toxigenic fungi and their associated mycotoxins for some Mediterranean crops. *European Journal of Plant Pathology* **109:** 645-667. (*lactis, poae*)

1281. **Logrieco, A., J. Chelkowski, A. Bottalico, and A. Visconti.** 1990. Further data on specific trichothecene production by *Fusarium* section *Sporotrichiella*. *Mycological Research* **94:** 587-589. (*tricinctum*)

1282. **Logrieco, A., B. Doko, A. Moretti, S. Frisullo, and A. Visconti.** 1998. Occurrence of fumonisin B_1 and B_2 in *Fusarium proliferatum* infected asparagus plants. *Journal of Agricultural and Food Chemistry* **46:** 5201-5204. (*proliferatum*)

1283. **Logrieco, A., A. Moretti, G. Castellá, M. Kostecki, P. Golinski, A. Ritieni, and J. Chelkowski.** 1998. Beauvericin production by *Fusarium* species. *Applied and Environmental Microbiology* **64:** 3084-3088. (*beomiforme, dlamini, equiseti, longipes, nygamai, oxysporum, poae, sambucinum, semitectum, subglutinans*)

1284. **Logrieco, A., A. Moretti, F. Fornelli, V. Fogliano, A. Ritieni, M. F. Caiaffa, G. Randazzo, A. Bottalico, and L. Macchia.** 1996. Fusaproliferin production by *Fusarium subglutinans* and its toxicity to *Artemia salina*, SF-9 insect cells, and IARC/LCL 171 human B lymphocytes. *Applied and Environmental Microbiology* **62:** 3378-3384. (*subglutinans*)

1285. **Logrieco, A., A. Moretti, A. Ritieni, A. Bottalico, and P. Corda.** 1995. Occurrence and toxigenicity of *Fusarium proliferatum* from preharvest maize ear rot, and associated mycotoxins, in Italy. *Plant Disease* **79:** 727-731. (*proliferatum*)

1286. **Logrieco, A., G. Mulè, and A. Bottalico.** 1994. Antagonistic activity in *Fusarium acuminatum*. *Journal of Phytopathology* **140:** 193-200. (*acuminatum*)

1287. **Logrieco, A., S. W. Peterson, and A. Bottalico.** 1991. Phylogenetic affinities of the species in *Fusarium*. *Experimental Mycology* **15:** 174-179. (*tricinctum*)

1288. **Logrieco, A., S. W. Peterson, and A. Bottalico.** 1995. Phylogenetic relationship within *Fusarium sambucinum* Fückel *sensu lato*, determined from ribosomal RNA sequences. *Mycopathologia* **129:** 153-158. (*sambucinum*, *torulosum*, *venenatum*)

1289. **Logrieco, A., A. Rizzo, R. Ferracane, and A. Ritieni.** 2002. Occurrence of beauvericin and enniatins in wheat affected by *Fusarium avenaceum* head blight. *Applied and Environmental Microbiology* **68:** 82-85. (*avenaceum*)

1290. **Longhi, S., M. Czjzek, V. Lamzin, A. Nicolas, and C. Cambillau.** 1997. Atomic resolution (1.0 Å) crystal structure of *Fusarium solani* cutinase: Stereochemical analysis. *Journal of Molecular Biology* **268:** 779-799. (*solani*)

1291. **Longhi, S., A. Nicolas, L. Creveld, M. Egmond, C. T. Verrips, J. de Vlieg, C. Martinez, and C. Cambillau.** 1996. Dynamics of *Fusarium solani* cutinase investigated through structural comparison among different crystal forms of its variants. *Proteins Structure Function and Genetics* **26:** 442-458. (*solani*)

1292. **Lopes, J. O., E. S. de Mello, and C. Klock.** 1995. Mixed intranasal infection caused by *Fusarium solani* and a zygomycete in a leukaemic patient. *Mycoses* **38:** 281-284. (*solani*)

1293. **Lopez-Franco, R., S. Bartnicki-Garcia, and C. E. Bracker.** 1994. Pulsed growth of fungal hyphal tips. *Proceedings of the National Academy of Sciences (USA)* **91:** 12228-12232. (*culmorum*)

1294. **Lopez-Franco, R., R. J. Howard, and C. E. Bracker.** 1995. Satellite Spitzenkörper in growing hyphal tips. *Protoplasma* **188:** 85-103. (*culmorum*)

1295. **Lori, G., V. Edel-Hermann, N. Gautheron, and C. Alabouvette.** 2004. Genetic diversity of pathogenic and nonpathogenic populations of *Fusarium oxysporum* isolated from carnation fields in Argentina. *Phytopathology* **94:** 661-668. (11, *oxysporum*)

1296. **Lori, G. A., J. V. Sanchez, and A. M. Stehr.** 1994. *Fusarium decemcellulare*: Causal agent of gall nuts in Australian cedar (*Toona ciliata*). *Fitopatologia Brasileira* **19:** 476-478. (*decemcellulare*)

1297. **Loser, K., and K. M. Weltring.** 1998. Induction of a polyubiquitin gene (*ubi1*) by potato phytoalexins and heat shock in *Gibberella pulicaris*. *Current Genetics* **34:** 404-409. (*sambucinum*)

1298. **Louie, T., F. El-Baba, M. Shulman, and V. Jimenez-Lucho.** 1994. Endogenous endophthalmitis due to *Fusarium*: Case report and review. *Clinical Infectious Diseases* **18:** 585-588. (*solani*)

1299. **Lowe, D. A., G. Romancik, and E. P. Elander.** 1986. Enzymatic hydrolysis of penicillin V to 6-aminopenicillanic acid by *Fusarium oxysporum*. *Biotechnology Letters* **8:** 151-156. (*oxysporum*)

1300. **Lui, L. H., and A. C. Kushalappa.** 2002. Response surface models to predict potato tuber infection by *Fusarium sambucinum* from duration of wetness and temperature, and dry rot lesion expansion from storage time and temperature. *International Journal of Food Microbiology* **76:** 19-25. (11)

1301. **Lukezic, F. L., and W. J. Kaiser.** 1966. Aero-biology of *Fusarium roseum* 'Gibbosum' associated with crown rot of boxed bananas. *Phytopathology* **56:** 545-548. (3, *semitectum*)

1302. **Lynch, P. T., H. A. Collin, and S. Isaac.** 1985. Isolation and regeneration of protoplasts from *Fusarium tricinctum* and *Fusarium oxysporum*. *Transactions of the British Mycological Society* **85:** 135-140. (*oxysporum*)

1303. **Lyrene, P. M., J. L. Dean, and N. I. James.** 1977. Inheritance of resistance to pokkah boeng in sugar cane crosses. *Phytopathology* **67:** 689-692. (*sacchari*)

1304. **Mabbayad, M. O., and A. K. Watson.** 2000. Rejection of *Fusarium pallidoroseum* as a biological control agent of *Mimosa invisa* in upland rice. *Biocontrol Science and Technology* **10:** 255-265. (*semitectum*)

1305. **Mach, R. L., C. M. Kullnig-Gradinger, A. H. Farnleitner, G. Reischer, A. Adler, and C. P. Kubicek.** 2004. Specific detection of *Fusarium langsethiae* and related species by DGGE and ARMS-PCR of a β-tubulin (*tub1*) gene fragment. *International Journal of Food Microbiology* **95:** 333-339. (*poae*)

1306. **Machado, C. M. M., B. O. Oishi, A. Pandey, and C. R. Soccol.** 2004. Kinetics of *Gibberella fujikuroi* growth and gibberellic acid production by solid-state fermentation in a packed-bed column bioreactor. *Biotechnology Progress* **20:** 1449-1453. (*fujikuroi*)

1307. **MacKenzie, S. E., M. E. Savard, B. A. Blackwell, J. D. Miller, and J. W. ApSimon.** 1998. Isolation of a new fumonisin from *Fusarium moniliforme* grown in liquid culture. *Journal of Natural Products* **61:** 367-369. (*verticillioides*)

1308. **MacNish, G. C.** 1977. Fungi associated with subterranean clover seeds and burrs collected from root-rot affected areas in Western Australia. *Journal of the Australian Institute of Agricultural Science* **43:** 157-159. (*avenaceum*)

1309. **Macris, B. J., and P. Markakis.** 1981. Characterization of extracellular β-D-galactosidase (e.c. 3.2.1.23) from *Fusarium moniliforme* grown in whey. *Applied and Environmental Microbiology* **41:** 956-958. (*verticillioides*)

1310. **Madan, M.** 1978. Trace element studies on four species of *Fusarium*. *Microbios* **22:** 161-172. (2)

1311. **Madar, Z., M. Kimchi, and Z. Solel.** 1996. Fusarium canker of Italian cypress. *European Journal of Forest Pathology* **26:** 107-112. (*compactum*)

1312. **Maddison, W. P.** 1997. Gene trees in species trees. *Systematic Biology* **46:** 523-536. (9)

1313. **Madhosingh, C.** 1990. *Fusarium graminearum* DNA induces variations in soilborne *Fusarium culmorum* plant pathogen. *Journal of Environmental Science and Health Part B* **25:** 801-816. (*culmorum*)

1314. **Madhosingh, C.** 1992. Interspecific hybrids between *Fusarium oxysporum lycopersici* and *Fusarium graminearum* by mycelial anastomoses. *Journal of Phytopathology* **136:** 113-123. (5)

1315. **Madhosingh, C.** 1994. Production of intraspecific hybrids of *Fusarium oxysporum* f. sp. *radicis-lycopersici* and *Fusarium oxysporum* f. sp. *lycopersici* by protoplast fusions. *Journal of Phytopathology* **142:** 301-309. (*oxysporum*)

1316. **Madhosingh, C., and W. Orr.** 1977. Hydroxymethyl glutaryl coenzyme A reductase in *Fusarium oxysporum*. *Agricultural and Biological Chemistry* **41:** 1519-1522. (*oxysporum*)

1317. **Madhosingh, C., and W. Orr.** 1978. Hydroxymethyl glutaryl coenzyme A reductase e.c. 1.1.1.34 purification and properties of the enzymes from *Fusarium oxysporum*. *Biochimica et Biophysica Acta* **523:** 283-296. (*oxysporum*)

1318. **Madhosingh, C., and W. Orr.** 1981. Sterol ester hydrolase e.c. 3.1.1.13 in *Fusarium oxysporum*. *Lipids* **16:** 125-132. (*oxysporum*)

1319. **Magnoli, C. E., M. A. Saenz, S. M. Chiacchiera, and A. M. Dalcero.** 1999. Natural occurrence of *Fusarium* species and fumonisin-production by toxigenic strains isolated from poultry feeds in Argentina. *Mycopathologia* **145:** 35-41. (*dlamini*, *napiforme*)

1320. **Mahaisavariya, P., A. Chaiprasert, A. Sivayathorn, and S. Khemngern.** 1999. Deep fungal and higher bacterial skin infections in Thailand: Clinical manifestations and treatment regimens. *International Journal of Dermatology* **38:** 279-284. (*compactum*)

1321. **Maher, A., A. Bassiouny, T. J. Bucci, M. K. Moawad, and D. S. Hendawy.** 1982. Tonsillomycosis: A myco-histopathological study. *Journal of Laryngology and Otology* **96:** 229-240. (*poae*)

1322. **Maia, M. M. D., A. Heasley, M. M. C. de Morais, E. H. M. Melo, M. A. Morais, Jr., W. M. Ledingham, and J. L. Lima-Filho.** 2001. Effect of culture conditions on lipase production by *Fusarium solani* in batch fermentation. *Bioresource Technology* **76:** 23-27. (*solani*)

1323. **Makropoulou, M., P. Christakopoulos, C. Tsitsimpikou, D. Kekos, F. N. Kolisi, and B. J. Macris.** 1998. Factors affecting the specificity of β-glucosidase from *Fusarium oxysporum* in enzymatic synthesis of alkyl-β-D-glucosides. *International Journal of Biological Macromolecules* **22:** 97-101. (*oxysporum*)

1324. **Malonek, S., M. C. Rojas, P. Hedden, P. Gaskin, P. Hopkins, and B. T. Tudzynski.** 2004. The NADPH-cytochrome P450 reductase gene from *Gibberella fujikuroi* is essential for gibberellin biosynthesis. *Journal of Biological Chemistry* **279:** 25075-25084. (*fujikuroi*)

1325. **Malz, A., M. N. Grell, C. Thrane, F. J. Maier, P. Rosager, A. Felk, K. S. Albertsen, S. Salomon, L. Bohn, W. Schäfer, and H. Giese.** 2005. Identification of a gene cluster responsible for the biosynthesis of aurofusarin in the *Fusarium graminearum* species complex. *Fungal Genetics and Biology* **42:** 420-433. (*graminearum*)

1326. **Mamma, D., P. Christakopoulos, D. Koullas, D. Kekos, B. J. Macris, and E. Koukios.** 1995. An alternative approach to the bioconversion of sweet sorghum carbohydrates to ethanol. *Biomass and Bioenergy* **8:** 99-103. (*oxysporum*)

1327. **Mandeel, Q. A.** 1996. A survey of *Fusarium* species in an arid environment of Bahrain. IV. Prevalence of *Fusarium* species in various soil groups using several isolation techniques. *Cryptogamie Mycologie* **17:** 149-163. (*compactum*, *equiseti*)

1328. **Mandeel, Q. A., J. A. Abbas, and A. M. Saeed.** 1995. Survey of *Fusarium* species in an arid environment of Bahrain: II. Spectrum of species on five isolation media. *Sydowia* **47:** 223-239. (*oxysporum*)

1329. **Mandeel, Q., A. Al-Laith, and L. Mohsen.** 1999. Survey of *Fusarium* species in an arid environment of Bahrain. V. Antimicrobial activity of some local and international *Fusarium* species. *Pharmaceutical Biology* **37:** 181-187. (*dimerum*)

1330. **Mandeel, Q., and R. Baker.** 1991. Mechanisms involved in biological control of Fusarium wilt of cucumber with strains of nonpathogenic *Fusarium oxysporum*. *Phytopathology* **81:** 462-469. (*oxysporum*)

1331. **Manicom, B. Q., and R. P. Baayen.** 1993. Restriction fragment length polymorphisms in *Fusarium oxysporum* f. sp. *dianthi* and other fusaria from *Dianthus* species. *Plant Pathology* **42:** 851-857. (*oxysporum*, *redolens*)

1332. **Manicom, B. Q., M. Bar Joseph, J. M. Kotze, and M. M. Becker.** 1990. A restriction fragment length polymorphism probe relating vegetative compatibility groups and pathogenicity in *Fusarium oxysporum* f. sp. *dianthi*. *Phytopathology* **80:** 336-339. (*oxysporum*)

1333. **Mannesse, M. L. M., R. C. Cox, B. C. Koops, H. M. Verheij, G. H. de Haas, M. R. Egmond, H. T. W. M. van der Hijden, and J. de Vlieg.** 1995. Cutinase from *Fusarium solani pisi* hydrolyzing triglyceride analogues: Effect of acyl chain length and position in the substrate molecule on activity and enantioselectivity. *Biochemistry* **34:** 6400-6407. (*solani*)

1334. **Mansuetus, A. S. B., G. N. Odvody, R. A. Frederiksen, and J. F. Leslie.** 1997. Biological species of *Gibberella fujikuroi* (*Fusarium* section *Liseola*) recovered from sorghum in Tanzania. *Mycological Research* **101:** 815-820. (9)

1335. **Manzo, S. K., and L. E. Claflin.** 1984. Survival of *Fusarium moniliforme* hyphae and conidia in grain sorghum (*Sorghum bicolor*) stalks. *Plant Disease* **68:** 866-867. (*thapsinum*)

1336. **Marasas, W. F. O., L. W. Burgess, R. Y. Anelich, S. C. Lamprecht, and D. J. van Schalkwyk.** 1988. Survey of *Fusarium* species associated with plant debris in South African soils. *South African Journal of Botany* **54:** 63-71. (*armeniacum*, *camptoceras*, *compactum*, *pseudonygamai*)

1337. **Marasas, W. F. O., K. Jaskiewicz, F. S. Venter, and D. J. van Schalkwyk.** 1988. *Fusarium moniliforme* contamination of maize in esophageal cancer areas in Transkei, South Africa. *South African Medical Journal* **74:** 110-114. (*verticillioides*)

1338. **Marasas, W. F. O., T. S. Kellerman, W. C. A. Gelderblom, J. A. W. Coetzer, P. G. Thiel, and J. J. van der Lugt.** 1988. Leukoencephalomalacia in a horse induced by fumonisin B_1 isolated from *Fuarium moniliforme*. *Onderstepoort Journal of Veterinary Research* **55:** 197-204. (*verticillioides*)

1339. **Marasas, W. F. O., L. Leistner, G. Hofman, and C. Eckardt.** 1979. Occurrence of toxigenic strains of *Fusarium* in maize and barley in Germany. *European Journal of Applied Microbiology and Biotechnology* **7:** 289-305. (*avenaceum*, *oxysporum*)

1340. **Marasas, W. F. O., J. D. Miller, R. T. Riley, and A. Visconti.** 2001. Fumonisins – Occurrence, toxicology, metabolism and risk assessment, p. 332-359. *In* B. A. Summerell, J. F. Leslie, D. Backhouse, W. L. Bryden, and L. W. Burgess (eds.), Fusarium*: Paul E. Nelson Memorial Symposium.* APS Press, St. Paul, Minnesota. (*proliferatum, thapsinum*)

1341. **Marasas, W. F. O., P. E. Nelson, and T. A. Toussoun.** 1984. *Toxigenic* Fusarium *Species: Identity and Mycotoxicology.* The Pennsylvania State University Press, University Park, Pennsylvania. (8, 9, *acuminatum, anthophilum, avenaceum, bulbicola, culmorum, decemcellulare, equiseti, graminearum, heterosporum, lateritium, merismoides, oxysporum, poae, proliferatum, sambucinum, semitectum, solani, sporotrichioides, subglutinans, tricinctum, verticillioides*)

1342. **Marasas, W. F. O., P. E. Nelson, and T. A. Toussoun.** 1985. *Fusarium dlamini*, a new species of *Fusarium* from southern Africa. *Mycologia* **77:** 971-975. (8, *dlamini*)

1343. **Marasas, W. F. O., P. E. Nelson, T. A. Toussoun, and P. S. van Wyk.** 1986. *Fusarium polyphialidicum*, a new species of *Fusarium* from South Africa. *Mycologia* **78:** 678-682. (*polyphialidicum*)

1344. **Marasas, W. F. O., C. J. Rabie, A. Lübben, P. E. Nelson, T. A. Toussoun, and P. S. van Wyk.** 1987. *Fusarium napiforme*, a new species from millet and sorghum in southern Africa. *Mycologia* **79:** 910-914. (4, 8, *napiforme*)

1345. **Marasas, W. F. O., C. J. Rabie, A. Lübben, P. E. Nelson, T. A. Toussoun, and P. S. van Wyk.** 1988. *Fusarium nygamai* from millet in southern Africa. *Mycologia* **80:** 263-266. (*pseudonygamai*)

1346. **Marasas, W. F. O., J. P. Rheeder, S. C. Lamprecht, K. A. Zeller, and J. F. Leslie.** 2001. *Fusarium andiyazi sp. nov.*, a new species from sorghum. *Mycologia* **93:** 1203-1210. (4, 7, 8, 9, 11, *andiyazi, thapsinum*)

1347. **Marasas, W. F. O., J. P. Rheeder, A. Logrieco, P. S. van Wyk, and J. H. Juba.** 1998. *Fusarium nelsonii* and *F. musarum*: Two new species in Section *Arthrosporiella* related to *F. camptoceras. Mycologia* **90:** 505-513. (4, *camptoceras, musarum, nelsonii*)

1348. **Marasas, W. F. O., R. T. Riley, K. A. Hendricks, V. L. Stevens, T. W. Sadler, J. Gelineau van Waes, S. A. Missmer, J. Cabrera, O. Torres, W. C. A. Gelderblom, J. Allegood, C. Martinez, J. Maddox, J. D. Miller, L. Starr, M. C. Sullards, A. V. Roman, K. A. Voss, E. Wang, A. H. Merrill, Jr.** 2004. Fumonisins disrupt sphingolipid metabolism, folate transport, and neural tube development in embryo culture and *in vivo*: A potential risk factor for human neural tube defects among populations consuming fumonisin-contaminated maize. *Journal of Nutrition* **134:** 711-716. (*verticillioides*)

1349. **Marasas, W. F. O., P. G. Thiel, C. J. Rabie, P. E. Nelson, and T. A. Toussoun.** 1986. Moniliformin production in *Fusarium* section *Liseola. Mycologia* **78:** 242-247. (*anthophilum, proliferatum, subglutinans, succisae*)

1350. **Marasas, W. F. O., P. G. Thiel, E. W. Sydenham, C. J. Rabie, A. Lubben, and P. E. Nelson.** 1991. Toxicity and moniliformin production by four recently described species of *Fusarium* and two uncertain taxa. *Mycopathologia* **113:** 191-197. (*beomiforme, dlamini, napiforme, nygamai, pseudonygamai*)

1351. **Marasas, W. F. O., B. Yagen, E. Sydenham, S. Combrinck, and P. G. Thiel.** 1987. Comparative yields of T-2 toxin and related trichothecenes from five toxicologically important strains of *Fusarium sporotrichioides. Applied and Environmental Microbiology* **53:** 693-696. (*sporotrichioides*)

1352. **Marek, E. T., C. L. Schardl, and D. A. Smith.** 1989. Molecular transformation of *Fusarium solani* with an antibiotic resistance marker having no fungal DNA homology. *Current Genetics* **15:** 421-428. (*solani*)

1353. **Marin, D. H., T. B. Sutton, S. M. Blankenship, and W. H. Swallow.** 1996. Pathogenicity of fungi associated with crown rot of bananas in Latin America on Grande Naine and disease-resistant hybrid bananas. *Plant Disease* **80:** 525-528. (*semitectum*)

1354. **Marin, S., X. Albareda, A. J. Ramos, and V. Sanchis.** 2001. Impact of environment and interactions of *Fusarium verticillioides* and *Fusarium proliferatum* with *Aspergillus parasiticus* on fumonisin B_1 and aflatoxins on maize grain. *Journal of the Science of Food and Agriculture* **81:** 1060-1068. (*verticillioides*)

1355. **Marin, S., N. Magan, J. Serra, A. J. Ramos, R. Canela, and V. Sanchis.** 1999. Fumonisin B_1 production and growth of *Fusarium moniliforme* and *Fusarium proliferatum* on maize, wheat, and barley grain. *Journal of Food Science* **64:** 921-924. (*verticillioides*)

1356. **Marin, S., V. Sanchis, F. Arnau, A. J. Ramos, and N. Magan.** 1998. Colonization and competitiveness of *Aspergillus* and *Penicillium* species on maize grain in the presence of *Fusarium moniliforme* and *Fusarium proliferatum. International Journal of Food Microbiology* **45:** 107-117. (*verticillioides*)

1357. **Marin, S., V. Sanchis, A. J. Ramos, I. Vinas, and N. Magan.** 1998. Environmental factors, in vitro interactions, and niche overlap between *Fusarium moniliforme, F. proliferatum*, and *F. graminearum, Aspergillus* and *Penicillium* species from maize grain. *Mycological Research* **102:** 831-837. (*verticillioides*)

1358. **Marin, S., V. Sanchis, A. Teixido, R. Saenz, A. J. Ramos, I. Vinas, and N. Magan.** 1996. Water and temperature relations and microconidial germination of *Fusarium moniliforme* and *Fusarium proliferatum* from maize. *Canadian Journal of Microbiology* **42:** 1045-1050. (*proliferatum, verticillioides*)

1359. **Markell, S. G., and L. J. Francl.** 2003. Fusarium head blight inoculum: Species prevalence and *Gibberella zeae* spore type. *Plant Disease* **87:** 814-820. (*poae*)

1360. **Marlatt, M. L., J. C. Correll, P. Kaufmann, and P. E. Cooper.** 1996. Two genetically distinct populations of *Fusarium oxysporum* f. sp. *lycopersici* race 3 in the United States. *Plant Disease* **80:** 1336-1342. (11, *oxysporum*)

1361. **Marley, P. S., D. A. Aba, J. A. Y. Shebayan, R. Musa, and A. Sanni.** 2004. Integrated management of *Striga hermonthica* in sorghum using a mycoherbicide and host plant resistance in the Nigerian Sudano-Sahelian savanna. *Weed Research* **44:** 157-162. (*oxysporum*)

1362. **Marley, P. S., S. M. Ahmed, J. A. Y. Shebayan, and S. T. O. Lagoke.** 1999. Isolation of *Fusarium oxysporum* with potential for biocontrol of the witchweed (*Striga hermonthica*) in the Nigerian savanna. *Biocontrol Science and Technology* **9:** 159-163. (*oxysporum*)

1363. **Marley, P. S., and J. A. Y. Shebayan.** 2005. Field assessment of *Fusarium oxysporum* based mycoherbicide for control of *Striga hermonthica* in Nigeria. *BioControl* **50:** 389-399. (*oxysporum*)

1364. **Marlida, Y., N. Saari, Z. Hassan, and S. Radu.** 2000. Raw starch-degrading enzyme from newly isolated strains of endophytic fungi. *World Journal of Microbiology and Biotechnology* **16:** 573-578. (*sambucinum*)

1365. **Marlida, Y., N. Saari, S. Radu, and F. Abu-Bakar.** 2000. Production of an amylase-degrading raw starch by *Gibberella pulicaris*. *Biotechnology Letters* **22:** 95-97. (*sambucinum*)

1366. **Marriott, A. C., S. A. Archer, and K. W. Buck.** 1984. Mitochondrial DNA in *Fusarium oxysporum* is a 46.5 kilobase pair circular molecule. *Journal of General Microbiology* **130:** 3001-3009. (*oxysporum*)

1367. **Martin-Udiroz, M., M. P. Madrid, and M. I. G. Roncero.** 2004. Role of chitin synthase genes in *Fusarium oxysporum*. *Microbiology* **150:** 3175-3187. (*oxysporum*)

1368. **Martinez, C., P. de Geus, M. Lauwereys, G. Matthyssens, and C. Cambillau.** 1992. *Fusarium solani* cutinase is a lipolytic enzyme with a catalytic serine accessible to solvent. *Nature* **356:** 615-618. (*solani*)

1369. **Martinez, C., P. de Geus, P. Stanssens, M. Lauwereys, and C. Cambillau.** 1993. Engineering cysteine mutants to obtain crystallographic phases with a cutinase from *Fusarium solani pisi*. *Protein Engineering* **6:** 157-165. (*solani*)

1370. **Martinez, M. J., M. T. Alconda, F. Guillen, C. Vazquez, and F. Reyes.** 1991. Pectic activities from *Fusarium oxysporum* f. sp. *melonis*: Purification and characterization of an exopolygalacturonase. *FEMS Microbiology Letters* **81:** 145-150. (*oxysporum*)

1371. **Martinez, M. J., C. Vazquez, R. Lahoz, and F. Reyes.** 1986. Role of uronic acids present in phytopathogenic fungi as inducers of polygalacturonases during autolysis. *Mycopathologia* **93:** 33-38. (*oxysporum*)

1372. **Martino, E., S. Cerminara, L. Prandi, B. Fubini, and S. Perotto.** 2004. Physical and biochemical interactions of soil fungi with asbestos fibers. *Environmental Toxicology and Chemistry* **23:** 938-944. (*oxysporum*)

1373. **Martino, P., R. Gastaldi, R. Raccah, and C. Girmenia.** 1994. Clinical patterns of *Fusarium* infections in immunocompromised patients. *Journal of Infection* **28 (Suppl. 1):** 7-15. (1, *chlamydosporum*)

1374. **Martins, E. R., and C. Kemmelmeier.** 1993. Zearalenone production in *Fusarium graminearum* variants after treatment with nitrosoguanidine. *Revista de Microbiologia* **24:** 248-254. (*graminearum*)

1375. **Maruyama, Y., and K. Onodera.** 1979. Production and some properties of invertase isozymes of *Fusarium oxysporum*. *Journal of General and Applied Microbiology* **25:** 361-366. (*oxysporum*)

1376. **Marzluf, G. A.** 1981. Regulation of nitrogen metabolism and gene expression in fungi. *Microbiological Reviews* **45:** 437-461. (2)

1377. **Mateo, J. J., R. Mateo, and M. Jimenez.** 2002. Accumulation of type A trichothecenes in maize, wheat and rice by *Fusarium sporotrichioides* isolates under diverse culture conditions. *International Journal of Food Microbiology* **72:** 115-123. (*sporotrichioides*)

1378. **Mathre, D., ed.** 1997. *Compendium of Barley Diseases*, 2nd ed. APS Press, St. Paul, Minnesota. (*graminearum*)

1379. **Mathur, S., and R. Utkhede.** 2004. Molecular detection of *Fusarium subglutinans*, the causal organism of internal fruit rot in greenhouse peppers. *Canadian Journal of Plant Pathology* **26:** 514-521. (*subglutinans*)

1380. **Mathur, S. K., S. B. Mathur, and P. Neergaard.** 1975. Detection of seed-borne fungi in sorghum and location of *Fusarium moniliforme* in the seed. *Seed Science and Technology* **3:** 683-690. (*semitectum*)

1381. **Mathur, S. K., R. Nath, and S. B. Mathur.** 1973. Seedborne fungi of pearl millet (*Pennisetum typhoides*) and their significance. *Seed Science and Technology* **1:** 811-820. (*semitectum*)

1382. **Matsuda, T., and T. Matsumoto.** 1986. Disseminated Hyalohyphomycosis in a leukemic patient. *Archives of Dermatology* **122:** 1171-1175. (*solani*)

1383. **Matsuda, Y., T. Beppu, and K. Arima.** 1978. Crystallization and positional specificity of hyperperoxidation of *Fusarium oxysporum* lipoxygenase e.c. 1.13.11.12. *Biochimica et Biophysica Acta* **530:** 439-450. (*oxysporum*)

1384. **Matsuda, Y., T. Beppu, and K. Arima.** 1979. Possible mechanism of oxygen activation in lineolate peroxidation by *Fusarium oxysporum* lipoxygenase e.c. 1.13.11.12. *Agricultural and Biological Chemistry* **43:** 1179-1186. (*oxysporum*)

1385. **Matsui, Y., and M. Watanabe.** 1988. Quantitative analysis of fusaric acid in the cultural filtrate and soybean plants inoculated with *Fusarium oxysporum* var. *redolens*. *Journal of Rakuno Gakuen University Natural Science* **13:** 159-168. (*redolens*)

1386. **Matsumoto, G., J. Wuchiyama, Y. Shingu, M. Kimura, K. Yoneyama, and I. Yamaguchi.** 1999. The trichothecene biosynthesis regulatory gene from the type B producer *Fusarium* strains: Sequence of *tri6* and its expression in *Escherichia coli*. *Bioscience Biotechnology and Biochemistry* **63:** 2001-2004. (*graminearum*)

1387. **Matsushima, T.** 1971. *Microfungi of the Solomon Islands and Papua-New Guinea*. Nippon Printing Publication Co., Kobe, Japan. (*proliferatum*)

1388. **Matuo, T.** 1972. Taxonomic studies of phytopathogenic fungi in Japan. *Review of Plant Protection Research* **5:** 34-45. (8)

1389. **Matuo, T., and W. C. Snyder.** 1972. Host virulence and the *Hypomyces* stage of *Fusarium solani* f. sp. *pisi*. *Phytopathology* **62:** 731-735. (*solani*)

1390. **Matuo, T., and W. C. Snyder.** 1973. Use of morphology and mating populations in the identification of special forms in *Fusarium solani*. *Phytopathology* **63:** 562-565. (*solani*)

1391. **Mauch-Mani, B., and A. J. Slusarenko.** 1994. Systemic acquired resistance in *Arabidopsis thaliana* induced by a predisposing infection with a pathogenic isolate of *Fusarium oxysporum*. *Molecular Plant-Microbe Interactions* **7:** 378-383. (*oxysporum*)

1392. **Mayayo, E., I. Pujol, and J. Guarro.** 1999. Experimental pathogenicity of four opportunist *Fusarium* species in a murine model. *Journal of Medical Microbiology* **48:** 363-366. (*oxysporum, proliferatum, solani, verticillioides*)

1393. **Mayr, E.** 1940. Speciation phenomena in birds. *American Naturalist* **74:** 249-278. (9)

1394. **Mayr, E.** 1963. *Animal Species and Evolution*. Harvard University Press: Cambridge, Massachusetts. (9)

1395. **McAllister, C. B., J. M. Garcia-Garrido, I. Garcia-Romera, A. L. Godeas, and J. A. Ocampo.** 1997. Interaction between *Alternaria alternata* or *Fusarium equiseti* and *Glomus mosseae* and its effects on plant growth. *Biology and Fertility of Soils* **24:** 301-305. (*equiseti*)

1396. **McCallum, B. D., A. Tekauz, and J. Gilbert.** 2001. Vegetative compatibility among *Fusarium graminearum* (*Gibberella zeae*) isolates from barley spikes in southern Manitoba. *Canadian Journal of Plant Pathology* **23:** 83-87. (*graminearum*)

1397. **McCallum, B. D., A. Tekauz, and J. Gilbert.** 2004. Barrage zone formation between vegetatively incompatible *Fusarium graminearum* (*Gibberella zeae*) isolates. *Phytopathology* **94:** 432-437. (5, *graminearum*)

1398. **McCormick, S. P., and N. J. Alexander.** 2002. *Fusarium Tri8* encodes a trichothecene C-3 esterase. *Applied and Environmental Microbiology* **68:** 2959-2964. (*sporotrichioides*)

1399. **McCormick, S. P., N. J. Alexander, S. E. Trapp, and T. M. Hohn.** 1999. Disruption of *TRI101*, the gene encoding trichothecene 3-O-acetyltransferase, from *Fusarium sporotrichioides*. *Applied and Environmental Microbiology* **65:** 5252-5256. (*sporotrichioides*)

1400. **McCormick, S. P., and T. M. Hohn.** 1997. Accumulation of trichothecenes in liquid cultures of a *Fusarium sporotrichioides* mutant lacking a functional trichothecene C-15 hydroxylase. *Applied and Environmental Microbiology* **63:** 1685-1688. (*sporotrichioides*)

1401. **McCormick, S. P., T. M. Hohn, and A. E. Desjardins.** 1996. Isolation and characterization of *Tri3*, a gene encoding 15-O-acetyltransferase from *Fusarium sporotrichioides*. *Applied and Environmental Microbiology* **62:** 353-359. (*sporotrichioides*)

1402. **McDonald, B. A., R. E. Pettway, R. S. Chen, J. M. Boeger, and J. P. Martinez.** 1995. The population genetics of *Septoria tritici* (teleomorph *Mycosphaerella graminicola*). *Canadian Journal of Botany* **73(Suppl. 1):** S292-S301. (9)

1403. **McGee, D. C., and A. W. Kellock.** 1974. *Fusarium avenaceum*, a seedborne pathogen of subterranean clover roots. *Australian Journal of Agricultural Research* **25:** 549-557. (*avenaceum*)

1404. **McGinnis, M. R., L. Pasarell, D. A. Sutton, A. W. Fothergill, and C. R. Cooper, Jr.** 1998. *In vitro* activity of voriconazole against selected fungi. *Medical Mycology* **36:** 239-242. (*oxysporum*)

1405. **McGinnis, M. R., L. C. Severo, R. Kalil, and P. T. Fallèiro.** 1994. Endocarditis caused by *Fusarium pallidoroseum*. *Journal de Mycologie Medicale* **4:** 45-47. (*semitectum*)

1406. **McGovern, R. J.** 1994. First report of corky dry rot of cantaloupe caused by *Fusarium semitectum* in Florida. *Plant Disease* **78:** 926. (*semitectum*)

1407. **McInnes, A. G., J. A. Walter, D. G. Smith, J. L. C. Wright, and L. C. Vining.** 1976. Biosynthesis of bikaverin in *Fusarium oxysporum*: Use of ^{13}C NMR with homonuclear ^{13}C decoupling to locate adjacent ^{13}C labels. *Journal of Antibiotics* **29:** 1050-1057. (*oxysporum*)

1408. **McKenzie, F., and G. S. Taylor.** 1983. *Fusarium* populations in British UK soils relative to different cropping practices. *Transactions of the British Mycological Society* **80:** 409-414. (*dimerum*)

1409. **McMahan, G., W. Yeh, M. N. Marshall, M. Olsen, S. Sananikone, J. Y. Wu, D. E. Block, and J. S. van der Gheynst.** 2001. Characterizing the production of a wild-type and benomyl-resistant *Fusarium lateritium* for biocontrol of *Eutypa lata* on grapevine. *Journal of Industrial Microbiology and Biotechnology* **26:** 151-155. (*lateritium*)

1410. **McMullen, M. P., and R. W. Stack.** 1983. *Fusarium* species associated with grassland soils. *Canadian Journal of Botany* **61:** 2530-2538. (*dimerum, heterosporum, oxysporum, redolens, subglutinans*)

1411. **McMullen, M. V. P., and R. W. Stack.** 1983. Effect of isolation techniques and media on the differential isolation of *Fusarium* species. *Phytopathology* **73:** 458-462. (3)

1412. **McMullen, M. P., and R. W. Stack.** 1984. The effects of surface mining and reclamation on *Fusarium* populations of grassland soils. *Reclamation and Revegetation Research* **2:** 253-266. (*heterosporum, oxysporum, redolens, subglutinans, verticillioides*)

1413. **Mebalds, M. I.** 1987. Mycoflora of *Medicago trunca-tula, M. rugosa* and *M. littoralis* seed produced in Victoria, Australia. *Seed Science and Technology* **15:** 175-183. (*acuminatum, avenaceum, torulosum*)

1414. **Medentsev, A. G., & V. K. Akimenko.** 1992. Mechanism of phytotoxic action of naphthoquinone pigments of the fungus *Fusarium decemcellulare*. *Phytochemistry* **31:** 77-80. (*decemcellulare*)

1415. **Meek, I. B., A. W. Peplow, C. Ake, Jr., T. D. Phillips, and M. N. Beremand.** 2003. *Tri1* encodes the cytochrome P450 monooxygenase for C-8 hydroxylation during trichothecene biosynthesis in *Fusarium sporotrichioides* and resides upstream of another new *Tri* gene. *Applied and Environmental Microbiology* **69:** 1607-1613. (*sporotrichioides*)

1416. **Mehta, A., and P. Mehta.** 1985. Production of pectolytic and cellulolytic enzymes by *Fusarium oxysporum* and *Fusarium moniliforme* under different cultivation conditions. *Folia Microbiologica* **30:** 42-50. (*oxysporum*)

1417. **Mehta, B. J., and E. Cerda-Olmedo.** 2001. Intersexual partial diploids of *Phycomyces*. *Genetics* **158:** 635-641. (6)

1418. **Melcher, G. P., D. A. McGough, A. W. Fothergill, C. Norris, and M. G. Rinaldi.** 1993. Disseminated hyalohyphomycosis caused by a novel human pathogen, *Fusarium napiforme*. *Journal of Clinical Microbiology* **31:** 1461-1467. (*napiforme*)

1419. **Melcion, D., B. Cahagnier, and D. Richard-Molard.** 1997. Study of the biosynthesis of fumonisins B_1, B_2 and B_3 by different strains of *Fusarium moniliforme*. *Letters in Applied Microbiology* **24:** 301-305. (*verticillioides*)

1420. **Melo, E. P., C. M. L. Carvalho, M. R. Aires-Barros, S. M. B. Costa, and J. M. S. Cabral.** 1998. Deactivation and conformational changes of cutinase in reverse micelles. *Biotechnology and Bioengineering* **58:** 380-386. (*solani*)

1421. **Menkir, A., G. Ejeta, L. G. Butler, A. Melakeberhan, and W. H. Warren.** 1996. Fungal invasion of kernels and grain mold damage assessment in diverse sorghum germ plasm. *Plant Disease* **80:** 1399-1402. (*graminearum*)

1422. **Mercer, P. C., and N. V. Hardwick.** 1991. Control of seed-borne diseases of linseed. *Aspects of Applied Biology* **78:** 71-77. (*avenaceum*)

1423. **Mercier, J., J. Makhlouf, and R. A. Martin.** 1991. *Fusarium avenaceum*, a pathogen of stored broccoli. *Canadian Plant Disease Survey* **71:** 161-162. (*avenaceum*)

1424. **Mergoum, M., J. P. Hill, and J. S. Quick.** 1998. Evaluation of resistance to winter wheat to *Fusarium acuminatum* by inoculation of seedling roots with single, germinated conidia. *Plant Disease* **82:** 300-302. (*acuminatum*)

1425. **Merrill, A. H., Jr., G. van Echten, E. Wang, and K. Sandhoff.** 1993. Fumonisin B_1 inhibits sphingosine (sphinganine) N-acyltransferase and *de novo* sphingolipid biosynthesis in cultured neurons *in situ*. *Journal of Biological Chemistry* **268:** 27299-27306. (*verticillioides*)

1426. **Merz, W. G., J. E. Karp, M. Hoagland, M. Jett-Goheen, J. M. Junkins, and A. Hood.** 1988. Diagnosis and successful treatment of Fusariosis in the compromised host. *Journal of Infectious Diseases* **158:** 1046-1055. (*solani*)

1427. **Mes, J. J., M. A. Haring, and B. J. C. Cornelissen.** 2000. *Foxy*: An active family of short interspersed nuclear elements from *Fusarium oxysporum*. *Molecular and General Genetics* **263:** 271-280. (*oxysporum*)

1428. **Mes, J. J., and J. van Doorn, E. J. A. Roebroeck, E. van Egmond, J. van Aartrijk, and P. M. Boonekamp.** 1994. Restriction fragment length polymorphisms, races and vegetative compatibility groups within a worldwide collection of *Fusarium oxysporum* f.sp. *gladioli*. *Plant Pathology* **43:** 362-370. (*oxysporum*)

1429. **Mes, J. J., R. Wit, C. S. Testerink, F. de Groot, M. A. Haring, and B. J. C. Cornelissen.** 1999. Loss of avirulence and reduced pathogenicity of a γ-irradiated mutant of *Fusarium oxysporum* f. sp. *lycopersici*. *Phytopathology* **89:** 1131-1137. (*oxysporum*)

1430. **Messiaen, C. M.** 1959. La systématique du genre *Fusarium* selon Snyder et Hansen. *Rev. Pathol. Veg. Entomol. Agric. Fr.* **38:** 253-266. (8)

1431. **Messiaen, C. M., and R. Cassini.** 1968. Recherches sur la fusarioses. IV. La systématique des *Fusarium*. *Annals Epiphyt.* **19:** 387-454. (8)

1432. **Mesterhazy, A.** 1981. The role of aggressiveness of *Fusarium graminearum* isolates in inoculation tests on wheat in seedling state. *Acta Phytopathologica Academiae Scientiarum Hungaricae* **16:** 281-292. (*graminearum*)

1433. **Metzenberg, R. L.** 1985. Killing mites without killing *Neurospora*. Neurospora *Newsletter* **32:** 22. (4)

1434. **Miao, V. P., S. F. Covert, and H. D. VanEtten.** 1991. A fungal gene for antibiotic resistance in a dispensable ("B") chromosome. *Science* **254:** 1773-1776. (7, 11, *solani*)

1435. **Michailides, T. J., and D. P. Morgan.** 1994. Dynamics of *Blastophaga psenes* populations, availability of caprifigs, and fig endosepsis caused by *Fusarium moniliforme*. *Phytopathology* **84:** 1254-1263. (*lactis*)

1436. **Michailides, T. J., and D. P. Morgan.** 1998. Spread of endosepsis in Calimyrna fig orchards. *Phytopathology* **88:** 637-647. (*lactis*)

1437. **Michailides, T. J., D. P. Morgan, and R. Klamm.** 1994. Comparison of three methods for determining fig endosepsis caused by *Fusarium moniliforme* and other molds in Caprifigs and calimyrna figs. *Plant Disease* **78:** 44-50. (*lactis*)

1438. **Michailides, T. J., D. P. Morgan, and K. V. Subbarao.** 1996. An old disease still a dilemma for California growers. *Plant Disease* **80:** 828-841. (*lactis*, *ramigenum*)

1439. **Miedaner, T., and C. Reinbrecht.** 2001. Trichothecene content of rye and wheat genotypes inoculated with a deoxynivalenol- and a nivalenol-producing isolate of *Fusarium culmorum*. *Journal of Phytopathology* **149:** 245-251. (*culmorum*)

1440. **Migheli, Q., T. Berio, and M. L. Gullino.** 1993. Electrophoretic karyotypes of *Fusarium* spp. *Experimental Mycology* **17:** 329-337. (*oxysporum*, *verticillioides*)

1441. **Migheli, Q., T. Berio, M. L. Gullino, and A. Garibaldi.** 1995. Electrophoretic karyotype variation among pathotypes of *Fusarium oxysporum* f. sp. *dianthi*. *Plant Pathology* **44:** 308-315. (*oxysporum*, *verticillioides*)

1442. **Migheli, Q., O. Friard, D. del Tedesco, M. R. Musso, and M. L. Gullino.** 1996. Stability of transformed antagonistic *Fusarium oxysporum* strains *in vitro* and in soil microcosms. *Molecular Ecology* **5:** 641-649. (*oxysporum*)

1443. **Migheli, Q., R. Lauge, J. M. Daviere, C. Gerlinger, F. Kaper, T. Langin, and M.-J. Daboussi.** 1999. Transposition of the autonomous *fot1* element in the filamentous fungus *Fusarium oxysporum*. *Genetics* **151:** 1005-1013. (*oxysporum*)

1444. **Migheli, Q., C. Steinberg, J. M. Daviere, C. Olivain, C. Gerlinger, N. Gautheron, C. Alabouvette, and M.-J. Daboussi.** 2000. Recovery of mutants impaired in pathogenicity after transposition of *impala* in *Fusarium oxysporum* f. sp. *melonis*. *Phytopathology* **90:** 1279-1284. (*oxysporum*)

1445. **Mihlan, M., V. Homann, T. W. D. Liu, and B. Tudzynski.** 2003. AREA directly mediates nitrogen regulation of gibberellin biosynthesis in *Gibberella fujikuroi*, but its activity is not affected by NMR. *Molecular Microbiology* **47:** 975-991. (5, *fujikuroi*)

1446. **Miller, J. D., R. Greenhalgh, Y. Wang, and M. Lu.** 1991. Trichothecene chemotypes of three *Fusarium* species. *Mycologia* **83:** 121-130. (*crookwellense*)

1447. **Miller, J. D., and S. MacKenzie.** 2000. Secondary metabolites of *Fusarium venenatum* strains with deletions in the *Tri5* gene encoding trichodiene synthetase. *Mycologia* **92:** 764-771. (*venenatum*)

1448. **Miller, J. D., M. E. Savard, A. W. Schaafsma, K. A. Seifert, and L. M. Reid.** 1995. Mycotoxin production by *Fusarium moniliforme* and *Fusarium proliferatum* from Ontario and occurrence of fumonisin in the 1993 corn crop. *Canadian Journal of Plant Pathology* **17:** 233-239. (*proliferatum*, *verticillioides*)

1449. **Miller, S. S., D. M. P. Chabot, T. Ouellet, L. J. Harris, and G. Fedak.** 2004. Use of a *Fusarium graminearum* strain transformed with green fluorescent protein to study infection in wheat (*Triticum aestivum*). *Canadian Journal of Plant Pathology* **26:** 453-463. (*graminearum*)

1450. **Minuto, A., G. Minuto, Q. Migheli, M. Mocioni, and M. L. Gullino.** 1997. Effect of antagonistic *Fusarium* spp. and of different commercial biofungicide formulations on Fusarium wilt of basil (*Ocimum basilicum* L.). *Crop Protection* **16:** 765-769. (*oxysporum*)

1451. **Mirete, S., C. Vazquez, G. Mulè, M. Jurado, and M. T. Gonzalez-Jaen.** 2004. Differentiation of *Fusarium verticillioides* from banana fruits by IGS and EF-1α sequence analyses. *European Journal of Plant Pathology* **110:** 515-523. (*verticillioides*)

1452. **Mirocha, C. J., H. K. Abbas, and R. F. Vesonder.** 1990. Absence of trichothecenes in toxigenic isolates of *Fusarium moniliforme*. *Applied and Environmental Microbiology* **56:** 520-525. (*verticillioides*)

1453. **Mirocha, C. J., S. V. Pathre, and T. S. Robison.** 1981. Comparative metabolism of zearalenone in and transmission into bovine milk. *Food Cosmetics and Toxicology* **19:** 25-30. (*graminearum*)

1454. **Mirocha, C. J., R. A. Pawlosky, K. Chatterjee, S. Watson, and W. Hayes.** 1983. Analysis of *Fusarium* toxins in various samples implicated in biological warfare in Southeast Asia. *Journal of the Association of Official Analytical Chemists* **66:** 1485-1499. (1)

1455. **Mishra, P. K., R. T. V. Fox, and A. Culham.** 2003. Development of a PCR-based assay for rapid and reliable identification of pathogenic Fusaria. *FEMS Microbiology Letters* **218:** 329-332. (7, *oxysporum*, *sambucinum*)

1456. **Mishra, P. K., R. T. V. Fox, and A. Culham.** 2003. Inter-simple sequence repeat and aggressiveness analyses revealed high genetic diversity, recombination and long-range dispersal in *Fusarium culmorum*. *Annals of Applied Biology* **143:** 291-301. (*culmorum*)

1457. **Mishra, P. K., J. P. Tewari, R. M. Clear, and T. K. Turkington.** 2004. Molecular genetic variation and geographical structuring in *Fusarium graminearum*. *Annals of Applied Biology* **145:** 299-307. (*graminearum*)

1458. **Mitchell, R. G., J. Zwolinski, N. Jones, and T. Coutinho.** 2004. The effect of applying prophylactic measures on the post-planting survival of *Pinus patula* in South Africa. *Southern African Forestry Journal* **200:** 51-58. (*circinatum*)

1459. **Mitra, J. P. K. Mukherjee, S. P. Kale, and N. B. K. Murthy.** 2001. Bioremediation of DDT in soil by genetically improved strains of soil fungus *Fusarium solani*. *Biodegradation* **12:** 235-245. (*solani*)

1460. **Mohammedi, I., B. Gachot, M. Grossin, C. Marchie, M. Wolff, and F. Vachon.** 1995. Overwhelming myocarditis due to *Fusarium oxysporum* following bone marrow transplantation. *Scandinavian Journal of Infectious Diseases* **27:** 643-644. (*oxysporum*)

1461. **Mohr, P., C. Tamm, W. Zürcher, and M. Zehnder.** 1984. Sambucinol and sambucoin. Two new metabolites of *Fusarium sambucinum* possessing modified trichothecene structures. *Helvetica Chimica Acta* **67:** 406-412. (*sambucinum*)

1462. **Molnar, A., L. Sulyok, and L. Hornok.** 1990. Parasexual recombination between vegetatively incompatible strains in *Fusarium oxysporum*. *Mycological Research* **94:** 393-398. (5)

1463. **Momol, E. A., and H. C. Kistler.** 1992. Mitochondrial plasmids do not determine host range in crucifer-infecting strains of *Fusarium oxysporum*. *Plant Pathology* **41:** 103-112. (*oxysporum*)

1464. **Mondal, S. N., and M. Hyakumachi.** 1998. Carbon loss and germinability, viability, and virulence of chlamydospores of *Fusarium solani* f. sp. *phaseoli* after exposure to soil at different pH levels, temperatures, and matric potentials. *Phytopathology* **88:** 148-155. (*solani*)

1465. **Mondal, S. N., K. Kageyama, and M. Hyakumachi.** 1995. Germinability, viability, and virulence of chlamydospores of *Fusarium solani* f. sp. *phaseoli* as affected by the loss of endogenous carbon. *Phytopathology* **85:** 1238-1244. (*solani*)

1466. **Monds, R. D., M. G. Cromey, D. R. Lauren, M. di Menna, and J. Marshall.** 2005. *Fusarium graminearum, F. cortderiae* and *F. pseudograminearum* in New Zealand: Molecular phylogenetic analysis, mycotoxin chemotypes and coexistence of species. *Mycological Research* **109:** 410-420. (*graminearum*)

1467. **Monnet, D., D. Vidal, and O. Creach.** 1988. Influence of metabolic and physical factors on production of diacetoxyscirpenol by *Fusarium sambucinum* Fückel. *Applied and Environmental Microbiology* **54:** 2167-2169. (*sambucinum*)

1468. **Montes-Belmont, R., I. Mendez-Ramirez, H. E. Flores-Moctezuma, and R. A. Nava-Juarez.** 2003. Impact of planting dates and climatic factors on the incidence and severity of sorghum grain mold in Morelos, Mexico. *Plant Disease* **87:** 1139-1143. (*thapsinum*)

1469. **Moore, D., and L. N. Frazer.** 2002. *Essential Fungal Genetics*. Springer, New York. (4)

1470. **Moore, N. Y., K. G. Pegg, I. W. Buddenhagen, and S. Bentley.** 2001. Fusarium wilt of banana, p. 212-224. *In* B. A. Summerell, J. F. Leslie, D. Backhouse, W. L. Bryden, and L. W. Burgess (eds.), *Fusarium: Paul E. Nelson Memorial Symposium*. APS Press, St. Paul, Minnesota. (11)

1471. **Moore, W. C.** 1943. New and interesting plant diseases. *Transactions of the British Mycological Society* **26:** 20-23. (3)

1472. **Moosavi-Jorf, S. A., A. Alizadeh, J. Hayati, and Z. Banihashemi.** 1999. Investigation on *Fusarium* species associated with date palm in Khuzestan province. *Iranian Journal of Plant Pathology* **35:** 32-34. (*succisae*)

1473. **Moretti, A., A. Belisario, A. Tafuri, A. Ritieni, L. Corazza, and A. Logrieco.** 2002. Production of beauvericin by different races of *Fusarium oxysporum* f. sp. *melonis*, the Fusarium wilt agent of muskmelon. *European Journal of Plant Pathology* **108:** 661-666. (*oxysporum*)

1474. **Moretti, A., A. Logrieco, A. Bottalico, A. Ritieni, V. Fogliano, and G. Randazzo.** 1997. Diversity in beauvericin and fusaproliferin production by different populations of *Gibberella fujikuroi* (*Fusarium* section *Liseola*). *Sydowia*: **30:** 44-56. (*fujikuroi*)

1475. **Moretti, A., A. Logrieco, A. Bottalico, A. Ritieni, G. Randazzo, and P. Corda.** 1995. Beauvericin production by *Fusarium subglutinans* from different geographical areas. *Mycological Research* **99:** 282-286. (*subglutinans*)

1476. **Moretti, A., G. Mulè, A. Susca, M. T. González-Jaén, and A. Logrieco.** 2004. Toxin profile, fertility and AFLP analysis of *Fusarium verticillioides* from banana fruits. *European Journal of Plant Pathology* **110:** 601-609. (11, *verticillioides*)

1477. **Morgan, M. K., S. J. Bursian, G. E. Rottinghaus, G. A. Bennett, J. A. Render, and R. J. Aulerich.** 1998. Subacute and reproductive effects in mink from exposure to *Fusarium fujikuroi* culture material (M-1214) containing known concentrations of moniliformin. *Archives of Environmental Contamination and Toxicology* **35:** 513-517. (*fujikuroi*)

1478. **Morgan, P.** 1983. Fusarium stem rot of Douglas fir seedlings. *Plant Disease* **67:** 441-442. (*avenaceum*)

1479. **Moricca, S., A. Ragazzi, T. Kasuga, and K. R. Mitchelson.** 1998. Detection of *Fusarium oxysporum* f. sp. *vasinfectum* in cotton tissue by polymerase chain reaction. *Plant Pathology* **47:** 486-494. (*oxysporum*)

1480. **Moschini, R. C., M. R. Carranza, and M. A. Carmona.** 2004. Meteorological-based predictions of wheat head blight epidemics in the Southern Argentinian Pampas region. *Cereal Research Communications* **32:** 45-52. (11)

1481. **Moss, M. O., and J. M. Frank.** 1985. Influence of the fungicide tridemorph on T-2 toxin production by *Fusarium sporothrichioides*. *Transactions of the British Mycological Society* **84:** 585-590. (*sporotrichioides*)

1482. **Moss, M. O., and J. E. Smith, eds.** 1984. *The Applied Mycology of Fusarium*. Cambridge University Press, New York. (9)

1483. **Motta, E., T. Annesi, and V. Balmas.** 1996. Seedborne fungi in Norway spruce: Testing methods and pathogen control by seed dressing. *European Journal of Forest Pathology* **26:** 307-314. (*dlamini*)

1484. **Mouyna, I., J. L. Renard, and Y. Brygoo.** 1996. DNA polymorphism among *Fusarium oxysporum* f. sp. *elaeidis* populations from oil palm, using a repeated and dispersed sequence "Palm". *Current Genetics* **30:** 174-180. (*oxysporum*)

1485. **Mpofu, S. I., and K. Y. Rashid.** 2001. Vegetative compatibility groups within *Fusarium oxysporum* f. sp. *lini* from *Linum usitatissimum* (flax) wilt nurseries in western Canada. *Canadian Journal of Botany* **79:** 836-843. (*oxysporum*)

1486. **Mselle, J.** 1999. Fungal keratitis as an indicator of HIV infection in Africa. *Tropical Doctor* **29:** 133-135. (1, *solani*)

1487. **Mselle, J.** 2001. Use of topical clotrimazole in human keratomycosis. *Ophthalmologica* **215:** 357-360. (*solani*)

1488. **Mueller-Stoever, D., J. Kroschel, H. Thomas, and J. Sauerborn.** 2002. Chlamydospores of *Fusarium oxysporum* Schlecht. f. sp. *orthoceras* (Appel and Wollenw.) Bilai as inoculum for wheat-flour-kaolin granules to be used for the biological control of *Orobanche cumana* Wallr. *European Journal of Plant Pathology* **108:** 221-228. (*oxysporum*)

1489. **Mueller-Stoever, D., H. Thomas, J. Sauerborn, and J. Kroschel.** 2004. Two granular formulations of *Fusarium oxysporum* f. sp. *orthoceras* to mitigate sunflower broomrape *Orobanche cumana*. *BioControl* **49:** 595-602. (*oxysporum*)

1490. **Muhitch, M. J., S. P. McCormick, N. J. Alexander, and T. M. Hohn.** 2000. Transgenic expression of the *TRI101* or *PDR5* gene increases resistance of tobacco to the phytotoxic effects of the trichothecene 4,15-diacetoxyscirpenol. *Plant Science* **157:** 201-207. (*sporotrichioides*)

1491. **Mulè, G., A. Logrieco, G. Stea, and A. Bottalico.** 1997. Clustering of trichothecene-producing *Fusarium* strains determined from 28S ribosomal DNA sequences. *Applied and Environmental Microbiology* **63:** 1843-1846. (*camptoceras*, *musarum*)

1492. **Mulè, G., A. Susca, G. Stea, and A. Moretti.** 2004. Specific detection of the toxigenic species *Fusarium proliferatum* and *F. oxysporum* from asparagus plants using primers based on calmodulin gene sequences. *FEMS Microbiology Letters* **230:** 235-240. (*oxysporum*, *proliferatum*)

1493. **Mulè, G., A. Susca, G. Stea, and A. Moretti.** 2004. A species-specific PCR assay based on the calmodulin partial gene for identification of *Fusarium verticillioides*, *F. proliferatum* and *F. subglutinans*. *European Journal of Plant Pathology* **110:** 495-502. (*compactum*, *culmorum*, *proliferatum*, *subglutinans*, *verticillioides*)

1494. **Mulfinger, S., L. Niessen, and R. F. Vogel.** 2000. PCR based quality control of toxigenic *Fusarium* spp. in brewing malt using ultrasonication for rapid sample preparation. *Advances in Food Science* **22:** 38-46. (7)

1495. **Mullins, E. D., X. Chen, P. Romaine, R. Raina, D. M. Geiser, and S. Kang.** 2001. *Agrobacterium*-mediated transformation of *Fusarium oxysporum*: An efficient tool for insertional mutagenesis and gene transfer. *Phytopathology* **91:** 173-180. (*oxysporum*)

1496. **Munaut, F., and F. van Hove.** 2004. Mating type determination in *Fusarium phyllophilum* and *Fusarium ramigenum*. *Proceedings of the Second International Symposium on* Fusarium *Head Blight (Orlando, Florida)* **2:** 572. (*ramigenum*)

1497. **Munkvold, G. P., and W. M. Carlton.** 1997. Influence of inoculation method on systemic *Fusarium moniliforme* infection of maize plants grown from infected seeds. *Plant Disease* **81:** 211-216. (*verticillioides*)

1498. **Munkvold, G. P., R. L. Hellmich, and L. G. Rice.** 1999. Comparison of fumonisin concentrations in kernels of transgenic Bt maize hybrids and nontransgenic hybrids. *Plant Disease* **83:** 130-138. (*verticillioides*)

1499. **Munkvold, G. P., R. L. Helmich, and W. B. Showers.** 1997. Reduced Fusarium ear rot and symptomless infection in kernels of maize genetically engineered for European corn borer resistance. *Phytopathology* **87:** 1071-1077. (*proliferatum*, *subglutinans*, *verticillioides*)

1500. **Munkvold, G. P., and J. J. Marois.** 1993. Efficacy of natural epiphytes and colonizers of grapevine pruning wounds for biological control of *Eutypa* dieback. *Phytopathology* **83:** 624-629. (*lateritium*)

1501. ☑**Munkvold, G. P., D. C. McGee, and W. M. Carlton.** 1997. Importance of different pathways for maize kernel infection by *Fusarium moniliforme*. *Phytopathology* **87:** 209-217. (*verticillioides*)

1502. **Murillo, I., L. Cavallarin, and B. S. Segundo.** 1998. The development of a rapid PCR assay for detection of *Fusarium moniliforme*. *European Journal of Plant Pathology* **104:** 301-311. (*verticillioides*)

1503. **Murphy, C. A., J. A. Cameron, S. J. Huang, and R. T. Vinopal.** 1996. *Fusarium* polycaprolactone depolymerase is cutinase. *Applied and Environmental Microbiology* **62:** 456-460. (*solani*)

1504. **Murphy, C. A., J. A. Cameron, S. J. Huang, and R. T. Vinopal.** 1998. A second polycaprolactone depolymerase from *Fusarium*, a lipase distinct from cutinase. *Applied Microbiology and Biotechnology* **50:** 692-696. (*solani*)

1505. **Murray, M. G., and W. F. Thompson.** 1980. Rapid isolation of high molecular weight plant DNA. *Nucleic Acids Research* **8:** 4321-4325. (7)

1506. **Musser, S. M., M. L. Gay, E. P. Mazzola, and R. D. Plattner.** 1996. Identification of a new series of fumonisins containing 3-hydroxypyridine. *Journal of Natural Products* **59:** 970-972. (*verticillioides*)

1507. **Musser, S. M., and R. D. Plattner.** 1997. Fumonisin composition in cultures of *Fusarium moniliforme*, *Fusarium proliferatum* and *Fusarium nygamai*. *Journal of Agricultural and Food Chemistry* **45:** 1169-1173. (*nygamai*, *proliferatum*)

1508. **Myc, A., T. Vanhecke, J. J. Landers, T. Hamouda, and J. R. Baker, Jr.** 2002. The fungicidal activity of novel nanoemulsion (X8W60PC) against clinically important yeast and filamentous fungi. *Mycopathologia* **155:** 195-201. (*oxysporum*)

1509. **Mylyk, O. M.** 1975. Heterokaryon incompatibility genes in *Neurospora crassa* detected using duplication-producing chromosome rearrangements. *Genetics* **80:** 197-124. (5)

1510. **Myoken, Y., T. Sugata, T. Kyo, and M. Fujihara.** 1995. Oral *Fusarium* infection in a granulocytopenic patient with acute myelogenous leukemia: a case report. *Journal of Oral Pathology and Medicine* **24:** 237-240. (*verticillioides*)

1511. **Möller, E. M., J. Chełkowski, and H. H. Geiger.** 1999. Species-specific PCR assays for the fungal pathogens *Fusarium moniliforme* and *Fusarium subglutinans* and their application to diagnose maize ear rot disease. *Journal of Phytopathology* **147:** 497-508. (7, *subglutinans*, *verticillioides*)

1512. **Müller, F. M. C., K. E. Werner, M. Kasai, A. Francesconi, S. J. Chanock, and T. J. Walsh.** 1998. Rapid extraction of genomic DNA from medically important yeasts and filamentous fungi by high-speed cell disruption. *Journal of Clinical Microbiology* **36:** 1625-1629. (*solani*)

1513. **Nagao, T., Y. Shimada, A. Sugihara, A. Murata, S. Komemushi, and Y. Tominaga.** 2001. Use of thermostable *Fusarium heterosporum* lipase for production of structured lipid containing oleic and palmitic acids in organic solvent-free system. *Journal of the American Oil Chemists' Society* **78:** 167-172. (*heterosporum*)

1514. **Nagao, T., Y. Shimada, A. Sugihara, and Y. Tominaga.** 1994. Cloning and nucleotide sequence of cDNA encoding a lipase from *Fusarium heterosporum*. *Journal of Biochemistry* **116:** 536-540. (*heterosporum*)

1515. **Nagao, T., Y. Shimada, A. Sugihara, and Y. Tominaga.** 1998. C-terminal peptide of *Fusarium heterosporum* lipase is necessary for its increasing thermostability. *Journal of Biochemistry* **124:** 1124-1129. (*heterosporum*)

1516. **Nagao, T., Y. Shimada, A. Sugihara, and Y. Tominaga.** 2000. Amino acid residues contributing to stabilization of *Fusarium heterosporum* lipase. *Journal of Bioscience and Bioengineering* **89:** 446-450. (*heterosporum*)

1517. **Nagao, T., Y. Shimada, A. Sugihara, and Y. Tominaga.** 2002. Increase in stability of *Fusarium heterosporum* lipase. *Journal of Molecular Catalysis. B. Enzymatic* **17:** 125-132. (*heterosporum*)

1518. **Nagpal, A., M. P. Valley, P. F. Fitzpatrick, and A. M. Orville.** 2004. Crystallization and preliminary analysis of active nitroalkane oxidase in three crystal forms. *Acta Crystallographica D. Biological Crystallography* **60:** 1456-1460. (*oxysporum*)

1519. **Nagy, R. and L. Hornok.** 1994. Electrophoretic karyotype differences between two subspecies of *Fusarium acuminatum*. *Mycologia* **86:** 203-208. (*acuminatum*, *armeniacum*, *sporotrichioides*)

1520. **Nagy, R., E. Taborhegyi, A. Witner, and L. Hornok.** 1995. Mini-chromosomes in *Fusarium sporotrichioides* are mosaics of dispersed repeats and unique sequences. *Microbiology* **141:** 713-719. (*sporotrichioides*)

1521. **Nahalkova, J., and J. Fatehi.** 2003. Red fluorescent protein (DsRed2) as a novel reporter in *Fusarium oxysporum* f. sp. *lycopersici*. *FEMS Microbiology Letters* **225:** 305-309. (*oxysporum*)

1522. **Naik, D. M., I. N. Nawa, and R. H. Raemaekers.** 1982. Absence of an effect from internally seedborne *Fusarium moniliforme* on emergence, plant growth and yield of maize (*Zea mays*). *Seed Science and Technology* **10:** 347-356. (*verticillioides*)

1523. **Nakahara, K., and H. Shoun.** 1996. N-terminal processing and amino acid sequence of two isoforms of nitric oxide reductase cytochrome P450nor from *Fusarium oxysporum*. *Journal of Biochemistry* **120:** 1082-1087. (*oxysporum*)

1524. **Nakahara, K., T. Tanimoto, K.-I. Hatano, K. Usuda, and H. Shoun.** 1993. Cytochrome P-450 55A1 (P-450dNIR) acts as nitric oxide reductase employing NADH as the direct electron donor. *Journal of Biological Chemistry* **268:** 8350-8355. (*oxysporum*)

1525. **Nakai, T., W. Yoshikawa, H. Nakamura, and H. Yoshida.** 1992. The three dimensional structure of guanine-specific ribonuclease F_1 in solution determined by NMR spectroscopy and distance geometry. *European Journal of Biochemistry* **208:** 41-51. (*verticillioides*)

1526. **Nakamura, C., N. Kodo, T. Shimizu, and N. Mori.** 1997. Construction of physical maps of mitochondrial DNAs in *Fusarium* species causing Fusarium head blight. *Cereal Research Communications* **25:** 321-324. (*culmorum*, *graminearum*)

1527. **Nakamura, T., K. Mitsuoka, M. Sugano, K. Tomita, and T. Murayama.** 1985. Effects of auxin and gibberellin on conidial germination and elongation of young hyphae in *Gibberella fujikuroi* and *Penicillium notatum*. *Plant and Cell Physiology* **26:** 1433-1438. (*fujikuroi*)

1528. **Namba, H., F. Nakatsubo, and T. Higuchi.** 1983. Degradation of β-1 linked dilignols by *Fusarium solani* M-13-1. *Wood Research* **69:** 52-60. (*solani*)

1529. **Namiki, F., M. Matsunaga, M. Okuda, I. Inoue, K. Nishi, Y. Fujita, and T. Tsuge.** 2001. Mutation of an arginine biosynthesis gene causes reduced pathogenicity in *Fusarium oxysporum* f. sp. *melonis*. *Molecular Plant-Microbe Interactions* **14:** 580-584. (*oxysporum*)

1530. **Namiki, F., T. Shiomi, T. Kayamura, and T. Takashi.** 1994. Characterization of the *formae speciales* of *Fusarium oxysporum* causing wilts of cucurbits by DNA fingerprinting with nuclear repetitive DNA sequences. *Applied and Environmental Microbiology* **60:** 2684-2691. (*oxysporum*)

1531. **Namiki, F., T. Shiomi, K. Nishi, T. Kayamura, and T. Takashi.** 1998. Pathogenic and genetic variation in the Japanese strains of *Fusarium oxysporum* f. sp. *melonis*. *Phytopathology* **88:** 804-810. (*oxysporum*)

1532. **Narendra, D. V., and M. V. N. Setty.** 1979. Seed mycoflora of sugarcane and their importance in nurseries. *Seed Research* **7:** 145-150. (*sacchari*)

1533. **Naseema, A., R. Praveena, R. R. Nam, and C. K. Peethambaran.** 2004. *Fusarium pallidoroseum* for management of water hyacinth. *Current Science* **86:** 770-771. (*semitectum*)

1534. **Nash, S. N., and W. C. Snyder.** 1962. Quantitative estimations by plate counts of propagules of the bean rot *Fusarium* in field soils. *Phytopathology* **73:** 458-462. (2)

1535. **Navas-Cortes, J. A., B. Hau, and R. M. Jimenez-Diaz.** 2000. Yield loss in chickpeas in relation to development of Fusarium wilt epidemics. *Phytopathology* **90:** 1269-1278. (11)

1536. **Nazareth, S., and S. Mavinkurve.** 1986. Degradation of ferulic acid via 4-vinylguaiacol by *Fusarium solani*. *Canadian Journal of Microbiology* **32:** 494-497. (*solani*)

1537. **Nazareth, S. W., and C. V. Bruschi.** 1994. Electrophoretic karyotype of *Fusarium solani*. *Microbiology* **140:** 1373-1375. (*solani*)

1538. **Neish, G. A.** 1987. Neotypification of *Fusarium tricinctum*. *Canadian Journal of Botany* **65:** 589-591. (*tricinctum*)

1539. **Neish, G. A., E. R. Farnworth, and H. Cohen.** 1982. Zearalenone and trichothecene production by some *Fusarium* species associated with Canadian grains. *Canadian Journal of Plant Pathology* **4:** 191-194. (*equiseti, poae, sporotrichioides*)

1540. **Nelson, P. E.** 1981. Life cycle and epidemiology of *Fusarium oxysporum*, p. 51-80. *In* M. E. Mace, A. A. Bell, and C. H. Beckman (eds.), *Fungal Wilt Diseases of Plants*. Academic Press, New York City, New York. (*oxysporum*)

1541. **Nelson, P. E., L. W. Burgess, and B. A. Summerell.** 1990. Some morphological and physiological characters of *Fusarium* species in sections *Liseola* and *Elegans* and similar new species. *Mycologia* **82:** 99-106. (*anthophilum, beomiforme, napiforme, nygamai, oxysporum, proliferatum, subglutinans, verticillioides*)

1542. **Nelson, P. E., R. J. Cole, T. A. Toussoun, J. W. Dorner, and R. M. Windingstad.** 1990. *Fusarium* species recovered from waste peanuts associated with sandhill crane mortality. *Mycologia* **82:** 562-565. (*compactum*)

1543. **Nelson, P. E., A. E. Desjardins, and R. D. Plattner.** 1993. Fumonisins, mycotoxins produced by *Fusarium* species: Biology, chemistry and significance. *Annual Review of Phytopathology* **31:** 233-252. (*verticillioides*)

1544. **Nelson, P. E., R. K. Horst, and S. S. Woltz.** 1981. Fusarium diseases of ornamental plants, p. 121-128. *In* P. E. Nelson, T. A. Toussoun, and R. J. Cook (eds.), Fusarium*: Diseases, Biology and Taxonomy*. Pennsylvania State University Press, University Park, Pennsylvania. (*oxysporum*)

1545. **Nelson, P. E., J. H. Juba, and C. A. Clark.** 1995. An unusual strain of *Fusarium lateritium* affecting sweet potato. *Mycologia* **87:** 507-509. (*denticulatum*)

1546. **Nelson, P. E., J. H. Juba, P. F. Ross, and L. G. Rice.** 1994. Fumonisin production by *Fusarium* species on solid substrates. *Journal of AOAC International* **77:** 522-525. (*proliferatum, verticillioides*)

1547. **Nelson, P. E., B. W. Pennypacker, T. A. Toussoun, and R. K. Horst.** 1975. Fusarium stub dieback of carnations. *Phytopathology* **65:** 575-581. (*graminearum*)

1548. **Nelson, P. E., R. D. Plattner, D. D. Shackelford, and A. E. Desjardins.** 1992. Fumonisin B_1 production by *Fusarium* species other than *F. moniliforme* in section *Liseola* and some related species. *Applied and Environmental Microbiology* **58:** 984-989. (*anthophilum, beomiforme, dlamini, napiforme, nygamai, proliferatum, subglutinans, succisae*)

1549. **Nelson, P. E., T. A. Toussoun, and L. W. Burgess.** 1987. Characterization of *Fusarium beomiforme sp. nov. Mycologia* **79:** 884-889. (9, *beomiforme*)

1550. **Nelson, P. E., T. A. Toussoun, and R. J. Cook, eds.** 1981. Fusarium*: Diseases Biology and Taxonomy*. Pennsylvania State University Press, University Park, Pennsylvania. (9, *oxysporum, pseudograminearum, solani*)

1551. **Nelson, P. E., T. A. Toussoun, and W. F. O. Marasas.** 1983. Fusarium *Species: An Illustrated Manual for Identification*. Pennsylvania State University Press, University Park, Pennsylvania. (P, 2, 8, 9, 11, 12, *anthophilum, camptoceras, chlamydosporum, equiseti, lactis, merismoides, poae, proliferatum, redolens, sacchari, sambucinum, scirpi, semitectum, solani, sporotrichioides, subglutinans, succisae, tricinctum, udum, venenatum, verticillioides*)

1552. **Nemec, S.** 1987. *Fusarium solani* association with branch and trunk cankers on citrus weakened by cold weather in Florida, USA. *Mycopathologia* **97:** 143-150. (*solani*)

1553. **Nemec, S.** 1995. Stress-related compounds in xylem fluid of blight-diseased citrus containing *Fusarium solani* naphthazarin toxins and their effects on the host. *Canadian Journal of Microbiology* **41:** 515-524. (*solani*)

1554. **Nemec, S., S. Jabaji-Hare, and P. M. Charest.** 1991. ELISA and immunocytochemical detection of *Fusarium solani*-produced naphthazarin toxins in citrus trees in Florida. *Phytopathology* **81:** 1497-1503. (*solani*)

1555. **Nemec, S., D. Phelps, and R. Baker.** 1989. Effects of dihydrofusarubin and isomarticin from *Fusarium solani* on carbohydrate status and metabolism of rough lemon seedling. *Phytopathology* **79:** 700-705. (*solani*)

1556. **Neumann, M. J., D. Backhouse, D. A. Carter, B. A. Summerell, and L. W. Burgess.** 2004. Genetic structure of populations of *Fusarium proliferatum* in soils associated with *Livistona mariae* palms in Little Palm Creek, Northern Territory, Australia. *Australian Journal of Botany* **52:** 543-550. (*proliferatum*)

1557. **Newsham, K. K.** 2003. Response of saprotrophic fungal communities to declining SO_2 pollution in the natural environment. *Pedobiologia* **47:** 77-84. (*avenaceum*)

1558. **Nicaudet, P., G. Beaurain, J. Leibowitch, and J. F. Bach.** 1980. *In vitro* characteristics on human lymphocyte functions of a new immuno-modulatory agent, a cyclic peptide cyclomunine. *Clinical and Experimental Immunology* **42:** 315-323. (*equiseti*)

1559. **Nicholson, R. L., L. F. Bauman, and H. L. Warren.** 1976. Association of *Fusarium moniliforme* with brown midrib maize. *Plant Disease Reporter* **60:** 908-910. (*verticillioides*)

1560. **Nicolas, A., M. Egmond, C. T. Verrips, J. de Vlieg, S. Longhi, C. Cambillau, and C. Martinez.** 1996. Contribution of cutinase serine 42 side chain to the stabilization of the oxyanion transition state. *Biochemistry* **35:** 398-410. (*solani*)

1561. **Nicoletti, R., F. Raimo, and E. Cozzolino.** 2002. *In vitro* evaluation of fungal antagonists of *Phytophthora nicotianae*. *Plant Protection Science* **38 (Special Issue 2):** 634-637. (*chlamydosporum*)

1562. **Nicosia, M. G. L. D., C. Brocard-Masson, S. Demais, A. H. Van, M.-J. Daboussi, and C. Scazzocchio.** 2001. Heterologous transposition in *Aspergillus nidulans*. *Molecular Microbiology* **39:** 1330-1344. (*oxysporum*)

1563. **Niessen, L., H. Schmidt, and R. F.Vogel.** 2004. The use of *tri5* gene sequences for PCR detection and taxonomy of trichothecene-producing species in the *Fusarium* section *Sporotrichiella*. *International Journal of Food Microbiology* **95:** 305-319. (*poae, sporotrichioides*)

1564. **Nik, W. Z., and D. G. Parberry.** 1977. Studies of seedborne fungi of tropical pasture legume species. *Australian Journal of Agricultural Research* **28:** 821-841. (*acuminatum*)

1565. **Ninet, B., I. Jan, O. Bontems, B. Lechenne, O. Jousson, D. Lew, J. Schrenzel, R. G. Panizzon, and M. Monod.** 2005. Molecular identification of *Fusarium* species in onychomycoses. *Dermatology* **210:** 21-25. (*oxysporum, proliferatum, solani*)

1566. **Nirenberg, H. I.** 1976. Untersuchungen über die morphologische und biologische Differenzierung in der *Fusarium* Sektion *Liseola*. *Mitteilungen aus der Biologischen Bundesanstalt Für Land- und Forstwirtschaft (Berlin – Dahlem)* **169:** 1-117. (2, 8, *anthophilum, bulbicola, fujikuroi, phyllophilum, proliferatum, succisae*)

1567. **Nirenberg, H. I.** 1990. Recent advances in the taxonomy of *Fusarium*. *Studies in Mycology* **32:** 91-101. (*crookwellense, semitectum*)

1568. **Nirenberg, H. I.** 1995. Morphological differentiation of *Fusarium sambucinum* Fückel *sensu stricto*, *F. torulosum* (Berk. & Curt.) Nirenberg *comb. nov.*, and *F. venenatum* Nirenberg *sp. nov.* *Mycopathologia* **129:** 131-141. (8, *sambucinum, torulosum, venenatum*)

1569. **Nirenberg, H., and T. Aoki.** 1997. *Fusarium nisikadoi*, a new species from Japan. *Mycoscience* **38:** 329-333. (8, *nisikadoi*)

1570. **Nirenberg, H. I., and K. O'Donnell.** 1998. New *Fusarium* species and combinations within the *Gibberella fujikuroi* species complex. *Mycologia* **90:** 434-458. (8, 9, 11, *acutatum, babinda, begoniae, bulbicola, circinatum, concentriccum, denticulatum, guttiforme, lactis, phyllophilum, pseudocircinatum, pseudonygamai, ramigenum, subglutinans*)

1571. **Nirenberg, H. I., K. O'Donnell, J. Kroschel, A. P. Andrianaivo, J. M. Frank, and W. Mubatanhema.** 1998. Two new species of *Fusarium*: *Fusarium brevicatenulatum* from the noxious weed *Striga asiatica* in Madagascar and *Fusarium pseudoanthophilum* from *Zea mays* in Zimbabwe. *Mycologia* **90:** 459-464. (8, 9, *brevicatenulatum, pseudoanthophilum*)

1572. **Niture, S. K., and A. Pant.** 2004. Purification and biochemical characterization of polygalacturonase II produced in semi-solid medium by a strain of *Fusarium moniliforme*. *Microbiological Research* **159:** 305-314. (*verticillioides*)

1573. **Niture, S. K., A. Pant, and A. R. Kumar.** 2001. Active site characterization of the single endo-polygalacturonase produced by *Fusarium moniliforme* NCIM 1276. *European Journal of Biochemistry* **268:** 832-840. (*verticillioides*)

1574. **Nixon, K. C., and Q. D. Wheeler.** 1990. An amplification of the phylogenetic species concept. *Cladistics* **6:** 211-223. (9)

1575. **Njiti, V. N., T. W. Doubler, R. J. Suttner, L. E. Gray, P. T. Gibson, and D. A. Lightfoot.** 1998. Resistance to soybean sudden death syndrome and root colonization by *Fusarium solani* f. sp. *glycine* in near-isogenic lines. *Crop Science* **38:** 472-477. (*solani*)

1576. **Noble, J. A., S. A. Crow, D. G. Ahearn, and F. A. Kuhn.** 1997. Allergic fungal sinusitis in the southeastern USA: Involvement of a new agent *Epicoccum nigrum* Ehrenb. ex Schlecht. 1824. *Journal of Medical and Veterinary Mycology* **35:** 405-409. (*solani*)

1577. **Nogawa, M., T. Kageyama, A. Nakatani, G. Taguchi, M. Shimosaka, and M. Okazaki.** 1996. Cloning and characterization of mycovirus double-stranded RNA from the plant pathogenic fungus, *Fusarium solani* f. sp. *robiniae*. *Bioscience Biotechnology and Biochemistry* **60:** 784-788. (*solani*)

1578. **Nogawa, M., A. Nakatani, K. Gonda, M. Shimosaka, and M. Okazaki.** 1996. Replication of double-stranded RNA in mycovirus from the plant pathogenic fungus, *Fusarium solani*. *FEMS Microbiology Letters* **137:** 45-49. (*solani*)

1579. **Nogawa, M., M. Shimosaka, T. Kageyama, and M. Okazaki.** 1993. A double-stranded RNA mycovirus from the plant pathogenic fungus, *Fusarium solani* f. sp. *robiniae*. *FEMS Microbiology Letters* **110:** 153-157. (*solani*)

1580. **Nolan, L. M., P. A. Harnedy, P. Turner, A. B. Hearne, and C. O'Reilly.** 2003. The cyanide hydratase enzyme of *Fusarium lateritium* also has nitrilase activity. *FEMS Microbiology Letters* **221:** 161-165. (*lateritium*)

1581. **Nonomura, T., Y. Matsuda, M. Takasugi, T. Ootani, T. Hasegawa, K. Miyajima, T. Hatasa, and H. Toyoda.** 2001. A monitoring system for green fluorescence protein gene-transformed *Fusarium oxysporum* in melon seedlings. *Journal of General Plant Pathology* **67:** 273-280. (*oxysporum*)

1582. **Nonomura, T., H. Tajima, Y. Kitagawa, N. Sekiya, K. Shitomi, M. Tanaka, K. Maeda, Y. Matsuda, and H. Toyoda.** 2003. Distinguishable staining with neutral red for GFP-marked and GFP-nonmarked *Fusarium oxysporum* strains simultaneously colonizing root surfaces. *Journal of General Plant Pathology* **69:** 45-48. (*oxysporum*)

1583. **Noriega-Cantú, D. H., D. Teliz, G. Mora-Aguilera, J. Rodriguez-Alcazar, E. Zavaleta-Mejia, G. Otero-Colinas, and C. L. Campbell.** 1999. Epidemiology of mango malformation in Guerrero, Mexico, with traditional and integrated management. *Plant Disease* **83:** 223-228. (*mangiferae*)

1584. **Norred, W. P., C. W. Bacon, R. D. Plattner, and R. F. Vesonder.** 1991. Differential cytotoxicity and mycotoxin content among isolates of *Fusarium moniliforme*. *Mycopathologia* **115:** 37-43. (*verticillioides*)

1585. **Norris, D. M.** 1980. Degradation of ^{14}C-labeled lignins and ^{14}C-labeled aromatic acids by *Fusarium solani*. *Applied and Environmental Microbiology* **40:** 376-380. (*solani*)

1586. **Nowrousian, M., U. Kueck, K. Loser, and K. M. Weltring.** 2000. The fungal *acl1* and *acl2* genes encode two polypeptides with homology to the N- and C-terminal parts of the animal ATP citrate lyase polypeptide. *Current Genetics* **37:** 189-193. (*sambucinum*)

1587. **Nuero, O. M.** 1995. Production of chitinase by *Fusarium* species. *Current Microbiology* **30:** 287-289. (*subglutinans*)

1588. **Nyvall, R. F., J. A. Percich, and C. J. Mirocha.** 1999. Fusarium head blight of cultivated and natural wild rice (*Zizania palustris*) in Minnesota caused by *Fusarium graminearum* and associated *Fusarium* spp. *Plant Disease* **83:** 159-164. (*anthophilum, graminearum, subglutinans*)

1589. **Ocamb, C., M. J. Juzwik, and F. B. Martin.** 2002. *Fusarium* spp. and *Pinus strobus* seedlings: Root disease pathogens and taxa associated with seed. *New Forests* **24:** 67-79. (*polyphialidicum, proliferatum*)

1590. **O'Day, D. M., P. L. Akrabawi, L. D. Richmond, B. R. Jones, and Y. Clayton.** 1979. An animal model of *Fusarium solani* endophthalmitis. *British Journal of Ophthalmology* **63:** 277-280. (*solani*)

1591. **O'Day, D. M., W. A. Ray, S. Head, R. D. Robinson, and T. E. Williams.** 1991. Influence of corticosteroid on experimentally induced keratomycosis. *Archives of Ophthalmology* **109:** 1601-1604. (*solani*)

1592. **O'Donnell, K.** 1992. Ribosomal DNA internal transcribed spacers are highly divergent in the phytopathogenic ascomycete *Fusarium sambucinum* (*Gibberella pulicaris*). *Current Genetics* **22:** 213-220. (*sambucinum*)

1593. **O'Donnell, K.** 2000. Molecular phylogeny of the *Nectria haematococca-Fusarium solani* species complex. *Mycologia* **92:** 919-938. (8, 9)

1594. **O'Donnell, K., and E. Cigelnik.** 1997. Two divergent intragenomic rDNA ITS2 types within a monophyletic lineage are nonorthologous. *Molecular Phylogenetics and Evolution* **7:** 103-116. (9)

1595. **O'Donnell, K., E. Cigelnik, and H. H. Casper.** 1998. Molecular, phylogenetic, morphological, and mycotoxin data support reidentification of the Quorn mycoprotein fungus as *Fusarium venenatum*. *Fungal Genetics and Biology* **23:** 57-67. (9, *graminearum, venenatum*)

1596. **O'Donnell, K., E. Cigelnik, and H. I. Nirenberg.** 1998. Molecular systematics and phylogeography of the *Gibberella fujikuroi* species complex. *Mycologia* **90:** 465-493. (7, 8, 9, *anthophilum, babinda, beomiforme, brevicatenulatum, circinatum, dlamini, fujikuroi, nygamai, pseudoanthophilum, pseudocircinatum, pseudonygamai, succisae*)

1597. **O'Donnell, K., and L. E. Gray.** 1995. Phylogenetic relationships of the soybean sudden death syndrome pathogen *Fusarium solani* f. sp. *phaseoli* inferred from rDNA sequence data and PCR primers for its identification. *Molecular Plant-Microbe Interactions* **8:** 709-716. (*solani*)

1598. **O'Donnell, K., H. C. Kistler, E. Cigelnik, and R. C. Ploetz.** 1998. Multiple evolutionary origins of the fungus causing Panama disease of banana: Concordant evidence from nuclear and mitochondrial gene genealogies. *Proceedings of the National Academy of Sciences (USA)* **95:** 2044-2049. (8, 9, *oxysporum*)

1599. **O'Donnell, K., H. C. Kistler, B. K. Tacke, and H. H. Casper.** 2000. Gene genealogies reveal global phylogeographic structure and reproductive isolation among lineages of *Fusarium graminearum*, the fungus causing wheat scab. *Proceedings of the National Academy of Sciences (USA)* **97:** 7905-7910. (9, 11, *graminearum*)

1600. **O'Donnell, K., H. I. Nirenberg, T. Aoki, and E. Cigelnik.** 2000. A multigene phylogeny of the *Gibberella fujikuroi* species complex: Detection of additionally phylogenetically distinct species. *Mycoscience* **41:** 61-78. (8, 9, *circinatum*)

1601. **O'Donnell, K., T. J. Ward, D. M. Geiser, H. C. Kistler, and T. Aoki.** 2004. Genealogical concordance between the mating type locus and seven other nuclear genes supports formal recognition of nine phylogenetically distinct species within the *Fusarium graminearum* clade. *Fungal Genetics and Biology* **41:**600-623. (6, 7, 8, 9, 11, *graminearum*)

1602. **Ofoegbu, C. A. E.** 1993. Root rot disease of cocoyam [*Xanthosoma sagittifolium* (L.) Schott] in the South Eastern state of Nigeria. *Discovery and Innovation* **5:** 373-376. (*solani*)

1603. **Ogawa, J. M., and H. English.** 1955. The efficiency of a quantitative spore collector using the cyclone method. *Phytopathology* **45:** 239-240. (3)

1604. **Ogiso, H., M. Fujinaga, H. Saito, T. Takehara, and S. Yamanaka.** 2002. Physiological races and vegetative compatibility groups of *Fusarium oxysporum* f. sp. *lactucae* isolated from crisphead lettuce in Japan. *Journal of General Plant Pathology* **68:** 292-299. (*oxysporum*)

1605. **Ohara, T., I. Inoue, F. Namiki, H. Kunoh, and T. Tsuge.** 2004. *REN1* is required for development of microconidia and macroconidia, but not of chlamydospores, in the plant pathogenic fungus *Fusarium oxysporum*. *Genetics* **166:** 113-124. (*oxysporum*)

1606. **Ohara, T., and T. Tsuge.** 2004. *FoSTUA*, encoding a basic helix-loop-helix protein, differentially regulates development of three kinds of asexual spores, macroconidia, microconidia, and chlamydospores, in the fungal plant pathogen *Fusarium oxysporum*. *Eukaryotic Cell* **3:** 1412-1422. (*oxysporum*)

1607. **Ohshima, S., M. Yanagisawa, A. Katoh, T. Fujii, T. Sano, S. Matsukuma, T. Furumai, M. Fujiu, K. Watanabe, K. Yokose, M. Arisawa, and T. Okuda.** 1994. *Fusarium merismoides* Corda NR 6356, the source of the protein kinase C inhibitor, azepinostatin: Taxonomy, yield improvement, fermentation and biological activity. *Journal of Antibiotics* **47:** 639-647. (*merismoides*)

1608. **Oji, E. O., and Y. M. Clayton.** 1982. The role of econazole in the management of oculomycosis. *International Ophthamology* **4:** 137-142. (*solani*)

1609. **Okada, H., S. Hamatani, M. Kondo, T. Imai, S. Itoh, K. Isobe, and S. Onishi.** 2000. Successful treatment of disseminated *Fusarium* infection in an infant with leukemia. *International Journal of Hematology* **72:** 494-498. (*verticillioides*)

1610. **Okawa, Y., and T. Yamaguchi.** 1977. Studies on sterol ester hydrolase from *Fusarium oxysporum*. Part I. Partial purification and properties. *Journal of Biochemistry* **81:** 1209-1216. (*oxysporum*)

1611. **Okiror, M. A.** 2002. Genetics of resistance to *Fusarium udum* in pigeonpea [*Cajanus cajan* (L.) Millsp.]. *Indian Journal of Genetics and Plant Breeding* **62:** 218-220. (*udum*)

1612. **Okiror, M. A., and P. M. Kimani.** 1997. Pathogenic variation of *Fusarium udum* of pigeonpea. *Indian Journal of Genetics and Plant Breeding* **57:** 186-192. (*udum*)

1613. **Okubara, P. A., A. E. Blechl, S. P. McCormick, N. J. Alexander, R. Dill-Macky, and T. M. Hohn.** 2003. Engineering deoxynivalenol metabolism in wheat through the expression of a fungal trichothecene acetyltransferase gene. *Theoretical and Applied Genetics* **106:** 74-83. (*sporotrichioides*)

1614. **Okuda, C., M. Ito, Y. Sato, K. Oka, and M. Hotchi.** 1987. Disseminated cutaneous *Fusarium* infection with vascular invasion in a leukemic patient. *Journal of Medical and Veterinary Mycology* **25:** 177-186. (*solani*)

1615. **Okuda, M., K. Ikeda, F. Namiki, K. Nishi, and T. Tsuge.** 1998. *Tfo1*: An *Ac*-like transposon from the plant pathogenic fungus *Fusarium oxysporum*. *Molecular and General Genetics* **258:** 599-607. (*oxysporum*)

1616. **Olivain, C., and C. Alabouvette.** 1999. Process of tomato root colonization by a pathogenic strain of *Fusarium oxysporum* f. sp. *lycopersici* in comparison with a nonpathogenic strain. *New Phytologist* **141:** 497-510. (*oxysporum*)

1617. **Olivain, C., S. Trouvelot, M. N. Binet, C. Cordier, A. Pugin, and C. Alabouvette.** 2003. Colonization of flax roots and early physiological responses of flax cells inoculated with pathogenic and nonpathogenic strains of *Fusarium oxysporum*. *Applied and Environmental Microbiology* **69:** 5453-5462. (*oxysporum*)

1618. **Olive, L. S.** 1954. Cross-karyogamy and segregation in a homothallic fungus. *Bulletin of the Torrey Botanical Club* **81:** 95-97. (6)

1619. **Olivieri, F., M. E. Zanetti, C. R. Oliva, A. A. Covarrubias, and C. A. Casalongue.** 2002. Characterization of an extracellular serine protease of *Fusarium eumartii* and its action on pathogenesis related proteins. *European Journal of Plant Pathology* **108:** 63-72. (*solani*)

1620. **Oller-Lopez, J. L., J. Avalos, A. F. Barrero, and J. E. Oltra.** 2003. Improved GA_1 production by *Fusarium fujikuroi*. *Applied Microbiology and Biotechnology* **63:** 282-285. (*fujikuroi*)

1621. **Olszak, M.** 1994. Aetiology of sour cherry fungal diseases in Poland. III. Pathogenicity of the isolated fungi. *Journal of Fruit and Ornamental Plant Research* **2:** 165-184. (*avenaceum, equiseti, poae*)

1622. **O'Mara, J. K., and G. P. Munkvold.** 1997. Fungal colonization of alfalfa stubble following harvest. *Journal of the Iowa Academy of Science* **104:** 1-3. (*acuminatum, sambucinum, sporotrichioides*)

1623. **O'Neil, C. E., M. L. McCants, J. E. Salvaggio, and S. B. Lehrer.** 1986. *Fusarium solani*: Prevalence of skin reactivity and antigenic allergenic analysis. *Journal of Allergy and Clinical Immunology* **77:** 842-849. (*solani*)

1624. **O'Neil, C. E., M. A. Reed, L. Aukrust, and B. T. Butcher.** 1983. The antigenic composition of aqueous cotton dust extracts. *International Archives of Allergy and Applied Immunology* **72:** 294-298. (*solani*)

1625. **O'Neil, C. E., M. A. Reed, J. M. Hughes, B. T. Butcher, and S. B. Lehrer.** 1987. *Fusarium solani*: Evidence for common antigenic-allergenic determinants with other *fungi imperfecti*. *Clinical Allergy* **17:** 127-134. (*solani*)

1626. **Onyike, N. B. N., and P. E. Nelson.** 1992. *Fusarium* spp. associated with sorghum grain from Nigeria, Lesotho, and Zimbabwe. *Mycologia* **84:** 452-458. (*napiforme, nygamai, subglutinans*)

1627. **Onyike, N. B. N., and P. E. Nelson.** 1993. The distribution of *Fusarium* species in soils planted to millet and sorghum in Lesotho, Nigeria and Zimbabwe. *Mycopathologia* **121:** 105-114. (*longipes, polyphialidicum*)

1628. **Onyike, N. B. N., P. E. Nelson, and W. F. O. Marasas.** 1991. *Fusarium* spp. associated with millet grain from Nigeria, Lesotho and Zimbabwe. *Mycologia* **83:** 708-712. (*napiforme, nygamai, subglutinans*)

1629. **Ooka, J. J., and T. Kommedahl.** 1977. Wind and rain dispersal of *Fusarium moniliforme* in corn fields. *Phytopathology* **67:** 1023-1026. (3, *verticillioides*)

1630. **Opperman, L., and F. C. Wehner.** 1994. Survey of fungi associated with grass roots in virgin soils on the Springbok Flats. *South African Journal of Botany* **60:** 67-72. (*nygamai, oxysporum*)

1631. **Oren, L., S. Ezrati, D. Cohen, and A. Sharon.** 2003. Early events in the *Fusarium verticillioides*-maize interaction characterized by using a green fluorescent protein-expressing transgenic isolate. *Applied and Environmental Microbiology* **69:** 1695-1701. (*verticillioides*)

1632. **Ortoneda, M., J. Guarro, M. P. Madrid, Z. Caracuel, M. I. G. Roncero, E. Mayayo, and A. di Pietro.** 2004. *Fusarium oxysporum* as a multihost model for the genetic dissection of fungal virulence in plants and mammals. *Infection and Immunity* **72:** 1760-1766. (*oxysporum*)

1633. **Ospina-Giraldo, M. D., E. Mullins, and S. Kang.** 2003. Loss of function of the *Fusarium oxysporum SNF1* gene reduces virulence on cabbage and *Arabidopsis*. *Current Genetics* **44:** 49-57. (*oxysporum*)

1634. **Osuchowski, M. F., G. L. Edwards, and R. P. Sharma.** 2005. Fumonisin B_1-induced neurodegeneration in mice after intracerebroventricular infusion is concurrent with disruption of sphingolipid metabolism and activation of proinflammatory signaling. *Neurotoxicology* **26:** 211-221. (*verticillioides*)

1635. **Owen, D., and D. Adams.** 2001. Impact of pitch canker on ornamental Monterey pines in Santa Cruz County, California, U.S., 1987-2000. *Journal of Arboriculture* **27:** 298-305. (*circinatum*)

1636. **Oyarzun, P. J., J. Postma, A. J. K. Luttikholt, and A. E. Hoogland.** 1994. Biological control of foot and root rot in pea caused by *Fusarium solani* with nonpathogen *Fusarium* *oxysporum* isolates. *Canadian Journal of Botany* **72:** 843-852. (*oxysporum*)

1637. **Oyeka, C. A., and H. C. Gugnani.** 1992. Skin infections due to *Hendersonula toruloidea*, *Scytalidium hyalinum*, *Fusarium solani* and dermatophytes in cement factory workers. *Journal de Mycologie Medicale* **2:** 197-201. (*solani*)

1638. **Pagliarusco, A., L. Tomazzoli, G. Amalfitano, L. Polonelli, and L. Bonomi.** 1995. Mycotic keratitis by *Fusarium moniliforme*. *Acta Ophthalmologica Scandinavica* **73:** 560-562. (*verticillioides*)

1639. **Palyusik, M., and E. M. Moran.** 1994. Porcine pulmonary edema with hydrothorax: A review. *Journal of Environmental Pathology, Toxicology and Oncology* **13:** 63-66. (*verticillioides*)

1640. **Panagiotou, G., and P. Christakopoulos.** 2004. NADPH-dependent D-aldose reductases and xylose fermentation in *Fusarium oxysporum*. *Journal of Bioscience and Bioengineering* **97:** 299-304. (*oxysporum*)

1641. **Panagiotou, G, P. Christakopoulos, S. G. Villas-Boas, and L. Olsson.** 2005. Fermentation performance and intracellular metabolite profiling of *Fusarium oxysporum* cultivated on a glucose-xylose mixture. *Enzyme and Microbial Technology* **36:** 100-106. (*oxysporum*)

1642. **Panagiotou, G., D. Kekos, B. J. Macris, and P. Christakopoulos.** 2002. Purification and characterization of NAD^+-dependent xylitol dehydrogenase from *Fusarium oxysporum*. *Biotechnology Letters* **24:** 2089-2092. (*oxysporum*)

1643. **Panagiotou, G., E. Topakas, L. Economou, D. Kekos, B. J. Macris, and P. Christakopoulos.** 2003. Induction, purification, and characterization of two extracellular α-L-arabinofuranosidases from *Fusarium oxysporum*. *Canadian Journal of Microbiology* **49:** 639-644. (*oxysporum*)

1644. **Panagiotou, G., S. G. Villas-Boas, P. Christakopoulos, J. Nielsen, and L. Olsson.** 2005. Intracellular metabolite profiling of *Fusarium oxysporum* converting glucose to ethanol. *Journal of Biotechnology* **115:** 425-434. (*oxysporum*)

1645. **Pandey, K. K., and J. P. Upadhyay.** 1999. Comparative study of chemical, biological and integrated approach for management of Fusarium wilt of pigeonpea. *Journal of Mycology and Plant Pathology* **29:** 214-216. (*udum*)

1646. **Pandey, M. K., D. K. Chakrabarti, and S. Kumar.** 2003. Analyzing mango (*Mangifera indica*) malformation in relation to the host age. *Indian Journal of Agricultural Sciences* **73:** 395-396. (11, *mangiferae*)

1647. **Pandey, R. N., S. E. Pawar, and C. R. Bhatia.** 1996. Inheritance of wilt resistance in pigeonpea. *Indian Journal of Genetics and Plant Breeding* **56:** 305-308. (*udum*)

1648. **Papst, C., H. F. Utz, A. E. Melchinger, J. Eder, T. Magg, D. Klein, and M. Bohn.** 2005. Mycotoxins produced by *Fusarium* spp. in isogenic Bt vs. non-Bt maize hybrids under European corn borer pressure. *Agronomy Journal* **97:** 219-224. (*graminearum, verticillioides*)

1649. **Parguey-Leduc, A.** 1964. Developpement d'une Nectriale: *Gibberella pulicaris*. *Compte Rendus Herbd. Seances Acad.* **258:** 2141-2144. (10)

1650. **Park, J. J., and F. S. Chu.** 1996. Partial purification and characterization of an esterase from *Fusarium sporotrichioides*. *Natural Toxins* **4:** 108-116. (*solani*)

1651. **Park, J. J., and F. S. Chu.** 1996. Immunochemical studies of an esterase from *Fusarium sporotrichioides*. *Food and Agricultural Immunology* **8:** 41-49. (*solani*)

1652. **Park, S.-Y., H. Shimizu, S.-I. Adachi, A. Nakagawa, I. Tanaka, K. Nakahara, H. Shoun, E. Obayashi, H. Nakamura, T. Iizuka, and Y. Shiro.** 1997. Crystal structure of nitric oxide reductase from denitrifying fungus *Fusarium oxysporum*. *Nature Structural Biology* **4:** 827-832. (*oxysporum*)

1653. **Pascoe, I. G.** 1990. *Fusarium* morphology I: Identification and characterization of a third conidial type, the mesoconidium. *Mycotaxon* **37:** 121-160. (12)

1654. **Pasquali, M., A. Acquadro, V. Balmas, Q. Migheli, M. L. Gullino, and A. Garibaldi.** 2004. Development of PCR primers for a new *Fusarium oxysporum* pathogenic on Paris daisy (*Argyranthemum frutescens* L.). *European Journal of Plant Pathology* **110:** 7-11. (*oxysporum*)

1655. **Pasquali, M., F. Dematheis, G. Gilardi, M. L. Gullino, and A. Garibaldi.** 2005. Vegetative compatibility groups of *Fusarium oxysporum* f. sp. *lactucae* from lettuce. *Plant Disease* **89:** 237-240. (*oxysporum*)

1656. **Pasquali, M., L. Marena, E. Fiora, P. Piatti, M. L. Gullino, and A. Garibaldi.** 2004. Real-time polymerase chain reaction for identification of a highly pathogenic group of *Fusarium oxysporum* f. sp. *chrysanthemi* on *Argyranthemum frutescens* L. *Journal of Plant Pathology* **86:** 53-59. (*oxysporum*)

1657. **Pathre, S. V., W. B. Gleason, Y. W. Lee, and C. J. Mirocha.** 1986. The structure of fusarochromanone, a new mycotoxin from *Fusarium roseum graminearum*. *Canadian Journal of Chemistry* **64:** 1308-1311. (*graminearum*)

1658. **Patil, P. J., and D. N. Padule.** 2000. Effects of grain mold fungi on seed germination and seedling vigor index of sorghum seeds var. CSH-9 in western Maharashtra. *Seed Research* **28:** 190-192. (*culmorum*)

1659. **Patino, B., S. Mirete, M. T. Gonzalez-Jaen, G. Mulè, M. T. Rodriguez, and C. Vazquez.** 2004. PCR detection assay of fumonisin-producing *Fusarium verticillioides* strains. *Journal of Food Protection* **67:** 1278-1283. (*verticillioides*)

1660. **Patino, B., M. L. Posada, M. T. Gonzalez-Jaen, A. Martinez del-Pozo, and C. Vazquez.** 1997. Control of polygalacturonase synthesis in *Fusarium oxysporum* f. sp. *radicis-lycopersici*. *Canadian Journal of Microbiology* **43:** 1084-1090. (*oxysporum*)

1661. **Paul, P. A., S. M. El-Allaf, P. E. Lipps, and L. V. Madden.** 2004. Rain splash dispersal of *Gibberella zeae* within wheat canopies in Ohio. *Phytopathology* **94:** 1342-1349. (*graminearum*)

1662. **Paulitz, T. C.** 1996. Diurnal release of ascospores by *Gibberella zeae* in inoculated wheat plots. *Plant Disease* **80:** 674-678. (3, 10, *graminearum*)

1663. **Paulitz, T. C., P. Dutilleul, S. H. Yamasaki, W. G. D. Fernando, and W. L. Seaman.** 1999. A generalized two-dimensional Gaussian model of disease foci of head blight of wheat caused by *Gibberella zeae*. *Phytopathology* **89:** 74-83. (11)

1664. **Paz, E., J. C. Cochrane, and V. W. Cochrane.** 1984. Spore germination and carbon metabolism in *Fusarium solani*. 6. Ethanol metabolism and the biosynthesis of amino acids. *Experimental Mycology* **8:** 1-12. (*solani*)

1665. **Pearson, C. A. S., J. F. Leslie, and F. W. Schwenk.** 1986. Variable chlorate resistance in *Macrophomina phaseolina* from corn, soybean and soil. *Phytopathology* **76:** 646-649. (4)

1666. **Pecchia, S.** 1999. Lisianthus crown and stem rot caused by *Fusarium avenaceum* in central Italy. *Plant Disease* **83:** 304. (*avenaceum*)

1667. **Pedersen, P. B., and A. Broadmeadow.** 2000. Toxicological studies on *Thermomyces lanuginosus* xylanase expressed by *Fusarium venenatum*, intended for use in food. *Food Additives and Contaminants* **17:** 739-747. (*venenatum*)

1668. **Pedersen, P. B., and J. D. Miller.** 1999. The fungal metabolite culmorin and related compounds. *Natural Toxins* **7:** 305-309. (*graminearum*)

1669. **Pegg, K. G., S. L. Willingham, R. G. O'Brien, A. W. Cooke, and L. M. Coates.** 2002. Base rot of golden passionfruit caused by a homothallic strain of *Fusarium solani*. *Australasian Plant Pathology* **31:** 305-306. (*solani*)

1670. **Peplow, A. W., I. B. Meek, M. C. Wiles, T. D. Phillips, and M. N. Beremand.** 2003. *Tri16* is required for esterification of position C-8 during trichothecene mycotoxin production by *Fusarium sporotrichioides*. *Applied and Environmental Microbiology* **69:** 5935-5940. (*sporotrichioides*)

1671. **Peplow, A. W., A. G. Tag, G. F. Garifullina, and M. N. Beremand.** 2003. Identification of new genes positively regulated by *Tri10* and a regulatory network for trichothecene mycotoxin production. *Applied and Environmental Microbiology* **69:** 2731-2736. (*sporotrichioides*)

1672. **Pereira, P. T., J. D. Arrabaca, and M. T. Amaral-Collaco.** 1996. Isolation, selection and characterization of a cyanide-degrading fungus from an industrial effluent. *International Biodeterioration and Biodegradation* **37:** 45-52. (*oxysporum*)

1673. **Pereira, P. T., M. Monteiro de Carvalho, J. D. Arrabaca, M. T. Amaral-Collaco, and J. C. Roseiro.** 1997. Alternative respiratory system and formamide hydrolyase activity as the key components of the cyanide-resistance mechanism in *Fusarium oxysporum*. *Canadian Journal of Microbiology* **43:** 929-936. (*oxysporum*)

1674. **Pereira, P., A. S. Pires, and J. C. Roseiro.** 1999. The effect of culture aging, cyanide concentration and induction time on formamide hydrolyase activity of *Fusarium oxysporum* CCMI 876. *Enzyme and Microbial Technology* **25:** 736-744. (*oxysporum*)

1675. **Pereiro, M., Jr., J. Labandeira, and J. Toribio.** 1999. Plantar hyperkeratosis due to *Fusarium verticillioides* in a patient with malignancy. *Clinical and Experimental Dermatology* **24:** 175-178. (*verticillioides*)

1676. **Pereiro, M., Jr., E. Pereiro, J. Toribio, and M. Pereiro-Miguens.** 1997. Superficial white toenail onchomycosis due to *Fusarium oxysporum*: A case report and review of the literature. *Journal de Mycologie Medicale* **7:** 219-222. (*oxysporum*)

1677. **Pereyra, S. A., R. Dill-Macky, and A. L. Sims.** 2004. Survival and inoculum production of *Gibberella zeae* in wheat residue. *Plant Disease* **88:** 724-730. (*graminearum*)

1678. **Perez-Artes, E., M. I. G. Roncero, and R. M. Jimenez-Diaz.** 1995. Restriction fragment length polymorphism analysis of the mitochondrial DNA of *Fusarium oxysporum* f. sp. *ciceris*. *Journal of Phytopathology* **143:** 105-109. (*oxysporum*)

1679. **Perez-Artes, E., and M. Tena.** 1989. Pectic enzymes from two races of *Fusarium oxysporum* f. sp. *ciceri*: enzyme production in culture and enzymatic activity on isolated chickpea cell walls. *Journal of Phytopathology* **124:** 39-51. (*oxysporum*)

1680. **Perez-Artes, E., and M. Tena.** 1990. Purification and characterization of pectic enzymes from two races of *Fusarium oxysporum* f. sp. *ciceri* differing in virulence to chickpea *Cicer arietinum* L. *Physiological and Molecular Plant Pathology* **37:** 107-124. (*oxysporum*)

1681. **Perez-Espinosa, A., T. Roldan-Arjona, and M. Ruiz-Rubio.** 2001. Pantothenate synthetase from *Fusarium oxysporum* f. sp. *lycopersici* is induced by α-tomatine. *Molecular Genetics and Genomics* **265:** 922-929. (*oxysporum*)

1682. **Perkins, D. D.** 1962. Preservation of *Neurospora* stock cultures with anhydrous silica gel. *Canadian Journal of Microbiology* **8:** 591-594. (4)

1683. **Perkins, D. D.** 1977. Behavior of *Neurospora sitophila* mating-type alleles in heterozygous duplications after introgression into *Neurospora crassa*. *Experimental Mycology* **1:** 166-172. (5)

1684. **Perkins, D. D.** 1987. Mating type switching in filamentous ascomycetes. *Genetics* **115:** 215-216. (6)

1685. **Perkins, D. D.** 1994. How should the interfertility of interspecies crosses be designated? *Mycologia* **86:** 758-761. (9)

1686. **Perkins, D. D., A. Radford, and M. S. Sachs.** 2001. *The* Neurospora *Compendium: Chromosomal Loci*. Academic Press, San Diego, California. (5, 7)

1687. **Perkowski, J., E. Foremska, and D. Latus-Zietkiewicz.** 1989. The yields of diacetoxyscirpenol produced by *Fusarium sambucinum* cultures isolated from potato tubers and their toxicity to brine shrimps (*Artemia salina*). *Mycotoxin Research* **5:** 61-67. (*sambucinum*)

1688. **Perkowski, J., I. Kiecana, J. Stachowiak, and T. Basinski.** 2003. Natural occurrence of scirpentriol in cereals infected by *Fusarium* species. *Food Additives and Contaminants* **20:** 572-578. (*sporotrichioides*)

1689. **Petersen, S. B., P. Fojan, E. I. Petersen, and M. T. N. Petersen.** 2001. The thermal stability of the *Fusarium solani pisi* cutinase as a function of pH. *Journal of Biomedicine and Biotechnology* **1:** 62-69. (*solani*)

1690. **Peterson, S. W., and A. Logrieco.** 1991. Ribosomal RNA sequence variation among interfertile strains of some *Gibberella* spp. *Mycologia* **83:** 397-402. (*fujikuroi, proliferatum, sambucinum, verticillioides*)

1691. **Pethybridge, S. J., F. S. Hay, C. R. Wilson, L. J. Sheriff, and G. W. Leggett.** 2001. First report of *Fusarium crookwellense* causing tip blight on cones of hop. *Plant Disease* **85:** 1208. (*crookwellense*)

1692. **Petmy, J. L., M. F. Temfack, L. Kaptue, J. Mbuagbauw, C. P. Raccurt, and C. Hennequin.** 2002. Invasive *Fusarium* infection in AIDS patients: Report of a case and review of the literature. *Journal de Mycologie Medicale* **12:** 146-148. (*verticillioides*)

1693. **Pettersson, H.** 1991. Nivalenol production by *Fusarium poae*. *Mycotoxin Research* **7A:** 26-30. (*poae*)

1694. **Phillips, N. J., J. T. Goodwin, A. Fraiman, R. J. Cole, and D. G. Lynn.** 1989. Characterization of the *Fusarium* toxin equisetin: The use of phenylboronates in structure assignment. *Journal of the American Chemical Society* **111:** 8223-8231. (*equiseti*)

1695. **Phillott, A. D., and C. J. Parmenter.** 2001. The distribution of failed eggs and the appearance of fungi in artificial nests of green (*Chelonia mydas*) and loggerhead (*Caretta caretta*) sea turtles. *Australian Journal of Zoology* **49:** 713-718. (*solani*)

1696. **Phillott, A. D., C. J. Parmenter, and C. J. Limpus.** 2004. Occurrence of mycobiota in eastern Australian sea turtle nests. *Memoirs of the Queensland Museum* **49:** 701-703. (*oxysporum, solani*)

1697. **Pieper, R., A. Haese, W. Schroeder, and R. Zocher.** 1995. Arrangement of catalytic sites in the multifunctional enzyme enniatin synthetase. *European Journal of Biochemistry* **230:** 119-126. (*sambucinum, scirpi*)

1698. **Pieper, R., H. Kleinkauf, and R. Zocher.** 1992. Enniatin synthetatses from different Fusaria exhibiting distinct amino acid specificities. *Journal of Antibiotics* **45:** 1273-1277. (*lateritium*)

1699. **Pinto-Sousa, A. M., J. M. S. Cabral, and M. R. Aires-Barros.** 1996. Stability of a *Fusarium solani pisi* recombinant cutinase in phosphatidylcholine reversed micelles. *Biotechnology Letters* **18:** 583-586. (*solani*)

1700. **Pirotta, R., and G. Riboni.** 1879. Studii sul latte. *Arch. Lab. Bot. Crittogam. Pavia* **2:** 316-317. (*lactis*)

1701. **Pirttila, A. M., L. M. McIntyre, G. A. Payne, and C. P. Woloshuk.** 2004. Expression profile analysis of wild-type and *fcc1* mutant strains of *Fusarium verticillioides* during fumonisin biosynthesis. *Fungal Genetics and Biology* **41:** 647-656. (*verticillioides*)

1702. **Plattner, R. D., and B. E. Branham.** 1994. Labeled fumonisins: Production and use of fumonisin B_1 containing stable isotopes. *Journal of AOAC International* **77:** 525-532. (*verticillioides*)

1703. **Plattner, R. D., A. E. Desjardins, J. F. Leslie, and P. E. Nelson.** 1996. Identification and characterization of strains of *Gibberella fujikuroi* mating population A (*Fusarium moniliforme*) with rare fumonisin phenotypes. *Mycologia* **87:** 416-424. (*verticillioides*)

1704. **Plattner, R. D., and P. E. Nelson.** 1994. Production of beauvericin by a strain of *Fusarium proliferatum* isolated from corn fodder for swine. *Applied and Environmental Microbiology* **60:** 3894-3896. (*proliferatum*)

1705. **Ploetz, R. C.** 1990. Variability in *Fusarium oxysporum* f. sp. *cubense*. *Canadian Journal of Botany* **68:** 1357-1363. (*oxysporum*)

1706. **Ploetz, R. C., ed.** 1990. *Fusarium Wilt of Banana*. APS Press, St. Paul, Minnesota. (1)

1707. **Ploetz, R. C.** 1991. Sudden wilt of passionfruit in southern Florida caused by *Nectria haematococca*. *Plant Disease* **75:** 1071-1073. (*solani*)

1708. **Ploetz, R. C.** 1994. Distribution and prevalence of *Fusarium subglutinans* in mango trees affected by malformation. *Canadian Journal of Botany* **72:** 7-9. (*mangiferae*)

1709. **Ploetz, R. C.** 2001. Malformation: A unique and important disease of mango, *Mangifera indica* L., p. 233-247. *In* B. A. Summerell, J. F. Leslie, D. Backhouse, W. L. Bryden, and L. W. Burgess (eds.), Fusarium*: Paul E. Nelson Memorial Symposium.* APS Press, St. Paul, Minnesota. (*mangiferae*)

1710. **Ploetz, R., A. Vazquez, and D. Benscher.** 1996. First report of *Fusarium decemcellulare* as a pathogen of mango in the United States. *Plant Disease* **80:** 1207. (*decemcellulare*)

1711. **Ploetz, R., Q. Zheng, A. Vazquez, and M. A. A. Sattar.** 2002. Current status and impact of mango malformation in Egypt. *International Journal of Pest Management* **48:** 279-285. (*mangiferae*)

1712. **Plyler, T. R., G. W. Simone, D. Fernandez, and H. C. Kistler.** 1999. Rapid detection of the *Fusarium oxysporum* lineage containing the Canary Island date palm wilt pathogen. *Phytopathology* **89:** 407-413. (*oxysporum*)

1713. **Plyler, T. R., G. W. Simone, D. Fernandez, and H. C. Kistler.** 2000. Genetic diversity among isolates of *Fusarium oxysporum* f. sp. *canariensis*. *Plant Pathology* **49:** 155-164. (5, *oxysporum*)

1714. **Pocalyko, D. J., and M. Tallman.** 1998. Effects of amphipaths on the activity and stability of *Fusarium solani pisi* cutinase. *Enzyme and Microbial Technology* **22:** 647-651. (*solani*)

1715. **Podila, G. K., M. B. Dickman, and P. E. Kolattukudy.** 1988. Transcriptional activation of a cutinase gene in isolated fungal nuclei by plant cutin monomers. *Science* **242:** 922-925. (*solani*)

1716. **Podila, G. K., E. Rosen, M. J. D. San Francisco, and P. E. Kolattukudy.** 1995. Targeted secretion of cutinase in *Fusarium solani* f. sp. *pisi* and *Colletotrichum gloeosporioides*. *Phytopathology* **85:** 238-242. (*solani*)

1717. **Polizzi, A., C. Siniscalchi, A. Mastromarino, and S. C. Sacca.** 2004. Effect of voriconazole on a corneal abscess caused by *Fusarium*. *Acta Ophthalmologica Scandinavica* **82:** 762-764. (*solani*)

1718. **Pomazi, A., A. Wittner, M. Pesti, and L. Hornok.** 1994. A PCR-generated simple RFLP-probe differentiates three distinct groups with *Fusarium oxysporum* f. sp. *pisi*. *Acta Phytopathologica et Entomologica Hungarica* **29:** 203-213. (*oxysporum*)

1719. **Pontecorvo, G. C.** 1953. The genetics of *Aspergillus nidulans*. *Advances in Genetics* **5:** 141-238. (6)

1720. **Porcelli, F., and S. Frisullo.** 1998. The fungus *Nectria aurantiicola* Berk et Ber. (*Fusarium larvarum* Fückel) a biocontrol agent of the armored scale *Suturaspis archangelskyae* (Lindinger) in Apulia (southern Italy). *Entomologica* **32:** 109-119. (10)

1721. **Porter, J. K., C. W. Bacon, W. P. Norred, E. M. Wray, G. A. Kuldau, A. E. Glenn, and J. F. Leslie.** 2002. Mycotoxins from fungal-infected sorghum: *Claviceps* vs. *Fusarium* and the *Striga* connection, p. 229-235. *In* J. F. Leslie (ed.), *Sorghum and Millets Pathology*. Iowa State Press, Ames, Iowa. (*thapsinum*)

1722. **Porter, J. K., C. W. Bacon, E. M. Wray, and W. M. Hagler, Jr.** 1995. Fusaric acid in *Fusarium moniliforme* cultures, corn, and feeds toxic to livestock and the neurochemical effects in the brain and pineal gland of rats. *Natural Toxins* **3:** 91-100. (*verticillioides*)

1723. **Porter, J. K., C. W. Bacon, E. M. Wray, G. A. Kuldau, and J. F. Leslie.** 2000. Fusaric acid (FA), other alkylpyridinecarboxylic acids (APAs), and fumonisin (FB_1) by *Fusarium thapsinum* and *F. moniliforme*: GC/MS analysis, p. 41-47. *In* W. J. de Koe, R. A. Samson, H. P. van Egmond, J. Gilbert, and M. Sabino (eds.), *Mycotoxins and Phycotoxins in Perspective at the Turn of the Millennium*. W. J. de Koe, Wageningen, The Netherlands. (*thapsinum*)

1724. **Posada, M. L., B. Patino, S. Mirete, M. C. Munoz, C. Vazquez, and M. T. Gonzalez-Jaen.** 2001. Comparative analysis of polygalacturonases in isolates of seven species of *Fusarium* from *Pinus pinea*. *Mycological Research* **105:** 100-104. (*oxysporum*)

1725. **Postma, J., and A. J. G. Luttikholt.** 1996. Colonization of carnation stems by a nonpathogenic isolate of *Fusarium oxysporum* and its effect on *Fusarium oxysporum* f. sp. *dianthi*. *Canadian Journal of Botany* **74:** 1841-1851. (*oxysporum*)

1726. **Postma, J., and H. Rattink.** 1992. Biological control of Fusarium wilt of carnation with a nonpathogenic isolate of *Fusarium oxysporum*. *Canadian Journal of Botany* **70:** 1199-1205. (*oxysporum*)

1727. **Pouleur, S., C. Richard, J. G. Martin, and H. Antoun.** 1992. Ice nucleation activity in *Fusarium acuminatum* and *Fusarium avenaceum*. *Applied and Environmental Microbiology* **58:** 2960-2964. (*acuminatum, avenaceum*)

1728. **Powell, W. A., and H. C. Kistler.** 1990. *In vivo* rearrangement of foreign DNA by *Fusarium oxysporum* produces linear self-replicating plasmids. *Journal of Bacteriology* **172:** 3163-3171. (*oxysporum, solani*)

1729. **Power, R., H., and K. Verhoeff.** 1984. Dieback of passion fruit (*Passiflora edulis* f. *flavicarpa*) in Surinam. *Phytopathologische Zeitschrift* **110:** 336-345. (*solani*)

1730. **Prado, M. M., A. Prado-Cabrero, R. Fernandez-Martin, and J. Avalos.** 2004. A gene of the opsin family in the carotenoid gene cluster of *Fusarium fujikuroi*. *Current Genetics* **46:** 47-58. (*fujikuroi*)

1731. **Prakash, O. M.** 2000. Integrated management of mango seedling (planting material) diseases: Present status. *Biological Memoirs* **26:** 37-43. (*mangiferae*)

1732. **Prasad, M., and S. K. Chaudhary.** 1974. *In vitro* production of fusaric acid and its impact on growth and sporulation in *Fusarium oxysporum* f. *udum*. *Phytopathologische Zeitschrift* **80:** 279-282. (*udum*)

1733. **Prasad, R. D., R. Rangeshwaran, S. V. Hegde, and C. P. Anuroop.** 2002. Effect of soil and seed application of *Trichoderma harzianum* on pigeonpea wilt caused by *Fusarium udum* under field conditions. *Crop Protection* **21:** 293-297. (*udum*)

1734. **Preece, T. F., T. R. Pettitt, and D. T. Biggs.** 1994. *Fusarium heterosporum* growing on ergots (*Claviceps purpurea*) in spikelets of common cord-grass (*Spartina anglica*) in the Isle of Wight. *Mycologist* **8:** 9-11. (*heterosporum*)

1735. **Prelusky, D. B., P. M. Scott, H. L. Trenholm, and G. A. Lawrence.** 1990. Minimal transmission of zearalenone to milk of dairy cows. *Journal of Environmental Science and Health Part B. Pesticides, Food Contaminants and Agricultural Wastes* **25:** 87-104. (*graminearum*)

1736. **Priest, M. J., and D. B. Letham.** 1996. Vascular wilt of *Phoenix canariensis* in New South Wales caused by *Fusarium oxysporum*. *Australasian Plant Pathology* **25:** 110-113. (11)

1737. **Proctor, R. H. D. W. Brown, R. D. Plattner, and A. E. Desjardins.** 2003. Co-expression of 15 contiguous genes delineates a fumonisin biosynthetic gene cluster in *Gibberella moniliformis*. *Fungal Genetics and Biology* **38:** 237-249. (*verticillioides*)

1738. **Proctor, R. H., T. M. Hohn, and S. P. McCormick.** 1997. Restoration of wild-type virulence to *Tri5* disruption mutants of *Gibberella zeae* via gene reversion and mutant complementation. *Microbiology* **143:** 2583-2591. (*graminearum*)

1739. **Proctor, R. H., T. M. Hohn, S. P. McCormick, and A. E. Desjardins.** 1995. *Tri6* encodes an unusual zinc finger protein involved in regulation of trichothecene biosynthesis in *Fusarium sporotrichioides*. *Applied and Environmental Microbiology* **61:** 1923-1930. (*sporotrichioides*)

1740. **Proctor, R. H., R. D. Plattner, D. W. Brown, J.-A. Seo, and Y.-W. Lee.** 2004. Discontinuous distribution of fumonisin biosynthetic genes in the *Gibberella fujikuroi* species complex. *Mycological Research* **108:** 815-822. (*anthophilum, fujikuroi, globosum, nygamai, oxysporum, proliferatum, verticillioides*)

1741. **Prom, L. K.** 2004. The effects of *Fusarium thapsinum*, *Curvularia lunata*, and their combination on sorghum germination and seed mycoflora. *Journal of New Seeds* **6:** 39-49. (*thapsinum*)

1742. **Prom, L. K., R. D. Waniska, A. I. Kollo, and W. L. Rooney.** 2003. Response of eight sorghum cultivars inoculated with *Fusarium thapsinum*, *Curvularia lunata*, and a mixture of the two fungi. *Crop Protection* **22:** 623-628. (*thapsinum*)

1743. **Prompers, J. J., A. Groenewegen, R. C. van Schaik, H. A. M. Pepermans, and C. W. Hilbers.** 1997. ^{1}H, ^{13}C, and ^{15}N resonance assignments of *Fusarium solani pisi* cutinase and preliminary features of the structure in solution. *Protein Science* **6:** 2375-2384. (*solani*)

1744. **Puhalla, J. E.** 1984. A visual indicator of heterokaryosis in *Fusarium oxysporum* f. sp. *apii* from celery. *Canadian Journal of Botany* **62:** 540-545. (5)

1745. **Puhalla, J. E.** 1985. Classification of *Fusarium oxysporum* on the basis of vegetative compatibility. *Canadian Journal of Botany* **63:** 179-183. (2, 5, 11, *oxysporum*)

1746. **Puhalla, J. E., and P. T. Spieth.** 1983. Heterokaryosis in *Fusarium moniliforme*. *Experimental Mycology* **7:** 328-335. (2, *verticillioides*)

1747. **Puhalla, J. E., and P. T. Spieth.** 1985. A comparison of heterokaryosis and vegetative incompatibility among varieties of *Gibberella fujikuroi* (*Fusarium moniliforme*). *Experimental Mycology* **9:** 39-47. (*verticillioides*)

1748. **Pujol, I., J. Guarro, J. Gene, and J. Sala.** 1997. *In vitro* antifungal susceptibility of clinical and environmental *Fusarium* spp. strains. *Journal of Antimicrobial Chemotherapy* **39:** 163-167. (*avenaceum, chlamydosporum, dimerum, equiseti, graminearum, oxysporum, proliferatum, semitectum, solani, subglutinans, verticillioides*)

1749. **Purss, G. S.** 1969. The relationship between strains of *Fusarium graminearum* causing crown rot of various gramineous hosts and stalk rot of maize in Queensland. *Australian Journal of Agricultural Research* **20:** 257-264. (*pseudograminearum*)

1750. **Purss, G. S.** 1971. Pathogenic specialization in *Fusarium graminearum*. *Australian Journal of Agricultural Research* **22:** 553-561. (*pseudograminearum*)

1751. **Pöggeler, S., and U. Kück.** 2000. Comparative analysis of mating-type loci from *Neurospora crassa* and *Sordaria macrospora*: Identification of novel transcribed ORFs. *Molecular and General Genetics* **263:** 292-301. (6)

1752. **Qian, X.-M., J. C. du Preez, and S. G. Kilian.** 1994. Factors affecting gibberellic acid production by *Fusarium moniliforme* in solid-state cultivation on starch. *World Journal of Microbiology and Biotechnology* **10:** 93-99. (*fujikuroi*)

1753. **Quarta, A. G. Mita, M. Haidukowski, A. Santino, G. Mulè, and A. Visconti.** 2005. Assessment of trichothecenes chemotypes of *Fusarium culmorum* occurring in Europe. *Food Additives and Contaminants* **22:** 309-315. (*culmorum*)

1754. **Quimby, P. C., Jr., N. K. Zidack, C. D. Boyette, and W. E. Grey.** 1999. A simple method for stabilizing and granulating fungi. *Biocontrol Science and Technology* **9:** 5-8. (*oxysporum*)

1755. **Raad, I., and R. Hachem.** 1995. Treatment of central venous catheter-related fungemia due to *Fusarium oxysporum*. *Clinical Infectious Diseases* **20:** 709-711. (*oxysporum*)

1756. **Rabie, C. J. and A. Lübben.** 1984. The mycoflora of sorghum: *Sorghum caffrorum* malt. *South African Journal of Botany* **3:** 251-255. (*subglutinans*)

1757. **Rabie, C. J., A. Lübben, A. I. Louw, E. B. Rathbone, P. S. Steyn, and R. Vleggaar.** 1978. Moniliformin, a mycotoxin from *Fusarium fusarioides*. *Journal of Agricultural and Food Chemistry* **26:** 375-379. (*chlamydosporum*)

1758. **Rabie, C. J., W. F. O. Marasas, P. G. Thiel, A. Lübben, and R. Vleggaar.** 1982. Moniliformin production and toxicity of different *Fusarium* species from southern Africa. *Applied and Environmental Microbiology* **43:** 517-521. (*avenaceum, chlamydosporum, semitectum*)

1759. **Rabie, C. J., E. W. Sydenham, P. G. Thiel, A. Lübben, and W. F. O. Marasas.** 1986. T-2 toxin production by *Fusarium acuminatum* isolated from oats and barley. *Applied and Environmental Microbiology* **52:** 594-596. (*acuminatum, armeniacum*)

1760. **Rabodonirina, M., M. A. Piens, M. F. Monier, E. Gueho, D. Fiere, and M. Mojon.** 1994. *Fusarium* infections in immunocompromised patients: Case reports and literature review. *European Journal of Clinical Microbiology and Infectious Diseases* **13:** 152-161. (1, *chlamydosporum, solani*)

1761. **Rai, B., and R. S. Upadhyay.** 1982. *Gibberella indica*: The perfect state of *Fusarium udum*. *Mycologia* **74:** 343-346. (*udum*)

1762. **Rai, B., and R. S. Upadhyay.** 1983. Competitive saprophytic colonization of pigeon pea substrate by *Fusarium udum* in relation to environmental factors, chemical treatments and microbial antagonism. *Soil Biology and Biochemistry* **15:** 187-192. (*udum*)

1763. **Raillo, A.** 1935. The diagnostic estimation of morphological and cultural characters of the species of the genus *Fusarium*. *Publ. All-Union Acad. Agric. Sciences, Series 2 (Leningrad)* **7:** 1-98. (8)

1764. **Raillo, A.** 1936. Systematics and methods for the determination of the species of the genus *Fusarium*. *Acta Bot. Inst. Acad. Sciences USSR, Series 2* **3:** 803-857. (8)

1765. **Raillo, A.** 1950. *Fungi of the genus* Fusarium. Pub. State Agr. Lit., Moscow, USSR. (8)

1766. **Rajendran, S., K. Parasakthy, P. D. Deepalakshmi, and S. N. Devaraj.** 1996. Oxidative stress in rat liver and lung induced by furanoterpenoids isolated from *Fusarium-solani*-infected sweet potatoes. *Indian Journal of Experimental Biology* **34:** 57-60. (*solani*)

1767. **Raju, N. B.** 1980. Meiosis and ascospore genesis in *Neurospora*. *European Journal of Cell Biology* **23:** 208-223. (10)

1768. **Raju, N. B.** 1994. Ascomycete spore killers: Chromosomal elements that distort genetic ratios among the products of meiosis. *Mycologia* **86:** 461-473. (10)

1769. **Raju, N. B., and J. F. Leslie.** 1992. Cytology of recessive sexual-phase mutants from wild strains of *Neurospora crassa*. *Genome* **35:** 816-825. (6)

1770. **Raju, N. B., and D. D. Perkins.** 1991. Expression of meiotic drive elements *spore killer-2* and *spore killer-3* in asci of *Neurospora tetrasperma*. *Genetics* **129:** 25-38. (10)

1771. **Raju, N. B., and D. D. Perkins.** 1994. Diverse programs of ascus development in pseudohomothallic species of *Neurospora*, *Gelasinospora*, and *Podospora*. *Developmental Genetics* **15:** 104-118. (6)

1772. **Ramirez, M. L., S. N. Chulze, and N. Magan.** 2004. Impact of osmotic and matric water stress on germination, growth, mycelial water potentials and endogenous accumulation of sugars and sugar alcohols in *Fusarium graminearum*. *Mycologia* **96:** 470-478. (*graminearum*)

1773. **Ramsdale, M.** 2001. Fungi with a sense of time: Molecular genetics of temporal organization in *Neurospora crassa*. *The Mycologist* **15:** 10-15. (4)

1774. **Randazzo, G., V. Fogliano, A. Ritieni, L. Mannina, E. Rossi, A. Scarallo, and A. L. Segre.** 1993. Proliferin, a new sesterterpene from *Fusarium proliferatum*. *Tetrahedron* **49:** 10883-10896. (*proliferatum*)

1775. **Rao, M. N., A. A. Kembhavi, and A. Pant.** 1996. Role of lysine, tryptophan and calcium in the β-elimination activity of a low-molecular-mass pectate lyase from *Fusarium moniliforme*. *Biochemical Journal* **319:** 159-164. (*verticillioides*)

1776. **Rao, N. N. R., and M. S. Pavgi.** 1976. A mycoparasite on *Sclerospora graminicola*. *Canadian Journal of Botany* **54:** 220-223. (*semitectum*)

1777. **Rao, V. G., and K. I. M. Varghese.** 1991. On two fungal rots of cacao. *World Journal of Microbiology and Biotechnology* **7:** 428-430. (*decemcellulare*)

1778. **Rao, V. P., and R. P. Thakur.** 1988. *Fusarium semitectum* var. *majus* - a potential biocontrol agent of ergot (*Claviceps fusiformis*) of pearl millet. *Indian Phytopathology* **41:** 567-574. (*semitectum*)

1779. **Rapp, P.** 1995. Production, regulation, and some properties of lipase activity from *Fusarium oxysporum* f. sp. *vasinfectum*. *Enzyme and Microbial Technology* **17:** 832-838. (*oxysporum*)

1780. **Rataj-Guranowska, M., A. Paszewski, and I. Walkowiak.** 1991. *Fusarium oxysporum* forms heterokaryons with *Fusarium redolens*. *Journal of Phytopathology* **132:** 294-302. (*redolens*)

1781. **Rataj-Guranowska, M., and B. Wolko.** 1991. Comparisons of *Fusarium oxysporum* and *Fusarium oxysporum* var. *redolens* by analyzing the isozyme and serological patterns. *Journal of Phytopathology* **132:** 287-293. (*redolens*)

1782. **Raybould, A. F., A. J. Gray, and R. T. Clarke.** 1998. The long-term epidemic of *Claviceps purpurea* on *Spartina anglica* in Poole Harbour: Pattern of infection, effects on seed production and the role of *Fusarium heterosporum*. *New Phytologist* **138:** 497-505. (*heterosporum*)

1783. **Rebell, G.** 1981. Fusarium infections in human and veterinary medicine, p. 210-220. *In* P. E. Nelson, T. A. Toussoun, and R. J. Cook (eds.), Fusarium*: Diseases, Biology and Taxonomy*. Pennsylvania State University Press, University Park, Pennsylvania. (1, 9, *solani*)

1784. **Recorbet, G., C. Steinberg, C. Olivain, V. Edel, S. Trouvelot, E. Dumas-Gaudot, S. Gianinazzi, and C. Alabouvette.** 2003. Wanted: Pathogenesis-related marker molecules for *Fusarium oxysporum*. *New Phytologist* **159:** 73-92. (*oxysporum*)

1785. **Reddy, A. S., and S. M. Reddy.** 1983. Elaboration of mycotoxins by fungi associated with til (*Sesamum indicum* L.). *Current Science* **52:** 613-614. (*scirpi*)

1786. **Reddy, M. V., and T. N. Raju.** 1997. Evaluation of pigeonpea (*Cajanus cajan*) varieties of resistance to wilt caused by *Fusarium udum* and sterility mosaic disease in a perennial system. *Indian Journal of Agricultural Sciences* **67:** 437-439. (*udum*)

1787. **Reed, J. E., P. T. Nordquist, and J. E. Partridge.** 1983. Fungal colonization of stalks and roots of grain sorghum (*Sorghum bicolor*) during the growing season. *Plant Disease* **67:** 417-420. (*thapsinum*)

1788. **Regalado, V., F. Perestelo, A. Rodriguez, A. Carnicero, F. J. Sosa, G. de la Fuente, and M. A. Falcon.** 1999. Activated oxygen species and two extracellular enzymes: Laccase and aryl-alcohol oxidase, novel for the lignin-degrading fungus *Fusarium proliferatum*. *Applied Microbiology and Biotechnology* **51:** 388-390. (*proliferatum*)

1789. **Regalado, V., A. Rodriguez, F. Perestelo, A. Carnicero, G. de la Fuente, and A. Falcon.** 1997. Lignin degradation and modification by the soil-inhabiting fungus *Fusarium proliferatum*. *Applied and Environmental Microbiology* **63:** 3716-3718. (*proliferatum*)

1790. **Reid, L. M., A. T. Bolton, R. I. Hamilton, T. Woldemariam, and D. E. Mather.** 1992. Effect of silk age on resistance of maize to *Fusarium graminearum*. *Canadian Journal of Plant Pathology* **14:** 293-298. (*graminearum*)

1791. **Reid, L. M., D. E. Mather, R. I. Hamilton, and A. T. Bolton.** 1992. Diallel analysis of resistance in maize to *Fusarium graminearum* infection via the silk. *Canadian Journal of Plant Science* **72:** 915-923. (*graminearum*)

1792. **Reis, E. M.** 1990. Survival of perithecia of *Gibberella zeae* on naturally infected wheat kernels under field conditions. *Fitopatologia Brasileira* **15:** 254-255. (*graminearum*)

1793. **Reis, E. M.** 1990. Effects of rain and relative humidity on the release of ascospores and on the infection of wheat heads by *Gibberella zeae*. *Fitopatologia Brasileira* **15:** 339-343. (*graminearum*)

1794. **Reischer, G. H., M. Lemmens, A. Farnleitner, A. Adler, and R. L. Mach.** 2004. Quantification of *Fusarium graminearum* in infected wheat by species specific real-time PCR applying a TaqMan probe. *Journal of Microbiological Methods* **59:** 141-146. (*graminearum*)

1795. **Remotti, P. C., and H. J. M. Loffler.** 1996. The involvement of fusaric acid in the bulb rot of gladiolus *Journal of Phytopathology* **144:** 405-411. (*oxysporum*)

1796. **Rensing, L., and P. Ruoff.** 2002. Temperature effect on entrainment, phase shifting, and amplitude of circadian clocks and its molecular bases. *Chronobiology International* **19:** 807-864. (4)

1797. **Rep, M., H. C. van der Does, and B. J. C. Cornelissen.** 2005. *Drifter*, a novel hAT-like transposon in *Fusarium oxysporum* is activated during starvation. *Fungal Genetics and Biology* **42:** 546-553. (*oxysporum*)

1798. **Rep, M., H. C. van der Does, M. Meijer, R. van Wijk, P. M. Houterman, H. L. Dekker, C. G. de Koster, and B. J. C. Cornelissen.** 2004. A small, cysteine-rich protein secreted by *Fusarium oxysporum* during colonization of xylem vessels is required for *I-3*-mediated resistance in tomato. *Molecular Microbiology* **53:** 1373-1383. (*oxysporum*)

1799. **Reuben, A., E. Anaissie, P. E. Nelson, R. Hashem, C. Legrand, D. H. Ho, and G. P. Bodey.** 1989. Antifungal susceptibility of 44 clinical isolates of *Fusarium* spp. determined by using a broth microdilution method. *Antimicrobial Agents and Chemotherapy* **33:** 1647-1649. (*oxysporum, semitectum, solani, verticillioides*)

1800. **Reynoso, M. M., A. M. Torres, and S. N. Chulze.** 2004. Fusaproliferin, beauvericin and fumonisin production by different mating populations among the *Gibberella fujikuroi* species complex isolated from maize. *Mycological Research* **108:** 154-160. (*proliferatum, subglutinans, verticillioides*)

1801. **Reynoso, M. M., A. M. Torres, M. L. Ramirez, M. I. Rodriguez, S. N. Chulze, and N. Magan.** 2002. Efficacy of antioxidant mixtures on growth, fumonisin production and hydrolytic enzyme production by *Fusarium verticillioides* and *F. proliferatum in vitro* on maize-based media. *Mycological Research* **106:** 1093-1099. (*proliferatum, verticillioides*)

1802. **Rheeder, J. P., W. F. O. Marasas, and P. E. Nelson.** 1996. *Fusarium globosum*, a new species from corn in southern Africa. *Mycologia* **88:** 509-513. (8, *globosum*)

1803. **Rheeder, J. P., W. F. O. Marasas, P. G. Thiel, E. W. Sydenham, G. S. Shephard, and D. J. van Schalkwyk.** 1992. *Fusarium moniliforme* and fumonisins in corn in relation to human esophageal cancer in Transkei. *Phytopathology* **82:** 353-357. (*verticillioides*)

1804. **Rheeder, J. P., W. F. O. Marasas, and P. S. van Wyk.** 1990. Fungal associations in corn kernels and effects on germination. *Phytopathology* **80:** 131-134. (*graminearum, verticillioides*)

1805. **Rheeder, J. P., P. S. van Wyk, and W. F. O. Marasas.** 1990. *Fusarium* spp. from Marion and Prince Edward Islands sub-Antarctic. *South African Journal of Botany* **56:** 482-486. (*heterosporum, merismoides*)

1806. **Richard, C., J. G. Martin, and S. Pouleur.** 1996. Ice nucleation activity identified in some phytopathogenic *Fusarium* species. *Phytoprotection* **77:** 83-92. (*oxysporum, tricinctum*)

1807. **Richardson, K. E., and P. B. Hamilton.** 1987. Preparation of scirpentriol and triacetoxyscirpenol in good yield from cultures of *Fusarium sambucinum* NRRL 13495. *Applied and Environmental Microbiology* **53:** 457-459. (*sambucinum*)

1808. **Richardson, K. E., and P. B. Hamilton.** 1987. Preparation of 4,15-diacetoxyscirpenol from cultures of *Fusarium sambucinum* NRRL 13495. *Applied and Environmental Microbiology* **53:** 460-462. (*sambucinum*)

1809. **Richardson, K. E., G. E. Toney, C. A. Haney, and P. B. Hamilton.** 1989. Occurrence of scirpentriol and its seven acetylated derivatives in culture extracts of *Fusarium sambucinum* NRRL 13495. *Journal of Food Protection* **52:** 871-876. (*sambucinum*)

1810. **Richardson, M. D., and C. W. Bacon.** 1995. Catabolism of 6-methoxy-benzoxazolinone and 2-benzoxazolinone by *Fusarium moniliforme*. *Mycologia* **87:** 510-517. (*verticillioides*)

1811. **Richey, M. G., E. T. Marek, C. L. Schardl, and D. A. Smith.** 1989. Transformation of filamentous fungi with plasmid DNA by electroporation. *Phytopathology* **79:** 844-847. (*solani*)

1812. **Rippon, J. W., R. A. Larson, D. M. Rosenthal, and J. Clayman.** 1988. Disseminated cutaneous and peritoneal hyalohyphomycosis caused by *Fusarium* sp. three cases and review of the literature. *Mycopathologia* **101:** 105-112. (*oxysporum, verticillioides*)

1813. **Ritieni, A., V. Fogliano, G. Randazzo, A. Scarallo, A. Logrieco, A. Moretti, L. Mannina, and A. Bottalico.** 1995. Isolation and characterization of fusaproliferin, a new toxic metabolite from *Fusarium proliferatum*. *Natural Toxins* **3:** 17-20. (*proliferatum*)

1814. **Robin, J. B., R. C. Arffa, I. Avni, and N. A. Rao.** 1986. Rapid visualization of three common fungi using fluorescein-conjugated lectins. *Investigative Ophthalmology and Visual Science* **27:** 500-506. (*solani*)

1815. **Robson, G. D., M. G. Wiebe, and A. P. J. Trinci.** 1991. Involvement of calcium in the regulation of hyphal extension and branching in *Fusarium graminearum* A 3-5. *Experimental Mycology* **15:** 263-272. (*venenatum*)

1816. **Robson, G. D., M. G. Wiebe, and A. P. J. Trinci.** 1991. Low calcium concentrations induce increased branching in *Fusarium graminearum*. *Mycological Research* **95:** 561-565. (*venenatum*)

1817. **Rodier, M. H., B. El-Moudni, C. Kauffman-Lacroix, and J. L. Jacquemin.** 1997. Purification of an intracellular metallopeptidase of M-r 45 000 in *Fusarium moniliforme*. *Mycological Research* **101:** 678-682. (*verticillioides*)

1818. **Rodriguez-Galvez, E., and K. Mendgen.** 1995. The infection process of *Fusarium oxysporum* in cotton root tips. *Protoplasma* **189:** 61-72. (*oxysporum*)

1819. **Rodriguez-Herrera, R., W. L. Rooney, D. T. Rosenow, and R. A. Frederiksen.** 2000. Inheritance of grain mold resistance in grain sorghum without a pigmented testa. *Crop Science* **40:** 1573-1578. (*thapsinum*)

1820. **Roebroeck, E. J. A., M. J. W. Jansen, and J. J. Mes.** 1991. A mathematical model describing the combined effect of exposure time and temperature of hot water treatments on survival of *Gladiolus* cormels. *Annals of Applied Biology* **119:** 89-96. (3)

1821. **Rogers, L. M., M. A. Flaishman, and P. E. Kolattukudy.** 1994. Cutinase gene disruption in *Fusarium solani* f. sp. *pisi* decreases its virulence on pea. *Plant Cell* **6:** 935-945. (*solani*)

1822. **Rohnert, U., I. Heiser, S. Nemec, R. Baker, W. Osswald, and E. F. Elstner.** 1998. Diaphorase-mediated oxygen activation and uncoupling of mitochondrial electron transport by naphthazarin toxins produced by *Fusarium solani*. *Journal of Plant Physiology* **153:** 684-692. (*solani*)

1823. **Rohrbach, K. G., and J. B. Pfeiffer.** 1976. Susceptibility of pineapple cultivars to fruit diseases incited by *Penicillium funiculosum* and *Fusarium moniliforme*. *Phytopathology* **66:** 1386-1390. (*guttiforme*)

1824. **Rohrbach, K. G., and G. Taniguchi.** 1984. Effects of temperature, moisture, and stage of inflorescence development on infection of pineapple *Ananas comosus* by *Penicillium funiculosum* and *Fusarium moniliforme* var. *subglutinans*. *Phytopathology* **74:** 995-1000. (*guttiforme*)

1825. **Rojas, M. C., O. Urrutia, C. Cruz, P. Gaskin, B. Tudzynski, and P. Hedden.** 2004. Kaurenolides and fujenoic acids are side products of the gibberellin P450-1 monooxygenase in *Gibberella fujikuroi*. *Phytochemistry* **65:** 821-830. (*fujikuroi*)

1826. **Roldan-Arjona, T., A. Perez-Espinosa, and M. Ruiz-Rubio.** 1999. Tomatinase from *Fusarium oxysporum* f. sp. *lycopersici* defines a new class of saponinases. *Molecular Plant-Microbe Interactions* **12:** 852-861. (*oxysporum*)

1827. **Romano, C., C. Miracco, and E. M. Difonzo.** 1998. Skin and nail infections due to *Fusarium oxysporum* in Tuscany, Italy. *Mycoses* **41:** 433-437. (*oxysporum*)

1828. **Romano, C., L. Presenti, and L. Massai.** 1999. Interdigital intertrigo of the feet due to therapy-resistant *Fusarium solani*. *Dermatology* **199:** 177-179. (*solani*)

1829. **Romero, M. C., M. L. Salvioli, M. C. Cazau, and A. M. Arambarri.** 2002. Pyrene degradation by yeasts and filamentous fungi. *Environmental Pollution* **117:** 159-163. (*solani*)

1830. **Rondon, G. A., and M. Materan.** 1990. Onoto and mammon – new hosts of *Fusarium decemcellulare* in Venezuela. *Fitopatologia Venezolana* **3:** 27-28. (*decemcellulare*)

1831. **Rosa, R. H., Jr., D. Miller, and E. C. Alfonso.** 1994. The changing spectrum of fungal keratitis in South Florida. *Ophthalmology* **101:** 1005-1013. (*oxysporum*)

1832. **Rose, F. L., J. Koke, R. Koehn, and D. Smith.** 2001. Identification of the etiological agent for necrotizing scute disease in the Texas tortoise. *Journal of Wildlife Diseases* **37:** 223-228. (*semitectum*)

1833. **Rosen, R. T., and J. D. Rosen.** 1982. Presence of four *Fusarium* mycotoxins and synthetic material in "Yellow Rain": Evidence for the use of chemical weapons in Laos. *Biomedical Mass Spectrometry* **9:** 443-450. (1)

1834. **Rosewich, U. L., R. E. Pettway, T. Katan, and H. C. Kistler.** 1999. Population genetic analysis corroborates dispersal of *Fusarium oxysporum* f. sp. *radicis-lycopersici* from Florida to Europe. *Phytopathology* **89:** 623-630. (5, 11, *oxysporum*)

1835. **Rossi, V., L. Languasco, E. Pattori, and S. Giosue.** 2002. Dynamics of airborne *Fusarium* macroconidia in wheat fields naturally infected by head blight. *Journal of Plant Pathology* **84:** 53-64. (3, 11)

1836. **Rosslein, L., C. Tamm, W. Zurcher, A. Riesen, and M. Zehnder.** 1988. Sambucinic acid, a new metabolite of *Fusarium sambucinum*. *Helvetica Chimica Acta* **71:** 588-595. (*sambucinum*)

1837. **Rossman, A. Y.** 1996. Morphological and molecular perspectives on systematics of the Hypocreales. *Mycologia* **88:** 1-19. (10, *solani*)

1838. **Rossman, A. Y., G. J. Samuels, C. T. Rogerson, and R. Lowen.** 1999. Genera of Bionectriaceae, Hypocreaceae and Nectriaceae (Hypocreales, Ascomycetes). *Studies in Mycology* **42:** 1-248. (10, *solani*)

1839. **Rothe, A., M. Seibold, T. Hoppe, H. Seifert, A. Engert, C. Caspar, M. Karthaus, G. Fatkenheuer, U. Bethe, K. Tintelnot, and O. A. Cornely.** 2004. Combination therapy of disseminated *Fusarium oxysporum* infection with terbinafine and amphotericin B. *Annals of Hematology* **83:** 394-397. (*oxysporum*)

1840. **Roux, J., E. T. Steenkamp, W. F. O. Marasas, M. J. Wingfield, and B. D. Wingfield.** 2001. Characterization of *Fusarium graminearum* from *Acacia* and *Eucalyptus* using β-tubulin and histone gene sequences. *Mycologia* **93:** 704-711. (*graminearum*)

1841. **Roy, K. W.** 1997. *Fusarium solani* on soybean roots: Nomenclature of the causal agent of sudden death syndrome and identity and relevance of *F. solani* form B. *Plant Disease* **81:** 259-266. (*solani*)

1842. **Royer, J. C., L. M. Christianson, W. T. Yoder, G. A. Gambetta, A. V. Klotz, C. L. Morris, H. Brody, and S. Otani.** 1999. Deletion of the trichodiene synthase gene of *Fusarium venenatum*: Two systems for repeated gene deletions. *Fungal Genetics and Biology* **28:** 68-78. (*venenatum*)

1843. **Royer, J. C., D. L. Moyer, S. G. Reiwitch, M. S. Madden, E. B. Jensen, S. H. Brown, C. C. Yonker, J. A. Johnstone, E. J. Golightly, W. T. Yoder, and J. R. Shuster.** 1995. *Fusarium graminearum* A 3/5 as a novel host for heterologous protein production. *Bio/Technology* **13:** 1479-1483. (*venenatum*)

1844. **Ruan, Y., V. Kotraiah, and D. C. Straney.** 1995. Flavonoids stimulate spore germination in *Fusarium solani* pathogenic on legumes in a manner sensitive to inhibitors of cAMP-dependent protein kinase. *Molecular Plant-Microbe Interactions* **8:** 929-938. (*solani*)

1845. **Ruan, Y., and D. C. Straney.** 1994. *In vitro* transcription from the *Nectria haematococca PDA1* promoter in an homologous extract reflects *in vivo* pisatin-responsive regulation. *Current Genetics* **27:** 46-53. (*solani*)

1846. **Ruan, Y., and D. C. Straney.** 1996. Identification of elements in the *PDA1* promoter of *Nectria haematococca* necessary for a high level of transcription *in vitro*. *Molecular and General Genetics* **250:** 29-38. (*solani*)

1847. **Rubidge, T.** 1986. Survey of *Fusarium* species for plasmid-like DNA and some evidence for its occurrence in a strain of *Fusarium merismoides*. *Transactions of the British Mycological Society* **87:** 463-466. (*merismoides*)

1848. **Ruiz-Roldan, M. C., A. di Pietro, M. D. Huertas-Gonzalez, and M. I. G. Roncero.** 1999. Two xylanase genes of the vascular wilt pathogen *Fusarium oxysporum* are differentially expressed during infection of tomato plants. *Molecular and General Genetics* **261:** 530-536. (*oxysporum*)

1849. **Ruiz-Roldan, M. C., A. di Pietro, and M. I. G. Roncero.** 1997. Purification and characterization of an acidic endo-β-1,4-xylanase from the tomato vascular pathogen *Fusarium oxysporum* f. sp. *lycopersici*. *FEMS Microbiology Letters* **148:** 75-82. (*oxysporum*)

1850. **Rupe, J. C.** 1989. Frequency and pathogenicity of *Fusarium solani* recovered from soybeans with sudden death syndrome. *Plant Disease* **73:** 581-584. (*solani*)

1851. **Rupe, J. C., J. C. Correll, J. C. Guerber, C. M. Becton, E. F. Gbur, Jr., M. S. Cummings, and P. A. Yount.** 2001. Differentiation of the sudden death syndrome pathogen of soybean, *Fusarium solani* f. sp. *glycines*, from other isolates of *F. solani* based on cultural morphology, pathogenicity, and mitochondrial DNA restriction fragment length polymorphisms. *Canadian Journal of Botany* **79:** 829-835. (*solani*)

1852. **Ryan, F. J., G. W. Beadle, and E. L. Tatum.** 1943. The tube method of measuring the growth rate of *Neurospora*. *American Journal of Botany* **30:** 784-789. (4, 11)

1853. **Rynkiewicz, M. J., D. E. Cane, and D. W. Christianson.** 2001. Structure of trichodiene synthase from *Fusarium sporotrichioides* provides mechanistic inferences on the terpene cyclization cascade. *Proceedings of the National Academy of Sciences USA* **98:** 13543-13548. (*sporotrichioides*)

1854. **Rypniewski, W. J., C. Dambmann, C. von der Osten, M. Dauter, and K. S. Wilson.** 1995. Structure of inhibited trypsin from *Fusarium oxysporum* at 1.55 Å. *Acta Crystallographica Section D Biological Crystallography* **51:** 73-84. (*oxysporum*)

1855. **Rypniewski, W. R., S. Hastrup, C. Betzel, M. Dauter, Z. Dauter, G. Papendorf, S. Branner, and K. S. Wilson.** 1993. The sequence and X-ray structure of the trypsin from *Fusarium oxysporum*. *Protein Engineering* **6:** 341-348. (*oxysporum*)

1856. **Ryu, D., C. Munimbazi, and L. B. Bullerman.** 1999. Fumonisin B_1 production by *Fusarium moniliforme* and *Fusarium proliferatum* as affected by cycling temperatures. *Journal of Food Protection* **62:** 1456-1460. (*verticillioides*)

1857. **Sabo, J., T. Duric, and S. Jasnic.** 2002. *Fusarium* fungi as a pathogen causing hop wilt. *Plant Protection Science* **38 (Special Issue 2):** 308-310. (*culmorum*)

1858. **Saccardo, P. A.** 1882-1925. *Sylloge Fungorum omnium hucusque cognitorum*. Padua, Italy. (*succisae*)

1859. **Sage, R. F.** 2002. How terrestrial organisms sense, signal, and respond to carbon dioxide. *Integrative and Comparative Biology* **42:** 469-480. (6)

1860. **Saha, B. C.** 2001. Purification and characterization of an extracellular β-xylosidase from a newly isolated *Fusarium verticillioides*. *Journal of Industrial Microbiology and Biotechnology* **27:** 241-245. (*verticillioides*)

1861. **Saha, B. C.** 2001. Xylanase from a newly isolated *Fusarium verticillioides* capable of utilizing corn fiber xylan. *Applied Microbiology and Biotechnology* **56:** 762-766. (*verticillioides*)

1862. **Saha, B. C.** 2002. Production, purification and properties of xylanase from a newly isolated *Fusarium proliferatum*. *Process Biochemistry* **37:** 1279-1284. (*proliferatum*)

1863. **Saha, B. C.** 2003. Purification and properties of an extracellular β-xylosidase from a newly isolated *Fusarium proliferatum*. *Bioresource Technology* **90:** 33-38. (*proliferatum*)

1864. **Saito, H., and T. Matuo.** 1982. Heterothallism in *Fusarium lateritium* f. sp. *mori*. *Transactions of the Mycological Society of Japan* **23:** 73-78. (*lateritium*)

1865. **Sajise, C. E.** 1988. Influence of cultivar inoculum density and plant age on the incidence of Fusarium root and stem rot in cowpea. *Annals of Tropical Research* **10:** 9-15. (*solani*)

1866. **Sakai, Y., N. Yoshida, A. Isogai, Y. Tani, and N. Kato.** 1995. Purification and properties of fructosyl lysine oxidase from *Fusarium oxysporum* S-1F4. *Bioscience Biotechnology and Biochemistry* **59:** 487-491. (*oxysporum*)

1867. **Salama, A. M., and A. G. Mishricky.** 1973. Seed transmission of maize wilt fungi with special reference to *Fusarium moniliforme* Sheld. *Phytopathologische Zeitschrift* **77:** 356-362. (3)

1868. **Salas, B., B. J. Steffenson, H. H. Casper, B. Tacke, L. K. Prom, T. G. Fetch, Jr., and P. B. Schwarz.** 1999. *Fusarium* species pathogenic to barley and their associated mycotoxins. *Plant Disease* **83:** 667-674. (*poae*, *sporotrichioides*)

1869. **Salch, Y. P., and M. N. Beremand.** 1993. *Gibberella pulicaris* transformants: State of transforming DNA during asexual and sexual growth. *Current Genetics* **23:** 343-350. (*sambucinum*)

1870. **Saleh, O. I.** 1997. Wilt, root rot and seed diseases of groundnut in El-Minia governorate, Egypt. *Egyptian Journal of Phytopathology* **25:** 1-18. (*compactum*)

1871. **Salerno, M. I., S. Gianinazzi, C. Arnould, and V. Gianinazzi-Pearson.** 2004. Ultrastructural and cell wall modifications during infection of *Eucalyptus viminalis* roots by a pathogenic *Fusarium oxysporum* strain. *Journal of General Plant Pathology* **70:** 145-152. (*oxysporum*)

1872. **Salerno, M. I., S. Gianinazzi, C. Arnould, and V. Gianinazzi-Pearson.** 2004. Cell interactions between a non-pathogenic *Fusarium oxysporum* strain and root tissues of *Eucalyptus viminalis*. *Journal of General Plant Pathology* **70:** 153-158. (*oxysporum*)

1873. **Sallam, L. A. R., A. M. El-Refai, and H. A. El-Minofi.** 2005. Physiological and biochemical improvement of the enzyme side-chain degradation of cholesterol by *Fusarium solani*. *Process Biochemistry* **40:** 203-206. (*solani*)

1874. **Sallam, L. A. R., A. H. El-Refai, N. Naim, and H. A. El-Minofi.** 1987. Side-chain degradation of sterols by some local microorganisms. *Egyptian Journal of Microbiology* **22:** 339-348. (*solani*)

1875. **Salleh, B.** 1994. Current status and control of plant diseases caused by *Fusarium* in Malaysia, p. 23-26. *In* M. A. Rifai, E. S. Scott, F. C. Quebral, and O. S. Dharmaputra (eds.), Biotrop Special Publication no. 54. Biotrop, Bogor, Indonesia. (11)

1876. **Salleh, B., and R. N. Strange.** 1988. Toxigenicity of some Fusaria associated with plant and human diseases in the Malaysian Peninsula, Malaysia. *Journal of General Microbiology* **134:** 841-848. (*longipes*, *nygamai*, *redolens*)

1877. **Samac, D. A., and S. A. Leong.** 1988. Two linear plasmids in mitochondria of *Fusarium solani* f. sp. *cucurbitae*. *Plasmid* **19:** 57-67. (*solani*)

1878. **Samac, D. A., and S. A. Leong.** 1989. Disease development in *Cucurbita maxima* squash infected with *Fusarium solani* f. sp. *cucurbitae*. *Canadian Journal of Botany* **67:** 3486-3489. (*solani*)

1879. **Samac, D. A., and S. A. Leong.** 1989. Characterization of the termini of linear plasmids from *Nectria haematococca* and their use in construction of an autonomously replicating transformation vector. *Current Genetics* **16:** 187-194. (*solani*)

1880. **Sampedro, I., E. Aranda, J. Martín, J. M. García-Garrido, I. García-Romera, and J. A. Ocampo.** 2004. Saprobic fungi decrease plant toxicity caused by olive mill residues. *Applied Soil Ecology* **26:** 149-156. (*oxysporum*)

1881. **Samuels, G. J., and I. C. Hallet.** 1983. *Microdochium stoveri* and *Monographella stoveri*, new combinations for *Fusarium stoveri* and *Micronectriella stoveri*. *Transactions of the British Mycological Society* **81:** 473-483. (10)

1882. **Samuels, G. J., H. I. Nirenberg, and K. A. Seifert.** 2001. Perithecial species of *Fusarium*, p. 1-14. *In* B. A. Summerell, J. F. Leslie, D. Backhouse, W. L. Bryden, and L. W. Burgess (eds.), Fusarium*: Paul E. Nelson Memorial Symposium*. APS Press, St. Paul, Minnesota. (9, 10)

1883. **Sanchez, L. E., J. V. Leary, and R. M. Endo.** 1976. Heterokaryosis in *Fusarium oxysporum* f. sp. *lycopersici*. *Journal of General Microbiology* **93:** 220-227. (*oxysporum*)

1884. **Sanchez-Fernandez, R., J. Avalos, and E. Cerda-Olmedo.** 1997. Inhibition of gibberellin biosynthesis by nitrate in *Gibberella fujikuroi*. *FEBS Letters* **413:** 35-39. (*fujikuroi*)

1885. **Sanchez-Murillo, I. R., M. de la Torre Martinez, J. Aguirre-Linares, and A. Herrera-Estrella.** 2004. Light regulated asexual reproduction in *Paecilomyces fumosoroseus*. *Microbiology* **150:** 311-319. (4)

1886. **Sander, A., U. Beyer, and R. Amberg.** 1998. Systemic *Fusarium oxysporum* infection in an immunocompetent patient with an adult respiratory distress syndrome (ARDS) and extracorporal membrane oxygenation (ECMO). *Mycoses* **41:** 109-111. (*oxysporum*)

1887. **Sands, D. C., E. J. Ford, R. V. Miller, B. K. Sally, M. K. McCarthy, T. W. Anderson, M. B. Weaver, C. T. Morgan, A. L. Pilgeram, and L. C. Darlington.** 1997.

Characterization of a vascular wilt of *Erythroxylum coca* caused by *Fusarium oxysporum* f. sp. *erythroxyli forma specialis nova*. *Plant Disease* **81:** 501-504. (*oxysporum*)

1888. **Sangalang, A. E., D. Backhouse, and L. W. Burgess.** 1995. Survival and growth in culture of four *Fusarium* species in relation to occurrence in soils from hot climatic regions. *Mycological Research* **99:** 529-533. (*acuminatum*)

1889. **Sangalang, A. E., L. W. Burgess, D. Backhouse, J. Duff, and M. Wurst.** 1995. Mycogeography of *Fusarium* species in soils from tropical, arid and Mediterranean regions of Australia. *Mycological Research* **99:** 523-528. (*acuminatum, anthophilum, avenaceum, chlamydosporum, dimerum, longipes, napiforme, nygamai, polyphialidicum*)

1890. **Sangalang, A. E., B. A. Summerell, L. W. Burgess, and D. Backhouse.** 1995. Characterization of *Fusarium avenaceum* ssp. *aywerte* and *Fusarium avenaceum* ssp. *nurragi*. *Mycological Research* **99:** 287-290. (8, 9, *avenaceum, aywerte, nurragi*)

1891. **Sanson, D. R., D. G. Corley, C. L. Barnes, S. Searles, E. O. Schlemper, M. S. Tempesta, and G. E. Rottinghaus.** 1989. New mycotoxins from *Fusarium sambucinum*. *Journal of Organic Chemistry* **54:** 4313-4318. (*sambucinum*)

1892. **Saparrat, M. C. N., M. J. Martinez, H. A. Tournier, M. N. Cabello, and A. M. Arambarri.** 2000. Production of ligninolytic enzymes by *Fusarium solani* strains isolated from different substrata. *World Journal of Microbiology and Biotechnology* **16:** 799-803. (*solani*)

1893. **Saremi, H., and L. W. Burgess.** 2000. Effect of soil temperature on distribution and population dynamics of *Fusarium* species. *Journal of Agricultural Science and Technology* **2:** 119-125. (*acuminatum*)

1894. **Saremi, H., L. W. Burgess, and D. Backhouse.** 1999. Temperature effects on the relative abundance of *Fusarium* species in a model plant-soil ecosystem. *Soil Biology and Biochemistry* **31:** 941-947. (11, *equiseti*)

1895. **Sarquis, M. I. D. M., and C. D. M. Borba.** 1997. *Fusarium* species in sandy soil from Ipanema Beach, Rio de Janeiro, Brazil. *Journal of Basic Microbiology* **37:** 425-429. (*sacchari*)

1896. **Sasaki, O., and S. Tani.** 1997. Chemical and biological control of land-dwelling leech, *Haemadipsa japonica*. *Medical Entomology and Zoology* **48:** 303-309. (*solani*)

1897. **Satou, M., M. Ichinoe, F. Fukumoto, N. Tezuka, and S. Horiuchi.** 2001. Fusarium blight of kangaroo paw (*Anigozanthos* spp.) caused by *Fusarium chlamydosporum* and *Fusarium semitectum*. *Journal of Phytopathology* **149:** 203-206. (*chlamydosporum, semitectum*)

1898. **Satyaprasad, K, G. L. Bateman, and E. Ward.** 2000. Comparisons of isolates of *Fusarium avenaceum* from white lupin and other crops by pathogenicity tests, DNA analyses, and vegetative compatibility tests. *Journal of Phytopathology* **148:** 211-219. (*avenaceum*)

1899. **Saubois, A., E. P. Laforet, M. C. Nepote, and M. L. Wagner.** 1999. Mycological evaluation of a sorghum grain of Argentina, with emphasis on the characterization of *Fusarium* species. *Food Microbiology* **16:** 435-445. (*subglutinans*)

1900. **Saunders, R., P. S. J. Cheetham, and R. Hardman.** 1986. Microbial transformation of crude Fenugreek (*Trigonella foenumgraecum*) steroids. *Enzyme and Microbial Technology* **8:** 549-555. (*solani*)

1901. **Saunders, R. P., R. Hardman, and P. S. J. Cheetham.** 1985. The use of a hydrophobic resin as a product reservoir in steroid transformations. *Biotechnology and Bioengineering* **27:** 825-831. (*solani*)

1902. **Saupe, S. J.** 2000. Molecular genetics of heterokaryon incompatibility in filamentous ascomycetes. *Microbiology and Molecular Biology Reviews* **64:** 489-502. (5)

1903. **Saupe, S. J., and N. L. Glass.** 1997. Allelic specificity at the *het-c* heterokaryon incompatibility locus of *Neurospora crassa* is determined by a highly variable domain. *Genetics* **146:** 1299-1309. (5)

1904. **Savard, M. E. and J. D. Miller.** 1992. Characterization of fusarin F, a new fusarin from *Fusarium moniliforme*. *Journal of Natural Products* **55:** 64-70. (*verticillioides*)

1905. **Savard, M. E., J. D, Miller, B. Salleh, and R. N. Strange.** 1990. Chlamydosporol, a new metabolite from *Fusarium chlamydosporum*. *Mycopathologia* **110:** 177-181. (*chlamydosporum*)

1906. **Savir, H., E. Henig, and N. Lehrer.** 1978. Exogenous mycotic infections of the eye and adnexa. *Annals of Ophthalmology* **10:** 1013-1018. (*solani*)

1907. **Scala, F., G. Cristinzio, F. Marziano, and C. Noviello.** 1981. Endo-polygalacturonase zymograms of *Fusarium* species. *Transactions of the British Mycological Society* **77:** 587-592. (*oxysporum*)

1908. **Schaafsma, A. W., M. E. Savard, R. Clear, and J. Dexter.** 2004. Methods and issues regarding detection of deoxynivalenol, *Fusarium* damaged kernels, and *Fusarium* spp. in commercial grain in Canada. *Canadian Journal of Plant Pathology* **26:** 443-452. (*graminearum*)

1909. **Scheetz, R. W., H. A. Whelan, and J. C. Wriston.** 1971. Purification and properties of L-asparaginase (e.c. 3.5.1.1) from *Fusarium tricinctum*. *Archives of Biochemistry and Biophysics* **142:** 184-189. (*tricinctum*)

1910. **Schilling, A. G., E. M. Möller, and H. H. Geiger.** 1996. Polymerase chain reaction-based assays for species-specific detection of *Fusarium culmorum, F. graminearum* and *F. avenaceum*. *Phytopathology* **86:** 515-522. (7)

1911. **Schlub, R. L., L. Lockwood, and H. Komada.** 1981. Colonization of soybean (*Glycine max*) seeds and plant tissue by *Fusarium* species in soil. *Phytopathology* **71:** 693-697. (*subglutinans*)

1912. **Schmale, D. G., III, and G. C. Bergstrom.** 2004. Spore deposition of the ear rot pathogen, *Gibberella zeae*, inside corn canopies. *Canadian Journal of Plant Pathology* **26:** 591-595. (3, *graminearum*)

1913. **Schmale, D. G., III, and T. R. Gordon.** 2003. Variation in susceptibility to pitch canker disease, caused by *Fusarium circinatum*, in native stands of *Pinus muricata*. *Plant Pathology* **52:** 720-725. (*circinatum*)

1914. **Schmale, D. G., III, D. A. Shah, and G. C. Bergstrom.** 2005. Spatial patterns of viable spore deposition of *Gibberella zeae* in wheat fields. *Phytopathology* **95:** 472-479. (3, *graminearum*)

1915. **Schmidt, H., A. Adler, A. Holst-Jensen, S. S. Klemsdal, A. Logrieco, R. L. Mach, H. I. Nirenberg, U. Thrane, M.**

Torp, R. F. Vogel, T. Yli-Mattila, and L. Niessen. 2004. An integrated taxonomic study of *Fusarium langsethiae*, *Fusarium poae* and *Fusarium sporotrichioides* based on the use of composite datasets. *International Journal of Food Microbiology* **95:** 341-349. (*poae*)

1916. **Schmidt, H., L. Niessen, and R. F. Vogel.** 2004. AFLP analysis of *Fusarium* species in the section *Sporotrichiella* – Evidence for *Fusarium langsethiae* as a new species. *International Journal of Food Microbiology* **95:** 297-304. (*poae*)

1917. **Schmidt, R., P. Zajkowski, and J. Wink.** 1995. Toxicity of *Fusarium sambucinum* Fückel *sensu lato* to brine shrimp. *Mycopathologia* **129:** 173-175. (*sambucinum, venenatum*)

1918. **Schneider, E. F., and W. L. Seaman.** 1974. Development of a cellular body during differentiation of conidial chlamydospores in *Fusarium sulphureum*. *Canadian Journal of Microbiology* **20:** 1205-1208. (*sambucinum*)

1919. **Schneider, E. F., and W. L. Seaman.** 1974. Development of conidial chlamydospores of *Fusarium sulphureum* in distilled water. *Canadian Journal of Microbiology* **20:** 247-254. (*sambucinum*)

1920. **Schneider, E. F., and W. L. Seaman.** 1977. Ontogeny of lipid bodies in the endoplasmic reticulum of *Fusarium sulphureum*. *Canadian Journal of Microbiology* **23:** 190-196. (*sambucinum*)

1921. **Schneider, R. W., and W. E. Pendery.** 1983. Stalk rot of corn: Mechanism of predisposition by an early season water stress. *Phytopathology* **73:** 863-871. (*verticillioides*)

1922. **Schoental, R.** 1994. Mycotoxins in food and the plague of Athens. *Journal of Nutritional Medicine* **4:** 83-85. (1)

1923. **Schreuder, W., S. C. Lamprecht, and G. Holz.** 2000. Race determination and vegetative compatibility grouping of *Fusarium oxysporum* f. sp. *melonis* from South Africa. *Plant Disease* **84:** 231-234. (*oxysporum*)

1924. **Schreuder, W., S. C. Lamprecht, W. F. O. Marasas, and F. J. Calitz.** 1995. Pathogenicity of three *Fusarium* species associated with asparagus decline in South Africa. *Plant Disease* **79:** 177-181. (*proliferatum*)

1925. **Schroeder, J. J., H. M. Crane, J. H. Xia, D. C. Liotta, and A. H. Merrill, Jr.** 1994. Disruption of sphingolipid metabolism and stimulation of DNA synthesis by fumonisin B_1: A molecular mechanism for carcinogenesis associated with *Fusarium moniliforme*. *Journal of Biological Chemistry* **269:** 3475-3481. (*verticillioides*)

1926. **Schroers, H.-J., R. P. Baayen, J. P. Meffert, J. de Gruyter, M. Hooftman, and K. O'Donnell.** 2004. *Fusarium foetens*, a new species pathogenic to *Begonia elatior* hybrids (*Begonia* × *hiemalis*) and the sister taxon of the *Fusarium oxysporum* species complex. *Mycologia* **96:** 393-406. (*foetens*)

1927. **Schröter, J.** 1874. Uber *Peronospora violacea* Berkeley und einige verwandte *Peronospora*-Arten. *Hedwigia* **12:** 180. (*succisae*)

1928. **Schutt, C., and D. Netzly.** 1991. Effects of apiforol and apigeninidin on growth of selected fungi. *Journal of Chemical Ecology* **17:** 2261-2266. (*graminearum*)

1929. **Schweigkofler, W., K. O'Donnell, and M. Garbelotto.** 2004. Detection and quantification of airborne conidia of *Fusarium circinatum*, the causal agent of pine pitch canker, from two California sites by using a real-time PCR approach combined with a simple spore trapping method. *Applied and Environmental Microbiology* **70:** 3512-3520. (*circinatum*)

1930. **Scott, P. M., H. K. Abbas, C. J. Mirocha, G. A. Lawrence, and D. Weber.** 1987. Formation of moniliformin by *Fusarium sporotrichioides* and *Fusarium culmorum*. *Applied and Environmental Microbiology* **53:** 196-197. (*culmorum, sporotrichioides*)

1931. **Secor, G. A., and N. C. Gudmestad.** 1999. Managing fungal diseases of potato. *Canadian Journal of Plant Pathology* **21:** 213-221. (*solani*)

1932. **Seefelder, W., M. Gossmann, and H. U. Humpf.** 2002. Analysis of fumonisin B_1 in *Fusarium proliferatum*-infected asparagus spears and garlic bulbs from Germany by liquid chromatography-electrospray ionization mass spectrometry. *Journal of Agricultural and Food Chemistry* **50:** 2778-2781. (*proliferatum*)

1933. **Segal, B. H., T. J. Walsh, J. M. Liu, J. D. Wilson, and K. J. Kwon-Chung.** 1998. Invasive infection with *Fusarium chlamydosporum* in a patient with aplastic anemia. *Journal of Clinical Microbiology* **36:** 1772-1776. (*chlamydosporum*)

1934. **Seifert, K. A.** 1995. Notes on the typification of *Gibberella zeae*. *Sydowia* **48:** 83-89. (*graminearum*)

1935. **Seifert, K. A.** 2001. *Fusarium* and anamorph generic concepts, p. 15-28. *In* B. A. Summerell, J. F. Leslie, D. Backhouse, W. L. Bryden, L. W. Burgess (eds.), Fusarium*: Paul E. Nelson Memorial Symposium.* APS Press, St. Paul, Minnesota. (8, 9)

1936. **Seifert, K. A., T. Aoki, R. P. Baayen, D. Brayford, L. W. Burgess, S. Chulze, W. Gams, D. Geiser, D., J. de Gruyter, J. F. Leslie, A. Logrieco, W. F. O. Marasas, H. I. Nirenberg, K. O'Donnell, J. P. Rheeder, G. J. Samuels, B. A. Summerell, U. Thrane, and C. Waalwijk.** 2003. The name *Fusarium moniliforme* should no longer be used. *Mycological Research* **107:** 643-644. (8, 9, *verticillioides*)

1937. **Sekhon, A. S., L. Kaufman, N. Moledina, R. C. Summerbell, A. A. Padhye, E. A. Ambrosie, and T. Panter.** 1995. An exoantigen test for the rapid identification of medically significant *Fusarium* species. *Journal of Medical and Veterinary Mycology* **33:** 287-289. (*solani*)

1938. **Sekhon, A. S., A. A. Padhye, A. K. Garg, H. Ahmad, and N. Moledina.** 1994. *In vitro* sensitivity of medically significant *Fusarium* species to various antimycotics. *Chemotherapy* **40:** 239-244. (*equiseti, oxysporum*)

1939. **Sella, L., C. Castiglioni, S. Roberti, R. D'Ovidio, and F. Favaron.** 2004. An endo-polygalacturonase (PG) of *Fusarium moniliforme* escaping inhibition by plant polygalacturonase-inhibiting proteins (PGIPs) provides new insights into the PG-PGIP interaction. *FEMS Microbiology Letters* **240:** 117-124. (*verticillioides*)

1940. **Semangun, H.** 1992. *Host Index of Plant Diseases in Indonesia.* Gadjah Mada University Press, Yogyakarta, Indonesia. (11)

1941. **Seo, J.-A., J.-C. Kim, and Y.-W. Lee.** 1996. Isolation and characterization of two new type C fumonisins produced by

Fusarium oxysporum. *Journal of Natural Products* **59:** 1003-1005. (*oxysporum*)

1942. **Seo, J.-A., J.-C. Kim, and Y.-W. Lee.** 1999. *N*-Acetyl derivatives of type C fumonisins produced by *Fusarium oxysporum*. *Journal of Natural Products* **62:** 355-357. (*oxysporum*)

1943. **Seo, J.-A., R. H. Proctor, and R. D. Plattner.** 2001. Characterization of four clustered and coregulated genes associated with fumonisin biosynthesis in *Fusarium verticillioides*. *Fungal Genetics and Biology* **34:** 155-165. (*verticillioides*)

1944. **Seta, S., M. Gonzalez, and G. Lori.** 2004. First report of walnut canker caused by *Fusarium incarnatum* in Argentina. *Plant Pathology* **53:** 248. (*semitectum*)

1945. **Seth, P. K., and T. R. Shandilya.** 1978. Mycoflora associated with the deterioration of the cultivated mushrooms (*Agaricus bisporus*) in storage. *Indian Journal of Mycology and Plant Pathology* **8:** 209-210. (*semitectum*)

1946. **Sewram, V., N. Mshicileli, G. S. Shephard, H. F. Vismer, J. P. Rheeder, Y.-W. Lee, J. F. Leslie, and W. F. O. Marasas.** 2005. Production of fumonisin B and C analogs by several *Fusarium* species. *Journal of Agricultural and Food Chemistry* **53:** 4861-4866. (*oxysporum*, *verticillioides*)

1947. **Sewram, V., T. W. Nieuwoudt, W. F. O. Marasas, G. S. Shephard, and A. Ritieni.** 1999. Determination of the *Fusarium* mycotoxins, fusaproliferin and beauvericin by high-performance liquid chromatography-electrospray ionization mass spectrometry. *Journal of Chromatography A* **858:** 175-185. (*subglutinans*)

1948. **Sewram, V., T. W. Nieuwoudt, W. F. O. Marasas, G. S. Shephard, and A. Ritieni.** 1999. Determination of the mycotoxin moniliformin in cultures of *Fusarium subglutinans* and in naturally contaminated maize by high-performance liquid chromatography-atmospheric pressure chemical ionization mass spectrometry. *Journal of Chromatography A* **848:** 185-191. (*subglutinans*)

1949. **Shabana, Y. M., J. P. Cuda, and R. Charudattan.** 2003. Combining plant pathogenic fungi and the leaf-mining fly, *Hydrellia pakistanae*, increases damage to hydrilla. *Journal of Aquatic Plant Management* **41:** 76-81. (*culmorum*)

1950. **Shabana, Y. M., J. P. Cuda, and R. Charudattan.** 2003. Evaluation of pathogens as potential biocontrol agents of hydrilla. *Journal of Phytopathology* **151:** 607-613. (*culmorum*)

1951. **Shabana, Y. M., D. Mueller-Stoever, and J. Sauerborn.** 2003. Granular Pesta formulation of *Fusarium oxysporum* f. sp. *orthoceras* for biological control of sunflower broomrape: Efficacy and shelf-life. *Biological Control* **26:** 189-201. (*oxysporum*)

1952. **Shaoxi, W., G. Ningro, and L. Guixia.** 1996. A rare case of *Fusarium verticillioides* facial granuloma successfully treated by itraconazole. *Journal de Mycologie Medicale* **6:** 88-90. (*verticillioides*)

1953. **Shaykh, M., C. Soliday, and P. E. Kolattukudy.** 1977. Proof for the production of cutinase by *Fusarium solani* f. *pisi* during penetration into its host *Pisum sativum*. *Plant Physiology* **60:** 170-172. (*solani*)

1954. **Sheldon, J. L.** 1904. A corn mold (*Fusarium moniliforme* n. sp.). Agricultural Experiment Station of Nebraska, 17th Annual Report, pp. 23-32. (8)

1955. **Shelton, B. G., K. H. Kirkland, W. D. Flanders, and G. K. Morris.** 2002. Profiles of airborne fungi in buildings and outdoor environments in the United States. *Applied and Environmental Microbiology* **68:** 1743-1753. (3)

1956. **Shephard, G. S., V. Sewram, T. W. Nieuwoudt, W. F. O. Marasas, and A. Ritieni.** 1999. Production of the mycotoxins fusaproliferin and beauvericin by South African isolates in the *Fusarium* section *Liseola*. *Journal of Agricultural and Food Chemistry* **47:** 5111-5115. (*globosum*, *proliferatum*, *subglutinans*, *verticillioides*)

1957. **Shephard, G. S., L. van der Westhuizen, P. G. Thiel, W. C. A. Gelderblom, W. F. O. Marasas, and D. J. van Schalkwyk.** 1996. Disruption of sphingolipid metabolism in non-human primates consuming diets of fumonisin-containing *Fusarium moniliforme* culture material. *Toxicon* **34:** 527-534. (*verticillioides*)

1958. **Sherbakoff, C. D.** 1915. Fusaria of potatoes. *Memoirs of the Cornell University Agricultural Experimental Station* **6:** 87-270. (*sporotrichioides*)

1959. **Shetty, P. H., and R. V. Bhat.** 1997. Differential production of fumonisin B_1 in maize inbreds and hybrids in laboratory. *Cereal Research Communications* **25:** 1011-1015. (*verticillioides*)

1960. **Shi, J., W. C. Mueller, and C. H. Beckman.** 1992. Vessel occlusion and secretory activities of vessel contact cells in resistant or susceptible cotton plants infected with *Fusarium oxysporum* f. sp. *vasinfectum*. *Physiological and Molecular Plant Pathology* **40:** 133-147. (*oxysporum*)

1961. **Shi, J. H., Y. Suzuki, B. D. Lee, S. Nakai, and M. Hosomi.** 2002. Isolation and characterization of the ethynylestradiol-biodegrading microorganism *Fusarium proliferatum* strain HNS-1. *Water Science and Technology* **45:** 175-179. (*proliferatum*)

1962. **Shier, W. T., H. K. Abbas, and F. A. Badria.** 1995. Complete structures of the sphingosine analog mycotoxins fumonisin B_1 and AAL toxin T-A: Absolute configuration of the side chains. *Tetrahedron Letters* **36:** 1571-1574. (*verticillioides*)

1963. **Shiflett, A. M., J. Enkerli, and S. F. Covert.** 2002. *Nht2*, a *copia* LTR retrotransposon from a conditionally dispensable chromosome in *Nectria haematococca*. *Current Genetics* **41:** 99-106. (*solani*)

1964. **Shillingford, C. A., and J. B. Sinclair.** 1980. Macerating enzyme production by *Colletotrichum musae* and *Fusarium semitectum*, incitants of banana fruit decay. *Phytopathologische Zeitschrift* **97:** 127-135. (*semitectum*)

1965. **Shim, W.-B., J. E. Flaherty, and C. P. Woloshuk.** 2003. Comparison of fumonisin B_1 biosynthesis in maize germ and degermed kernels by *Fusarium verticillioides*. *Journal of Food Protection* **66:** 2116-2122. (*verticillioides*)

1966. **Shim, W.-B, and C. P. Woloshuk.** 1999. Nitrogen repression of fumonisin B_1 biosynthesis in *Gibberella fujikuroi*. *FEMS Microbiology Letters* **177:** 109-116. (*verticillioides*)

1967. **Shim, W.-B., and C. P. Woloshuk.** 2001. Regulation of fumonisin B_1 biosynthesis and conidiation in *Fusarium verticillioides* by a cyclin-like (C-Type) gene, *FCC1*. *Applied and Environmental Microbiology* **67:** 1607-1612. (*verticillioides*)

1968. **Shimizu, S.** 2003. Biochemical and applied studies of vitamin production by microorganisms. *Vitamins* **77:** 131-145. (*oxysporum*)

1969. **Shimizu, S., M. Kataoka, K. Honda, and K. Sakamoto.** 2001. Lactone-ring-cleaving enzymes of microorganisms: Their diversity and applications. *Journal of Biotechnology* **92:** 187-194. (*oxysporum*)

1970. **Shimosaka, M., M. Kumehara, X.-Y. Zhang, M. Nogawa, and M. Okazaki.** 1996. Cloning and characterization of a chitosanase gene from the plant pathogenic fungus *Fusarium solani*. *Journal of Fermentation and Bioengineering* **82:** 426-431. (*solani*)

1971. **Shimosaka, M., M. Nogawa, Y. Onho, and M. Okazaki.** 1993. Chitosanase from the plant pathogenic fungus, *Fusarium solani* f. sp. *phaseoli*: Purification and some properties. *Bioscience Biotechnology and Biochemistry* **57:** 231-235. (*solani*)

1972. **Shiro, Y., M. Fujii, T. Iizuka, S.-I. Adachi, K. Tsukamoto, K. Nakahara, and S. Hirofumi.** 1995. Spectroscopic and kinetic studies on reaction of cytochrome P450nor with nitric oxide: Implication for its nitric oxide reduction mechanism. *Journal of Biological Chemistry* **270:** 1617-1623. (*oxysporum*)

1973. **Shiu, P. K. T., and N. L. Glass.** 1999. Molecular characterization of *tol*, a mediator of mating-type-associated vegetative incompatibility in *Neurospora crassa*. *Genetics* **151:** 545-555. (5)

1974. **Shoun, H., Y. Sudo, Y. Seto, and T. Beppu.** 1983. Purification and properties of a cytochrome P-450 of a fungus *Fusarium oxysporum*. *Journal of Biochemistry* **94:** 1219-1230. (*oxysporum*)

1975. **Shukla, R., S. Chand, and A. K. Srivastava.** 2005. Batch kinetics and modeling of gibberellic acid production by *Gibberella fujikuroi*. *Enzyme and Microbial Technology* **36:** 492-497. (*fujikuroi*)

1976. **Shukla, R., A. K. Srivastava, and S. Chand.** 2003. Bioprocess strategies and recovery processes in gibberellic acid fermentation. *Biotechnology and Bioprocess Engineering* **8:** 269-278. (1, *fujikuroi*)

1977. **Siddiqui, Z. A., and I. Mahmood.** 1995. Some observations on the management of the wilt disease complex of pigeonpea by treatment with vesicular arbuscular fungus and biocontrol agents for nematodes. *Bioresource Technology* **54:** 227-230. (*udum*)

1978. **Sidhu, G. S.** 1984. Genetics of *Gibberella fujikuroi*. V. Spore killer alleles in *G. fujikuroi*. *Journal of Heredity* **75:** 237-238. (10)

1979. **Sidhu, G. S.** 1986. Genetics of *Gibberella fujikuroi*. VIII. Vegetative compatibility groups. *Canadian Journal of Botany* **64:** 117-121. (*verticillioides*)

1980. **Sidhu, G. S., and J. M. Webster.** 1979. A study of heterokaryosis and its influence on virulence in *Fusarium oxysporum lycopersici*. *Canadian Journal of Botany* **57:** 548-555. (*oxysporum*)

1981. **Silbernagel, M. J. and L. J. Mills.** 1990. Genetic and cultural control of Fusarium root rot in bush snap beans. *Plant Disease* **74:** 61-66. (*solani*)

1982. **Silva-Hanlin, D. M. W., M. Menezes, R. T. Hanlin, and E. Ramalho-Neto.** 1999. Ribosomal DNA sequencing data reveals low genetic variability among *Fusarium oxysporum* f. sp. *cubense* isolates. *Fitopatologia Brasileira* **24:** 534-539. (*oxysporum*)

1983. **Simpson, D. R., J. M. Withers, M. G. Wiebe, G. D. Robson, and A. P. J. Trinci.** 1998. Mutants with general growth rate advantages are the predominant morphological mutants to be isolated from the Quorn production plant. *Mycological Research* **102:** 221-227. (*venenatum*)

1984. **Simpson, E. B., S. L. Ross, S. E. Marchetti, and J. C. Kennell.** 2004. Relaxed primer specificity associated with reverse transcriptases encoded by the p*FOXC* retroplasmids of *Fusarium oxysporum*. *Eukaryotic Cell* **3:** 1589-1600. (*oxysporum*)

1985. ☑**Singh, A., R. C. Kuhad, and M. Kumar.** 1995. Xylanase production by a hyperxylanolytic mutant of *Fusarium oxysporum*. *Enzyme and Microbial Technology* **17:** 551-553. (*oxysporum*)

1986. **Singh, A., P. K. R. Kumar, and K. Schuegerl.** 1992. The oxygen requirement of *Fusarium oxysporum* for the fermentation of D-xylose and D-glucose. *Biotechnology and Applied Biochemistry* **16:** 296-302. (*oxysporum*)

1987. **Singh, A., P. K. R. Kumar, and K. Schuegerl.** 1992. D-Xylose fermentation and catabolism in *Fusarium oxysporum*. *Biochemistry International* **27:** 831-839. (*oxysporum*)

1988. **Singh, A., and K. Schuegerl.** 1992. Induction and regulation of D-xylose catabolizing enzymes in *Fusarium oxysporum*. *Biochemistry International* **28:** 481-488. (*oxysporum*)

1989. **Singh, I. P., V. Dhar, and R. G. Chaudhary.** 1998. Inheritance of resistance to Fusarium wilt in pigeonpea. *Indian Journal of Agricultural Sciences* **68:** 729-731. (*udum*)

1990. **Singh, K. G.** 1980. A checklist of host and diseases in Malaysia. Ministry of Agriculture, Kuala Lumpur, Malaysia. (11)

1991. **Singh, R. S., and Y. L. Nene.** 1965. Malachite green in synthetic medium for the isolation of *Fusarium* spp. from plant tissues. *Naturwissenschaften* **52:** 94. (2)

1992. **Singh, S. B., D. L. Zink, J. D. Polishook, A. W. Dombrowski, S. J. Darkin-Rattray, D. M. Schmatz, and A. M. Goetz.** 1996. Apicidins: Novel cyclic tetrapeptides as coccidiostats and antimalarial agents from *Fusarium pallidoroseum*. *Tetrahedron Letters* **37:** 8077-8080. (*semitectum*)

1993. **Sinha, O. K., and M. N. Khare.** 1977. Control of seedborne *Macrophomina phaseolina* and *Fusarium equiseti* by hot water treatment of cowpea seeds. *Seed Research* **5:** 21-22. (3)

1994. **Sitepu, D., and H. R. Wallace.** 1984. Biological control of *Sclerotinia sclerotiorum* in lettuce by *Fusarium lateritium*. *Australian Journal of Experimental Agriculture and Animal Husbandry* **24:** 272-276. (*lateritium*)

1995. **Skadsen, R. W., and T. M. Hohn.** 2004. Use of *Fusarium graminearum* transformed with *gfp* to follow infection patterns in barley and *Arabidopsis*. *Physiological and Molecular Plant Pathology* **64:** 45-53. (*graminearum*)

1996. **Skoglund, L. G., and W. M. Brown.** 1988. Effects of tillage regimes and herbicides on *Fusarium* spp. associated with corn stalk rot. *Canadian Journal of Plant Pathology* **10:** 332-338. (*verticillioides*)

1997. **Skov, J., M. Lemmens, and H. Giese.** 2004. Role of a *Fusarium culmorum* ABC transporter (FcABC1) during infection of wheat and barley. *Physiological and Molecular Plant Pathology* **64:** 245-254. (*culmorum*)

1998. **Skovgaard, K., L. Bodker, and S. Rosendahl.** 2002. Population structure and pathogenicity of members of the *Fusarium oxysporum* complex isolated from soil and root necrosis of pea (*Pisum sativum* L.). *FEMS Microbiology Ecology* **42:** 367-374. (*oxysporum*)

1999. **Skovgaard, K., H. I. Nirenberg, K. O'Donnell, and S. Rosendahl.** 2001. Evolution of *Fusarium oxysporum* f. sp. *vasinfectum* races inferred from multigene genealogies. *Phytopathology* **91:** 1231-1237. (*oxysporum*)

2000. **Smith, A. G., C. I. Bustamante, and C. Wood.** 1993. Disseminated cutaneous and vascular invasion by *Fusarium moniliforme* in a fatal case of acute lymphocytic leukemia. *Mycopathologia* **122:** 15-20. (*verticillioides*)

2001. ☑**Smith, A. G., A. G. Muhvich, K. H. Muhvich, and C. Wood.** 1989. Fatal *Fusarium solani* infections in baby sharks. *Journal of Medical and Veterinary Mycology* **27:** 83-92. (*solani*)

2002. **Smith, D. A., J. M. Harrer, and T. E. Cleveland.** 1981. Simultaneous detoxification of phytoalexins by *Fusarium solani* f. sp. *phaseoli*. *Phytopathology* **71:** 1212-1215. (*solani*)

2003. **Smith, D. A., P. J. Kuhn, J. A. Bailey, and R. S. Burden.** 1980. Detoxification of phaseolidin by *Fusarium solani* f. sp. *phaseoli*. *Phytochemistry* **19:** 1673-1676. (*solani*)

2004. **Smith, D. A., H. E. Wheeler, S. W. Banks, and T. E. Cleveland.** 1984. Association between lowered kievitone hydratase activity and reduced virulence virulence to bean in variants of *Fusarium solani* f. sp. *phaseoli*. *Physiological Plant Pathology* **25:** 135-148. (*solani*)

2005. **Smith, J. S., J. Fotso, J. F. Leslie, X. Wu, D. van der Velde, and R. A. Thakur.** 2004. Characterization of bostrycoidin: an analytical analog of zearalenone. *Journal of Food Science* **69:** 227-232. (*verticillioides*)

2006. **Smith, M. L., O. C. Micali, S. P. Hubbard, N. Mir-Rashed, D. J. Jacobson, and N. L. Glass.** 2000. Vegetative compatibility in the *het-6* region of *Neurospora crassa* is mediated by two linked genes. *Genetics* **155:** 1095-1104. (5)

2007. **Smith, M. N., and C. R. Olien.** 1978. Pathological factors affecting survival of winter barley following controlled freeze tests. *Phytopathology* **68:** 773-777. (*avenaceum*)

2008. **Smith-White, J. L., L. V. Gunn, and B. A. Summerell.** 2001. Analysis of diversity within *Fusarium oxysporum* populations using molecular and vegetative compatibility grouping. *Australasian Plant Pathology* **30:** 153-157. (*oxysporum*)

2009. **Snijders, C. H. A.** 1990. Systemic fungal growth of *Fusarium culmorum* in stems of winter wheat. *Journal of Phytopathology* **129:** 133-140. (*culmorum*)

2010. **Snijders, C. H. A., and J. Perkowsji.** 1990. Effects of head blight caused by *Fusarium culmorum* on toxin content and weight of wheat kernels. *Phytopathology* **80:** 566-570. (*culmorum*)

2011. **Snijders, C. H. A., R. A. Samson, E. S. Hoekstra, T. Ouellet, J. D. Miller, P. C. E. M. de Rooij van der Goes, A. J. M. Barr, A. E. J. Dubois, and H. F. Kauffman.** 1996. Analysis of *Fusarium* causing dermal toxicosis in marram grass planters. *Mycopathologia* **135:** 119-128. (*culmorum*)

2012. **Snyder, W. C., and H. N. Hansen.** 1940. The species concept in *Fusarium*. *American Journal of Botany* **27:** 64-67. (8, 9, *redolens*, *solani*)

2013. **Snyder, W. C., and H. N. Hansen.** 1941. The species concept in *Fusarium* with reference to section *Martiella*. *American Journal of Botany* **28:** 738-742. (8)

2014. **Snyder, W. C. and H. N. Hansen.** 1941. The effect of light on taxonomic characters in *Fusarium*. *Mycologia* **33:** 580-591. (4)

2015. **Snyder, W. C., and H. N. Hansen.** 1945. The species concept in *Fusarium* with reference to *Discolor* and other sections. *American Journal of Botany* **32:** 657-666. (8)

2016. **Snyder, W. C., and H. N. Hansen.** 1946. Control of culture mites by cigarette paper barriers. *Mycologia* **38:** 455-462. (4)

2017. **Snyder, W. C., and H. N. Hansen.** 1947. Advantages of natural media and environments in the culture of fungi. *Phytopathology* **37:** 420-421. (2, 4)

2018. **Snyder, W. C., and H. N. Hansen.** 1954. Variation and speciation in the genus *Fusarium*. *Annals of the New York Academy of Sciences* **60:** 16-23. (8, 9)

2019. **Sobek, E. A., and G. P. Munkvold.** 1999. European corn borer (Lepidoptera: Pyralidae) larvae as vectors of *Fusarium moniliforme*, causing kernel rot and symptomless infection of maize kernels. *Journal of Economic Entomology* **92:** 503-509. (*verticillioides*)

2020. **Solfrizzo, M., and A. Visconti.** 1996. Simultaneous high-performance liquid chromatographic determination of visoltricin, acuminatopyrone, and chlamydosporols in *Fusarium* cultures on maize. *Journal of Chromatography A* **730:** 69-73. (*chlamydosporum*)

2021. **Soliday, C. L., M. B. Dickman, and P. E. Kolattukudy.** 1989. Structure of the cutinase gene and detection of promoter activity in the 5' flanking region by fungal transformation. *Journal of Bacteriology* **171:** 1942-1951. (*solani*)

2022. **Soliday, C. L., W. H. Flurkey, and T. W. Okita.** 1984. Cloning and structure determination of complementary DNA for cutinase, an enzyme involved in fungal penetration for plants. *Proceedings of the National Academy of Sciences USA* **81:** 3939-3943. (*solani*)

2023. **Soliday, C. L., and P. E. Kolattukudy.** 1983. Primary structure of the active site region of fungal cutinase, an enzyme involved in phytopathogenesis. *Biochemical and Biophysical Research Communications* **114:** 1017-1022. (*solani*)

2024. **Soni, G. L., and I. S. Bhatia.** 1979. Studies on the extracellular cellulase from *Fusarium oxysporum*. *Indian Journal of Biochemistry and Biophysics* **16:** 223-226. (*oxysporum*)

2025. **Soni, G. L., and I. S. Bhatia.** 1981. Studies on pectinases from *Fusarium oxysporum*. *Indian Journal of Experimental Biology* **19:** 547-550. (*oxysporum*)

2026. **Soonthornpoct, P., L. E. Trevathan, and D. Ingram.** 2000. The colonization of maize seedling roots and rhizosphere by *Fusarium* spp. in Mississippi in two soil types under conventional tillage and no-tillage systems. *Phytoprotection* **81:** 97-106. (*verticillioides*)

2027. **Soriano, J. M., and S. Dragacci.** 2004. Occurrence of fumonisins in foods. *Food Research International* **37:** 985-1000. (*verticillioides*)

2028. **Souheil, H., A. Vey, P. Thuet, and J.-P. Trilles.** 1999. Pathogenic and toxic effects of *Fusarium oxysporum* (Schlecht.) on survival and osmoregulatory capacity of *Penaeus japonicus* (Bate). *Aquaculture* **178:** 209-224. (*oxysporum*)

2029. **Speeleveld, E., B. Gordts, H. W. van Landuyt, C. de Vroey, and C. Raes-Wuytack.** 1996. Susceptibility of clinical isolates of *Fusarium* to antifungal drugs. *Mycoses* **39:** 37-40. (*oxysporum*)

2030. **Spencer, G. F., A. E. Desjardins, and R. D. Plattner.** 1990. 5-2-Carboxy-6-hydroxy-7-methoxybenzofuran, a fungal metabolite of xanthotoxin. *Phytochemistry* **29:** 2495-2498. (*sambucinum*)

2031. **Srdic, N., S. Radulovic, Z. Nonkovic, S. Velimirovic, L. Cvetkovic, and I. Vico.** 1993. Two cases of exogenous endophthalmitis due to *Fusarium moniliforme* and *Pseudomonas* species as associated aetiological agents. *Mycoses* **36:** 441-444. (*subglutinans*)

2032. **Staben, C., and C. Yanofsky.** 1990. *Neurospora crassa a* mating type region. *Proceedings of the National Academy of Sciences (USA)* **87:** 4917-4921. (6)

2033. **Stack, R. W.** 1989. A comparison of the inoculum potential of ascospores and conidia of *Gibberella zeae*. *Canadian Journal of Plant Pathology* **11:** 137-142. (*graminearum*)

2034. **Stack, R. W., R. K. Horst, P. E. Nelson, and R. W. Langhans.** 1976. Differential susceptibility to Fusarium stub dieback in carnation cultivars. *Journal of the American Society for Horticultural Science* **101:** 654-657. (*graminearum*)

2035. **Stahl, D. J., A. Theuerkauf, R. Heitefuss, W. Schaefer.** 1994. Cutinase of *Nectria haematococca* (*Fusarium solani* f. sp. *pisi*) is not required for fungal virulence or organ specificity on pea. *Molecular Plant-Microbe Interactions* **7:** 713-725. (*solani*)

2036. **Stalhammar, C. M.** 1976. An electron microscope study of spore germination in *Fusarium sambucinum* var. *coeruleum*. *Transactions of the British Mycological Society* **67:** 83-90. (*torulosum*)

2037. **Starkey, D. E., T. J. Ward, K. L. O'Donnell, D. M. Geiser, G. Kuldau, R. M. Clear, L. R. Gale, H. C. Kistler, and T. Aoki.** 2005. Delineation of species boundaries within *Fusarium graminearum*, the causative agent of Fusarium Head Blight. *Fungal Genetics Newsletter* **52 (Suppl.):** 71. (*graminearum*)

2038. **Steenkamp, E., H. Britz, T. Coutinho, B. Wingfield, W. F. O. Marasas, and M. J. Wingfield.** 2000. Molecular characterization of *Fusarium subglutinans* associated with mango malformation. *Molecular Plant Pathology* **1:** 187-193. (*mangiferae*)

2039. **Steenkamp, E. T., T. A. Coutinho, A. E. Desjardins, B. D. Wingfield, W. F. O. Marasas, and M. J. Wingfield.** 2001. *Gibberella fujikuroi* mating population E is associated with maize and teosinte. *Molecular Plant Pathology* **2:** 215-221. (8, *subglutinans*)

2040. **Steenkamp, E. T., B. D. Wingfield, T. A. Coutinho, M. J. Wingfield, and W. F. O. Marasas.** 1999. Differentiation of *Fusarium subglutinans* f. sp. *pini* by histone gene sequence data. *Applied and Environmental Microbiology* **65:** 3401-3406. (9, *circinatum*, *proliferatum*)

2041. **Steenkamp, E. T., B. D. Wingfield, T. A. Coutinho, K. A. Zeller, M. J. Wingfield, W. F. O. Marasas, and J. F. Leslie.** 2000. PCR-based identification of *MAT-1* and *MAT-2* in the *Gibberella fujikuroi* species complex. *Applied and Environmental Microbiology* **66:** 4378-4382. (6, 7, 9, 11)

2042. **Steenkamp, E. T., B. D. Wingfield, A. E. Desjardins, W. F. O. Marasas, and M. J. Wingfield.** 2002. Cryptic speciation in *Fusarium subglutinans*. *Mycologia* **94:** 1032-1043. (8, *subglutinans*)

2043. **Steinberg, C., V. Edel, N. Gautheron, C. Abadie, T. Vallaeys, and C. Alabouvette.** 1997. Phenotypic characterization of natural populations of *Fusarium oxysporum* in relation to genotypic characterization. *FEMS Microbiology Ecology* **24:** 73-85. (*oxysporum*)

2044. **Steinberg, C., J. M. Whipps, D. Wood, J. Fenlon, and C. Alabouvette.** 1999. Mycelial development of *Fusarium oxysporum* in the vicinity of tomato roots. *Mycological Research* **103:** 769-778. (*oxysporum*)

2045. **Stephens, C. T., R. M. de Vries, and K. C. Sink.** 1989. Evaluation of *Asparagus* species for resistance to *Fusarium oxysporum* f. sp. *asparagi* and *Fusarium moniliforme*. *Hortscience* **24:** 365-368. (*verticillioides*)

2046. **Stevenson, W. R., R. Loria, G. D. Franc, and D. P. Weingartner.** 2001. *Compendium of Potato Diseases*, 2nd ed. APS Press, St. Paul, Minnesota. (*sambucinum*)

2047. **Stewart, A., and J. W. Deacon.** 1995. Vital fluorochromes as tracers for fungal growth studies. *Biotechnic and Histochemistry* **70:** 57-65. (*oxysporum*)

2048. **Stewart, D. W., L. M. Dwyer, and L. M. Reid.** 1998. Aspects of maize modeling in eastern Canada. *Canadian Journal of Soil Science* **78:** 421-429. (11)

2049. **Stewart, D. W., L. M. Reid, R. W. Nicol, and A. W. Schaafsma.** 2002. A mathematical simulation of growth of *Fusarium* in maize ears after artificial inoculation. *Phytopathology* **92:** 534-541. (11, *verticillioides*)

2050. **Stone, J. M., J. E. Heard, T. Asai, and F. M. Ausubel.** 2000. Simulation of fungal-mediated cell death by fumonisin B_1 and selection of fumonisin B_1-resistant (*fbr*) *Arabidopsis* mutants. *Plant Cell* **12:** 1811-1822. (*verticillioides*)

2051. **Storer, A. J., D. L. Wood, and T. R. Gordon.** 2004. Twig beetles, *Pifyophthorus* spp. (Coleoptera: Scolytidae), as vectors of the pitch canker pathogen in California. *Canadian Entomologist* **136:** 685-693. (*circinatum*)

2052. **Stover, R. H.** 1959. Studies on Fusarium wilt of bananas. IV. Clonal differentiation among wild type isolates of *F. oxysporum* f. *cubense*. *Canadian Journal of Botany* **37:** 245-255. (5)

2053. **Strand, L. L., M. E. Corden, and D. L. Macdonald.** 1976. Characterization of 2-endopolygalacturonase e.c. 3.2.1.15 isozymes produced by *Fusarium oxysporum* f. sp. *lycopersici*. *Biochimica et Biophysica Acta* **429:** 870-883. (*oxysporum*)

2054. **Straney, D., Y. Ruan, and J. He.** 1994. *In vitro* transcription and binding analysis of promoter regulation by a host-specific signal in a phytopathogenic fungus. *Antonie van Leeuwenhoek* **65:** 183-189. (*solani*)

2055. **Straney, D. C., and H. D. VanEtten.** 1994. Characterization of the *PDA1* promoter of *Nectria haematococca* and identification of a region that binds a pisatin-responsive DNA binding factor. *Molecular Plant-Microbe Interactions* **7:** 256-266. (*solani*)

2056. **Sturtevant, A. H.** 1913. The linear arrangement of six sex-linked factors in *Drosophila*, as shown by their mode of association. *Journal of Experimental Zoology* **14:** 43-59. (7)

2057. **Styer, R. C., and D. J. Cantliffe.** 1984. Infection of two endosperm mutants of sweet corn by *Fusarium moniliforme* and its effect on seedling vigor. *Phytopathology* **74:** 189-194. (*verticillioides*)

2058. **Subbarao, K. V., and T. J. Michailides.** 1992. A reevaluation of *Fusarium moniliforme* var. *fici*, the causal agent of fig endosepsis. *Mycological Research* **96:** 766-768. (*lactis*)

2059. **Subbarao, K. V., and T. J. Michailides.** 1993. Virulence of *Fusarium* species causing fig endosepsis in cultivated and wild caprifigs. *Phytopathology* **83:** 527-533. (*dimerum, lactis*)

2060. **Subbarao, K. V., and T. J. Michailides.** 1995. Effects of temperature on isolates of *Fusarium moniliforme* causing fig endosepsis and *Aspergillus niger* causing smut. *Phytopathology* **85:** 662-668. (*lactis*)

2061. **Subbarao, K. V., T. J. Michailides, and D. P. Morgan.** 1993. Effects of osmotic potential and temperature on growth of two pathogens of figs and a biocontrol agent. *Phytopathology* **83:** 1454-1459. (*lactis*)

2062. **Subramanian, C. V.** 1971. *Hyphomycetes – an Account of Indian Species, except Cercosporae*. Indian Council for Agricultural Research, New Delhi. (*semitectum*)

2063. **Subramanian, C. V., and D. J. Bhat.** 1983. Developmental morphology of ascomycetes. 9. *Calonectria rigidiuscula*. *Cryptogamie Mycologie* **4:** 269-282. (*decemcellulare*)

2064. **Suga, H., L. R. Gale, and H. C. Kistler.** 2004. Development of VNTR markers for two *Fusarium graminearum* clade species. *Molecular Ecology Notes* **4:** 468-470. (*graminearum*)

2065. **Suga, H., T. Hasegawa, H. Mitsui, K. Kageyama, and M. Hyakumachi.** 2000. Phylogenetic analysis of the phytopathogenic fungus *Fusarium solani* based on the rDNA-ITS region. *Mycological Research* **104:** 1175-1183. (8, 9, *solani*)

2066. **Suga, H., S. Ikeda, M. Taga, K. Kageyama, and M. Hyakumachi.** 2002. Electrophoretic karyotyping and gene mapping of seven *formae speciales* in *Fusarium solani*. *Current Genetics* **41:** 254-260. (*solani*)

2067. **Suga, H., K. Oyabu, M. Ito, K. Kageyama, and M. Hyakumachi.** 2000. Detection of intron-like sequences in the small subunit rDNA 3' region of *Fusarium solani*. *Mycological Research* **104:** 782-787. (*solani*)

2068. **Sugiura, Y., J. R. Barr, D. B. Barr, J. W. Brock, C. M. Elie, Y. Ueno, D. G. Patterson, Jr., M. E. Potter, and E. Reiss.** 1999. Physiological characteristics and mycotoxins of human clinical isolates of *Fusarium* species. *Mycological Research* **103:** 1462-1468. (*solani*)

2069. **Sugiura, Y., K. Fukasaku, T. Tanaka, Y. Matsui, and Y. Ueno.** 1993. *Fusarium poae* and *Fusarium crookwellense*, fungi responsible for the natural occurrence of nivalenol in Hokkaido. *Applied and Environmental Microbiology* **59:** 3334-3338. (*poae*)

2070. **Sugiura, Y., K. Y. Sugita, S. Kumagai and E. Reiss.** 2003. Experimental murine hyalohyphomycosis with soil-derived isolates of *Fusarium solani*. *Medical Mycology* **41:** 241-247. (*solani*)

2071. **Suihko, M. L.** 1983. The fermentation of different carbon sources by *Fusarium oxysporum*. *Biotechnology Letters* **5:** 721-724. (*oxysporum*)

2072. **Suihko, M. L., and T. M. Enari.** 1981. Production of ethanol from D-glucose and D-xylose by different *Fusarium* strains. *Biotechnology Letters* **3:** 723-728. (*oxysporum*)

2073. **Suihko, M. L., I. Suomalainen, and T. M. Enari.** 1983. D-xylose catabolism in *Fusarium oxysporum*. *Biotechnology Letters* **5:** 525-530. (*oxysporum*)

2074. **Suleman, P., A. M. Tohamy, A. A. Saleh, M. A. Madkour, and D. C. Straney.** 1996. Variation in sensitivity to tomatine and rishitin among isolates of *Fusarium oxysporum* f. sp. *lycopersici*, and strains not pathogenic on tomato. *Physiological and Molecular Plant Pathology* **48:** 131-144. (*oxysporum*)

2075. **Sulzenbacher, G., M. Schulein, and G. J. Davies.** 1997. Structure of the endoglucanase I from *Fusarium oxysporum*: Native, cellobiose, and 3,4-epoxybutyl β-D-cellobioside-inhibited forms, at 2.3 Å resolution. *Biochemistry* **36:** 5902-5911. (*oxysporum*)

2076. **Summerbell, R. C., S. E. Richardson, and J. Kane.** 1988. *Fusarium proliferatum* as an agent of disseminated infection in an immunosuppressed patient. *Journal of Clinical Microbiology* **26:** 82-87. (*proliferatum*)

2077. **Summerell, B. A., and L. W. Burgess.** 1988. Stubble management practices and the survival of *Fusarium graminearum* Group 1 in wheat stubble residues. *Australasian Plant Pathology* **17:** 88-93. (2, *pseudograminearum*)

2078. **Summerell, B. A., L. W. Burgess, D. Backhouse, S. Bullock, and L. J. Swan.** 2001. Natural occurrence of perithecia of *Gibberella coronicola* on wheat plants with crown rot in Australia. *Australasian Plant Pathology* **30:** 353-356. (10, *pseudograminearum*)

2079. **Summerell, B. A., L. W. Burgess, S. Bullock, D. Backhouse, and D. T. Nguyen.** 1998. Occurrence of perithecia of *Gibberella fujikuroi* mating population A (*Fusarium moniliforme*) on maize stubble in northern Vietnam. *Mycologia* **90:** 890-895. (10, *verticillioides*)

2080. **Summerell, B. A., L. W. Burgess, and T. A. Klein.** 1989. The impact of stubble management practices on crown rot of wheat. *Australian Journal of Experimental Agriculture* **29:** 1015-1019. (*pseudograminearum*)

2081. **Summerell, B. A., L. W. Burgess, and P. G. Nixon.** 1990. Crown and stem canker of waratah caused by *Cylindrocarpon destructans*. *Australasian Plant Pathology* **19:** 13-15. (2)

2082. **Summerell, B. A., and J. F. Leslie.** 2004. Genetic diversity and population structure of plant-pathogenic species in the genus *Fusarium*, p. 205-221. *In* M. Gillings and A. Holmes (eds.), *Plant Microbiology*. BIOS Scientific Publishers, New York. (*oxysporum*)

2083. **Summerell, B. A., J. F. Leslie, D. Backhouse, W. L. Bryden, and L. W. Burgess, eds.** 2001. Fusarium*: Paul E. Nelson Memorial Symposium*. APS Press, St. Paul, Minnesota. (P, 9)

2084. **Summerell, B. A., and C. A. Rugg.** 1992. Vascular wilt of *Helichrysum* species caused by *Fusarium oxysporum*. *Australasian Plant Pathology* **21:** 18-19. (*oxysporum*)

2085. **Summerell, B. A., C. A. Rugg, and L. W. Burgess.** 1993. Mycogeography of *Fusarium*: A survey of *Fusarium* species in forest and woodland soils in north Queensland, Australia. *Mycological Research* **97:** 1015-1019. (*avenaceum, longipes, solani*)

2086. **Summerell, B. A., C. A. Rugg, and L. W. Burgess.** 1995. Characterization of *Fusarium babinda* sp. nov. *Mycological Research* **99:** 1345-1348. (*babinda*)

2087. **Summerell, B. A., B. Salleh, and J. F. Leslie.** 2003. A utilitarian approach to *Fusarium* identification. *Plant Disease* **87:** 117-128. (P, 11, 12)

2088. **Sunder, S. S.** 1998. Vegetative compatibility, biosynthesis of GA_3 and virulence of *Fusarium moniliforme* isolates from bakanae disease of rice. *Plant Pathology* **47:** 767-772. (5, *fujikuroi*)

2089. **Sutherland, J. B., A. L. Pometto, III, and D. L. Crawford.** 1983. Lignocellulose degradation by *Fusarium* spp. *Canadian Journal of Botany* **61:** 1194-1198. (*solani*)

2090. **Sutherland, M. L., and G. F. Pegg.** 1992. The basis of host recognition in *Fusarium oxysporum* f. sp. *lycopersici*. *Physiological and Molecular Plant Pathology* **40:** 423-436. (*oxysporum*)

2091. **Swan, L. J., D. Backhouse, and L. W. Burgess.** 2000. Surface soil moisture and stubble management practice effects on the progress of infection of wheat by *Fusarium pseudograminearum*. *Australian Journal of Experimental Agriculture* **40:** 693-698. (*pseudograminearum*)

2092. **Swart, K., D. van Heemst, M. Slakhorst, F. Debets, and C. Heyting.** 2001. Isolation and characterization of sexual sporulation mutants of *Aspergillus nidulans*. *Fungal Genetics and Biology* **33:** 25-35. (6)

2093. **Swift, C. E., E. R. Wickliffe, and H. F. Schwartz.** 2002. Vegetative compatibility groups of *Fusarium oxysporum* f. sp. *cepae* from onion in Colorado. *Plant Disease* **86:** 606-610. (*oxysporum*)

2094. **Sydenham, E. W., W. F. O. Marasas, P. G. Thiel, G. S. Shephard, and J. J. Nieuwenhuis.** 1991. Production of mycotoxins by selected *Fusarium graminearum* and *F. crookwellense* isolates. *Food Additives and Contaminants* **8:** 31-41. (*crookwellense*)

2095. **Sydenham, E. W., G. S. Shephard, S. Stockenstrom, J. P. Rheeder, W. F. O. Marasas, and M. J. van der Merwe.** 1997. Production of fumonisin B analogues and related compounds by *Fusarium globosum*, a newly described species from corn. *Journal of Agricultural and Food Chemistry* **45:** 4004-4010. (*globosum*)

2096. **Szathmary, C. I., C. J. Mirocha, M. Palyusik, and S. V. Pathre.** 1976. Identification of mycotoxins produced by species of *Fusarium* and *Stachybotrys* obtained from Eastern Europe. *Applied and Environmental Microbiology* **32:** 579-584. (*poae, sporotrichioides*)

2097. **Szecsi, A., and A. Dobrovolszky.** 1985. Phylogenetic relationships among *Fusarium* spp. measured by DNA reassociation. *Mycopathologia* **89:** 89-94. (*oxysporum*)

2098. **Szecsi, A., and A. Dobrovolszky.** 1985. Genetic distance in the fungal genus *Fusarium* measured by comparative computer analysis of DNA thermal denaturation profiles. *Mycopathologia* **89:** 95-100. (*oxysporum*)

2099. **Tag, A. G., G. F. Garifullina, A. W. Peplow, C. Ake, Jr., T. D. Phillips, T. M. Hohn, and M. N. Beremand.** 2001. A novel regulatory gene, *Tri10*, controls trichothecene toxin production and gene expression. *Applied and Environmental Microbiology* **67:** 5294-5302. (*sambucinum, sporotrichioides*)

2100. **Tag, A., J. Hicks, G. Garifullina, C. Ake, Jr., T. D. Phillips, M. Beremand, and N. Keller.** 2000. G-protein signaling mediates differential production of toxic secondary metabolites. *Molecular Microbiology* **38:** 658-665. (*sporotrichioides*)

2101. **Taga, M., M. Murata, and H. Saito.** 1998. Comparison of different karyotyping methods in filamentous ascomycetes: A case study of *Nectria haematococca*. *Mycological Research* **102:** 1355-1364. (*solani*)

2102. **Taga, M., C. Waalwijk, W. G. Flier, and G. H. J. Kema.** 2003. Cytological karyotyping of somatic chromosomes from *Phytophthora infestans*, *Mycosphaerella graminicola*, and *Fusarium* spp. *Fungal Genetics Newsletter* **50 (Suppl.):** 149. (7)

2103. **Takaya, N.** 2002. Dissimilatory nitrate reduction metabolisms and their control in fungi. *Journal of Bioscience and Bioengineering* **94:** 506-510. (*oxysporum*)

2104. **Takaya, N., S. Kuwazaki, Y. Adachi, S. Suzuki, T. Kikuchi, H. Nakamura, Y. Shiro, and H. Shoun.** 2003. Hybrid respiration in the denitrifying mitochondria of *Fusarium oxysporum*. *Journal of Biochemistry* **133:** 461-465. (*oxysporum*)

2105. **Takaya, N., and H. Shoun.** 2000. Nitric oxide reduction, the last step in denitrification by *Fusarium oxysporum*, is obligatorily mediated by cytochrome P450nor. *Molecular and General Genetics* **263:** 342-348. (*oxysporum*)

2106. **Takaya, N., S. Suzuki, S. Kuwazaki, H. Shoun, F. Maruo, M. Yamaguchi, and K. Takeo.** 1999. Cytochrome P450nor, a novel class of mitochondrial cytochrome P450 involved in nitrate respiration in the fungus *Fusarium oxysporum*. *Archives of Biochemistry and Biophysics* **372:** 340-346. (*oxysporum*)

2107. **Takaya, N., H. Uchimura, Y. Lai, and H. Shoun.** 2002. Transcriptional control of nitric oxide reductase gene (*CYP55*) in the fungal denitrifier *Fusarium oxysporum*. *Bioscience, Biotechnology and Biochemistry* **66:** 1039-1045. (*oxysporum*)

2108. **Takehara, T., and K. Kuniyasu.** 1995. Use of nitrate non-utilizing mutants in ecological studies of Fusarium diseases: III. Growth, benomyl sensitivity, pathogenicity, and stability of *nit* mutants of *Fusarium oxysporum* compared to wild-type strains. *Annals of the Phytopathological Society of Japan* **61:** 541-548. (*oxysporum*)

2109. **Takehara, T., K. Kuniyasu, M. Mori, and H. Hagiwara.** 2003. Use of a nitrate non-utilizing mutant and selective media to examine population dynamics of *Fusarium oxysporum* f. sp. *spinaciae* in soil. *Phytopathology* **93:** 1173-1181. (*oxysporum*)

2110. **Takeuchi, M., M. Nakajima, T. Ogita, M. Inukai, K. Kodama, K. Furuya, H. Nagaki, and T. Haneishi.** 1989. Fosfonochlorin: A new antibiotic with spheroplast forming activity. *Journal of Antibiotics* **42:** 198-205. (*tricinctum*)

2111. **Talbot, N. J., P. Vincent, and H. G. Wildman.** 1996. The influence of genotype and environment on the physiological and metabolic diversity of *Fusarium compactum*. *Fungal Genetics and Biology* **20:** 254-267. (*compactum*)

2112. **Tan, M.-K., and L. M. Niessen.** 2003. Analysis of rDNA ITS sequences to determine genetic relationships among, and provide a basis for simplified diagnosis of, *Fusarium* species causing crown rot and head blight of cereals. *Mycological Research* **107:** 811-821. (*avenaceum, oxysporum, pseudograminearum, sambucinum, solani, torulosum, tricinctum*)

2113. **Tan, M.-K., S. Simpfendorfer, D. Backhouse, and G. M. Murray.** 2004. Occurrence of Fusarium head blight (FHB) in southern NSW in 2000: Identification of causal fungi and determination of putative chemotype of *Fusarium graminearum* isolates by PCR. *Australasian Plant Pathology* **33:** 385-392. (*graminearum*)

2114. **Tang, Y.-X., Z.-H. Sun, L. Hua, C.-F. Lv, X.-F. Guo, and J. Wang.** 2002. Kinetic resolution of DL-pantolactone by immobilized *Fusarium moniliforme* SW-902. *Process Biochemistry* **38:** 545-549. (*verticillioides*)

2115. **Tantaoui, A., M. Ouinten, J. P. Geiger, and D. Fernandez.** 1996. Characterization of a single clonal lineage of *Fusarium oxysporum* f. sp. *albedinis* causing Bayoud disease of date palm in Morocco. *Phytopathology* **86:** 787-792. (*oxysporum*)

2116. **Tarekegn, G., N. W. McLaren, and W. J. Swart.** 2004. Relationship between grain development stage and sorghum cultivar susceptibility to grain mold. *African Plant Protection* **10:** 53-62. (*thapsinum*)

2117. **Tatum, J. H., and R. A. Baker.** 1983. Naphthoquinones produced by *Fusarium solani* isolated from citrus. *Phytochemistry* **22:** 543-548. (*solani*)

2118. **Tatum, J. H., R. A. Baker, and R. E. Berry.** 1985. Three further naphthoquinones produced by *Fusarium solani*. *Phytochemistry* **24:** 3019-3022. (*solani*)

2119. **Tatum, J. H., R. A. Baker, and R. E. Berry.** 1987. Naphthoquinones and derivatives from *Fusarium*. *Phytochemistry* **26:** 795-798. (*solani*)

2120. **Tatum, J. H., R. A. Baker, and R. E. Berry.** 1989. Metabolites of *Fusarium solani*. *Phytochemistry* **28:** 283-284. (*solani*)

2121. **Taylor, J. W., D. J. Jacobson, S. Kroken, T. Kasuga, D. M. Geiser, D. S. Hibbett, and M. C. Fisher.** 2000. Phylogenetic species recognition and species concepts in fungi. *Fungal Genetics and Biology* **31:** 21-32. (9)

2122. **Tee, R. D., D. J. Gordon, J. A. Welch, and A. J. N. Taylor.** 1993. Investigation of possible adverse allergic reactions to mycoprotein ("Quorn"). *Clinical and Experimental Allergy* **23:** 257-260. (*venenatum*)

2123. **Teichert, S., B. Schoenig, S. Richter, and B. Tudzynski.** 2004. Deletion of the *Gibberella fujikuroi* glutamine synthetase gene has significant impact on transcriptional control of primary and secondary metabolism. *Molecular Microbiology* **53:** 1661-1675. (*fujikuroi*)

2124. **Tekauz, A., B. McCallum, N. Ames, and J. M. Fetch.** 2004. Fusarium head blight of oat – Current status in western Canada. *Canadian Journal of Plant Pathology* **26:** 473-479. (*graminearum, poae*)

2125. **Templeton, A. R.** 1989. The meaning of species and speciation: A genetic perspective, p. 3-27. *In* D. Otte and J. A. Endler, eds., *Speciation and its Consequences*. Sinauer Associates, Sunderland, Massachusetts. (9)

2126. **Temporini, E. D., and H. D. VanEtten.** 2004. An analysis of the phylogenetic distribution of the pea pathogenicity genes of *Nectria haematococca* MPVI supports the hypothesis of their origin by horizontal transfer and uncovers a potentially new pathogen of garden pea: *Neocosmospora boniensis*. *Current Genetics* **46:** 29-36. (*solani*)

2127. **Tesso, T., L. E. Claflin, and M. R. Tuinstra.** 2004. Estimation of combining ability for resistance to Fusarium stalk rot in grain sorghum. *Crop Science* **44:** 1195-1199. (*thapsinum*)

2128. **Teunissen, H. A. S., Y. Boer, J. Verkooijen, B. J. C. Cornelissen, and M. A. Haring.** 2003. A near-isogenic *Fusarium oxysporum* f. sp. *lycopersici* strain with a novel combination of avirulence characteristics. *Phytopathology* **93:** 1360-1367. (*oxysporum*)

2129. **Teunissen, H. A. S., M. Rep, P. M. Houterman, B. J. C. Cornelissen, and M. A. Haring.** 2003*b*. Construction of a mitotic linkage map of *Fusarium oxysporum* based on *Foxy*-AFLPs. *Molecular Genetics and Genomics* **269:** 215-226. (*oxysporum*)

2130. **Teunissen, H. A. S., J. Verkooijen, B. J. C. Cornelissen, and M. A. Haring.** 2002. Genetic exchange of avirulence determinants and extensive karyotype rearrangements in parasexual recombinants of *Fusarium oxysporum*. *Molecular Genetics and Genomics* **268:** 298-310. (*oxysporum*)

2131. **Tezuka, N., and T. Makino.** 1991. Biological control of Fusarium wilt of strawberry by nonpathogenic *Fusarium oxysporum* isolated from strawberry. *Annals of the Phytopathological Society of Japan* **57:** 506-511. (*oxysporum*)

2132. **Thiel, P. G., W. F. O. Marasas, E. W. Sydenham, G. S. Shephard, W. C. A. Gelderblom, and J. J. Nieuwenhuis.** 1991. Survey of fumonisin production by *Fusarium* spp. *Applied and Environmental Microbiology* **57:** 1089-1093. (*nygamai, proliferatum*)

2133. **Thiel, P. G., C. J. Meyer, and W. F. O. Marasas.** 1982. Natural occurrence of moniliformin together with deoxynivalenol and zearalenone in Transkeian South Africa corn. *Journal of Agricultural and Food Chemistry* **30:** 308-312. (*subglutinans*)

2134. **Thomas, H., A. Heller, J. Sauerborn, and D. Mueller-Stoever.** 1999. *Fusarium oxysporum* f. sp. *orthoceras*, a potential mycoherbicide, parasitizes seeds of *Orobanche cumana* (sunflower broomrape): A cytological study. *Annals of Botany* **83:** 453-458. (*oxysporum*)

2135. **Thomas, H., J. Sauerborn, D. Mueller-Stoever, A. Ziegler, J. S. Bedi, and J. Kroschel.** 1998. The potential of *Fusarium oxysporum* f. sp. *orthoceras* as a biological control agent for *Orobanche cumana* in sunflower. *Biological Control* **13:** 41-48. (*oxysporum*)

2136. **Thomas, J. L., and M. O. Moss.** 1990. The loss of biological activity of the preservative bronopol associated with *Fusarium solani*. *International Biodeterioration* **26:** 327-336. (*solani*)

2137. **Thomas, M. D., and I. W. Buddenhagen.** 1980. Incidence and persistence of *Fusarium moniliforme* in symptomless maize (*Zea mays*) kernels and seedlings in Nigeria. *Mycologia* **72:** 882-887. (*verticillioides*)

2138. **Thomas, P. A., D. J. Abraham, C. M. Kalavathy, and J. Rajasekaran.** 1988. Oral itraconazole therapy for mycotic keratitis. *Mycoses* **31:** 271-279. (*solani*)

2139. **Thrall, P. H., J. D. Bever, J. D. Mihail, and H. M. Alexander.** 1997. The population dynamics of annual plants and soilborne fungal pathogens. *Journal of Ecology* **85:** 313-328. (11)

2140. **Thrane, U.** 1988. Screening for fusarin C production by European isolates of *Fusarium* species. *Mycotoxin Research* **4:** 2-10. (*avenaceum*, *crookwellense*, *culmorum*, *graminearum*, *poae*, *sporotrichioides*, *tricinctum*)

2141. **Thrane, U.** 1990. Grouping *Fusarium* section *Discolor* isolates by statistical analysis of quantitative high performance liquid chromatographic data on secondary metabolite production. *Journal of Microbiological Methods* **12:** 23-40. (*culmorum*)

2142. **Thrane, U.** 2001. Developments in the taxonomy of *Fusarium* species based on secondary metabolites, p. 29-49. *In* B. A. Summerell, J. F. Leslie, D. Backhouse, W. L. Bryden, and L. W. Burgess (eds.), Fusarium*: Paul E. Nelson Memorial Symposium.* APS Press, St. Paul, Minnesota. (9, 11, 12)

2143. **Thrane, U., A. Adler, P. E. Clasen, F. Galvano, W. Langseth, A. Logrieco, K. F. Nielsen, and A. Ritieni.** 2004. Diversity in metabolite production by *Fusarium langsethiae*, *Fusarium poae*, and *Fusarium sporotrichioides*. *International Journal of Food Microbiology* **95:** 257-266. (*poae*)

2144. **Thrane, U., and U. Hansen.** 1995. Chemical and physiological characterization of taxa in the *Fusarium sambucinum* complex. *Mycopathologia* **129:** 183-190. (*sambucinum*, *torulosum*, *venenatum*)

2145. **Tibayrenc, M.** 1999. Toward an integrated genetic epidemiology of parasitic protozoa and other pathogens. *Annual Review of Genetics* **33:** 449-477. (9)

2146. **Tiedt, L. R., and W. J. Jooste.** 1988. Aberrant conidiogenesis in a *Fusarium subglutinans* isolate. *Transactions of the British Mycological Society* **90:** 527-530. (*subglutinans*)

2147. **Tiedt, L. R., and W. J. Jooste.** 1988. Ultrastructure of collarette formation in *Fusarium* section *Liseola* and some taxonomic implications. *Transactions of the British Mycological Society* **90:** 531-536. (*fujikuroi*, *succisae*, *verticillioides*)

2148. **Tiedt, L. R., and W. J. Jooste.** 1992. Ultrastructural aspects of conidiogenesis of *Fusarium* spp. in the section *Liseola*. *Mycological Research* **96:** 187-193. (*verticillioides*)

2149. **Tiedt, L. R., W. J. Jooste, and A. V. L. Hamilton.** 1986. Ultrastructural aspects of conidiogenesis of *Fusarium sacchari* f. sp. *subglutinans*. *Transactions of the British Mycological Society* **87:** 237-242. (*subglutinans*)

2150. **Tinline, R. D., G. B. Wildermuth, and D. T. Spurr.** 1988. Inoculum density of *Cochliobolus sativus* in soil and common root rot of wheat cultivars in Queensland. *Australian Journal of Agricultural Research* **39:** 569-577. (*chlamydosporum*)

2151. **Tio, M., L. W. Burgess, P. E. Nelson, and T. A. Toussoun.** 1977. Techniques for the isolation, culture, and preservation of the Fusaria. *Australasian Plant Pathology Society Newsletter* **6:** 11-13. (2, 4)

2152. **Tiribelli, M., F. Zaja, C. Fili, T. Michelutti S. Prosdocimo, A. Candoni, and R. Fanin.** 2002. Endogenous endophthalmitis following disseminated fungemia due to *Fusarium solani* in a patient with acute myeloid leukemia. *European Journal of Haematology* **68:** 314-317. (*solani*)

2153. **To-Anun, C., H. Nelson, and S. Ouchi.** 1995. Electrophoretic karyotyping of *Fusarium oxysporum*. *Annals of the Phytopathological Society of Japan* **61:** 350-356. (*oxysporum*)

2154. **Togawa, M.** 1992. Effects of sterilization methods, plant varieties and leaf stages on conidia and perithecia formation in the genus *Fusarium* in CLA culture. *Transactions of the Mycological Society of Japan* **33:** 385-393. (2)

2155. **Togawa, M., and A. Nomura.** 1998. Dieback of Atemoya caused by *Fusarium decemcellulare* Brick. *Annals of the Phytopathological Society of Japan* **64:** 217-220. (*decemcellulare*)

2156. **Tokai, T., M. Fujimura, H. Inoue, T. Aoki, K. Ohta, T. Shibata, I. Yamaguchi, and M. Kimura.** 2005. Concordant evolution of trichothecene 3-O-acetyltransferase and an rDNA species phylogeny of trichothecene-producing and non-producing Fusaria and other ascomycetous fungi. *Microbiology* **151:** 509-519. (*decemcellulare*, *fujikuroi*, *oxysporum*, *solani*)

2157. **Tomasini, A., C. Fajardo, and J. Barrios-Gonzalez.** 1997. Gibberellic acid production using different solid-state fermentation systems. *World Journal of Microbiology and Biotechnology* **13:** 203-206. (1)

2158. **Tomimori-Yamashita, J., M. M. Ogawa, S. H. Hirata, O. Fischman, N. S. Michalany, H. K. Yamashita, and M. M. A. Alchorne.** 2002. Mycetoma caused by *Fusarium solani* with osteolytic lesions on the hand: Case report. *Mycopathologia* **153:** 11-14. (*solani*)

2159. **Topakas, E., and P. Christakopoulos.** 2004. Production and partial characterization of alkaline feruloyl esterases by *Fusarium oxysporum* during submerged batch cultivation. *World Journal of Microbiology and Biotechnology* **20:** 245-250. (*oxysporum*)

2160. **Topakas, E., P. Christakopoulos, and C. B. Faulds.** 2005. Comparison of mesophilic and thermophilic feruloyl esterases: Characterization of their substrate specificity for methyl phenylalkanoates. *Journal of Biotechnology* **115:** 355-366. (*oxysporum*)

2161. **Topakas, E., H. Stamatis, P. Biely, D. Kekos, B. J. Macris, and P. Christakopoulos.** 2003. Purification and characterization of a feruloyl esterase from *Fusarium oxysporum* catalyzing esterification of phenolic acids in ternary water-organic solvent mixtures. *Journal of Biotechnology* **102:** 33-44. (*oxysporum*)

2162. **Topakas, E., H. Stamatis, M. Mastihubova, P. Biely, D. Kekos, B. J. Macris, and P. Christakopoulos.** 2003. Purification and characterization of a *Fusarium oxysporum* feruloyl esterase (FoFAE-I) catalyzing transesterification of phenolic acid esters. *Enzyme and Microbial Technology* **33:** 729-737. (*oxysporum*)

2163. **Torp, M., and W. Langseth.** 1999. Production of T-2 toxin by a *Fusarium* resembling *Fusarium poae*. *Mycopathologia* **147:** 89-96. (*poae*)

2164. **Torp, M., and H. I. Nirenberg.** 2004. *Fusarium langsethiae* sp. nov. on cereals in Europe. *International Journal of Food Microbiology* **95:** 247-256. (*poae*)

2165. **Torres, A., S. M. Li, S. Roussos, and M. Vert.** 1996. Screening of microorganisms for biodegradation of poly(lactic acid) and lactic acid-containing polymers. *Applied and Environmental Microbiology* **62:** 2393-2397 (*verticillioides*)

2166. **Toth, A., I. Barna-Vetro, A. Gyongyosi, A. Pomazi, A. Szecsi, and L. Hornok.** 1993. Prevalence and mycotoxin production of *Fusarium* species isolated from wheat grains in Hungary. *Acta Phytopathologica et Entomologica Hungarica* **28:** 3-12. (*poae*)

2167. **Toth, B., A. Mesterhazy, P. Nicholson, J. Teren, and J. Varga.** 2004. Mycotoxin production and molecular variability of European and American isolates of *Fusarium culmorum*. *European Journal of Plant Pathology* **110:** 587-599. (*culmorum*)

2168. **Toussoun, T. A., and P. E. Nelson.** 1975. Variation and speciation in the Fusaria. *Annual Review of Phytopathology* **13:** 11-13. (4)

2169. **Toussoun, T. A., and P. E. Nelson.** 1976. *A Pictorial Guide to the Identification of* Fusarium *Species*, 2nd ed. Pennsylvania State University Press, University Park, Pennsylvania. (8)

2170. **Toyota, K., T. Tsuge, and M. Kimura.** 1992. Potential application of genetic transformants of *Fusarium oxysporum* f. sp. *raphani* for assessing fungal autecology. *Soil Biology and Biochemistry* **24:** 489-494. (*oxysporum*)

2171. **Trail, F., and R. Common.** 2000. Perithecial development by *Gibberella zeae*: A light microscopy study. *Mycologia* **92:** 130-138. (10)

2172. **Trail, F., I. Gaffoor, and S. Vogel.** 2005. Ejection mechanics and trajectory of the ascospores of *Gibberella zeae* (anamorph *Fusarium graminearum*). *Fungal Genetics and Biology* **42:** 528-533. (10, *graminearum*)

2173. **Trail, F., and H. Xu.** 2002. Purification and characterization of mannitol dehydrogenase and identification of the corresponding cDNA from the head blight fungus, *Gibberella zeae* (*Fusarium graminearum*). *Phytochemistry* **61:** 791-796. (10)

2174. **Trail, F, H. Xu, R. Loranger, and D. Gadoury.** 2002. Physiological and environmental aspects of ascospore discharge in *Gibberella zeae* (anamorph *Fusarium graminearum*). *Mycologia* **94:** 181-189. (10)

2175. **Trail, F., J.-R. Xu, P. San Miguel, R. G. Halgren, and H. C. Kistler.** 2003. Analysis of expressed sequence tags from *Gibberella zeae* (anamorph *Fusarium graminearum*). *Fungal Genetics and Biology* **38:** 187-197. (10, *graminearum*)

2176. **Trimboli, D. S.** 1972. *Celosia* collar rot. *Australasian Plant Pathology Society Newsletter* **1:** 10. (*lateritium*)

2177. **Trimboli, D. S., and L. W. Burgess.** 1983. Reproduction of *Fusarium moniliforme* basal stalk rot and root rot of grain sorghum (*Sorghum bicolor*) in the greenhouse. *Plant Disease* **67:** 891-894. (*thapsinum*)

2178. **Trimboli, D. S., and L. W. Burgess.** 1985. Fungi associated with basal stalk rot and root rot of dryland grain sorghum in New South Wales. *Plant Protection Quarterly* **1:** 3-9. (*nygamai*)

2179. **Trinci, A. P. J.** 1992. Myco-protein – a twenty-year overnight success story. *Mycological Research* **96:** 1-13. (*venenatum*)

2180. **Trinci, A. P. J., G. D. Robson, and M. G. Wiebe.** 2001. Evolution of *Fusarium venenatum* A 3/5 in continuous flow (chemostat) culture, p. 161-175. *In* B. A. Summerell, J. F. Leslie, D. Backhouse, W. L. Bryden, and L. W. Burgess (eds.), Fusarium*: Paul E. Nelson Memorial Symposium.* APS Press, St. Paul, Minnesota. (*venenatum*)

2181. **Trouvelot, S., C. Olivain, G. Recorbet, Q. Migheli, and C. Alabouvette.** 2002. Recovery of *Fusarium oxysporum* Fo47 mutants affected in their biocontrol activity after transposition of the *Fot1* element. *Phytopathology* **92:** 936-945. (*oxysporum*)

2182. **Tschanz, A. T., R. K. Horst, and P. E. Nelson.** 1976. The effect of environment on sexual reproduction of *Gibberella zeae*. *Mycologia* **68:** 327-340. (*graminearum*)

2183. **Tudzynski, B.** 2005. Gibberellin biosynthesis in fungi: Genes, enzymes, evolution, and impact on biotechnology. *Applied Microbiology and Biotechnology* **66:** 597-611. (*fujikuroi*)

2184. **Tudzynski, B., P. Hedden, E. Carrera, and P. Gaskin.** 2001. The P450-4 gene of *Gibberella fujikuroi* encodes *ent*-kaurene oxidase in the gibberellin biosynthesis pathway. *Applied and Environmental Microbiology* **67:** 3514-3522. (*fujikuroi*)

2185. **Tudzynski, B., V. Homann, B. Teng, and G. A. Marzluf.** 1999. Isolation, characterization and disruption of the *areA* nitrogen regulatory gene of *Gibberella fujikuroi*. *Molecular and General Genetics* **261:** 106-114. (5, *fujikuroi*)

2186. **Tudzynski, B., S. Liu, and J. M. Kelly.** 2000. Carbon catabolite repression in plant pathogenic fungi: Isolation and characterization of the *Gibberella fujikuroi* and *Botrytis cinerea creA* genes. *FEMS Microbiology Letters* **184:** 9-15. (*fujikuroi*)

2187. **Tudzynski, B., M. Mihlan, M. C. Rojas, P. Linnemannstoens, P. Gaskin, and P. Hedden.** 2003. Characterization of the final two genes of the gibberellin biosynthesis gene cluster of *Gibberella fujikuroi*: *des* and *P450-3* encode GA4 desaturase and the 13-hydroxylase, respectively. *Journal of Biological Chemistry* **278:** 28635-28643. (*fujikuroi*)

2188. **Tudzynski, B., M. C. Rojas, P. Gaskin, and P. Hedden.** 2002. The gibberellin 20-oxidase of *Gibberella fujikuroi* is a multifunctional monooxygenase. *Journal of Biological Chemistry* **277:** 21246-21253. (*fujikuroi*)

2189. **Tuite, J. F.** 1969. *Plant Pathological Methods: Fungi and Bacteria*. Burgess Publishing, Minneapolis, Minnesota. (2)

2190. **Turbek, C. S., D. Li, G. H. Choi, C. L. Schardl, and D. A. Smith.** 1990. Induction and purification of kievitone hydratase from *Fusarium solani* f. sp. *phaseoli*. *Phytochemistry* **29:** 2841-2846. (*solani*)

2191. **Turbek, C. S., D. A. Smith, and C. L. Schardl.** 1992. An extracellular enzyme from *Fusarium solani* f. sp. *phaseoli* which catalyses hydration of the phytoalexin phaseollidin. *FEMS Microbiology Letters* **94:** 187-190. (*solani*)

2192. **Turgeon, B. G., and O. C. Yoder.** 2000. Proposed nomenclature for mating type genes of filamentous ascomycetes. *Fungal Genetics and Biology* **31:** 1-5. (6)

2193. **Turkington, T. K., R. M. Clear, P. A. Burnett, S. K. Patrick, D. D. Orr, and K. Xi.** 2002. Fungal plant pathogens infecting barley and wheat seed from Alberta, 1995-1997. *Canadian Journal of Plant Pathology* **24:** 302-308. (*avenaceum*)

2194. **Turnbull, W. H., A. R. Leeds, and G. D. Edwards.** 1990. Effect of mycoprotein on blood lipids. *American Journal of Clinical Nutrition* **52:** 646-650. (*venenatum*)

2195. **Twigg, L. E., and L. V. Socha.** 2001. Defluorination of sodium monofluoroacetate by soil microorganisms from central Australia. *Soil Biology and Biochemistry* **33:** 227-234. (*oxysporum*)

2196. **Tzatzarakis, M. N., A. M. Tsatsakis, E. Charvalos, and D. Vakalounakis.** 2001. Comparison of *in vitro* activities of amphotericin, clotrimazole, econazole, miconazole, and nystatin against *Fusarium oxysporum*. *Journal of Environmental Science and Health. Part B. Pesticides Food Contaminants and Agricultural Wastes* **B36:** 331-340. (*oxysporum*)

2197. **Uchimura, H., H. Enjoji, T. Seki, A. Taguchi, N. Takaya, and H. Shoun.** 2002. Nitrate reductase-formate dehydrogenase couple involved in the fungal denitrification by *Fusarium oxysporum*. *Journal of Biochemistry* **131:** 579-586. (*oxysporum*)

2198. **Uchiyama, N., Y. Shindo, H. Mikoshiba, and H. Saito.** 1987. Cutaneous infection caused by *Fusarium solani* and *Fusarium oxysporum*. *Journal of Dermatology* **14:** 471-475. (*oxysporum, solani*)

2199. **Ullstrup, A. J.** 1971. Hypersusceptibility of high lysine corn to kernel and ear rots. *Plant Disease Reporter* **55:** 1046. (*verticillioides*)

2200. **Umemura, K., S. Tanino, T. Nagatsuka, J. Koga, M. Iwata, K. Nagashima, and Y. Amemiya.** 2004. Cerebroside elicitor confers resistance to Fusarium disease in various plant species. *Phytopathology* **94:** 813-818. (*oxysporum*)

2201. **Upadhyay, R. S.** 1992. Ecology and biological control of *Fusarium udum* in relation to soil fungistasis of antagonistic microorganisms. *Journal of Plant Protection in the Tropics* **9:** 1-9. (*udum*)

2202. **Upadhyay, R. S., B. Rai, and R. C. Gupta.** 1981. *Fusarium udum*: Parasitic on *Aspergillus luchuensis* and *Syncephalastrum racemosum*. *Plant and Soil* **63:** 407-414. (*udum*)

2203. **Urban, M., S. Daniels, E. Mott, and K. Hammond-Kosack.** 2002. Arabidopsis is susceptible to the cereal ear blight fungal pathogens *Fusarium graminearum* and *Fusarium culmorum*. *Plant Journal* **32:** 961-973. (*culmorum, graminearum*)

2204. **Urban, M., E. Mott, T. Farley, and K. Hammond-Kossack.** 2003. The *Fusarium graminearum MAP1* gene is essential for pathogenicity and development of perithecia. *Molecular Plant Pathology* **4:** 347-359. (10, *graminearum*)

2205. **Urrutia, O., P. Hedden, and M. C. Rojas.** 2001. Monooxygenases involved in GA12 and GA14 synthesis in *Gibberella fujikuroi*. *Phytochemistry* **56:** 505-511. (*fujikuroi*)

2206. **Uzura, A., T. Katsuragi, and Y. Tani.** 2001. Stereoselective oxidation of alkylbenzenes by fungi. *Journal of Bioscience and Bioengineering* **91:** 217-221. (*verticillioides*)

2207. **Vaidya, R. J., S. L. A. Macmil, P. R. Vyas, L. V. Ghetiya, K. J. Thakor, and H. S. Chatpar.** 2003. Biological control of Fusarium wilt of pigeonpea, *Cajanus cajan* (L.) Millsp., with chitinolytic *Alcaligenes xylosoxydans*. *Indian Journal of Experimental Biology* **41:** 1469-1472. (*udum*)

2208. **Vakalounakis, D. J., and G. A. Fragkiadakis.** 1999. Genetic diversity of *Fusarium oxysporum* isolates from cucumber: Differentiation by pathogenicity, vegetative compatibility and RAPD fingerprinting. *Phytopathology* **89:** 161-168. (5, *oxysporum*)

2209. **Vakalounakis, D. J., Z. Wang, G. A. Fragkiadakis, G. N. Skaracis, and D. B. Li.** 2004. Characterization of *Fusarium oxysporum* isolates obtained from cucumber in China by pathogenicity, VCG, and RAPD. *Plant Disease* **88:** 645-649. (*oxysporum*)

2210. **Valenstein, P., and W. A. Schell.** 1986. Primary intranasal *Fusarium* infection potential for confusion with rhinocerebral zygomycosis. *Archives of Pathology and Laboratory Medicine* **110:** 751-754. (*oxysporum*)

2211. **van der Hoeven, E., J. M. Mitali, and D. Gindrat.** 1979. Evaluation of microorganisms antagonistic to *Phomopsis sclerotiodes* by laboratory methods. *Bulletin de la Societe Botanique Suisse* **89:** 145-153. (*solani*)

2212. **van der Molen, G. E., J. M. Labavitch, and J. E. Devay.** 1986. *Fusarium* induced vascular gels from banana, *Musa acuminata*, roots: A partial chemical characterization. *Physiologia Plantarum* **66:** 298-302. (*oxysporum*)

2213. **van der Schaft, P. H., N. ter Burg, S. van den Bosch, and A. M. Cohen.** 1992. Fed-batch production of 2-heptanone by *Fusarium poae*. *Applied Microbiology and Biotechnology* **36:** 709-711. (*poae*)

2214. **van Dijk, E., W. H., H. W. van den Berg, and A. J. Landwehr.** 1980. *Fusarium solani* infection of a hypertensive leg ulcer in a diabetic. *Mykosen* **23:** 603-606. (*solani*)

2215. **van Eck, W. H.** 1976. Ultrastructure of forming and dormant chlamydospores of *Fusarium solani* in soil. *Canadian Journal of Microbiology* **22:** 1634-1642. (*solani*)

2216. **van Eck, W. H.** 1978. Lipid body content and persistence of chlamydospores of *Fusarium solani* f. sp. *pisi* in soil. *Canadian Journal of Microbiology* **24:** 65-69. (*solani*)

2217. **van Eck, W. H., and B. Schippers.** 1976. Ultrastructure of developing chlamydospores of *Fusarium solani* f. *cucurbitae in vitro*. *Soil Biology and Biochemistry* **8:** 1-6. (*solani*)

2218. **van Gemeren, I. A., W. Musters, C. A. M. J. J. van den Hondel, and C. T. Verrips.** 1995. Construction and heterologous expression of a synthetic copy of the cutinase cDNA from *Fusarium solani pisi*. *Journal of Biotechnology* **40:** 155-162. (*solani*)

2219. **van Maanen, A., and X. M. Xu.** 2003. Modeling plant disease epidemics. *European Journal of Plant Pathology* **109:** 669-682. (11)

2220. **van Rensburg, J. C. J., N. Labuschagne, and S. Nemec.** 2001. Occurrence of *Fusarium*-produced naphthazarins in citrus trees and sensitivity of rootstocks to isomarticin in relation to citrus blight. *Plant Pathology* **50:** 258-265. (*solani*)

2221. **van Staden, J., and R. I. D. Nicholson.** 1989. Cytokinins and mango malformation II. The cytokinin complement produced by *Fusarium moniliforme* and the ability of the fungus to incorporate 8-^{14}C-labeled adenine into cytokinins. *Physiological and Molecular Plant Pathology* **35:** 423-432. (*mangiferae*)

2222. **van Wyk, P. S., O. Los, and W. F. O. Marasas.** 1988. Pathogenicity of a new *Fusarium* sp. from crown rot on wheat in South Africa. *Phytophylactica* **20:** 73-76. (*longipes*)

2223. **van Wyk, P. S., O. Los, G. D. C. Pauer, and W. F. O. Marasas.** 1987. Geographic distribution and pathogenicity of *Fusarium* species associated with crown rot of wheat in the Orange Free State, South Africa. *Phytophylactica* **19:** 271-274. (*armeniacum, longipes, scirpi*)

2224. **van Wyk, P. S., D. J. Scholtz, and O. Los.** 1986. A selective medium for the isolation of *Fusarium* spp. from soil debris. *Phytophylactica* **18:** 67-69. (2)

2225. **van Wyk, P. S., D. J. Scholtz, and W. F. O. Marasas.** 1988. Protection of maize seedlings by *Fusarium moniliforme* against infection by *Fusarium graminearum* in the soil. *Plant and Soil* **107:** 251-258. (*verticillioides*)

2226. **van Wyk, P. S., E. Venter, M. J. Wingfield, and W. F. O. Marasas.** 1988. Enteroblastic first macroconidia in *Fusarium crookwellense*. *Canadian Journal of Botany* **66:** 1364-1366. (*crookwellense*)

2227. **van Wyk, P. S., E. Venter, M. J. Wingfield, and W. F. O. Marasas.** 1988. Nuclear division and septation in macroconidia of *Fusarium crookwellense*. *South African Journal of Botany* **54:** 118-122. (*crookwellense*)

2228. **van Wyk, P. S., M. J. Wingfield, and W. F. O. Marasas.** 1988. Delimitation of *Fusarium crookwellense* macroconidia. *Transactions of the British Mycological Society* **91:** 611-618. (*crookwellense*)

2229. **van Wyk, P. S., M. J. Wingfield, W. F. O. Marasas, and J. L. Bosman.** 1991. Development of microconidia in *Fusarium* section *Sporotrichiella*. *Mycological Research* **95:** 284-289. (*chlamydosporum, sporotrichioides, tricinctum*)

2230. **VanEtten, H. D., and H. G. Kolmark.** 1977. Modifying the growth habit of the filamentous fungus *Fusarium solani* to facilitate replica plating procedures. *Canadian Journal of Botany* **55:** 848-851. (*solani*)

2231. **VanEtten, H. D., D. C. Straney, S. F. Covert, and H. C. Kistler.** 2001. The genetics of *Nectria haematococca* mating population VI with special emphasis on its conditionally dispensable chromosomes: A source of habitat-specific genes, p. 97-112. *In* B. A. Summerell, J. F. Leslie, D. Backhouse, W. L. Bryden, and L. W. Burgess (eds.), Fusarium*: Paul E. Nelson Memorial Symposium*. APS Press, St. Paul, Minnesota. (7, 11)

2232. **Vannacci, G., and C. Cristani.** 1998. Characterization of chlorate-resistant sectors from isolates of *Fusarium moniliforme* and *F. proliferatum*. *Journal of Microbiological Methods* **31:** 175-184. (*verticillioides*)

2233. **Varma, A., and B. K. Tandan.** 1996. Pathogenicity of three entomogenous fungi against insect pests of sugarcane. *Journal of Biological Control* **10:** 87-91. (*oxysporum*)

2234. **Vazquez, C., B. Patino, and M. J. Martinez.** 1993. Purification and characterization of an exopolygalacturonase produced by *Fusarium oxysporum* f. sp. *radicis-lycopersici*. *FEMS Microbiology Letters* **110:** 191-196. (*oxysporum*)

2235. **Vazquez, C., F. Reyes, and M. J. Martinez.** 1993. Comparative study of pectic activities from different *formae speciales* of *Fusarium oxysporum*. *Letters in Applied Microbiology* **16:** 210-213. (*oxysporum*)

2236. **Vazquez, F., and de Figueroa, L. I. C.** 1996. Protoplast formation in *Fusarium* species. *Biotechnology Techniques* **10:** 93-98. (*verticillioides*)

2237. **Velluti, A., S. Marin, L. Bettucci, A. J. Ramos, and V. Sanchis.** 2000. The effect of fungal competition on colonization of maize grain by *Fusarium moniliforme*, *F. proliferatum* and *F. graminearum* and on fumonisin B_1 and zearalenone formation. *International Journal of Food Microbiology* **59:** 59-66. (*verticillioides*)

2238. **Venter, S. L., D. J. Theron, P. J. Steyn, D. I. Ferreira, and A. Eicker.** 1992. Relationship between vegetative compatibility and pathogenicity of isolates of *Fusarium oxysporum* f. sp. *tuberosi* from potato. *Phytopathology* **82:** 858-862. (*oxysporum*)

2239. **Verdin, A., A. L. H. Sahraoui, R. Newsam, G. Robinson, and R. Durand.** 2005. Polycyclic aromatic hydrocarbons storage by *Fusarium solani* in intracellular lipid vesicles. *Environmental Pollution* **133:** 283-291. (*solani*)

2240. **Verma, J., and S. V. Gangal.** 1994. Studies of *Fusarium solani* Cross-reactivity among *Fusarium* species. *Allergy* 1994; **49:** 330-336. (*equiseti, solani*)

2241. **Verma, J., and S. V. Gangal.** 1994. *Fusarium solani*: Immunochemical characterization of allergens. *International Archives of Allergy and Immunology* **104:** 175-183. (*solani*)

2242. **Verma, J., S. Pasha, and S. V. Gangal.** 1994. Purification and characterization of Fus SI-3596, a 65 kd allergen of *Fusarium solani*. *Molecular and Cellular Biochemistry* **131:** 157-166. (*solani*)

2243. **Verma, J., B. P. Singh, S. Sridhara, S. N. Gaur, and N. Arora.** 2003. Purification and characterization of a cross-reactive 45-kD major allergen of *Fusarium solani*. *International Archives of Allergy and Immunology* **130:** 193-199. (*solani*)

2244. **Verma, J., S. Sridhara, D. Rai, and S. V. Gangal.** 1998. Isolation and immunobiochemical characterization of a major allergen (65 kDa) from *Fusarium equiseti*. *Allergy* **53:** 311-315. (*equiseti, solani, verticillioides*)

2245. **Verma, J., S. Sridhara, B. P. Singh, S. Pasha, S. V. Gangal, and N. Arora.** 2001. *Fusarium solani* major allergen peptide IV-1 binds IgE but does not release histamine. *Clinical and Experimental Allergy* **31:** 920-927. (*solani*)

2246. **Vesonder, R. F.** 1986. Monilifomin produced by cultures of *Fusarium moniliforme* var. *subglutinans* isolated from swine feed. *Mycopathologia* **95:** 149-154. (*subglutinans*)

2247. **Vesonder, R. F., A. Logrieco, A. Bottalico, C. Altomare, and S. W. Peterson.** 1995. *Fusarium* species associated with banana fruit rot and their potential toxigenicity. *Mycotoxins Research* **11:** 78-83. (*concentricum*)

2248. **Vesonder, R. F., and W. Wu.** 1998. Correlation of moniliformin, but not fumonisin B_1 levels, in culture materials of *Fusarium* isolates to acute death in ducklings. *Poultry Science* **77:** 67-72. (*proliferatum*)

2249. **Vesonder, R. F., W. Wu, D. Weisleder, S. H. Gordon, T. Krick, W. Xie, H. K. Abbas, and C. E. McAlpin.** 2000. Toxigenic strains of *Fusarium moniliforme* and *Fusarium proliferatum* isolated from dairy cattle feed produce fumonisins, moniliformin and a new $C_{21}H_{38}N_2O_6$ metabolite phytotoxic to *Lemna minor* L. *Journal of Natural Toxins* **9:** 103-112. (*verticillioides*)

2250. **Vidhyasekaran, P., K. Sethuraman, K. Rajappan, and K. Vasumathi.** 1997. Powder formulations of *Pseudomonas fluorescens* to control pigeonpea wilt. *Biological Control* **8:** 166-171. (*udum*)

2251. **Vigier, B. L. M. Reid, K. A. Seifert, and D. W. Stewart.** 1997. Distribution and prediction of *Fusarium* species associated with maize ear rot in Ontario. *Canadian Journal of Plant Pathology* **19:** 60-65. (11)

2252. **Vignutelli, A., C. Wasternack, K. Apel, and H. Bohlmann.** 1998. Systemic and local induction of an *Arabidopsis* thionin gene by wounding and pathogens. *Plant Journal* **14:** 285-295. (*oxysporum*)

2253. **Vilich, V., B. Loehndorf, R. A. Sikora, and A. Friebe.** 1999. Metabolism of benzoxazolinone allelochemicals of *Zea mays* by *Fusarium subglutinans*. *Mycological Research* **103:** 1529-1532. (*subglutinans*)

2254. **Viljoen, A., M. J. Wingfield, T. R. Gordon, and W. F. O. Marasas.** 1997. Genotypic diversity in a South African population of the pitch canker fungus *Fusarium subglutinans* f. sp. *pini*. *Plant Pathology* **46:** 590-593. (*circinatum*)

2255. **Villalba, F., M.-H. Lebrun, A. Hua-Van, M.-J. Daboussi, and M. C. Grosjean-Cournoyer.** 2001. Transposon *impala*, a novel tool for gene tagging in the rice blast fungus *Magnaporthe grisea*. *Molecular Plant-Microbe Interactions* **14:** 308-315. (*oxysporum*)

2256. **Viscoli, C., E. Castagnola, C. Moroni, A. Garaventa, G. Manno, and C. Savioli.** 1990. Infection with *Fusarium* spp. in two children with neuroblastoma. *European Journal of Clinical Microbiology and Infectious Diseases* **9:** 773-776. (*solani, verticillioides*)

2257. **Visconti, A., L. A. Blais, J. W. ApSimon, R. Greenhalgh, and J. D. Miller.** 1992. Production of enniatins by *Fusarium acuminatum* and *Fusarium compactum* in liquid culture: Isolation and characterization of three new enniatins B_2, B_3 and B_4. *Journal of Agricultural and Food Chemistry* **40:** 1076-1082. (*compactum*)

2258. **Visconti, A., F. Minervini, M. Solfrizzo, C. Bottalico, and G. Lucivero.** 1992. Toxicity of some *Fusarium* section *Sporotrichiella* strains in relation to mycotoxin production. *Applied and Environmental Microbiology* **58:** 769-772. (*poae, sporotrichioides*)

2259. **Visconti, A., C. J. Mirocha, A. Logrieco, A. Bottalico, and M. Solfrizzo.** 1989. Mycotoxins produced by *Fusarium acuminatum*. Isolation and characterization of acuminatin: A new trichothecene. *Journal of Agricultural and Food Chemistry* **37:** 1348-1351. (*compactum*)

2260. **Visconti, A., and M. Solfrizzo.** 1994. Isolation, characterization, and biological activity of visoltricin, a novel metabolite of *Fusarium tricinctum*. *Journal of Agricultural and Food Chemistry* **42:** 195-199. (*tricinctum*)

2261. **Visconti, A., and M. Solfrizzo.** 1995. Visoltricin, a novel biologically active compound produced by *Fusarium tricinctum*. *Food Additives and Contaminants* **12:** 515-519. (*tricinctum*)

2262. **Visconti, A., M. Solfrizzo, A. Fruchier, and J. W. ApSimon.** 1994. Acuminatopyrone: Revised structure and production by *Fusarium chlamydosporum* and *Fusarium tricinctum*. *Journal of Natural Products* **57:** 695-699. (*chlamydosporum, tricinctum*)

2263. **Vismer, H. F., W. F. O. Marasas, J. P. Rheeder, and J. J. Joubert.** 2002. *Fusarium dimerum* as a cause of human eye infections. *Medical Mycology* **40:**399-406. (1, *dimerum*)

2264. **Vismer, H. F., P. W. Snijman, W. F. O. Marasas, and D. J. van Schalkwyk.** 2004. Production of fumonisins by *Fusarium verticillioides* strains on solid and in a defined liquid medium – Effects of L-methionine and inoculum. *Mycopathologia* **158:** 99-106. (*verticillioides*)

2265. **von Arx, J. A.** 1984. Notes on *Monographella* and *Microdochium*. *Transactions of the British Mycological Society* **83:** 373-374. (*dimerum*)

2266. **Vos, P., R. Hogers, M. Bleeker, M. Reijans, T. van de Lee, M. Hornes, A. Frijters, J. Pot, J. Peleman, M. Kuiper, and M. Zabeau.** 1995. AFLP: A new technique for DNA fingerprinting. *Nucleic Acids Research* **23:** 4407-4414. (7)

2267. **Voss, K. A., F. I. Meredith, and C. W. Bacon.** 2003. Effect of baking and frying on the *in vivo* toxicity to rats of cornmeal containing fumonisins. *Journal of Agricultural and Food Chemistry* **51:** 5546-5551. (*verticillioides*)

2268. **Voss, K. A., W. P. Norred, R. D. Plattner, and C. W. Bacon.** 1989. Hepatotoxicity and renal toxicity in rats of corn samples associated with field cases of equine leukoencephalomalacia. *Food and Chemical Toxicology* **27:** 89-96. (*verticillioides*)

2269. **Voss, K. A., R. D. Plattner, R. T. Riley, F. I. Meredith, and W. P. Norred.** 1998. *In vivo* effects of fumonisin B_1-producing and fumonisin B_1-nonproducing *Fusarium moniliforme* isolates are similar: Fumonisins B_2 and B_3 cause hepato- and nephrotoxicity in rats. *Mycopathologia* **141:** 45-58. (*verticillioides*)

2270. **Voss, K. A., R. T. Riley, W. P. Norred, C. W. Bacon, F. I. Meredith, P. C. Howard, R. D. Plattner, T. F. X. Collins, D. K. Hansen, and J. K. Porter.** 2001. An overview of rodent toxicities: liver and kidney effects of fumonisins and *Fusarium moniliforme*. *Environmental Health Perspectives* **109 (Suppl. 2):** 259-266. (*verticillioides*)

2271. **Vujanovic, V., C. Hamel, S. Jabaji-Hare, and M. St. Arnaud.** 2002. Development of a selective myclobutanil agar (MBA) medium for the isolation of *Fusarium* species from asparagus fields. *Canadian Journal of Microbiology* **48:** 841-847. (2)

2272. **Waalwijk, C., and R. P. Baayen.** 1995. Identification of races of *Fusarium oxysporum* f. sp. *dianthi* by DNA fingerprinting and vegetative compatibility. *Bulletin OEPP* **25:** 117-124. (*oxysporum*)

2273. **Waalwijk, C., R. P. Baayen, J. R. A. de Koning, and W. Gams.** 1996. Ribosomal DNA analyses challenge the status of *Fusarium* sections *Liseola* and *Elegans*. *Sydowia* **48:** 90-104. (9)

2274. **Waalwijk, C., P. Kastelein, I. de-Vries, Z. Kerenyi, T. van der Lee, T. Hesselink, J. Kohl, and G. Kema.** 2003. Major changes in *Fusarium* spp. in wheat in the Netherlands. *European Journal of Plant Pathology* **109:** 743-754. (*culmorum, proliferatum*)

2275. **Waalwijk, C., R. van der Heide, I. de Vries, T. van der Lee, C. Schoen, G. Costrel de Corainville, I. Hauser-**

Hahn, P. Kastelein, J. Kohl, P. Lonnet, T. Demarquet, and G. H. J. Kema. 2004. Quantitative detection of *Fusarium* species in wheat using TaqMan. *European Journal of Plant Pathology* **110:** 481-494. (*avenaceum, culmorum, graminearum, poae*)

2276. **Waalwijk, C., T. van der Lee, I. de Vries, T. Hesselink, J. Arts, and G. H. J. Kema.** 2004. Synteny in toxigenic *Fusarium* species: The fumonisin gene cluster and the mating type region as examples. *European Journal of Plant Pathology* **110:** 533-544. (*proliferatum, verticillioides*)

2277. **Wadhwani, K., and K. Srivastava.** 1984. Fungi from *otis media* of agricultural field workers. *Mycopathologia* **88:** 155-160. (*verticillioides*)

2278. **Wadia, K. D. R., and C. Manoharachary.** 1980. Three post harvest diseases of *Zizyphus mauritiana* fruits from India. *Plant Disease* **64:** 323-324. (*decemcellulare*)

2279. **Wahab, S., B. Lal, Z. Jacob, V. C. Pandey, and O. P. Srivastava.** 1979. Studies on a strain of *Fusarium solani* isolated from a case of mycotic keratitis. *Mycopathologia* **68:** 31-38. (*solani*)

2280. **Wahab, S., R. N. Tandon, Z. Jacob, O. P. Srivastava, V. C. Pandey, and B. Lal.** 1981. Comparative effect of amphotericin B on *Rhizoctonia* sp., *Paecilomyces lilacinus* and *Fusarium solani in vitro* and in experimental keratitis. *Indian Journal of Experimental Biology* **19:** 458-461. (*solani*)

2281. **Walker, J. R. L., and B. G. Taylor.** 1983. Metabolism of phloroglucinol by *Fusarium solani*. *Archives of Microbiology* **134:** 123-126. (*solani*)

2282. **Wallbridge, A.** 1981. Fungi associated with crown-rot disease of boxed bananas from the Windward Islands during a two-year survey. *Transactions of the British Mycological Society* **77:** 567-577. (*semitectum*)

2283. **Wallbridge, A., and J. A. Pinegar.** 1975. Fungi associated with crown-rot disease of bananas from St. Lucia in the Windward Islands. *Transactions of the British Mycological Society* **64:** 247-254. (*semitectum*)

2284. **Waller, J. M., and D. Brayford.** 1990. Fusarium diseases in the tropics. *Tropical Pest Management* **36:** 181-194. (11)

2285. **Walther, T. C., and J. C. Kennell.** 1999. Linear mitochondrial plasmids of *F. oxysporum* are novel, telomere-like retroelements. *Molecular Cell* **4:** 229-238. (*oxysporum*)

2286. **Wang, B., C. L. Brubaker, and J. J. Burdon.** 2004. *Fusarium* species and Fusarium wilt pathogens associated with native *Gossypium* populations in Australia. *Mycological Research* **108:** 35-44. (1, 8, *oxysporum*)

2287. **Wang, B., and S. N. Jeffers.** 2002. Effects of cultural practices and temperature on Fusarium root and crown rot of container-grown hostas. *Plant Disease* **86:** 225-231. (*hostae*)

2288. **Wang, E., W. P. Norred, C. W. Bacon, R. T. Riley, and A. H. Merrill, Jr.** 1991. Inhibition of sphingolipid biosynthesis by fumonisins: Implications for diseases associated with *Fusarium moniliforme*. *Journal of Biological Chemistry* **266:** 14486-14490. (*verticillioides*)

2289. **Wang, Y. Z., and J. D. Miller.** 1988. Effects of *Fusarium graminearum* metabolites on wheat tissue in relation to Fusarium Head Blight resistance. *Journal of Phytopathology* **122:** 118-125. (*graminearum*)

2290. **Waniska, R. D., R. T. Venkatesha, A. Chandrashekar, S. Krishnaveni, F. P. Bejosano, J. Jeoung, J. Jayaraj, S. Muthukrishnan, and G. H. Liang.** 2001. Antifungal proteins and other mechanisms in the control of sorghum stalk rot and grain mold. *Journal of Agricultural and Food Chemistry* **49:** 4732-4742. (*thapsinum*)

2291. **Ward, T. J., J. P. Bielawski, H. C. Kistler, E. Sullivan, and K. O'Donnell.** 2002. Ancestral polymorphisms and adaptive evolution in the trichothecene mycotoxin gene cluster of phytopathogenic *Fusarium*. *Proceedings of the National Academy of Sciences (USA)* **99:** 9278-9283. (9, 11, *crookwellense, graminearum*)

2292. **Warfield, C. Y., and R. M. Davis.** 1996. Importance of the husk covering on the susceptibility of corn hybrids to Fusarium ear rot. *Plant Disease* **80:** 208-210. (*verticillioides*)

2293. **Warfield, C. Y., and D. G. Gilchrist.** 1999. Influence of kernel age on fumonisin B_1 production in maize by *Fusarium moniliforme*. *Applied and Environmental Microbiology* **65:** 2853-2856. (*verticillioides*)

2294. **Warnock, D. W.** 1995. Fungal complications of transplantation: Diagnosis, treatment and prevention *Journal of Antimicrobial Chemotherapy* **36 (Suppl. B):** 73-90. (*oxysporum*)

2295. **Warren, H. L.** 1978. Comparison of normal and high lysine maize inbreds for resistance to kernel rot caused by *Fusarium moniliforme*. *Phytopathology* **68:** 1331-1335. (*verticillioides*)

2296. **Wasmann, C. C., and H. D. VanEtten.** 1996. Transformation-mediated chromosome loss and disruption of a gene for pisatin demethylase decrease the virulence of *Nectria haematococca* on pea. *Molecular Plant-Microbe Interactions* **9:** 793-803. (*solani*)

2297. **Watanabe, T., and T. Umehara.** 1977. The perfect state of the causal fungus of bakanae disease of rice plants recollected at Toyama, Japan. *Nippon Kingakukai Kaiho* **18:** 136-142. (10, *fujikuroi*)

2298. **Webster, R. K., and P. S. Gunnell.** 1992. *Compendium of Rice Diseases*. APS Press, St. Paul, Minnesota. (3)

2299. **Weidemann, G. J., and G. E. Templeton.** 1988. Control of Texas Gourd, *Cucurbita texana* with *Fusarium solani* f. sp. *cucurbitae*. *Weed Technology* **2:** 271-274. (*solani*)

2300. **Weidemann, G. J., and G. E. Templeton.** 1988. Efficacy and soil persistence of *Fusarium solani* f. sp. *cucurbitae* for control of Texas Gourd, *Cucurbita texana*. *Plant Disease* **72:** 36-38. (*solani*)

2301. **Weideman, H., and F. C. Wehner.** 1993. Greenhouse evaluation of *Trichoderma harzianum* and *Fusarium oxysporum* for biological control of citrus root rot in soils naturally and artificially infested with *Phytophthora nicotianae*. *Phytophylactica* **25:** 101-105. (*oxysporum*)

2302. **Weissgold, D. J., S. E. Orlin, M. E. Sulewski, W. C. Frayer, and R. C. Eagle, Jr.** 1998. Delayed onset fungal keratitis after endophthalmitis. *Ophthalmology* **105:** 258-262. (*solani*)

2303. **Weltring, K. M., and M. Altenburger.** 1998. Metabolism of the phytoalexin rishitin by *Gibberella pulicaris* is highly reduced in liquid culture. *Zeitschrift für Naturforschung Section C Journal of Biosciences* **53:** 806-810. (*sambucinum*)

2304. **Weltring, K. M., K. Loser, and J. Weimer.** 1998. Genetic instability of rishitin metabolism and tolerance and virulence on potato tubers of a strain of *Gibberella pulicaris*. *Journal of Phytopathology* **146:** 393-398. (*sambucinum*)

2305. **Weltring, K. M., J. Wessels, and R. Geyer.** 1997. Metabolism of the potato saponins α-chaconine and α-solanine by *Gibberella pulicaris*. *Phytochemistry* **46:** 1005-1009. (*sambucinum*)

2306. **Weltring, K. M., J. Wessels, and G. F. Pauli.** 1998. Metabolism of the tomato saponin α-tomatine by *Gibberella pulicaris*. *Phytochemistry* **48:** 1321-1328. (*sambucinum*)

2307. **Wene, E. G., and A. A. Antonpoulos.** 1988. D-xylose fermentation by *Fusarium oxysporum* strains. *Biomass* **17:** 13-20. (*oxysporum*)

2308. **Wheeler, M. H., R. D. Stipanovic, and L. S. Puckhaber.** 1999. Phytotoxicity of equisetin and epi-equisetin isolated from *Fusarium equiseti* and *F. pallidoroseum*. *Mycological Research* **103:** 967-973. (*equiseti, semitectum*)

2309. **Wheeler, M. S., M. R. McGinnis, W. A. Schell, and D. H. Walker.** 1981. *Fusarium* infection in burned patients. *American Journal of Clinical Pathology* **75:** 304-311. (*oxysporum*)

2310. **White, D. G.** 1978. Correlation of corn stalk rot reactions caused by *Diplodia maydis* and *Gibberella zeae*. *Plant Disease Reporter* **62:** 1016-1018. (*graminearum*)

2311. **White, D. G., ed.** 1999. *Compendium of Corn Diseases*, 3rd ed. APS Press, St. Paul, Minnesota. (*graminearum, proliferatum, subglutinans*)

2312. **Whitehead, D. S., A. Coddington, and B. G. Lewis.** 1992. Classification of races by DNA polymorphism analysis and vegetative compatibility grouping in *Fusarium oxysporum* f. sp. *pisi*. *Physiological and Molecular Plant Pathology* **41:** 295-305. (*oxysporum*)

2313. **Wick, R. L., and D. B. Schroeder.** 1982. *In vitro* production of pectolytic and cellulolytic enzymes by *Fusarium tricinctum*. *Mycologia* **74:** 460-466. (*tricinctum*)

2314. **Wicklow, D. T.** 1988. Patterns of fungal association within maize kernels harvested in North Carolina, USA. *Plant Disease* **72:** 113-115. (*verticillioides*)

2315. **Wicklow, D. T., B. W. Horn, O. L. Shotwell, C. W. Hesseltine, and R. W. Caldwell.** 1988. Fungal interference with *Aspergillus flavus* infection and aflatoxin contamination of maize grown in a controlled environment. *Phytopathology* **78:** 68-74. (*verticillioides*)

2316. **Wiebe, M. G.** 2002. Siderophore production by *Fusarium venenatum* A3/5. *Biochemical Society Transactions* **30:** 696-698. (*venenatum*)

2317. **Wiebe, M. G.** 2002. Myco-protein from *Fusarium venenatum*: A well-established product for human consumption. *Applied Microbiology and Biotechnology* **58:** 421-427. (*venenatum*)

2318. **Wiebe, M. G.** 2004. Quorn mycoprotein – Overview of a successful fungal product. *The Mycologist* **18:** 17-20. (*venenatum*)

2319. **Wiebe, M. G., G. D. Robson, B. Cunliffe, A. P. J. Trinci, and S. G. Oliver.** 1992. Nutrient-dependent selection of morphological mutants of *Fusarium graminearum* A3/5 isolated from long-term continuous flow cultures. *Biotechnology and Bioengineering* **40:** 1181-1189. (*venenatum*)

2320. **Wiebe, M. G., G. D. Robson, J. R. Shuster, and A. P. J. Trinci.** 1999. pH regulation of recombinant glucoamylase production in *Fusarium venenatum* JeRS 325, a transformant with a *Fusarium oxysporum* alkaline (trypsin-like) protease promoter. *Biotechnology and Bioengineering* **64:** 368-372. (*venenatum*)

2321. **Wiebe, M. G., G. D. Robson, J. Shuster, and A. P. J. Trinci.** 2000. Growth-rate-independent production of recombinant glucoamylase by *Fusarium venenatum* JeRS 325. *Biotechnology and Bioengineering* **68:** 245-251. (*venenatum*)

2322. **Wiebe, M. G., G. D. Robson, J. Shuster, and A. P. J. Trinci.** 2001. Evolution of a recombinant (gucoamylase-producing) strain of *Fusarium venenatum* A3/5 in chemostat culture. *Biotechnology and Bioengineering* **73:** 146-156. (*venenatum*)

2323. **Wiebe, M. G., G. D. Robson, and A. P. J. Trinci.** 1989. Effect of choline on the morphology, growth and phospholipid composition of *Fusarium graminearum*. *Journal of General Microbiology* **135:** 2155-2162. (*venenatum*)

2324. **Wiebe, M. G., A. P. J. Trinci, B. Cunliffe, G. D. Robson, and S. G. Oliver.** 1991. Appearance of morphological colonial mutants in glucose-limited continuous flow cultures of *Fusarium graminearum* A3-5. *Mycological Research* **95:** 1284-1288. (*venenatum*)

2325. **Wiese, M. V., ed.** 1987. *Compendium of Wheat Diseases*, 2nd ed. APS Press, St. Paul, Minnesota. (*graminearum*)

2326. **Wietor-Orlandi, E. A., and D. A. Smith.** 1985. Metabolism of the phytoalexin phaseolin isoflavan by *Fusarium solani* f. sp. *phaseoli*. *Physiological Plant Pathology* **27:** 197-208. (*solani*)

2327. **Wikler, K., T. R. Gordon, S. L. Clark, M. J. Wingfield, and H. Britz.** 2000. Potential for outcrossing in an apparently asexual population of *Fusarium circinatum*, the causal agent of pitch canker disease. *Mycologia* **92:** 1085-1090. (*circinatum*)

2328. **Wikler, K., A. J. Storer, W. Newman, T. R. Gordon, and D. L. Wood.** 2003. The dynamics of an introduced pathogen in a native Monterey pine (*Pinus radiata*) forest. *Forest Ecology and Management* **179:** 209-221. (*circinatum*)

2329. **Wilhelmus, K. R., N. M. Robinson, R. A. Font, M. B. Hamill, and D. B. Jones.** 1988. Fungal keratitis in contact lens wearers. *American Journal of Ophthalmology* **106:** 708-714. (*solani*)

2330. **Willeke, U., and W. Barz.** 1982. Catabolism of 5-hydroxyisoflavones by fungi of the genus *Fusarium*. *Archives of Microbiology* **132:** 266-269. (*solani*)

2331. **Willeke, U., K. M. Weltring, W. Barz, and H. D. VanEtten.** 1983. Degradation of the isoflavone biochanin A by isolates of *Nectria haematococca* (*Fusarium solani*). *Phytochemistry* **22:** 1539-1542. (*solani*)

2332. **Willemsen, M. J., A. L. de Coninck, P. J. E. Coremans, P. M. A. Marichal, and D. I. Roseeuw.** 1986. Parasitic invasion of *Fusarium oxysporum* in an arterial ulcer in an otherwise healthy patient. *Mykosen* **29:** 248-252. (*oxysporum*)

2333. **Williams, G., F. Billson, R. Husain, S. A. Howlader, N. Islam, and K. McClellan.** 1987. Microbiological diagnosis of suppurative keratitis in Bangladesh. *British Journal of Ophthalmology* **71:** 315-321. (*solani*)

2334. **Williams, K. J., J. I. Dennis, C. Smyl, and H. Wallwork.** 2002. The application of species-specific assays based on the polymerase chain reaction to analyze Fusarium crown rot of durum wheat. *Australasian Plant Pathology* **31:** 119-127. (7, *acuminatum*, *pseudograminearum*)

2335. **Wilson, A., D. Simpson, E. Chandler, P. Jennings, and P. Nicholson.** 2004. Development of PCR assays for the detection and differentiation of *Fusarium sporotrichioides* and *Fusarium langsethiae*. *FEMS Microbiology Letters* **233:** 69-76. (*poae*, *sporotrichioides*)

2336. **Wilson, J. F., L. Garnjobst, and E. L. Tatum.** 1961. Heterokaryon incompatibility in *Neurospora crassa* microinjection studies. *American Journal of Botany* **48:** 299-305. (5)

2337. **Wilson, J. P.** 2002. Fungi associated with the stalk rot complex of pearl millet. *Plant Disease* **86:** 833-839. (*semitectum*)

2338. **Wilson, J. P., H. H. Casper, and D. M. Wilson.** 1995. Effect of delayed harvest on contamination of pearl millet grain with mycotoxin-producing fungi and mycotoxins. *Mycopathologia* **132:** 27-30. (*semitectum*)

2339. **Wilson, J. P., W. W. Hanna, D. M. Wilson, R. W. Beaver, and H. H. Casper.** 1993. Fungal and mycotoxin contamination of pearl millet grain in response to environmental conditions in Georgia. *Plant Disease* **77:** 121-124. (*semitectum*)

2340. **Wilson, R. A., and H. N. Arst, Jr.** 1998. Mutational analysis of AREA, a transcriptional activator mediating nitrogen metabolite repression in *Aspergillus nidulans* and a member of the "Streetwise" GATA family of transcription factors. *Microbiology and Molecular Biology Reviews* **62:** 586-596. (5)

2341. **Wilson, W. J., C. L. Strout, T. Z. DeSantis, J. L. Stilwell, A. V. Carrano, and G. L. Andersen.** 2002. Sequence-specific identification of 18 pathogenic microorganisms using microarray technology. *Molecular and Cellular Probes* **16:** 119-127. (7)

2342. **Windels, C. E.** 2000. Economic and social impacts of Fusarium head blight: Changing farms and rural communities in the Northern Great Plains. *Phytopathology* **90:** 17-21. (1, 9)

2343. **Windels, C. E., P. M. Barnes, and T. Kommedahl.** 1988. Five-year preservation of *Fusarium* species on silica gel and soil. *Phytopathology* **78:** 107-109. (4)

2344. **Windels, C. E., P. M. Barnes, and T. Kommedahl.** 1993. *Fusarium* species stored for silica gel and soil for ten years. *Mycologia* **85:** 21-23. (4)

2345. **Windels, C. E., C. J. Mirocha, H. K. Abbas, and W. Xie.** 1989. Perithecium production in *Fusarium graminearum* populations and lack of correlation with zearalenone production. *Mycologia* **81:** 272-277. (10, *graminearum*)

2346. **Windingstad, R. M., R. J. Cole, P. E. Nelson, T. J. Roffe, R. R. George, and J. W. Dorner.** 1989. Fusarium mycotoxins from peanuts suspected as a cause of sandhill crane mortality. *Journal of Wildlife Diseases* **25:** 38-46. (*compactum*)

2347. **Wineland, G. O.** 1924. An ascigerous stage and synonomy for *Fusarium moniliforme*. *Journal of Agricultural Research* **28:** 909-922. (8)

2348. **Wing, N., W. L. Bryden, D. R. Lauren, and L. W. Burgess.** 1994. Toxigenicity of *Fusarium* species and subspecies in section *Gibbosum* from different regions of Australia. *Mycological Research* **97:** 1441-1446. (*acuminatum*, *armeniacum*, *longipes*, *scirpi*)

2349. **Wing, N., L. W. Burgess, and W. L. Bryden.** 1995. Cultural degeneration in two *Fusarium* species and its effects on toxicity and cultural morphology. *Mycological Research* **99:** 615-620. (4, *armeniacum*, *circinatum*)

2350. **Wing, N., D. R. Lauren, W. L. Bryden, and L. W. Burgess.** 1993. Toxicity and trichothecene production by *Fusarium acuminatum* subsp. *acuminatum* and *Fusarium acuminatum* subsp. *armeniacum*. *Natural Toxins* **1:** 229-234. (*acuminatum*, *armeniacum*)

2351. **Winn, R. M., C. Gil-Lamaignere, A. Maloukou, and E. Roilides.** 2003. Interactions of human phagocytes with molds *Fusarium* spp. and *Verticillium nigrescens* possessing different pathogenicity. *Medical Mycology* **41:** 503-509. (*oxysporum*, *solani*)

2352. **Wolf, J. C., J. R. Lieberman, and C. J. Mirocha.** 1972. Inhibition of F-2 (zearalenone) biosynthesis and perithecium production in *Fusarium roseum graminearum*. *Phytopathology* **62:** 937-939. (10, *graminearum*)

2353. **Wolf, J. C., and C. J. Mirocha.** 1973. Regulation of sexual reproduction in *Gibberella zeae* (*Fusarium roseum graminearum*) by zearalenone. *Canadian Journal of Microbiology* **19:** 725-734. (10, *graminearum*)

2354. **Wolf, J. C., and C. J. Mirocha.** 1977. Control of sexual reproduction in *Gibberella zeae* (*Fusarium roseum graminearum*). *Applied and Environmental Microbiology* **33:** 546-550. (10, *graminearum*)

2355. **Wollenweber, H. W.** 1913. Studies on the *Fusarium* problem. *Phytopathology* **3:** 24-50. (*redolens*)

2356. **Wollenweber, H. W.** 1916-1935. *Fusaria autographice delineate*. Berlin: Selbstverlag. 1200 pls. (*scirpi*)

2357. **Wollenweber, H. W.** 1931. *Fusarium - Monographie Fungi parasitici et saprophytici. Zeitschrift Parisitenk.* **3:** 269-516. (*succisae*)

2358. **Wollenweber, H. W., and O. A. Reinking.** 1925. Aliquot *Fusaria* tropicalia nova vel revisa. *Phytopathology* **15:** 155-169. (*camptoceras*, *sacchari*)

2359. **Wollenweber, H. W, and O. A. Reinking.** 1935. *Die Fusarien, ihre Beschreibung, Schadwirkung und Bekampfung*. Verlag Paul Parey, Berlin, Germany. (F, 8, 9, *acuminatum*, *chlamydosporum*, *compactum*, *culmorum*, *decemcellulare*, *dimerum*, *equiseti*, *lactis*, *lateritium*, *longipes*, *merismoides*, *nygamai*, *phyllophilum*, *poae*, *sambucinum*, *scirpi*, *semitectum*, *sporotrichioides*, *subglutinans*, *udum*, *venenatum*)

2360. **Woloshuk, C. P., and P. E. Kolattukudy.** 1986. Mechanism by which contact with plant cuticle triggers cutinase gene expression in the spores of *Fusarium solani* f. sp. *pisi*. *Proceedings of the National Academy of Sciences USA* **83:** 1704-1708. (*solani*)

2361. **Wong, G. K., S. Griffith, I. Kojima, and A. L. Demain.** 1998. Antifungal activities of rapamycin and its derivatives, prolylrapamycin, 32-desmethylrapamycin, and 32-desmethoxyrapamycin. *Journal of Antibiotics* **51:** 487-491. (*oxysporum*)

2362. **Woo, S. L., A. Zoina, G. del Sorbo, M. Lorito, B. Nanni, F. Scala, and C. Noviello.** 1996. Characterization of *Fusarium oxysporum* f. sp. *phaseoli* by pathogenic races VCGs, RFLPs, and RAPD. *Phytopathology* **86:** 966-973. (*oxysporum*)

2363. **Woodward, V. W., J. R. de Zeeuw, and A. M. Srb.** 1954. The separation and isolation of particular biochemical mutants of *Neurospora* by differential germination of conidia, followed by filtration and selective plating. *Proceedings of the National Academy of Sciences (USA)* **40:** 192-200. (4)

2364. **Woudt, L. P.** 1995. Identification of *Fusarium oxysporum* f. sp. *cyclaminis* by DNA fingerprinting and PCR. *Bulletin OEPP* **25:** 109-112. (*oxysporum*)

2365. **Woudt, L. P., A. Neuvel, A. Sikkema, W. A. J. de Milliano, C. L. Campbell, and J. F. Leslie.** 1995. *Fusarium* from cyclamens. *Phytopathology* **85**: 1348-1355. (5, 11, *oxysporum*)

2366. **Wright, G. F. K., D. I. Guest, D. L. S. Wimalajeewa, and R. van Heeswijck.** 1996. Characterization of *Fusarium oxysporum* isolated from carnation in Australia based on pathogenicity, vegetative compatibility and random amplified polymorphic DNA (RAPD) assay. *European Journal of Plant Pathology* **102:** 451-457. (*oxysporum*)

2367. **Wu, T.-G., V. V. Keasler, B. M. Mitchell, and K. R. Wilhelmus.** 2004. Immunosuppression affects the severity of experimental *Fusarium solani* keratitis. *Journal of Infectious Diseases* **190:** 192-198. (*solani*)

2368. **Wu, X., J. F. Leslie, R. A. Thakur, and J. S. Smith.** 2003. Preparation of a fusaproliferin standard from the culture of *Fusarium subglutinans* E-1583 by high performance liquid chromatography. *Journal of Food and Agricultural Chemistry* **51:** 383-388. (*subglutinans*)

2369. **Wu, Z., Y. Tsumura, G. Blomquist, and X.-R. Wang.** 2003. 18s rRNA gene variation among common airborne fungi, and the development of specific oligonucleotide probes for the detection of fungal isolates. *Applied and Environmental Microbiology* **69:** 5389-5397. (7)

2370. **Wu, Z., X.-R. Wang, and G. Blomquist.** 2002. Evaluation of PCR primers and PCR conditions for specific detection of common airborne fungi. *Journal of Environmental Monitoring* **4:** 377-382. (7)

2371. **Xie, W., C. J. Mirocha, R. J. Pawlosky, Y. Wen, and X. Xu.** 1989. Biosynthesis of fusachromanone and its monoacetyl derivative by *Fusarium equiseti*. *Applied and Environmental Microbiology* **55:** 794-797. (*equiseti*)

2372. **Xu, F., E. J. Golightly, P. Schneider, R. M. Berka, K. M. Brown, J. A. Johnstone, D. H. Baker, C. C. Fuglsang, S. H. Brown, A. Svendsen, and A. V. Klotz.** 2000. Expression and characterization of a recombinant *Fusarium* spp. galactose oxidase. *Applied Biochemistry and Biotechnology* **88:** 23-32. (*venenatum*)

2373. **Xu, J.-R., and J. F. Leslie.** 1996. A genetic map of *Fusarium moniliforme* (*Gibberella fujikuroi* mating population A). *Genetics* **143:** 175-189. (6, 7, 10, *verticillioides*)

2374. **Xu, J.-R., K. Yan, M. B. Dickman, and J. F. Leslie.** 1995. Electrophoretic karyotypes distinguish the biological species of *Gibberella fujikuroi* (*Fusarium* section *Liseola*). *Molecular Plant-Microbe Interactions* **8:** 74-84. (7, *fujikuroi*, *thapsinum*, *verticillioides*)

2375. **Xue, A. G., K. C. Armstrong, H. D. Voldeng, G. Fedak, and C. Babcock.** 2004. Comparative aggressiveness of isolates of *Fusarium* spp. causing head blight on wheat in Canada. *Canadian Journal of Plant Pathology* **26:** 81-88. (*graminearum*)

2376. **Yajima, H., M. Takao, S. Yasuhira, J.-H. Zhao, C. Ishii, H. Inoue, and A. Yasui.** 1995. A eukaryotic gene encoding an endonuclease that specifically repairs DNA damaged by ultraviolet light. *EMBO Journal* **14:** 2393-2399. (4)

2377. **Yamaguchi, K., M. Kida, M. Arita, and M. Takahashi.** 1992. Induction of systemic resistance by *Fusarium oxysporum* MT0062 in Solanaceous crops. *Annals of the Phytopathological Society of Japan* **58:** 16-22. (*oxysporum*)

2378. **Yamaguchi, K., T. Sano, M. Arita, and M. Takahashi.** 1992. Biocontrol of Fusarium wilt of tomato and Verticillium wilt of eggplant by non-pathogenic *Fusarium oxysporum* MT0062. *Annals of the Phytopathological Society of Japan* **58:** 188-194. (*oxysporum*)

2379. **Yamamoto, K., Y. Tsuji, H. Kumagai, and T. Tochikura.** 1986. Induction and purification of α-L-fucosidase from *Fusarium oxysporum*. *Agricultural and Biological Chemistry* **50:** 1689-1696. (*oxysporum*)

2380. **Yamauchi, N., J. Shimazu, M. Satou, S. Horiuchi, and T. Shirakawa.** 2004. Physiological races and vegetative compatibility groups of butterhead lettuce isolates of *Fusarium oxysporum* f. sp. *lactucae* in Japan. *Journal of General Plant Pathology* **70:** 308-313. (*oxysporum*)

2381. **Yan, K., and M. B. Dickman.** 1996. Isolation of a β-tubulin gene from *Fusarium moniliforme* that confers cold-sensitive benomyl resistance. *Applied and Environmental Microbiology* **62:** 3053-3056. (*verticillioides*)

2382. **Yanase, H., A. Sakamoto, K. Okamoto, K. Kita, and Y. Sato.** 2000. Degradation of the metal-cyano complex tetracyanonickelate (II) by *Fusarium oxysporum* N-10. *Applied Microbiology and Biotechnology* **53:** 328-334. (*oxysporum*)

2383. **Yates, I. E., J. W. Arnold, D. M. Hinton, W. Basinger, and R. R. Walcott.** 2003. *Fusarium verticillioides* induction of maize seed rot and its control. *Canadian Journal of Botany* **81:** 422-428. (*verticillioides*)

2384. **Yates, I. E., C. W. Bacon, and D. M. Hinton.** 1997. Effects of endophytic infection by *Fusarium moniliforme* on corn growth and cellular morphology. *Plant Disease* **81:** 723-728. (*verticillioides*)

2385. **Yates, I. E., K. L. Hiett, D. R. Kapczynski, W. Smart, A. E. Glenn, D. M. Hinton, C. W. Bacon, R. Meinersmann, S. Liu, and A. J. Jaworski.** 1999. GUS transformation of the maize fungal endophyte *Fusarium moniliforme*. *Mycological Research* **103:** 129-136. (*verticillioides*)

2386. **Yates, I. E., and A. J. Jaworski.** 2000. Differential growth of *Fusarium moniliforme* relative to tissues from 'Silver Queen', a sweet maize. *Canadian Journal of Botany* **78:** 472-480. (*verticillioides*)

2387. **Yildiran, S. T., S. Komurcu, M. A. Saracli, A. Gonlum, C. Beyan, H. Gun, and A. Yalcin.** 1998. *Fusarium* fungaemia in severely neutropenic patients. *Mycoses* **41:** 467-469. (*verticillioides*)

2388. **Yli-Mattila, T., R. L. Mach, I. A. Alekhina, S. A. Bulat, S. Koskinen, C. M. Kullnig-Gradinger, C. P. Kubicek,**

and S. S. Klemsdal. 2004. Phylogenetic relationship of *Fusarium langsethiae* to *Fusarium poae* and *Fusarium sporotrichioides* as inferred by IGS, ITS, β-tubulin sequences and UP-PCR hybridization analysis. *International Journal of Food Microbiology* **95:** 267-285. (*poae*)

2389. **Yli-Mattila, T., S. Paavanen-Huhtala, S. A. Bulat, I. A. Alekhina, and H. I. Nirenberg.** 2002. Molecular, morphological and phylogenetic analysis of the *Fusarium avenaceum/F. arthrosporioides/F. tricinctum* species complex: A polyphasic approach. *Mycological Research* **106:** 655-669. (*acuminatum, avenaceum*)

2390. **Yli-Mattila, T., S. Paavanen-Huhtala, P. Parikka, P. Konstantinova, and T. Y. Gagkaeva.** 2004. Molecular and morphological diversity of *Fusarium* species in Finland and northwestern Russia. *European Journal of Plant Pathology* **110:** 573-585. (*sporotrichioides*)

2391. **Yoder, O. C.** 1979. Experience with the Applegate-Nelson-Metzenberg method of mutant enrichment in high sorbose medium. Neurospora *Newsletter* **26:** 23-24. (4)

2392. **Yoder, W. T., and L. M. Christianson.** 1998. Species-specific primers resolve members of *Fusarium* section *Fusarium*. Taxonomic status of the edible "Quorn" fungus reevaluated. *Fungal Genetics and Biology* **23:** 68-80. (7, 9, *graminearum, sambucinum, venenatum*)

2393. **Yonli, D., H. Traore, D. E. Hess, A. A. Abbasher, and I. J. Boussim.** 2004. Effect of growth medium and method of application of *Fusarium oxysporum* on infestation of sorghum by *Striga hermonthica* in Burkina Faso. *Biocontrol Science and Technology* **14:** 417-421. (*oxysporum*)

2394. **Yoshida, H., M. Iizuka, N. Norioka, S. Norioka, and F. Sakiyama.** 1998. Cloning and sequencing of cDNA encoding ribonuclease F_1 from *Fusarium moniliforme*. *Biochemistry and Molecular Biology International* 45: 555-560. (*verticillioides*)

2395. **Yoshida, H., and N. Tamiya.** 1971. Acid phophatases (e.c. 3.1.3.2) from *Fusarium moniliforme*: Purification and enzymatic properties. *Journal of Biochemistry* **69:** 525-534. (*verticillioides*)

2396. **Yoshida, N., T. Fukushima, H. Saito, M. Shimosaka, and M. Okazaki.** 1989. Cellulase and xylan-degrading enzymes of the plant pathogenic fungus *Fusarium oxysporum* SUF850. *Agricultural and Biological Chemistry* **53:** 1829-1836. (*oxysporum*)

2397. **Young, C. N., and A. M. Myers.** 1979. Opportunistic fungal infection by *Fusarium oxysporum* in a renal transplant patient. *Sabouraudia* **17:** 219-224. (*oxysporum*)

2398. **Young, N. A., K. J. Kwon-Chung, T. T. Kubota, A. E. Jennings, and R. I. Fisher.** 1978. Disseminated infection by *Fusarium moniliforme* during treatment for malignant lymphoma. *Journal of Clinical Microbiology* **7:** 589-594. (*verticillioides*)

2399. **Young, T. R., and T. A. Kucharek.** 1977. Succession of fungal communities in roots and stalks of hybrid field corn grown in Florida, USA. *Plant Disease Reporter* **61:** 76-80. (*verticillioides*)

2400. **Ypema, H. L., P. A. van de Pol, and G. J. Bollen.** 1987. Black rot of stentlings of roses, a disease caused by various soil fungi. *Scientia Horticulturae* **33:** 269-280. (*redolens*)

2401. **Yue, Q., C. W. Bacon, and M. D. Richardson.** 1998. Biotransformation of 2-benzoxazolinone and 6-methoxy-benzoxazolinone by *Fusarium moniliforme Phytochemistry* **48:** 451-454. (*verticillioides*)

2402. **Yun, S.-H., T. Arie, I. Kaneko, O. C. Yoder, and B. G. Turgeon.** 2000. Molecular organization of mating type loci in heterothallic, homothallic, and asexual *Gibberella/Fusarium* species. *Fungal Genetics and Biology* **31:** 7-20. (6, 9, *graminearum, oxysporum*)

2403. **Zan, G. H., and J. L. Brewbaker.** 1999. Seed quality of isogenic endosperm mutants in sweet corn. *Maydica* **44:** 271-277. (*verticillioides*)

2404. **Zapater, R. C., A. A. de Arrechea, and V. H. Guevara.** 1972. Keratomycosis caused by *Fusarium dimerum*. *Sabouraudia* **10:** 274-275. (*dimerum*)

2405. **Zare, R., and D. Ershad.** 1997. *Fusarium* species isolated from cereals in Gorgan area. *Iranian Journal of Plant Pathology* **33:** 1-4. (*scirpi*).

2406. **Zauberman, G., and M. Schiffmann-Nadel.** 1977. Physiological response of avocado fruit to infection by different *Fusarium* species. *Phytopathologische Zeitschrift* **89:** 359-365. (*sambucinum*)

2407. **Zedan, H. H., O. M. El-Tayeb, and M. Abdel Aziz.** 1976. Microbiological transformations of steroids. 3. Transformation of progesterone to androstadienedione by the mycelia and by nongerminating spores of two fungi. *Planta Medica* **30:** 360-368. (*solani*)

2408. **Zedan, H. H., and R. Plourde.** 1973. Steroid 1-dehydrogenation and side chain degradation enzymes in the life cycle of *Fusarium solani*. *Biochimica Biophysica Acta* **326:** 103-115. (*solani*)

2409. **Zeller, K. A., R. L. Bowden, and J. F. Leslie.** 2003. Diversity of epidemic populations of *Gibberella zeae* from small quadrats in Kansas and North Dakota. *Phytopathology* **93:** 874-880. (6, 7, 9, 11, *graminearum*)

2410. **Zeller, K. A., R. L. Bowden, and J. F. Leslie.** 2004. Population differentiation and recombination in wheat scab populations of *Gibberella zeae* in the United States. *Molecular Ecology* **13:** 563-571. (6, 7, 9, 11, *graminearum*)

2411. **Zeller, K. A., J. E. Jurgenson, and J. F. Leslie.** 2001. Simultaneous mapping of multiple *vic* loci in *Gibberella fujikuroi* Mating Population A (*Fusarium verticillioides*). *Phytopathology* **91:** s99. (5)

2412. **Zeller, K. A., B. A. Summerell, S. Bullock, and J. F. Leslie.** 2003. *Gibberella konza* (*Fusarium konzum*) *sp. nov.*, a new biological species within the *Gibberella fujikuroi* species complex from prairie grasses. *Mycologia* **95:** 943-954. (4, 6, 7, 8, 9, 10, *anthophilum, konzum*)

2413. **Zhang, J., B. D. Bruton, and C. L. Biles.** 1999. *Fusarium solani* endo-polygalacturonase from decayed muskmelon fruit: Purification and characterization. *Physiological and Molecular Plant Pathology* **54:** 171-186. (*solani*)

2414. **Zhang, J., C. R. Howell, J. L. Starr, and M. H. Wheeler.** 1996. Frequency of isolation and the pathogenicity of *Fusarium* species associated with roots of healthy cotton seedlings. *Mycological Research* **100:** 747-752. (*nygamai*)

2415. **Zhang, L., T. Kudo, N. Takaya, and H. Shoun.** 2002. The B′ helix determines cytochrome P450nor specificity for the electron donors NADH and NADPH. *Journal of Biological Chemistry* **277:** 33842-33847. (*oxysporum*)

2416. **Zhang, Y., and D. A. Smith.** 1983. Concurrent metabolism of the phytoalexins phaseolin kievitone and phaseolin isoflavan by *Fusarium solani* f. sp. *phaseoli*. *Physiological Plant Pathology* **23:** 89-100. (*solani*)

2417. **Zheng, Q., and R. C. Ploetz.** 2002. Genetic diversity in the mango malformation pathogen and development of a PCR assay. *Plant Pathology* **51:** 208-216. (*mangiferae*)

2418. **Zhou, Z., N. Takaya, A. Nakamura, M. Yamaguchi, K. Takeo, and H. Shoun.** 2002. Ammonia fermentation, a novel anoxic metabolism of nitrate by fungi. *Journal of Biological Chemistry* **277:** 1892-1896. (*oxysporum*)

2419. **Zhou, Z., N. Takaya, M. A. C. Sakairi, and H. Shoun.** 2001. Oxygen requirement for denitrification by the fungus *Fusarium oxysporum*. *Archives of Microbiology* **175:** 19-25. (*oxysporum*)

2420. **Zikmundova, M., K. Drandarov, L. Bigler, M. Hesse, and C. Werner.** 2002. Biotransformation of 2-benzoxazolinone and 2-hydroxy-1,4-benzoxazin-3-one by endophytic fungi isolated from *Aphelandra tetragona*. *Applied and Environmental Microbiology* **68:** 4863-4870. (*sambucinum*)

2421. **Zimmermann, W., and E. Seemüller.** 1984. Degradation of raspberry suberin by *Fusarium solani* f. sp. *pisi* and *Armillaria mellea*. *Phytopathologische Zeitschrift* **110:** 192-199. (*solani*)

2422. **Zocher, R., U. Keller, and H. Kleinkauf.** 1982. Enniatin synthetase, a novel type of multifunctional enzyme catalyzing depsi-peptide synthesis in *Fusarium oxysporum*. *Biochemistry* **21:** 43-48. (*oxysporum*)

2423. **Zocher, R., and H. Kleinkauf.** 1978. Biosynthesis of enniatin B partial purification and characterization of the synthesizing enzyme and studies of the biosynthesis. *Biochemical and Biophysical Research Communications* **81:** 1162-1167. (*oxysporum*)

2424. **Zonneveld, B. J. M.** 1988. Effect of carbon dioxide on fruiting in *Aspergillus nidulans*. *Transactions of the British Mycological Society* **91:** 625-630. (6)

2425. **Zuniga, T. L., T. A. Zitter, T. R. Gordon, D. T. Schroeder, and D. Okamoto.** 1997. Characterization of pathogenic races of *Fusarium oxysporum* f. sp. *melonis* causing Fusarium wilt of melon in New York. *Plant Disease* **81:** 592-596. (*oxysporum*)

Index

B

C

H

I

J

K

L

M

O

P

Q

R

S

T

Printed and bound by CPI Group (UK) Ltd, Croydon, CR0 4YY